Physics of Sun and Star Spots

IAU SYMPOSIUM No. 273

COVER ILLUSTRATION:
Left: George Ellery Hale

Hale in the central hall of the National Academy of Sciences, viewing a solar image projected on a circular drum by the coelostat telescope located in the dome above. Over the drum swings the Foucault pendulum, showing the earth's rotation. This exhibit was designed and constructed under Hale's supervision for permanent display in the Academy's new building, dedicated on April 28, 1924.

George Hale discovered magnetic field in sunspots in 1908 using 60 foot solar tower at Mt. Wilson Observatory by applying the principle of Zeeman Splitting. The Zeeman splitting of the spectral line into several components happens in presence of magnetic field. This discovery marked the presence of extraterrestrial magnetism. (credit: Caltech Archives)

Cover designed by John Hodgson II

INTERNATIONAL ASTRONOMICAL UNION
UNION ASTRONOMIQUE INTERNATIONALE

Physics of Sun and Star Spots

PROCEEDINGS OF THE 273th SYMPOSIUM OF THE INTERNATIONAL ASTRONOMICAL UNION HELD IN VENTURA, CALIFORNIA, USA AUGUST 22–26, 2010

Edited by

Debi Prasad Choudhary
California State University Northridge, 18111 Nordhoff St., Northridge, CA, 91330, CA

Klaus G. Strassmeier
Leibniz-Institute for Astrophysics, An der Sternwarte 16, D-14482 Potsdam, Germany

Shaftesbury Road, Cambridge CB2 8EA, United Kingdom

One Liberty Plaza, 20th Floor, New York, NY 10006, USA

477 Williamstown Road, Port Melbourne, VIC 3207, Australia

314–321, 3rd Floor, Plot 3, Splendor Forum, Jasola District Centre, New Delhi – 110025, India

103 Penang Road, #05–06/07, Visioncrest Commercial, Singapore 238467

Cambridge University Press is part of Cambridge University Press & Assessment, a department of the University of Cambridge.

We share the University's mission to contribute to society through the pursuit of education, learning and research at the highest international levels of excellence.

www.cambridge.org
Information on this title: www.cambridge.org/9780521760621

First published 2011

A catalogue record for this publication is available from the British Library

ISBN 978-0-521-76062-1 Hardback

Table of Contents

Oral Presentations

Poster Presentations

THE ORGANIZING COMMITTEE

Acknowledgements

The symposium is sponsored and supported by the IAU Divisions II (Sun and Heliosphere) and IV (Stars).

The Local Organizing Committee operated under the auspices of the Department of Physics and Astronomy, California State University, Northridge.
Funding by the following is gratefully acknowledged:
International Astronomical Union
National Science Foundation, USA (Grants: ATM-0548260 and AST-0968672)
Living With a Star Program of NASA (Grant: NASA LWS NNX10AQ67G)
California State University, Northridge

1. Kamel Yassin
2. Tersi Arias
3. Arlt Rainer
4. Carsten Denker
5. Na Deng
6. Koshiaki Kato
7. V.Bommier
8. Adriana Valio
9. Klaus Strassmeier
10. Antonio Lanza
11. Douglas Brown
12. Angie Cookson
13. Matthias Rempel
14. Durgesh Tripathi
15. Maurizio Ternullo
16. Guillaume Aulanier
17. Krisztián Vida
18. Isroil Sattarov
19. Katalin Olah
20. Jacquelynne Milingo
21. J. C. Pandey
22. Kiran Jain
23. Debi P. Choudhary
24. Hari O Vats
25. Alexander Kosovichev
26. Andrea Dupree
27. Nagendra Kumar
28. Arnab R. Choudhuri
29. Moira Jardine
30. Sanjay Gusain
31. Srikrishna Tripathy
32. Somasekhar Bagare
33. Sami Solanki
34. KR Sivaraman
35. Eugene Avrett
36. Yuan Yuan
37. Rolf Schlichenmeier
38. Jeongwoo Lee
39. Shuo Wang
40. Laurent Gizon
41.
42. Yixuan Li
43.
44. Oleg Kochukhov
45. Toshifumi Shimizu
46. Gary Chapman
47. Mark Giampapa
48. John Hodgson II
49. Brigitte Schmeider
50. Marcelo Lopez Fuentes
51. Rudolf Komm
52.
53. Heidi Korhonen
54. James Neff
55. Alex Brown
56.
57.
58. David Montes
59. Mikhail Demidov
60. Daniel Gomez
61. Sreejith Padinshatteeri
62. Etienne Pariat
63. Etienne Pariat
64. Sanjiv Kumar Twari
65. Judit Murakzy
66. Fraser Watson
67.
68. Andrs Ludmány
69. HP Singh
70. Andrew Gascoyne
71.
72. Zsolt Kővári
73. J. C. Pandey
74. Shaun Bloomfield
75. Irina Kitiashvili
76. Manolis K. Georgoulis
77.
78. SP Rajaguru
79. Valentin Martinez Pillet
80. Thomas Wiegelmann
81.
82.
83. Svetlana Berdyugina
84.
85.
86. Haimin Wang
87. Jose Carlos del Toro Iniesta
88. Jorn Warnecke
89. Matteo Cantiello
90. Thorsten Carroll
91. Lokesh Bharati
92. Qiang Hu

Top: Adam Kowalski talks to Mark. Bottom: Ichimoto San brings out result from his computer.

Top: Arnab Choudhuri explains to Manolis Georgoulis. Bottom: Dalda and Lucia at banquet table.

Top: Rachel Osten talks Star Spots to Steven Saar, behind them Jim Klimchuk makes a point.
Bottom: Finally Brigitte gets it!

Banquet Top: Tsuneta San's table. Bottom: Klaus and the Hungarian/Catania connection.

Top: Jose Carlos del Toro Iniesta talks about discovery of magnetic field in Sunspots by George Hale. Bottom: Gary Chapman and Debi Prasad Choudhary.

Abramenko, Valentyna, BBSO avi@bbso.njit.edu
Andic, Aleksandra, BBSO aandic@bbso.njit.edu
Arias, Tersi, California State University Northridge tersi.arias.25@csun.edu
Arlt, Rainer, Astrophysikalisches Institut Potsdam rarlt@aip.de
Aulanier, Guillaume, Paris Observatory, LESIA guillaume.aulanier@obspm.fr
Avrett, Eugene, Harvard-Smithsonian Center for Astrophysics avrett@cfa.harvard.edu
Ayres, Thomas, University of Colorado thomas.ayres@colorado.edu
Bagare, Somashekhar, IIAP bagare@iiap.res.in
Barnes, Graham, NWRA/CoRA graham@cora.nwra.com
Barnes, Sydney, Lowell Observatory barnes@lowell.edu
Berdyugina, Svetlana, Kiepenhuer Institut fuer Sonnenphysik sveta@kis.uni-freiburg.de
Bharti, Lokesh, Max-Planck-Institute for Solar System Research lokesh_bharti@yahoo.co.in
Bloomfield, David, Trinity College Dublin shaun.bloomfield@tcd.ie
Boisse, Isabelle, IAP iboisse@iap.fr
Bommier, Vronique, LERMA, Observatoire de Paris v.bommier@obspm.fr
Braun, Doug, NorthWest Research Assoc. dbraun@cora.nwra.com
Brown, Alexander, University of Colorado alexander.brown@colorado.edu
Cadavid, Ana, California State University Northridge ana.cadavid@csun.edu
Cantiello, Matteo, Argelander Institute for Astronomy cantiello@astro.uni-bonn.de
Carroll, Thorsten, Astrophysikalisches Institut Potsdam tcarroll@aip.de
Chapman, Gary, San Fernando Observatory, CSU Northridge gchapman@csun.edu
Chatterjee, Piyali, Nordita piyalic@nordita.org
Choudhary, Debi Prasad, California State University Northridge debiprasad.choudhary@csun.edu
Choudhuri, Arnab, Indian Institute of Science arnab@physics.iisc.ernet.in
Christian, Damian, California State University Northridge damian.christian@csun.edu
Cookson, Angela, San Fernando Observatory, CSU Northridge angie.cookson@csun.edu
Crouch, Ashley, NWRA/CoRA ash@cora.nwra.com
Del Toro Iniesta, Jose Carlos, IAA (CSIC) jti@iaa.es
Demidov, Mikhail, Institute of Solar-Terrestrial Physics demid@iszf.irk.ru
Deng, Na, California State University Northridge nd7@njit.edu
Denker, Carsten, Astrophysikalisches Institut Potsdam cdenker@aip.de
Dobias, Jan, California State University Northridge jan.dobias@csun.edu
Dupree, Andrea, CfA dupree@cfa.harvard.edu
Dwivedi, Vidya, APS University, India vidya_charan2000@yahoo.com
Featherstone, Nicholas, JILA, University of Colorado feathern@solarz.colorado.edu
Galsgaard, Klaus, Niels Bohr Institute kg@nbi.ku.dk
Gascoyne, Andrew, University of Sheffield app07adg@sheffield.ac.uk
Georgoulis, Manolis, RCAAM, Academy of Athens manolis.georgoulis@academyofathens.gr
Giampapa, Mark, National Solar Observatory giampapa@noao.edu
Gizon, Laurent, Max Planck Institute for Solar System Research gizon@mps.mpg.de
Golovin, Alexander, MAO NASU, Ukraine golovin.alex@gmail.com
Gomez, Daniel, IAFE dgomez@df.uba.ar
Gosain, Sanjay, Udaipur Solar Observatory sgosaiw@gmail.com
Gritcyk, Pavel, Sternberg Astronomical Institute, Lomonosov Moscow State University funt@inbox.ru
Hastie, Morag, MMT Observatory mhastie@mmto.org
Hazarika, A. B. Rajib, Diphu Govt. College,Diphu,Assam, India drabrh_dgc51632@rediffmail.com
Hodgson, John, California State University Northridge john.hodgson.71@my.csun.edu
Hu, Qiang, CSPAR/UAHuntsville qh0001@uah.edu
Ichimoto, Kiyoshi, Kyoto University ichimoto@kwasan.kyoto-u.ac.jp
Jain, Kiran, National Solar Observatory kjain@noao.edu
Jardine, Moira, University of St Andrews mmj@st-andrews.ac.uk
Jeffers, Sandra, University of Utrecht j.v.jeffers@uu.nl
Karak, Bidya, IISC, India bidya_karak@physics.iisc.ernet.in
Kato, Yoshiaki, ISAS/JAXA kato.yoshiaki@isas.jaxa.jp
Kemel, Koen, Nordita koen@nordita.org
Khelfi, Khaled, CRAAG, Algiers, Algeria k.khelfi@craag.dz
Kitiashvili, Irina, Stanford University irinasun@standord.edu
Kleint, Lucia, Institute of Astronomy, ETH Zurich kleintl@astro.phys.ethz.ch
Klimchuk, James, NASA GSFC james.a.klimchuk@nasa.gov
Kochukhov, Oleg, Uppsala University oleg.kochukhov@gysast.uu.se
Komm, Rudolf, National Solar Observatory komm@naoa.edu
Korhonen, Heidi, European Southern Observatory hkorhone@eso.org
Kosovichev, Alexander, Stanford University sasha@sun.stanford.edu
Kővri, Zsolt, Konkoly Observatory kovari@konkoly.hu
Kowalski, Adam, University of Washington kowalski@astro.washington.edu
Kuhn, Jeff, IfA/UH/Maui kuhn@ifa.hawaii.edu
Kumar, Nagendra, M.M.H. College nagendrakgk@rediffmail.com
Kuttickat, Raju, Indian Institute of Astrophysics kpr@iiap.es.in
Lanza, Antonio Francesco, INAF-Osservatorio Astrofisico di Catania nlanza@oact.inaf.it
Lee, Jeongwoo, New Jersey Institute of Technology leej@njit.edu
Lefevre, Laure, Royal Observatory of Belgium laure.lefevre@oma.be
Li, Yixuan, New Jersey Institute of Technology yl89@njit.edu
Lindborg, Marjaana, University of Helsinki marjaana.lindborg@helsinki.fi
Lockwood, George, Lowell Observatory gwl@lowell.edu
Lopez Fuentes, Marcelo, Instituto de Astronomia y Fisica del Espacio lopezf@iafe.uba.ar
Ludmny, Andrs, Heliophysical Observatory ludmany@tigris.unideb.hu
Lukicheva, Maria, St.Petersburg University marija@peterlink.ru
MacDonald, Gordon, California State University Northridge gordon.macdonald.31@my.csun.edu
Mackay, Duncan, University of St Andrews duncan@mcs.st-and.ac.uk
Marschall, Laurence, Gettysburg College marschal@gettysburg.edu
Martin, Sara, Helio Research, USA sara@helioresearch.org
Milingo, Jacquelynne, Gettysburg College jmilingo@gettysburg.edu
Montes, David, UCM, Universidad Complutense Madrid dmg@astrax.fis.ucm.es
Morin, Julien, DIAS jmorin@cp.dias.ie
Murakozy, Judit, Konkoly Observatory murakozyj@puma.unideb.hu
Murray, Sophie, Trinity College Dublin somurray@tcd.ie
Nakatsukasa, Ken, California State University Northridge ken.nakatsukasa.918@my.csun.edu
Neff, James, College of Charleston neffj@cofc.edu
Nelson, Nicholas, JILA, University of Colorado at Boulder nnelson@lcd.colorado.edu

Nordlund, Aake, Niels Bohr Institute aake@nbi.dk
Nunez, Marlon, Universidad de Malaga mnuez@uma.es
Olah, Katalin, Konkoly Observatory olah@konkoly.hu
Olshevsky, Vyacheslav, Center for Turbulence Research sya@stanford.edu
Orozco Surez, David, National Astronomical Observatory of Japan d.orozco@nao.ac.jp
Osten, Rachel, STScI osten@stsci.edu
Padinhatteeri, Sreejith, ISRO Satellite Center sreejith.p@gmail.com
Pandey, Jeewan, ARIES, Nainital jeewan.pandey@gmail.com
Parchevsky, Konstantin, Stanford University kparchevsky@solar.stanford.edu
Pariat, Etienne, Observatoire de Paris etienne.pariat@obspm.fr
Pasqua, Antonio, University of Manchester, UK pasqua.antonio@yahoo.com
Penn, Matt, National Solar Observatory mpenn@noao.edu
Pillet, Valentin, Instituto de Astrofisica de Canarias vmp@iac.es
Preminger, Dora, San Fernando Observatory, CSU Northridge dora.preminger@csun.edu
Priest, Eric, St. Andrews University eric@mcs.st-and.ac.uk
Rajaguru, Paul, Indian Institute of Astrophysics rajaguru@iiap.res.in
Rempel, Matthias, HAO/NCAR rempel@ucar.edu
Ren, Deqing, California State University Northridge ren.deqing@csun.edu
Rhodes, Edward, University of Southern California erhodes@usc.edu
Saar, Steven, SAO saar@cfa.harvard.edu
Sainz Dalda, Alberto, Stanford-Lockheed Institute for Space Research asdalda@stanford.edu
Sattarov, Isroil, Tashkent State Pedagogical University isattar@astrin.uzsci.net
Savanov, Igor isavanov@rambler.ru
Schlichenmaier, Rolf, KIS schliche@kis.uni-freiburg.de
Schmieder, Bridgitte, Observatoire de Paris brigitte.schmieder@obspm.fr
Schrijver, Karel, Lockheed Martin ATC schryver@lmsal.com
Sennhauser, Christian, ETH Zurich csennhau@astro.phys.ethz.ch
Shimizu, Toshifumi, ISAS/JAXA shimizu@solar.isas.jaxa.jp
Singh, Harinder, University of Delhi singh@iucaa.emet.in
Sinha, Krishnanand, ARIES ksinha2000@hotmail.com
Sivaraman, Koduvayur, Indian Institute of Astrophysics, Bangalore kr_sivaraman@yahoo.com
Solanki, Sami, Max-Planck-Institute for Solar System Research solanki-office@mps.mpg.de
Still, Martin, NASA Ames Research Center martin.still@nasa.gov
Strassamier, Klaus, Astrophysical Institute Potsdam kstrassmeier@aip.de
Ternullo, Maurizio, INAF - Osservatorio Astrofisico Catania mternullo@oact.inaf.it
Tiwari, Sanjiv, Udaipur, Solar Observatory stiwari@prl.res.in
Tripathi, Durgesh, University of Cambridge d.tripathi@damtp.cam.ac.uk
Tripathi, Shreekrishna, University of California, Los Angeles tripathi@physics.ucla.edu
Tsuneta, Saku, NAOJ saku.tsuneta@nao.ac.jp
Valio, Adriana, Mackenzie University adrivalia@gmail.com
Vats, Ho, Astronomy Astrophysics Division vats@pri.res.in
Vida, Krisztin, Konkoly Observatory vidakris@konkoly.hu
Wang, Haimin, New Jersey Institute of Technology haimin.wang@njit.edu
Wang, Shuo, New Jersey Institute of Technology sw84@njit.edu
Warnecke, Jrn, Nordita joern@nordita.org
Watson, Fraser, University of Glasgow f.watson@astrolgla.ac.uk
White, Stephen, AFRL stephen.white@kirtland.af.mil
Wiegelmann, Thomas, MPS wiegelmann@mps.mpg.de
Yassin, Kemal, California State University Northridge kemal.yassin.233@my.csun.edu
Yuan, Yuan, New Jersey Institute of Technology yy46@njit.edu
Yurchyshyn, Vasyl, BBSO vayur@bbso.njit.edu
Zolotova, Nadezhda, Institute of Physics, St.Petersburg State University ned@geo.phys.spbu.ru

Dear Colleagues,

Recalling the observational history of sunspots led to the germ of an idea for this symposium. Even though the Chinese had records of sunspots going back hundreds of years, 2009 marked the four hundredth anniversary of their first viewing through a telescope by Galileo Galilei and Thomas Harriot. In addition, 2008 celebrated the one hundredth anniversary of George Ellery Hale's discovery of the magnetic nature of sunspots at the Mount Wilson Observatory (Los Angeles), marking the first detection of magnetism outside the Earth. It seemed natural to have a celebration of Hale's discovery in Los Angeles, but the original idea needed the impetus of a current scientific theme.

In spite of one hundred years of observational and theoretical research on solar magnetic fields, understanding the mechanisms that govern the origin and decay of sunspots is far from complete. Indeed, the delay of the onset of solar cycle 24 came as a complete surprise to the scientific community. While sunspots have been the subject of detailed studies, spots on other stars cannot yet receive the same level of scrutiny. Combining the two fields of research is mutually beneficial since solar investigations can gain perspective from the long-term evolution of stellar magnetism, and stellar research can gain insight into the root origin of spots. We hope that these proceedings not only reflect the present state of knowledge but contribute to furthering the cross-fertilization of ideas between the solar and stellar research communities. The oral presentations of the symposium were recorded and will be made available on the symposium website http://www.csun.edu/physicsandastronomy/IAUS273/. They are specially useful to follow the discussions at the end of each presentation.

It was a pleasure to welcome to beautiful Ventura, near Los Angeles, over 140 scientists from all over the world. Both the excellent scientific program designed by the SOC and the relaxed setting characterized by spectacular sunsets over the Pacific Ocean, lead to a stimulating and welcoming atmosphere for the exchange of ideas. The smooth and successful running of the event depended on the efficient planning and professionalism of four members of the LOC: Angie Cookson, John Hodgson, Debbie Klevens and Dora Preminger. To them goes our deep appreciation.

We are also extremely grateful for the essential financial support from the following agencies: The National Science Foundation in the Atmospheric Sciences and the Astronomy divisions; the NASA Living with a Star Program; the College of Science and Mathematics and the Office of Research and Sponsored Projects at California State University Northridge. On the final day of the symposium, forty participants took the two and a half hour trip to the Mount Wilson Observatory. There they toured the research-active 60-foot and 150-foot solar towers, as well as the historic Hooker 100-inch telescope and the Snow solar telescope. Walking through the facilities designed and used by Hale was, for some, a brush with the past, bringing deeper significance to their present day work.

Ana Cristina Cadavid, Debi Prasad Choudhary, Klaus G. Strassmeier
March 2011

The Physics of Sun and Star Spots
Proceedings IAU Symposium No. 273, 2010
D.P. Choudhary & K.G. Strassmeier, eds.

doi:10.1017/S174392131101492X

Cosmic magnetic fields in the Sun: Current Outstanding Problems (Invited Review)

Eric Priest

Mathematics Institute, University of St Andrews,
North Haugh, St Andrews, KY16 9SS, UK
email: eric@mcs.st-and.ac.uk

Abstract. In the Sun there has been much progress towards answering fundamental problems with profound implications for the behaviour of cosmic magnetic fields in other stars. A review is given here of such problems, including identifying some of the outstanding questions that remain. In the solar interior, the main dynamo operates at the base of the convection zone, but its details have not been identified. In the solar surface, recent observations have revealed many new and surprising properties of magnetic fields, but understanding the key processes of flux emergence, fragmentation, merging and cancellation is rudimentary. Sunspots have until very recently been an enigma. In the atmosphere, there are many new ideas for coronal heating and solar wind acceleration, but the mechanisms have not yet been pinned down. Also, the detailed mechanisms for solar flares and coronal mass ejections remain controversial. In future, new generations of space and ground-based measurements and computational modelling should enable a definitive physical understanding of these puzzles.

Keywords. Magnetic fields, MHD, sun: corona, sunspots, sun: photosphere.

1. Introduction

There has been a revolution in solar physics over the past 10 years, with many advances that have far-reaching implications for other stars, where similar processes are operating but under different parameter regimes. Major progress has been made on fundamental questions about the nature of the Sun, such as the dynamo generation of its magnetic field, sunspot structure, coronal heating, solar wind acceleration, solar flare and coronal mass ejection origin, but as yet no definitive answers have been given. Here I give a brief overview of this *new Sun*, referring to the review talks that follow and leaving the listeners to make their own connections to similar phenomena on other stars.

These advances have arisen from a combination of ground-based observations, theoretical and computational modelling, and especially space observations from the following satellites: the Yohkoh mission (1992–2002) which revealed the dynamic nature of the corona; the SoHO mission (1995–...) which viewed the interior and atmosphere; the TRACE satellite (1998–2010) which showed the fine-scale nature of the corona; RHESSI (2002–...) which has been studying high-energy processes in solar flares. More recently, the Stereo spacecraft (2006–...) has built up stereoscopic images of coronal mass ejections from two locations, and the Hinode satellite (2006–...) has studied the connections between photosphere and corona, as described in the talk by Saku Tsuneta (Fig. 1a).

In future, we expect to learn much from the Sunrise mission (see talk by Sami Solanki) and from the Solar Dynamics Observatory (launched in February, 2010), which includes EVE (extreme ultraviolet variability experiment), HMI (heliospheric and magnetic

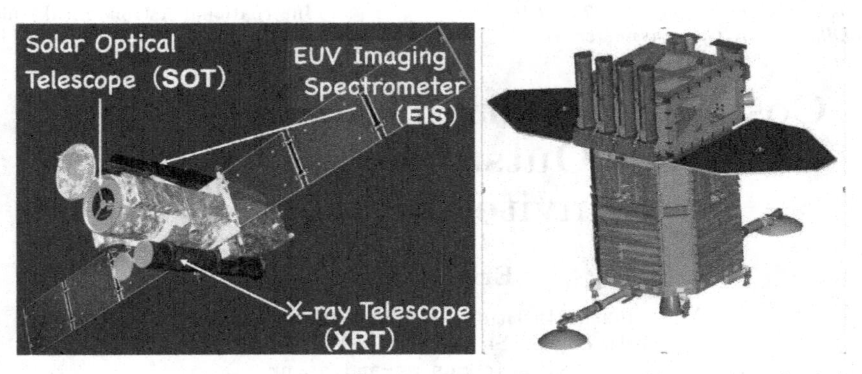

Figure 1. The Hinode satellite (left) and Solar Dynamics Observatory (right).

imager) and AIA (atmospheric imaging assembly). AIA in particular is a super-TRACE instrument, which has the same spatial resolution as TRACE but for the full disc and with images at 8 temperatures (from 20,000 K to 20 MK) every 10 seconds (Fig. 1b).

So what is the current status of understanding and what are the main questions about the solar interior, photosphere and corona?

2. The Solar interior

The solar interior has been opened up to study by the advent of helioseismology, whereby the properties of global and local oscillations have been observed by ground-based networks (GONG and Bison) as well as the MDI instrument on SoHO. The deduced internal temperature structure agrees with the standard model to within 1%, but the internal rotation structure has been a big surprise. At the solar surface, the equator rotates more rapidly than the poles, and so the internal rotation was expected to be constant on cylinders with the dynamo generated throughout the convection zone.

However, the seismology result was that rotation is instead constant on cones in the convection zone and possesses a strong shear layer at the base of the convection zone, termed the *tachocline*. It is now believed that the main solar dynamo responsible for active regions and sunspots is generated at the tachocline, although the details are uncertain.

The main questions about the solar interior, therefore, are:

What is the detailed internal flow structure, especially near the poles and below the convection zone, as well as the deep-seated nature of the meridional flow?

For dynamo theory, there is uncertainty about the validity of mean-field theory and the alpha effect. Also, is the main dynamo generated at the tachocline or is it generated by a Babcock-Leighton effect at the solar surface? Furthermore, the nature of the connection between the tachocline and the radiative interior and convection zone is unclear. (See the talk by Arnab Choudhuri for clarification of these issues.)

Robust, reliable predictions of the solar cycle are also needed. The current solar minimum, the lowest for 200 years, was a big surprise to most of us (except for Matt Penn, who will be talking later). When and how large will the next solar maximum be?

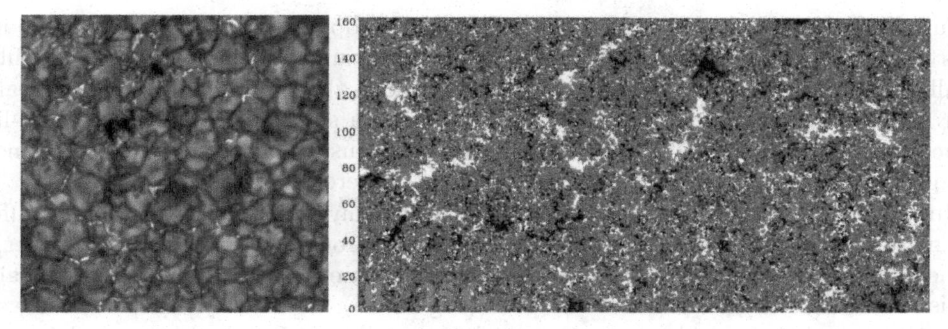

Figure 2. (a) A close-up of the photosphere from the Swedish Solar Telescope (courtesy M. Carlsson and the Swedish Academy of Sciences). (b) The line-of-sight photospheric magnetic field from Hinode (courtesy B. Lites).

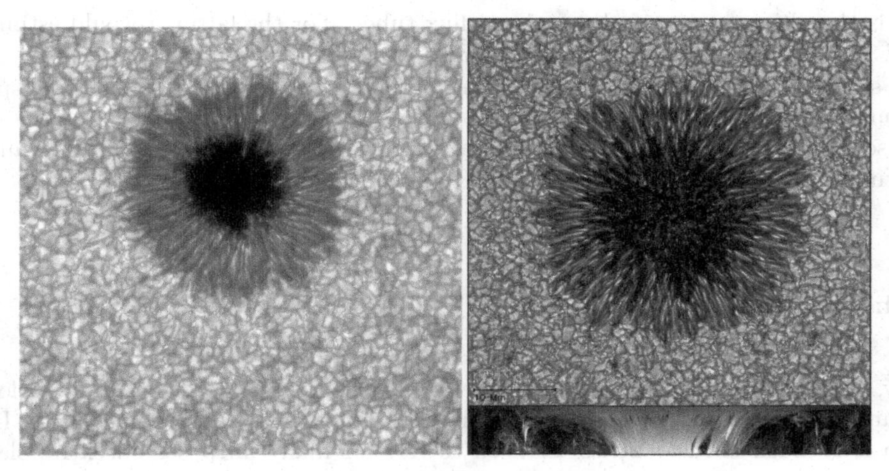

Figure 3. Images of sunspots from: (a) the Swedish Solar Telescope (courtesy L. Rouppe van Der Voordt and Swedish Academy of Sciences); (b) numerical computation (courtesy M. Rempel).

3. The Photosphere

The standard picture was until recently that outside active regions the photospheric magnetic field is vertical and concentrated at the edges of supergranule cells (of width 15–30 Mm by comparison with the solar radius of 700 Mm). Magnetic flux emerged mainly as ephemeral active regions in the supergranule cell interiors and then fragmented as it was carried to the cell boundaries. It then moved along the cell boundaries and either merged with like-polarity elements or cancelled with those of opposite polarity. Such flux was reprocessed every 14 hours (Hagenaar *et al.* 2003).

However, observations of the photosphere from the Swedish Solar Telescope on La Palma has revealed bright points at the edges of granules that represent intense magnetic flux concentrations (Figure 2a). Furthermore, the Hinode satellite has measured the line-of-site field at low thresholds and deduced that there is a huge amount of complex vertical magnetic flux in the interior of superganules in the boundaries between individual granules (Figure 2b). In addition, a big surprise was its discovery of ubiquitous horizontal fields in the photosphere on the edges of granules (see talk by Saku Tsuneta).

Turning to sunspots, again the Swedish Solar Telescope has produced remarkable images with 0.03 arcsec pixels, which show fine-scale granular behaviour and incredibly small scales in the penumbra (Figure 3a). Furthermore, the Hinode satellite has revealed that the penumbral magnetic field varies strongly on a small scale and has a comb-like structure, with low-lying field lines closing back down just outside the sunspot and more oblique field lines arching much higher in the atmosphere.

Until recently, the nature of the penumbra was a mystery that several quite different models tried to explain. However, a new breakthrough has come from a series of numerical computations by Matthias Rempel of magnetoconvection, with remarkably realistic-looking images of model sunspots (Figure 3b). Previously, the nature of, say umbral dots, penumbral filaments, Evershed flow and moat flow had been explained by separate models, but Rempel *et al.* (2009) give a unified explanation for them all in terms of the natural effects of convection in an inclined magnetic field (see talk by Rempel).

The main questions for the Quiet Sun are: how much flux is inside supergranules and how much is at the edge? What is the effect of the fine-scale structure on coronal heating? What is the ultimate size of photospheric flux tubes? For the latter, I would estimate a size of 30 km, based on the photospheric magnetic diffusivity.

For sunspots, what is the nature of the outer penumbra and the subphotospheric structure and are the penumbral field lines held down by granular flux pumping?

These are tackled in the talks by Carsten Denker, Laurent Gizon, Kioshi Ichimoto, Valentin Martinez, Matthias Rempel and Rolf Schlichenmaier.

4. The Corona

4.1. *Chromospheric and coronal magnetic fields*

Firstly, how can we measure solar magnetic fields reliably? Talks by José Carlos del Toro and Hongqi Zhang will address this in the photosphere and chromosphere. In the corona it is much harder and so we need to rely for many purposes on extrapolations from the photosheric field. Potential extrapolations are usually very poor: – for example, a comparison of coronal holes and regions of open magnetic fields in global potential extrapolations is highly disappointing. Nevertheless, nonlinear force-free extrapolations both of local and global fields are much more helpful (see talk by Thomas Wiegelmann).

The global coronal magnetic field evolves in response to the emergence of new flux, magnetic diffusion, meridional flow and differential rotation. These have been incorporated into a model which impressively predicts the location of most quiescent prominences outside active regions (Mackay & van Ballegooijen 2006). It also predicts the eruptions of many coronal mass ejections (see talk by Duncan Mackay).

4.2. *Coronal Heating*

The nature of coronal heating remains a mystery. Close *et al.* (2004) constructed magnetic field lines in the Quiet Sun from SoHO MDI magnetograms and found that each photospheric source connects on average to 8 others. Most of them close down close to the surface and a few reach large heights. More surprisingly, when they followed the motions of the photospheric sources and recalculated the connectivity, they found that the time for all the field lines on the Quiet Sun to reconnect is only 1.5 hours. In other words, there is an incredible amount of reconnection going on continually in the corona and this is probably responsible for its heating.

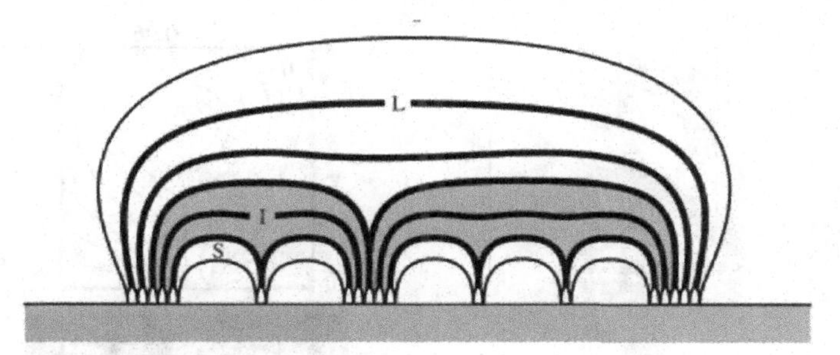

Figure 4. The Coronal Tectonics Model.

The topology of coronal fields is highly complex, but an invaluable method for estimating the topology is to calculate the *skeleton* of the field. In two dimensions, a plane is separated up into different regions by separatrix curves that intersect at an X-type null point. In three dimensions, a volume is separated into topologically different regions by separatrix surfaces, which meet in a *separator* – a special field line that usually begins and ends at a null point. The skeleton then consists of a web of separatrix surfaces. Parnell *et al.* (2009) have developed sophisticated techniques to calculate such a skeleton for any magnetic structure, whether produced in a numerical experiment or arising as a result of a photospheric extrapolation.

The *Coronal Tectonics Model* (Priest *et al.* 2002) was developed in order to take account of the magnetic carpet and to refine Parker's braiding model for coronal heating (Figure 4). Each coronal loop that is currently observed in practice goes down through the solar surface in many tiny sources. Thus, the flux from each source is separated by a separatrix surface. The idea of the model is then that the corona is filled with myriads of current sheets forming and dissipating impulsively at separators and also at *Quasi-Separatrix Layers* where the mapping gradient is large rather than being infinite (as it is at a separatrix) (see Priest & Démoulin (1995); Démoulin *et al.* (1996); Aulanier *et al.* (2006)).

A key question is how to identify the mechanism or mechanisms that are heating the corona. Toshifumi Shimizu will describe new observations of active regions, and Daniel Gomez will consider the role of turbulence, while Klaus Glasgaard and Guillaume Aulanier will describe heating at null points, separators and quasi-separatrix layers.

4.3. *Solar flares*

By comparing the structure of active regions with the location of solar flares, some authors suggest they occur at separators (Longcope *et al.* 2007), while others suggest the location is a quasi-separatrix layer (Aulanier *et al.* 2006).

Four suggestions have been proposed for the cause of the magnetic eruption that produces a coronal mass ejection and two-ribbon solar flare. These are kink instability, magnetic nonequilibrium, torus instability and breakout (Figure 5). A key question is therefore to determine for realistic active regions and flare situations (Figure 6) which of the four possibilities possesses the correct physics (see talk by Guillaume Aulanier).

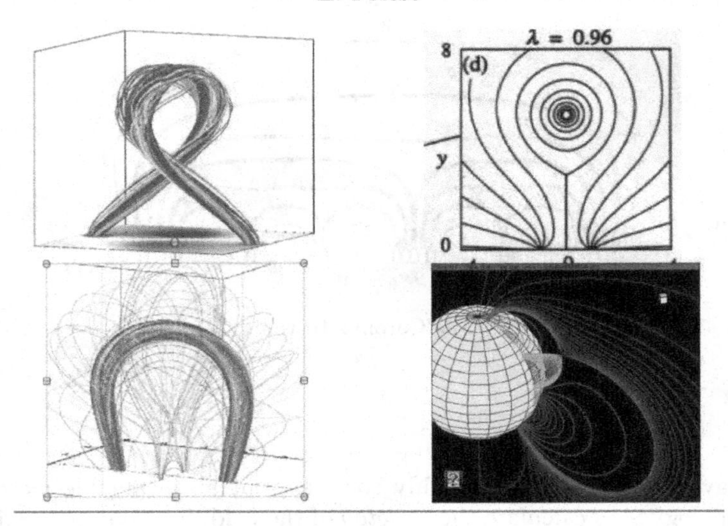

Figure 5. Four possible models for the eruption of coronal magnetic fields in a two-ribbon solar flare or coronal mass ejection.

Figure 6. An image from AIA on the Solar Dynamics Observatory of the active solar corona during the occurrence of several solar flares (courtesy K Schrijver).

5. Conclusions

Solar physics is currently in a golden age of great vitality as we home in on explanations for many of the fundamental questions in the field. These include the nature of the solar interior and the dynamo process, the behaviour of photospheric magnetic fields and sunspots, the heating of the corona and the origin of solar flares.

In this endeavour computational experiments and space observations provide a key role. It will be fascinating to hear the insights of the array of outstanding speakers that have been assembled for this meeting.

References

Aulanier, G., Pariat, E., Démoulin, P., & Devore, C. R. 2006, *Solar Phys.*, 238, 347

Close, R., Parnell, C., Mackay, D. H., & Priest, E. R. 2004, *Solar Phys.*, 212, 251

Démoulin, P., Henoux, J. C., Priest, E. R., & Mandrini, C. H. 1996, *Astron. Astrophys*, 308, 643

Hagenaar, H. J., Schrijver, C. J., & Title, A. M. 2003, *Astrophys. J.*, 584, 1107

Longcope, D. W., Beveridge, C., Qiu, J., Ravindra, B., Barnes, G., & Dasso, S. 2007, *Solar Phys.*, 244, 45S

Mackay, D. H. & van Ballegooijen, A. A., 2006, *Astrophys. J.*, 641, 577
Parnell, C. E., Haynes, A. L., & Galsgaard, K., 2009, *Journal of Geophysical Research*, 115, 2102
Priest, E. R. & Démoulin, P., 1995, *Journal of Geophysical Research*, 23, 443
Priest, E. R., Heyvaerts, J., & Title, A. M., 2002, *Astrophys. J.*, 576, 533
Rempel, M., Schüssler, M., Cameron, R. H., & Knölker, M., 2009, *Science*, 325, 171

The Physics of Sun and Star Spots
Proceedings IAU Symposium No. 273, 2010
D.P. Choudhary & K.G. Strassmeier, eds.

doi:10.1017/S1743921311014931

3D numerical MHD modeling of sunspots with radiation transport

Matthias Rempel
High Altitude Observatory, National Center for Atmospheric Research,
P.O. Box 3000, Boulder, CO 80307, USA
email: rempel@ucar.edu

Abstract. Sunspot fine structure has been modeled in the past by a combination of idealized magneto-convection simulations and simplified models that prescribe the magnetic field and flow structure to a large degree. Advancement in numerical methods and computing power has enabled recently 3D radiative MHD simulations of entire sunspots with sufficient resolution to address details of umbral dots and penumbral filaments. After a brief review of recent developments we focus on the magneto-convective processes responsible for the complicated magnetic structure of the penumbra and the mechanisms leading to the driving of strong horizontal outflows in the penumbra (Evershed effect). The bulk of energy and mass is transported on scales smaller than the radial extent of the penumbra. Strong horizontal outflows in the sunspot penumbra result from a redistribution of kinetic energy preferring flows along the filaments. This redistribution is facilitated primarily through the Lorentz force, while horizontal pressure gradients play only a minor role. The Evershed flow is strongly magnetized: While we see a strong reduction of the vertical field, the horizontal field component is enhanced within filaments.

Keywords. MHD, radiative transfer, Sun: sunspots, Sun: photosphere, Sun: magnetic fields

---

1. Introduction

Sun and starspots play a central role for our understanding of solar and stellar magnetism. Since direct observations of magnetic field in stellar convection zones are very limited (e.g. helioseismic inversions), sun and starspots provide a window to understand the magnetism of stellar interiors, provided we understand in detail how starspots form, evolve and decay. Sunspots are a multi scale problem with respect to spatial and temporal scales. While their typical size is > 20 Mm, fine structure is observed at the smallest scales currently observable of about 200 km (e.g. high resolution ground based observations with the *SST*, Scharmer *et al.* (2002), or space based observations with *HINODE*, see e.g. Ichimoto *et al.* (2007)). Properly resolving the scales currently observed requires grid resolutions of 20 km or less. Details of the penumbra evolve over time scales of hours, while the life time of sunspots and the evolution of the adjacent moat happens on time scales of days to weeks. On the other hand typical numerical time steps are of the order of 0.1 sec (assuming already that the very fast Alfvén velocities found above a sunspot umbra were removed from the system, see e.g. appendix of Rempel *et al.* (2009a)). As a consequence a well resolved realistic numerical simulation of an entire sunspot requires billions of grid points and hundred thousands of time steps. Since the essential physics that need to be considered (MHD, 3D radiative transfer, realistic equation of state) are well known and have been included in MHD codes for more than 2 decades, the primary challenge for addressing sunspot structure is linked to robust efficient numerical schemes and availability of powerful computing resources. The latter became finally available on the scale needed, which allowed for substantial progress within a time frame of only a few years. In the following sections we will very briefly summarize recent progress and a few aspects of sunspot structure for which numerical simulations provided substantial insight.

2. Recent developments

We limit the following discussion entirely to MHD simulations that include 3D radiative transfer and a realistic equation of state. Here, progress in applications related to sunspots started with the work by Schüssler & Vögler (2006) who presented a MHD simulation of a sunspot umbra showing the development of magneto-convection in form of umbral dots. The simulation revealed that almost field free upflow plumes can form within an initially monolithic umbra, which transport energy through overturning convection. Heinemann *et al.* (2007) focused on a narrow slab through the center of sunspot. Their setup allowed to model the transition from umbra toward penumbra and granulation, while keeping the computational expense at a moderate level. Their simulation showed the formation

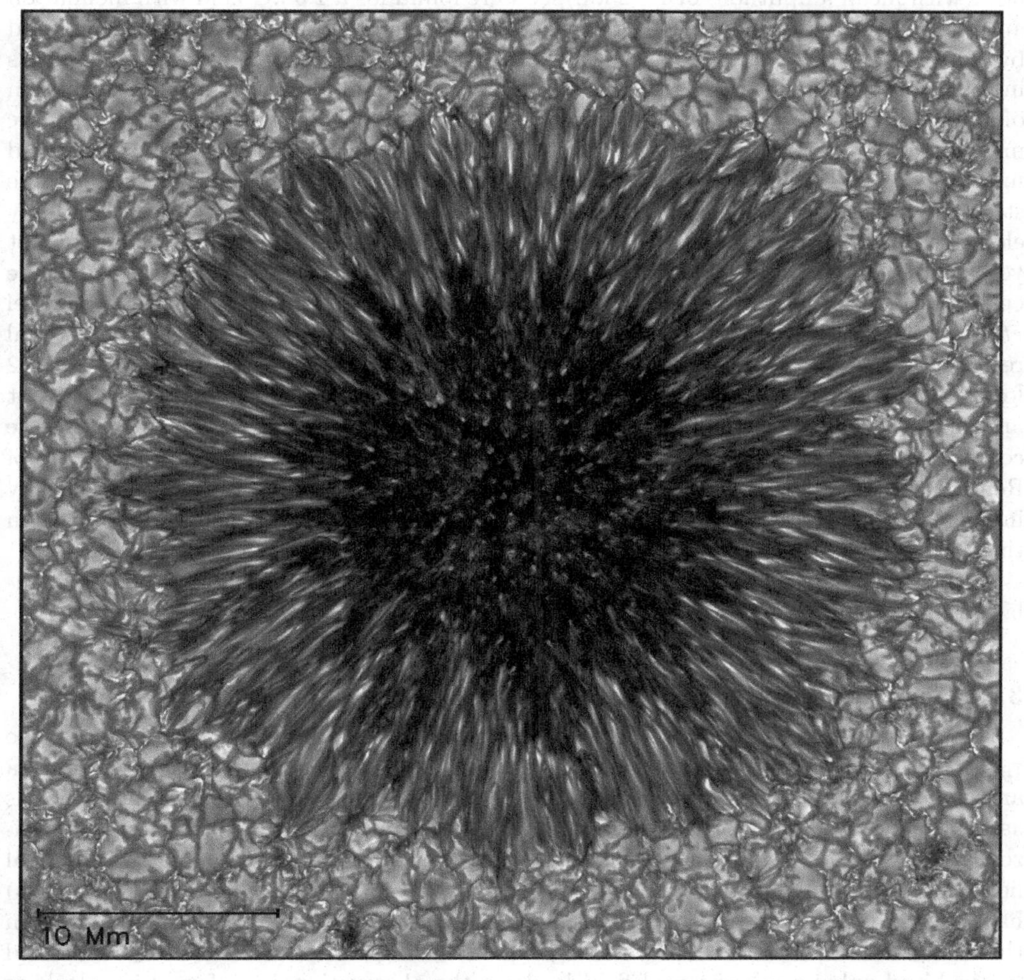

Figure 1. Numerical sunspot model. The domain size is 49 × 49 × 6.1 Mm, the grid resolution 16×16×12 km (3072×3072×512 grid points). The simulation was performed with grey radiative transfer, displayed is the bolometric intensity for a vertical ray in the range from $0.25 - 1.5 I_\odot$.

of short filaments with dark lanes at the umbra/granulation interface, however, umbral dots were not present. The filaments were propagating inward during the formation phase and showed outflow along their axis. Based on this simulation Scharmer *et al.* (2008) interpreted the Evershed flow as convective flow component along filaments. In a very similar setup with an initially larger segment of a sunspot Rempel *et al.* (2009a) were able to produce umbral dots and filaments of about 3 Mm length, showing the smooth transition from central to peripheral umbral dots and filaments in the inner penumbra. The latter show the presence of a bright head propagating inward and detaching from the filament during the formation phase and up to 3 Mm long dark lanes with evidence of twisting motions. Along filaments, outflows of a few km/s were present, and on a larger scale a moat flow, previously also reported in Heinemann *et al.* (2007). Overall these simulations clearly stress the common magneto-convective origin of umbral dots and penumbral filaments, but a realistic outer penumbra with strong outflows was not present. Kitiashvili *et al.* (2009) conducted a study of how the magnetic field strength and inclination angle influence the formation of horizontal flows in magneto-convection. Fast flows with mean amplitudes of $1-2$ km/sec were found for a 1.5 kG field with inclination angle of 85 degrees. The first comprehensive simulation of entire sunspots was presented by Rempel *et al.* (2009b). Using a setup including a pair of opposite polarity sunspots in a 98×49 Mm wide and 6.1 Mm deep domain, this simulation showed the formation of extended (up to 10 Mm wide) penumbrae with horizontal outflows of up to 6 km/sec mean- and 14 km/sec peak flow speeds. At the same time this simulation contained umbral dots and a substantial moat region, connecting all the aspects of sunspot fine structure in one comprehensive simulation run. While the penumbra contained radially elongated convection cells, the intensity image (see Fig. 1 in Rempel *et al.* (2009b)) did not yet show the typical radial structure of narrow filaments known from observations. The currently best resolved sunspot simulation is presented in Fig. 1. Compared to Rempel *et al.* (2009b) the horizontal resolution is doubled (from 32 to 16 km), while the vertical resolution is increased from 16 to 12 km, leading to a total grid size of $3072 \times 3072 \times 512$ for a domain of the size $49 \times 49 \times 6.1$ Mm. Here we focused again on a single spot, but artificially enhanced the field inclination through the top boundary condition to generate conditions that are comparable to the region in between the opposite polarity spots of Rempel *et al.* (2009b). While the photospheric appearance of the sunspot is substantially improved compared to previous simulations, we did not find fundamental differences in the underlying magneto-convection process.

In the following section we highlight a few central aspects we learned from the simulations summarized above.

3. Magnetic fine structure of penumbra

Figure 2 summarizes the magnetic fine structure of umbra and penumbra at the $\tau_{\rm Ross} = 1$ level for the simulation presented in Fig. 1. A common element in umbra and penumbra is a strong reduction of the vertical magnetic field component, which is associated with umbral dots, peripheral umbral dots and penumbral filaments. The horizontal magnetic field component shows a different behavior. A reduction of the horizontal field strength is only present in the umbra, whilst the penumbra filaments (flow channels) have an enhanced horizontal magnetic field strength throughout, which is evident from the correlations shown in the bottom right panel of Fig. 2. The combination of reduced vertical and enhanced horizontal field leads to the observationally inferred interlocking comb structure with strong variation of the field inclination angle and strong horizontal flows in the component with nearly horizontal field (see e.g. Solanki (2003)) . The combined effect of the weakening of B_z and strengthening of B_R withing flow channels results in an overall reduction of $|B|$ in the inner, and increase of $|B|$ in the outer, penumbra. Recently Ichimoto *et al.* (2007) found this trend in the observed $V_R - |B|$ correlation, even though positive values of the correlation were not observed. An Evershed flow in regions with enhanced field strength in the outer penumbra was proposed by Tritschler *et al.* (2007) based on zero-crossings of the NCP.

Overall the numerical simulation presented here supports a strongly magnetized penumbra and Evershed flow. While the underlying structure of the magneto-convection is well captured by the "gappy" penumbra model of Scharmer & Spruit (2006) and Spruit & Scharmer (2006), we do not see any support for the claim that these gaps are close to field free. While the latter is a good approximation for the vertical field component alone,

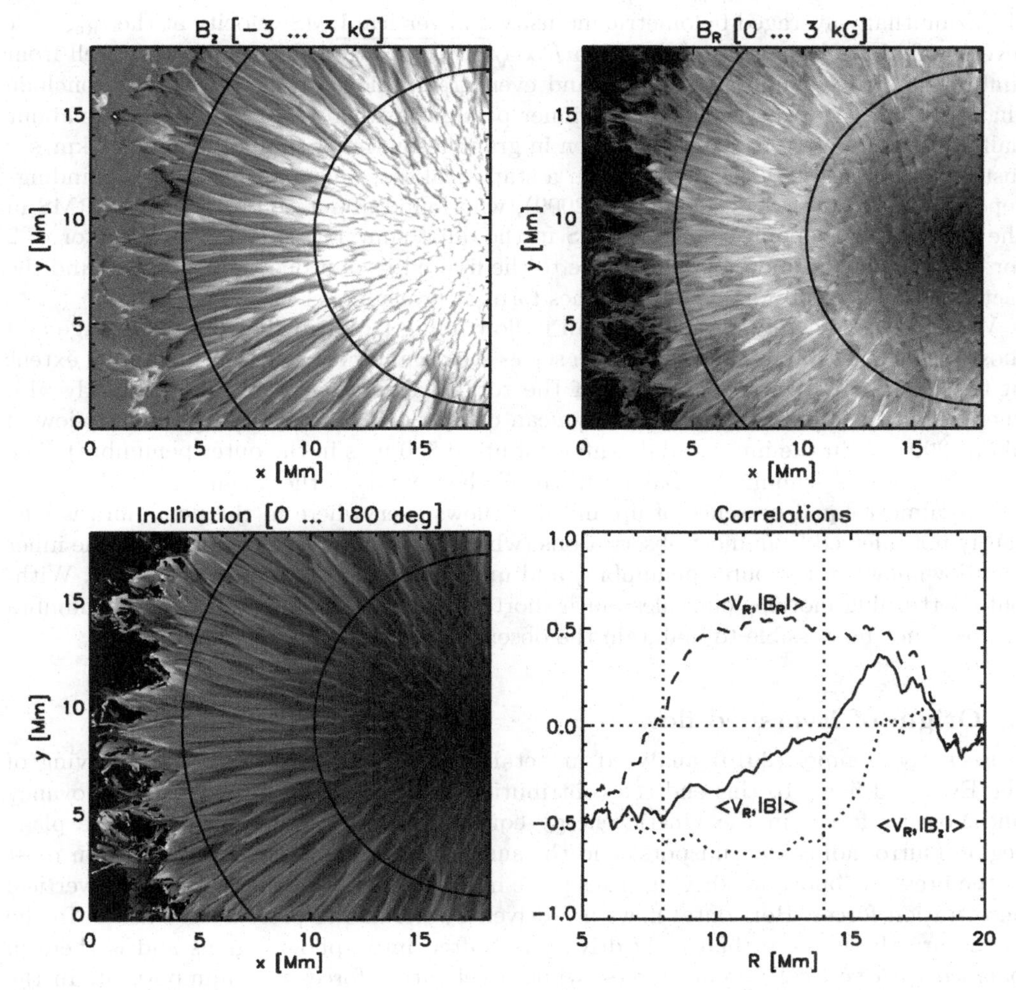

Figure 2. Magnetic fine structure of sunspot at $\tau = 1$ level. The top panels show vertical (± 3 kG) and radial field strength ($0 \ldots 3$ kG). The bottom left panel shows the inclination, with black indicating vertical outward, grey horizontal, and white vertical inward directed field. We set the inclination to 0 for regions with less than 500 G field strength (dark color outside the penumbra). The bottom right shows correlations between horizontal flow velocity and magnetic field strength fluctuations as a function of the spot radius. The vertical dotted lines correspond to the dark concentric circles shown in the other three panels.

horizontal field is actually enhanced compared to the background field. Recently Rempel (2010) showed that the horizontal field originates from the strong subsurface shear profile of the Evershed flow, which leads to an induction term $B_z \partial_z v_R$ with a strength of about $5 - 10$ G/sec a few 100 km beneath $\tau = 1$. Relative contributions from numerical magnetic diffusivity are only of a few percent. A recent convergence study (Rempel, in preparation) from 96×32 up to 16×12 km resolution (Fig. 1 presents the highest resolution case) strongly supports the robustness of this result.

4. Convective energy transport

All numerical simulations performed to date (most of them are summarized in Sect. 2) point toward a common magneto-convective origin of sunspot fine structure. Comparing the azimuthally averaged bolometric intensity and vertical RMS velocity at the $\tau_{\rm Ross} = 1$ level we find a relationship of the form $I \propto \sqrt{v_z^{RMS}(\tau = 1)}$, which holds very well from umbral dots to penumbral filaments and even granulation. From this we can conclude that the vertical RMS velocity in the inner penumbra with $I \approx 0.7\, I_\odot$ should be about half of the convective RMS velocity seen in granulation, i.e. it should be about 1 km/sec instead of 2 km/sec. This is, at least in a statistical sense, compatible with the findings reported in Franz & Schlichenmaier (2009), who found about 500 m/s vertical RMS in the penumbra vs about 1 km/sec RMS in the quiet sun. The shortfall by a factor of 2 for both penumbra and quiet sun is due to the limited resolution of observations and the fact that the typically used spectral lines form at levels higher than $\tau_{\rm Ross} = 1$.

We find throughout the penumbra an upflow filling factor in the $40 - 60\%$ range and most of the mass flux is turning over on scales significantly shorter than the radial extent of the penumbra. Only about 15% of the total overturning mass flux (similarly also energy flux) is found in the azimuthal mean component (corresponding to an upflow of about 200 m/s in the inner and downflow of up to 500 m/s in the outer penumbra). The mass flux of the penumbra is balanced within the bounds of the penumbra.

The almost equal presence of up and downflows everywhere in the penumbra is currently not inferred from most observations (which see a preference of upflows in the inner and downflows in the outer penumbra) and might require even higher resolution. Without overturning motions on scales much shorter than the radial extent of the penumbra it would not be possible to maintain the observed brightness of about $0.7\, I_\odot$.

5. Origin of Evershed flow

Recently Rempel (2010) analyzed in detail the processes underlying the driving of the Evershed flow. To this end the contributions from acceleration, pressure, buoyancy and Lorentz forces in the kinetic energy equation were compared between the plage region (surrounding the sunspots) and the sunspot penumbra. In the plage region most of the pressure/buoyancy driving takes place in downflows and is in balance with vertical acceleration forces. Horizontal flows are driven by horizontal pressure gradients. In the penumbra the pressure/buoyancy driving is shifted into upflow regions and is there in balance with vertical Lorentz forces, while acceleration forces are unimportant. In the horizontal direction acceleration forces are in balance with the horizontal Lorentz force component, while horizontal pressure gradients do not play a major role throughout most of the penumbra. The Lorentz force facilitates the energy exchange between motions in the vertical and horizontal direction, while the net work done by Lorentz forces remains negative (sink of kinetic energy). Some aspects of this picture (pressure driving in upflows, deflection and guiding of motions by magnetic field) have been captured to some extent in simplified flux tube models such as Montesinos & Thomas (1997), Schlichenmaier *et al.* (1998b) and Schlichenmaier *et al.* (1998a). However, the presence of vigorous convection leads to a situation in which the outflowing mass is continuously replaced through upflows extending along filaments – a situation which is inherently opposite to the concept of a flux tube. Overall the Evershed flow is best characterized as convective flow (Scharmer *et al.* 2008), although, compared to field free convection, notable differences exist with respect to the underlying driving forces.

6. Conclusions

Numerical simulations have been advanced to the point where they provide a consistent and unifying picture of the magneto-convective processes underlying the energy transport, magnetic fine structure and origin of large scale flows in sunspots. The bulk of energy and mass is transported by overturning convection with scales substantially shorter than the radial extent of the penumbra. The filamentation of the penumbra and the driving of large scale outflows is strongly linked to the presence of anisotropic convective flows. We find the above picture being converged with respect to numerical resolution. We explored the range from 96×32 km to 16×12 km and found that most aspects are already well described with 48×24 km resolution. For the currently accessible resolution range, flows within filaments are mostly laminar; whether a possible transition to turbulent flows at higher resolution could change results remains an open question (due to the rather strong magnetic field turbulence might remain suppressed even at higher resolutions than currently affordable). Initial state and boundary conditions at top and bottom have a strong influence on the global magnetic structure and stability (i.e. they can make the difference between having or not having a penumbra), however the details of sunspot fine structure as discussed here are influenced to a much lesser degree. While the overall radial extent of the penumbra is subject to boundary conditions, the details of filamentation energy transport and driving mechanism of Evershed flow are not.

Acknowledgements

The National Center for Atmospheric Research is sponsored by the National Science Foundation. Computing time was provided by the National Center for Atmospheric Research, the Texas Advanced Computing Center (TACC), the National Institute for Computational Sciences (NICS) and the NASA High End Computing Program.

References

Franz, M. & Schlichenmaier, R. 2009, *A & A*, 508, 1453
Heinemann, T., Nordlund, Å., Scharmer, G. B., & Spruit, H. C. 2007, *ApJ*, 669, 1390
Ichimoto, K., Shine, R. A., Lites, B., Kubo, M., Shimizu, T., Suematsu, Y., Tsuneta, S., Katsukawa, Y., Tarbell, T. D., Title, A. M., Nagata, S., Yokoyama, T., & Shimojo, M. 2007, *Publ. Astron. Soc. Jpn.*, 59, 593
Kitiashvili, I. N., Kosovichev, A. G., Wray, A. A., & Mansour, N. N. 2009, *ApJL*, 700, L178
Montesinos, B. & Thomas, J. H. 1997, *Nature*, 390, 485
Rempel, M. 2010, *ApJ*, submitted
Rempel, M., Schüssler, M., & Knölker, M. 2009a, *ApJ*, 691, 640
Rempel, M., Schüssler, M., Cameron, R. H., & Knölker, M. 2009b, *Science*, 325, 171
Scharmer, G. B., Gudiksen, B. V., Kiselman, D., Löfdahl, M. G., & Rouppe van der Voort, L. H. M. 2002, *Nature*, 420, 151
Scharmer, G. B., Nordlund, Å., & Heinemann, T. 2008, *ApJL*, 677, L149
Scharmer, G. B. & Spruit, H. C. 2006, *A & A*, 460, 605

Schlichenmaier, R., Jahn, K., & Schmidt, H. U. 1998a, *ApJ*, 493, L121
—. 1998b, *A & A*, 337, 897
Schüssler, M. & Vögler, A. 2006, *ApJL*, 641, L73
Solanki, S. K. 2003, *A & A*, 11, 153
Spruit, H. C. & Scharmer, G. B. 2006, *A & A*, 447, 343
Tritschler, A., Müller, D. A. N., Schlichenmaier, R., & Hagenaar, H. J. 2007, *ApJL*, 671, L85

The Physics of Sun and Star Spots
Proceedings IAU Symposium No. 273, 2010
D.P. Choudhary & K.G. Strassmeier, eds.

doi:10.1017/S1743921311014943

Rapid changes of sunspot structure associated with solar eruptions

Haimin Wang and Chang Liu

Space Weather Research Lab, New Jersey Institute of Technology, Newark, NJ 07102, USA
email: haimin.wang@njit.edu

Abstract. In this paper we summarize the studies of flare-related changes of photospheric magnetic fields. When vector magnetograms are available, we always find an increase of transverse field at the polarity inversion line (PIL). We also discuss 1 minute cadence line-of-sight MDI magnetogram observations, which usually show prominent changes of magnetic flux contained in the flaring δ spot region. The observed limb-ward flux increases while disk-ward flux decreases rapidly and irreversibly after flares. These observations provides evidences, either direct or indirect, for the theory and prediction of Hudson, Fisher & Welsch (2008) that the photospheric magnetic fields would respond to coronal field restructuring and turn to a more horizontal state near the PIL after eruptions. From the white-light observations, we find that at flaring PIL, the structure becomes darker after an eruption, while the peripheral penumbrae decay. Using high-resolution Hinode data, we find evidence that only dark fibrils in the "uncombed" penumbral structure disappear while the bright grains evolve to G-band bright points after flares.

Keywords. Sun: activity, sun: flares, sun: coronal mass ejections (CMEs), sun: magnetic fields

1. Introduction

Solar eruptions, including flares, filament eruptions, and coronal mass ejections (CMEs) have been understood as the result of magnetic reconnection in the solar corona (e.g., Kopp & Pneuman 1976; Antiochos *et al.* 1999). Although surface magnetic field evolution (such as new flux emergence and shear motion) play important roles in building energy and triggering eruption, most models of flares and CMEs have the implication that photospheric magnetic fields do not have rapid, irreversible changes associated with the eruptions. The key reason behind this assumption is that the solar surface, where the coronal magnetic fields are anchored, has much higher density and gas pressure than the corona. Recently, we note the work by Hudson, Fisher & Welsch (2008, hereafter HFW08), who quantitatively assessed the back reaction on the solar surface and interior resulting from the coronal field evolution required to release energy, and made the prediction that after flares, the photospheric magnetic fields become more horizontal. This is one of the very few models that specifically predict that flares can be accompanied by rapid and irreversible changes of photospheric magnetic fields. Fisher *et al.* (2010) elaborated this work further.

On the observational side, earlier studies were inconclusive on the flare-related changes of photospheric magnetic field topology. Wang (1992) and Wang *et al.* (1994) showed impulsive changes of vector fields after flares including some unexpected patterns such as increase of magnetic shear along the PIL, while mixed results were also reported (e.g., Hagyard *et al.* 1999; Chen *et al.* 1994). It is not until recently that rapid and permanent changes of photospheric magnetic fields, mainly the line-of-sight component, are observed

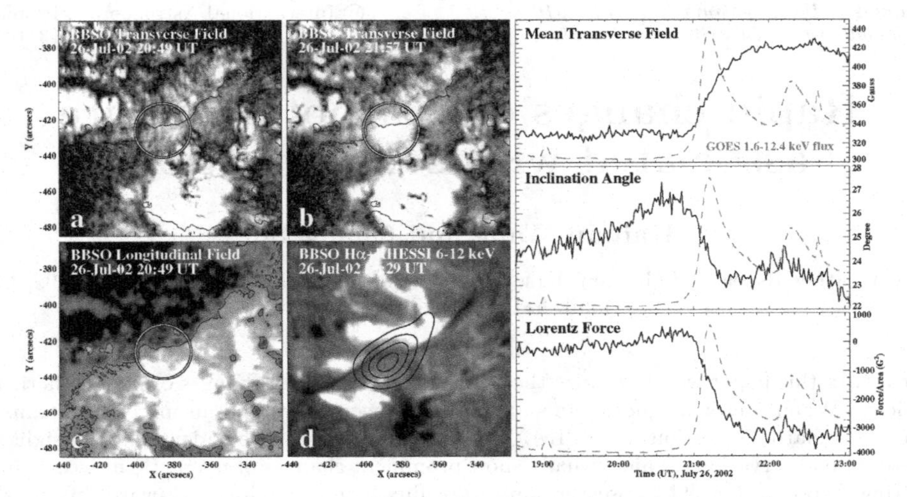

Figure 1. Time profiles of transverse field, inclination angle, and Lorentz force per unit area within a white circled region (in *a–c*) at the PIL (black line in *a–c*) for the 2002 July 26 M8.7 flare (see an Hα image in *d* overplotted with RHESSI soft X-ray source), calculated using BBSO vector magnetograms. In *GOES* 10 soft X-ray flux (dashed line), the flare started at 20:51 UT, peaked at 21:12 UT, and ended at 21:29 UT.

to consistently appear in major flares and considered as indicative of flare energy release (Kosovichev & Zharkova 2001; Sudol & Harvey 2005). In particular, a number of papers of our group have been devoted to the finding of sudden unbalanced magnetic flux change (Spirock *et al.* 2002; Wang *et al.* 2002; Yurchyshyn *et al.* 2004; Wang *et al.* 2004a; Wang 2006) and a new phenomenon of sunspot white-light structure change (Wang *et al.* 2004b; Wang *et al.* 2005; Liu *et al.* 2005; Deng *et al.* 2005; Chen *et al.* 2007; Jing *et al.* 2008; Li *et al.* 2010) associated with flares. Liu *et al.* (2005) proposed a reconnection picture where the active region field collapses inward after flares as signified by HFW08.

Wang & Liu (2010) recently examined the observations in a systematic fashion and compare them quantitatively with the prediction of HFW08. Meanwhile, several conflicting concepts in our earlier papers can be reconciled with new physical understanding.

2. Observations and Results

The most straightforward way to determine changes of vector fields associated with flares is to monitor the time sequence of vector magnetograms. In order to detect any rapid and subtle variation, however, magnetogram observations with high cadence (a few minutes), high resolution ($1''$), and high polarization accuracy are required, and it is understood that *HMI/SDO* will be able to provide unprecedented data with these characteristics. Yet in the past two decades there are some vector magnetograms available that can tackle this topic with certain limitations. All the results obtained thus far point to the conclusion that transverse magnetic field strength increases at the PIL after major flares (namely, the fields there turn more horizontal), which provides direct and strong observational support for the theory of HFW08. Following HFW08, we further quantify the change of Lorentz force per unit area in the vertical direction using the formula:

$$\delta f_z = (B_z \delta B_z - B_x \delta B_x - B_y \delta B_y)/4\pi \ , \tag{2.1}$$

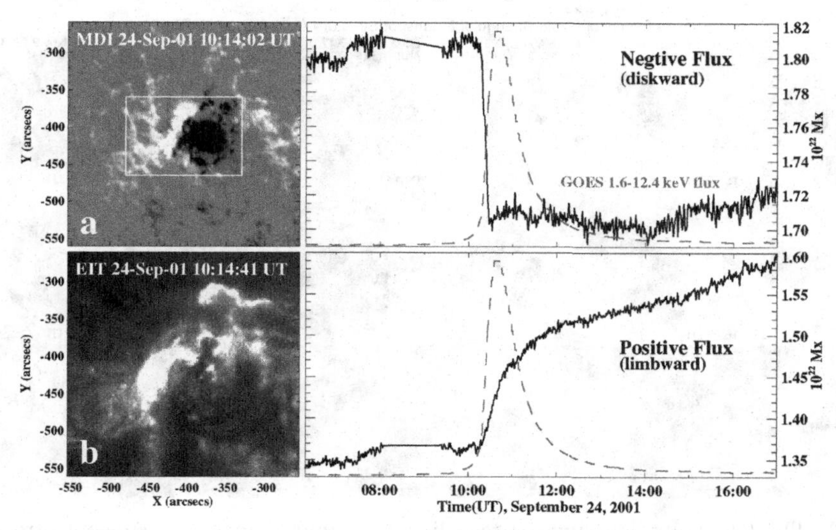

Figure 2. Time profiles of negative and positive MDI line-of-sight magnetic fields within a boxed region (in *a*) covering the entire δ spot for the 2001 September 24 X2.6 flare, seen in an EIT image (*b*). The flare started at 09:32 UT, peaked at 10:38 UT, and ended at 11:09 UT.

and present an example of our re-analysis of the vector field observations associated with the 2002 July 26 M8.7 flare in Figure 1. For the compact region at the flaring PIL, the results unambiguously show the following. First, the mean transverse field strength increases 90 G in about one hour ensuing from the rapid rising of the flare soft X-ray emission at 20:51 UT. Second, the inclination angle decreases 3° accordingly, which indicates that magnetic field lines there turn to a more horizontal direction as predicted by HFW08. Third, the change of the Lorentz force per unit area is ~ -5000 G^2, integrating which over the analyzed area of $\sim 3.2 \times 10^{18}$ cm^2 yields a downward net Lorentz force in the order of 1.6×10^{22} dynes, comparable to what is expected by HFW08. Similar results have been found for other flares as listed in Table 1 of Wang & Liu (2010).

Considering that vector magnetograms covering major flares with sufficient cadence and quality are rare and that reliable detection of rapid changes in the observed magnetic signals would ideally require stable observing conditions, we use the line-of-sight magnetograms ($\sim 2''$ pixel^{-1}) measured with *SOHO*/MDI to study more events, taking advantage of its long-sequence and seeing-free data set. We surveyed all the X-class flares satisfying our event selection criteria, and analyzed 18 events. Except for one case, all events exhibit an increase of limbward flux and a decrease of diskward flux of active region magnetic fields after flares with an order of magnitude of 10^{20} Mx (no obvious changes can be detected for one of the two polarities in some flares). Figure 2 shows the field changes associated with the 2001 September 24 X2.6 flare at NOAA AR 09632, which are calculated for the entire δ spot region. It can be clearly seen that the limbward/diskward (positive/negative) fluxes increase/decrease for a similar amount right after the flare impulsive phase. The change-over time for positive and negative fluxes are 40 and 10 minutes, respectively. As a matter of fact, these behaviors of line-of-sight fields also imply that the active region magnetic fields become more horizontal after flares, which was indicated by Wang (2006) based on limited sample events.

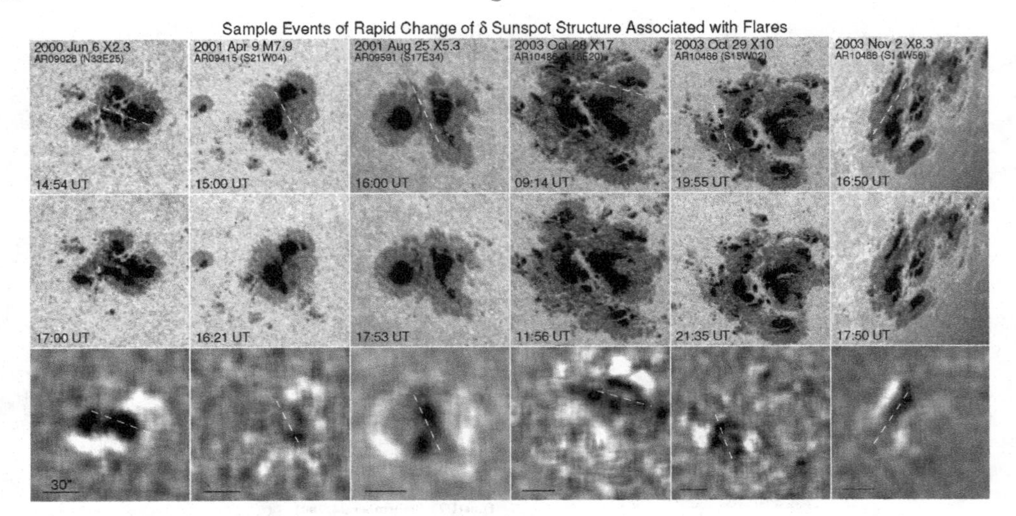

Figure 3. TRACE white-light images revealing the rapid change of δ sunspot structure associated with six major flares. The top, middle, and bottom rows show the pre-flare images, post-flare images, and the difference images between them, respectively. The white feature in the difference image indicates the region of penumbral decay, while the dark feature indicates the region of central umbral/penumbral enhancement. For each event, the white dashed line denotes the approximate flaring PIL and the black line represents a scale of $30''$ (adapted from Liu *et al.* 2005).

From more recent investigation, we began to appreciate the consistent pattern of magnetic field changes associated with flares using simple white-light observations (Wang *et al.* 2004b; Liu *et al.* 2005; Deng *et al.* 2005). The most outstanding changes are the decay of penumbral structure in the outer sides of the δ spot and the enhancement of sunspot structure near the flaring PILs. Figure 3 demonstrates examples of sunspot structure change. The difference between post- and pre-flare images always shows a dark patch at the flaring magnetic polarity inversion line surrounded by a bright ring, corresponding to the enhancement of center sunspot structure and decay of peripheral penumbra. These examples were discussed in detail by Liu *et al.* (2005), where we showed that these rapid changes were associated with the flares and were irreversible.

It is recently found that the sunspot penumbra has two components forming an "uncombed" structure, i.e., a more inclined dark component added to the bright grains that have a more vertical field structure (e.g., Langhans *et al.*, 2005). Therefore, it is natural to ask how do the two components evolve in the above-mentioned penumbral decay. We investigated the X6.5 flare on 2006 December 6 and show the evolution of the G-band images in Figure 4. The penumbral decay is obvious in the region marked by a white eclipse. Examining the time-lapse movie, it is evident that the penumbral decay may make the dark fibrils in the penumbrae disappear, while the bright grains evolve to the G-band bright points corresponding to vertical magnetic flux tubes.

3. Summary and Discussion

Synthesizing the research of flare-related rapid and irreversible changes in both vector and line-of-sight magnetic fields, we have revealed a pattern that magnetic fields near the PIL become more horizontal, while peripheral penumbral fields become more vertical.

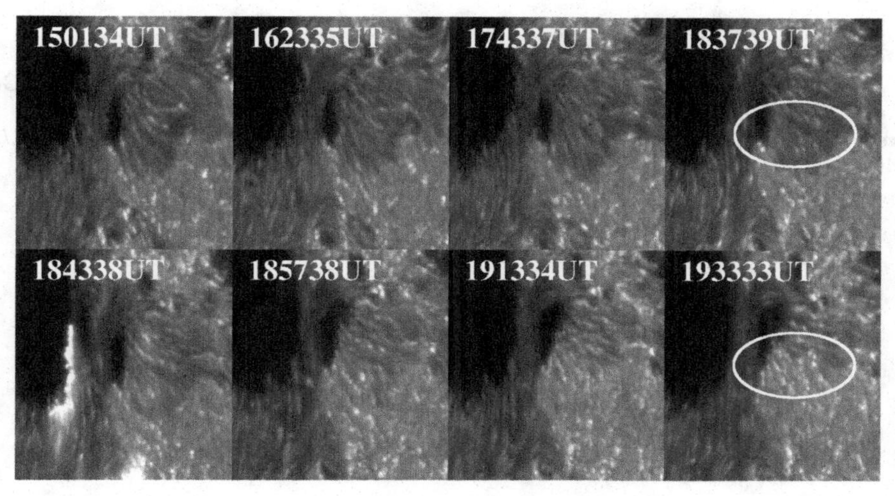

Figure 4. Time sequence of G-band images as observed by Hinode/SOT covering the X6.5 flare on 2006 December 6. The white eclipse marks the most prominent region of penumbral decay. The while-light flare is evident in the frame at 18:43:38 UT. The field of view is 43″ × 48″.

The change-over time lies between ~10 minutes to 1 hour, and all the changes are co-temporal with the flare initiation. We discussed the results in the context of HFW08. However, as the result of the discussion in this Symposium, the fields near the PIL to become compressed after flare/CME may be due to relaxation of the magnetic stresses (Klimchuk, 1990). The Figure 2 in that paper may also explain the observed penumbral decay.

References

Antiochos, S. K., DeVore, C. R., & Klimchuk, J. A., 1999, *ApJ*, 510, 485
Chen, J., Wang, H., Zirin, H., & Ai, G., 1994, *Solar Phys.*, 154, 261
Chen, W., Liu, C., & Wang, H., 2007, *ChJAA*, 7, 733
Deng, N., Liu, C., Yang, G., & Wang, H., Denker C., 2005, *ApJ*, 623, 1195
Fisher, G. H., Bercik, D. J., Welsch, B. T., & Hudson, H. S., 2010, arXiv:1006.5247
Hagyard *et al.*, 1999, *Solar Phys.*, 184, 133
Hudson, H. S., Fisher, G. H., & Welsch, B. T., 2008, *ASP Conference Series*, 383, 221
Jing, J., Wiegelmann, T., Suematsu, Y., Kubo, M., & Wang, H., 2008, *ApJL*, 676, L81
Klimchuck, J. A., 1990, *ApJ*, 354, 745
Kopp, R. A. & Pneuman, G. W., 1976, *Solar Phys.*, 50, 85
Kosovichev, A. G. & Zharkova, V. V. 2001, *ApJL*, 550, L105
Langhans, K., Scharmer, G. B., Kiseman, D., Lofdahl, M. G., & Berger, T. E., 2005, *A & A*, 436, 1087
Li, Y., Jing, J., Fan, Y., & Wang, H., 2010, this proceeding
Liu, C., Deng, N., Liu, Y., Falconer, D., Goode, P. R., Denker, C., & Wang, H., 2005, *ApJ*, 622, 722
Spirock, T. J., Yurchyshyn, V. B., & Wang, H., 2002, *ApJ*, 572, 1072
Sudol, J. J. & Harvey, J. W., 2005, *ApJ*, 635, 647
Wang, H., 1992, *Solar Phys.*, 140, 85
Wang, H., 2006, *ApJ*, 649, 490
Wang, H. & Liu, C., 2010, *ApJL*, 716, L195
Wang, H., Ewell, M. W., Zirin, H., & Ai, G. 1994, *ApJ*, 424, 436
Wang, H., Liu, C., Qiu, J., Deng, N., Goode, P. R., & Denker, C., 2004b, *ApJL*, 601, L195

Wang, H., Liu, C., Zhang, H., & Deng, Y, 2005, *ApJ*, 627, 1031
Wang, H., Qiu, J., Jing, J., Spirock, T. J., & Yurchyshyn, V., 2004a, *ApJ*, 605, 931
Wang, H., Spirock, T. J., Qiu, J., Ji, H., Yurchyshyn, V., Moon, Y. J., Denker, C., & Goode, P. R., 2002, *ApJ*, 576, 497
Yurchyshyn, V. B., Wang, H., Abramenko, V., Spirock, T. J., & Krucker, S., 2004, *ApJ*, 605, 546

The Physics of the Sun and Star Spots
Proceedings IAU Symposium No. 273, 2010
D.P. Choudhary and K. Strassmeier

doi:10.1017/S1743921311014955

Helicity of the solar magnetic field

Sanjiv Kumar Tiwari
Udaipur Solar Observatory, Physical Research Laboratory, Dewali, Bari Road,
Udaipur - 313 001, India
email: stiwari@prl.res.in

Abstract. Helicity measures complexity in the field. Magnetic helicity is given by a volume integral over the scalar product of magnetic field **B** and its vector potential **A**. A direct computation of magnetic helicity in the solar atmosphere is not possible due to unavailability of the observations at different heights and also due to non-uniqueness of **A**. The force-free parameter α has been used as a proxy of magnetic helicity for a long time. We have clarified the physical meaning of α and its relationship with the magnetic helicity. We have studied the effect of polarimetric noise on estimation of various magnetic parameters. Fine structures of sunspots in terms of vertical current (J_z) and α have been examined. We have introduced the concept of signed shear angle (SSA) for sunspots and established its importance for non force-free fields. We find that there is no net current in sunspots even in presence of a significant twist, showing consistency with their fibril-bundle nature. The finding of existence of a lower limit of SASSA for a given class of X-ray flare will be very useful for space weather forecasting. A good correlation is found between the sign of helicity in the sunspots and the chirality of the associated chromospheric and coronal features. We find that a large number of sunspots observed in the declining phase of solar cycle 23 do not follow the hemispheric helicity rule whereas most of the sunspots observed in the beginning of new solar cycle 24 do follow. This indicates a long term behaviour of the hemispheric helicity patterns in the Sun. The above sums up my PhD thesis.

Keywords. Sun: atmosphere, Sun: magnetic fields, Sun: sunspots

---

1. Introduction

Magnetic helicity is a physical quantity that measures the degree of linkage and twistedness in the magnetic field lines (Moffatt 1978). It is given as

$$H_m = \int \mathbf{A} \cdot \mathbf{B} \, dV \tag{1.1}$$

The term magnetic helicity was introduced by Elsasser (1956) and many of its important characteristics were studied by Woltjer (1958); Taylor (1974); Berger & Field (1984) etc.

The handedness associated with the field is defined by 'chirality'. Helicity is closely related to chirality. If the twist on the surface is clockwise, the chirality is negative and the field bears dextral chirality. The sunspot twist direction is decided by the curvature of sunspot whirls (Martin 1998; Tiwari 2009). If the twist is counterclockwise (when we go from sunspot center towards outside), the chirality is sinistral and sign of helicity is positive. Reverse is true for the dextral chirality. These definitions of chirality have been used to study the hemispheric patterns of the active regions (Tiwari *et al.* 2008, 2010b).

One of the main motivations of the thesis (Tiwari 2009) was to use the helicity or related parameters to help in predicting the severity of the solar flares. If done so, this would contribute in improving the space-weather forecasting. We have found the parameter signed shear angle (SSA) to be very useful in this context (Tiwari *et al.* 2010a).

Some of the important results of my Ph.D. thesis are summarized very briefly in the following sections.

2. Estimating magnetic parameters

Physical meaning of α. We arrive at the following depiction of α (for details, please see Appendix A of Tiwari *et al.* (2009a)):

$$\alpha = 2\,\frac{d\phi}{dz} \tag{2.1}$$

From Equation 2.1, it is clear that the α gives twice the degree of twist per unit axial length. If we take one complete rotation of flux tube i.e., $\phi = 2\pi$, and loop length $\lambda \approx 10^9$ meters, then

$$\alpha = \frac{2 \times 2\pi}{\lambda} \tag{2.2}$$

comes out to be of the order of 10^{-8} per meter.

Correlation between sign of H_m *and that of* α. Vector potential in terms of scalar potential ϕ can be expressed as (for details, please see Appendix B of Tiwari *et al.* (2009a))

$$\mathbf{A} = \mathbf{B}\alpha^{-1} + \nabla\phi \tag{2.3}$$

which is valid only for constant α. Using this relation in Equation 1, we get magnetic helicity as

$$\begin{aligned} H_m &= \int (\mathbf{B}\alpha^{-1} + \nabla\phi)\cdot\mathbf{B}\; dV \\ &= \int B^2\alpha^{-1}dV + \int (\mathbf{B}\cdot\nabla)\phi\; dV \sim (\int(\phi\;\mathbf{B})\cdot\mathbf{n}\; dS) \end{aligned} \tag{2.4}$$

showing that the force free parameter α has the same sign as that of the magnetic helicity iff $\mathbf{n}\cdot\mathbf{B} = \mathbf{0}$ i.e., no field lines cross the boundary, which is not the case with the Sun.

A direct method for calculating global α. We prefer to use the second moment of minimization (Tiwari *et al.* 2009a) leading to the following expression:

$$\alpha_g = \frac{\sum(\frac{\partial B_y}{\partial x} - \frac{\partial B_x}{\partial y})B_z}{\sum B_z^2}. \tag{2.5}$$

This formula gives a single global value of α in a sunspot and is the similar to $\alpha_{av}^{(2)}$ of Hagino & Sakurai (2004). We do not use direct mean (0th order moment) as it leads to singularities at neutral lines where $B_z \sim 0$. First order moment will also lead to singularities when flux is balanced.

Estimating the effect of polarimetric noise in the measurement of field parameters. Using the analytical bipole method (Low 1982), non-potential force-free field components B_x, B_y & B_z in a plane have been generated. We calculate the synthetic Stokes profiles for each B, γ and ξ in a grid of 100 x 100 pixels, using the He-Line Information Extractor "HELIX" code (Lagg *et al.* 2004). We add random noise of 0.5 % of the continuum intensity I_c (Ichimoto *et al.* 2008) to the polarimetric profiles as observed in SOT/SP aboard Hinode. In addition, we also study the effect of adding a noise of 2.0% level to Stokes profiles as a worst case scenario. We add 100 realizations of the noise of the orders mentioned above to each pixel and invert the corresponding 100 noisy profiles using the "HELIX" code. The effect of polarimetric noise in the derivation of vector fields and other parameters such as α_g and magnetic energy is found to be very small. We have done similar investigations as a second step using real data (Gosain *et al.* 2010). As a third step we plan to use MHD simulated data to check the inversion codes including the effect of optical depth corrugation.

3. Global twist of sunspot magnetic fields

Introduction of signed shear angle (SSA). To emphasize the sign of shear angle we introduce the signed shear angle (SSA) for the sunspots as follows: choose an initial reference azimuth for a current-free field (obtained from the observed line of sight field). Then move to the observed field azimuth from the reference azimuth through an acute angle. If this rotation is counter-clockwise, then assign a positive sign for the SSA. A negative sign is given for clockwise rotation. This sign convention will be consistent with the sense of azimuthal field produced by a vertical current. This sign convention is also consistent with the sense of chirality (Tiwari *et al.* 2009b). The SSA is computed from the following formula (Tiwari *et al.* 2010a):

$$SSA = \tan^{-1}\left(\frac{B_{yo}B_{xp} - B_{yp}B_{xo}}{B_{xo}B_{xp} + B_{yo}B_{yp}}\right) \tag{3.1}$$

where B_{xo}, B_{yo} and B_{xp}, B_{yp} are observed and potential transverse components of sunspot magnetic fields respectively. A spatial average of the SSA (SASSA) gives the global twist of sunspot magnetic fields at observed height irrespective of the force-free nature of the field and shape of sunspots (Venkatakrishnan & Tiwari 2009).

Fine structures in terms of J_z and α. Local J_z and α patches of opposite signs are present in the umbra of each sunspot. The amplitude of the spatial variation of local α in the umbra is typically of the order of the global α of the sunspot. We find that the local α is distributed as alternately positive and negative filaments in the penumbra. The amplitude of azimuthal variation of the local α in the penumbra is approximately an order of magnitude larger than that in the umbra. The contributions of the local positive and negative currents and α in the penumbra cancel each other giving almost no contribution for their global values for whole sunspot. The data sets used in the analysis are taken from ASP/DLSP and Hinode (SOT/SP). See for details: Tiwari *et al.* (2009b). Most of the data sets we studied are observed during the declining minimum phase of solar cycle 23. All except 5, out of 43 sunspots observed, follow the reverse twist hemispheric rule, while 5 follow the conventional helicity rule. Also, α_g has same sign as the SASSA and therefore the same sign of the photospheric chirality of the sunspots, but the magnitudes of SASSA and α_g are not well correlated. This lack of correlation could be due to a variety of reasons: (a) departure from the force-free nature (b) even for the force-free fields, α is the gradient of twist variation whereas SASSA is purely an angle. The missing link is the scale length of variation of twist.

4. Net current in sunspots

Expression for net current. We consider a long straight flux bundle surrounded by a region of "field free" plasma following Parker (1996). Parker (1996) assumed azimuthal symmetry as well as zero radial component B_r, of the magnetic field. For realistic sunspot fields, we have already seen the ubiquitous fine structure of the radial magnetic field. Hence, we need to relax both these assumptions.

The vertical component of the electric current density consists of two terms, viz. $-\frac{1}{\mu_0 r}\frac{\partial B_r}{\partial \psi}$ and $\frac{1}{\mu_0 r}\frac{\partial (rB_\psi)}{\partial r}$. We will call the first term as the "pleat current density",

j_p and the second term as the "twist current density", j_t. The total current I_z within a distance ϖ from the center is then given by

$$I_z(\varpi) = \int_0^{2\pi} d\psi \int_0^{\varpi} rdr(j_p + j_t) \tag{4.1}$$

The ψ integral over j_p vanishes, while the second term yields

$$I_z(\varpi) = \frac{\varpi}{\mu_0} \int_0^{2\pi} d\psi B_\psi(\varpi, \psi) \tag{4.2}$$

which gives the net currents within a circular region of radius ϖ. The transverse vector can be expressed in cylindrical geometry as

$$B_r = \frac{1}{r}(xB_x + yB_y) \tag{4.3}$$

$$B_\psi = \frac{1}{r}(-yB_x + xB_y) \tag{4.4}$$

The azimuthal field B_ψ is then used in Equation 4.2 for obtaining the value for the total vertical current within a radius ϖ.

No net current: an evidence for fibril bundle nature of sunspot magnetic field? As expected from the trend in Figure 3 of Venkatakrishnan & Tiwari (2009), the net current shows evidence for a rapid decline after reaching a maximum. Similar trends were seen in other sunspots. This can be interpreted as evidence for the neutralization of the net current. Table 1 of Venkatakrishnan & Tiwari (2009) shows the summary of results for all the sunspots analyzed. Along with the power law index δ of B_ψ decrease, we have also shown the average deviation of the azimuth from the radial direction ("twist angle $= tan^{-1}(B_\psi/B_r)$"), as well as the SASSA. The average deviation of the azimuth is well correlated with the SASSA for nearly circular sunspots, but is not correlated with SASSA for more irregularly shaped sunspots. Thus, SASSA is a more general measure of the global twist of sunspots, irrespective of their shape.

As is well known for astrophysical plasmas, that the plasma distorts the magnetic field and the curl of this distorted field produces a current by Ampere's law (Parker 1979). Parker's (1996) expectation of net zero current in a sunspot was basically motivated by the concept of a fibril structure for the sunspot field. However, he also did not rule out the possibility of vanishing net current for a monolithic field where the azimuthal component of the vector field in a cylindrical geometry declines faster than $1/\varpi$. While it is difficult to detect fibrils using the Zeeman effect, notwithstanding the superior resolution of SOT on *Hinode*, the stability and accuracy of the measurements have allowed us to detect the faster than $1/\varpi$ decline of the azimuthal component of the magnetic field, which in turn can be construed as evidence for the confinement of the sunspot field by the external plasma. The resulting pattern of curl **B** appears as a sharp decline in the net current at the sunspot boundary. Although the existence of a global twist in the absence of a net current is possible for a monolithic sunspot field (Baty 2000), a fibril model of the sunspot field can accommodate a global twist even without a net current (Parker 1996). A sunspot, made up of a bundle of magnetically isolated current free fibrils, can be given an overall torsion without inducing a global current. For details and more discussions please see Venkatakrishnan & Tiwari (2009); Tiwari (2009, 2010).

5. Relationship between the SASSA of active regions and associated GOES X-ray flux

We find an upper limit of peak X-ray flux for a given value of SASSA can be given for different classes of X-ray flares. Figures 5(a) and 5(b) of Tiwari *et al.* (2010a) represent scatter plots between the peak GOES X-ray flux and interpolated SASSA and mean weighted shear angle (MWSA: Wang (1992)) values for that time, respectively. The cubic spline interpolation of the sample of the SASSA and the MWSA values has been done to get the SASSA and MWSA exactly at the time of peak flux of the X-ray flare. For details kindly see Tiwari *et al.* (2010a). We find that the SASSA, apart from its helicity sign related studies, can also be used to predict the severity of the solar flares. However to establish these lower limits of SASSA for different classes of X-ray flares, we need more cases to study. The SASSA already gives a good indication of its utility from the present four case studies using 115 vector magnetograms from Hinode (SOT/SP). Once the vector magnetograms are routinely available with higher cadence, the lower limit of SASSA for each class of X-ray flare can be established by calculating the SASSA in a series of vector magnetograms. This will provide the inputs to space weather models. Also, SASSA has shown a good correlation with the free magnetic energy computed by Jing *et al.* (2010).

The other non-potentiality parameter MWSA studied in Tiwari *et al.* (2010a) does show a similar trend as that of the SASSA. The magnitudes of MWSA, however, do not show consistent threshold values as related with the peak GOES X-ray flux of different classes of solar flares. One possible reason for this behavior may be explained as follows: The MWSA weights the strong transverse fields e.g., penumbral fields. From the recent studies (Su *et al.* 2009; Tiwari *et al.* 2009b; Tiwari 2009; Venkatakrishnan & Tiwari 2009, 2010) it is clear that the penumbral field contains complicated structures with opposite signs of vertical current and vertical component of the magnetic tension forces. Although the amplitudes of the magnetic parameters are found high in the penumbra, they do not contribute to their global values because they contain opposite signs, which cancel out in the averaging process (Tiwari 2009; Tiwari *et al.* 2009b). On the other hand, the MWSA adds those high values of shear and produces a pedestal that might mask any relation between the more relevant global non-potentiality and the peak X-ray flux. Whereas the SASSA perhaps gives more relevant value of the shear after cancelation of the penumbral contribution.

6. Solar cycle dependence

Helicity hemispheric rule. We compare the behaviour of magnetic helicity sign of AR's observed in the beginning of 24th solar cycle with some AR's observed in the declining phase of 23rd solar cycle. We find that the majority of active regions in the beginning of solar cycle 24 do follow the hemispheric helicity rule whereas those observed in the declining phase of solar cycle 23 do not (Tiwari, 2009; Tiwari, 2010).

Sign of magnetic helicity at different heights in the solar atmosphere. A good correlation has been found among the sign of helicity in the associated features observed at photospheric, chromospheric and coronal heights without solar cycle dependence (Tiwari *et al.* 2008; Tiwari, 2009; Tiwari *et al.* 2010b; Tiwari, 2010).

7. Conclusions

The magnetic field parameters can be derived very accurately using the recent data available (e.g. from *Hinode* (SOT/SP)) and advanced inversion codes. The SASSA is the best measure of the global magnetic twist of sunspot magnetic fields at observed height, irrespective of the force-free nature and the shape of sunspots. The sunspots with significant twist and no net currents show consistency with the fibril bundle nature of the sunspots. The study of evolution of SASSA of sunspots showed threshold values for different classes of X-ray flares. This is an important discovery which was being sought after for many decades. The magnetic helicity sign of active regions studied, has good correlation with the sign of chirality of associated features observed at chromospheric and coronal heights. The majority of sunspots observed in the declining phase of solar cycle 23 follow a reverse hemispheric helicity rule, whereas most of the AR's emerged in the beginning of solar cycle 24 follow the conventional helicity rule. This result indicates that revisiting the hemispheric helicity rule using data sets of several years is required.

Acknowledgements

The presentation of this paper in the IAU Symposium 273 was possible due to partial support from the National Science Foundation grant numbers ATM 0548260, AST 0968672 and NASA - Living With a Star grant number 09-LWSTRT09-0039. I am grateful to Professor P. Venkatakrishnan for his invaluable and patient guidance during my PhD thesis. Hinode is a Japanese mission developed and launched by ISAS/JAXA, with NAOJ as domestic partner and NASA and STFC (UK) as international partners. It is operated by these agencies in co-operation with ESA and NSC (Norway).

References

Baty, H. 2000, *A&A*, 360, 345
Berger, M. A. & Field, G. B. 1984, *Journal of Fluid Mechanics*, 147, 133
Elsasser, W. M. 1956, *Rev. Mod. Phys.*, 28, 135
Fan, Y. & Gibson, S. E. 2004, *ApJ*, 609, 1123
Gosain, S., Tiwari, S. K., & Venkatakrishnan, P. 2010, *ApJ*, 720, 1281
Hagino, M. & Sakurai, T. 2004, *PASJ*, 56, 831
Ichimoto, K., *et al.* 2008, *Solar Phys.*, 249, 233
Jing, J., Tan, C., Yuan, Y., Wang, B., Wiegelmann, T., Xu, Y., & Wang, H. 2010, *ApJ*, 713, 440
Lagg, A., Woch, J., Krupp, N., & Solanki, S. K. 2004, *A&A*, 414, 1109
Low, B. C. 1982, *Solar Phys.*, 77, 43
Martin, S. F. 1998, in Astronomical Society of the Pacific Conference Series, Vol. 150, IAU Colloq. 167: New Perspectives on Solar Prominences, ed. D. F. Webb, B. Schmieder, & D. M. Rust, 419
Moffatt, H. K. 1978, Magnetic field generation in electrically conducting fluids, ed. H. K. Moffatt
Parker, E. N. 1979, Cosmical magnetic fields: Their origin and their activity (Oxford, Clarendon Press; New York, Oxford University Press, 1979)
Parker, E. N. 1996, *ApJ*, 471, 485
Su, J. T., Sakurai, T., Suematsu, Y., Hagino, M., & Liu, Y. 2009, *Astrophys. J. Lett.*, 697, L103
Taylor, J. B. 1974, *Physical Review Letters*, 33, 1139
Tiwari, S. K. 2009, Ph.D. thesis, Udaipur Solar Observatory/Physical Research Laboratory, Mohanlal Sukhadia University, Udaipur
Tiwari, S. K. 2010, Helicity of the Solar Magnetic Field: An application to predicting the severity of solar flares (*Lambert Academic Publishing*, 2010, ISBN: 978-3-8383-9771-9)
Tiwari, S. K., Joshi, J., Gosain, S., & Venkatakrishnan, P. 2008, in Turbulence, Dynamos, Accretion Disks, Pulsars and Collective Plasma Processes, ed. S. S. Hasan, R. T. Gangadhara, & V. Krishan, *Astrophysics and Space Science Proceedings*, 329
Tiwari, S. K., Venkatakrishnan, P., & Gosain, S. 2010a, *ApJ*, 721, 622
Tiwari, S. K., Venkatakrishnan, P., Gosain, S., & Joshi, J. 2009a, *ApJ*, 700, 199
Tiwari, S. K., Venkatakrishnan, P., & Sankarasubramanian, K. 2009b, *Astrophys. J. Lett.*, 702, L133

Tiwari, S. K., Venkatakrishnan, P., & Sankarasubramanian, K. 2010b, in Magnetic Coupling between the Interior and Atmosphere of the Sun, ed. by S. S. Hasan and R. J. Rutten; *Astrophysics and Space Science Proceedings.* Published by Springer Berlin Heidelberg; ISBN 978-3-642-02858-8 (Print) 978-3-642-02859-5 (Online), pp. 443-447

Venkatakrishnan, P. & Tiwari, S. K. 2009, *Astrophys. J. Lett.*, 706, L114

Venkatakrishnan, P. & Tiwari, S. K. 2010, *A&A*, 516, L5

Wang, H. 1992, *Solar Phys.*, 140, 85

Woltjer, L. 1958, *ApJ*, 128, 384

The Physics of Sun and Star Spots
Proceedings IAU Symposium No. 273, 2010
D.P. Choudhary & K.G. Strassmeier, eds.

doi:10.1017/S1743921311014967

Origin of solar magnetism

Arnab Rai Choudhuri
Department of Physics, Indian Institute of Science, Bangalore-560012
email: arnab@physics.iisc.ernet.in

Abstract. The most promising model for explaining the origin of solar magnetism is the flux transport dynamo model, in which the toroidal field is produced by differential rotation in the tachocline, the poloidal field is produced by the Babcock–Leighton mechanism at the solar surface and the meridional circulation plays a crucial role. After discussing how this model explains the regular periodic features of the solar cycle, we come to the questions of what causes irregularities of solar cycles and whether we can predict future cycles. Only if the diffusivity within the convection zone is sufficiently high, the polar field at the sunspot minimum is correlated with strength of the next cycle. This is in conformity with the limited available observational data.

Keywords. Solar magnetism, dynamo, tachocline, convection zone

---

1. Introduction

The dynamo process taking place in the convection zone of the Sun is believed to be the source of solar magnetism. A solar dynamo model needs to explain two things. Firstly, it has to explain why the sunspot cycle is roughly periodic. Secondly, it has to explain what causes the variabilities of the cycles. Why are not all the cycles identical? We shall address both these issues in this review. However, we would like to emphasize that this is *not* meant to be a comprehensive review of the solar dynamo. Limitation of space forces us to concentrate only on certain very limited aspects of the solar dynamo problem. The readers are referred to the excellent reviews by Ossendrijver (2003) and Charbonneau (2005) for discussions of other aspects.

2. Some basic considerations

Polarities of sunspot pairs indicate a toroidal magnetic field underneath the Sun's surface. On the other hand, the magnetic field of the Earth is of poloidal nature. Parker (1955b) — in perhaps the most important paper on dynamo theory ever written — proposed that the sunspot cycle is produced by an oscillation between toroidal and poloidal components, just as we see an oscillation between kinetic and potential energies in a simple harmonic oscillator. This was a truly extraordinary suggestion because almost nothing was known about the Sun's poloidal field at that time. Babcock & Babcock (1955) were the first to detect the weak poloidal field having a strength of about 10 G near the Sun's poles. Only from mid-1970s we have systematic data of the Sun's polar fields. Fig. 1 shows the polar fields of the Sun plotted as a function of time, with the sunspot number plotted below. It is clear that the sunspot number, which is a proxy of the toroidal field, is maximum at a time when the polar field is nearly zero. On the other hand, the polar field is maximum when the sunspot number is nearly zero. This clearly shows an oscillation between the toroidal and poloidal components, as envisaged by Parker (1955b).

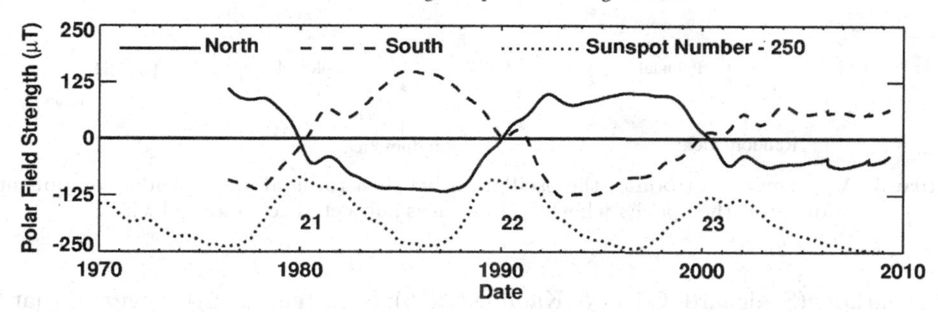

Figure 1. The polar field of the Sun as a function of time (on the basis of the Wilcox Solar Observatory data) with the sunspot number shown below. From Hathaway (2010).

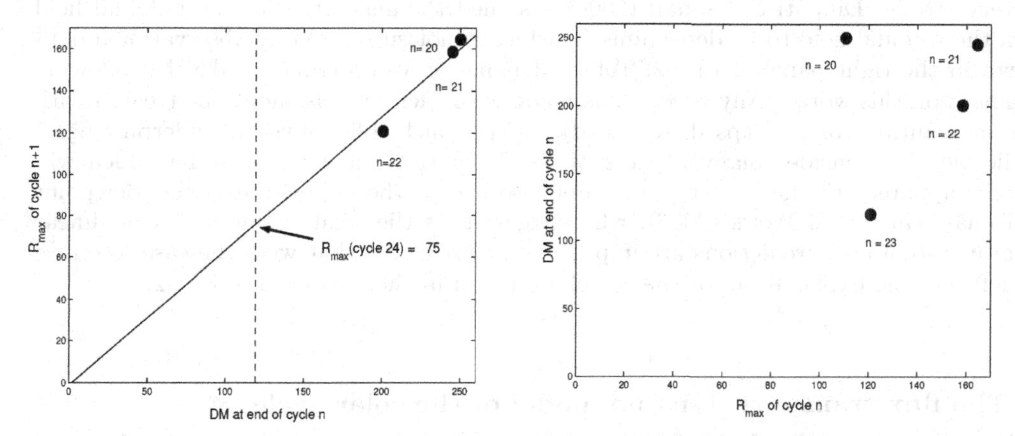

Figure 2. The left panel shows a plot of the strength of cycle $n + 1$ against the polar field at the end of cycle n. The right panel shows a plot of the polar field at the end of cycle n against the strength of the cycle n. From Choudhuri (2008).

We note another important thing in Fig. 1. The polar field at the end of cycle 22 was weaker than the polar field in the previous minimum. We see that this weaker polar field was followed by the cycle 23 which was weaker than the previous cycle. Does this mean that there is a correlation between the polar field during a sunspot minimum and the next sunspot cycle? In the left panel of Fig. 2, we plot the polar field in the minimum along the horizontal axis and the strength of the next cycle along the vertical axis. Although there are only 3 data points so far, they lie so close to a straight line that one is tempted to conclude that there is a real correlation. On the other hand, the right panel of Fig. 2, which has the cycle strength along the horizontal axis and the polar field at the end of that cycle along the vertical axis, has points which are scattered around. Choudhuri, Chatterjee & Jiang (2007) proposed the following to explain these observations. While an oscillation between toroidal and poloidal components takes place, the system gets random kicks at the epochs indicated in Fig. 3. Then the poloidal field and the next toroidal field should be correlated, as suggested by the left panel of Fig. 2. On the other hand, the random kick ensures that the toroidal field is not correlated with the poloidal field coming after it, as seen in the right panel of Fig. 2.

If there is really a correlation between the polar field at the minimum and the next cycle, then one can use the polar field to predict the strength of the next cycle (Schatten *et al.* 1978). Since the polar field in the just concluded minimum has been rather weak,

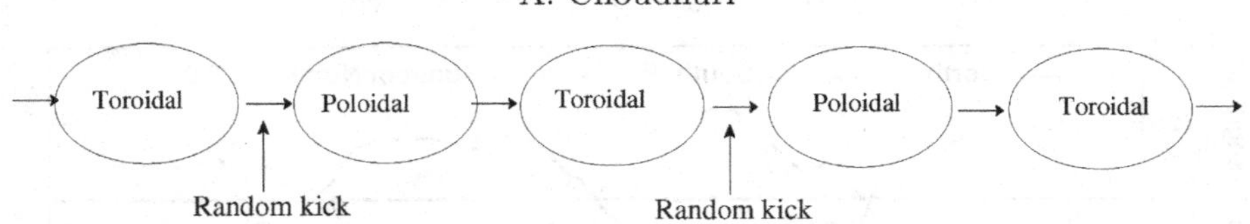

Figure 3. A schematic cartoon of the oscillation between toroidal and poloidal components, indicating the epochs when the system is subjected to random kicks.

several authors (Svalgaard, Cliver & Kamide (2005); Schatten (2005)) suggested that the coming cycle 24 will be rather weak. Very surprisingly, the first theoretical prediction based on a dynamo model made by Dikpati & Gilman (2006) is that the cycle 24 will be very strong. Dikpati & Gilman (2006) assumed the generation of the poloidal field from the toroidal field to be deterministic, which is not supported by observational data shown in the right panel of Fig. 2. Tobias, Hughes & Weiss (2006) make the following comment on this work: "Any predictions made with such models should be treated with extreme caution (or perhaps disregarded), as they lack solid physical underpinnings." While we also consider many aspects of the Dikpati–Gilman work wrong which will become apparent to the reader, we cannot also accept the opposite extreme viewpoint of Tobias, Hughes & Weiss (2006), who suggest that the solar dynamo is a nonlinear chaotic system and predictions are impossible or useless. If that were the case, then we are left with no explanation for the correlation seen in the left panel of Fig. 2.

3. The flux transport dynamo model of the solar cycle

We now come to the theoretical question as to what causes the oscillation between the toroidal and poloidal components. Then in §4 we shall discuss how the random kicks shown in Fig. 3 arise.

The differential rotation of the Sun stretches any poloidal field line to produce a toroidal field, primarily in the tachocline where the differential rotation is strongest. Then this toroidal field rises due to magnetic buoyancy and produces sunspots (Parker (1955a)). It is fairly easy to see how the toroidal field can be produced from the poloidal field. The production of the poloidal field from the toroidal field is more non-trivial. The original idea of Parker (1955b) and Steenbeck, Krause & Rädler (1966) was that the toroidal field is twisted by the helical turbulence in the convection zone giving rise to a field in the poloidal plane. This is, however, possible only if the toroidal field is not stronger than its equipartition value. Simulations based on the thin flux tube equation (Spruit (1981); Choudhuri (1990)) suggest that the toroidal field is likely to be as strong as 10^5 G at the bottom of the convection zone (Choudhuri (1989); D'Silva & Choudhuri (1993)) and cannot be twisted by helical turbulence. We therefore invoke the alternate idea proposed by Babcock (1961) and Leighton (1969). We know that bipolar sunspots on the solar surface have tilts with respect to latitude lines, the leading sunspot appearing nearer the equator. Suppose we consider a pair in which the positive polarity sunspot is nearer the equator. When this pair decays, more positive polarity is spread around in lower latitudes and more negative polarity in the higher latitudes. This causes a poloidal magnetic field. A sunspot pair which forms from the toroidal field, therefore, acts as a conduit through which a conversion from the toroidal field to the poloidal field takes place. In a nutshell, the differential rotation acting on the poloidal field produces the toroidal field. The toroidal field, in turn, gives rise to the poloidal field by the

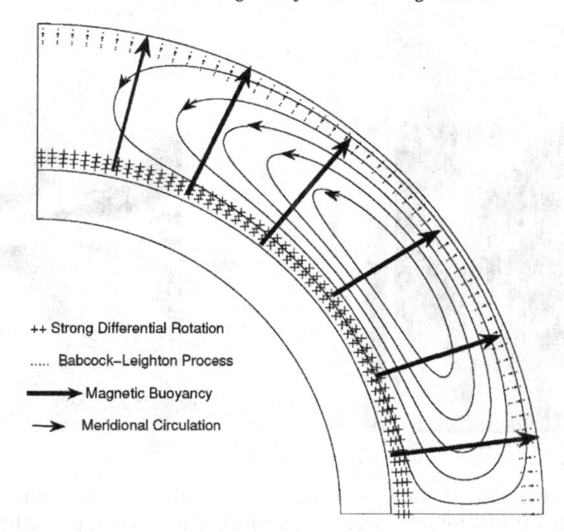

Figure 4. A cartoon explaining how the flux transport dynamo works.

Babcock–Leighton mechanism, thereby completing the cycle. However, a theoretical dynamo model with differential rotation and Babcock–Leighton mechanism alone gives rise to belts of toroidal field propagating poleward, in contradiction to the observation that the sunspots belts propagate equatorward. So we need something else. Choudhuri, Schüssler & Dikpati (1995), who pointed out this difficulty, also proposed a solution. They realized that the meridional circulation of the Sun can turn things around and can make the dynamo wave propagate in the correct direction.

Fig. 4 schematically shows the dynamo in which the meridional circulation plays a key role. Such a dynamo is often called a *flux transport dynamo*. First two-dimensional models of the flux transport dynamo were constructed by Choudhuri, Schüssler & Dikpati (1995) and Durney (1995), although the basic ideas were given in an early paper by Wang, Sheeley & Nash (1991). We now write down the basic equations of the flux transport dynamo. The axisymmetric magnetic field is represented in the form

$$\mathbf{B} = B(r,\theta,t)\mathbf{e}_\phi + \nabla \times [A(r,\theta,t)\mathbf{e}_\phi], \tag{1}$$

where $B(r,\theta,t)$ and $A(r,\theta,t)$ respectively correspond to the toroidal and poloidal components. They evolve according to the equations

$$\frac{\partial A}{\partial t} + \frac{1}{s}(\mathbf{v}.\nabla)(sA) = \eta_p \left(\nabla^2 - \frac{1}{s^2}\right) A + \alpha B, \tag{2}$$

$$\frac{\partial B}{\partial t} + \frac{1}{r}\left[\frac{\partial}{\partial r}(r v_r B) + \frac{\partial}{\partial \theta}(v_\theta B)\right] = \eta_t \left(\nabla^2 - \frac{1}{s^2}\right) B + s(\mathbf{B}_p.\nabla)\Omega + \frac{1}{r}\frac{d\eta_t}{dr}\frac{\partial}{\partial r}(rB), \tag{3}$$

where $s = r\sin\theta$ and the other symbols have the usual meanings. Since nothing much can be done analytically, these equations have to be solved numerically. We have developed a code *Surya*, which we make available to anybody upon request. The right panel of Fig. 5, taken from Chatterjee, Nandy & Choudhuri (2004), shows the theoretical butterfly diagram superposed on a time-latitude plot indicating how B_r evolves at the solar surface.

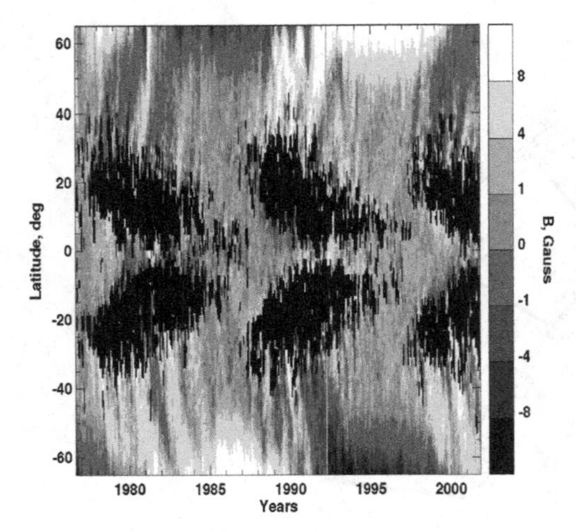

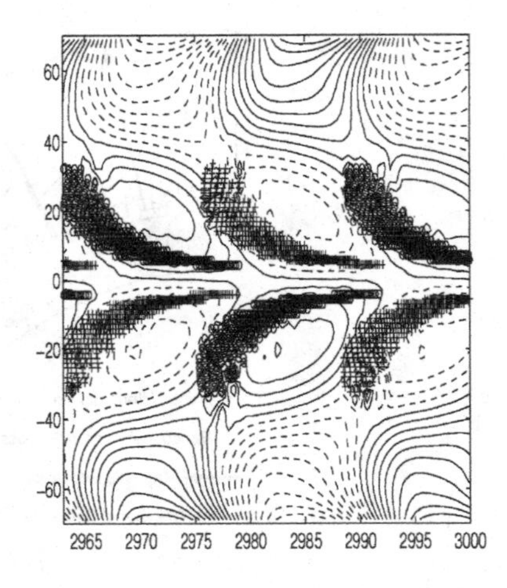

Figure 5. Butterfly diagram of sunspots superposed on the time-latitude plot of B_r. The observational plot is shown on the left. The comparable theoretical plot obtained by the dynamo model of Chatterjee, Nandy & Choudhuri (2004) is on the right.

The equatorward propagating meridional circulation at the bottom of the convection zone forces the sunspot belts to migrate equatorward, whereas the poleward propagating meridional circulation at the surface advects the poloidal field poleward. The right panel of Fig. 5 has to be compared with the comparable plot of observational data shown in the left panel. Given the fact that this was one of the first efforts of matching observational data in such detail, the fit is not too bad.

The original flux transport dynamo model of Choudhuri, Schüssler & Dikpati (1995) led to two offsprings: a high diffusivity model and a low diffusivity model. The diffusion times in these two models are of the order of 5 years and 200 years respectively. The high diffusivity model has been developed by a group working in IISc Bangalore (Choudhuri, Nandy, Chatterjee, Jiang, Karak), whereas the low diffusivity model has been developed by a group working in HAO Boulder (Dikpati, Charbonneau, Gilman, de Toma). The differences between these models have been systematically studied by Jiang, Chatterjee & Choudhuri (2007) and Yeates, Nandy & Mckay (2008). Both these models are capable of giving rise to oscillatory solutions resembling solar cycles. However, when we try to study the variabilities of the cycles, the two models give completely different results. We need to introduce fluctuations to cause variabilities in the cycles. In the high diffusivity model, fluctuations spread all over the convection zone in about 5 years. On the other hand, in the low diffusivity model, fluctuations essentially remain frozen during the cycle period. Thus the behaviours of the two models are totally different on introducing fluctuations.

4. Modelling irregularities of solar cycles

Let us now finally come to the question as to what produces the variabilities of cycles and whether we can predict the strength of a cycle before its advent. Some processes in nature can be predicted and some not. We can easily calculate the trajectory of a projectile by using elementary mechanics. On the other hand, when a dice is thrown, we cannot predict which side of the dice will face upward when it falls. Is the solar dynamo more like the trajectory of a projectile or more like the throw of a dice? Our point of view is that the solar dynamo is not a simple unified process, but a complex combination

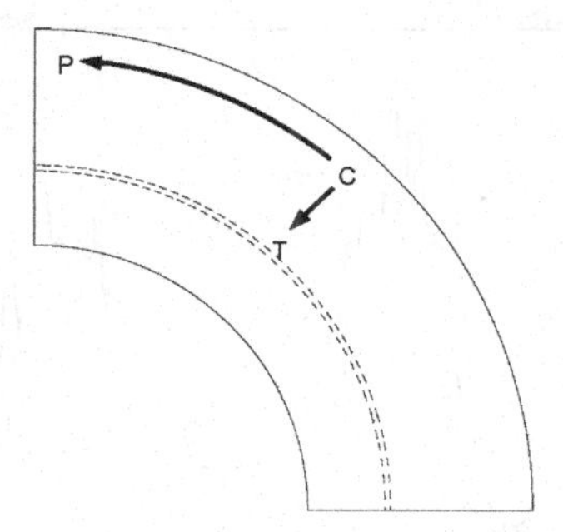

Figure 6. A sketch indicating how the poloidal field produced at C during a maximum gives rise to the polar field at P during the following minimum and the toroidal field at T during the next maximum. From Jiang, Chatterjee & Choudhuri (2007).

of several processes, some of which are predictable and others not. Let us look at the processes which make up the solar dynamo.

The flux transport dynamo model combines three basic processes. (i) The strong toroidal field is produced by the stretching of the poloidal field by differential rotation in the tachocline. (ii) The toroidal field generated in the tachocline gives rise to active regions due to magnetic buoyancy and then the decay of tilted bipolar active regions produces the poloidal field by the Babcock–Leighton mechanism. (iii) The poloidal field is advected by the meridional circulation first to high latitudes and then down to the tachocline, while diffusing as well. We believe that the processes (i) and (iii) are reasonably ordered and deterministic. In contrast, the process (ii) involves an element of randomness due to the following reason. The poloidal field produced from the decay of a tilted bipolar region by the Babcock–Leighton process depends on the tilt. While the average tilt of bipolar regions at a certain latitude is given by Joy's law, we observationally find quite a large scatter around this average. Presumably the action of the Coriolis force on the rising flux tubes gives rise to Joy's law (D'Silva & Choudhuri (1993)), whereas convective buffeting of the rising flux tubes in the upper layers of the convection zone causes the scatter of the tilt angles (Longcope & Choudhuri (2002)). This scatter in the tilt angles certainly introduces a randomness in generation process of the poloidal field from the toroidal field. Choudhuri, Chatterjee & Jiang (2007) identified it as the main source of irregularity in the dynamo process, which is in agreement with Fig. 3. It may be noted that Choudhuri (1992) was the first to suggest that the randomness in the poloidal field generation process is the source of fluctuations in the dynamo.

The poloidal field gets built up during the declining phase of the cycle and becomes concentrated near the poles during the minimum. The polar field at the solar minimum produced in a theoretical mean field dynamo model is some kind of 'average' polar field during a typical solar minimum. The observed polar field during a particular solar minimum may be stronger or weaker than this average field. The theoretical dynamo model has to be updated by feeding the information of the observed polar field in an appropriate way, in order to model particular cycles. Choudhuri, Chatterjee & Jiang (2007) proposed to model this in the following way. They ran the dynamo code from a minimum to the next minimum in the usual way. After stopping the code at the minimum, the poloidal field of the theoretical model was multiplied by a constant factor everywhere above $0.8R_\odot$ to bring it in agreement with the observed poloidal field. Since some of the poloidal field at the bottom of the convection zone may have been produced in the still earlier cycles, it is left unchanged by not doing any updating below $0.8R_\odot$. Only the poloidal field produced in the last cycle which is concentrated in the upper layers gets

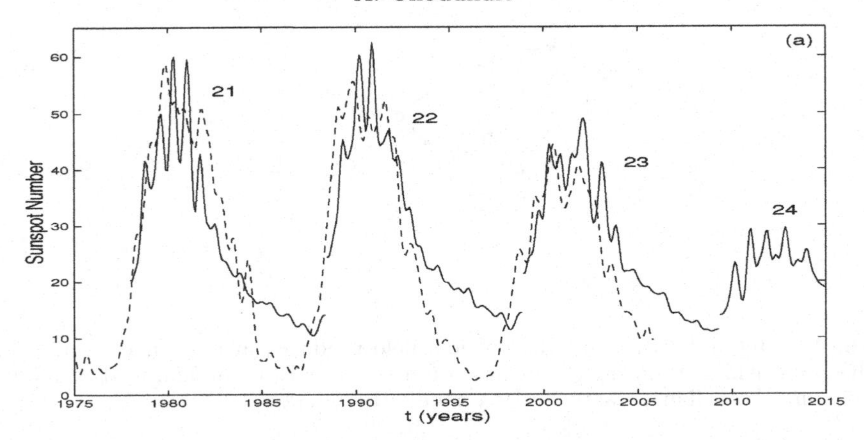

Figure 7. The theoretical monthly sunspot number (solid line) for the last few years as well as the upcoming next cycle, plotted along with the observational data (dashed line) for the last few years. From Choudhuri, Chatterjee and Jiang (2007).

updated. After this updating, we run the code till the next minimum, when the code is again stopped and the same procedure is repeated. Our solutions are now no longer self-generated solutions from a theoretical model alone, but are solutions in which the random aspect of the dynamo process has been corrected by feeding the observational data of polar fields into the theoretical model.

Before presenting the results obtained with this procedure, we come to the question how the correlation between the polar field at the minimum and the strength of the next cycle as seen in the left panel of Fig. 2 may arise. This was first explained by Jiang, Chatterjee & Choudhuri (2007). The Babcock–Leighton process would first produce the poloidal field around the region C in Fig. 6. Then this poloidal field will be advected to the polar region P by meridional circulation and will also diffuse to the tachocline T. In the high diffusivity model, this diffusion will take only about 5 years and the toroidal field of the next cycle will be produced from the poloidal field that has diffused to T. If the poloidal field produced at C is strong, then both the polar field at P at the end of the cycle and the toroidal field at T for the next cycle will be strong (and vice versa). We thus see that the polar field at the end of a cycle and the strength of the next cycle will be correlated in the high diffusivity model. But this will not happen in the low diffusivity model where it will take more than 100 years for the poloidal field to diffuse from C to T and the poloidal field will reach the tachocline only due to the advection by meridional circulation taking a time of about 20 years. If we believe that the 3 data points in the left panel of Fig. 2 indicate a real correlation, then we have to accept the high diffusivity model!

Finally the solid line in Fig. 7 shows the sunspot number calculated from our high diffusivity model (Choudhuri, Chatterjee & Jiang (2007)). Since systematic polar field measurements are available only from the mid-1970s, the procedure outlined above could be applied only from that time. It is seen from Fig. 7 that our model matches the last three cycles (dashed line) reasonably well and predicts a weak cycle 24.

5. Conclusion

It seems that the flux transport dynamo model explains many aspects of solar magnetism and probably will stand the test of time as the appropriate theoretical model for the solar cycle. In the recent past, two versions of the flux transport dynamo model have been developed in considerable detail: the high diffusivity version and the low diffusivity version. In the high diffusivity version, the poloidal field created at the surface by the Babcock–Leighton process is transported to the tachocline by diffusion. On the other hand, this transport is affected by the meridional circulation in the low diffusivity version of the flux transport dynamo. Several authors (Chatterjee, Nandy & Choudhuri (2004); Chatterjee & Choudhuri (2006); Jiang, Chatterjee & Choudhuri (2007); Goel & Choudhuri (2009); Hotta & Yokoyama (2010a, 2010b); Karak & Choudhuri (2011)) have given several independent arguments in support of the high diffusivity model. This model has also been applied to provide theoretical explanations of such things as the helicity of active regions (Choudhuri, Chatterjee & Nandy (2004)), the torsional oscillations (Chakraborty, Choudhuri & Chatterjee (2009)) and the Maunder minimum (Choudhuri & Karak (2009)). However, a crucial test of the high diffusivity model lies ahead. Since the high diffusivity model makes the polar field at a minimum correlated with the strength of the next cycle and the polar field during the immediate past minimum was weak, an inescapable conclusion of the high diffusivity model is that the upcoming cycle will be weak (Choudhuri, Chatterjee & Jiang (2007); Jiang, Chatterjee & Choudhuri (2007)). While the early indications are that this will indeed be the case, we need to wait for a couple of years to be absolutely sure. If cycle 24 turns out to be strong as predicted by Dikpati & Gilman (2006), then there will be no way of reconciling that fact with the high diffusivity model and the model will have to be discarded. On the other hand, if cycle 24 turns out to be weak, that will provide a further corroboration of the high diffusivity model.

References

Babcock, H. W. 1961, *ApJ*, 133, 572
Babcock, H. W. & Babcock, H. D. 1955, *ApJ*, 121, 349
Chakraborty, S., Choudhuri, A. R., & Chatterjee, P. 2009, *Phys. Rev. Lett.*, 102, 041102
Charbonneau, P. 2005 *Living Rev. Solar Phys.*, 2, 2
Chatterjee, P. & Choudhuri, A. R. 2006, *Solar Phys.*, 239, 29
Chatterjee, P., Nandy, D., & Choudhuri, A. R. 2004, *A&A*, 427, 1019
Choudhuri, A. R. 1989, *Solar Phys.*, 123, 217
Choudhuri, A. R. 1990, *A&A*, 239, 335
Choudhuri, A. R. 1992, *A&A*, 253, 277
Choudhuri, A. R. 2008, *J. Astrophys. Astron.*, 29, 41
Choudhuri, A. R., Chatterjee, P., & Jiang, J. 2007, *Phys. Rev. Lett.*, 98, 131103
Choudhuri, A. R., Chatterjee, P., & Nandy, D. 2004, *ApJ*, 615, L57
Choudhuri, A. R. & Karak, B. B. 2009, *RAA*, 9, 953
Choudhuri, A. R., Schüssler, M., & Dikpati, M. 1995, *A&A*, 303, L29
Dikpati, M. & Gilman, P. A. 2006, *ApJ*, 649, 498
D'Silva, S. & Choudhuri, A. R. 1993, *A&A*, 272, 621
Durney, B. R. 1995, *Solar Phys.*, 160, 213
Goel, A. & Choudhuri, A. R. 2009, *RAA*, 9, 115
Hathaway, D. H. 2010, *Living Rev. Solar Phys.*, 7, 1
Hotta, H. & Yokoyama, T. 2010a, *ApJ*, 709, 1009
Hotta, H. & Yokoyama, T. 2010b, *ApJ*, 714, L308
Jiang, J., Chatterjee, P., & Choudhuri, A. R. 2007, *MNRAS*, 381, 1527
Karak, B. B. & Choudhuri, A. R. 2011, *MNRAS*, 410, 1503
Leighton, R. B. 1969, *ApJ*, 156, 1
Longcope, D. W. & Choudhuri, A. R. 2002, *Solar Phys.*, 205, 63
Ossendrijver, M. A. J. H. 2003, *A&AR*, 11, 287
Parker, E. N. 1955a, *ApJ*, 121, 491

Parker, E. N. 1955b, *ApJ*, 122, 293
Schatten, K. 2005, *Geo. Res. Lett.*, 32, L21106
Schatten, K. H., Scherrer, P. H., Svalgaard, L., & Wilcox, J. M. 1978, *Geo. Res. Lett.*, 5, 411
Spruit, H. C. 1981, *A&A*, 98, 155
Steenbeck, M., Krause, F., & Rädler, K. H. 1966, *Z. Naturforsch.*, 21, 369
Svalgaard, L. & Cliver, E. W., Kamide Y. 2005, *Geo. Res. Lett.*, 32, L01104
Tobias, S., Hughes, D., & Weiss, N. 2006, *Nature*, 442, 26.
Wang, Y.-M., Sheeley, N. R., & Nash, A. G. 1991, *ApJ*, 383, 431
Yeates, A. R., Nandy, D., & Mackay, D. H. 2008, *ApJ*, 673, 544

The Physics of Sun and Star Spots
Proceedings IAU Symposium No. 273, 2010
D.P. Choudhary & K.G. Strassmeier, eds.

doi:10.1017/S1743921311014979

Diagnostics for spectropolarimetry and magnetography

Jose Carlos del Toro Iniesta[1] **and Valentín Martínez Pillet**[2]

[1]Instituto de Astrofísica de Andalucía (CSIC),
Apdo. de Correos 3004, E-18080, Granada, Spain
email: jti@iaa.es

[2]Instituto del Astrofísica de Canarias,
Vía Láctea, s/n, E-38200, La Laguna, Spain
email: vmp@iac.es

Abstract. An assessment on the capabilities of modern spectropolarimeters and magnetographs is in order since most of our astrophysical results rely upon the accuracy of the instrumentation and on the sensitivity of the observables to variations of the sought physical parameters. A contribution to such an assessment will be presented in this talk where emphasis will be made on the use of the so-called response functions to gauge the probing capabilities of spectral lines and on an analytical approach to estimate the uncertainties in the results in terms of instrumental effects. The Imaging Magnetograph eXperiment (IMaX) and the Polarimetric and Helioseismic Imager (PHI) will be used as study cases.

Keywords. Sun: magnetic fields, polarization, radiative transfer, instrumentation: polarimeters, instrumentation: spectrographs

1. Introduction

Modern solar spectropolarimeters and magnetographs are vectorial because all four Stokes parameters of the light spectrum are measured. Longitudinal magnetography (i.e., Stokes $I \pm V$) can be interesting for some specific applications, but the partial analysis is usually included (if possible) as a particular case of the more general, full-Stokes polarimetry. Some of these modern instruments have been recently or are currently in operation (e.g., the Tenerife Infrared Polarimeter, TIP, Martínez Pillet *et al.* 1999, Collados *et al.* (2007); the Diffraction-Limited Spectro-Polarimeter, DLSP, Sankarasubramanian *et al.* 2004; the spectropolarimeter, SP, Lites *et al.* 2001, for the *Hinode* mission, Kosugi *et al.* 2007; CRISP, Narayan *et al.* 2008; the Visible Imaging Polarimeter, VIP, Beck *et al.* 2010; the Imaging Magnetograph eXperiment, IMaX, Martínez Pillet *et al.* 2010, for the *Sunrise* mission, Bartol *et al.* 2010; and the Helioseismic and Magnetic Imager, HMI, Graham *et al.* 2003, for the *Solar Dynamics Observatory* mission, Title 2000), some other are being designed and built for near future operation and missions (e.g., the Polarimetric and Helioseismic Imager, SO/PHI, [formerly called VIM, Martínez Pillet, 2006] for the *Solar Orbiter* mission, Marsch *et al.* 2005). Assessing their capabilities in terms of their accuracy for retrieving the solar line-of-sight (LOS) velocity ($v_{\rm LOS}$) and vector magnetic field (of components B, γ, and ϕ) is in order since such an analysis can diagnose how far reaching is our current and near-future understanding of the solar atmosphere. The diagnostics is relevant both for the design of new instruments in order to maximize their performances and for the analysis of uncertainties in data coming from currently operating devices. Certainly, no fully general assessment can be devised that includes all possible polarimeters and a family of them should be considered in each specific study. Here we restrict our analysis to those spectropolarimeters and magnetographs whose polarization modulator consists of two nematic liquid crystal variable retarders (LCVRs). Hopefully, the discussion presented in this invited contribution helps further diagnostics of other instruments.

2. Rules for improving the measurements

Since we only measure photons, every inference we can make out of the observations naturally depends on photometric accuracy. Assuming that systematic errors are under control (ideally absent) two are, therefore, the pillars which the quality of measurements rests upon: the signal-to-noise ratio (S/N) and the minimum variations, δS_i, that the Stokes parameters exhibit after a perturbation in the solar physical quantities.† If the latter are larger than the uncertainties in the Stokes signals due to noise, the measurements are useful. Otherwise, they are not. One should, then, design new instruments so that S/N and δS_i are maximized and results from current instruments are more accurate wherever these quantities are larger.

2.1. *Increasing the signal-to-noise ratio*

As a first tool for improving S/N, modern polarimeters introduce image accumulation of N_a individual exposures. Besides, every Stokes parameter is obtained from N_p polarization modulation states, so that a total of $N_p N_a$ individual frames contribute to a given Stokes parameter image. If $\overline{\sigma}_i$ stands for the individual frame contribution to σ_i, the uncertainty in S_i then is

$$\sigma_i = \overline{\sigma}_i \sqrt{N_p N_a}, \tag{2.1}$$

where we have assumed photon noise.

According to Martínez Pillet *et al.* (1999) and to Del Toro Iniesta & Collados (2000),

$$\overline{\sigma}_i = \frac{\sigma}{\epsilon_i}, \tag{2.2}$$

where σ is the noise-induced uncertainty for each individual exposure and ϵ_i is the so-called polarimetric efficiency for Stokes S_i. Then, it is easy to see that, if s/n denotes the signal-to-noise ratio of each individual exposure,

$$(S/N)_i = (s/n)\, \epsilon_i \sqrt{N_p N_a}. \tag{2.3}$$

Equation (2.3) tells us that the larger the polarimetric efficiencies and/or the larger the number of individual exposures, the larger the signal-to-noise ratio for each Stokes parameter. N_p is often (advisably) kept to its minimum value of 4 in order to preserve integrity in the Stokes analysis in a minimum time. This can only be done, however, when the polarization modulator permits it as, indeed, in our LCVR-based polarimeters, but it seldom exceeds 6 or 8. The number of accumulations is usually traded-off with the solar dynamic time scales, in order not to blur information on time-evolving solar features with a too long effective exposure time. Then, optimization of polarimetric measurements basically lies in maximization of polarimetric efficiencies. According to Del Toro Iniesta & Collados (2000), an ideal polarimeter wishing to have equal signal-to-noise ratios for

† We shall hereafter denote by $\mathbf{S} = (S_1, S_2, S_3, S_4)$ the vector of Stokes parameters, usually called (I, Q, U, V).

Stokes S_2, S_3, and S_4 can reach maximum efficiencies given by

$$\epsilon_1 = 1, \ \epsilon_{2,3,4} = 1/\sqrt{3}. \tag{2.4}$$

Since we usually speak of only one ("the") signal-to-noise ratio of the observations, we are implicitly meaning $(S/N)_1$, that is, the signal-to-noise ratio for the intensity. Therefore, after Eq. (2.3), we can write

$$(S/N)_i = \frac{\epsilon_i}{\epsilon_1} (S/N), \tag{2.5}$$

so that, if we have for instance $S/N = 10^3$, then $(S/N)_{2,3,4} \leqslant 577$, according to Eqs. (2.4). It is important to point out that we are speaking about single wavelength samples. Every observational quantity involving several samples can certainly improve $(S/N)_{2,3,4}$ above this limit. Although simple, the result in Eq. (2.5) has not ever been brought to the attention of the community as far as we know, and is paramount to assessing observational accuracies: polarimetry imposes an extra penalty in terms of S/N as compared to normal spectroscopy or photometry. Such a penalty roots in the differential character of polarimetric measurements. A discussion on how optimum polarimetric efficiencies can be reached (at least theoretically) with LCVR-based polarimeters is deferred to Sec. 3.

2.2. *Maximizing the spectral line sensitivities*

As explained in the beginning of Sect. 2, the other ingredient for improving measurements quality is the spectral line sensitivity. Increasing δS_i can only be achieved by carefully selecting the line. Fortunately, the tools for such a selection are at our disposal as well. As explained by Ruiz Cobo & Del Toro Iniesta (1994; see references to pioneering work over there), the sensitivity of Stokes profiles to perturbations in the solar physical quantities are directly given by the response functions (RFs). We insist on the importance of perturbations: we can only discern different LOS velocities or field strengths in two structures provided the modification in the Stokes profiles are large enough (that is, larger than the threshold imposed by noise) in one of the structures as compared to the other. It is perturbation theory the technique that enables us to evaluate how large δS_i are for given variations in $v_{\rm LOS}$, B, γ, and/or ϕ. RFs are defined such that, for every single wavelength,

$$\delta S_{i_x} = \int_{-\infty}^{+\infty} R_{i_x} \, \delta x \, {\rm d}\tau, \tag{2.6}$$

where x is an index representing either one of the physical quantities of interest, τ stands for the optical depth, and R_{i_x} is the response function of S_i to perturbations in x. Naturally, and within a linear approximation, $\delta S_i = \sum \delta S_{i_x}$, the sum being extended to all the quantities. Equation (2.6) paves the way for estimating detection thresholds for the different physical quantities. For example, the detectable two-sigma field strength, $\delta B_{\rm min}$, would be such that $\delta S_{i_{B_{\rm min}}} = 2\sigma_i$.

Analytic expressions for RFs are available under the Milne-Eddington (ME) approximation (Orozco Suárez & Del Toro Iniesta, 2007) that are useful even for more accurate estimates of uncertainty levels (Del Toro Iniesta, Orozco Suárez, & Bellot Rubio, 2010) accounting for details of the specific technique used to retrieve given quantities. But purely phenomenological approaches are also valid to establish real rankings of spectral lines according to their ability for inferring velocities, magnetic fields, and so on (Cabrera Solana *et al.*, 2005). Finally, a further approach to determine which particular line is more useful for being used with a given instrument has recently been provided by simulations. MHD simulations are a modern and useful tool to elaborate educated guesses of instrument behavior since real observations can be computationally reproduced. This has been the way, for instance, how Orozco Suárez *et al.* (2010) have been able to gather evidence in favor of the Fe I line at 525.02 nm against that at 525.06 nm, which was originally foreseen for the IMaX instrument (see details in Martínez Pillet *et al.* 2010).

2.3. *Detection thresholds*

Scientific requirements on given physical quantities can be translated into instrument requirements, provided an inference technique to retrieve that quantity is known. This is an advisable exercise that helps for trading-off the many instrumental parameters that must be taken into account during the design phase. As a first example, imagine we are going to infer LOS velocities through a Fabry-Pérot spectrometer and the Fourier tachometer technique. Such a technique involves four Stokes S_1 samples that combined, according to Fernandes (1992), give

$$v_{\rm LOS} = \frac{2c\,\delta\lambda}{\pi\lambda_0} \arctan \frac{I_{-9} + I_{-3} - I_{+3} - I_{+9}}{I_{-9} - I_{-3} - I_{+3} + I_{+9}}, \tag{2.7}$$

where c stands for the speed of light, $\delta\lambda$ is the étalon spectral resolution, λ_0 is the central wavelength of the line, and $I_{\pm i}$ represent the Stokes S_1 (Stokes I) samples at the given wavelengths in picometers.

Error propagation in Eq. (2.7) can be shown to give the LOS-velocity expected uncertainties in terms of the étalon roughness, $\sigma_{\delta\lambda}$, of the thermal and voltage instabilities of the spectrometer, σ_T and σ_V, and on the noise of the observations, σ_1. Without entering into details of the (easy but) lengthy calculations, the variance of the retrieved velocities can be written as

$$\sigma^2_{v_{\rm LOS}} = f(v_{\rm LOS}, \delta\lambda)\sigma^2_{\delta\lambda} + g(v_{\rm LOS}, \lambda_0, \delta\lambda, I_i, s_i)(k_T^2\sigma_T^2 + k_V^2\sigma_V^2) + h(\lambda_0, \delta\lambda, I_i)\sigma_1^2, \tag{2.8}$$

where f, g, and h are given functions of the specified variables, s_i represent the Stokes S_1 profile derivatives with respect to wavelength at the sample wavelengths, and k_T and k_V are the temperature and voltage calibration constants for tuning the étalon, respectively. Assume now, for example, that $\lambda_0 = 617.3$ nm and $\delta\lambda = 100$ mÅ as for the SO/PHI instrument. Then, an étalon roughness leading to $\sigma_{\delta\lambda} = 1$ mÅ produces $\sigma_{v_{\rm LOS}} = 1\ {\rm m\,s^{-1}}$ for velocities of 100 ms^{-1} (and is linear in $v_{\rm LOS}$); pure photon noise with $S/N = 10^3$ induces $\sigma_{v_{\rm LOS}} = 7\ {\rm m\,s^{-1}}$ or, in other words, a scientific requirement on $v_{\rm LOS}$ stability of 1 m s^{-1} (that can be of interest for low-l, global helioseismology) demands a stability of 0.55 mK in temperature or 42 mV in voltage! Only state-of-the-art technology can aim at such thermal stabilities in a space environment, but the voltage requirement is very stringent as well since $LiNbO_3$ étalons are tuned with voltages of the order of 10^3 V.

Take now as a second example the inference of longitudinal and transverse field strengths through the magnetograph equations

$$B_{\rm lon} \equiv k_{\rm lon}\frac{V_{\rm s}}{I_{\rm c}} \text{ and } B_{\rm tran} \equiv k_{\rm tran}\sqrt{\frac{L_{\rm s}}{I_{\rm c}}}, \tag{2.9}$$

where $k_{\rm lon}$ and $k_{\rm tran}$ are calibration constants, $V_{\rm s}$ and $L_{\rm s}$ are the circular and linear polarization magnetographic signals,

$$V_{\rm s} \equiv \frac{1}{n_\lambda}\sum_{j=1}^{n_\lambda} a_j S_{4,j}, \ L_{\rm s} \equiv \frac{1}{n_\lambda}\sum_{j=1}^{n_\lambda}\sqrt{S^2_{2,j} + S^2_{3,j}}, \tag{2.10}$$

where $a_j = 1$ or -1 depending on whether the sample is to the blue (including the zero shift) or the red side of the central wavelength of the line, with n_λ being the number of wavelength samples within the spectral line. Since $n_\lambda = 4$ for the IMaX instrument, the uncertainties in V_s and L_s are necessarily a factor 2 smaller than $\sigma_{2,3,4}$ because information from four independent wavelengths is averaged for building the magnetograms. In such a case, one can estimate photon-noise-induced uncertainties of $\sigma_{B_{\rm lon}} = 4.8$ G and $\sigma_{B_{\rm lon}} = 80$ G.

3. Maximizing the polarimetric efficiencies

Once we know the effect of noise in the final inferences made with the instrument and how to improve S/N, that in the end turns out to maximizing efficiencies, let us check whether or not real polarimeters can (theoretically) achieve or (practically) approach the optimum polarimetric efficiencies of ideal instruments.

A rule of thumb in polarimetry is to put the polarization modulator as early in the optical path as possible so as to minimize the influence of the remaining optics: after the modulator, light is encoded and, no matter the path, can finally be analyzed properly before reaching the detector. This property has not been demonstrated, however, for polarimetric efficiencies so far. In other words, can polarimeters preserve the polarimetric efficiencies regardless of the retardations and changes of phase induced by the optics between the modulator and the analyzer? This is not a trivial question because intermediate optics might change the polarimeter's Mueller matrix in a way that would modify the efficiencies; indeed, not all polarimeters can reach the optimum efficiencies. Martínez Pillet *et al.* (2004) pointed out that nematic-LCVR-based polarimeters can theoretically achieve optimum efficiencies for both vector and longitudinal magnetography. This fact is easy to understand as we are going to demonstrate.

Assume we have two nematic LCVRs of retardances ρ and τ, respectively for the first and the second one to be reached by light. Such retardances can be changed at will by simply modifying the tuning voltage of the devices. If the optical axis of the first LCVR is put at 0° with respect to the positive S_2 direction (X axis, for instance), the second LCVR has its axis at $\pi/4$, and the linear analyzer is at 0°, then the rows of the modulation matrix (Del Toro Iniesta & Collados, 2000) are

$$O_{ij} = (1, \cos\tau_i, \sin\rho_i \sin\tau_i, -\cos\rho_i \sin\tau_i), \tag{3.1}$$

where index $i = 1, 2, 3, 4$ corresponds to each of the four measurements. A polarimeter having these four elements equal to the efficiencies in Eqs. (2.4) is an optimum one. Since all four $O_{i1} = 1$, Stokes S_1 can reach its maximum efficiency. At least four different solutions can also be found to equations resulting from making the other three components equal to $1/\sqrt{3}$. Therefore, the remaining Stokes parameters can also reach their maximum efficiencies. It is also easy to understand that the best longitudinal analysis ($S_1 \mp S_4$) can be obtained by tuning the retardance of the first LCVR to 0° and that of the second to $\pm\pi/2$.

Let us see now what is the effect of an étalon in between the modulator and the analyzer as in the IMaX or SO/PHI instruments. The most general way of modeling the polarization properties of such a device is by assuming it behaves like a retarder oriented at an angle θ and with a retardance δ. Now the Mueller matrix of the polarimeter gets modified because the Mueller matrix of the étalon, $\mathbf{M}_3$, has to be inserted between those of the LCVRs, $\mathbf{M}_1$ and $\mathbf{M}_2$, and that of the analyzer, $\mathbf{M}_4$. The Mueller matrix of the system is then $\mathbf{M} = \mathbf{M}_4\mathbf{M}_3\mathbf{M}_2\mathbf{M}_1$ and the modulation matrix turns modified to $O_{ij} = \mathbf{M}_{1j}(\tau_i, \rho_i)$. Since $M_{11} = 1$, $O_{i1} = 1$, $\forall i$. If we proceed now as before by equating $O_{i2,3,4} = 1/\sqrt{3}$, we obtain trascendental equations. Fortunately, they can be shown to have a solution and, therefore, optimum efficiencies can also be reached theoretically with these *real* instruments.

Besides the étalon, several mirrors may be needed in the design in between the modulator and the analyzer. Then, a Mueller matrix representing all the mirrors has to be inserted between $\mathbf{M}_3$ and $\mathbf{M}_4$. The effect of such an insertion can be demonstrated to be a global multiplication of all the modulation matrix elements by a constant factor. Therefore, the solutions for the transcendental equations are the same and, again, optimum efficiencies can theoretically be reached. Of course, real instruments may have modulation matrices that slightly differ from the optimum ones and calibration is always necessary.

4. Summary and conclusions

The accuracy in line-of-sight velocity and vector magnetic fields inferred from observations roots in photometric accuracy and, hence, in S/N. The instruments have, therefore, to be designed so that they collect, with the best polarimetric efficiencies, as many photons as possible in wavelengths of spectral lines that are as much sensitive as possible to these physical quantities.

In this contribution we have gathered rules for increasing the S/N, for finding out the more sensitive spectral lines to given quantities by means of the response functions, and for deducing detectability thresholds imposed by noise. The optimization of the signal-to-noise ratio of the observations goes necessarily through the maximization of polarimetric efficiencies and we have also shown that the optimum theoretical efficiencies can be reached with nematic-LCVR-based spectropolarimeters and magnetographs like IMaX and SO/PHI.

Acknowledgements

This work has been partially funded by the Spanish MICINN, through project AYA2009-14105-C06, and Junta de Andalucía, through project P07-TEP-02687.

References

Bartol, P. *et al.* 2010, *Solar Phys.*, in press

Lites, B. W., Elmore, D. F., & Streander, K. V. 2001, in *ASP Conf. Ser.*, 236, M. Sigwarth (ed.), 33

Cabrera Solana, D., Bellot Rubio, L. R., & Del Toro Iniesta, J. C. 2005, *ApJ*, 439, 687

Collados, M., Lagg, A., Díaz Garcí A, J. J., Hernández Suárez, E., López López, R., Páez Mañá, E., & Solanki, S. K. 2007, in *The Physics of Chromospheric Plasmas*, APS Conf. Ser., 368, 611

Del Toro Iniesta, J. C. & Collados, M. 2000, *Appl. Optics*, 39, 1637

Del Toro Iniesta, J. C., Orozco Suárez, D., & Bellot Rubio, L. R. 2010, *AJ*, 711, 312

Fernandes, D. N. 1992, PhD Thesis, Standford University

Graham, J. D., Norton, A., López Ariste, A., Lites, B., Socas-Navarro, H., & Tomczyk, S. 2003, *APS Conf. Ser.*, 307, 131

Kosugi, T. *et al.* 2007, *Sol. Phys.*, 243, 3

Landi Degl'Innocenti, E. & Landi Degl'Innocenti, M. 1977, *Astron. Astrophys*, 56, 111

Beck, C., Bellot Rubio, L. R., Kentischer, T. J., Tritschler, A., & Del Toro Iniesta, J. C. 2010, *Astron. Astrophys.*, 520, A115

Marsch, E., Marsden, R., Harrison, R., Wimmer-Schweingruber, R., & Fleck, B. 2005, *Adv. in Space Res.*, 36, 1360

Martínez Pillet, V., Collados, M., & Sánchez Almeida, J., *et al.* 1999, in High Resolution Solar Physics: Theory, Observations, and Techniques, T. R. Rimmele, K. S. Balasubramaniam, & R. R. Radick (eds.), *APS Conf. Ser.*, 183, 264

Martínez Pillet, V. 2006, in *Proc. Second Solar Orbiter Workshop*, R. Marsden & L. Conroy (eds.) ESA SP-641

Martínez Pillet, V., *et al.* 2004, *Proc. of SPIE*, 5487, 1152

Martínez Pillet, V., *et al.* 2010, *Solar Phys.*, in press

Mein, P. 1971, *Solar Phys.*, 20, 3

Narayan, G., Scharmer, G. B., Hillberg, T., Lofdahl, M., van Noort, M., Sutterlin, P., & Lagg, A. 2008, *12th European Solar Physics Meeting, Freiburg, Germany*

Orozco Suárez, D., Bellot Rubio, L. R., Martínez Pillet, V., Bonet, J. A., Vargas Domínguez, S., & del Toro Iniesta, J. C. 2010, *Astron. Astrophys.*, 522, A101

Orozco Suárez, D. & Del Toro Iniesta, J. C. 2007, *Astron. Astrophys*, 462, 1137

Ruiz Cobo, B. & Del Toro Iniesta, J. C. 1994, *Astron. Astrophys*, 283, 129

Sankarasubramanian, K., *et al.* 2004, *Proc. SPIE*, 5171, 207

Title, A. 2000, *Bull. Am. Astron. Soc.*, 32, 839

The Physics of Sun and Star Spots
Proceedings IAU Symposium No. 273, 2010
D.P. Choudhary & K.G. Strassmeier, eds.

doi:10.1017/S1743921311014980

Heating of coronal active regions

Daniel O. Gómez[1,2]

[1]Instituto de Astronomía y Física del Espacio, C.C. 67 - Suc. 28, (1428) Buenos Aires, Argentina - email: gomez@iafe.uba.ar

[2]Departamento de Física, Facultad de Ciencias Exactas y Naturales (UBA), Ciudad Universitaria, (1428) Buenos Aires, Argentina

Abstract. Recent observations of coronal loops in solar active regions show that their heating must be a truly dynamic process. Even though it seems clear that the energy source is the magnetic field that confines the coronal plasma, the details of how it dissipates are still a matter of debate. In this presentation we review the theoretical models of coronal heating, which have been traditionally clasified as DC or AC depending on the electrodynamic response of the loops to the photospheric driving motions.

Also, we show results from numerical simulations of the internal dynamics of coronal loops within the framework of the reduced MHD approximation. These simulations indicate that the application of a stationary velocity field at the photospheric boundary leads to a turbulent stationary regime after several photospheric turnover times. Once this turbulent regime is set, both DC and AC stresses dissipate at faster rates as a result of a direct energy cascade.

Keywords. Sun: corona, Sun: X-rays, magnetic fiels, MHD, turbulence.

---

1. Introduction

Theoretical models of coronal heating have been traditionally classified into two broad categories, according to the time scales involved in the driving at the loop bases: (a) AC or wave models, for which the energy is provided by waves at the Sun's photosphere, with timescales much faster than the time it takes an Alfven wave to cross the loop; (b) DC or stress models, which assume that energy dissipation takes place by magnetic stresses driven by slow footpoint motions (compared to the Alfven wave crossing time) at the Sun's photosphere. Although these scenarios seem mutually exclusive, two common factors prevail: (i) the ultimate energy source is the kinetic energy of the subphotospheric velocity field, (ii) the existence of fine scale structure is essential to speed up the dissipation mechanisms invoked.

Review articles on coronal heating (Narain & Ulmschneider 1990; Gómez 1990; Zirker 1993; Narain & Ulmschneider 1996) explore the theoretical models in further detail. More recent reviews can be found in Mandrini *et al.* (2000); Demoulin *et al.* (2003); Aschwanden (2004), where observational tests of the models are also described.

A natural candidate for the dissipation of the energy provided by subphotospheric motions is Joule heating, but the typical time scale to dissipate coronal magnetic stresses at the length scale of the driving motions is exceedingly long. This time scale can be estimated as $l^2/\eta \approx 10^6$ years (l is a typical length scale an η is the plasma resistivity). Most of the theories of coronal heating invoke different mechanisms to speed up energy dissipation (Parker 1972, 1988; Heyvaerts & Priest 1992; van Ballegooijen 1986; Mikić *et al.* 1989; Longcope & Sudan 1994; Hendrix & van Hoven 1996; Galsgaard & Nordlund 1996; Gudiksen & Nordlund 2002).

One of the proposed scenarios to speed up dissipation is the assumption that the magnetic and velocity fields of the coronal plasma are in a turbulent state (Gómez & Ferro

Fontán 1988, 1992; Heyvaerts & Priest 1992). On these models, turbulent fluctuations of both fields are predominantly in the directions perpendicular to the main magnetic field. In a turbulent regime, energy is transferred from photospheric motions to the magnetic field and then cascades toward small scales due to nonlinear interactions, until highly structured electric currents are formed. The development of fine scales (i.e. the drastic reduction of l) to enhance the dissipation of either waves or DC currents is a natural outcome of turbulence models.

2. Reduced MHD

To theoretically describe the dynamics of coronal loops in solar (or stellar) active regions, we assume these loops to be relatively homogeneous bundles of fieldlines, with their footpoints deeply rooted into the photosphere. Subphotospheric convective motions move individual fieldlines around, generating magnetic stresses in the coronal portion of the loop. More specifically, we consider a simplified model of a coronal magnetic loop with length L and cross section $2\pi l_p \times 2\pi l_p$, where l_p is the lengthscale of typical subphotospheric motions. For elongated loops, *i.e.* such that $2\pi l_p << L$, it seems reasonable to neglect toroidal effects. The main magnetic field $\mathbf{B_0}$ is assumed to be uniform and parallel to the axis of the loop (the z axis). The planes at $z = 0$ and $z = L$ correspond to the photospheric footpoints.

Under these symplifying assumptions, we are able to use the reduced MHD approximation (Strauss 1976)), according to which the plasma moves incompressibly in planes perpendicular to the axial field $\mathbf{B_0}$, and the transverse component of the magnetic field is small compared to $\mathbf{B_0}$. The very high electric conductivity (frozen field) allows photospheric motions to easily drive magnetic stresses in the corona (Parker 1972), since the field lines twist and bend due to these motions, generating tranverse components of velocity $\mathbf{u}$ and magnetic field $\mathbf{b}$. Therefore:

$$\mathbf{B} = B_0\mathbf{z} + \mathbf{b}(x, y, z, t) \ , \quad \mathbf{b} \cdot \mathbf{z} = 0 \tag{2.1}$$

$$\mathbf{u} = \mathbf{u}(x, y, z, t), \quad \mathbf{u} \cdot \mathbf{z} = 0 \tag{2.2}$$

Since both $\mathbf{b}$ and $\mathbf{u}$ are two-dimensional and divergence-free fields, they can be represented by scalar potentials:

$$\mathbf{b} = \nabla \times (a\mathbf{z}) = \nabla a(x, y, z, t) \times \mathbf{z} \tag{2.3}$$

$$\mathbf{u} = \nabla \times (\psi\mathbf{z}) = \nabla\psi(x, y, z, t) \times \mathbf{z} \tag{2.4}$$

where ∇ indicates derivatives in the x, y plane. The reduced MHD equations for the stream function $\psi\mathbf{z}$ and the vector potential $a\mathbf{z}$ are:

$$\partial_t a = v_A \partial_z \psi + [\psi, a] + \eta \nabla^2 a \tag{2.5}$$

$$\partial_t w = v_A \partial_z j + [\psi, w] - [a, j] + \nu \nabla^2 w \tag{2.6}$$

where $w = -\nabla^2 \psi$ is the fluid vorticity, $j = -\nabla^2 a$ is the current density, $v_A = B_0/\sqrt{4\pi\rho}$ is the Alfvén velocity and ρ is the plasma density.

Eqn 2.5 describes the advection of the potential a and Eqn 2.6 corresponds to the evolution of vorticity w. The terms $v_A \partial_z$ represent the coupling between neighboring z =constant planes and describe parallel wave propagation. The ∇^2 terms represent dissipative effects, the constants η and ν being the resistivity and viscosity coefficients. The nonlinear terms are those represented by the Poisson brackets ($[A, B] = \partial_x A \partial_y B - \partial_y A \partial_x B$). Their role is to couple normal modes in such a way that energy, and other ideal invariants, can be transferred between them.

3. RMHD turbulence

We numerically explore the feasibility of a turbulent scenario, describing the internal dynamics of coronal loops through the RMHD approximation. We assume periodicity for the lateral boundary conditions, and specify the velocity fields at the $z = 0$ and $z = L$ photospheric boundaries. In particular, we assume $\psi(z = 0) = 0$ and $\psi(z = L) = \Psi(x, y)$ where the stream function $\Psi(x, y)$ describes stationary and incompressible footpoint motions on the photospheric plane. We specify the Fourier components of $\Psi(x, y)$ as $\Psi_{\mathbf{k}} = \Psi_0$ inside the ring $3 < l_p|k| < 4$ on the Fourier plane, and $\Psi_{\mathbf{k}} = 0$ elsewhere, to simulate a stationary and isotropic pattern of photospheric granular motions of diameters between $2\pi l_p/4$ and $2\pi l_p/3$. The strength Ψ_0 is proportional to a typical photospheric velocity $V_p \approx 1\ km.s^{-1}$. The typical timescale associated to these driving motions, is the eddy turnover time, which is defined as $t_p = l_p/V_p \approx 10^3\ sec$. We choose a narrowband and non-random forcing to make sure that the broadband energy spectra and the signatures of intermittency that we obtained (see below) are exclusively determined by the nonlinear nature of the MHD equations.

To transform (2.5)-(2.6) into their dimensionless form, we choose l_p and L as the units for transverse and longitudinal distances ($l_p \approx 10^3\ km$ and $L \approx\ 10^4 - 10^5\ km$) and $t_A \equiv L/v_A$ as the time unit ($t_A \approx 10 - 100\ sec$).

In Figure 1 we show the results obtained from a simulation extending from $t = 0$ to $t = 100t_A$. The upper panel shows the magnetic (E_B, thin trace) and kinetic (E_U, dotted line) energies, as well as the total energy ($E = E_U + E_B$, thick trace). We can see that after about ten Alfven times, the energy reaches a stationary regime, since the work done by footpoint motions statistically (i.e. in time average) reaches an equilibrium with the dissipative processes (electric resistivity and fluid viscosity). In this stationary regime most of the energy is magnetic, while kinetic energy is only about 5% of the total. In the lower panel, we show the dissipation rate (D, thick trace) and the incoming Poynting flux (P, thin trace), showing that their time averages are approximately equal.

Gómez & Ferro Fontán (1988) (also Gómez & Ferro Fontán (1992)) showed that this stationary equilibrium corresponds to a turbulent regime (see also **?**). The associated energy cascade bridges the gap between the large spatial scales where energy is injected, to the much smaller scales where it dissipates (see Dmitruk & Gómez (1997)). The dependence of this mean dissipation rate $\epsilon =< D >=< P >$ ($< \ldots >$: time average)

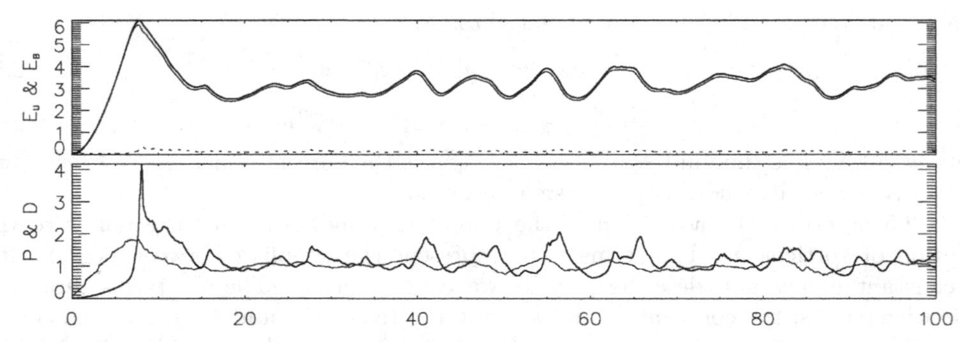

Figure 1. Time series obtained from simulations externally driven by a stationary velocity field. **Upper panel**: Magnetic energy (thin), kinetic energy (dotted) and total energy (thick). **Lower panel**: Energy dissipation rate (thick) and Poynting flux (thin).

with the physical parameters of the loop is (Dmitruk & Gómez 1999)

$$\epsilon \propto \frac{\rho l_p^2}{t_A^3}\left(\frac{t_A}{t_p}\right)^{\frac{3}{2}} \tag{3.1}$$

The kinetic and magnetic Reynolds numbers in our simulations ($Re = l_p^2/(\eta t_A) = Rm$) were carefully chosen to guarantee a proper resolution of the smallest spatial structures (i.e. the largest wavenumbers), where dissipation becomes dominant.

Another feature which can readily be observed in Figure 1 is the spiky nature of these time series, which is caused by the ubiquitous presence of intermittency in turbulent regimes. Dmitruk *et al.* (1998) (see also Georgoulis *et al.* (1998)) associated each of these spikes of energy dissipation with Parker's *nanoflares* (see Parker (1988)) and studied the statistical properties of these dissipation events. The main result from this statistical study (see also Gómez & Dmitruk (2008)) is that the number of dissipation events (or nanoflares) as a function of their individual energies $N(E)$ follows a power law $N(E) \approx E^{-3/2}$, which is remarkably comparable to the result obtained for larger dissipation events (such as microflares and flares), gathering a large number of observational studies and reported by Aschwanden (2004).

4. Wave propagation and dissipation

To study the response of coronal loops to waves being pumped at their footpoints, we performed RMHD simulations applying a time-periodic velocity field at $z = L$, with a time frequency w_0. In Figure 2 we show the results obtained for a relatively slow frequency $\omega_0 = 0.1$ (i.e. a wave period equal to ten loop Alfven times). In Figure 3 we show the results for the case $\omega_0 = 1.0$. Note that both the energy and dissipation rate reach maximum levels much larger than the previous case, which is to be expected for a resonant mode.

We repeat this type of RMHD simulations for a range of wave frequencies and compute the asymptotic value for the dissipation rate, which is shown in Figure 4. This curve displays the expected behavior of an externally driven resonant system. The particular case of externally driven coronal loops was addressed analytically by Inverarity & Priest (1995). The full line in Figure 4 shows the analytical result for the viscosity used in our simulations, i.e. $\nu = \eta = 4.10^{-3}$. The dotted trace corresponds to the ideal case.

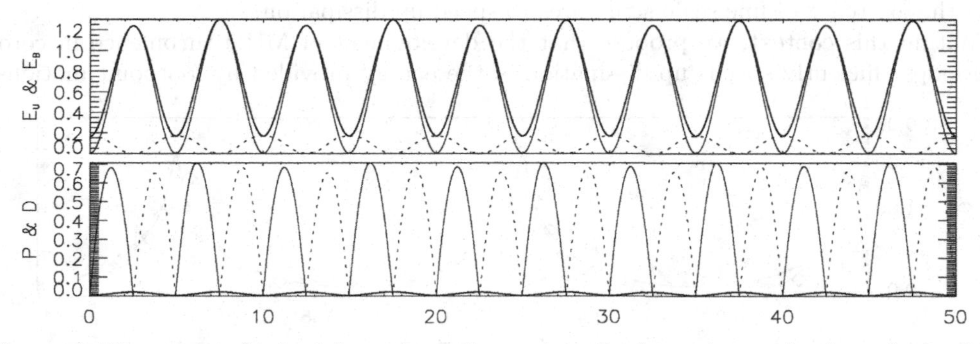

Figure 2. Time series obtained from simulations externally driven by a pulsating velocity field at the frequency $\omega_0 = 0.1$. **Upper panel**: Magnetic energy (thin), kinetic energy (dotted) and total energy (thick). **Lower panel**: Energy dissipation rate (thick), positive (thin) and negative (dotted) Poynting flux.

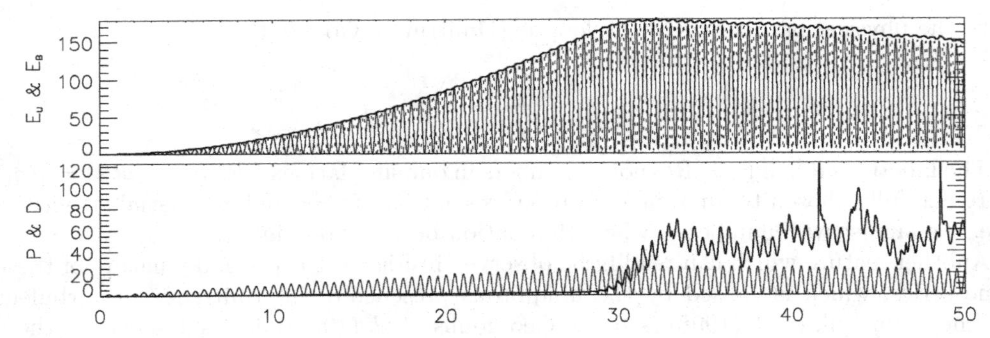

Figure 3. Time series obtained from simulations externally driven by a pulsating velocity field at the frequency $\omega_0 = 1.0$. **Upper panel**: Magnetic energy (thin), kinetic energy (dotted) and total energy (thick). **Lower panel**: Energy dissipation rate (thick), positive (thin) and negative (dotted) Poynting flux.

If the wave propagates in a turbulent flow, we would expect its dissipation to be enhanced with respect to the case of propagation in a laminar medium. We therefore performed the following combined simulations: drive the loop with the stationary motions described in §3 until reaching a stationary turbulent regime, and then add the wave pumping boundary motion at a given frequency ω_0 described in this section. In Figure 5, we show the asymptotic dissipation rate as a function of the external pumping frequency. The result shown in Figure 5 shows both a broadening and a reduction of the resonant peaks, as one would expect for waves propagation in a more dissipative medium. Our next step was to quantify this enhanced dissipation by fitting the analytical result with ν_{eff} as free parameter. The best fit to the numerical results shown in Figure 5 corresponds to

$$\nu_{eff} \approx 0.03 \tag{4.1}$$

which is much larger than the flow viscosity used in our simulations $\nu = \eta = 4.10^{-3}$.

5. Conclusions

We briefly summarized our current understanding on the problem of coronal heating (see also Aulanier (2011)). Even though there are several theoretical models for coronal heating proposed in the literature, which can be broadly classified as DC or AC, we identified two key assumptions that are common to almost all of these models. Namely, that the ultimate energy source is the kinetic energy of the subphotospheric convection, and the existence of fine scale structures to speed up dissipation.

Within this context, we propose that the development of MHD turbulence in coronal loops efficiently speeds up dissipation of the energy provided by footpoint motions.

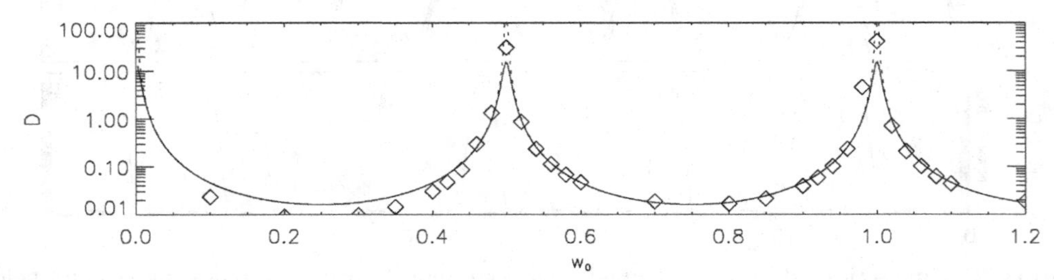

Figure 4. Asymptotic dissipation rate (log-scale) vs. the wave frequency ω_0. Diamonds correspond to numerical results, the full trace corresponds to the analytical curve for $\nu = 4.10^{-3}$ and the dotted trace corresponds to the ideal limit.

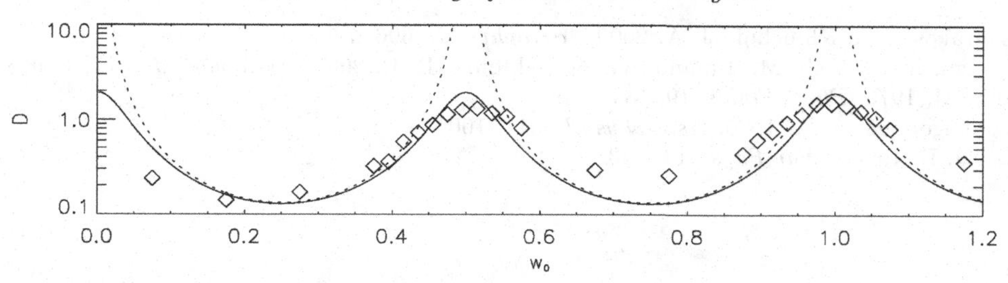

Figure 5. Asymptotic dissipation rate (log-scale) vs. the wave frequency ω_0. Diamonds correspond to numerical results with turbulent background, the full trace corresponds to the analytical curve for $\nu_{eff} = 0.03$ and the dotted trace corresponds to the ideal limit.

Turbulent fluctuations of the velocity and magnetic field within the loop, contribute to transfer energy toward small scales due to nonlinear interactions, until highly structured electric currents are formed. Numerical simulations of the RMHD equations show that even slow and stationary footpoint motions develop a turbulent regime after a few photospheric turnover times. Our simulations also show that Alfven waves propagating into coronal loops dissipate much faster in the presence of a turbulent regime.

In summary, the development of fine scales to enhance the dissipation of either waves or DC currents is a natural outcome of turbulence models. Furthermore, the intermittent nature of turbulent dissipation is fully consistent with the highly dynamic behavior displayed in recent observations of coronal loops in solar active regions (see for instance Patsourakos & Klimchuk (2009)).

References

Aschwanden, M. J. 2004, in *Physics of the Solar Corona. An Introduction*, Springer-Verlag, Berlin.
Aulanier, G. 2011, in "Proc. IAU Symp. 273: Physics of the Sun and Star spots", in press.
Demoulin, P., van Driel-Gesztelyi, L., Mandrini, C. H., Klimchuk, J. A., & Harra, L. 2003, *Astrophys. J.*, 586, 592.
Dmitruk, P. & Gómez, D. O. 1997, *Astrophys. J.*, 484, L83.
Dmitruk, P., Gómez, D. O., & DeLuca, E. 1998, *Astrophys. J.*, 505, 974.
Dmitruk, P. & Gómez, D. O. 1999, *Astrophys. J.*, 527, L63.
Galsgaard, K. & Nordlund, A. 1996, *J. Geophys. Res.*, 101, 13445.
Georgoulis, M., Velli, M., & Einaudi, G. 1998, *Astrophys. J.*, 497, 957.
Gómez, D. O. 1990, *Fund. Cosmic Phys.*, 14, 361.
Gómez, D. O. & Ferro Fontán, C. 1988, *Solar Phys.*, 116 33.
Gómez, D. O. & Ferro Fontán, C. 1992, *Astrophys. J.*, 394, 662.
Gómez, D. O. & Dmitruk, P. 2008, in "Proc. IAU Symp. 247: Waves and Oscillations in the Solar Atmosphere", (Eds. R. Erdelyi & C.A. Mendoza-Briceño), 269.
Gudiksen, B. V. & Nordlund, A. 2002, *Astrophys. J.*, 572, L113.
Hendrix, D. L. & van Hoven, G. 1996, *Astrophys. J.*, 467, 887.
Heyvaerts, J. & Priest, E. R. 1992, *Astrophys. J.*, 390, 297.
Inverarity, G. W. & Priest, E. R. 1995, *Astron. Astrophys*, 302, 567.
Longcope, D. W. & Sudan, R. N. 1994, *Astrophys. J.*, 437, 491.
Mandrini, C. H., Demoulin, P., & Klimchuk, J. A. 2000, *Astrophys. J.*, 530, 999.
Mikić, Z., Schnack, D. D., & van Hoven, G. 1989, *Astrophys. J.*, 338, 1148.
Narain, U. & Ulmschneider, P. 1990, *Space Sci. Rev.*, 54, 377.
Narain, U. & Ulmschneider, P. 1996, *Space Sci. Rev.*, 75, 453.
Parker, E. N. 1972, *Astrophys. J.*, 174, 499.
Parker, E. N. 1988, *Astrophys. J.*, 330, 474.

Patsourakos, S. & Klimchuk, J. A. 2009, *Astrophys. J.*, 696, 760.
Rappazzo, A. F., Velli, M., Einaudi, G., & Dahlburg, R. B. 2008, *Astrophys. J.*, 677, 1348.
Strauss, H. 1976, *Phys. Fluids*, 19, 134
van Ballegooijen, A. A. 1986, *Astrophys. J.*, 311, 1001.
Zirker, J. B. 1993, *Solar Phys.*, 148, 43.

The Physics of Sun and Star Spots
Proceedings IAU Symposium No. 273, 2010
D.P. Choudhary & K.G. Strassmeier, eds.

doi:10.1017/S1743921311014992

Automated sunspot detection and the evolution of sunspot magnetic fields during solar cycle 23

Fraser Watson and Lyndsay Fletcher

Department of Physics and Astronomy
Kelvin Building, University of Glasgow, Glasgow, UK
email: f.watson@astro.gla.ac.uk

Abstract. The automated detection of solar features is a technique which is relatively underused but if we are to keep up with the flow of data from spacecraft such as the recently launched Solar Dynamics Observatory, then such techniques will be very valuable to the solar community. Automated detection techniques allow us to examine a large set of data in a consistent way and in relatively short periods of time allowing for improved statistics to be carried out on any results obtained. This is particularly useful in the field of sunspot study as catalogues can be built with sunspots detected and tracked without any human intervention and this provides us with a detailed account of how various sunspot properties evolve over time. This article details the use of the Sunspot Tracking And Recognition Algorithm (STARA) to create a sunspot catalogue. This catalogue is then used to analyse the magnetic fields in sunspot umbrae from 1996-2010, taking in the whole of solar cycle 23.

Keywords. Sun: evolution, sun: magnetic fields, sun: photosphere, sunspots, techniques: image processing

1. Introduction

To examine the magnetic fields measured in sunspots it is useful to have large datasets as there are vast differences between a simple sunspot surrounded by quiet sun and a sunspot which is in the centre of a complex active region. The large dataset was assembled by using an automated sunspot detection algorithm developed by Watson *et al.* (2009). The Sunspot Tracking And Recognition Algorithm (STARA) is a quick and reliable way to process a large number of solar images and has been tested on images from a variety of sources including ground based observatories (such as Kanzelhöhe Observatory, see http://www.kso.ac.at/sonnenbeobachtung/spot_rec_en.php for details on how the algorithm is being used), the MDI instrument on SOHO and the HMI instrument on the SDO satellite.

The data used in this article were level 1.8 data recorded by the MDI instrument (Scherrer *et al.* 1995) and are taken from the launch of the instrument in 1996 through to early 2010 which covers the whole of solar cycle 23. We use both white light continuum observations and magnetograms which allows us to detect the sunspots and determine their magnetic properties at the same time.

2. Creating the Catalogue

To ensure that the magnetic fields of sunspots could be measured at the same time as their detection, the times of measurement had to be as close as possible. Due to the cadence of the MDI measurements, this was only the case once per day at 00:00UT giving a dataset of around 5000 continuum and magnetogram images to process. Processing these images takes 30-40 hours on a single processor depending on the number of sunspots present. To extract the sunspots from the data, techniques from the field of mathematical morphology were used (see Matheron (1975) and Serra (1982) for more detail).

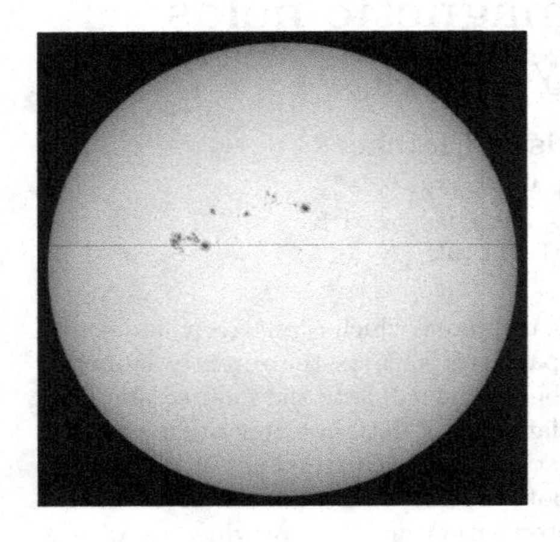

Figure 1. An example solar continuum image from MDI. The black horizontal line passes through 2 sunspots.

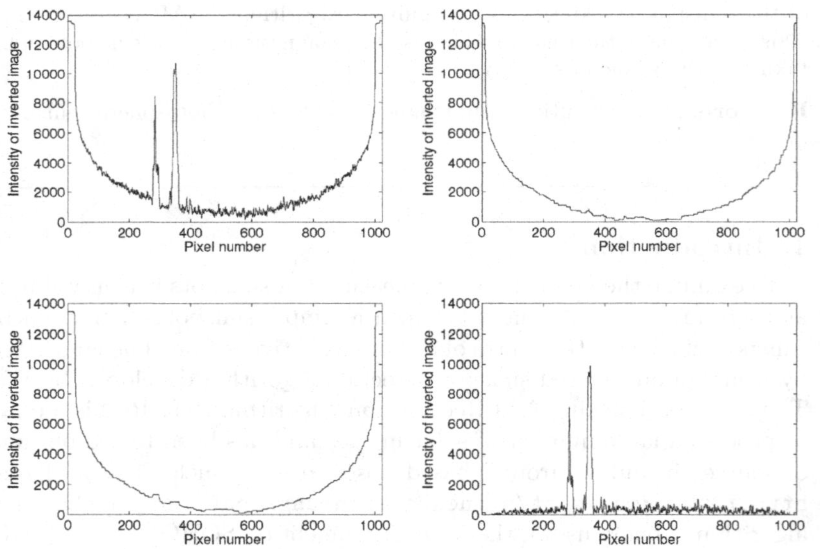

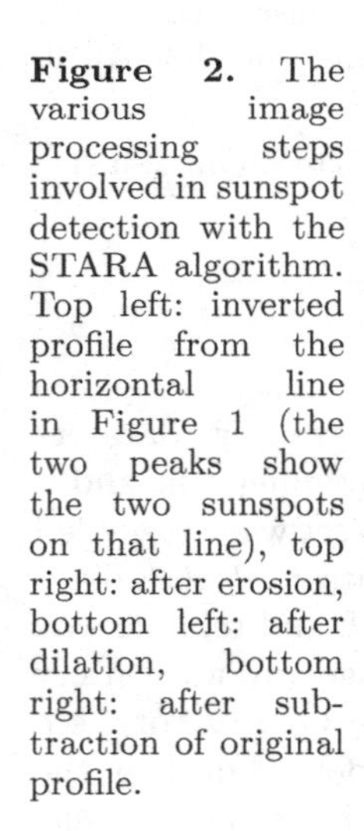

Figure 2. The various image processing steps involved in sunspot detection with the STARA algorithm. Top left: inverted profile from the horizontal line in Figure 1 (the two peaks show the two sunspots on that line), top right: after erosion, bottom left: after dilation, bottom right: after subtraction of original profile.

Figure 1 and Figure 2 shows the various steps involved in detecting sunspots using the STARA code and the process works as follows (note that this example is given in 2D for simplicity looking at the two sunspots along the dark line in Figure 1 but the same applies to the full 3D Sun case) :

- Invert the image so that the sunspots appear as bright peaks on a darker background. This is shown in the top left panel of Figure 2.
- A 'top-hat' transform is then applied which consists of an erosion and a dilation (explanations of these terms can be found in Serra, 1982).
- Subtract the profile after this transform from the original to obtain the bottom right panel of Figure 2.
- Apply a threshold to give locations of the sunspots.

The full 3D case works in the same way but rather than operating on a 2D 'U' shaped profile, a 3D bowl shaped profile is used. Further processing is then applied to separate the umbra and penumbra of the sunspots. To complete this process on a single image takes around 4 seconds. The sunspot locations are then superimposed on a magnetogram taken at almost the same time so that the magnetic fields can be recorded. This is repeated until all of the images have been processed.

3. Magnetic fields in sunspots from 1996–2010

As the magnetic fields present in sunspots had already been measured, we could easily look at the trends present over the length of the catalogue.

Figure 3 shows the maximum magnetic field detected in the umbra of all sunspots present on a given day. It is assumed that the field in the sunspot umbrae are in the local vertical direction and so a cosine correction is applied to the MDI line of sight magnetic field. To minimise the effects of this, only sunspots with a value of $\mu > 0.95$ were used (where μ is the cosine of the angle between the local solar vertical and the observers line of sight). Also, if there is a day with no sunspots, that day is omitted from the plot.

We can immediately see that there is a large variation in the sunspot fields even over short timescales of a few weeks and that the majority of maximum umbral fields fall between 1500 and 3500 gauss.

Penn & Livingston (2006) looked at this long term trend using the McMath-Pierce telescope on Kitt Peak and measured the magnetic field in the darkest observed part of sunspots from around 1993 to the present. This was done by measuring the Zeeman splitting present in the Fe I line to infer a magnetic field strength at the location of the measurement. When their whole data set is taken into account they find a trend of decreasing sunspot magnetic field strength of about 52 Gauss per year which is obtained by binning the data by year and looking at the mean of each bin along with the standard error on the mean. In Figure 4 we show the same treatment of the data from the STARA algorithm. The best fitting straight line to the data is shown with a dashed line and there are two sets of error bars present. The thin error bars show the standard deviation of all the data in that bin and the thick error bar only takes into account data which would meet the criteria for being a sunspot by Penn & Livingston (2006). This is because their data excludes pores, some of which can be as large as 10" in size and we observe sunspots that are of that order in size. These sunspots are the primary reason for the

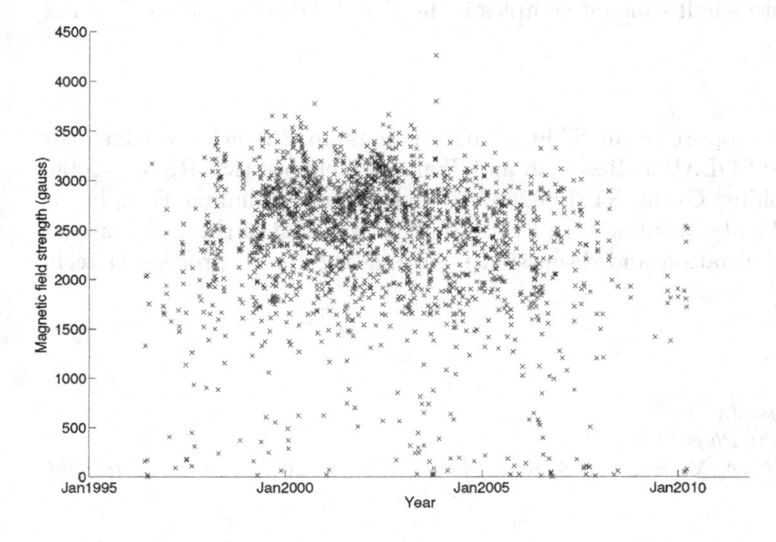

Figure 3. Maximum sunspot umbra fields from 1996 to 2010 as measured by the STARA algorithm.

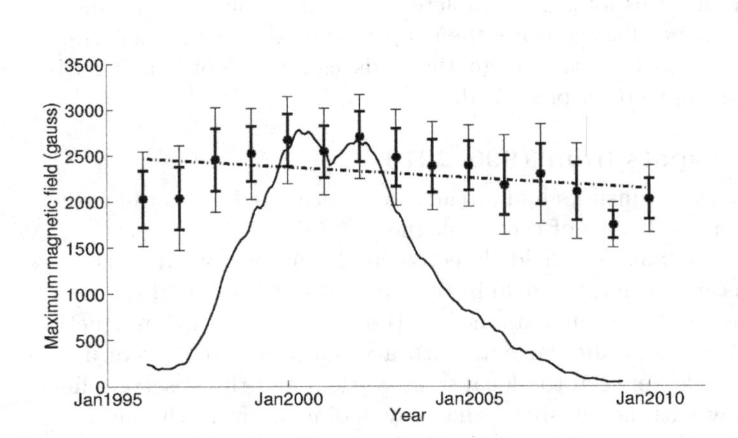

Figure 4. The data from Figure 3 have been binned by year and the mean of each bin plotted. The dashed line is the best fit straight line to those points. The gradient corresponds to a trend of -23.6 gauss per year. Thin error bars correspond to the standard deviation of all data in Figure 3 whereas thick error bars only take into account the data which met the criteria for being a sunspot in the Penn & Livingston (2006) article. The solid line is the scaled international sunspot number over the same time period shown for reference.

fields in Figure 3 that lie below 1500 gauss as they correspond to small sunspots that do not yet have magnetic field strengths comparable to larger, fully formed spots.

We observe the same decreasing trend as Penn & Livingston (2006) but with a shallower gradient of 23.6 gauss per year. However, it could be argued that there is a slight cyclic variation - unfortunately we have no cycle 22 data with which to investigate this. The data from the STARA algorithm also has a much larger spread making the errors larger in this result. However, the advantage of making measurements in this way is that it allows for a completely automated system that always processes data in the same way. From this data we cannot say for certain that a long term trend exists until data from the new solar cycle is obtained and a scaled plot of the international sunspot number is included in Figure 4 to show that the mean magnetic field is increasing and decreasing along with the solar activity. A change of 600G over the solar cycle, as suggested by Penn & Livingston (2006) would cause a change in the mean umbral radius as a relationship has been shown by Kopp & Rabin (1992) and Schad & Penn (2010) but observations by Penn & McDonald (2007) could not uncover this in the data. Mathew *et al.* (2007) also suggests that the size distribution of sunspots, although constant over the solar cycle, could introduce a bias into small sunspot samples if the size distribution of spots is not calculated.

Acknowledgements

FW acknowledges the support of an STFC PhD studentship. LF acknowledges the support of the EC-funded SOLAIRE Research and Training Network (MTRN-CT-2006-035484), the STFC (Rolling Grant ST/F002637/1) and the Leverhulme Foundation (Grant F/00 179/AY). We also acknowledge Prof. Stephen Marshall of the University of Strathclyde for useful information and discussion regarding the image processing techniques used.

References

Hale, G. E. 1908, *Astrophys. J.*, 315

Kopp, G. & Rabin, D., *Solar Phys.*, 141, 253

Mathew, S. K., Martínez Pillet, V., Solanki, S. K. & Krivova, N. A., 2007, *Astron. Astrophys*, 465, 291

Matheron, G. 1975, *Random Sets and Integral Geometry,* Wiley, New York

Penn, M. J. & Livingston, W. 2006, *Astrophys. J.*, 649, L45

Penn, M. J. & MacDonald, R. K. D., 2007, *Astrophys. J.*, 662, L123

Schad, T. A. & Penn, M. J., 2010, *Solar Phys.*, 262, 19

Scherrer, P. H., Bogart, R. S., Bush, R. I., Hoeksema, J. T., Kosovichev, A. G., Schou, J. & others 1995, *Solar Phys..*, 162, 129

Serra, J. 1982, *Image Analysis and Mathematical Morphology,* Academic Press, London

Solanki, S. K. 2003, *A&AR,* 11, 153

Watson, F., Fletcher, L., Dalla, S., & Marshall, S. 2009, *Solar Phys.*, 260, 5

The Physics of Sun and Star Spots
Proceedings IAU Symposium No. 273, 2010
D.P. Choudhary & K.G. Strassmeier, eds.

doi:10.1017/S1743921311015006

On the manifestation in the Sun-as-a-star magnetic field measurements of the quiet and active regions

Mikhail Demidov

Institute of Solar-Terresrial Physics, P.O. Box 291, 664033 Irkutsk, Russia
email: demid@iszf.irk.ru

Abstract. The best way to test the stellar magnetic field mapping codes is to apply them, with some changes, to the Sun, where high-precision disk-integrated and disk-resolved observations are available for a long time. Data sets of the full-disk magnetograms and the solar mean magnetic fields (SMMF) measurements are provided, for example, by the J.M.Wilcox Solar observatory (WSO) and by the Sayan Solar observatory (SSO). In the second case the measurements in the Stokes-meter mode simultaneously in many spectral lines are available. This study is devoted to analysis of the SSO quasi-simultaneous full-disk magnetograms and SMMF measurements. Changes of the SMMF signal with rotation of the surface large-scale magnetic fields are demonstrated. Besides, by deleting of selected pixels with active regions (AR) from the maps their contribution to the integrated SMMF signal is evaluated. It is shown that in some cases the role of AR can be rather significant.

Keywords. Sun: photosphere, Sun: magnetic fields

1. Introduction and Motivation

Observations of the solar magnetic fields with different spatial (angular) resolution are important. No doubts that the dominant tendency in present experimental solar physics is to achieve extremely high resolution up to the parts of arc second. This holds for the new space missions (Hinode, SDO), and for the largest ground-based telescopes (GREGOR, NST, ATST, etc.) as well. But establishment of this fact does not mean at all that observations with moderate or even with low spatial resolution, including integral observations of the Sun-as-a-star, became unimportant. There are a lot of applications where spectral and polarimetric low-resolution observations with high sensitivity and high accuracy are very demanded. Among the examples are the problems connected with solar-terrestrial physics. Indeed, it is sufficient to have the full-disk magnetograms with a resolution of no more than several dozens of arc seconds for calculations of the magnetic field structures in the corona and in the interplanetary media. Even more, the best correlation of interplanetary magnetic field (IMF) structures is found with the Sun-as-a-star, or, in other words, with the solar mean magnetic field (SMMF) measurements.

The other "ecological niche" of such kind observations belong to solar-stellar relations. A significant progress in observations of the magnetic fields on the stars has been achieved recently (Donati *et al.*, 2008; Strassmeier, 2009). Quite reliable measurements of the extremely weak magnetic fields even on the solar-type stars are available now (Petit *et al.*, 2008). And a lot of efforts are undertaken in the attempts to construct the surface distribution of stellar magnetic fields. Naturally, the best way to justify such methods of stellar mapping, at least for the case of solar twin stars, is to apply them to solar data. Indeed, the Sun is the only celestial body where disk-integrated and

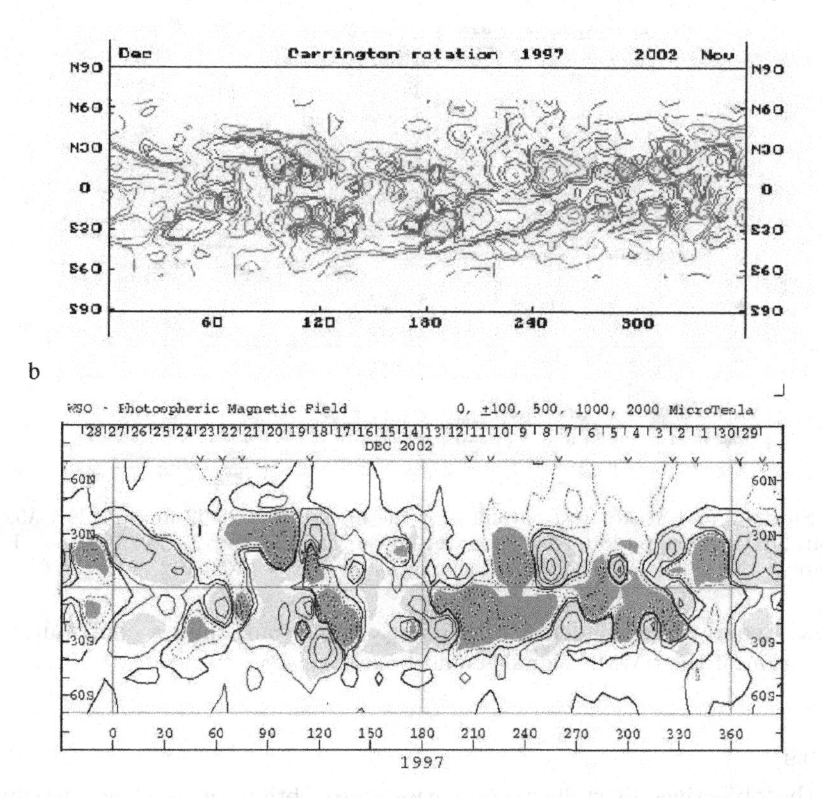

Figure 1. Large-scale solar magnetic fileds synoptic maps for the Carrington rotation 1997, constructed from SSO (top panel) and WSO (bottom panel) observations.

disk-resolved observations of magnetic fields are available simultaneously. And observations with sub- arc second spatial resolution are not necessary in this case. Observations with a resolution of about some arc minutes are sufficient.

At present time there are only two observatories in the world which provide regularly quasi simultaneous observations of the magnetic field of the Sun-as-a-star and full-disk magnetograms. It is the J.M.Wilcox Solar observatory (WSO) at Stanford, USA (Scherrer *et al.*, 1977), and the Sayan Solar observatory (SSO) in Russia (Demidov *et al.*, 2002). The advantage of the SSO observations with the Solar Telescope of Operative Predictions (STOP) is that they are made in the Stokes-meter mode, when the Stokes I and Stokes V parameters for many spectral lines are registered simultaneously. Usually, several lines in the vicinity of Fe I λ525.02 nm are observed.

One of the essential questions of the interpretation of SMMF measurements is the origin of such signals. Understanding that is important for the pure solar physics, and for stellar physics as well. It was shown earlier by Scherrer *et al.* (1972) and Kotov *et al.* (1977), that the main source of the SMMF is the rather weak large-scale back-ground magnetic field, covering the most part of the solar surface, and the role of active regions is small. However, it is known, that the SMMF strength is weak during the periods of minimal solar activity, when the number of active regions (AR) is small, and strong enough during the epochs of maximal activity. Besides, the model simulations of the temporal evolution of SMMF, regarding bipolar AR as sources of the flux (Sheeley *et al.*,

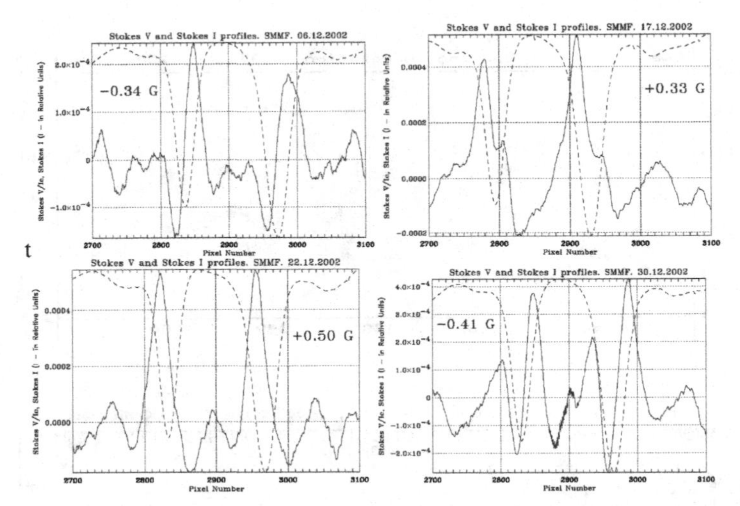

Figure 2. Stokes I and Stokes V/I_c profiles for the lines Fe I λ525.02 nm and Fe I λ525.06 nm for the sevaral days of the period, corresponding to the Carrington rotation 1997 (see Figure 1). The numbers in the panels are the SMMF strengths for the Fe I λ525.02 nm.

1985), show a good correspondence with reality. Thus sometimes a contribution of AR in the formation of the SMMF signals could be significant.

2. Results

One of the objectives of stellar observatios is to obtain the surface distribution of the magnetic fields. In the case of the Sun such distributions are presented in synoptic maps. Examples of such maps for the Carrington rotation (CR) 1997 are shown in Figure 1: the top panel is constructed on the basis of the STOP SSO measurements, and the bottom one is the WSO map (http://soi.stanford.edu/data). It is seen that despite of the differences in the spatial resolution used (about 3 arc min in the WSO and about 100// in the SSO), the two maps are similar.

During the rotation of the Sun, different magnetic field structures are visible in the central zone of the solar disk (Demidov *et al.*, 2002), and, as a consequence, different SMMF strengths and Stokes profiles are registered. Some examples of Stokes V and Stokes I profiles for the pair of lines Fe I λ525.02 nm and Fe I λ525.06 nm for the different phases of the CR 1997 are shown in Figure 2.

To explore the role of AR in the formation of the SMMF signal, the following experiment was made. First of all, the real observed disk-integrated Stokes V profiles and magnetic field strengths B for the line Fe I λ525.02 nm were compared with the zonal Stokes V profiles and stregths B, calculated for different circular zones of solar disk using all corresponding points (pixels) of the magnetograms. Then the same procedure was repeated after deleting those pixels where AR are situated. For different days, different results were obtained depending on the number of deleted pixels and their position on the solar disk. Examples for two days with different results are shown in Figures 3 and 4. Figure 3 shows the magnetograms for 7 March 2001 (left panel) and 29 March 2001 (right panel), where pixels (by number N = 25 in the first case and N = 60 in the second), corresponding to AR, are deleted. Figure 4 shows the impact of deleting for the these days. Two top panels show the correlation coefficents between integrated and zonal Stokes V profiles, while the bottom panels show the behaviour of the magnetic field strengths B.

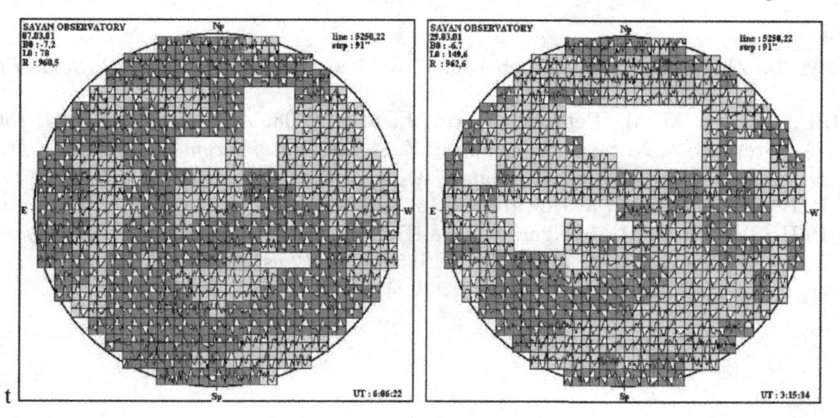

Figure 3. SSO magnetograms for 7 March 2001 (left panel) and 29 March 2001 (right panel) with deleted active regions pixels.

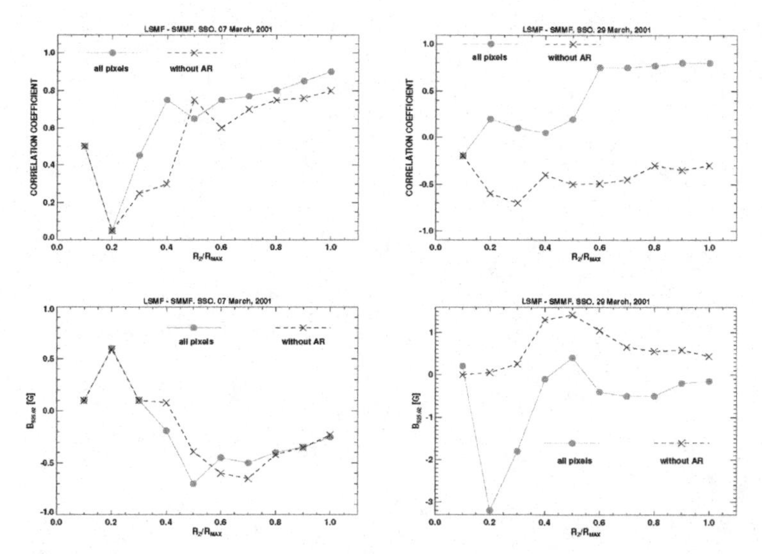

Figure 4. Top panels: comparison of the correlation coefficients between disk-integrated and zonal Stokes V profiles of the line Fe I λ525.02 nm for the cases when all pixels are taken into account (solid lines), and when active region pixels are deleted (dash lines). Bottom panels: the same as for the top panels, but for the magnetic field strengths B.

It is easy to see that the influence of the deleted AR is almost negligible for the first day, but is rather essential for the second one.

Acknowledgements

The presentation of this paper in the IAU Symposium 273 was possible due to partial support from the National Scince Foundation grant numbers ATM 0548260, AST 0968672 and NASA - Living With a Star grant number 09-LWSTRT09-0039, and Russian Foundation for Basic Researches travel grant 10-02-08457-3. The author acknowledges Dr. H. Balthasar for reading the manuscript and discussions. This work has been supported by the Deutsche Forschungsgemeinschaft (DFG) grant BA 1875/6-1.

References

Demidov, M. L., Zhigalov, V. V., Peshcherov, V. S., & Grigoryev, V. M. 2002, *Solar Phys.* 209, 217.
Donati, J.-F., Jardine, M. M., Petit, P., Morin, J., *et al.* 2008, *ASP Conf. Ser.* 384, 156.
Kotov, V. A., Stepanyan, N. S., & Sherbakova, Z. A. 1977, *Izv. Krymsk. Astrofiz. Obs.* 56, 75.
Petit, P., Dintrans, B., Solanki, S. K., Donati, J.-F., *et al.* 2008, *MNRAS.* 388, 80.
Scherrer, P. H., Wilcox, J. M., & Howard, R. 1972, *Solar Phys.* 22, 418.
Scherrer, P. H., Wilcox, J. M., Svalgaard, L., & Duvall, T. L., *et al.* 1977, *Solar Phys.* 54, 353.
Sheeley, N. R., DeVore, C. R., & Boris, J. P. 1985, *Solar Phys.* 98, 219.
Strassmeier, K. G. 2009, *Proc. IAU Symp.N.259.* 4, 363.

The Physics of Sun and Star Spots
Proceedings IAU Symposium No. 273, 2010
D. P. Choudhary & K. G. Strassmeier, eds.

doi:10.1017/S1743921311015018

Starspots, cycles, and magnetic fields

Steven H. Saar[1]

[1]Smithsonian Astrophysical Obs.,
60 Garden Street, Cambridge, MA 02138, USA
email: saar@cfa.harvard.edu

Abstract. I make a perhaps slightly foolhardy attempt to synthesize a semi-coherent scenario relating cycle characteristics, starspots, and the underlying magnetic fields with stellar properties such as mass and rotation. Key to this attempt is to first study single dwarfs; differential rotation plays a surprising role.

Keywords. Stars: magnetic fields, stars: spots, stars: rotation, stars: coronae, stars: activity, stars: late-type, stars: evolution

1. Introduction

Starspots are one of the most visible manifestations of stellar magnetism, and their periodic variation is often taken evidence for magnetic cycles. The most successful models of cyclic dynamos involve the interaction of differential rotation (DR) and helical turbulence to effect a polarity reversal and drive the cycle (in mean-field terms, $\alpha\Omega$ or $\alpha^2\Omega$ dynamos). However, a recent broad-based survey of surface DR measurements (the angular shear $\Delta\Omega$) in cool stars Barnes *et al.* (2005) showed virtually *no* dependence on rotation: $\Delta\Omega \propto \Omega^{0.15}$. Fundamentally, it is difficult to explain the clear, strong increase in magnetic activity with rotation (by a factor of $\sim 10^4$ in coronal emission, for example) if DR has virtually no Ω dependence (it seems unlikely that the α effect can have such a strong Ω dependence to compensate; Durney *et al.* (1993). Thus we are left with the profoundly puzzling situation of a magnetic dynamo whose products show strong rotational dependence, while its component parts (DR, α effect) show little such dependence.

I would like to propose a different way of looking at the data which is physically motivated, and in the end recovers an important, and in some aspects, surprising role for DR in magnetic field generation and cycles. The basic idea is an outgrowth of work presented last year Saar (2009), extended by additional data. I then see how this new view of DR may help to better understand the behavior of cycle periods, amplitudes, and starspot amplitudes and distributions (see also Saar *et al.* 2010, these proceedings).

2. DR Data and Analysis

The Barnes *et al.* (2005) DR sample is quite heterogeneous, including evolved stars, binaries, and a wide range of spectral types. Saar (2009) proposed that the sample was *too* broad, mixing categories (e.g., evolved and main sequence, binaries and single) which would more safely be analyzed separately, to better isolate the various parameters which may affect DR. I continue this line of thinking here, updating the Saar (2009) sample with more recent results e.g., (Morin *et al.* 2008; Lanza *et al.* 2009; Katsova *et al.* 2010). The sample is composed of single (or wide binary) dwarfs, later than F5 (B−V> 0.46). The latter restriction avoids concerns that hotter stars may have quite different magnetic generation processes (possibly due to the extreme thinness of their convection zones) as suggested by lack of any rotation-activity relationship Böhm-Vitense *et al.* (2002). Where multiple measurement exist (some stars may have variable DR; Jeffers *et al.* (2007)), I use the highest DR value as indicative of the maximum strength of the dynamo for a given set of stellar conditions. I also require stars with rotation period ($P_{\rm rot}$) measurements, to avoid sin$i$ uncertainties, and systematics from activity-rotation based estimates (unfortunately, this rules out most DR values determined from the Fourier transform method, as most of these stars have only $v \sin i$ data; Reiners & Schmitt (2003).

Several important caveats should be noted. Although the DR sample is pruned, the data *themselves* are also heterogeneous, with several quite different detection methods used. Timing methods (measuring drifts in $P_{\rm rot}$ from photometry or Ca II HK) give lower limits to DR. There is also the general worry that $P_{\rm rot}$ derived from photometry/HK can be difficult to interpret properly when the star is highly spotted Eaton *et al.* (1996); Jeffers (2005). Doppler imaging methods can give better DR values, but are subject to extrapolation uncertainties when features are only visible in a limited range of latitudes, and systematics in latitude placement. Fourier methods in principle derive DR over the whole surface, but are confused by spots (especially at the poles).

With these issues in mind, I analyze the expanded Saar (2009) sample for trends. Figure 1 shows a clear relationship $\Delta\Omega \propto \Omega^{0.68}$ ($\sigma_{\rm fit}$ = 0.246 dex) for $\Omega \leqslant 3$ d^{-1} ($\approx 12\Omega_\odot$), already quite different from the Barnes *et al.* (2005) result ($\propto \Omega^{0.15}$), but very similar to the old HK-based result ($\propto \Omega^{0.7}$; (Donahue *et al.* 1996, Donahue *et al.* 1996). There is also reasonable agreement with detailed 3-D hydrodynamic models, which find $\Delta\Omega \propto \Omega^{0.4}$ (Brown *et al.* 2008, Brown *et al.* 2008). The relationship disappears for $\Omega > 3$ d^{-1} ($\approx 12\Omega_\odot$), but a new, color dependent one emerges: $\log_{10} \Omega \propto -2.27$(B−V) (Fig. 1. right; $\sigma_{\rm fit}$ = 0.129 dex; throughout, the smallest $\Delta\Omega$ value is neglected from the fits as an outlier). This suggests DR in rapid rotators may depend on $T_{\rm eff}$ (cf. (Barnes *et al.* 2005, Barnes *et al.* 2005)) or the convective turnover timescale τ_C, which also produce reasonable fits.

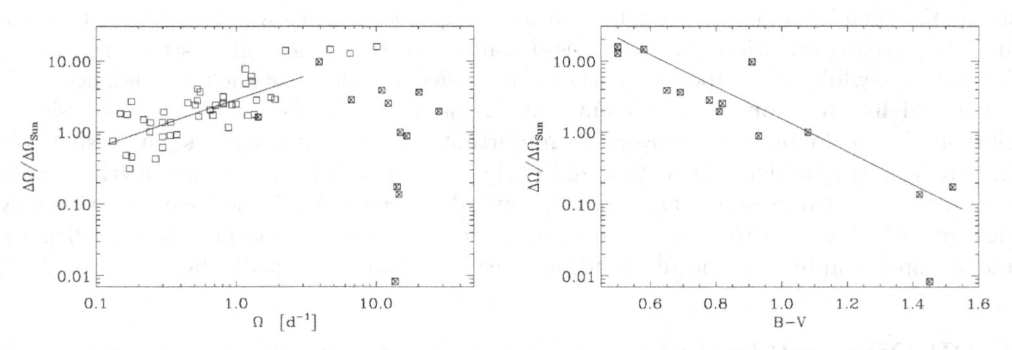

Figure 1. [Left]: Surface DR $\Delta\Omega$ (normalized to the solar value) vs. Ω for the sample, saturated activity (SA) stars (see Fig. 2, right) are crossed. A least squares fit $\Delta\Omega \propto \Omega^{0.68}$ (solid) is shown for $\Omega < 3$ d^{-1}. [Right]: Normalized $\Delta\Omega$ vs. B-V color for stars with $\Omega > 3$ d^{-1}.

The latter fit suggests one physically motivated way to combine the Ω and color relationships would be through the inverse Rossby number Ro^{-1} = $\tau_C\Omega$ (I take τ_C from Gunn *et al.* (1998). This idea has the added advantage of tying DR in with the mean-field dynamo number ($\propto$ Ro^{-2}) and with rotational evolution (Barnes & Kim 2010, Barnes & Kim 2010). When DR is plotted against Ro^{-1} (Fig. 2, left), the high rotation portion collapses onto a single relationship, and the low rotation portion is mostly reshuffled, yielding $\Delta\Omega \propto \mathrm{Ro}^{-1.00}$ ($\sigma_{\mathrm{fit}} = 0.244$ dex) for $\mathrm{Ro}^{-1} < 80$, and $\Delta\Omega \propto \mathrm{Ro}^{1.28}$ for $\mathrm{Ro}^{-1} > 80$($\sigma_{\mathrm{fit}} = 0.367$ dex). Thus, not only does $\Delta\Omega$ in single dwarfs appear to have a *significant* Ω dependence, it *reverses* its strong dependence at high Ro^{-1}.

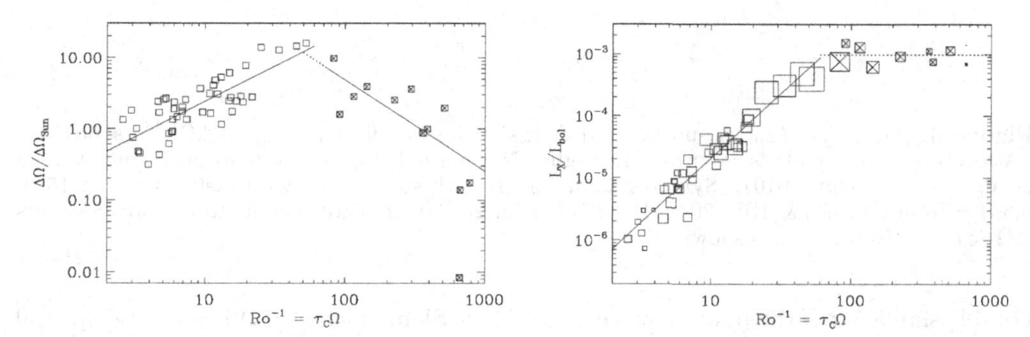

Figure 2. [Left]: Normalized $\Delta\Omega$ vs. $\mathrm{Ro}^{-1} = \tau_C\Omega$; symbols as in Fig. 1. Least square fits of $\Delta\Omega$ vs. $\mathrm{Ro}^{-1.00}$ ($\mathrm{Ro}^{-1} < 80$) and $\Delta\Omega$ vs. $\mathrm{Ro}^{1.28}$ ($\mathrm{Ro}^{-1} > 80$) are shown (solid). [Right]: L_X/L_{bol} vs. Ro^{-1}; symbols as in Fig. 1, with size scaling as $(\Delta\Omega)^{0.5}$. A fit with $L_X/L_{\mathrm{bol}} \propto \mathrm{Ro}^{-2.088}$ (solid) together with the SA level (dotted) is indicated.

The striking trend reversal in $\Delta\Omega$ finds an impressive parallel in the traditional rotation-activity relationship. It has long been known that various measures of magnetic activity reach a maximum level - "saturate" - above some critical rotation level. If one studies the fractional (relative to bolometric) X-ray luminosity L_X/L_{bol} vs. Ro^{-1} for the DR sample (X-ray data from ROSAT All Sky Survey; calibration from Hünsch *et al.* (1999), it is immediately apparent that saturated activity (=SA) occurs at the peak in DR: at $\mathrm{Ro}^{-1} \approx 80$ (Fig. 2, right). This can be even more strikingly demonstrated by directly plotting L_X/L_{bol} against DR (Fig. 3, left). The diagram splits into two radically different regimes, giving the appearance of a large number "7": a low activity regime showing $L_X/L_{\mathrm{bol}} \propto \Delta\Omega^{1.36}$ ($\sigma_{\mathrm{fit}} = 0.428$ dex) and a SA regime where coronal emission is independent of $\Delta\Omega$ over 3 orders of magnitude in rotational shear. Since L_X is proportional to magnetic flux over many decades (Pevtsov *et al.* (2003)), the implication is both clear and surprising: in the SA regime, a maximal amount of coronal/magnetic flux can be generated by stellar dynamos *independent of the differential rotation rate.* This would appear to be striking proof that *a fundamentally different kind of dynamo dominates the SA realm.* Hints of this have already arisen in the persistence of low latitude magnetic features on rapid rotators (e.g., Deluca *et al.* (1997)).

It is also informative to explore the time dependence of DR. I use the latest formulation of gyrochronology, which uses Ω and τ_C to derive stellar ages, t (Barnes & Kim 2010, Barnes & Kim 2010). When DR is plotted versus gyrochronological age, a gently decline of DR with time is seen (Fig. 3, right). When SA stars ($\mathrm{Ro}^{-1} > 80$) are excluded, the relationship tightens considerably, yielding $\Delta\Omega \propto t^{-0.57}$($\sigma_{\mathrm{fit}} = 0.232$ dex). The $\Delta\Omega$ in SA stars apparently increases, evolving up to this line. There is also a color/mass dependent effect: high mass stars appear on the declining sequence first, with lower mass stars joining later. The paucity of stars limits accuracy, but F8 stars appear to join the decline by t ~30 Myr, G1 by ~70 Myr, K2 by ~200 Myr, ~500 Myr for K5, and ~2 Gyr at M4. This roughly corresponds to when stars of these masses leave the SA regime. The

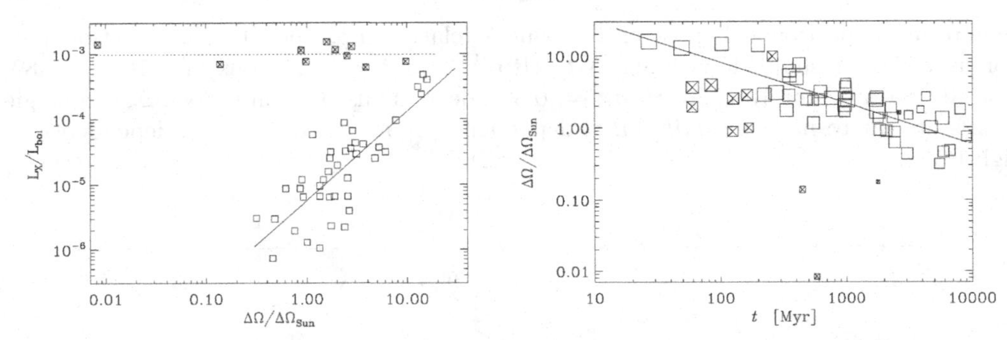

Figure 3. [Left]: $L_X/L_{\rm bol}$ vs. normalized $\Delta\Omega$, showing the fit $L_X/L_{\rm bol} \propto \Delta\Omega^{1.36}$ (solid) and SA level (dotted); symbols as in Fig. 1. [Right]: Normalized $\Delta\Omega$ vs. age t (from (Barnes & Kim 2010, Barnes & Kim 2010)). Symbols as in Fig. 1, with size scaling with stellar mass (1.15 to $0.3M_\odot$; from (Barnes & Kim 2010, Barnes & Kim 2010)). A power law fit to the non-SA stars $\Delta\Omega \propto t^{-0.57}$ (solid) is also shown.

notable similarity of the power law with the classic Skumanich (1972) $v \propto t^{-0.5}$ empirical spindown law, together with the mass dependent start of the spindown, strongly hint that *DR has an important role in controlling the rotational evolution of cool stars.* We will explore this further in an upcoming paper (Barnes & Saar 2010, in prep.)

3. Cycle Data and Analysis

Armed with this new understanding of the Ω and mass dependence of DR in single dwarfs, I now turn to the question of cycle properties. I adopt identical selection criteria as for the DR sample, and also insist vis. Saar & Brandenburg (1999) that the cycle be a relatively "clean", high quality detection (this is admittedly subjective in some cases). We adopt the cycle sample of Saar (2009), with some recent additions (e.g., (Oláh *et al.* 2009; Fares *et al.* 2009); Milingo *et al.* 2010, these proceedings). A quick look at the data reveals no strong trend with Ω (as found by, e.g., (Oláh *et al.* 2009). However, if one separates the secondary (= weaker amplitude) cycle period $P_{\rm cyc}(2)$, present in many moderate-to-active stars, and insists that they somehow be in a relationship separate from the main stellar cycle e.g, (Soon *et al.* 1993; Saar & Brandenburg 1999), a more complex, multitiered pattern emerges (Fig. 4). When cycle frequency $\omega_{\rm cyc}$ is plotted against Ω, three parallel tracks, separated by a factor of ~3.5, each with $\omega_{\rm cyc} \propto \Omega^{1.1}$. At $\Omega \approx 10\Omega_\odot$, very close to DR peak (Fig. 1, right), the relations reverse, showing $\omega_{\rm cyc} \propto \Omega^{-0.6}$ (Fig. 4). Only six of the 15 double cycle stars fail to have their $P_{\rm cyc}$ on separate branches. Thus the DR peak would appear to have left a mark in $\omega_{\rm cyc}$ as well.

Key to this interpretation of the cycle data is the reality of the $P_{\rm cyc}(2)$ as true, separate, polarity-reversing cycles (rather than just amplitude modulations of the main cycle). While this is not known to be true in general, there is evidence supporting this idea. First, $P_{\rm cyc}(2)$ are generally short, and recent Zeeman Doppler imaging evidence demonstrates that at least some short cycles *do* reverse polarity (Fares *et al.* 2009; Petit *et al.* 2009, Fares *et al.* 2009; Petit *et al.* 2009). Second, the amplitudes of the primary and secondary cycles show opposite trends with increasing Ro^{-1} Moss *et al.* (2008), suggesting they have different physical sources Böhm-Vitense (2007); possibly related to higher order dynamo modes Petit *et al.* (2009). The amplitudes of many cycles vary in time though Oláh *et al.* (2009), so assigning a single amplitude to each cycle is likely too simplistic.

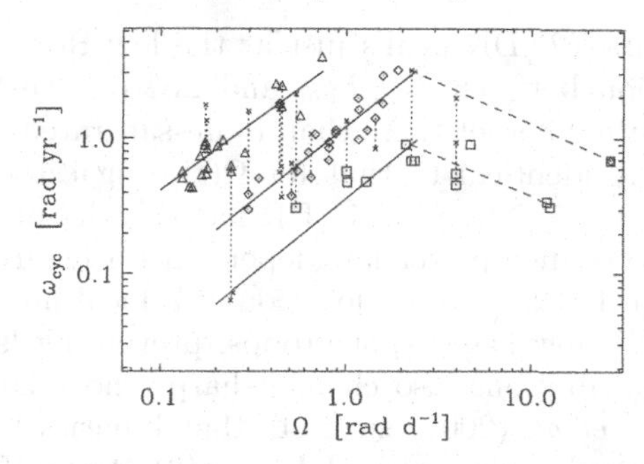

Figure 4. Cycle frequency $\omega_{\rm cyc}$ versus Ω for the dataset, triangles, diamonds, and squares represent stars assigned to the three "branches". Power law fits yield $\omega_{\rm cyc} \propto \Omega^{1.1}$ and $\sigma_{\rm fit} \approx$ 0.12 dex for each branch below $10\Omega_\odot$ (at DR peak and the start of mass-dependent drop in $\Delta\Omega$; see Fig. 1 left). The (less well defined) descending portions at high Ω are consistent with $\omega_{\rm cyc} \propto \Omega^{-0.6}$. A single fit to all the data yields $\omega_{\rm cyc} \propto \Omega^{0.15}$ with considerably higher scatter ($\sigma_{\rm fit}$ =0.33 dex; not shown).

4. Other Implications and Connections

The DR results for the sample adopted here show a strong, two regime Ω dependence. Yet Barnes *et al.* (2005), with a similar sample which, however, also included binaries and evolved stars, found much weaker Ω dependence, and lower overall average DR values (compare their Fig. 3 with Fig. 1 here). This suggests that binary and/or evolved stars must have weaker DR, which likely also has weaker Ω dependence than the present (dwarf, single) sample. A detailed investigation of this is beyond the scope of this paper, but is clearly worthwhile. I note that there has already been hints of a binary/single difference in how Ω evolves in clusters. If binaries do have lower DR, and DR is related to spindown (e.g., Fig. 3, right), this may explain the slower than expected rotation rates seen in some cluster binaries Meibom *et al.* (2006).

The sharp peak in DR may also have an effect on starspot distributions. There is a sharp drop in the maximum observed photometric spot amplitudes $A_{\rm max}$ at nearly the same Ro^{-1} ($\approx$ 100) as the DR peak (Saar *et al.* 2010, these proceedings). The σ width, skewness, and kurtosis of the spot amplitude distributions also drop sharply near the DR peak. The authors speculate that strong DR may strongly shear spot groups, increasing homogeneity, and thereby decreasing $A_{\rm max}$ and distribution moments.

Spectropolarimetric data has recently been starting to reveal large scale topologies in cool stars (Morin 2010, these proceedings). The DR peak corresponds roughly where dynamos transition to a more axisymmetric, poloidal geometry. An earlier transition from solar-like poloidal to strongly toroidal, non-axisymmetric geometry Petit *et al.* (2008) may, combined with the sharp $A_{\rm max}$ peak around Ro^{-1} $\sim$ 60 (Saar *et al.* 2010, these proceedings), may be the observational signature of the formation of powerful convection-zone dynamos and toroidal magnetic "wreathes" (Brown *et al.* 2010, Brown *et al.* 2010). Tachocline-driven dynamos may only be prominent at relatively low Ro^{-1}, since meridional flows needed to sustain them drop with Ω (Brown *et al.* 2008; Jouve *et al.* 2010).

5. Summary and Conclusions

To summarize the results: in single dwarfs, (1) DR has a strong, two part dependence on rotation, increasing as $\Delta\Omega \propto$ Ro^{-1} for coronally un-saturated stars and as Ro$^{1.38}$

for saturated activity stars. (2) DR peaks just at the low Ro^{-1} end of the SA branch. (3) The lack of correlation between L_X/L_{bol} and $\Delta\Omega$ on the SA branch implies the dynamo mechanism is quite different there than in un-saturated stars. (4) The age/mass dependence of DR (nearly identical to that for $\Omega(t)$) suggest DR plays an important role in rotational evolution. (5) The peak in DR has an echo in the multitiered pattern of ω_{cyc} values, which also change power law dependence roughly at the Ω and Ro that DR does. (6) The peak in DR is also the location of a local minimum in the maximum spot amplitudes; peak DR may shear spot groups, preventing larger $A_{\max}$ values. The moments of the A_{spot} distributions also change sharply near DR peak. (7) Comparing these results with Barnes *et al.* (2005) suggests that binaries likely have reduced DR, possibly with weaker dependence on Ω and Ro^{-1}. (8) Peak DR may correspond to a transition in dynamo symmetry from toroidal and non-axisymmetric (at lower Ro^{-1}) towards more poloidal and axisymmetric (at higher Ro^{-1}).

There is a clear need to study DR in binaries, and expand investigations to the pre- and post-MS. But it is also important to realize that the present results may well be revised, or even overturned, very quickly! COROT and Kepler will be soon be flooding us with vast mountains of new A_{spot}, P_{rot}, and $\Delta\Omega$ measurements. All these will surely help us better understand the complex workings of stellar dynamos and their myriad products... or at least provide much stimulating confusion!

Acknowledgments: This work was supported by a Solar Heliospheric Guest Investigator grant NNX10AF29G, and by Chandra grants GO8-9025A and GO0-11041A. I am indebted to S. Barnes, B. Brown, M. Browning, S. Meibom, A. Muñoz-Jaramillo, D. Nandy, K. Olah, and M. Rempel, among many others, for helpful conversations.

References

Böhm-Vitense, E. 2007, *Astrophys. J.*, 657, 486

Böhm-Vitense, E., Robinson, R., Carpenter, K., & Mena-Werth, J. 2002, *Astrophys. J.*, 569, 941

Barnes, J. R., Cameron, A. C., Donati, J.-F., James, D. J., Marsden, S. C., & Petit, P. 2005, *Mon. Not. Roy. Astron. Soc.*, 357, L1

Barnes, S. A. & Kim, Y.-C. 2010, *Astrophys. J.*, 721, 675

Brown, B. P., Browning, M. K., Brun, A. S., Miesch, M. S., & Toomre, J. 2010, *Astrophys. J.*, 711, 424

Brown, B. P., Browning, M. K., Brun, A. S., Miesch, M. S., & Toomre, J. 2008, *Astrophys. J.*, 689, 1354

Deluca, E. E., Fan, Y., & Saar, S. H. 1997, *Astrophys. J.*, 481, 369

Donahue, R. A., Saar, S. H., & Baliunas, S. L. 1996, *Astrophys. J.*, 466, 384

Durney, B. R., De Young, D. S., & Roxburgh, I. W. 1993, *Solar Phys.*, 145, 207

Eaton, J. A., Henry, G. W., & Fekel, F. C. 1996, *Astrophys. J.*, 462, 888

Fares, R., *et al.* 2009, *Mon. Not. Roy. Astron. Soc.*, 398, 1383

Gunn, A. G., Mitrou, C. K., & Doyle, J. G. 1998, *Mon. Not. Roy. Astron. Soc.*, 296, 150

Hünsch, M., Schmitt, J. H. M. M., Sterzik, M. F., & Voges, W. 1999, *A&AS*, 135, 319

Jeffers, S. V. 2005, *Mon. Not. Roy. Astron. Soc.*, 359, 729

Jeffers, S. V., Donati, J.-F., & Collier Cameron, A. 2007, *Mon. Not. Roy. Astron. Soc.*, 375, 567

Jouve, L., Brown, B. P., & Brun, A. S. 2010, *Astron. Astrophys*, 509, A32

Katsova, M. M., Livshits, M. A., Soon, W., Baliunas, S. L., & Sokoloff, D. 2010, *NewA*, 15, 274

Lanza, A. F., *et al.* 2009, *Astron. Astrophys*, 506, 255

Meibom, S., Mathieu, R. D., & Stassun, K. G. 2006, *Astrophys. J.*, 653, 621

Morin, J., *et al.* 2008, *Mon. Not. Roy. Astron. Soc.*, 384, 77

Moss, D., Saar, S. H., & Sokoloff, D. 2008, *Mon. Not. Roy. Astron. Soc.*, 388, 416

Oláh, K., *et al.* 2009, *Astron. Astrophys*, 501, 703

Petit, P., Dintrans, B., Morgenthaler, A., van Grootel, V., Morin, J., Lanoux, J., Aurière, M., & Konstantinova-Antova, R. 2009, *Astron. Astrophys*, 508, L9

Petit, P., *et al.* 2008, *Mon. Not. Roy. Astron. Soc.*, 388, 80

Pevtsov, A. A., Fisher, G. H., Acton, L. W., Longcope, D. W., Johns-Krull, C. M., Kankelborg, C. C., & Metcalf, T. R. 2003, *Astron. Astrophys*, 598, 1387

Reiners, A. & Schmitt, J. H. M. M. 2003, *Astron. Astrophys*, 398, 647

Saar, S. H. 2009, Stellar Dynamos as Revealed by Helio- and Asteroseismology, ASP Conf. Ser. 416, 375

Saar, S. H. & Brandenburg, A. 1999, *Astrophys. J.*, 524, 295

Skumanich, A. 1972, *Astrophys. J.*, 171, 565

Soon, W. H., Baliunas, S. L., & Zhang, Q. 1993, *Astrophys. J.* l, 414, L33

The Physics of Sun and Star Spots
Proceedings IAU Symposium No. 273, 2010
D.P. Choudhary & K.G. Strassmeier, eds.

doi:10.1017/S174392131101502X

The evolution of stellar surface activity and possible effects on exoplanets

Mark S. Giampapa[1]

[1]National Solar Observatory/NOAO
950 N. Cherry Ave., POB 26732, Tucson, Arizona USA 85726-6732
email: giampapa@noao.edu

Abstract. The evolution of stellar activity involves a complex interplay between the interior dynamo mechanism, the emergent magnetic field configurations and their coupling with stellar winds, the subsequent angular momentum evolution, and fundamental stellar parameters. The discussion of the evolution of surface activity will emphasize the main sequence phase, from the ZAMS to stars of solar-age. We will focus particularly on the evolution of the fractional area coverages of spots on the surfaces of solar-type stars. We fit an empirical relation to the fractional mean spot area coverage as a function of age for ages greater than the Pleiades of the form $\log(\text{MeanSpotCoverage}) = 0.90(\pm 0.26) - 1.03(\pm 0.10)\log(\text{Age})$, where Age is in Myr. In addition, we summarize the relative evolution of radiative emissions in various short wavelength bands that are associated with stellar magnetic field-related activity. Possible effects on young planetary atmospheres also are appropriate to consider given that stellar surface activity is the origin of the high-energy component of the ambient radiation and particle fields in which planetary atmosphere evolution occurs.

Keywords. Stars:spots, stars:activity, stars:planetary systems

---

1. Introduction

Spots and granulation are the most obvious inhomogeneities on the surface of the Sun. Spots are associated with strong magnetic fields and they are the most obvious manifestation of stellar surface 'activity', which we will define for the purposes of this discussion as the consequence of interactions of magnetic fields with the surrounding plasma. The granulation pattern is the result of outer convection. While we typically do not associate granulation with magnetic activity, the interaction of magnetic fields and granulation is seen in the form of systematic changes that are discernible through high spectral resolution observations of photospheric line profiles. For example, W. C. Livingston (National Solar Observatory) and collaborators find that the average amplitude of the natural asymmetry of absorption lines in the spectrum of the Sun-as-a-star, arising from the velocity-brightness correlation in upward moving granular cells, varies with the solar cycle. In particular, the asymmetry as measured by the mean bisector amplitude (see Gray 2008) is at a maximum near the time of solar minimum. It declines to a minimum near the time of solar maximum, consistent with what would occur if surface magnetic fields acted to reduce upflowing granulation velocities.

Stellar surface activity is exhibited in the spatially unresolved observations of stars primarily through line and continuum emission arising from chromospheres and coronae consisting of plasma that has been heated to temperatures well in excess of the stellar effective temperature. Brightness changes in the photometric light curves of late-type stars on rotational time scales, or longer time scales that are the stellar analogs of the 11-year sunspot cycle, also are associated with stellar surface features that are, in turn, delineated by magnetic fields. In the Sun, these brightness changes are seen in the variability of the

Total Solar Irradiance (TSI) and the Spectral Solar Irradiance (SSI). Transient activity such as flare outbursts and Coronal Mass Ejections (CMEs), and irradiance changes, also may influence the evolution of planetary atmospheres, initially in young exoplanet systems and, later, affect planetary climates in more mature systems.

We will discuss the nature and apparent sources of irradiance and brightness changes seen in the Sun and sun-like stars, respectively, and estimate the form of the evolution of spot coverage on solar-type stars combined with a review of the evolution of coronal and chromospheric emission in late-type dwarfs. Potential effects on planetary atmospheres will be briefly discussed. Recent reviews of the evolution of stellar activity include Guinan & Engle (2009); Engle, Guinan & Mizusawa, T. (2009); and, Preibisch & Feigelson (2005). The possible effects of host stars in an astrobiology context are discussed by Cuntz, Guinan & Kurucz (2010). This is not an exhaustive list but these papers and references therein should enable the interested reader to pursue this topic in more depth.

2. Convection, magnetic fields and irradiance variability

Given the observation of solar irradiance variations in phase with the solar cycle, and in view of the $\sim$ 4 million granules on the disk of the Sun at any given time, we might ask whether random variations in the number and size of turbulent cells combined with modulation of granule structures by the solar magnetic field could contribute to variations in heat flux. Conversely, in view of the modulation of the solar irradiance primarily due to magnetic structures such as faculae and spots, and the existence of modulation of stellar photometric light curves, also presumably by cool spots, we might justifiably ask why these brightness changes or irradiance variations are seen at all instead of the "blocked light" (in the case of spots) simply being re-radiated elsewhere.

Apparently, the answer lies in the enormous thermal inertia of the outer convection zone in the Sun and late-type stars. This same inertia tends to damp luminosity variations we might otherwise expect from structural changes deeper in the Sun. The light blocked by sunspots simply isn't re-radiated elsewhere on the Sun. Instead, the energy is stored in the convective envelope where it very slightly increases the potential and internal energy of the entire outer convection zone. The blocked heat flux remains stored in the Sun for hundreds of millennia instead of appearing elsewhere on the surface (Foukal *et al.* 2006). In faculae, the effect of their magnetic structure is to cause a slight localized depression in the photosphere, enabling radiation from lower and hotter layers to escape more easily. The predicted effects on changes in the solar radius are unobservably small.

Stellar 'Irradiance' Variability. Analogous to solar TSI is brightness variability in solar-type stars as observed in selected photometric bands spanning the visible range. The same kind of correlated changes in brightness with variations in chromospheric Ca II H & K core emission that are seen in the Sun during the course of the solar cycle also are observed in sun-like stars. However, unlike the Sun, anti-correlated brightness and chromospheric variability also is observed, as depicted in Fig. 1 from Radick *et al.* (1998).

Inspection of Fig. 1 reveals that the correlated and anti-correlated variability is associated with mean chromospheric emission level. The more active stars characterized by anti-correlated variability are typically younger, as can be seen from the superimposed 600 Myr Hyades isochrone. The correlation itself is thought to arise from the balance in fractional area coverage of bright facular features versus cool, dark spots so that in more active solar-type stars that are younger than the Sun, the net effect of the fractional area coverage of spots versus faculae is to produce a relative dimming with increasing chromospheric emission. Thus, a salient feature of the evolution of surface activity in sun-like stars is the change in the relative balance between dark spots and bright faculae

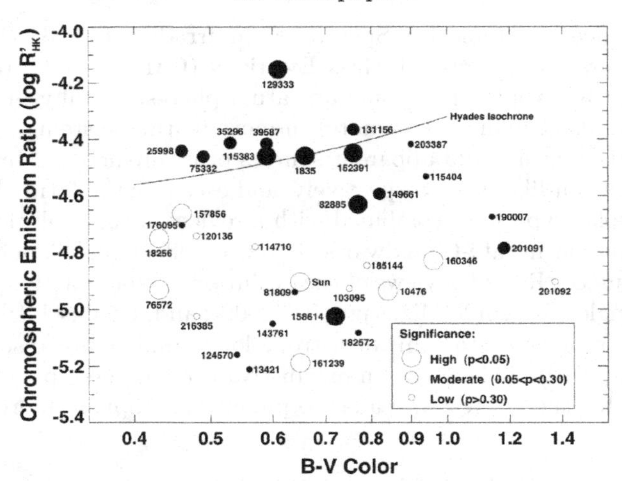

Figure 1. Correlated (open circles) and anti-correlated variability (filled circles) as a function of mean chromospheric emission level and color. Younger, more active stars exhibit anti-correlated variability while stars of solar-age with solar-like activity are characterized by variability that is directly correlated with chromospheric variations

in addition to the well known decline of the mean level of chromospheric emission with age.

3. The evolution of star spots

Using data from Messina (Rodonò & Guinan), Radick *et al.* (1995) and Saar, Barnes & Meibom (2010), I constructed Fig. 2 showing the logarithm of the 'Mean Spot Amplitude' as a function of age. The spot amplitude is really the amplitude of the photometric light curve, which is interpreted as the modulation of a star's V-band brightness by rotation carrying starspots across the line of sight. The vertical 'error bars' represent the one-sigma dispersion of spot amplitudes in the cluster.

Inspection of Fig. 2 shows that in the young clusters ranging from 30 Myr (IC 2391) to 100 Myr (Pleiades), spot amplitudes cluster around $\sim 10\%$. There is a sharp, linear (in the log) decline to around 1% by the age of the Hyades (600 Myr) and $\sim 0.1\%$ - 0.2% by the age of the Sun. However, there is a large gap in the available data at the intermediate age range between the Hyades and the Sun. Photometric monitoring of stars in an intermediate-age cluster such as NGC 752 (~ 2 Gyr) would be valuable in filling in this gap. Fitting the data in Fig. 2 with a linear function yields for the mean fractional area coverage of starspots at ages of the Pleiades or greater

$$\log(\mathrm{MeanSpotCoverage}) = 0.90(\pm 0.26) - 1.03(\pm 0.10)\log(\mathrm{Age}),$$

where Age is in units of Myr. The above fitted relation can be rewritten to a good approximation simply as (Starspot Area) $\approx 10/(\mathrm{Age})$, where, again, Age is in units of Myr.

The shape of the correlation in Fig. 2 is very much reminiscent of that for activity diagnostics as a function of rotation period, as shown in Fig. 3. The upper panel from Pizzolato *et al.* (2003) shows coronal X-ray luminosity as a function of rotation period (on a logarithmic scale) for individual objects. This radiative diagnostic can be compared to the lower graph of mean spot amplitude versus age. Of course, the extent of spots in the photospheres of solar-type and other late-type stars must be intimately related to

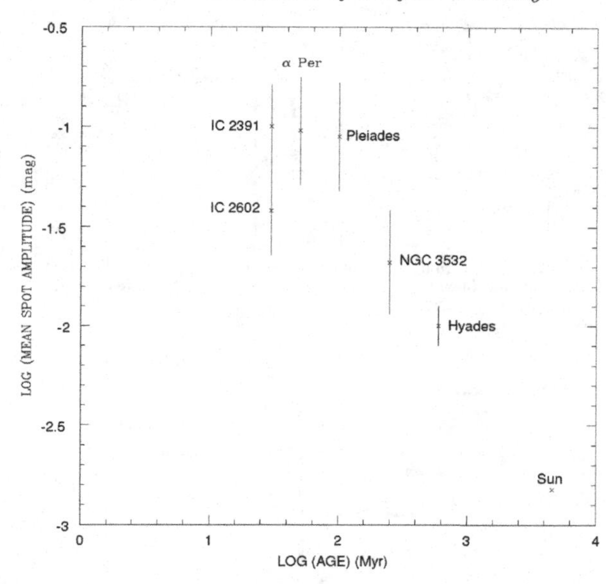

Figure 2. The evolution of the mean area coverage of star spots from the ages of young clusters to the age of the Sun. The vertical bars represent the one-sigma dispersion of spot amplitudes in the cluster

rotation via the internal dynamo. Stellar rotation rates are also a function of stellar age as a result of the interaction of stellar winds with extended magnetic field configurations combined with magnetic braking. The lower right panel in Fig. 3 shows the mean rotation period in the clusters and the Sun, and its correlation with age. The vertical 'error bars' represent the one-sigma dispersion in rotation periods for all the stars in their respective clusters with measured rotation periods, regardless of spectral type. Not surprisingly, the shape of this relation is very similar to the one for mean spot amplitude versus age. The similarity of the shape of the relations along with those of, say, L_x vs. rotation period, suggests that rotation determines total magnetic flux on stars, i.e., it is not just field strength that increases with rotation rates.

At this point we digress to note that the argument could be made that a decreasing amplitude of periodic variability in photometric light curves as a function of age could be interpreted as due to a more uniform spatial distribution of the same fractional area coverage of spots on the star as opposed to a real decrease in the total filling factor of spots. However, the concomitant decrease in mean levels of chromospheric and coronal emission along with the change from inversely to directly correlated brightness changes with chromospheric variability imply that the decline in photometric amplitudes with age is the result of a real decline in the filling factor of spots with stellar age.

Rotation rates decline more rapidly in the higher mass G dwarfs than in the lower mass M dwarfs. Thoughts as to why this is the case include decoupled spin-down of the outer convection zone, different dominant magnetic field configurations or simply a decrease in mass loss rates toward lower mass stars on the main sequence. The associated chromospheric and coronal emission also decline with decreasing rotation, with coronal X-ray emission depending sensitively on rotation and, hence, exhibiting a more rapid decline. The change (decline) in X-ray emission depends sensitively on rotation while lower-temperature chromospheric emission also decreases but not as rapidly. The rate of decline is more rapid at shorter wavelengths and becomes less rapid at UV wavelengths. It is these latter wavelengths that are more relevant to the photochemistry of the

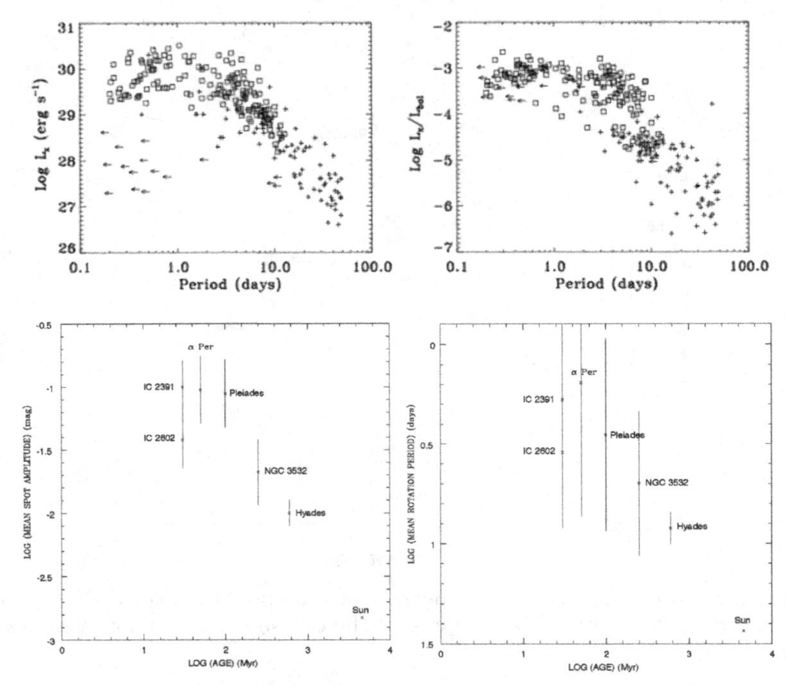

Figure 3. A comparison of the shapes of the mean spot fractional surface coverage versus age relation with the coronal activity and spot area coverage relations, respectively, each with rotation period. The similar shapes of these relations imply a common origin via the dynamo mechanism

atmospheres of planets, e.g., ozone chemistry is sensitive to UV emission. Hence, enhanced activity can have a more extended, longer-term influence in the UV compared to shorter wavelengths.

In this context, the recent study by Segura *et al.* (2010) merits a brief review. These authors examined the effect of a large UV flare event observed on the dM3.5e star, AD Leo, on an earth-like atmosphere for a planet located in the habitable zone at ~ 0.1 AU in a cool dwarf star of this type (Kaltenegger & Traub 2009). Segura *et al.* also assumed that an energetic proton event analogous to those that occur in major solar flares was associated with the UV event. The planet was assumed not to have a magnetic field that could shield it from energetic particles. They found that the enhanced UV and energetic particles incident on the earth-like atmosphere in the habitable zone destroyed $\sim 90\%$ of the ozone column for ~ 10 years before it recovered to its initial pre-flare state. The potential hazard for life is somewhat ambiguous because the pre-flare, 'quiescent' near-UV flux from this cool dwarf is relatively low compared to that of the Sun. However, it must be remembered that large flare events of this kind occur relatively frequently on a dwarf M flare star such as AD Leo. Moreover, evidence from X-ray light curves suggest that some degree of flaring is occurring continuously in this star (Giampapa *et al.* 1996) so that enhanced UV emission and energetic particle radiation likely characterizes the ambient radiative and particle environments in the habitable zones of active M dwarf stars.

Finally, recent work suggests that the nature of the outer convection zone in late-type stars not only plays an important role in the operative dynamo mechanism, the emergence of magnetic flux and the formation of spots but that it also influences the orbital inclination, or 'obliquities', of exoplanets. In particular, Winn *et al.* (2010) claim that tidal dissipation through interaction between the exoplanet and the mass in the outer convection zone acts to dampen obliquities to near zero in solar-type and cooler dwarfs. Conversely, in warmer stars with thin, low-mass convection zones, exoplanet systems exhibit a wide range of obliquities.

In conclusion, the enormous yield of starspot rotational modulation results emerging from the ultra-high precision photometry from the *Kepler* mission will undoubtedly increase our understanding of the relationship between spot characteristics and stellar properties on rotational and evolutionary time scales. We can also look forward to the development of techniques such as Zeeman Doppler Imaging and high precision spectropolarimetry combined with advanced modeling techniques that, together, will enable us to carry out 'solar physics' on stars.

References

Cuntz, M., Guinan, E. F., & Kurucz, R. L. 2010, *Solar and Stellar Variability: Impact on Earth and Planets*, Proceedings of the International Astronomical Union, IAU Symposium, 264, 419

Engle, S. G., Guinan, E. F., & Mizusawa, T. 2009, *Future Directions in Ultraviolet Spectroscopy: A Conference Inspired by the Accomplishments of the Far Ultraviolet Spectroscopic Explorer Mission*, AIP Conf. Proc., 1135, p. 221

Foukal, P., Frölich, C., Spruit, H., & Wigley, T. M. L. 2006, *Nature*, 443, 161

Giampapa, M. S., Rosner, R., Kashyap, V., Fleming, T. A., Schmitt, J.H.M.M., & Bookbinder, J. A.,1996, *Astrophys. J.*, 463, 707

Gray, D. F. 2008, *The Observation and Analysis of Stellar Photospheres* (Cambridge: Cambridge University Press)

Guinan, E. F & Engle, S. G. 2009, *The Ages of Stars*, Proceedings of the International Astronomical Union, IAU Symposium, 258, 395

Kaltenegger, L. & Traub, W. A. 2009, *Astrophys. J.*, 698, 519

Livingston, W. C. 2010, *private communication*

Messina, S., Rodonò, M., & Guinan, E. F. 2001, *Astron. Astrophys*, 366, 215

Pizzolato, N., Maggio, A., Micela, G., Sciortino, S., & Ventura, P. 2003, *Astron. Astrophys*, 397, 147

Preibisch, T. & Feigelson, E. D. 2005, *Astrophys. J. Suppl.*, 160, 390

Radick, R. R., Lockwood, G. W., Skiff, B. A., & Thompson, D. T. 1995, *Astrophys. J.*, 452, 332

Radick, R. R., Lockwood, G. W., Skiff, B. A., & Baliunas, S. L. 1998, *Astrophys. J. Suppl.*, 118, 239

Saar, S. H., Barnes, S., & Meibom, S. 2010, *BAAS*, 216, 833

Segura, A., Walkowicz, L., Meadows, V., Kasting, J., & Hawley, S. 2010, *Astrobiology*, in press

Winn, J. N., Fabrycky, D., Albrecht, S., & Johnson, J. A. 2010, *Astrophys. J. Lett.*, 718, L145

The Physics of Sun and Star Spots
Proceedings IAU Symposium No. 273, 2010
D.P. Choudhary & K.G. Strassmeier, eds.

doi:10.1017/S1743921311015031

Rotational modulation, shear, and cyclic activity in HII 1883

J. B. Milingo[1], S. H. Saar[2], L. A. Marschall[1], and J. R. Stauffer[3]

[1]Department of Physics, Gettysburg College, Gettysburg, PA 17325, USA
email: jmilingo@gettysburg.edu, marschal@gettysburg.edu

[2]Harvard-Smithsonian Center for Astrophysics, Cambridge, MA 02138, USA
email: saar@head.cfa.harvard.edu

[3]Spitzer Science Center, Pasadena, CA 91125, USA
email: stauffer@ipac.caltech.edu

Abstract. We present a 30 year compilation of V-band differential photometry for the Pleiades K dwarf HII 1883. HII 1883 has an average rotational period $\langle P_{rot} \rangle$ of $\sim 0.235d$ and displays rotational modulation due to non-uniform surface brightness as large as 0.2 magnitudes in V. Preliminary work yields a cycle period of $\sim 9yrs$ and rotational shear $\delta P_{rot}/\langle P_{rot} \rangle$ considerably less than solar. With such a long baseline of data available we can explore many aspects of the star's photometric variability. We present studies of the variation of the rotational modulation amplitude, $\langle V \rangle$, and P_{rot} over the cycle.

Keywords. Stars: activity, stars: late-type, stars: spots, stars: rotation

1. Introduction

Long-term photometry of spotted stars can be exploited to constrain our understanding of the underlying magnetic activity in late-type stars. Tying changes in integrated brightness to periodic changes in spot coverage, a long-running time series of photometry can be used to look for cycle periods analogous to the Sun's. Given that heavily spotted stars show rotational modulation due to overall spot coverage and asymmetry, if a time series is available that exceeds the rotational period of the star then seasons of observations can be examined for changes in the mean brightness of the star as well as changes in the rotation period. The rotational periods of spotted stars are determined from light curves that emerge as a result of the distribution of spots that exist at the time of observation. The period determined in this fashion depends on the latitude of the active regions producing the brightness modulation. Although the time scales associated with the emergence and decay of active regions must be considered here, changes in rotational period can be used as a limited measure of the surface differential rotation (Donahue *et al.* 1996).

This paper presents a nearly 30 year compilation of V band photometry for HII 1883. HII 1883 is one of three stars in a long-term study that Gettysburg College joined in 1992. HII 1883 is a (presumably) single K2 V Pleiades member that shows rotational modulation due to spot coverage (Stauffer 1984a; Stauffer *et al.* 1984b). With a V magnitude of 12.6 HII 1883 displays modulation as large as 0.2 magnitudes and a very rapid rotation rate. With published photometry dating back to 1980, the existing time series and continued monitoring of HII 1883 affords us the opportunity to explore various aspects of this active star's photometric variability.

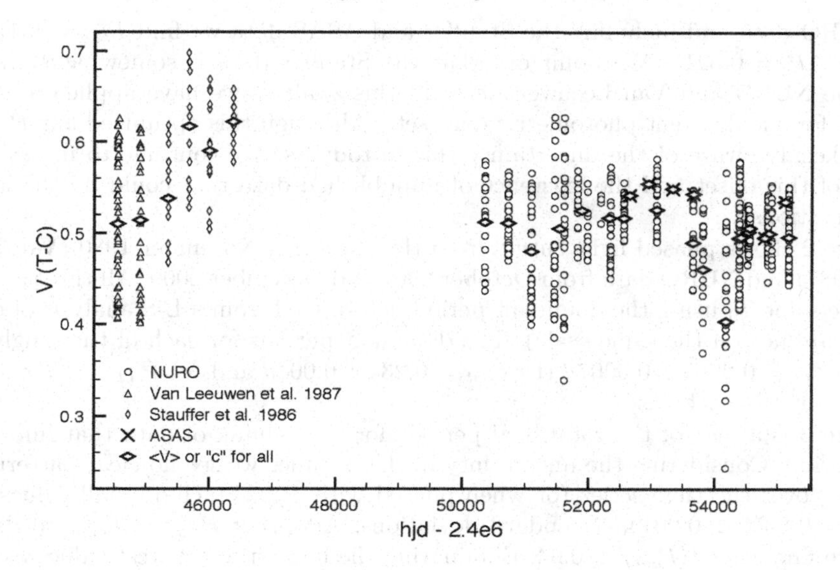

Figure 1. Compiled time series, differential V_{mag}.

2. Data

Figure 1 shows the compiled time series. The data from 1992-2009 was acquired using the Lowell Observatory 0.8 m telescope which, under an agreement with Northern Arizona University and the NURO Consortium, is operated 60% of the time as the National Undergraduate Research Observatory (NURO). As a NURO consortium member Gettysburg College is allocated 1-2 weeks of dedicated time each year to observe various targets including HII 1883. The NURO data is V band differential photometry using check and comparison stars from earlier studies by John Stauffer (Stauffer *et al.* 1987). The data from 1980 and 1981 is from Van Leeuwen *et al.* (1987), and data from the early-mid 1980's is from Stauffer *et al.* (1986). We have included additional data from the All Sky Automated Survey (ASAS) project (Pojmanski 2002) that stretches from 2003-2009. All data in the series is either standard or differential photometry using the comparison star from the NURO work. Horizontal diamonds in Fig. 1 mark the mean V of each chunk of data. The data from the ASAS project is noisy given the magnitude of HII 1883, hence these runs are simply represented by their mean values.

3. Analysis

Before it could be included in the time-series analysis, each run of data was examined to determine whether a light curve could be constructed. Breaks in the data considered the length of each run, how many observation points it contained, and possible contributions from the emergence and decay of different active regions. A value of $\langle V \rangle$ was determined for each chunk of data either by directly calculating the mean or by applying a fit to the phased light curve then taking the mean of that fit. This was done to mitigate any odd windowing of the observations.

A cycle period was determined by running a Lomb-Scargle (L-S) analysis on the $\langle V \rangle$ values. An initial rough analysis including just the Van Leeuwen and NURO data produced $P_{cyc} = 7.7 yrs$ with a L-S $FAP = 0.1315$. After a more careful examination of the NURO data and including the Stauffer and ASAS data we find $P_{cyc} = 9.1yrs$ with a L-S $FAP = 0.0215$. We point out that the Stauffer data is somewhat shifted from both the NURO and Van Leeuwen data. In this analysis we have applied an offset to account for the different photometric data sets. Although this optimized our P_{cyc} value, we are keenly aware of the uncertainty this introduces. We continue to investigate the nature of this offset and the existence of unpublished data that could fill the large gap in observations.

Figure 2 shows phased light curves from (left to right) November 1980 (Van Leeuwen *et al.* 1987) and NURO data from October 2007 and December 2009. All three are phased with the same T_0 using the dominant period determined from a L-S analysis of each run (y-axes are not on the same scale). The dominant periods for each of these light curves are (a) $P_{rot} = 0.2355 \pm 0.0007d$ (b) $P_{rot} = 0.234 \pm 0.004d$ and (c) $P_{rot} = 0.236 \pm 0.004d$; $FAP < 10^{-3}$ in each case.

Figure 3 is a plot of the rotational periods for each chunk of data that informed the cycle period. Considering the uncertainty in these values we see no clear pattern in P_{rot} emerging over the time series (or when phased with P_{cyc}). Including all values we find $\langle P_{rot} \rangle = 0.2352 \pm 0.0014d$ (standard deviation). Adopting $P_{max} - P_{min}$ to determine δP_{rot} we find $\delta P_{rot}/\langle P_{rot} \rangle \leqslant 0.04$. Eliminating the highest and lowest values as possible outliers we find $\langle P_{rot} \rangle = 0.2352 \pm 0.0006d$ and $\delta P_{rot}/\langle P_{rot} \rangle \leqslant 0.01$. We present both as upper limits considering the errors in P_{rot} overlap to a degree that makes our detection uncertain. Regardless our measure of differential rotation is considerably less than the solar value, measured as a star in Ca II HK, of 0.15 (Donahue *et al.* 1996).

Using the light curves from each data chunk, we examined the rotational modulation amplitudes as an alternate proxy, albeit muddy, for activity. Running a L-S analysis on the Van Leeuwen, Stauffer, and NURO data sets (no ASAS) we found $P_{cyc} \sim 2900d$ with a $FAP = 0.171$, only a weak indication of periodicity. There was a clear phase shift amongst the rotational light curves (see Figure 2). With all light curves phased to the same T_0 we ran a L-S analysis on the phase shifts and found $P_{cyc} \sim 2800d$ with a $FAP = 0.016$. With very little deviation in the P_{rot} values we interpret these phase shifts to be the march of active longitudes at play, but small changes in P_{rot} are tangled in here complicating the issue. With such a small spread in P_{rot} the strength of the dominant periodicity in the phase shifts is intriguing.

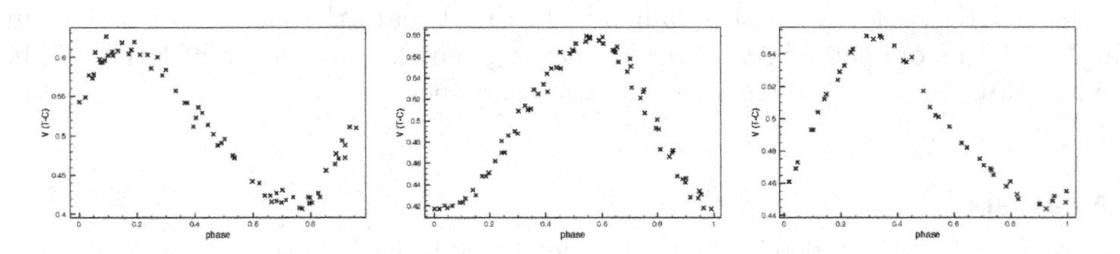

Figure 2. Light curves phased to the same T_0.

4. Future Work

Our work continues with HII 1883, as well as the other two Pleiades stars in this study (which are also rapid rotators). Continued monitoring at NURO and the addition of complimentary data, both archival (Krishnamurthi *et al.* (1998) for example) and

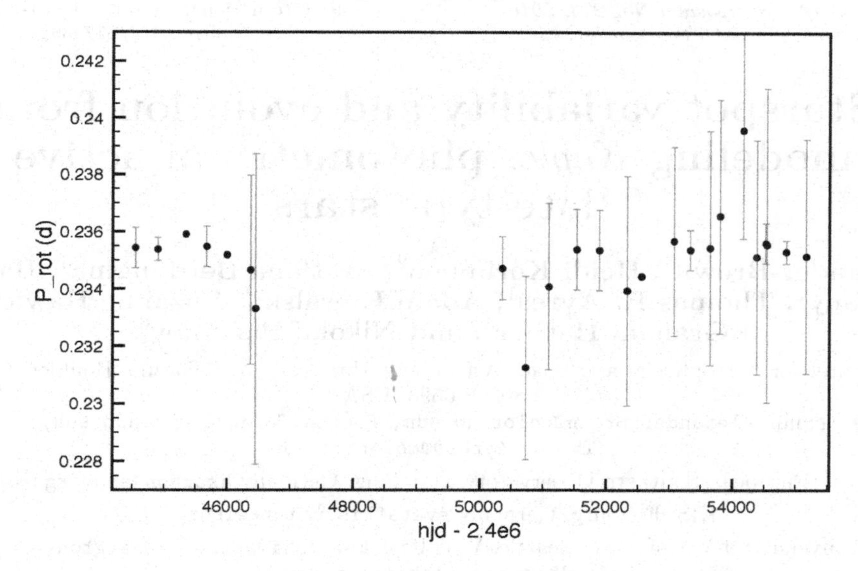

Figure 3. P_{rot} values throughout the time series.

newly acquired, will help paint a clearer picture of this active star. Many thanks to all, particularly Dr. Dana Backman, who have helped to move this project forward.

References

Donahue, R. A., Saar, S. H., & Baliunas, S. L. 1996, *Astrophys. J.*, 466, 384

Krishnamurthi, A., Terndrup, D. M., Pinsonneault, M. H., Sellgren, K., Stauffer, J. R., Schild, R., Backman, D. E., Beisser, K. B., Dahari, D. B., Dasgupta, A., Hagelgans, J. T., Seeds, M. A., Anand, R., Laaksonen, B. D., Marschall, L. A., & Ramseyer, T. 1998, *Astrophys. J.*, 493, 914

Pojmanski, G. 2002, *AcA*, 52, 397

Stauffer, J. R., Schild, R. A., Baliunas, S. L., & Africano, J. L. 1987, *Pub. Astron. Soc. Pac.*, 99, 471

Stauffer, J. R., Dorren, J. D., & Africano, J. L. 1986, *AJ*, 91, 1443

Stauffer, J. R. 1984, *Astrophys. J.*, 280, 189

Stauffer, J. R., Hartmann, L., Soderblom, D. R., & Burnham, N. 1984, *Astrophys. J.*, 280, 202

Van Leeuwen, F., Alphenaar, P., & Meys, J. J. M. 1987, *A&AS*, 67, 483

The Physics of Sun and Star Spots
Proceedings IAU Symposium No. 273, 2010
D.P. Choudhary & K.G. Strassmeier, eds.

doi:10.1017/S1743921311015043

Starspot variability and evolution from modeling *Kepler* photometry of active late-type stars

Alexander Brown[1], Heidi Korhonen[2], Svetlana Berdyugina[3], Barton Tofany[1], Thomas R. Ayres[1], Adam Kowalski[4], Suzanne Hawley[4], Graham Harper[5], and Nikolai Piskunov[6]

[1]Center for Astrophysics and Space Astronomy, University of Colorado, Boulder, CO 80309-0593, USA
email: Alexander.Brown@colorado.edu, Barton.Tofany@colorado.edu, Thomas.Ayres@@colorado.edu

[2]European Southern Observatoty, Garching, Germany hkorhon@eso.org

[3]KIS, Freiburg, Germany sveta@kis.uni-freiburg.de

[4]University of Washington, Seattle, WA, USA kowalski@astro.washington.edu, slh@astro.washington.edu

[5]Trinity College Dublin, Dublin, Ireland graham.harper@tcd.ie

[6]Dept. of Astronomy & Space Physics, Uppsala University, Box 515, SE-75120 Uppsala, Sweden
email: piskunov@astro.uu.se

Abstract. The *Kepler* satellite provides a unique opportunity to study the detailed optical photometric variability of late-type stars with unprecedentedly long (several year) continuous monitoring and sensitivity to very small-scale variations. We are studying a sample of over two hundred cool (mid-A - late-K spectral type) stars using *Kepler* long-cadence (30 minute sampling) observations. These stars show a remarkable range of photometric variability, but in this paper we concentrate on rotational modulation due to starspots and flaring. Modulation at the 0.1% level is readily discernable. We highlight the rapid timescales of starspot evolution seen on solar-like stars with rotational periods between 2 and 7 days.

Keywords. Stars:spots, stars:activity, stars:rotation, stars: late-type

---

1. Introduction

Starspots are the most blatant manifestations of photospheric magnetic activity patterns on late-type stars in the optical. Starspots are dark, and thus conspicuous, because the intense magnetic fields within them suppress the convective heat flux locally, thereby cooling the atmosphere above the spot. The optical photometric monitoring provided by the *Kepler* satellite (Koch *et al.* 2010) allows study of activity phenomena such as the growth, migration, and decay of starspots, differential rotation, activity cycles, and flaring. Better understanding of how magnetic activity and dynamos work on other stars should lead to a better understanding of and predictive capability for solar activity. Individual starspots have lifetimes ranging from a few months in tidally-locked binaries to a few weeks in single active dwarfs (Hussain 2002). On active stars multiple, independent activity cycles are seen with typical shorter periods of 3-4 years and longer periods of 8-10 years (Olah *et al.* 2009). Such rotational and activity cycle periods and spot lifetimes

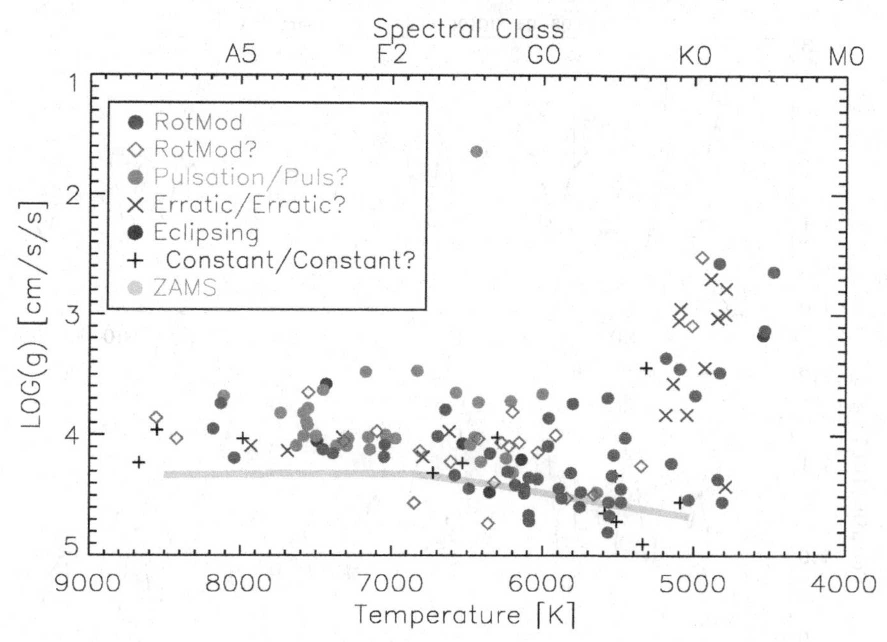

Figure 1. HR diagram for our *Kepler* Cycle 1 sample showing the dominant form of variability for each of these 135 stars. The effective temperatures and surface gravities are taken from the *Kepler Input Catalog.*

are well suited to study with the near-continuous *Kepler* photometry, particularly if long time series spanning the full mission lifetime ($\geqslant$ 3.5 years) can be obtained.

Three key observational attributes of the solar dynamo are: (1) its cyclic behavior, showing an ebb and flow of the spotted area over an $\sim$ 11-year time-scale; (2) migration of spots from mid-latitudes at the cycle onset to lower latitudes at the end (the "butterfly diagram"); and (3) preferred stable longitudes for spot emergence ("active longitudes"). Active stars, such as fast-rotating young dwarf stars and tidally-locked close binaries, clearly show enhanced magnetic activity compared with older, more slowly rotating dwarfs like the Sun. Their higher activity levels are manifested in intense X-ray, FUV, and Ca II H+K emissions, and a greater surface area coverage of spots (tens of %) than ever seen on the Sun (where sunspots typically cover only 0.01% of the surface). However, active star dynamos appear to differ substantially from the solar example: with cyclic behavior occurring on shorter time-scales and a strong tendency to show large spots at high latitudes, although still displaying preferences for active longitudes.

2. Target Selection Using *GALEX* FUV and NUV Imaging

Kepler's Field of View (FOV) covers 105 square degrees and is continuously pointed at the same area of sky in Cygnus/Lyra. However, while millions of potential stellar targets are available in the appropriate magnitude range (V = 9 - 15), only data for $\sim$ 150,000 stars is recorded because of telemetry limitations. Consequently, we developed a method using *GALEX* UV imaging to preselect stars likely to show high levels of stellar magnetic activity so that we could address our science goals of observing starspot evolution, activity patterns, differential rotation, and activity cycles on late-type stars. *GALEX* with a 1

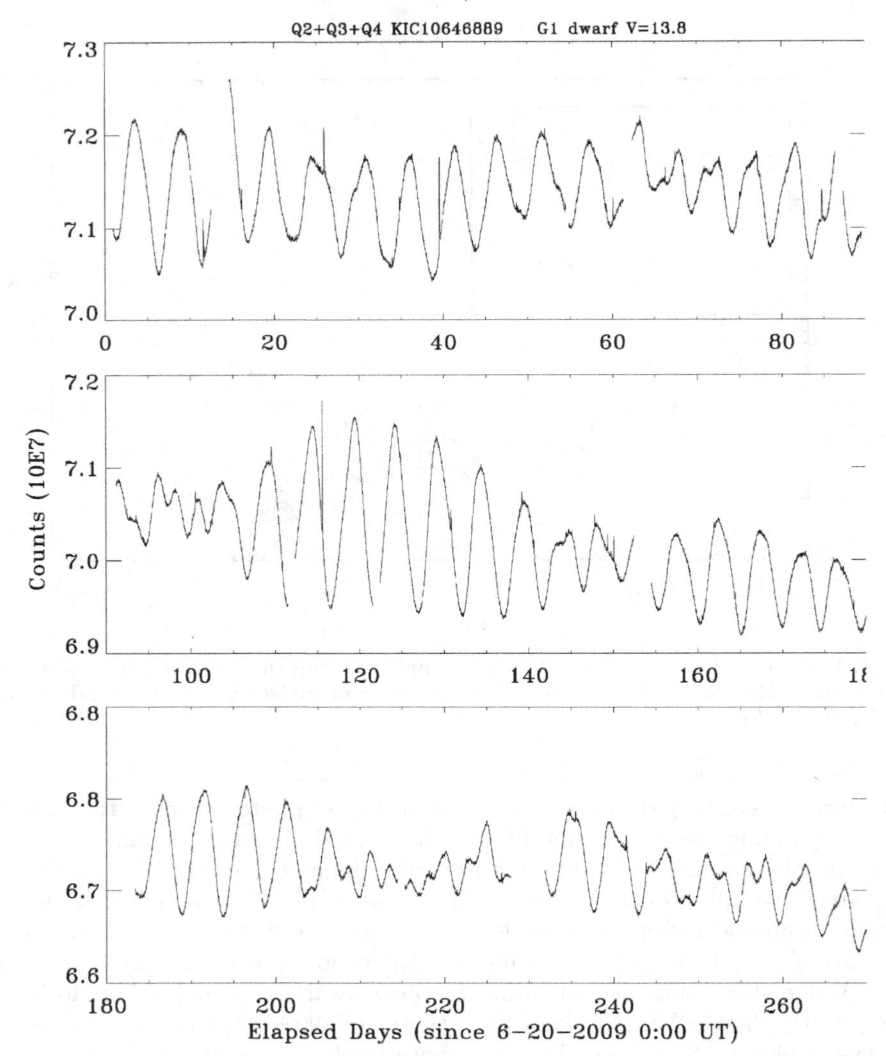

Figure 2. *Kepler* light-curve of the G1 dwarf KIC 10646889 over the 270 period covered by Kepler Quarters 2-3-4. Data for each Quarter is recorded by a different CCD. Prior to *Kepler* such time series were impossible.

square degree FOV is the only UV or X-ray satellite capable of surveying a significant fraction of the *Kepler* Field. Stellar activity is most obvious in the UV and X-ray spectral regions where the stellar photospheric radiation is weak. During *GALEX* Cycles 4 and 5 we imaged 28 square degrees of the *Kepler* Field using deep (up to 6 ksec) exposures. We used the FUV (1350-1800 Å) minus NUV (1800-2800 Å) color as a function of B-V to select active stars due to their FUV excess. We calibrated this method using nearly active and inactive stars. For inclusion in our *Kepler* sample stars had to — fall in our color-color selection box; be 3σ sources in both *GALEX* channels; have a stellar counterpart in the SDSS, 2MASS, or HST GSC catalogs; have no optical companion within 10 arcsec; have V= 9.0-15.0; and lie on the active *Kepler* CCD area.

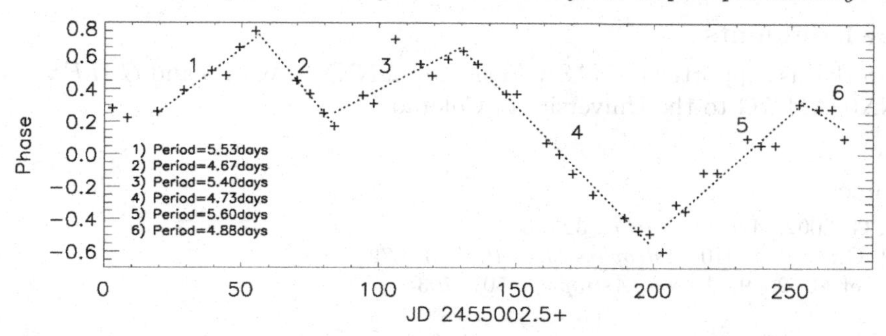

Figure 3. The early-G dwarf KIC 10646889 shows rapid spot migration in both longitude and latitude implying strong differential rotation. Alternatively, the active region "flip-flop" process might also be influencing the light-curve.

3. *Kepler* Observations

Kepler is a NASA satellite launched in March 2009 into an Earth-trailing orbit for thermal stability. It has a mission lifetime of at least 3.5 years and a primary science goal of detecting Earth-like planets through transit observations. *Kepler* has a vast array of 42 broadband optical (4300-8900 Å) CCDs that are 0.95 meters across at the focal plane of a 1.4 m telescope. Quarterly 90° rolls of the spacecraft maintain power to the spacecraft and aid thermal stability. Data downloads occur once a month.

Our GO Cycle 1 and 2 programs are obtaining 30 minute cadence photometry for a total of just over 200 stars with spectral types from mid-A to mid-K. We have currently received 9 months [3 quarters] of *Kepler* photometry for our Cycle 1 sample and data acquisition is on-going for our 185 Cycle 2 targets. Unprecedented near-continuous time series should result with durations of several years. An H-R diagram outlining the dominant types of variability is shown in Fig. 1.

In Fig. 2 we show 9 months of *Kepler* photometry for the G1 dwarf KIC 10646889, which has a mean rotational period is 5.16 days, which is rapid rotation for a solar-mass star. A strongly varying modulation with a maximum amplitude of 3% is seen due to starspots migrating relative to each other. Strong evidence for differential rotation is present (as illustrated in Fig. 3 where the apparent rotation period is measured for week-long subsamples of the data) — as dominant spots move to higher latitudes, longer apparent rotation periods are measured.

4. Conclusions

The *Kepler* data are of high quality and allow study of low amplitude variability. A wide diversity of types of variability is seen — including rotational modulation, pulsation, eclipses, flares, and erratic variability. Roughly one third of the stars shows strong rotational modulation, while a further third are dominated by pulsation. Pulsation is generally restricted to earlier A-F spectral types. White-light flares are common on rapidly-rotating solar-like stars. Erratic variability is typical for cooler giant stars. *Kepler* provides a unique, near-continuous record of stellar optical variability that allows detailed study of starspot modulation and the growth, decay and migration of starspots for stars with a wide range of fundamental properties. Rotational modulation periods range from a K dwarf with a 0.6 day period to a K giant with a 40 day period. The light-curve of the G1 dwarf KIC 10646889 illustrates the dramatic spot migration and strong inferred differential rotation that can be observed.

Acknowledgments

This work was supported by NASA *Kepler* grant NNX10AC51G and *GALEX* NNX08AU63G and NNX09AM47G to the University of Colorado.

References

Hussain, G. 2002, *Astron. Nachr.*, 323, 349
Koch, D. G., et al., 2010, *Astrophys. J. Lett.*, 713, L79
Olah, K., et al., 2009, *Astron. Astrophys*, 501, 703

Physics of Sun and Star Spots
Proceedings IAU Symposium No. 273, 2010
D. P. Choudhary & K. G. Strassmeier, eds.

doi:10.1017/S1743921311015055

The negative magnetic pressure effect in stratified turbulence

K. Kemel[1,2], A. Brandenburg[1,2], N. Kleeorin[3], and I. Rogachevskii[3]

[1]NORDITA, AlbaNova University Center, Roslagstullsbacken 23, SE-10691 Stockholm, Sweden
[2]Department of Astronomy, Stockholm University, SE–10691 Stockholm, Sweden
[3]Department of Mechanical Engineering, Ben-Gurion University of the Negev, POB 653, Beer-Sheva 84105, Israel

Abstract. While the rising flux tube paradigm is an elegant theory, its basic assumptions, thin flux tubes at the bottom of the convection zone with field strengths two orders of magnitude above equipartition, remain numerically unverified at best. As such, in recent years the idea of a formation of sunspots near the top of the convection zone has generated some interest. The presence of turbulence can strongly enhance diffusive transport mechanisms, leading to an effective transport coefficient formalism in the mean-field formulation. The question is what happens to these coefficients when the turbulence becomes anisotropic due to a strong large-scale mean magnetic field. It has been noted in the past that this anisotropy can also lead to highly non-diffusive behavior. In the present work we investigate the formation of large-scale magnetic structures as a result of a negative contribution of turbulence to the large-scale effective magnetic pressure in the presence of stratification. In direct numerical simulations of forced turbulence in a stratified box, we verify the existence of this effect. This phenomenon can cause formation of large-scale magnetic structures even from initially uniform large-scale magnetic field.

Keywords. Turbulence, MHD, sunspots

---

1. Introduction

The standard explanation for the appearance of strong magnetic fields at the solar surface involves the coherent rise of a tachocline-generated magnetic flux tube through the solar convection zone. Flux tube emergence simulations do give very promising results (Rempel *et al.* 2009), the paradigm looks elegant and is very textbook friendly, but some of its assumptions are problematic: the integrity of flux tubes, their rise, and even their very existence.

So far, numerical simulations have failed to produce the assumed thin magnetic flux tubes in the tachocline (Cattaneo *et al.* 2006; Parker 2009). Magnetic buoyancy as the driving transport mechanism through the convection zone can be dominated by downward pumping (Nordlund *et al.* 1992; Tobias *et al.* 1998). For the tubes to remain intact, strongly super-equipartition field strengths and strong twists are required (Fan 2001).

Thus, it seems not unreasonable to explore alternative mechanisms of formation of strong magnetic fields at the solar surface. A number of positive arguments has been put forward (Brandenburg 2005) in favour of formation of sunspots from local flux concentrations near the surface. In mean-field models, such magnetic instabilities have been produced by introducing a magnetic dependence of the thermal eddy diffusivity (Kitchatinov & Mazur 2000) and the viscous stress tensor (Brandenburg *et al.* 2010a).

Turbulence generally is associated with enhanced transport effects. However, it also exhibits non-diffusive behaviour, generating magnetic field on much larger scales than the driving scale. An example is turbulent dynamos that are able to produce large-scale magnetic fields (Brandenburg & Subramanian 2005). In a stratified layer, magnetic fields tend to become buoyantly unstable: assuming a constant temperature across the magnetic flux tube, pressure balance implies lower densities in regions of stronger magnetic field. Now one can wonder what the role of turbulent pressure is in such an equilibrium if the magnetic structures extend over several turbulent eddies. The turbulent pressure associated with the convective fluid motions is certainly not negligible and is strongly affected by the background magnetic field. The latter can be seen by evaluating the total turbulent dynamic pressure, here given for the isotropic case:

$$P_{\rm turb} = \tfrac{1}{3}\overline{\rho u^2} + \tfrac{1}{6}\overline{b^2}/\mu_0,$$

here $\boldsymbol{u}$ and $\boldsymbol{b}$ are the velocity and magnetic fluctuations, respectively, μ_0 the vacuum permeability and ρ the fluid density. Overbars indicate ensemble averaging. As shown in direct numerical simulations (Brandenburg *et al.* 2010a, hereafter referred to as BKR), in forced turbulence with imposed uniform large-scale magnetic field, the total turbulent energy is approximately conserved in this parameter regime.

$$\tfrac{1}{2}\overline{\rho u^2} + \tfrac{1}{2}\overline{b^2}/\mu_0 \equiv E_{\rm tot} \approx {\rm const.}$$

As a result, one finds a reversed feedback from the magnetic fluctuations on the turbulent pressure:

$$P_{\rm turb} = -\tfrac{1}{6}\overline{b^2}/\mu_0 + 2E_{\rm tot}/3$$

(Kleeorin *et al.* 1990; Rogachevskii & Kleeorin 2007, hereafter referred to as RK07). One can see that the effective mean magnetic pressure force is reduced and can, in a certain parameter range, be reversed. This effect would then counteract the aforementioned buoyancy instability in the presence of turbulence. RK07 have suggested that the reversed feedback instability could lead to the formation of magnetic flux concentrations near the solar surface.

Mean-field magnetohydromagnetic simulations by BKR confirmed the basic phenomenon of magnetic flux concentration by the effect of turbulence on the mean Lorentz force and for sufficient stratification, a linear instability was found. Direct numerical simulations (DNS) by BKR confirmed the reversed feedback phenomenology, but did not address the effect of stratification. This is one of the important additions of the more recent work of Brandenburg *et al.* (2010b), of which we report here the main highlights.

2. DNS model and analysis

DNS of forced turbulence were performed in a cubic computational domain of size L^3. For an isothermal equation of state and a constant vertical gravitational acceleration g, one finds an exponentially stratified density:

$$\rho = \rho_0 \exp\left(-z/H_\rho\right),$$

where $H_\rho = c_s^2/g$ is the constant density scale height, c_s is the isothermal sound speed and ρ_0 a normalisation factor. We choose $k_1 H_\rho = 1$, where $k_1 = 2\pi/L$, the smallest wavenumber. The density contrast then corresponds to $\exp 2\pi \approx 535$.

We solve the equations of compressible magneto-hydrodynamics in the form

$$\rho\frac{D\boldsymbol{U}}{Dt} = \boldsymbol{J}\times\boldsymbol{B} - c_s^2\nabla\ln\rho + \nabla\cdot(2\nu\rho\mathbf{S}) + \rho(\boldsymbol{f}+\boldsymbol{g}),$$

$$\frac{\partial\boldsymbol{A}}{\partial t} = \boldsymbol{U}\times\boldsymbol{B} + \eta\nabla^2\boldsymbol{A},$$

$$\frac{\partial \rho}{\partial t} = -\nabla \cdot \rho \boldsymbol{U},$$

where ν and η are kinematic viscosity and magnetic diffusivity, respectively. Furthermore, $\boldsymbol{B} = \boldsymbol{B}_0 + \nabla \times \boldsymbol{A}$ is the magnetic field consisting of a uniform mean field, $\boldsymbol{B}_0 = (0, B_0, 0)$, and a nonuniform part that is represented in terms of the magnetic vector potential $\boldsymbol{A}$, $\boldsymbol{J} = \nabla \times \boldsymbol{B}/\mu_0$ is the current density, and $\mathsf{S}_{ij} = \frac{1}{2}(U_{i,j} + U_{j,i}) - \frac{1}{3}\delta_{ij}\nabla \cdot \boldsymbol{U}$ is the traceless rate of strain tensor, where commas denote partial differentiation. The turbulence is driven with a forcing function $\boldsymbol{f}$ that consists of random plane non-polarized waves with an average wavenumber $k_{\rm f} = 5\,k_1$. The forcing strength is arranged such that the turbulent rms velocity, $u_{\rm rms} = \langle \boldsymbol{u}^2 \rangle^{1/2}$, is around $0.1\,c_s$. This value is small enough so that compressibility effects are weak.

In order to characterize our simulations, a set of dimensionless parameters is defined. The Reynolds number is given by $\mathrm{Re} = u_{\rm rms}/\nu k_{\rm f}$ and is of the order 120 in the simulations. The magnetic Prandtl number is $\pm = \nu/\eta$, although in reality this parameter is much smaller than unity; we choose here $\pm = 0.5 - 8$ in order to achieve higher values for the magnetic Reynolds number. The equipartition field strength $B_{\rm eq}$ is defined as a function of z and the imposed fields are normalised against $B_{\rm eq0}$, the equipartition strength in the middle of the domain:

$$B_{\rm eq}(z) = \left(\mu_0 \overline{\rho \boldsymbol{u}^2}\right)^{1/2}, \quad B_{\rm eq0} = (\mu_0 \rho_0)^{1/2} u_{\rm rms}.$$

The boundary conditions are stress-free perfect conductors at the top and bottom of the domain, periodicity in the horizontal direction. The simulations are performed with the PENCIL CODE†, which uses sixth-order explicit finite differences in space and a third-order accurate time stepping method (Brandenburg & Dobler 2002).

The contribution to the mean momentum density by the fluctuations is

$$\overline{\Pi}_{ij}^{f} = \overline{\rho u_i u_j} + \frac{1}{2}\delta_{ij}\overline{b^2} - \overline{b_i b_j},$$

where the overbars now indicate horizontal averages. The influence of the mean magnetic field can be found by subtracting the contributions that are present in the absence of a uniform background magnetic field and can be modelled by the following ansatz (RK07)

$$\overline{\Pi}_{ij}^{f,\overline{B}} - \overline{\Pi}_{ij}^{f,\overline{0}} = -\left(\frac{1}{2}\delta_{ij} q_p + e_i e_j q_e\right)\overline{\boldsymbol{B}}^2 + q_s \overline{B}_i \overline{B}_j,$$

which will then appear in the effective mean Lorentz force

$$\rho \boldsymbol{F}_i^M = -\nabla_j \left(\delta_{ij}\overline{\boldsymbol{B}}^2 + \overline{B}_i \overline{B}_j + \overline{\Pi}_{ij}^{f,\overline{B}} - \overline{\Pi}_{ij}^{f,\overline{0}}\right).$$

Thus the total effective magnetic pressure of the mean field is given by $1 - q_p\left(\overline{B}\right)\overline{\boldsymbol{B}}^2$ As $\overline{\boldsymbol{B}} \approx \left(0, \overline{B}, 0\right)$, and assuming no small-scale dynamo, we can determine q_p from

$$\rho\left(\overline{u_x^2} - \overline{u_{0x}^2}\right) + \frac{1}{2}\overline{\boldsymbol{b}^2} - \overline{b_x^2} = -\frac{1}{2} q_p \overline{\boldsymbol{B}}^2.$$

3. Results

We compare the effective magnetic pressure of the mean field with the turbulent kinetic energy density (Fig. 1a) and see that the resulting contribution of stratified turbulence

† `http://pencil-code.googlecode.com`

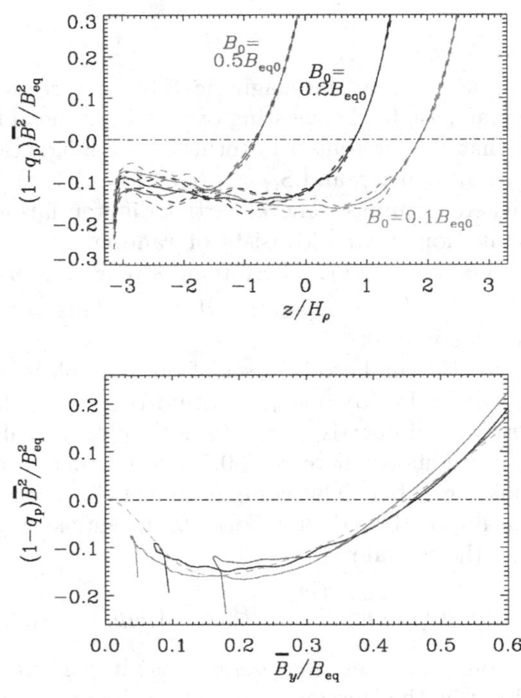

Figure 1. Normalized effective mean magnetic pressure for $B_0 = 0.1B_{\rm eq0}$, $B_0 = 0.2B_{\rm eq0}$, and $B_0 = 0.5B_{\rm eq0}$ using $Re = 120$. *Left*: as a function of depth *Right*: as function of the local value of the ratio of $B_0/B_{\rm eq}(z)$. Note that the curves for the different imposed field strengths collapse onto a single dependence and agree very well with the fit by BKR (dashed line). Adapted from Brandenburg *et al.* (2010b).

to the effective magnetic pressure is negative in a large part of the domain. Plotting the ratio of effective magnetic pressure to kinetic energy density as a function of the imposed uniform horizontal magnetic field divided by the equipartition field strength (Fig. 1b), collapses the observations from simulations with different imposed field strengths into a single dependence of q_p on $\overline{B_y}/B_{\rm eq}$. This result agrees with analytical calculations by RK07 and a fit based on simulations by BKR. The effect is fairly robust under an increase of the magnetic Prandtl number (Fig. 2a). While the reduced effective magnetic pressure is observed, the formation of local magnetic field concentrations as observed in mean-field simulations, has not yet been found in DNS (Fig. 2b).

4. Discussion

The DNS have shown that for an isothermal atmosphere with strong density stratification, the turbulent pressure is decreased due to a negative feedback from magnetic fluctuation generation, resulting in a negative effective mean magnetic pressure. The dependence on the ratio of imposed field to local equipartition field agrees with results obtained from analytic theory (RK07) and direct numerical simulations (BKR). The results are robust when changing the strength of the imposed field and the magnetic Prandtl number.

However, the simulations do not show any obvious signs of a large-scale instability that should result in magnetic flux concentrations, as was expected from mean-field calculations. A possible explanation for this discrepancy could be the simplicity of the ansatz

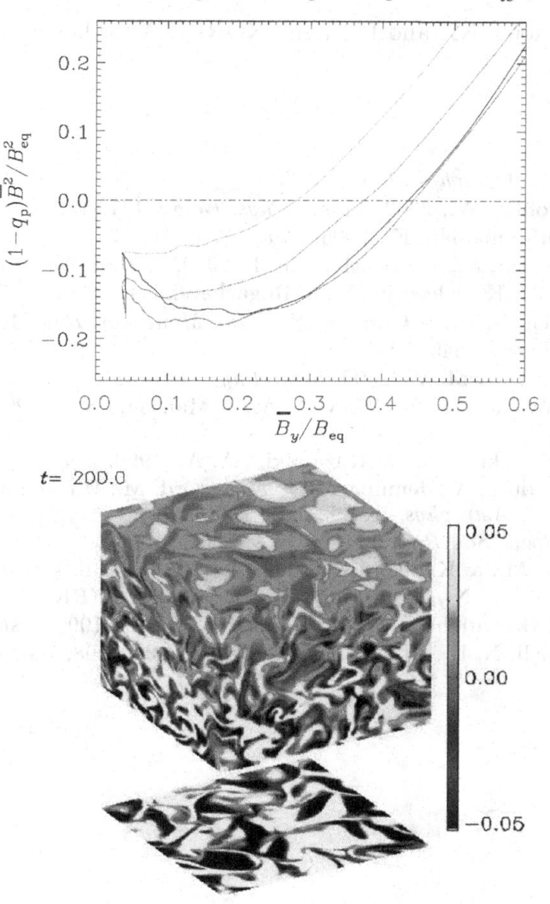

Figure 2. *Left*: Normalized effective mean magnetic pressure as a function of $B_0/B_{eq}(z)$ for $B_0 = 0.1B_{eq0}$, and varying magnetic Prandtl number,± = 0.05 (black), 2 (blue), 4 (red) and 8 (yellow). *Right*: Visualisation of $B_y - B_0$ on the periphery of the computational domain $B_0 = 0.1B_{eq0}$, and ± = 2.

for the effective Lorentz force. Indeed, early simulations of Tao *et al.* (1998) produced clear signs of flux separation into magnetised and unmagnetised regions in convection simulations at large aspect ratio. An increase of scale separation might alleviate the effect of higher order terms. Future work should incorporate this increase as well as a wider scan of the parameter regime and ultimately the inclusion of radiative transfer, as was done in simulations of Kitiashvili *et al.* (2010), which showed flux concentrations in the presence of a vertical field.

Acknowledgements

We acknowledge the allocation of computing resources provided by the Swedish National Allocations Committee at the Center for Parallel Computers at the Royal Institute of Technology in Stockholm and the National Supercomputer Centers in Linköping as well as the Norwegian National Allocations Committee at the Bergen Center for Computational Science. This work was supported in part by the European Research Council under the AstroDyn Research Project No. 227952 and the Swedish Research Council

Grant No. 621-2007-4064. NK and IR thank NORDITA for hospitality and support during their visits.

References

Brandenburg, A., 2005, *Astrophys. J.*, 625, 539
Brandenburg, A. & Dobler, W., 2002, *Comp. Phys. Comm.*147, 471
Brandenburg, A. & Subramanian, K., 2005, *Phys. Rep.*, 417, 1
Brandenburg, A., Kleeorin, N., & Rogachevskii, I., 2010, *Astron. Nachr.*, 331, 5 (BKR)
Brandenburg, A., Kemel, K., Kleeorin, N., & Rogachevskii, I., 2010, *arXiv:1005.5700*
Cattaneo, F., Brummell, N. H., & Cline, K. S., 2006, *Mon. Not. Roy. Astron. Soc.*, 365, 727
Fan, Y., 2001, *Astrophys. J.*, 546, 509
Kitchatinov, L. L. & Mazur, M. V., 2000, *Solar Phys.*, 191, 325
Kitiashvili, I. N., Kosovichev, A. G., Wray, A. A., & Mansour, N. N., 2010, *Astrophys. J.*, 719, 307
Kleeorin, N. I., Rogachevskii, I. V., & Ruzmaikin, A. A., 1990, *Sov. Phys.*, 70, 878
Nordlund, Å., Brandenburg, A., Jennings, R. L., Rieutord, M., Ruokolainen, J., Stein, R. F., & Tuominen, I., 1992, *Astrophys. J.*, 392, 647
Parker, E. N., 2009, *Space Sci. Rev.*, 144, 15
Rempel, M., Schüssler, M., & Knölker, M., 2009, *Astrophys. J.*, 691, 640
Rogachevskii, I. & Kleeorin, N., 2007, *Phys. Rev. E*, 76, 056307(RK07)
Tao, L., Weiss, N. O., Brownjohn, D. P., & Proctor, M. R. E., 1998, *Astrophys. J.*, 496, L39
Tobias, S. M., Brummell, N. H., Clune, T. L., & Toomre, J., 1998, *Astrophys. J.*, 502, L177

The Physics of Sun and Star Spots
Proceedings IAU Symposium No. 273, 2010
D. P. Choudhary & K. G. Strassmeier, eds.

doi:10.1017/S1743921311015067

Stellar activity, differential rotation, and exoplanets

A. F. Lanza

INAF-Osservatorio Astrofisico di Catania, Via S. Sofia, 78, 95123 Catania, Italy
email: nuccio.lanza@oact.inaf.it

Abstract. The photospheric spot activity of some of the stars with transiting planets discovered by the CoRoT space experiment is reviewed. Their out-of-transit light modulations are fitted by a spot model previously tested with the total solar irradiance variations. This approach allows us to study the longitude distribution of the spotted area and its variations versus time during the five months of a typical CoRoT time series. The migration of the spots in longitude provides a lower limit for the surface differential rotation, while the variation of the total spotted area can be used to search for short-term cycles akin the solar Rieger cycles. The possible impact of a close-in giant planet on stellar activity is also discussed.

Keywords. Stars: activity; stars: magnetic fields; stars: spots; stars: rotation; planetary systems

1. Introduction

Magnetic fields in the Sun and late-type stars produce active regions in the photosphere by affecting the transfer of energy and momentum in the outermost convective layers. These features consist of dark spots, bright faculae, and an enhanced network of magnetic flux tubes with radii down to about 100 km and kG field strength. Solar active regions can be studied in details thanks to a spatial resolution down to $\sim$ 100 km and a time resolution better than $\sim$ 1 s.

In distant stars, we lack spatial resolution (with the exception of the supergiant α Orionis, see Dupree (2011)), so we must apply indirect techniques to map their photospheres. The most successful is Doppler Imaging that produces a two-dimensional map of the surface from a sequence of high-resolution line profiles, sampling the different rotation phases of the star (see, e.g., Strassmeier (2009, 2011)). Its application requires that the rotational broadening of the spectral lines exceeds the macroturbulence by at least a factor of $4-5$, implying a minimum $v \sin i \sim 10-15$ km s^{-1} in dwarf stars. Therefore, late-type stars that rotate as slowly as the Sun cannot be imaged by this technique. For those stars, we derive the distribution of brightness inhomogeneities from the rotational modulation of the flux produced by cool spots and bright faculae that come into and out of view as the star rotates. By comparing successive rotations, it is possible to study the time evolution of the active regions.

Another method exploits the tiny light modulations produced by the occultation of starspots by a transiting giant planet moving in front of the disc of its star (e.g. Silva-Valio *et al.* (2010); Wolter *et al.* (2009)). It is a specialized version of the eclipse mapping technique, developed for the active components of close binary stars, e.g., Collier Cameron (1997); Lanza *et al.* (1998). Its application will be reviewed by Silva-Valio (2011). Stellar differential rotation and activity cycles can be studied also through techniques of time series analysis applied to sequences of seasonal optical photometry; see, e.g., Messina & Guinan (2003); Kolláth & Oláh (2009); Oláh (2011).

In this review, I shall briefly report on the modelling of the light curves of some planet-hosting stars as observed by the space experiment CoRoT.

2. Space-borne optical photometry

Space-borne optical photometry has provided time series for several late-type stars spanning from several days up to several months thanks to the space missions MOST (Walker *et al.* 2003), CoRoT (Auvergne *et al.* 2009), and Kepler (see, e.g., Ciardi *et al.* 2011).

MOST (the Microvariability and Oscillation of Stars satellite) has a telescope of 15 cm aperture that can observe a given target for up to $40 - 60$ days reaching a photometric precision of $50 - 100$ ppm (parts per million) on the brightest stars ($V \leqslant 4$). It has observed ϵ Eridani (Croll *et al.* 2006) and k^1 Ceti (Walker *et al.* 2007) whose light curves have been modelled to extract information on stellar differential rotation.

CoRoT (Convection, Rotation and Transits) is a space experiment devoted to asteroseismology and the search for extrasolar planets through the method of transits. With its 27-cm aperture telescope, it can observe up to 12,000 targets per field for intervals of 150 days searching for transit signatures in the light curves. Its white-band light curves have a sampling of 32 or 512 s and a bandpass ranging from 300 to 1100 nm. A photometric accuracy of ~ 100 ppm is achieved for a G or K-type star of $V \sim 12$ by integrating the flux over individual orbits of the satellite. CoRoT provides some chromatic information on the light variability of its brightest targets ($V < 14.5$).

Kepler, launched in March 2009, has a telescope of 95 cm aperture and is continuously monitoring $\sim 100,000$ dwarfs in a fixed field of view for a time interval of at least 3.5 years, searching for planetary transits. Its accuracy reaches ~ 30 ppm in 1 hour integration on a G2V target of $V \sim 12$.

Stars with transiting giant planets allow us to derive the inclination of the stellar rotation axis, which is important to modelling their light curves for stellar activity studies (see, e.g., Mosser *et al.* (2009)). Specifically, by fitting the transit light curve, we can measure the inclination of the planetary orbit along the line of sight, which is equal to the inclination of the stellar rotation axis if the spin and the orbital angular momentum are aligned. This hypothesis can be tested with the so-called Rossiter-McLaughlin effect, i.e., the apparent anomaly in the radial velocity of the star observed during the planetary transit that allows us to measure the angle between the projections of the stellar spin and the orbital angular momentum on the plane of the sky (Winn *et al.* 2005). Another advantage of stars with transiting planets is that their fundamental parameters have been well determined because accurate stellar masses and radii are needed to derive accurate planetary parameters from the transit modelling.

3. Light curve modelling

Spot models of the light curves of late-type stars observed by MOST and CoRoT have already been published, while the modelling of Kepler light curves has just started (Brown *et al.* 2011). MOST time series were exploited to measure stellar differential rotation by fitting the wide-band light modulation with a few circular spots with fixed contrast. Thanks to such assumptions, it was possible to derive the latitudes of the spots and measure the variation of the angular velocity vs. latitude (Croll *et al.* 2006; Walker *et al.* 2007). For a generalization of that approach with the CoRoT light curves, see Mosser *et al.* (2009).

In the Sun, active regions consist not only of cool spots but also of bright faculae whose contrast is maximum close to the limb and minimum at the disc centre. Moreover, the optical variability of the Sun is dominated by several active regions at the same time making a model based on a few spots poorly suitable to reproduce its active region pattern.

Lanza *et al.* (2007) used the time series of the Total Solar Irradiance (TSI, e.g., Fröhlich & Lean (2004)) as a proxy for the solar optical light curve to test different modelling approaches assuming that the active regions consist of dark spots and bright faculae with fixed contrasts and in a fixed area proportion. A model with a continuous distribution of active regions and the maximum entropy regularization, to warrant the uniqueness and stability of the solution, is the most suitable and reproduces the distribution of the area of the sunspot groups vs. longitude with a resolution better that $\sim 50°$, as well as the variation of the total spotted area vs. time. It allows a highly accurate reproduction of the TSI variations with a typical standard deviation of the residuals of $\sim 30 - 35$ ppm for time intervals of 14 days. However, the value of the faculae-to-spotted area ratio is a critical parameter because it affects the derived distribution of the active regions vs. longitude (see, Lanza *et al.* (2007), for a detailed discussion).

The maximum entropy spot model tested in the case of the Sun has been applied to CoRoT light curves to derive the distribution of the stellar active regions vs. longitude and the variation of their total area vs. time. In general, information on the latitudes of stellar active regions cannot be extracted from a one-dimensional data set such as an optical light curve. Moreover, since the inclination of the rotation axis of the stars with transiting planets is generally close to $90°$, it is impossible to constrain the spot latitudes because the duration of the transit of a spot across the stellar disc is independent of its latitude.

4. Results from CoRoT light curves

Lanza *et al.* (2009a) model the out-of-transit light curve of CoRoT-2, a G7V star with a giant planet with a mass of 3.3 Jupiter masses and an orbital period of 1.743 days (Alonso *et al.* 2008; Bouchy *et al.* 2008). Since the spot pattern is evolving rapidly, they model individual intervals of 3.15 days along a sequence of 142 days. The star has a light curve amplitude of 0.06 mag, i.e., about 20 times that of the Sun at the maximum of the 11-yr cycle. Solar-like contrasts for the spots and the faculae are adopted.

The distribution of the spotted area vs. longitude and time is plotted in Fig. 4 of Lanza *et al.* (2009a), here reproduced in Fig. 1. The active regions are mainly found within two active longitudes, initially separated by $\sim 180°$. The longitude initially around $0°$ does not migrate, i.e., it rotates with the same period as the adopted reference frame, while the other longitude, initially around $180°$, migrates backward, i.e., it rotates slower than the reference frame by ~ 0.9 percent. Individual active regions also migrate backward as they evolve, i.e., they rotate slower than the active longitude to which they belong, with a maximum difference of ≈ 2 percent. The relative migration of the active longitudes can be interpreted as a consequence of their different latitudes on a differentially rotating star, yielding a lower limit of 0.9 percent for the amplitude of the differential rotation (Lanza *et al.* 2009a). On the other hand, the backward migration of the individual active regions can be regarded as analogous to the braking of the rotation of solar active regions as they evolve because the relative amplitude of the angular velocity variation is remarkably similar e.g., (Zappalà & Zuccarello 1991; Schüssler & Rempel 2005).

Other authors have suggested a greater amplitude for the differential rotation of CoRoT-2, up to ~ 8 percent, from the migration of individual active regions (Fröhlich

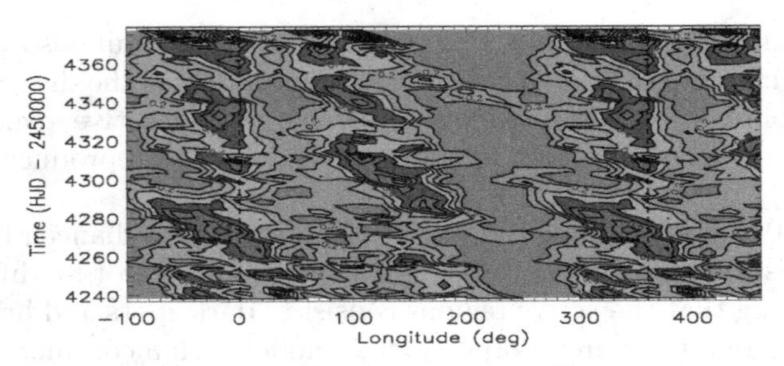

Figure 1. The isocontours of the spot filling factor vs. longitude and time for CoRoT-2 after Lanza *et al.* (2009a). The longitude reference frame rotates with the star with a period of 4.5221 days and the longitude increases in the same direction of the stellar rotation. The longitude scale has been extended beyond 0° and 360° (marked by the vertical dashed lines) to help follow the migration of the spots. In the electronic version, different colours indicate different relative filling factors of the starspots; from the minimum to the maximum: green, light green, light blue, blue, light pink, pink, and red.

et al. 2009; Huber *et al.* 2010). Savanov (2010) notices that the active regions appear in an alternate way in the two active longitudes, suggesting a short-term flip-flop phenomenon, reminiscent of the flip-flop cycles in some active stars that, however, have timescales of several years (Berdyugina & Tuominen 1998; Berdyugina 2005).

The variation of the total spotted area shows remarkable oscillations with a cycle of ~ 29 days (see Fig. 6 of Lanza *et al.* (2009a)). In the Sun, short-term oscillations of the total spotted area have been observed close to the maximum of some of the 11-yr cycles and are called Rieger cycles. They have periods around 160 days, i.e., about five times longer than the cycles observed in CoRoT-2 (Oliver *et al.* 1998; Krivova & Solanki 2002; Zaqarashvili *et al.* 2010). Lou (2000) suggests that they may be due to hydromagnetic Rossby-type waves trapped in the solar convection zone. Since the wave frequency is proportional to the rotation frequency of the star, the expected period is close to that observed in CoRoT-2 because this star rotates five times faster than the Sun.

The approach introduced for CoRoT-2 has been applied to other late-type planet-hosting stars, viz. CoRoT-4 (Lanza *et al.* 2011), CoRoT-6 (Lanza *et al.* 2010), and CoRoT-7 (Lanza et al. 2011). From the migration of their active longitudes, a lower limit for the amplitude of their latitudinal differential rotation has been obtained. The results are listed in Table 1, together with those derived from MOST photometry and those by Mosser *et al.* (2009) for two of the CoRoT asteroseismic targets. In Table 1, the columns from left to right list the name of the star, its effective temperature T_{eff}, its mean rotation period P_{rot}, the relative amplitude of the differential rotation $\Delta\Omega/\Omega$, and the references.

A comparison with the differential rotation amplitudes as derived from Doppler Imaging or the Fourier transform of the spectral line profiles (Reiners (2006)) shows that the values derived from the migration of the active longitudes are generally smaller than those expected in stars with the same effective temperature and rotation rates, by a typical factor of $2-3$. This suggests that their active regions are mostly localized at low latitudes, as in the case of the Sun.

5. The possible case for a magnetic star-planet interaction

The planets of CoRoT-2, CoRoT-4, and CoRoT-6 are hot Jupiters, i.e., giant planets orbiting within 0.15 AU from their host stars. They interact tidally and possibly magnetically with their stars, which may lead to observable effects on stellar activity (see,

Table 1. Stellar differential rotation from spot modelling of space-borne photometry.

Star	T_{eff} (K)	P_{rot} (days)	$\Delta\Omega/\Omega$	References
ϵ Eridani	4830	11.45	$0.11 \pm 0.03^*$	(Croll *et al.* 2006, Croll *et al.* (2006))
CoRoT-7	5275	23.64	0.058 ± 0.017	(Lanza *et al.* 2010b, Lanza *et al.* (2010b))
k^1 Ceti	5560	8.77	$0.09 \pm 0.006^*$	(Walker *et al.* 2007, Walker *et al.* (2007))
CoRoT-2	5625	4.52	$\sim 0.009 - 0.08$	(Lanza *et al.* 2009a, Lanza *et al.* (2009a)); (Fröhlich *et al.* 2009, Fröhlich *et al.*(2009))
HD 175726	6030	3.95	$\approx 0.40^*$	(Mosser *et al.* 2009, Mosser *et al.* (2009))
CoRoT-6	6090	6.35	0.12 ± 0.02	(Lanza *et al.* 2010a, Lanza *et al.* (2010a))
CoRoT-4	6190	9.20	0.057 ± 0.015	(Lanza *et al.* 2009b, Lanza *et al.* (2009b))
HD 181906	6360	2.71	$\approx 0.25^*$	(Mosser *et al.* 2009, Mosser *et al.* (2009))

Note:
*Value of the K coefficient estimated for a solar-like differential rotation law $P(\phi) = P_{eq}/(1 - K\sin^2\phi)$, where ϕ is the latitude, $P(\phi)$ the rotation period at latitude ϕ, and P_{eq} the rotation period at the equator.

e.g., Cuntz *et al.* 2000; Lanza 2008, 2009). Current evidence of star-planet magnetic interaction (hereafter SPMI) is limited to the modulation of the chromospheric flux with the orbital phase of the planet in a few stars and in some seasons (Shkolnik *et al.* (2005, 2008)). Evidence of a coronal flux enhacement is much more controversial (cf. Kashyap *et al.* 2008 and Poppenhaeger *et al.* 2010), although some possible cases have been presented (Saar *et al.* (2008); Pillitteri *et al.* (2010)). SPMI features in the photosphere have been proposed for τ Bootis (Walker *et al.* 2008), CoRoT-2 (Pagano *et al.* 2009), and, possibly, HD 192263 (Santos *et al.* 2003). The mean rotation of τ Boo is synchronized with the orbital motion of its giant planet, so a modulation of its optical flux with the orbital period of the planet cannot be unambiguously attributed to SPMI. Nevertheless, Walker *et al.* (2008) found an active region on the star which lasted for at least ~ 500 stellar rotations, i.e., 5 years, always leading the subplanetary meridian by $\sim 70°$. The persistence of such a feature strongly suggests a connection with the planet.

Lanza *et al.* (2009b) suggest a similar phenomenon in the other synchronous system CoRoT-4, finding an active region located at the subplanetary longitude that has persisted for ~ 70 days. Even more intriguing is the case of CoRoT-6, a non-synchronous system with a planetary orbital period of 8.886 days and a mean stellar rotation period of 6.35 days. Assuming a longitude reference frame rotating with the mean stellar rotation period, the maximum of the spot filling factor in several active regions occurs when they cross a meridian at $-200°$ from the subplanetary meridian. The probability of a chance occurrence is only ~ 0.8 percent (Lanza *et al.* (2010a)).

It is difficult to find a mechanism for the allegedly supposed influence of the planet on the formation of stellar active regions. Lanza (2008) conjectured that the reconnection of the stellar coronal field with the magnetic field of the planet may induce a longitudal dependence of the hydromagnetic dynamo action in the star, provided that some of the spot magnetic flux tubes come from the subphotospheric layers, as suggested by Brandenburg (2005). Nevertheless, further observations of the photospheric SPMI are needed firstly to confirm the reality of the phenomenon and secondly to derive its dependence on stellar and planetary parameters.

References

Aigrain, S., *et al.* 2008, *Astron. Astrophys*, 488, L43
Alonso, R., *et al.* 2008, *Astron. Astrophys*, 482, L21
Auvergne, M., *et al.* 2009, *Astron. Astrophys*, 506, 411

Berdyugina, S. V. 2005, *Living Reviews in Solar Physics*, 2, 8
Berdyugina, S. V. & Tuominen, I. 1998, *Astron. Astrophys*, 336, L25
Bouchy, F., *et al.* 2008, *Astron. Astrophys*, 482, L25
Brown, A., Korhonen, H., Berdyugina, S, *et al.*, 2011, these proceedings
Brandenburg, A. 2005, *Astrophys. J.*, 625, 539
Ciardi, D. R., van Braun, K., Bryden, G., *et al.* 2011, *Astronomical J.*, 141, 108
Collier Cameron, A. 1997, *MNRAS*, 287, 556
Croll, B., *et al.* 2006, *Astron. Astrophys*, 648, 607
Cuntz, M., Saar, S. H., & Musielak, Z. E. 2000, *Astrophys. J.*, 533, L151
Dupree, A. 2011, these proceedings
Fröhlich, H.-E., Küker, M., Hatzes, A. P., & Strassmeier, K. G. 2009, *Astron. Astrophys*, 506, 263
Fröhlich, C. & Lean, J. 2004, *A&AR*, 12, 273
Haas, M. R., *et al.* 2010, *Astrophys. J.*, 713, L115
Huber, K. F., Czesla, S., Wolter, U., & Schmitt, J. H. M. M. 2010, *Astron. Astrophys*, 514, A39
Kashyap, V. L., Drake, J. J., & Saar, S. H. 2008, *Astrophys. J.*, 687, 1339
Kolláth, Z. & Oláh, K. 2009, *Astron. Astrophys*, 501, 695
Krivova, N. A. & Solanki, S. K. 2002, *Astron. Astrophys*, 394, 701
Lanza, A. F. 2008, *Astron. Astrophys*, 487, 1163
Lanza, A. F. 2009, *Astron. Astrophys*, 505, 339
Lanza, A. F., Catalano, S., Cutispoto, G., Pagano, I., & Rodono, M. 1998, *Astron. Astrophys*, 332, 541
Lanza, A. F., Bonomo, A. S., & Rodonò, M. 2007, *Astron. Astrophys*, 464, 741
Lanza, A. F., *et al.* 2009a, *Astron. Astrophys*, 493, 193
Lanza, A. F., *et al.* 2009b, *Astron. Astrophys*, 506, 255
Lanza, A. F., *et al.* 2010, *Astron. Astrophys*, 520, A53
Lanza, A. F., *et al.* 2011, *Astron. Astrophys*, 525, A14
Lou, Y.-Q. 2000, *Astrophys. J.*, 540, 1102
Messina, S. & Guinan, E. F. 2003, *Astron. Astrophys*, 409, 1017
Mosser, B., Baudin, F., Lanza, A. F., *et al.* 2009, *Astron. Astrophys*, 506, 245
Oláh, K. 2011, these proceedings
Oliver, R., Ballester, J. L., & Baudin, F. 1998, *Nature*, 394, 552
Pagano, I., Lanza, A. F., Leto, G., *et al.* 2009, *Earth, Moon, and Planets*, 105, 373
Pillitteri, I., Wolk, S. J., Cohen, O. *et al.* 2010, *Astrophys. J.*, 722, 1216
Poppenhaeger, K., Robrade, J., & Schmitt, J. H. M. M. 2010, *Astron. Astrophys*, 515, A98
Reiners, A. 2006, *Astron. Astrophys*, 446, 267
Saar, S. H., Cuntz, M., Kashyap, V. L., & Hall, J. C. 2008, IAU Symposium, 249, 79
Santos, N. C., *et al.* 2003, *Astron. Astrophys*, 406, 373
Savanov, I. S. 2010, *Astron. Rep.*, 54, 437
Schüssler, M. & Rempel, M. 2005, *Astron. Astrophys*, 441, 337
Shkolnik, E., *et al.* 2005, *Astrophys. J.*, 622, 1075
Shkolnik, E., Bohlender, D. A., Walker, G. A. H., & Collier Cameron, A. 2008, *Astrophys. J.*, 676, 628
Silva-Valio, A. 2011, these proceedings
Silva-Valio, A., Lanza, A. F., Alonso, R., & Barge, P. 2010, *Astron. Astrophys*, 510, A25
Strassmeier, K. G., 2009, *A&AR*, 17, 251
Strassmeier, K. G., 2011, these proceedings
Walker, G., *et al.* 2003, *PASP*, 115, 1023
Walker, G. A. H., *et al.* 2007, *Astrophys. J.*, 659, 1611
Walker, G. A. H., *et al.* 2008, *Astron. Astrophys*, 482, 691
Winn, J. N., *et al.* 2005, *Astrophys. J.*, 631, 1215
Wolter, U., Schmitt, J. H. M. M., Huber, K. F., *et al.* 2009, *Astron. Astrophys*, 504, 561
Zappalà, R. A. & Zuccarello, F. 1991, *Astron. Astrophys*, 242, 480
Zaqarashvili, T. V., Carbonell, M., Oliver, R., & Ballester, J. L. 2010, *Astrophys. J.*, 709, 749

Discussion

K. Strassmeier: First of all, this is a very excellent idea to verify this with white-light photometry of the irradiance, but it does also ring a little bit of a warning bell because, for example, I presume – and please do correct me – that you assume a faculae component and a spot component at the same time. The time scales, the lifetime of these two features are markedly different; but you implicitly assume they are living the same time and actually give the same constant contribution. But if you have a different lifetimes, they would actually give a variable contribution.

lanza: Yes, you are right. Faculae are a possible source of problems in this kind of modeling. If the star is very active, probably the cool spots dominate the variation. When we consider the stars that have the same level of activity as the sun, faculae may be a problem; and also the longitudinal distribution of the spots may be affected. It is difficult for the Sun to have a good treatment of the faculae because, as you mentioned, they evolve on different timescales. The ratio between the faculae and the spotted area may change during the evolution of an active region. So this model is very simplified. It assumes a constant ratio, and that may affect both the evolution of the area and also that of the longitude positions.

A. Kosovichev: Antonino, interesting study. I'm wondering if your technique estimates the differential rotation for slow-rotating stars like the sun?

Lanza: We focus on stars with exoplanets because those are the targets with the better characterization, and we also have sometimes information on the inclination of the rotation axis. Certainly in the CoRoT-4 data bases, there are a lot of star that's rotate as slowly as the sun or even slower. But as far as I know, no planets have been found for those stars. And also another problem is that the time series is quite limited. So in 150 days, you have only a few rotations; and that may be difficult to trace the migration accurately. So we focused on stars that rotate faster than the sun because of these two limitations, but certainly it is a very interesting question, and it deserves more attention for the future.

The Physics of Sun and Star Spots
Proceedings IAU Symposium No. 273, 2010
D. P. Choudhary & K. G. Strassmeier, eds.

doi:10.1017/S1743921311015079

Study of stellar activity through transit mapping of starspots

Adriana Valio[1]

[1]Center for Radioastronomy and Astrophysics, Mackenzie University
São Paulo, Brazil
email: avalio@craam.mackenzie.br

Abstract. During the eclipse of a planet, spots and other features on the surface of the host star may be occulted. This will cause small variations in the light curve of the star. Detailed studies of these variations during planetary transits provide a wealth of information about the starspots properties such as size, position, temperature (i.e. intensity), and magnetic field. If observation of multiple transits is available, the spots lifetime can be estimated. Moreover it may also be possible to determine the stellar rotation and whether differential rotation is present. Here, the study is performed using a method that simulates the passage of a planet (dark disk) in front of a star with multiple spots of different sizes, intensities, and positions on its surface. The data variations in the light curve of the star are fit using this method, yielding the starspots properties. Results are presented for solar-like stars, such as the active star CoRoT-2a.

Keywords. Starspots, stellar activity, planetary transit, exoplanet

1. Introduction

Presently, there are almost 500 known planets orbiting other stars other than our Sun, discovered during the past 15 to 18 years. About one fifth of them (101 planets, or 20%) eclipses their host star. During one of these transits, the planet may pass in front of a spot group and cause a detectable signal in the light curve of the star, of the order of a few tenths of a percent.

A method that simulates planetary transits was developed by Silva (2003), where the planet is used as a probe to investigate the presence and characteristics of features on the surface of the star, such as spots. The physical parameters of these spots resulting from this study are size, intensity (or temperature), location on the surface of the star (longitude and latitude), and lifetime, among others. Stellar properties such as rotation (Silva-Valio 2008) and differential rotation can also be estimated.

Lightcurves of individual planetary transits are fit by the model of a star with several round spots on its surface, and the spots properties is obtained by a χ^2 minimization technique. The star is modeled as a 2-D matrix with a given limb darkening (linear or quadratic profiles), or a real image of the Sun can be used. On the otherhand, the planet is modeled as an opaque disk of radius r/R_s, where R_s is the stellar radius. Every 2 min (or the desired time interval), the position of the planet in its orbit is calculated. The orbit is assumed circular with radius a_{orb}/R_s and inclination angle i. By summing over all the pixels in the image, the lightcurve intensity is obtained at every time interval.

At any given transit, the number of spots on the stellar surface is variable. Each spot is characterized by 3 parameters – *Intensity*: measured with respect to stellar maximum

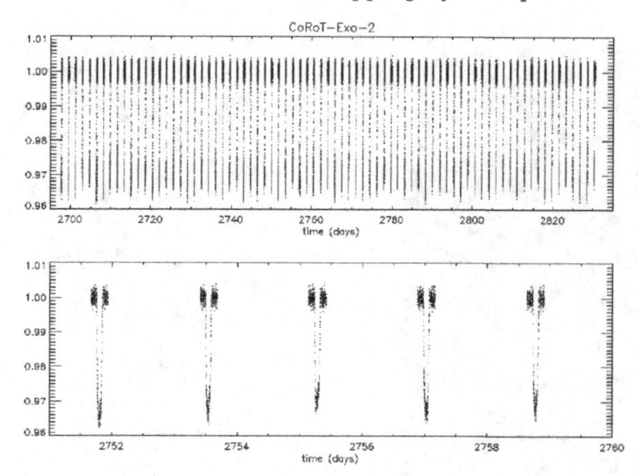

Figure 1. Top: Lightcurve of the 77 transits of CoRoT-2b. Bottom: Examples of 5 of these transits.

intensity (at disk center); *Size*: measured in units of planetary radius; and *Position*: Latitude (restricted to the transit band) and Longitude.

Here the technique of spot mapping is applied to the case study of CoRoT-2, the second planet-hosting-star found by the CoRoT satellite (Auvergne *et al.* 2009). CoRoT-2 is a G7V star of 0.97 M_{Sun} and 0.902 R_{Sun}, with an effective temperature of 5625 K. It is a fairly young star, with an estimated age of less than 0.5 Gyr. The planet CoRoT-2b is a hot Jupiter with $M\sin(i) = 3.31 M_{Jup}$, radius of $1.465 R_{Jup}$, orbiting the star every 1.743 day at a distance of only 6.7 R_s (approximately 0.03 A.U.) in and inclined orbit of 87.84° (Alonso *et al.* 2008). This data was previously analyzed by Lanza *et al.* (2009); Wolter *et al.* (2009) and Huber *et al.* (2010).

2. Transit mapping

Because CoRoT observes the same region of the sky for up to 150 days, a total of 77 transits of CoRoT-2b were observed, which are shown in the top panel of Figure 1. Due to its size, orbital radius, and inclination angle, the planet obscures a band on the stellar surface of 20° centered at −14.6° latitude.

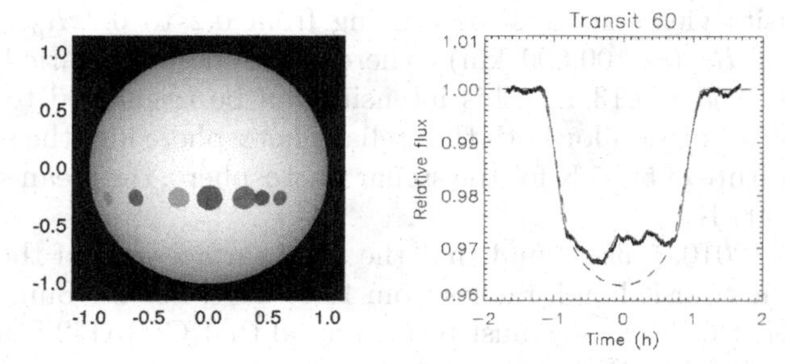

Figure 2. Left: modeled star with 7 spots. Right: Lightcurve of transit 60, with the bars representing the uncertainty of the CoRoT-2 data, $\sigma = 0.0006$. The solid gray line represents the best fit to the data, where 7 spots were considered. The dashed line is the result from a model of a spotless star.

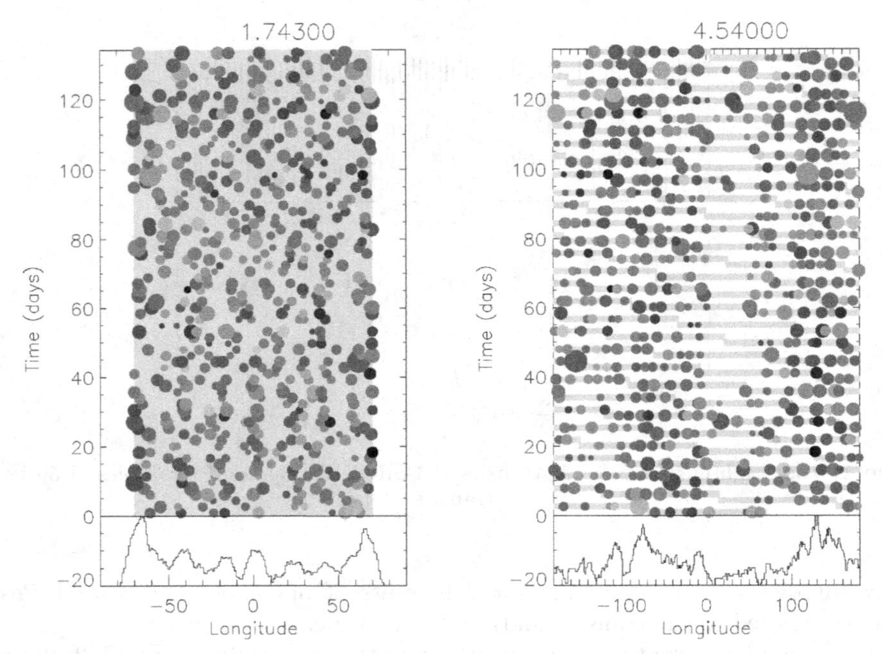

Figure 3. Left: Spot map of the stellar surface considering a reference frame fixed with respect to the observer, that is spots position (topocentric longitudes) at every transit. Right: Same as in left panel but in the rotating frame of the star. The curves on the bottom of each panel represent the time integrated flux deficit of the spots for each longitude.

The presence of spots on the stellar surface causes the transit to be less deep, because the region been eclipsed by the planet has a smaller brightness than otherwise (see right panel of Figure 2). Up to a maximum of 9 spots are considered in each transit, with their longitudes constrained between ±70° due to the very steep variation of the lightcurve during the times of ingress and egress of the planet. The radius, R, and intensity, I, of the spots are restricted between 0 and 1, in units of planet radius (R_p) and central stellar brightness (I_c). An example of the fit for the 60th transit with 7 spots is shown in Figure 2.

The analysis of the 77 transits has been presented by Silva-Valio *et al.* (2010a). A total of about 600 spots were modeled, where only spots with a flux deficit, defined as $F = (1 - I/I_c) \times (R/R_p)^2$, larger than 0.02 in relative units were considered. The fitting of the transits yields spots sizes ranging from 0.2 to 0.7 R_p, with an average value of 0.46 ± 0.11 R_p ($\sim$ 100,000 km), whereas the intensity varied from 0.3 to 0.8 I_c, with a mean of 0.55 ± 0.13 I_c. This intensity can be translated to temperature by assuming black-body emission for both the stellar photosphere and the spot. Considering an effective temperature of 5625 K for the stellar photosphere, the mean spot temperature found was 4700 ± 300 K.

Silva-Valio *et al.* (2010a) also found that the total surface area of the star covered by the spots, within the transit band, ranges from 10 to 20%. For the Sun, the spot covered area is less than 1%. However, one must bear in mind that CoRoT-2 is a young and very active star, with an age less than one tenth that of the Sun.

Maps of the stellar surface region occulted by the planet can be reproduced for each transit. Such map is shown in the left panel of Figure 3, where the best result for each transit is shown as filled circles representing spots of a certain relative size and

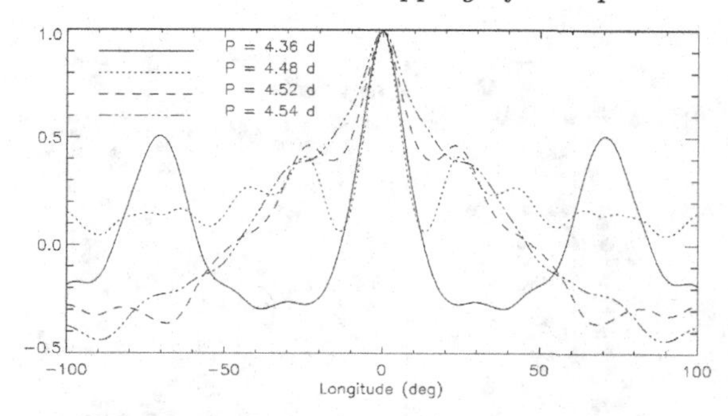

Figure 4. Auto–correlation functions of the flux deficit integrated in time for each longitude bin for stellar rotation periods of 4.36, 4.48, 4.52, and 4.54 days.

intensity placed at their fitted longitude for each transit in time (vertical axis). The shaded region in the background represents the visible stellar hemisphere (within ±70° longitudes) during each transit. The fitted longitudes shown in the left panel of Figure 3 are the topocentric longitude, measured in a fixed reference frame with respect to the observer, where zero longitude coincides with the projection of the line–of–sight at mid transit.

To study the temporal evolution of the spots, it is necessary to associate whether some of the spots are detected again in a posterior transit. For that, it is necessary to use a reference frame that rotates with the star. In this case, the spots will be identified by their common rotation longitude. The rotation longitude, β_{rot}, may be obtained from the topocentric longitude, β_{topo}, once the stellar rotation period, P_{star}, is known.

$$\beta_{rot} = \beta_{topo} - 360^\circ \frac{n\ P_{orb}}{P_{star}} \tag{2.1}$$

where $P_{orb} = 1.743$ day is the planet orbital period, P_{star} the stellar rotation period, and n the transit number. As a first estimate for the rotation period within the transit band, the average period of the star of 4.54 days obtained from the out–of–transit data is used (Alonso *et al.* 2008). A map of the spotted stellar surface in time (for every transit) versus rotational longitudes is shown in the right panel of Figure 3. The figure clearly shows that there is a longitude band within 0 and 40° where there are almost no spots visible during the total observation period of 134 days.

3. Differential rotation

The differential rotation profile of the stellar surface may be calculated from the measured rotation period within a given latitude band and the average rotation of the star. If we are to consider that the rotation period at −14.6° is equal to the mean rotation period of the star of 4.54 d, this implies that this star presents no differential rotation. However, this is an active star with many spots on its surface at any given time, therefore a magnetic dynamo process must be taking place within the star. Thus one possibility is that the rotation period within the transit band is different than 4.54 days, and what is seen here is something similar to the counterpart of a coronal hole observed on our Sun.

Coronal holes are regions of open magnetic field lines in the Sun, with lower temperature and density than the rest of the corona. They present smaller differential rotation, sometimes nonexistent, than the photosphere where the spots are located. Also from

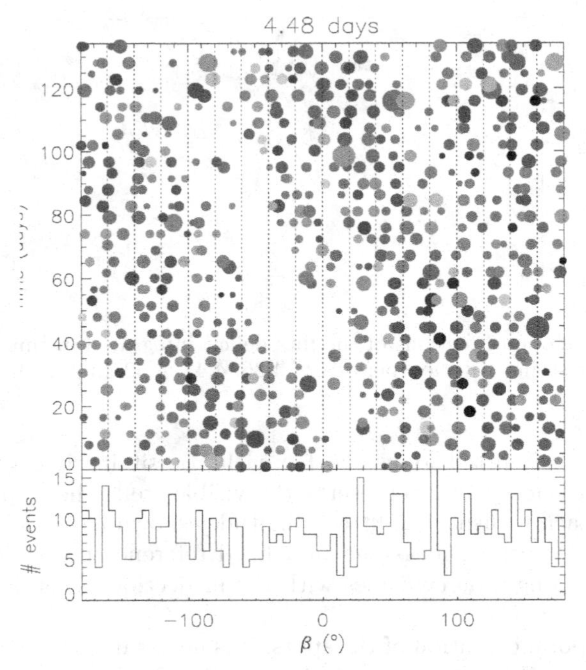

Figure 5. Same as Figure 3 for a stellar rotation period of 4.48 days within the transit latitudes.

solar observations, it is known that the photosphere below the coronal holes has no spot. Therefore, we will assume that the spotless photosphere within $0 - 40°$ longitudes of CoRoT-2 is the surface counterpart of a coronal hole that reaches near the equator of the star.

Thus the problem now is to determine the true rotation period of the transit band. For this, we calculated the stellar surface map for different rotation periods from 2 to 6 days, every 0.01 day. For each map (or rotation period), the flux deficit due to spots for a given longitude is summed over all transits. Next, the auto–correlation of the flux deficit as a function of longitude is calculated, and the curve with the smaller width is assumed to be the best estimate of the stellar rotation period within $-25°$ and $-5°$ latitudes. Examples of the auto–correlation functions for 4 rotation periods of the transit band are shown in Figure 4.

The auto–correlation function of the time integrated flux deficit was found to be the narrowest for a rotation period of 4.48 days (see Figure 4). Thus hereafter, the rotation period of the star within the transit latitudes of $-14.6 \pm 10°$ is assumed to be 4.48 days, and the corresponding stellar surface map is shown in Figure 5.

Now, the differential rotation of the star can be estimated, this is assumed to follow the same functional form as that of the Sun, where the rotational angular velocity, Ω, is given by $\Omega = A - B\cos^2\theta$, where θ is the co-latitude, and in the case of the Sun $A = 14.55°\ \mathrm{day}^{-1}$ and $B = 2.87°\ \mathrm{day}^{-1}$. For CoRoT-2 these constants are estimated by considering that the rotation period at co-latitude of $75.4° = 90° - 14.6°$ is 4.48 days, and that the rotation period averaged over all latitudes is 4.54 days. These two values yield $A = 80.5°\ \mathrm{day}^{-1}$ and $B = 2.4°\ \mathrm{day}^{-1}$, therefore:

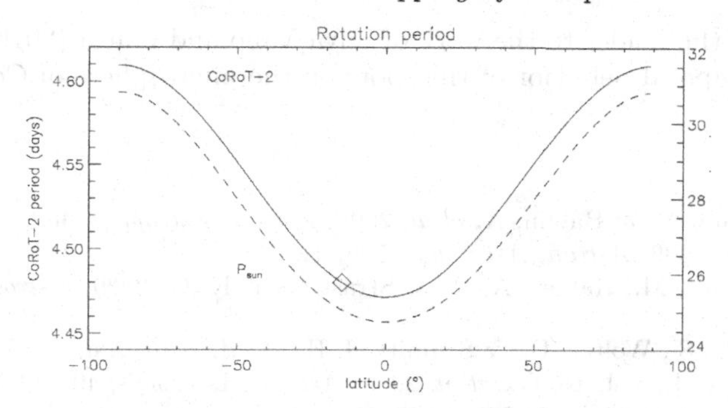

Figure 6. Differential rotation profile, that is, rotation period versus latitude, for CoRoT-2 (solid) and the Sun (dashed, scale on the right of the figure).

$$P_{CoRoT-2} = \frac{360°}{80.5 - 2.4\cos^2(90° - \alpha)} \text{days}, \tag{3.1}$$

where α is the latitude. The rotation period as a function of latitude is plotted in Figure 6. The solar differential rotation is also plotted as a dotted line. The transit latitude value is marked by the diamond, that is, the period of 4.48 days at $\alpha = -14.6°$.

4. Discussion and conclusions

A total of 77 consecutive transits were observed for the planet hosting star CoRoT-2. This was possible because the orbital period of the planet is only 1.743 day. By modeling small variations in the light curve observed during transits as due to star spots, it was possible to infer the presence of at about 600 spots on the stellar surface during these 134 days of observation.

The modelling of spots yields its physical characteristics such as size, temperature, location, surface area coverage, etc (Silva 2003; Silva-Valio *et al.* 2010a). If multiple transits are available, also the stellar rotation can be obtained (Silva-Valio 2008), and even stellar differential rotation such as the study presented here.

To estimate the differential rotation of CoRoT-2a, first it was necessary to determine the rotation period within the given latitude band. This was obtained by calculating the time integrated spot flux deficit for each longitude for several values of stellar rotation. Then, the auto–correlation of the flux deficit function with the narrowest peak was chosen as that representing the true rotation period within the $-14.6 \pm 10°$ latitudes. This procedure yields a value of 4.48 days for this rotation period.

The differential rotation profile obtained here shows that close to the equator, the star rotates faster, with a period of 4.47 d, while the longest period of 4.61 days is found at the poles. Using the values listed in Eq. (3.1) resulted in $\Delta\Omega = \Omega_{eq} - \Omega_{pole} = 0.042$ rad/d. This corresponds to $(P_{pole} - P_{eq})/ < P >= 3\%$, where $< P >$ is the mean rotation period. From out–of–transit data analysis, Lanza *et al.* (2009) estimated a lower limit to the amplitude of the relative differential rotation of only $\sim 0.7 - 1$ % for this same star, while from a model with three spots only, Fröhlich *et al.* (2009) find a mean differential rotation of 0.11 rad/d and a maximum differential rotation of 0.129 rad/d. As can be seen, the differential rotation estimated here is a intermediate value between both estimates.

Once the rotation period of the star within the transit latitude band is known, it is possible to identify the same spot on multiple transits and study their temporal evolution.

However, we refer the reader to the work of Silva-Valio and Lanza (2010b) for a detailed analysis of the temporal behavior of the spots on the photosphere of CoRoT-2.

References

Alonso, R., Auvergne, A., & Baglin, A. *et al.* 2008, *Astron. Astrophys*, 482, L21

Auvergne, M., *et al.* 2009, *Astron. Astrophys*, 506, 411

Fröhlich, H.-E., Küker, M., Hatzes, A. P., & Strassmeier, K. G. 2009, *Astron. Astrophys*, 506, 263

Huber, K. F., Czesla, S., Wolter, U., & Schmitt, J. H. M. M. 2010, *Astron. Astrophys*, 514, A39

Lanza, A. F., Pagano, I., & Leto, G., *et al.* 2009, *Astron. Astrophys*, 493, 193

Silva, A. R. V. 2003, *Astrophys. J.*, 585, L147

Silva-Valio, A. 2008, *Astrophys. J.*, 683, L179

Silva-Valio, A., Lanza, A. F., Alonso, R., & Barge, P. 2010, *Astron. Astrophys*, 510, A25

Silva-Valio, A. & and Lanza, A. F. 2010, *Astron. Astrophys* (submitted)

Wolter, U., Schmitt, J. H. M. M., Huber, K. F., Czesla, S., Müller, H. M., Guenther, E. W., & Hatzes, A. P. 2009, *Astron. Astrophys*, 504, 561

Discussion

DUPREE: I would like follow up on your detection of coronal holes and ask, A, what the lifetime was during the whole COROT point of 180 days which seems anomalous or very long compared to the sun? And can you follow up with Chandra observations or Hubble observations to actually detect modulation and the diminution, X-ray flux, and symmetries of lines that might indicate out flow?

VALIO: Yes, it lasts for the whole 140 days. Maybe at the end there might be some evidence that it's kind of not there anymore but yes, yes. And for the second comment, that's a great idea.

KITIASHVILI: Very interesting method you show. My question is about hot Jupiters which you observe like a transit across a star. No one [INAUDIBLE] models of hot Jupiters – hot Jupiters have very massive atmosphere. How will it effect the atmosphere on your results?

VALIO: Okay. For this one, I'm just modeling as a black – completely black disk, absolutely no emission. And of course you can change that and add a transparency like varying towards the edge of the planet, but I haven't done that. But I don't think that's going to be – unless it's very transparent, it's – I don't think it's going to matter much for the spot study. It will matter if you're analyzing the shape of the transit and things like that.

JARDINE: Do you have any plans to look at the transiting planets that are in highly inclined orbits because that will give you access to different latitudes in the star?

VALIO: Yes, yes, that's next. Hopefully try and measure the angle.

KLIMCHUK: Is it your technique sensitive enough that you could say something about possibly penumbrae?

VALIO: Maybe if they are big as I was hoping, I could use this for COROT-7 or other smaller planets. But the notes were the COROT, again I can't look in the vision transits.

But for Kepler, I did a very brief simulation. If you get a earth-size planet, then you get all this – like on the solar planet, not on this, there you have all the activity regions and the penumbra; and you do get the structures. So, yes, with this model planet, you can do that.

The Physics of Sun and Star Spots
Proceedings IAU Symposium No. 273, 2010
D. P. Choudhary & K. G. Strassmeier, eds.

doi:10.1017/S1743921311015080

Time series photometry and starspot properties

Katalin Oláh

Konkoly Observatory, Budapest, Hungary
email: olah@konkoly.hu

Abstract. Systematic efforts of monitoring starspots from the middle of the XXth century, and the results obtained from the datasets, are summarized with special focus on the observations made by automated telescopes. Multicolour photometry shows correlations between colour indices and brightness, indicating spotted regions with different average temperatures originating from spots and faculae. Long-term monitoring of spotted stars reveals variability on different timescales.

On the rotational timescale new spot appearances and starspot proper motions are followed from continuous changes of light curves during subsequent rotations. Sudden interchange of the more and less active hemispheres on the stellar surfaces is the so called flip-flop phenomenon. The existence and strength of the differential rotation is seen from the rotational signals of spots being at different stellar latitudes.

Long datasets, with only short, annual interruptions, shed light on the nature of stellar activity cycles and multiple cycles. The systematic and/or random changes of the spot cycle lengths are discovered and described using various time-frequency analysis tools. Positions and sizes of spotted regions on stellar surfaces are calculated from photometric data by various softwares. From spot positions derived for decades, active longitudes on the stellar surfaces are found, which, in case of synchronized eclipsing binaries can be well positioned in the orbital frame, with respect to, and affected by, the companion stars.

Keywords. Stars: activity, stars: spots, stars: rotation, stars: late-type

1. Introduction

Nearly a century ago, at about the same time when the magnetic field of the sunspots were first measured, a short note of Eberhard & Schwarzschild (1913) called the attention to the similarity of the solar activity and the newly measured stellar feature in the CaII H&K lines: *The same kind of eruptive activity that appears in sun-spots, flocculi and prominences, we probably also have to deal with in Arcturus and Aldebaran and in a very greatly magnified scale in σ Geminorum.* In the same time photographic monitoring of certain parts of the sky commenced in a few observatory. Later these plates were used to extend the databases of active stars to check their long term behaviour. Starting with data from the Harvard plate collection by Phillips & Hartmann (1978) (big dots and bars until 1970) to the most recent results of Messina (2008) (bars after 1990) - and dozens of published data in between (small dots) - a more than century long light curve of BY Dra is presented in Fig. 1. This figure shows clearly how short is a century for studying long term variations in stellar activity. However, results based on time series photometric data emerge as the databases are growing in time. These results are reviewed in the followings.

The origin of the photometric variability of active stars is due to the work of magnetic dynamos in their interiors. The description of the dynamo processes and how spots are formed is far beyond the possibility of this review, however, the reader finds ample papers to read more about the dynamo mechanism within this Proceedings.

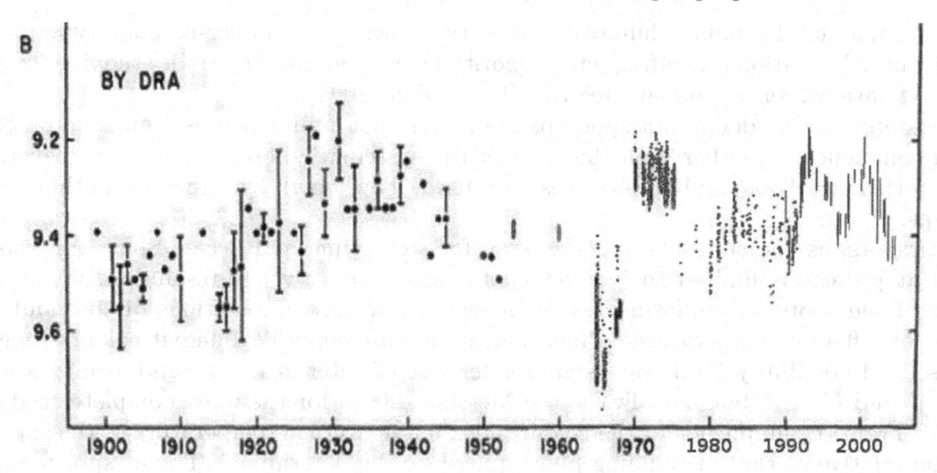

Figure 1. Long term light variation of BY Dra over a century, from photographic and photometric data.

2. Telescopes, filters and tools

Photometric observations of active stars began in the mid 1960's in different observatories after discovering the "rotational modulation" of the already known and studied flare stars. The systematic studying (day-by-day observations) of stellar activity needed cheap and dedicated instruments. Since the manpower is the biggest cost in operating a small telescope suitable for measuring the relatively bright (4-10 mag.) stars, fully automated telescopes were built for the task. The first one was the Phoenix 25cm APT which served the community for 23 years, its program consisted of several different types of variable stars. Dedicated mostly to active star research the Catania APT operated from 1992 till the recent years (see e.g. Strassmeier *et al.* (1997b)). At present, three APTs are gathering the data on active stars for more than a decade: the Wolfgang-Amadeus Vienna twin APT Strassmeier *et al.* (1997a) and APT of the Four-College Consortium. Rodonò *et al.* (2004) a few years ago summarized the use of robotic telescopes in stellar activity; the interested reader finds useful references in this paper. Space observatories (CoRoT, Kepler) are also gathering data on active stars in a much higher magnitude range than from the Earth, giving uninterrupted and very high quality data for hundreds of objects. The planned durations of these projects are, unfortunately, below the time scale of spot cycle variations.

The filter sets were different in the APTs of long run. The Phoenix 25cm and the Catania APT had Johnson UBV, while now Wolfgang (T6) has $uvby$, Amadeus (T7) has $V(RI)_C$, and the Four-College APT has $uvby$, $V(RI)_C$, TiO and Hα filter sets. This means, that the longest databases to study activity cycles are available only in one passband, i.e., in V combined with the very similar y. The datasets in the different colour indices $(B-V)$, $(V-I)_C$ are generally shorter and even less is the case when both blue and red colour index time series observations exist for one star. Even worse is the situation with the space observatories: at present they do not observe in any established colour system.

Rotational and cycle periods from the photometric data are searched for by different period finding algorithms. The most commonly applied one-dimensional Fourier transform however, is not well suited to find periods of rotational modulations because of the differential rotation which can be strong - although can give a hint about it. Fourier analysis may fail for double humped light curves; these cases Lafler-Kinman type procedure can help. Various time-frequency algorithms exist and are used for studying longer-term variations: for a summary see Kolláth & Oláh (2009).

Attempts for modeling starspot positions and spot temperatures from photometric data commenced together with the start of the systematic observations of active stars. Two other methods and approaches are used, developed by different authors and groups.

Starspots, as the sunspots, emerge, stay for some time and decay. Since our knowledge at present is limited to indirect spot maps with very low resolution, we do not know if the spots are uniform dark, cool regions, or are combinations of cool and hot areas of different temperatures. Photometric measurements in different colours suggest the second possibility. Active stars are redder when fainter in $V - I_C$ and usually also in $B - V$ and $U - B$, but not always, see Messina (2008) for the most complete study in UBV. From the amplitude of the colour index curve average temperature of the spotted region relative to the surrounding photosphere can be determined. These "spot temperatures" are warmer than those on the Sun and show variability on different timescales, therefore are more resembling to the active nests on the Sun than to the individual sunspots.

3. Starspot changes on rotational timescale

Systematic monitoring of active stars show that the measured light variations caused by spots on the rotating stellar surfaces show continuous, and sometimes sudden, changes. The origin of these observed changes are reflections of the changes in the spots' properties: positions, temperature, lifetime.

Fig. 2 shows the time-frequency behaviour of CoRoT-Exo-2a with the planet transits and starport changes. It is well seen that the signal of the transit is (naturally) very regular and several harmonics are present. Together with this regular pattern the rotational signal of the active (host) star is seen, with slow changes of the period around 4.5 days, due to the differential rotation. A temporary period doubling is also seen indicating two close periods by spots at different latitudes. Longer, slowly varying periods of about 28-29 and around 86 days are also seen. All periodicities are discussed in detail by Lanza *et al.* (2009). Fig. 2 shows the power of the time-frequency analysis in giving a first estimate of the observed periodic signals and a first guess about their origin.

3.1. *Differential rotation - quiet spot patterns*

On the stellar surfaces active regions with starspots are found at different latitudes, rotating with different periods. The signal of the differential rotation is usually present in the observed light curves appearing as several close periods in period search results. The bunch of resulting periods give only a guess about the existence and a limit of the strength of the differential rotation. For short period active stars, when enough data is available, seasonal mean rotational periods can be obtained. However, the errors should be calculated with care to see the reliability of the differences, if any. Only in lucky cases and with the help of corresponding Doppler imaging to the photometry is possible to derive the the strength and the sign (solar type or opposite) of the differential rotation. One good example is given by Oláh *et al.* (2003) for the active giant star UZ Lib, where a long-lasting equatorial and high latitude spotted regions were identified in Doppler images and their rotational periods were found from photometry.

Until recently the most straightforward, direct approach to derive differential rotation from photometric data failed, since the photometric precision (a few thousandths of a

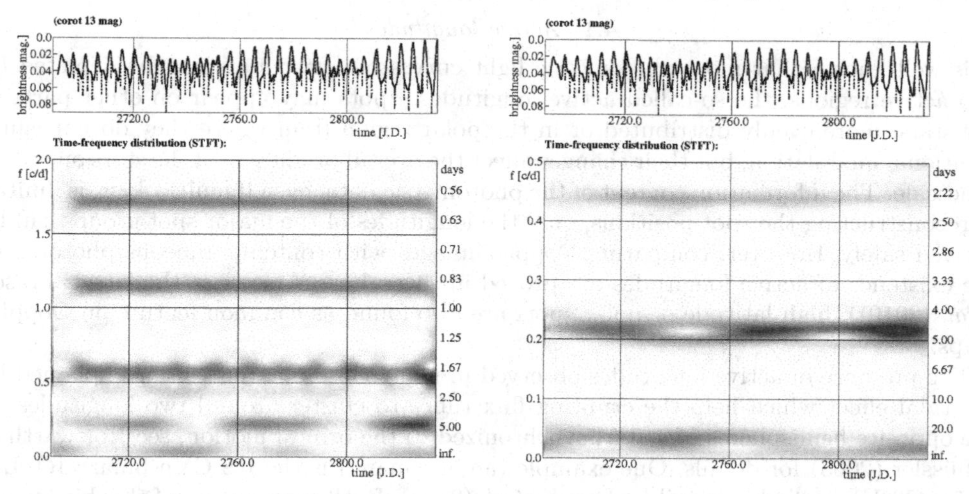

Figure 2. Analysis of the original dataset of CoRoT-Exo-2a (with some pre-filtering only). Left: time-frequency diagram from 0.5 days, right: the same from 2 days. On the left the periodicity of the planet transit (1.743 days) is seen with some cross-talk with the half of the star's rotational period. On the right only the signals from the star (rotational period of 4.5 days, its half and long periods) are seen.

magnitude) was too low to derive spot latitudes, which is strongly related to the inclination of the star and is affected by the limb darkening at different latitudes. However, the present, very high precision, uninterrupted datasets acquired from satellites make possible to get this important parameter directly from the light curves. Walker *et al.* (2007) derived accurate differential rotation parameter for κ^1 Cet from photometric data obtained by the MOST satellite.

3.2. *Spot emergence, decay and proper motion - fast changing spot patterns*

Spot emergences - sudden change in the light curve - and decays have already been measured for several active stars during their continuous monitoring. The effect is best seen in uninterrupted, satellite datasets. The period doubling shown in the time-frequency plot on Fig. 2 and the corresponding sudden change of the light curve suggests a restructuring of the spots on the stellar surface, possibly by new spot emergence(s) at different latitudes. The two periods reflect the differential rotation but additional spot proper motion (typical for newly emerged spots) could change the rotational period of the spots. In the same dataset traces of spot decay and emergence of new spots are also found by Lanza *et al.* (2009)). It is interesting to compare the results originating from two totally different methods using on the same dataset.

4. Long term variability of starspots

The behaviour of the spottedness of stars is particularly interesting on the decadal timescale. This includes the changes (or constancy) in position (longitude), size and temperature of the starspots. Only APTs using *the same instrument* with well-defined filter sets are able to follow the temperature and the closely related size variations of the starspots.

4.1. *Active longitudes*

It is well known that most photometric light curves of active stars can be depicted by two active regions, the so-called active longitudes. Spots may appear on other parts of the stars, more evenly distributed or in the polar region from where they do not cause rotational modulation, but their changes affect the overall brightness of the stars on longer timescale. The information content of the photometric data, even if multicolour, is limited in reconstructing the spot positions, only the longitudes of the major spot groups can be derived safely. However, comparing Doppler images with contemporaneous photometry the existence of active longitudes are proved in several cases (see e.g. the latest: Frasca *et al.* (2010)), high latitude or polar spots are also found as common feature on Doppler maps.

The presence of active longitudes observed in many close binaries can be explained by the tidal effect which help the erupting flux tubes to cluster around two longitudes on the opposite hemispheres of a star, synchronized to the orbital motion, see Holzwarth & Schüssler (2003) for details. One example (among many) is the RS CVn binary RT Lac (G5:+G9IV) studied in detail by Lanza *et al.* (2002). Both components of this binary are active and show enhanced spottedness at the substellar points and on their opposite sides.

4.2. *Flip-flop phenomenon*

Time series photometry shows, that the dominance of the activity can change between two active longitudes, which are usually separated by about 180°. This effect is called as flip-flop, and the elapsed time between two such change is the flip-flop period. Several active stars show this feature, a recent summary is found in Korhonen & Järvinen (2007). In lucky cases the flip-flop, i.e., the interchange between the more active hemisphere of a star is directly observed, as in the case of FK Com by Oláh *et al.* (2006). The long term homogeneous datasets are of vital importance in studying this strange behavior of magnetic activity. For theoretical background of the phenomenon see Korhonen & Elstner (2005) and the references in its Introduction.

4.3. *Activity cycles*

An important - maybe the most important - result of the time-series observations is the detection and tracking the magnetic cycles of different types of active stars. This study needs the most expensive factor of an astronomical research: *time*. Only one smooth sine-like change in the overall brightness of a star during a few (a dozen) years cannot be called cycle, which, by meaning, is a repetitive pattern, with at least a characteristic timescale (see the variability of BY Dra for more than 100 years in Fig. 1). The correct timescales of the activity cycles is of fundamental importance as an observational basis of dynamo modeling. Studying cyclic behaviour of different stars may help to generalize this feature, which then helps us to better understand the behaviour of our Sun, and that is of vital importance, by its full meaning.

Active stars are found in the galactic field as well as in young open clusters. An important study yet to be done, is to monitor the time behaviour of activity of stars in clusters which would allow to compare the cyclic (and also rotational) behaviour of coeval stars. This way we get rid of the age factor and most of the composition difference, so we can concentrate on the evolutionary effect, which depend on the mass of the stars. A pioneering long-term study is being done on stars in Pleiades by Milingo *et al.* (2011).

On Fig. 3, left, the time behaviour of the activity cycles of EI Eri is plotted based on 28 years of monitoring. In a way this picture is typical: most studied active stars show

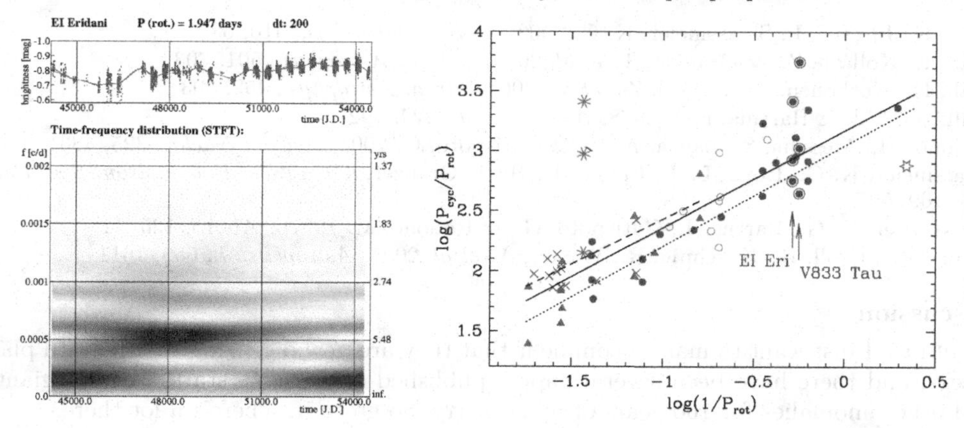

Figure 3. Left: Time-frequency plot of EI Eri, showing multiple and changing cycles. Right: the relation between rotational and cycle periods. The solid line shows the relation for all derived cycles except the long cycle of the Sun (stars), dashed line is from Baliunas *et al.* (1996), dotted line is the fit to the shortest cycles.

multiple and changing cycles, like this example. Dwarfs and giants, singles and binaries, young and evolved stars are among the studied 20 objects by Oláh *et al.* (2003), and all share the same characteristics. The right side of Fig. 3 shows the cycle length - rotational period relation ($P_{\rm cyc}/P_{\rm rot}$) originally given by Baliunas *et al.* (1996) derived from the Wilson sample: *"Theoretical considerations suggest that the ratio is the observational equivalent of the stellar dynamo number, D."*.

The two marked stars on Fig. 3, right, EI Eri and V833 Tau have very similar rotational periods of 1.95 and 1.79 days, respectively, but are much different in all physical parameters, EI Eri is a subgiant and is 10 times older than the mid-K dwarf V833 Tau. Yet, their short cycles are similar, but not the long ones. Whether it is due to the limited length of datasets remains to be seen by my successors, *in a century or more.*

5. Acknowledgements

This paper is dedicated to the memory of Prof. M. Rodonò. Thanks are due to K. Vida who designed Fig. 1., and to Z. Kolláth and Zs. Kővári for useful discussions. This work was supported by the Hungarian Research grants OTKA T-068626 and K-081421.

References

Baliunas, S., Nesme-Ribes, E., Sokoloff, D., Soon, W. 1996, *Astrophys. J.*, 460, 848

Eberhard, G. & Schwarzschild, M., 1913, *Astrophys. J.*, 38, 292

Frasca, A., Biazzo, K., Kővári, Zs. *et al.* 2010, *Astron. Astrophys.*, 518, A48

Holzwarth, V., Schüssler, M. 2003, *Astron. Astrophys.*, 405, 303

Kolláth, Z. & Oláh, K. 2009, *Astron. Astrophys.*, 501, 695

Korhonen, H. & Elstner, D. 2005, *Astron. Astrophys.*, 440, 1161

Korhonen, H., & Järvinen, S. 2007, *Binary Stars as Critical Tools & Tests in Contemporary Astrophysics* (Proc. IAU Symp. 240, eds.: W.I. Hartkopf, E.F. Guinan and P. Harmanec, Cambridge University Press) p. 453

Lanza, A. f., Pagano, I., Leto, G., Messina, S. *et al.* 2009, *Astron. Astrophys.*, 493, 193

Lanza, A. F., Catalano, S., Rodonò, M. *et al.* 2002, *Astron. Astrophys.*, 386, 583

Messina, S. 2008, *Astron. Astrophys.*, 480, 495

Milingo, J., Saar, S., Marschall, L., & Stauffer, J. 2011, *this proceedings*

Oláh, K. Jurcsik, J., Strassmeier, K. G. 2003, *Astron. Astrophys.*, 410, 685
Oláh, K. Kolláth, Z., & Granzer, T. *et al.* 2009, *Astron. Astrophys.*, 501, 703
Oláh, K., Korhonen, H., Kővári, Zs. *et al.*2006, *Astron. Astrophys.*, 452, 303
Phillips, M. J. & Hartmann, J. 1978, *Astrophys. J.*, 224, 192
Rodonò, M., Messina, S., Lanza, A. F., & Cutispoto, G. 2004, *Astron. Nachr.*, 325, 483
Strassmeier, K. G., Boyd, L. J., Epand, D. H., & Granzer, Th. 1997a, *Pub. Astron. Soc. Pac.*, 109, 697
Strassmeier, K. G., Bartus, J., Cutispoto, G., & Rodonò, M. 1997b, *A&AS*, 125, 11
Walker, G., Croll, B., Kuschnig, R., Walker, A. *et al.* 2007, *Astrophys. J.*, 659, 1611

Discussion

DUPREE: I just want to make a comment that they are now digitizing the Harvard plate stacks, and there have been several papers published on variable stars, variable giants, and other anomalies for 100 years of photometry. So go to it. There's a lot there.

OLÁH: Thank you.

STRASSMEIER: Well, I couldn't agree with you more, Katalin, as you probably know; but I do disagree on one point. You mentioned that, if the light curve gets precise enough, you can retrieve latitudes. I think this is wrong. You can have infinite precision light curves. You're not going to get latitude out of a one-dimensional light curve. It's, I think, mathematically impossible. If you assume a circle, a model, per se – if you assume that, then you get the latitude out from high-precision light curves if you assume square spots, you also get a latitude curve out. If you assume triangle spot, you also get a latitude out. So it's model dependent. Then, yes, but not in general if you let spot areas completely free, pixelize it, as we do in inversion, then you do not get a latitude out.

OLÁH: You are absolutely right, and I suggest to devote a special conference to the question. Why is it fair to suppose a circular spot? Because we see circular spots on the Sun. Circular spot is an assumption, and I fully agree with what you say when you suspect other spot shapes.

KŐVÁRI: *To Strassmeier* This is not fully true because once we have limb darkening, we'll have latitudinal information, too.

The Physics of Sun and Star Spots
Proceedings IAU Symposium No. 273, 2010
D. P. Choudhary & K. G. Strassmeier, eds.

doi:10.1017/S1743921311015092

Exploring the deep convection and magnetism of A-type stars

Nicholas A. Featherstone[1], Matthew K. Browning[2], Allan Sacha Brun[3] and Juri Toomre[1]

[1] JILA and Department of Astrophysical and Planetary Sciences, University of Colorado, Boulder, CO 80309-0440, USA
email: *feathern@lcd.colorado.edu*

[2] Canadian Institute for Theoretical Astrophysics, University of Toronto, Toronto, ON M5S3H8, Canada

[3] DSM/IRFU/SAp, CEA-Saclay and UMR AIM, CEA-CNRS-Universite Paris 7, 91191 Gif-sur-Yvette, France

Abstract. A-type stars have both a near-surface layer of fast convection that can excite acoustic modes and a deep zone of core convection whose properties may be probed with asteroseismology. Many A-type stars also exhibit large magnetic spots that are often attributed to surviving primordial fields of global scale in the intervening radiative zone. We have explored the potential for core convection in rotating A-type stars to build strong magnetic fields through dynamo action. These 3-D simulations using the ASH code provide guidance on the nature of differential rotation and magnetic fields that may be present in the deep interiors of these stars, thus informing the asteroseismic deductions now becoming feasible. Our models encompass the inner 30% by radius of a two solar mass A-type star, rotating at four times the solar rate and capturing the convective core and a portion of the overlying radiative envelope. Convection in these stars drives a strong retrograde differential rotation and yields a core that is prolate in shape. When dynamo action is admitted, the convection generates strong magnetic fields largely in equipartition with the dynamics. Remarkably, introducing a modest but large-scale external field threading the radiative envelope (which may be of primordial origin) can substantially alter the turbulent dynamics of the convective interior. The resulting convection involves a complex assembly of helical rolls that link distant portions of the core and stretch and advect magnetic field, ultimately yielding magnetic fields of super-equipartition strength.

Keywords. Astroseismology, A-type stars, differential rotation, deep convection

1. Introduction

The peculiar A stars (Ap) exhibit strong and variable spectral lines (relative to solar values) in Si and certain rare earth metals (e.g., Sr and Hg). Of the Si and Sr-Cr-Eu peculiarity classes, many possess equally variable and unusually strong magnetic fields with strengths ranging from a lower threshold of ~300 G up to 20,000 G (e.g., Mestel 1999; Aurière *et al.* 2007). When variable, these fields appear to change at the stellar rotation rate (e.g., Deutsch 1956; Mestel 1999), suggesting that the magnetic fields are "frozen in" to the radiative envelope of the star. A typical Ohmic decay time is much longer than the lifetime of an A star, and so these magnetic fields are likely the remnants of a primordial magnetic that threaded the star's natal molecular cloud (Cowling 1945).

The subsurface nature of these strong fields remains unknown and may depend on the dynamics of the star's convective core. Recently, 3-D numerical simulations have demonstrated that nuclear burning cores of A-type stars can drive vigorous convection capable of generating equipartition strength magnetic fields (Browning *et al.* 2004; Brun

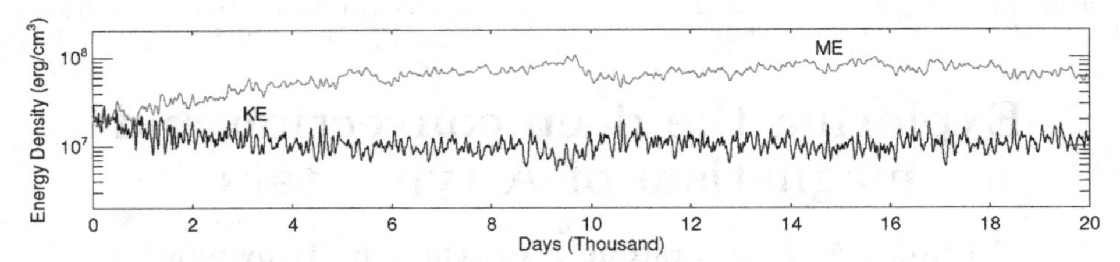

Figure 1. Evolution of volume-averaged energy densities following the imposition of an external mixed magnetic field spanning 20,000 days (or about 2,900 rotation periods). Magnetic energy ME (red) has grown in strength to become about ten-fold greater than the kinetic energy KE (black) and is thus super-equipartition in nature.

et al. 2005). If the surface fields are indicative of a global-scale magnetic field with roots in the core, then the primordial field must exert some influence on the core dynamo, and vice versa. We explore this possibility by modeling a core dynamo with a large-scale field of primordial origin threading the radiative envelope.

2. Modeling the A Star

We use the 3-D anelastic spherical harmonic (ASH) code (see Brun *et al.* 2004) to model convection and dynamo action in a main-sequence A-type star of $2M_\odot$ rotating at four times the solar rate (with a rotation period of seven days). Our computational domain encompasses the inner 0.3 of the A star by radius, the inner 0.15 of which is convectively unstable, with an overlying stable, radiative envelope comprising the remainder of the domain. We adopt a Prandtl number of 0.25 and a magnetic Prandtl number of 5 throughout the domain, achieving a Reynolds number following equilibration of ~136 and a magnetic Reynolds number of ~680. Further details of this model may be found in Featherstone *et al.* (2009).

Our fossil field system was initialized by augmenting the magnetic fields from a mature A-star dynamo simulation (case C4m) of Brun *et al.* (2005). Their dynamo was started using a well equilibrated hydrodynamic simulation from Browning *et al.* (2004) by adding a small dipole seed field to this system. Persistent dynamo action realized in that simulation yielded equipartion magnetic energies with respect to the convective flows.

The geometry most likely to support a primordial field against decay is likely to require a poloidal and toroidal component if the field is to remain stable over the lifetime of a star (e.g., Prendergast 1956; Braithwaite & Spruit 2004). We have adopted such a twisted field geometry for our model fossil field, placing a magnetic torus with amplitude 30 kG in the lower radiative zone. A poloidal field component consistent with a current threading through the center of our magnetic torus was then added to give some twist to the field, yielding a fossil field whose magnetic energy constituted a modest 10% increase to the magnetic energy of the overall system.

3. Discussion

The most striking effect arising from the inclusion of a fossil field in our core dynamo model is apparent in the temporal evolution of the system's (globally-averaged) energy densities. The inclusion of a fossil magnetic field in our system has led to a fivefold increase in the magnetic energy (ME) that gradually develops over approximately one magnetic diffusion time (~ 7000 days) following the imposition of the external field. As

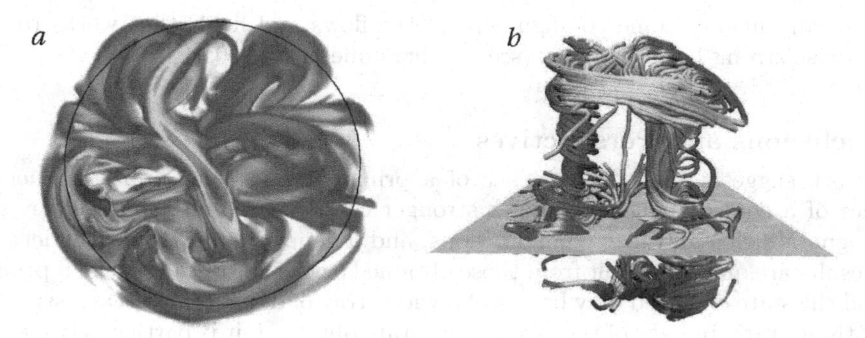

Figure 2. Convective and magnetic structures near day 15,000. (*a*) Magnetic magnetic energy density as realized in the equatorial region of the convection zone and region of overshooting; yellow/green tones indicate high values, and red tones low values. (*b*) Interior view of columnar convection in the core of the A-star visualized near day 15,000 using instantaneous streamlines. Streamlines are colored by velocity component along the rotation axis v_z. Blue (yellow) tones indicate northward (southward) motion; equatorial plane indicated by light blue.

the magnetic energy rises, the kinetic energy (KE) halves, yielding a ratio of ME/KE $\sim$ 10 that is maintained (with some variation) throughout the remainder of the simulation. Such a change in the energy balance is remarkable given that the imposed fossil field constituted but a small perturbation to the magnetic energy of the system. Moreover, one might anticipate that strong Lorentz forces would accompany such high magnetic field strengths, ultimately suppressing the very convective motions that spur their genesis, but this does not occur.

The increase in ME is largely associated with non-axisymmetric magnetic fields contained within the core. Typical magnetic field strengths have transitioned from $\sim$67 kG in the absence of a fossil field (based on rms values at mid-core) to $\sim$80 kG, and typical flow speeds are $\sim$20 m s^{-1} (vs. $\sim$29 m s^{-1} initially). While typical magnetic energies have not changed dramatically, the strong magnetic flux ropes that now pervade much of the core possess interior field strengths of $\sim$300 kG (blue regions Figure 2*a*). These regions of highly concentrated magnetic flux are largely responsible for the increase in ME seen in Figure 1. Core convection in the presence of such strong magnetic fields is now characterized by four to six evolving cylindrical rolls encircling the rotation axis and aligned with it. Similar to previous studies of rotating spherical convection using rigid boundaries (e.g., Busse 2002), where these cylindrical rolls intersect the spherical edge of the convective core, their motions "tilt" as they conform to the edge of the convective core. This behavior is visible in detailed rendering of the streamlines in Figure 2*b*. Pressure gradients created by this tilting drive flows along the axes of these rolls, imbuing neighboring rolls with opposite senses of helicity. Poleward axial flows from one roll thus link to the equatorward axial flows of their neighbors. As these rolls drive only a weak differential rotation, downflows from one set of rolls freely cross the rotation axis undeflected, connecting distant regions of the core which would otherwise remain relatively isolated from one another.

The imprint that this cylindrical roll-like convection leaves on the magnetic field is illustrated in Figure 2*a* showing magnetic energy in the equatorial region. The helical convection creates broad flux ropes of strong ($\sim$ 300 kG) field that extend from pole to pole encircling the periphery of the cylindrical rolls. Magnetic fields in the core thus exhibit a globally-connected topology reflective of the convective motions, with strong fields tending to exist on the periphery of regions of strong flow. Exceptions to this trend typically involve some co-alignment of the flows and fields that works to dimish the otherwise strong Lorentz forces (see Featherstone *et al.* 2009).

4. Conclusions and Perspectives

Our work suggests that the presence of a primordial field can subtly influence the dynamics of a core dynamo, yielding a stronger core dynamo characterized by global-scale magnetic fields and flow configurations, and a super-equipartition magnetic state. These results are quite different from those obtained earlier in the absence of a primordial field, and the state obtained may be one of a wide array of dynamo states accessible to the cores of these stars. In light of the strong field state obtained, it is particularly interesting to ask whether the strong mean fields generated near the edge of the convective core might become buoyant, ultimately contributing to the large star spots seen at the surface of the Ap and Bp stars.

5. Acknowledgements

The authors gratefully acknowledge Nicholas Nelson for presenting this work. This research is supported by NASA through Heliophysics Theory Program grants NNG05G124G and NNX08AI57G, with additional support for Featherstone through NASA GSRP program by award NNX07AP34H. The simulations were carried out with NSF PACI support of PSC, NCSA and TACC, and by NASA HEC support at Project Columbia. Browning was supported by NSF Astronomy and Astrophysics postdoctoral fellowship AST 05-02413, and by a Jeffery L. Bishop fellowship at CITA. Brun was partly supported by the Programme National Soleil-Terre of CNRS/INSU (France), and by the STARS2 grant from the European Research Council.

References

Aurière, M. *et al.*, 2007, *A&A* 475 1053
Braithwaite, J. & Spruit, H. C., 2004 *Nature* 431 819
Browning, M. K., Brun, A. S., & Toomre J., 2004 , *ApJ* 601 512
Brun, A. S. , Browning, M. K., & Toomre, J., 2005 *ApJ* 629 461
Brun A. S., Miesch, M. S., & Toomre, J., 2004 *ApJ* 614 1073
Busse, F. H., 2002 *Phys. Fluids* f 14 1301
Deutsch, A. J., 1956 *PASP* 68 92
Featherstone, N. A., Browning, M. K., Brun, A. S., & Toomre, J. 2009 *ApJ* 705 1000
Mestel, L., 1999 *Stellar Magnetism* (Oxford: Clarendon Press)
Prendergast, K. H., 1956 *ApJ* 123 498
Preston, G. W., 1971 *PASP* 83 571

Discussion

STRASSMEIER: You said there's 300 kilogauss in your peak and you still don't see flux surfacing. So, what sort of field strength or fluxes would you need in order that flux ropes can surface up all the way up the real surface?

FEATHERSTONE: That is a good question. To address the first part, we see these very strong fields. They tend to be in the convective roles. The simulation is sufficiently diffusive that we would probably need much stronger fields, order of magnitude stronger field in order to get a buoyancy instability. We have some newer methods to try to reduce the diffusion in these simulations that may be more effective in promoting some sort of magnetic stability and buoyancy that hasn't been tried in the A-type stars. I'll show some results in my talk Thursday in similar things we have been able to get in solar-type stars.

ARLT: It looks like, let me call it a one-scale dynamo. You have the motions, and you have the fields. Would you expect or do you have a feeling on, when you drive up the resolution, that you can actually get scale separation and eventually end up with the possibility of describing it by a mean field dynamo? It's hard to say?

FEATHERSTONE: Yeah, it is hard to say. I would say that, in this particular case, I don't know that you are going to get a strong scale separation between your convective motions and your dynamo scales if you will in the core. I should say partly because the convective motions are – are – in terms of a spherical harmonic degree are somewhere in a spherical harmonic degree of 10, 12, or 20, somewhere in this ballpark, which is very well-resolved in these simulations. So I don't know if your convective motions are going to get much smaller and your dynamo scales aren't going to get much larger, bigger.

ARLT: Then we would have an option of alpha-square-type turbulence which could give us these obliqueness of the things, but it doesn't look like this because you have revolved the motions already. So there's not much room for small-scale turbulence perhaps.

FEATHERSTONE: Yeah.

The Physics of Sun and Star Spots
Proceedings IAU Symposium No. 273, 2010
D.P. Choudhary & K.G. Strassmeier, eds.

doi:10.1017/S1743921311015109

Chemical spots and their dynamical evolution on HgMn stars

Heidi Korhonen[1], Swetlana Hubrig[2], Maryline Briquet[3], Federico González[4], and Igor Savanov[5]

[1]European Southern Observatory, Karl-Schwarzschild-Str 2, D-85748, Garching bei München, Germany
email: hkorhone@eso.org

[2]Astrophysikalisches Institut Potsdam, An der Sternwarte 16, D-14482 Potsdam, Germany

[3]Instituut voor Sterrenkunde, Katholieke Universiteit Leuven, Celestijnenlaan 200 D, B-3001 Leuven, Belgium

[4]Instituto de Ciencias Astronomicas, de la Tierra, y del Espacio (ICATE), 5400 San Juan, Argentina

[5]Institute of Astronomy, Russian Academy of Sciences, Pyatnitskaya 48, Moscow 119017, Russia

Abstract. Our recent studies of late B-type stars with HgMn peculiarity revealed for the first time the presence of fast dynamical evolution of chemical spots on their surfaces. These observations suggest a hitherto unknown physical process operating in the stars with radiative outer envelopes. Furthermore, we have also discovered existence of magnetic fields on these stars that have up to now been thought to be non-magnetic. Here we will discuss the dynamical spot evolution on HD 11753 and our new results on magnetic fields on AR Aur.

Keywords. Stars: atmospheres, chemically peculiar, early-type, magnetic fields, spots

1. Introduction

Recently, Kochukhov *et al.* (2007), using Doppler Imaging technique, reported a discovery of secular evolution of the mercury distribution on the surface of the HgMn star α And. However, this result was never verified by other studies due to the lack of observational data. Until very recently, the only other HgMn star with a published surface elemental distribution was AR Aur (Hubrig *et al.* 2006), where the discovered surface chemical inhomogeneities are related to the relative position of the companion star. The elements Y and Sr are strongly concentrated in an equatorial ring, which has a gap exactly on the area permanently facing the secondary.

A large number of spectra of a sample of HgMn stars were obtained with the CORALIE spectrograph at the 1.2 m Euler telescope on La Silla during a programme dedicated to a search for SPB-like pulsations in B-type stars. In Briquet *et al.* (2010) we published two sets of surface maps of HD 11753 based on these data. The maps are separated by approximately two months and are obtained from three different elements: Y II, Ti II and Sr II. The maps made from the Y II 4900 Å line (see Fig. 1) exhibit a high abundance region at phases 0.5–1.0 extending from the latitude 45° to the pole, with an extension to the equator around the phase 0.8. The Y II abundance distribution shows also a high latitude lower abundance spot around phases 0.2–0.4. Clear evolution in the surface features is present during the two months that separate the datasets. The lower abundance high latitude feature at phases 0.2–0.4 becomes more extended and less prominent in the

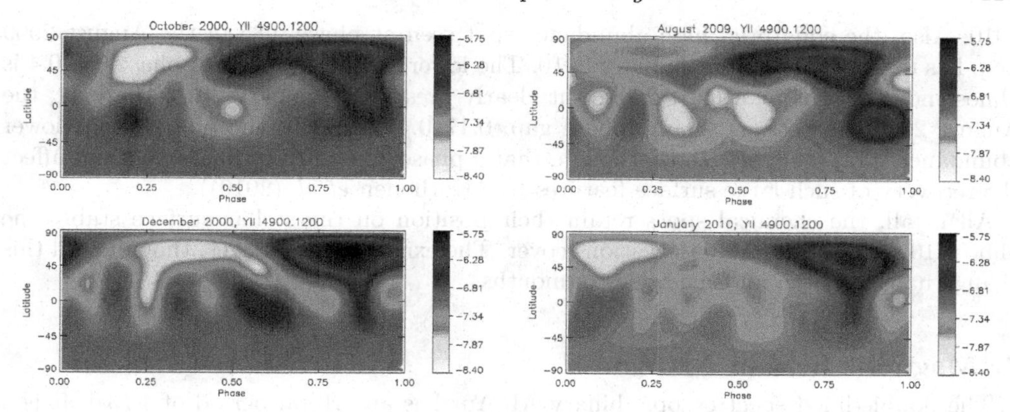

Figure 1. Chemical surface maps of HD 11753 from the Y II 4900 Å line at four different epochs. In the maps the abscissa is the longitude in phases and the ordinate is the latitude in degrees. Colour indicates different abundances, with darker denoting higher abundance.

second set, while the abundance of the high abundance spot at phases 0.6–1.0 gets more prominent with time.

We have also obtained measurements of the magnetic field strength with the moment technique using several elements in a circularly polarised high resolution spectrum of another HgMn star, AR Aur. These observations revealed the presence of a longitudinal magnetic field in both stellar components (Hubrig *et al.* 2010).

Here, we continue the investigation of HD 11753 using newer CORALIE data from 2009 and 2010, and discuss the magnetic field measurements of AR Aur.

2. Spot evolution in HD11753 for 2000–2010

HD 11753 is a single-lined spectroscopic binary with an effective temperature of 10612 K (Dolk *et al.* 2003). According to our observations the orbital period of the binary would be long, and the projected rotational velocity vsin i=13.5km/s (Briquet *et al.* 2010). After adding the new 2009 and 2010 datasets the rotational period is improved to P=9.531 d (Korhonen *et al.* 2010).

We have obtained Doppler images of HD 11753 from CORALIE spectra for four different epochs. In Doppler imaging spectroscopic observations at different rotational phases are used to measure the rotationally modulated distortions in the line-profiles. These distortions are produced by the inhomogeneous distribution of a surface characteristic, e.g., surface temperature or element abundance. Surface maps are constructed by combining all the observations from different phases and comparing them with synthetic model line-profiles. For accurate Doppler imaging the shape and changes of the line-profile have to be well defined. This requires high resolving power and high signal-to-noise-ratio.

The Doppler images of HD 11753 using the Y II 4900 Å line are shown in Fig. 1. The two first maps, both from 2000, have 65 days in between them. Clear temporal evolution of the chemical spots occurs even on such short timescales. The high abundance spot around the phase 0.75 gets more concentrated with time, and the lower abundance spot around the phase 0.25 more extended. These two maps were already published by Briquet *et al.* (2010), but the August 2009 and January 2010 maps are previously unpublished. These latter maps have approximately four and half months in between them, and again they show temporal evolution of the surface structures. The high abundance spot of phase 0.75 is at high latitudes much more extended in August 2009 than in January

2010. Also, the equatorial high abundance spot seen at phase 0.85 in the August 2009 map has disappeared before January 2010. The lower abundance spot of phase 0.0–0.4 is almost non-existent in August 2009, but clearly present in January 2010. However, the August 2009 dataset has a large phase gap, 0.17–0.47, close to the phase of the lower abundance spot. Our tests show, though, that a phase gap of 0.3 in phase does not affect the recovery of such large surface features (see Korhonen *et al.* (2010)).

All in all, the chemical spots retain their position on the stellar surface stably the almost 10 year period our observations cover. The exact shape changes, though, and this change happens even on time scales of months.

3. Magnetic field in AR Aur

The double-lined spectroscopic binary AR Aur has an orbital period of 4.13 d. It is a young system with an age of only 4×10^6 yr and its primary, showing HgMn peculiarity, is exactly on the Zero Age Main Sequence while the secondary is still contracting towards it. Variability of spectral lines associated with a large number of chemical elements was reported for the first time for the primary component of this eclipsing binary by Hubrig *et al.* (2006).

Doppler maps for the elements Mn, Sr, Y, and Hg using nine spectra of AR Aur observed at the European Southern Observatory with the UVES spectrograph at UT2 in 2005 were for the first time presented at the IAU Symposium 259 by Savanov *et al.* (2009). To prove the presence of a dynamical evolution of spots also on the surface of AR Aur, we obtained new spectroscopic data with the Coudé Spectrograph of the 2.0 m telescope of the Thüringer Landessternwarte and the SES spectrograph of the 1.2 m STELLA-I robotic telescope at the Teide Observatory. A number of SOFIN spectra of AR Aur were obtained in 2002 at the Nordic Optical Telescope, which we also used in our analysis. Our new results show secular evolution of the chemical spots on AR Aur (Hubrig *et al.* 2010). Fig. 2 shows Sr II 4215.5 Å line during three different epochs (late 2002, late 2005, and late 2008 – early 2009), but at the same rotational phase, ∼0.8. The shapes of the lineprofiles are clearly different indicating that also the Sr spots very likely changed their shape and abundance with time.

To pinpoint the mechanism responsible for the surface structure formation in HgMn stars, we carried out spectropolarimetric observations of AR Aur and investigated the presence of a magnetic field during a rotational phase of very good visibility of the spots of overabundant elements. The spectropolarimetric observations of AR Aur at the rotation phase 0.622 were obtained with the low-resolution camera of SOFIN ($R \approx 30\,000$) at the Nordic Optical Telescope. Since most elements are expected to be inhomogeneously distributed over the surface of the primary of AR Aur, magnetic field measurements were carried out for samples of Ti, Cr, Fe, and Y lines separately. Among the elements showing line variability, the selected elements have numerous transitions in the observed optical spectral region, allowing us to sort out the best samples of clean unblended spectral lines with different Landé factors.

Our magnetic field measurements, which are discussed in detail by Hubrig *et al.* (2010), were done using the formalism described by Mathys (1994). A longitudinal magnetic field at a level higher than 3σ of the order of a few hundred Gauss is detected in Fe II, Ti II, and Y II lines, while a quadratic magnetic field $\langle B\rangle$=8284 ± 1501 G at 5.5σ level was measured in Ti II lines. No crossover at 3σ confidence level was detected for the elements studied. Further, we detect a weak longitudinal magnetic field, $\langle B_z\rangle = -229 \pm 56$ G, in the secondary component using a sample of nine Fe II lines. The main limitation on the accuracy achieved in our determinations is set by the small number of lines that can

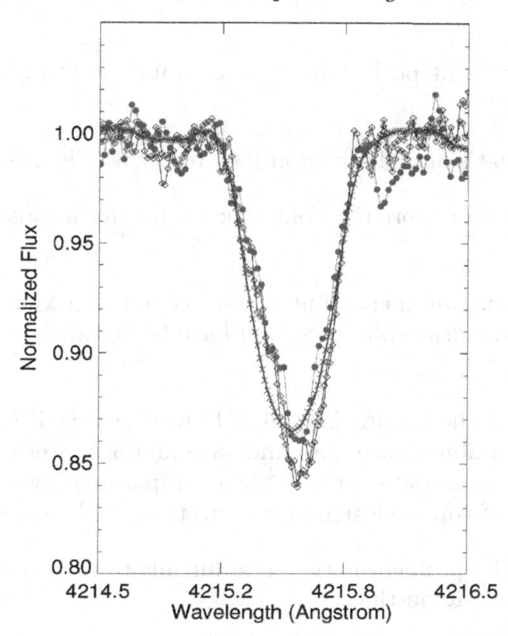

Figure 2. Line profiles of the Sr II 4215.5 Å line at the rotation phase ~0.8 for late 2002 (open diamonds), late 2005 (crosses), and late 2009 – early 2010 (filled circles) show clearly different shapes.

be used for magnetic field measurements. The diagnosis of the quadratic field is more difficult than that of the longitudinal magnetic field, and it depends much more critically on the number of lines that can be used for the analysis.

4. Summary

The fast dynamic evolution of the spots on HD 11753 implies hitherto unknown physical mechanism operating in the outer envelopes of late B-type stars with HgMn peculiarity and the detection of the magnetic field in AR Aur shatters the traditional view that HgMn stars do not exhibit magnetic fields. For the proper understanding of the nature of these stars we need accurate information on the element spot configuration and underlying magnetic fields in a sample of HgMn stars to determine a link between these properties and stellar fundamental parameters such as rotation rate, temperature, evolutionary state, stellar mass, multiplicity and orbital parameters. It is clear that time series of high resolution spectropolarimetric observations are needed to solve the puzzle these stars represent.

References

Briquet, M., Korhonen, H., González, J. F., Hubrig, S., & Hackman, T. 2010, *A&A*, 511, A71
Dolk, L., Wahlgren, G. M., & Hubrig, S. 2003, *A&A* 402, 299
Hubrig, S., González, J. F., & Savanov, I., *et al.* 2006, *MNRAS*, 371, 1953
Hubrig, S., Savanov I. & Ilyin, I., *et al.* 2010, *MNRAS*, 408, L61
Kochukhov, O., Adelman, S. J., Gulliver, A. F., & Piskunov, N. 2007, *Nature Physics*, 3, 526
Korhonen, H., González, J. F., & Briquet, M., *et al.* 2010, *A&A*, in prep.
Mathys G. 1994, *A&AS*, 108, 547
Savanov I. S., Hubrig S., González J. F., Schöller M. 2009, *IAU Symp. 259*, p. 401

Discussion

MORIN: Have you tried to fit both datasets with a unique Doppler map to assess variability?

KORHONEN: Oh, no, that's actually a good idea to try it. Thanks.

OLEG: Have you tried to use more than one spectral line for a construction of abundance maps?

KORHONEN: Yes, I have. But then – but I also, as you well know, I mean there's also radial stratification of the atmosphere. So you have to be careful which lines you choose to.

Q.: Well, actually, not in these stars. But what I meant was is, if you take one individual lines and reconstruct an abundance map and take another interim line and reconstruct another map from the same data set and then compare the two maps, that would be another way to see if – if you have significant effects related to specific lines?

KORHONEN: I agree. The problem is that it's difficult to find lines that are unblended; but yes, we are also working on that.

The Physics of Sun and Star Spots
Proceedings IAU Symposium No. 273, 2010
D.P. Choudhary & K.G. Strassmeier, eds.

doi:10.1017/S1743921311015110

Differential rotation on the young solar analogue V889 Herculis

Zsolt Kővári[1], Antonio Frasca[2], Katia Biazzo[3], Krisztián Vida[1], Ettore Marilli[2] and Ömür Çakırlı[4,5]

[1]Konkoly Observatory,
Budapest, XII. Konkoly Thege út 15-17., H-1121, Hungary
email: kovari@konkoly.hu, vidakris@konkoly.hu

[2]INAF, Osservatorio Astrofisico di Catania,
via S. Sofia, 78, 95123 Catania, Italy
email: antonio.frasca@oact.inaf.it, ettore.marilli@oact.inaf.it

[3]INAF, Osservatorio Astrofisico di Arcetri,
L.go E. Fermi, 5, 50125 Firenze, Italy
email: kbiazzo@arcetri.astro.it

[4]Ege University, Science Faculty, Astronomy and Space Sciences Department,
35100 Bornova, Izmir, Turkey
email: omur.cakirli@ege.edu.tr

[5]TÜBITAK National Observatory, Akdeniz University Campus, 07058 Antalya, Turkey

Abstract. V889 Herculis is one of the brightest single early-G type stars, a young Sun, that is rotating fast enough ($P_{\rm rot} = 1.337$ days) for mapping its surface by Doppler Imaging. The 10 FOCES spectra collected between 13-16 Aug 2006 at Calar Alto Observatory allowed us to reconstruct one single Doppler image for two mapping lines. The Fe I-6411 and Ca I-6439 maps, in a good agreement, revealed an asymmetric polar cap and several weaker features at lower latitudes. Applying the sheared-image method with our Doppler reconstruction we perform an investigation to detect surface differential rotation (DR). The resulting DR parameter, $\delta\Omega/\Omega \approx 0.009$ of solar type, is compared to previous studies which reported either much stronger shear or comparably weak DR, or just preferred rigid rotation. Theoretical aspects are also considered and discussed.

Keywords. Stars: activity, stars: imaging, stars: individual (V889 Her), stars: spots, stars: late-type

1. Introduction

The primary source of all the manifestations of the solar activity is the magnetic dynamo. However, the Sun is the only star, for which one of the conditions of a working dynamo, the surface differential rotation can be measured directly, so far. Today's advanced techniques can measure the solar surface and radial rotation profiles with high accuracy, but at the dawn of extensive solar observations, tracing of sunspots at different latitudes was the only tool to measure the surface shear on the Sun (e.g. Maunder & Maunder 1905). Similarly, on stars the detection of surface differential rotation is still a challenging observational task. Direct tracing of starspots is the usual way to observe surface shear, however, it requires a reliable surface reconstruction technique such as Doppler imaging.

Detecting surface differential rotation (DR) on stars of different types can provide essential observational input for the theoretical understanding of DR and dynamo. Indeed, investigating solar analogues, such as our target, V889 Her (HD 171488), provide us with

Table 1. Physical parameters of V889 Her taken from Frasca *et al.* (2010)

Parameter	Value
Spectral type	G2V
$T_{\rm eff}$ (K)	5750±130
$P_{\rm rot}$ (days)	1.33697±0.0002
$v\sin i$ (km s^{-1})	37.1±1.0
Inclination (°)	60±10
$\log g$	4.30 ± 0.15
Microturbulence (km s^{-1})	1.6
Macroturbulence $\xi_{\rm R} = \xi_{\rm T}$	3.0

useful information for further development of solar dynamo theory, which is still far from a thorough understanding.

The fast-rotating solar-type V889 Her is probably the brightest single early-G type star that is rotating fast enough for mapping its photosphere by means of Doppler imaging. The star is a member of the Local Association, a stream of young stars with ages ranging from 20 to 150 Myr (Montes *et al.* 2001). Strassmeier *et al.* (2003) made the first photometric and spectroscopic study of V889 Her and presented the first Doppler reconstruction showing a large polar spot and additional high-latitude features. Subsequent works by Marsden *et al.* (2006) and Jeffers & Donati (2008) based on Zeeman-Doppler imaging have also shown polar spottedness and found a strong solar-type differential rotation. However, Järvinen *et al.* (2008) reported instead a much weaker surface shear, while Huber *et al.* (2009) preferred rigid rotation. In a comprehensive study of the chromospheric and photospheric activity, including Doppler imaging, Frasca *et al.* (2010) concluded that V889 Her is a G2V type star. Hereafter, for further investigation of DR we adopt the reviewed astrophysical parameters given there (see Table 1).

2. Sheared-image reconstruction

In order to detect differential rotation, the Doppler reconstruction of Frasca *et al.* (2010) is repeated, but instead of assuming rigid rotation, latitude-dependency is allowed in the rotation profile. This technique known as 'sheared-image method' (Donati *et al.* 2000) incorporates a predefined solar-type differential rotation law in the Doppler imaging process. In practice, the image shear expressed by $\alpha = \delta\Omega/\Omega$ is kept fixed, and individual Doppler reconstructions are done over a reasonable range of values in the $P_{\rm rot} - \alpha$ parameter plane. Each of the resulting Doppler maps has a χ^2-value which describes the goodness-of-fit to the observed spectral line profiles. The best combination

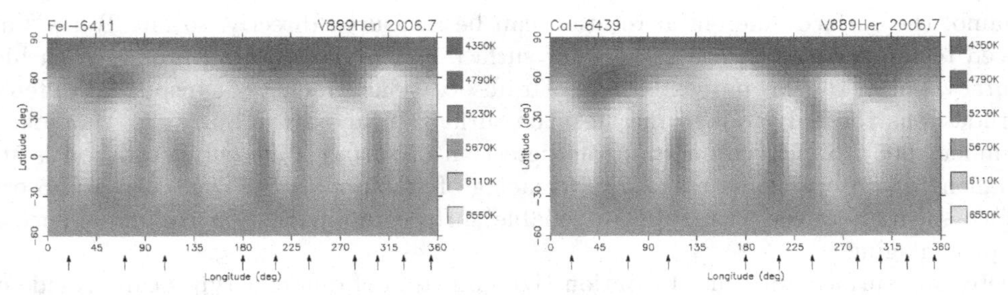

Figure 1. Doppler maps for the Fe I-6411 lines (*right*) and for the Ca I-6439 lines (*left*), taken from Frasca *et al.* (2010). The arrows below indicate the phases of spectroscopic observations.

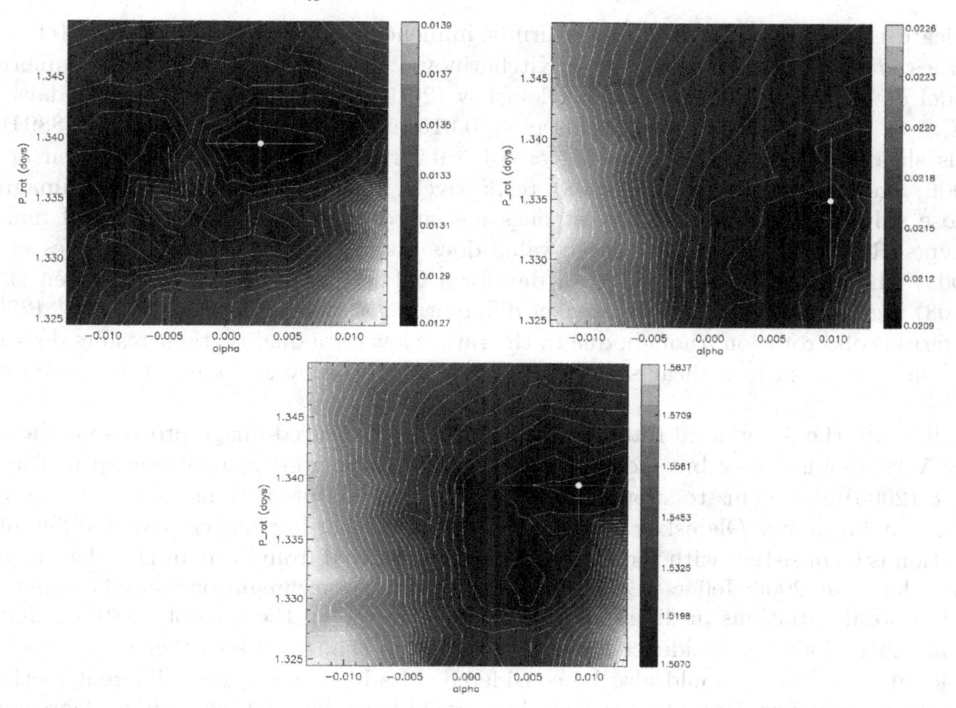

Figure 2. Resulting χ^2 maps of the sheared-image method over the $P_{\rm rot} - \alpha$ parameter plane. The best fit is found at $\alpha = 0.003 \pm 0.004$ and $P_{\rm rot} = 1.3395 \pm 0.0027$ days for the Fe I-6411 reconstructions (*top left panel*) and at $\alpha = 0.010 \pm 0.002$ and $P_{\rm rot} = 1.3350 \pm 0.0052$ days for the Ca I-6439 reconstructions (*top right panel*). Bottom panel shows the averaged χ^2 map, where minimum is found at $\alpha = 0.009 \pm 0.006$ and at $P_{\rm rot} = 1.3395 \pm 0.0028$ days.

of $P_{\rm rot}$ and α is then marked by the least χ^2 value. The ability of this method was demonstrated also on artificial data by Kővári & Weber (2004).

For the Doppler reconstructions we used the Doppler imaging code TEMPMAP by Rice *et al.* (1989). Doppler imaging was performed for the Fe I-6411, and Ca I-6439 mapping lines. The individual reconstructions shown in Fig. 1 revealed similar spot distributions, i.e., mainly cool polar spots with temperature contrasts of up to ≈1500 K with respect to the unspotted surface of ≈5800 K. Some low-latitude features are also recovered, however, with significantly weaker contrast ranging from ≈300 K (Ca I-6439) to a maximum of ≈500 K (Fe I-6411).

Fig. 2 displays the χ^2-maps of the trial-and-error process over the $P_{\rm rot} - \alpha$ parameter plane. The best fit minima have similar locations for the two independent image reconstructions. We find $\Omega(\theta) = 4.69 - 0.014 \sin^2\theta$ rad/day differential rotation law for the iron line and $\Omega(\theta) = 4.71 - 0.047 \sin^2\theta$ rad/day for the calcium line, while the combined χ^2 map suggests $\Omega_{\rm eq} = 4.69$ rad/day and $\delta\Omega = 0.042$ rad/day.

3. Discussion

Using the averaged χ^2 landscape of the two independent line reconstructions in Fig. 2 we find $\alpha = 0.009$ or equivalently $\delta\Omega = \Omega_{\rm eq} - \Omega_{\rm pole} = 0.042$ rad/day. Accordingly, the time the equator needs to lap the pole by one full rotation is about 150 days, i.e., of the same order of the solar value. Observations and theoretical model calculations indicate that differential rotation is firmly influenced by stellar temperature (cf. e.g., Barnes *et al.* 2005, Reiners 2006, & Kitchatinov & Rüdiger 1995). A recent numerical model developed by Kitchatinov & Olemskoy (2010) predicts $\delta\Omega \approx 0.075$ rad/days for a $T_{\rm eff} = 5800$ K dwarf, which yields $\alpha \approx 0.016$ at the angular velocity of V889 Her. This shear is much weaker than $\delta\Omega \approx 0.4 - 0.5$ rad/day obtained by Marsden *et al.* (2006) and by Jeffers & Donati (2008), respectively, both from Zeeman Doppler imaging. Those values are among the largest ones measured only for very few stars of mainly F-type (Reiners 2006). Such a large value does not fit the power law of Barnes *et al.* (2005) which suggests $\delta\Omega \approx 0.12$ rad/day for a G2-type star. Moreover, Järvinen *et al.* (2008) argues for a substantially weaker differential rotation. Indeed, Huber *et al.* (2009) preferred solid rotation, though, due to the rather low data quality, their results does not exclude a weak surface shear such as that was suggested by Järvinen *et al.* (2008) and by us.

All in all, the differential rotation derived from our sheared-image process for the G2-star V889 Her is below but close to the one from the empirical relationship in Barnes *et al.* (2005) and seems to support the theoretical model predictions (Küker & Rüdiger 2005; Kitchatinov & Olemskoy 2010) as well. On the other hand, such a weak differential rotation is inconsistent with the high surface shear derived from Zeeman Doppler imaging (Marsden *et al.* 2006; Jeffers & Donati 2008). This difference might partially be explained by temporal variations in differential rotation, however, in the case of V889 Her Jeffers *et al.* (2010) found no evidence for such a temporal change. If so, other effects such as rapid spot evolution should also be considered. In addition, the very different methods and data applied to detect the surface shear could have different, sometimes depreciated or unrecognized error sources, which can also explain such a disagreement (we refer T.A. Carroll's paper talk in this conference). For instance, when various spectral features used for imaging correspond to different surface layers, the detected shear would diverge.

Acknowledgements

ZsK and KV are supported by the Hungarian Science Research Program (OTKA) grants K-68626 K-81421 and by the Lendület Young Researchers' Program of the Hungarian Academy of Sciences. This work has been supported by the Italian *Ministero dell'Istruzione, Università e Ricerca* (MIUR) and by the *Regione Sicilia*.

References

Barnes, J. R., Collier Cameron, A., Donati, J.-F., James, D. J., Marsden, S. C., & Petit, P. 2005, *MNRAS* 357, L1

Donati, J.-F., Mengel, M., Carter, B. D., Marsden, S., Collier Cameron, A., & Wichmann, R.: 2000, *MNRAS* 316, 699

Frasca, A., Biazzo, K., Kővári, Zs., Marilli, E., & Cakirli, O. 2010, *A&A*, 518, A48

Huber, K. F., Wolter, U., Czesla, S., Schmitt, J. H. M. M., Esposito, M., Ilyin, I., & González-Pérez, J. N. 2009, *A&A*, 501, 715

Järvinen, S. P., Korhonen, H., Berdyugina, S. V., Ilyin, I., Strassmeier, K. G., Weber, M., Savanov, I., & Tuominen, I. 2008, *A&A*, 488, 1047

Jeffers, S. V. & Donati, J.-F. 2008, *MNRAS*, 390, 635

Jeffers, S. V., Donati, J.-F., Alecian, E., & Marsden, S. C. 2011, *MNRAS*, 411, 1301

Kitchatinov, L. L. & Olemskoy, S. V. 2011, *MNRAS*, 411, 1059

Kitchatinov, L. L. & Ruediger, G. 1995, *A&A*, 299, 446

Kővári, Zs., Weber, M. 2004, in: British-Hungarian N+N Workshop for Young Researchers on 'Computer Processing and Use of Satellite Data in Astronomy & Astrophysics' and 3rd Workshop of Young Researchers in Astronomy & Astrophysics, Eötvös University, 4–7 Feb,

2004, Budapest, Hungary, Publ. Astron. Dept. Eötvös University Vol. 14, E. Forgács-Dajka, K. Petrovay, R. Erdélyi (eds), p. 221

Küker, M. & Rüdiger, G. 2005, *Astron. Nachrichten*, 326, No. 3/4, 265

Marsden, S. C., Donati, J.-F., Semel, M., Petit, P., & Carter, B. D. 2006, *MNRAS*, 370, 468

Maunder, E. W. & Maunder, A. S. D. 1905, *MNRAS*, 65, 813

Montes, D., López-Santiago, J., Gálvez, M. C., *et al.* 2001, *MNRAS*, 328, 45

Reiners, A. 2006, *A&A*, 446, 267

Rice, J. B., Wehlau, W. H., & Khokhlova, V. L. 1989, *A&A*, 208, 179

Strassmeier, K. G., Pichler, T., Weber, M., & Granzer, T. 2003, *A&A*, 411, 595

The Physics of Sun and Star Spots
Proceedings IAU Symposium No. 273, 2010
D.P. Choudhary & K.G. Strassmeier, eds.

doi:10.1017/S1743921311015122

Long-term evolution of sunspot magnetic fields

Matthew J. Penn[1] and William Livingston[1]

[1]National Solar Observatory†, 950 N Cherry Av, Tucson AZ 85718
email: mpenn@nso.edu

Abstract. Independent of the normal solar cycle, a decrease in the sunspot magnetic field strength has been observed using the Zeeman-split 1564.8nm Fe I spectral line at the NSO Kitt Peak McMath-Pierce telescope. Corresponding changes in sunspot brightness and the strength of molecular absorption lines were also seen. This trend was seen to continue in observations of the first sunspots of the new solar Cycle 24, and extrapolating a linear fit to this trend would lead to only half the number of spots in Cycle 24 compared to Cycle 23, and imply virtually no sunspots in Cycle 25.

We examined synoptic observations from the NSO Kitt Peak Vacuum Telescope and initially (with 4000 spots) found a change in sunspot brightness which roughly agreed with the infrared observations. A more detailed examination (with 13,000 spots) of both spot brightness and line-of-sight magnetic flux reveals that the relationship of the sunspot magnetic fields with spot brightness and size remain constant during the solar cycle. There are only small temporal variations in the spot brightness, size, and line-of-sight flux seen in this larger sample. Because of the apparent disagreement between the two data sets, we discuss how the infrared spectral line provides a uniquely direct measurement of the magnetic fields in sunspots.

Keywords. Solar Cycle, Sunspot, FeI1564.8 nm

---

1. Introduction

Observations of the magnetic fields in sunspot umbrae have been carried out by Livingston at the National Solar Observatory's McMath-Pierce solar telescope atop Kitt Peak. These observations are made with a single-element detector and measure the intensity spectra of the 1564.8nm Fe I g=3 spectral line and nearby atomic and molecular absorption lines. While these observations began in the 1990's, the focus then was only on the larger sunspots visible on the solar disk. During the last 10 years these observations have become more synoptic in that all sunspots visible on the solar disk are observed in this way, from solar pores to the largest umbrae. (In the following text we use the term "spots" to refer to both sunspots with penumbrae and pores without penumbrae.) After fitting several spectral lines in the data, Livingston has compiled a table of the magnetic field strength at the darkest spot location, the continuum brightness at that location (normalized to nearby quiet Sun brightness), and the line depth of several OH molecular lines in the spectral field-of-view. It is important here to note (1) that no polarimetry is done, only intensity spectra are used, (2) that the 1564.8nm Fe I line is completely split (i.e. the Zeeman sigma components are shifted in wavelength more than their line widths) for the 1500 Gauss and larger magnetic fields seen in the spots, and (3) the splitting of the sigma components in the intensity spectrum measures the true magnetic field strength, not a vector component of the magnetic field.

† NSO is operated by AURA, Inc., under contract to the National Science Foundation

We reported in Penn & Livingston (2006) that a time series of this magnetic field data showed a decrease in the umbral magnetic field strength which was independent of the normal sunspot cycle. Also, the measurements revealed a threshold magnetic field strength of about 1500 Gauss, below which no dark pores formed. A linear extrapolation of the magnetic field trend suggested that the mean field strength would reach this threshold 1500 Gauss value in the year 2017. Furthermore, analysis of the umbral continuum brightness showed another linear trend, and extrapolation showed the umbral brightness would be equal to the quiet Sun brightness at about the same year. Finally, the molecular line depths showed a decreasing strength with time, and again the trend suggested that molecular absorption lines would disappear from the average sunspot umbra near 2017.

Recent observations spanning from the solar interior to the solar corona clearly show that solar Cycle 24 has started. Below the solar surface, helioseismic observations of the torsional oscillations have shown that the subsurface flow maxima migrated to latitudes of +/-23 degrees in February of 2009 coinciding with the flow latitude at the onset of the magnetic activity for solar Cycle 23 (). At the solar surface, the sunspot number is rising http://sidc.oma.be/sunspot-data/. The magnetic polarity of solar magnetic active regions has switched since Cycle 23, and the hemispheres now show new cycle magnetic flux consistent with Hale's polarity law http://www.nso.edu/press/cycle24.html. In the solar chromosphere the spectral Ca K index has shown an increase ftp://ftp.nso.edu/idl/cak_plot.gif. And in the solar corona, the radio emission from the Sun at 10.7 cm wavelength has begun to increase http://www.spaceweather.gc.ca/sx-6-eng.php, and the UV emission from the Sun has started to rise http://lasp.colorado.edu/lisird/sorce/sorce_ssi/ts.html. And finally of note, the Solar Cycle 24 Prediction Panel from the Space Weather Prediction Center has recognized that the minimum after solar Cycle 23 was reached in December 2008 http://www.swpc.noaa.gov/SolarCycle/.

If Cycle 24 has started, we are in the rise phase of the cycle; but where exactly in the cycle are we located? The helioseismic observations can tell us based on the latitude of the torsional oscillation bands. This gives us a phase indicator which is independent of the cycle duration or the amplitude of the activity peak for the cycle. We can extrapolate the latitudinal drift of the torsional bands () and then compare the current position with the position in Cycle 23. This calculation tells us that June 2010 in Cycle 24 corresponds with February 1998 in Cycle 23. It is instructive to examine the monthly sunspot numbers for those two months; for February 1998 that value was 40, and for June 2010 that value was 13 http://sidc.oma.be/sunspot-data/. Including the 5 months preceding these times, we find that for a 6 month period Cycle 24 has shown only 0.37 times the number of spots seen in Cycle 23. By correcting for the phase of the solar cycles, we are now seeing far fewer sunspots than we saw in the preceding cycle; solar Cycle 24 is producing an anomalously low number of dark spots and pores.

2. Recent observations

Figure 1 shows the observations of sunspot and pore magnetic fields from Livingston's data set. The total magnetic field strength at the darkest location in the umbra or pore is plotted against the date of the measurement. The raw measurements are shown as crosses. There is a large distribution of magnetic field strengths in spots visible on the solar photosphere, and there seems to be a lower threshold for the formation of dark spots, either pores or umbrae. No measurements show that the total magnetic field strength is less than about 1500 Gauss in a dark spot, and presumably magnetic regions with maximal field strengths less than this value do not undergo convective collapse. In

Figure 1 the annual bins of the measurements are shown as asterisks, and the standard deviation of the mean is shown as a vertical error bar on the asterisks.

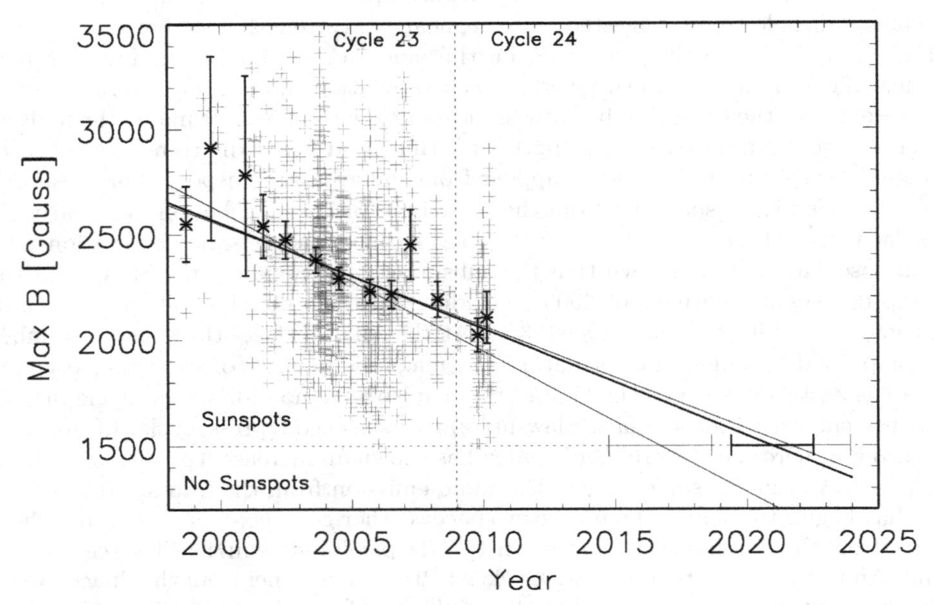

Figure 1. Measurements of the total magnetic field strength at the darkest location in umbrae and pores as a function of time. The crosses show the individual measurements, the asterisks show annual bins. Three linear fits are shown: the bottom fit line fits data from 1998-2006 as done in our 2006 paper. The top line fits all the data from Cycle 23, and the middle line fits all of the data.

Various linear fits are also shown in the Figure. The line to the left shows a linear fit from the work done by Penn & Livingston (2006); the extrapolated line shows an intercept with the 1500 Gauss value in 2017, and error bars of the computed intercept are also shown. The right-most line fits all of the data from Livingston's Cycle 23 observations, and the slight uptick in the magnetic field measurements from 2007 and 2008 move the 1500 Gauss intercept time out to 2022. The central line fits all of the data, including measurements from Cycle 24, and the intercept date now appears to be 2021, but it is within the error-bars from the fit to the Cycle 23 sunspot data. The linear fit to all of the data show a decrease of about 50 Gauss per year in the magnetic field strength at the darkest location in spots.

It is important to note that both sunspots and pores are included in this plot. Pores, lacking penumbra, often have magnetic fields less than 2000 Gauss, but always have magnetic fields stronger than 1500 Gauss. Secondly, the intercept of the mean magnetic field strength with this 1500 Gauss threshold does not imply that all sunspots will disappear by the year 2021; rather it implies that half of the sunspots which would normally appear on the surface of the Sun would be visible. Finally, the plot doesn't address the other magnetic fields on the Sun where field strengths are lower than 1500 Gauss; the temporal behavior of solar active network or quiet Sun magnetic fields may be different from the behavior shown by sunspots.

3. Searching for support in other data

The changes observed in sunspot brightness prompted an analysis of sunspot umbrae as observed in the synoptic data set from the National Solar Observatory Kitt Peak Vacuum Telescope (KPVT). In the first analysis of that data set, Penn & MacDonald (2007) selected isolated spots by hand. About 4000 sunspots and pores were examined, and a cyclic behavior was seen in the minimum brightness found in these dark spots in phase with the sunspot cycle; darker sunspots were more common during solar maximum, and brighter sunspots were more common during solar minimum. Strangely, no significant temporal change in the radius of the sunspot umbrae was seen. Using well-known wavelength scaling coefficients from Maltby (Maltby *et al.* 1986) the KPVT data showed a good correspondence with both a study using MDI data (Norton & Gilman 2004) and with the observations of sunspot intensities in the infrared by Livingston. At the time, the uptick in the magnetic field strengths seen by Livingston in 2007 and 2008 suggested that perhaps there was a solar-cycle dependence.

A more detailed analysis of the KPVT data set was performed by Tom Schad (Schad & Penn 2010) which included an automated sunspot selection procedure, resulting in the identification of over 13,000 dark spots, and an analysis of the brightness as well as the line-of-sight magnetogram data. This work showed that there were only small temporal changes in the spot intensities and magnetic field strengths. It also showed that two empirical sunspot relationships, the first between sunspot magnetic field strength and brightness, and the second between magnetic field strength and spot radius, both remained unchanged during the solar cycle. Both relationships did contain some scatter, but it was found that the temporal changes in the spot radius were consistent with the changes in magnetic fields and brightness. Finally, current work with the KPVT data set suggests that the twist of the sunspot magnetic fields does not vary significantly during the solar cycle. The horizontal pressure balance that spots achieve with the surrounding quiet Sun behaves the same way at all phases of the solar cycle.

Work from other authors have addressed some of these issues as well. Observations of the brightness of sunspots as measured with MDI showed no changes from 1998-2004 (Mathew *et al.* 2007) which is consistent with the observed KPVT data during this time interval. Measurements of the brightness of sunspot umbrae from the California State University Northridge San Fernando Observatory showed no changes during the interval from 1997-2004 (Wesolowski, Walton & Chapman 2008) although the brightness vs radius relationship from that data seems anomalous (Schad & Penn 2010). And most recently in these proceedings, measurements of the magnetic fields from sunspot umbrae near the center of the solar disk using MDI magnetograms Watson & Fletcher (2010) show a smaller decrease in the magnetic field strength, but that result is not significant compared to the standard errors of their fit.

Measuring the true magnetic field strength in the darkest sunspot or pore regions is known to be a difficult task since the brightness levels are low and the line depths are small (Liu, Norton & Scherrer 2007). Using simultaneous measurements of a large sunspot from Hinode and MDI, Moon *et al.* (2007) show that the MDI observations can underestimate the magnetic field strength by a factor of two. Imaging magnetographs have distinct advantages in terms of cadence of observations and the spatial integrity of the images, but spectrograph-based instruments which capture full line profiles in dark spots do have advantages in terms of accuracy.

It is also important to realize that the data obtained by Livingston using the infrared line at 1564.8nm with a Landé g-factor of 3.0 are measuring magnetic fields that are completely resolved. Using a conservative estimate (i.e. a large Doppler and instrumental

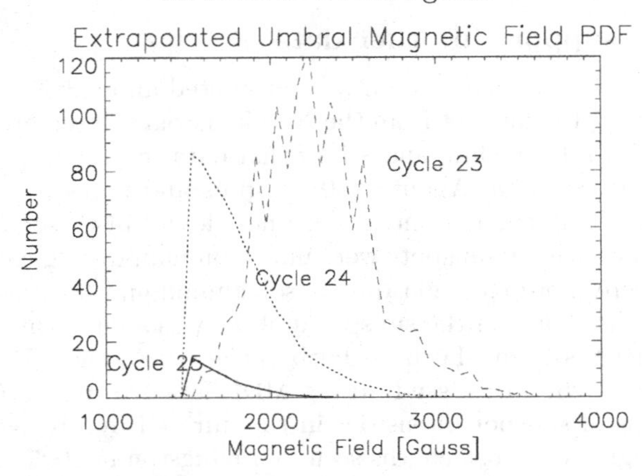

Figure 2. The magnetic probability distribution function (PDF) is show for the IR measurements of sunspots during Cycle 23. With the assumption discussed in the text, we can produce PDFs for Cycles 24 and 25. A simple scaling using the total number of spots suggests Cycle 24 will peak with a SSN of 66, and Cycle 25 will peak with a SSN of 7.

line width) the infrared spectral line can resolve fields with strengths greater than 750 Gauss. We can scale this value by the factor of g times lambda for many of the instruments used to study sunspot magnetic fields. The KPVT magnetic field resolution would be 2400 Gauss, and for MDI the resolved field strength is about 3600 Gauss. For the HMI and SOLIS spectrograph-based instruments, the magnetic field strength must be above 2200 Gauss to be fully resolved. If we examine the measurements in Figure 1 which have magnetic fields only above 2200 Gauss, the temporal trend is not apparent. Certainly magnetic fields can be determined for spots with fields below this magnetic resolution value, but there are assumptions, corrections and (in some cases) models which are used in that determination, and perhaps the scatter inherent in that process is enough to swamp the underlying temporal variation which is so apparent in the more direct infrared measurements.

The lack of significant brightness or radius variation of sunspots as seen with other instruments is more difficult to explain. While the infrared measurements suffer from less instrumental scattered light, and perhaps better ground-based seeing than ground-based visible observations, these advantages do not seem large enough to explain the lack of variation seen with other telescopes; it remains a mystery.

4. Implications and critical observations for the future

As suggested by Figure 1 a detailed analysis shows that the sunspots measured during the rise phase of Cycle 24 have the same shape in the distribution of magnetic field strengths as the spots seen during the decay phase of Cycle 23, but that the mean value of the distribution is reduced. This is a conservative conclusion from Livingston's observations. If we make three assumptions however, we see that there are more dramatic implications of these infrared observations. First we assume that the distribution of magnetic fields observed by Livingston from 1998-2008 is a good proxy for the true probability distribution function (PDF) for sunspot umbral magnetic fields for Cycle 23. Secondly, we assume that the magnetic threshold of 1500 Gauss represents a real physical limit for the formation of a dark spot (either a pore or a sunspot) on the solar photosphere. And finally, we assume that the mean of the magnetic field PDF continues to decrease linearly with time.

Figure 2 shows the computed magnetic PDF for the sunspots in cycles 24 and 25, using a linear decrease of the magnetic field of 65 Gauss per year and a duration of 11 years for each cycle. This is meant to represent an upper limit, and the magnetic change corresponds to the most steeply sloped line in Figure 1. We can see that the PDFs for Cycle 24 and Cycle 25 vary dramatically from that observed in Cycle 23. If we assume that the appearance time of sunspots during each cycle is similar, we can use the total number of spots in each cycle to compute the maximum activity level of that cycle, using the fact that Cycle 23 showed a peak smoothed sunspot number (SSN) of 130. The linear decrease of 65 Gauss per year predicts that Cycle 24 will peak with a smoothed SSN of 66, and Cycle 25 will peak with a smoothed SSN of 7. Using a value of 50 Gauss per year suggests a smoothed SSN of 87 for Cycle 24 and 20 for Cycle 25.

It is important to note that it is always risky to extrapolate linear trends; but the importance of the implications from making such an assumption justify its mention. Also of note is that while these PDFs are drawn from Livingston's observations, they are at best proxies for the true sunspot magnetic PDFs. While a sunspot with a magnetic field strength of 4200 Gauss was observed in Cycle 23 (NOAA 10930, Moon *et al.* 2007), it was not observed by Livingston and does not appear in this analysis. Thus the sunspot which appeared recently in Cycle 24 (NOAA 11092, August 2010) with a magnetic field strength of 3350 Gauss does not invalidate these assumptions. Certainly if a large number of sunspots with magnetic field strengths greater than 3000 Gauss do appear, then the extrapolated PDF will be shown to be erroneous. We will see in the coming months and years.

Umbral magnetic field measurements at 1564.8nm have been shown to reveal differences between the decay phase of Cycle 23 and the rise phase of Cycle 24, and they imply that the next two sunspot cycles might be very different from the last one. Observations with visible light magnetographs do not show significant support for these claims. Thus we feel it is essential to make synoptic observations using this very favorable infrared line to determine if these trends continue. It is essential to save the spectra and calibrations, and it would be very useful to make synoptic measurements of sunspots using temperature sensitive molecular lines such as the lines of OH near 1564.8nm.

References

Howe, R., Christensen-Dalsgaard, J., Hill, F., Komm, R., Schou, J. & Thompson, M. J. *Astrophys. J. Lett.* 701, L87

Liu, Y., Norton, A. A., & Scherrer, P. H. 2007, *Solar Phys.* 241, 185

Maltby, P., Avrett, E. H., Carlsson, M., Kjeldseth-Moe, O., Kurucz, R. L., & Loeser, R. 1986, *Astrophys. J.* 306, 284

Mathew, S. K., Martinez Pillet, V., Solanki, S. K., & Krivova, N. A. 2007, *Astron. Astrophys* 465, 291

Moon, Y.-J., Kim, Y.-H., Park, Y.-D., Ichimoto, K., Sakurai, T., Chae, J., Cho, K. S., Bong, S., Suematsu, Y., Tsuneta, S., Katsukawa, Y., Shimojo, M., Shimizu, T., Shine, R. A., Tarbell, T. D., Title, A. M., Lites, B., Kubo, M., Nagata, S., & Yokoyama, T. 2007, *Pub. Astron. Soc. Jap.* 59, 625.

Norton, A. A. & Gilman, P. A. 2004, *Astrophys. J.* 603, 348.

Penn, M. J. & Livingston, W. 2006, *Astrophys. J. Lett.*, 649, L45

Penn, M. J. & MacDonald, R. K. D.. 2007, *Astrophys. J. Lett.*, 662, L123

Schad, T. A. & Penn, M. J. 2010, *Solar Phys.* 262, 19
Watson, F. & Fletcher, L. 2010, *IAU Symposium 273, these proceedings*
Wesolowski, M. J., Walton, S. R., & Chapman, G. A. 2008, *Solar Phys.* 248, 141.

Discussion

KOSOVICHEV: I wonder if Livingston has the data for cycle 22 to confirm your trend?

PENN: Bill's coverage in cycle 22 is really limited. So I would really like to trust his observations after 2000.

GEORGOULIV: I was wondering if you have any additional information on how different or perhaps on how similar is the distribution of magnetic field strengths in the declining phase of the second cycle versus the rising phase of the next cycle? As we know for quite some time that the declining phases of a cycle often has surprises, that its very strong active regions. Don't forget that we had the Halloween period over the declining phase of 23. So I was wondering if there is anymore information on this apparently very complex correlation?

PENN: I must admit I have done a very bad job of looking through the literature to find out, but you beat me to some slides I didn't want to show. So here is a distribution for Cycle Number 23; and if we just move this on down using the mean decrease, here is what Bill has observed for Cycle 24 rise phase. So there is a really distinct difference between the two. So, yeah, I hope if you know of other work that has been done, I would like to see it, but it is pretty clear in his data.

OLAH: Just one comment, below 1500, what I think you have are magnetic concentrations that are not dark. So they are there. With the ASP Bruce and I were calling that azimuth centers and the typical field there would be 1400 or something, but I think it's the same that Bakers call the magnetic north. So there are magnetic structure, but they are just simply not dark. So it's a systematic effect have you there, but it's something there.

PENN: Exactly. We're looking at the high end of the distribution of magnetic fields of the sun when we look at sunspot umbra. So yeah, either the high end – if the high end is doing something strange, then we're being misled; but maybe it's following the mean.

SAAR: Could height formation differences account for some of the differences between IR and other spot measurements?

PENN: Right. The two power laws that I showed you are mostly accounted for by the height difference. So – but in terms of the long term variation with time, I'm not sure that can be accounted by the height differences.

STRASSMEIER: There is full disc H and K images or K-line images that Bill had actually monitored for ages as well and do these K images show the same trend. Don't you actually have information there for, say, the global field or proxy of the global field?

PENN: Right. And from what I have seen there is no long-term trend. It's a solar cycle variation in the data set that he has. We haven't seen a long-term trend like this in that data, no.

Unknown: Quick last question. Are we headed for a Maunder minimum, and is this going to solve global warming?

Penn: Well, according to the author of that book, "Red Hot Lies," I'm an outsider in solar physics; and I'm proposing this is a cause of global warming. I guess that is how I have been portrayed. So, yeah, we'll be lucky or unlucky in sunspots return, I guess.

The Physics of Sun and Star Spots
Proceedings IAU Symposium No. 273, 2010
D.P. Choudhary & K.G. Strassmeier, eds.

doi:10.1017/S1743921311015134

The formation of a penumbra as observed with the German VTT and SoHO/MDI

Rolf Schlichenmaier, Nazaret Bello González, and Reza Rezaei

Kiepenheuer Institut für Sonnenphysik,
Schöneckstr. 6, 79104 Freiburg, Germany
email: schliche@kis.uni-freiburg.de

Abstract. The generation of magnetic flux in the solar interior and its transport to the outer solar atmosphere will be in the focus of solar physics research for the next decades. One key-ingredient is the process of magnetic flux emergence into the solar photosphere, and the reorganization to form the magnetic phenomena of active regions like sunspots and pores.

On July 4, 2009, we observed a region of emerging magnetic flux, in which a proto-spot without penumbra forms a penumbra within some 4.5 hours. This process is documented by multi-wavelength observations at the German VTT: (a) imaging, (b) data with high resolution and temporal cadence acquired in Fe I 617.3 nm with the 2D imaging spectropolarimter GFPI, and (c) scans with the slit based spectropolarimeter TIP in Fe I 1089.6 nm. MDI contiuum maps and magnetograms are used to follow the formation of the proto-spot, and the subsequent evolution of the entire active region.

During the formation of the penumbra, the area and the magnetic flux of the spot increases. The additional magnetic flux is supplied by the adjacent region of emerging magnetic flux: As emerging bipole separate, the poles of the spot polarity migrate towards the spot, and finally merge with it. As more and more flux is accumulated, a penumbra forms. From inversions we infer maps for the magnetic field and the Doppler velocity (being constant along the line-of-sight). We calculate the magnetic flux of the forming spot and of the bipole footpoints that merge with the proto-spot. We witness the onset of the Evershed flow and the associated enhance of the field inclination as individual penumbral filaments form. Prior to the formation of individual penumbral sectors we detect the existence of 'counter' Evershed flows. These in-flows turn into the classical radial Evershed outflows as stable penumbra segments form.

Keywords. Magnetic fields, sunspots, Sun: photosphere, MHD, Sun: activity

---

1. Introduction

The solar photosphere exhibits magnetic features at a large range of scales. From these, sunspots are the largest. Simulations of flux emergence have made major progress recently (e.g., Cheung *et al.* 2008; Tortosa-Andreu & Moreno-Insertis 2009; Cheung *et al.* 2010). Yet, high-resolution spectropolarimetric observations of such formation process are rare and either lack spectropolarimetric data or high spatial resolution (Lites *et al.* 1998; Leka & Skumanich 1998; Yang *et al.* 2003). In this paper, we report on high resolution spectropolarimetric observations of the formation of a sunspot penumbra. This penumbra forms around a proto-spot, thereby transforming the proto-spot into a fully developed sunspot. In addition we use MDI data to learn about how the proto-spot formed and how the entire active region develops (Scherrer *et al.* 1995).

We analyze our measurements taken at the German VTT and attempt to understand how a penumbra forms. It is known (see e.g. the reviews of Solanki 2003; Bellot Rubio 2004, 2010; Borrero 2009; Schlichenmaier 2009) that a penumbra is a site of more inclined (with respect to the vertical) magnetic fields, as compared to the field inclination in umbrae and pores. A typical penumbral feature is its radial filamentation in white light images. A radially outward directed flow is present along these filaments – the Evershed flow. From Stokes profiles we know that this predominantly horizontal flow is magnetized (e.g., Bellot Rubio *et al.* 2003, 2004; Franz & Schlichenmaier 2009). This horizontal component is accompanied by a less inclined magnetic field component. Hence, the penumbra is a very peculiar phenomenon. And its structure is still not understood, although many model attempts have been made (e.g., Schlichenmaier *et al.* 1998; Schlichenmaier 2002; Scharmer & Spruit 2006; Heinemann *et al.* 2007; Ryutova *et al.* 2008; Rempel *et al.* 2009). One way to enhance our understanding is to study sunspot formation. Therefore we analyze our unique measurements as described in Schlichenmaier *et al.* (2010b,a) (hereafter paper I and II, respectively) taken at the German VTT and try to shed new light on the formation of the penumbra.

2. Observational setup and data acquisition

As detailed in paper I, the campaign consisted of a multi-instrument setup and involved KAOS (von der Lühe *et al.* 2003). Two imaging cameras were tuned to the G-band at 430 nm and to the Ca II K line core at 393 nm, respectively. These images were speckle reconstructed using KISIP (Wöger *et al.* 2008). This was enriched by two spectropolarimeters of different type: the Fabry-Pérot system GFPI (Puschmann *et al.* 2006; Bello González & Kneer 2008) and the slit-based TIP (Collados *et al.* 2007). With the GFPI, we scan Fe I 617.3 nm in 56 s to measure the maps of the Stokes parameters, and with TIP we observe Fe I 1089.6 nm, with an exposure time of 10 s per slit position.

For the GFPI and TIP data we perform inversions with SIR (Ruiz Cobo & del Toro Iniesta 1992; Bellot Rubio 2003) to infer physical parameters that are imprinted in the Stokes profiles by the Zeeman effect and other effects of radiative transport. To minimize the degrees of freedom, we assume that the magnetic and velocity fields are constant along the line-of-sight. Thereby, we correspondingly retrieve mean values, instead of attempting to resolve variations along the line-of-sight and laterally.

When scanning the spot with TIP, the spot image also moves in the GFPI field-of-view. We made scans spanning not more than $2''$ with TIP to assure that the spot stays within the $30''$ by $20''$ field-of-view of GFPI. Yet, between UT 11:43 and 11:59 we performed a scan with TIP covering the entire spot. With this scan we can demonstrate that the GFPI and the TIP not only measure maps of integrated Stokes profiles that look alike, but also produce physical maps that give consistent results for the two different lines.

3. Flux emergence and structure formation

The spot formation on July 4, 2009, takes place in the active region NOAA 11024. This active region rotated in onto the disk on June 29. At this stage, it was visible as a facular region without pores. Its magnetic signatures are seen in MDI magneto-grams, but only barely in MDI continuum images. On July 1, the region formed two pores (Fig. 1). These two pores – of opposite polarity, as we know from TIP observations at the German VTT– evolved significantly and disappeared by July 2. In the morning of July 4, at 08:30 UT, a proto-spot with two pronounced light bridges appeared close to the zero-meridian of the Sun (Fig. 1). In the following 4.5 hours a penumbra formed around the proto-spot. By 13:00 UT the seeing conditions did not allow for further observation. At that time the penumbra encircled some 220 degrees. No stable penumbra formed towards the site of magnetic flux emergence, i.e., in the direction of the opposite polarity of the active region.

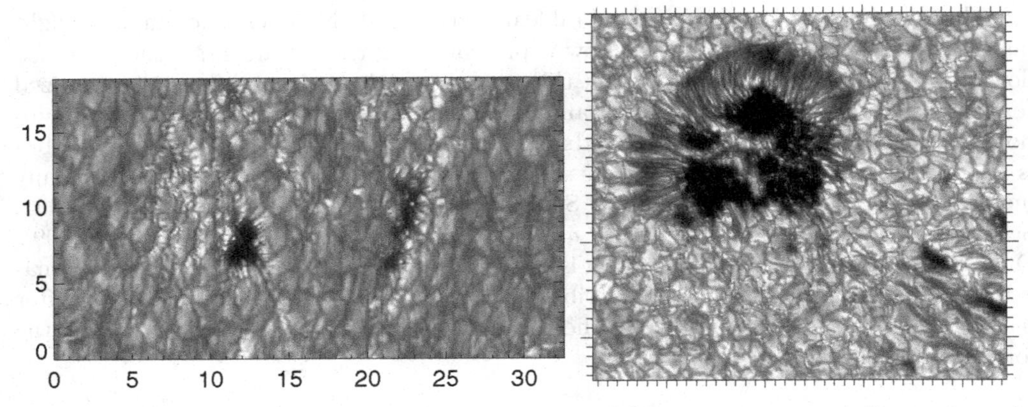

Figure 1. Left: On July 1st, 2009, the active region exhibits two pores at a heliocentric angle of $\theta \approx 70°$ (courtesy: T.A. Waldmann). Axis units are in arcsecs. From measurements with TIP we know, that they are of opposite polarity. During the course of the day the pores evolve and dissolve. Right: Snapshot of sunspot at $\theta = 28°$ on July 4 at 11:39, with tickmarks in arcsec.

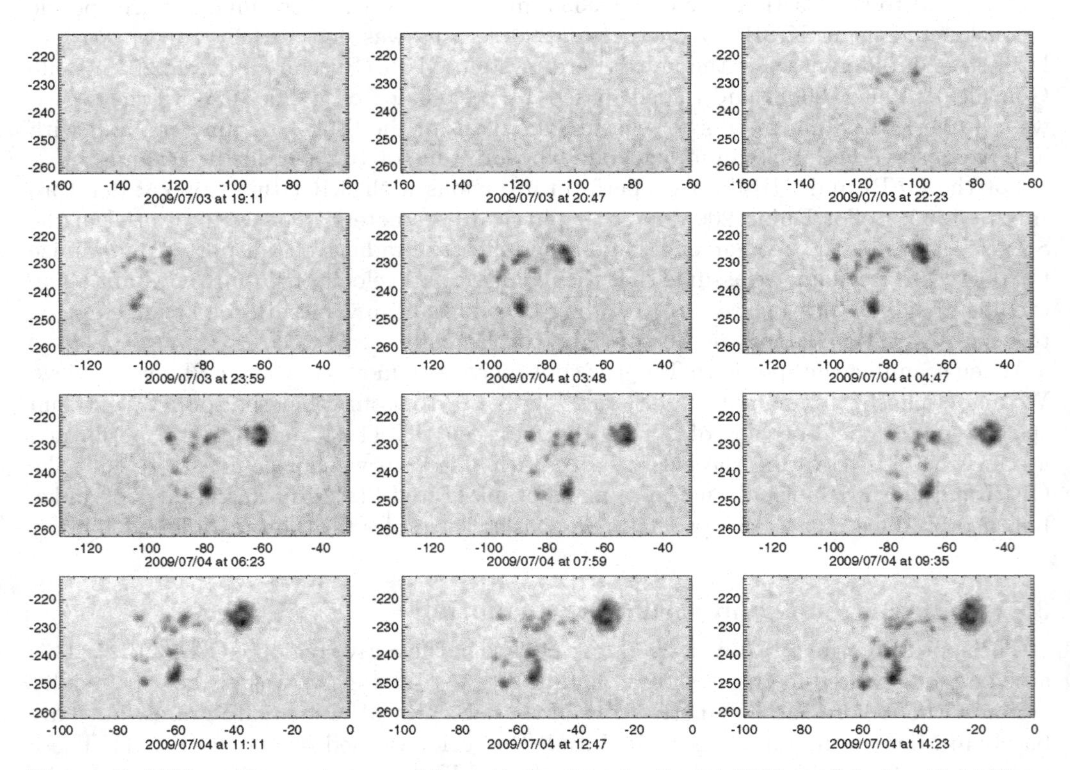

Figure 2. MDI continuum images of the active region starting at 19:11 on July 3 until 14:23 on July 4. The date and time are given below each image. The axis units are in arcsec and relative to disk center. X-axis varies to keep the active region in field-of-view.

Subsequently, the spot further increased in size. On July 5 the spot still showed a bridge. This light bridge disappeared by July 6. Also then, the penumbra did not fully encircle the umbra, but had a gap of 30 degrees toward the opposite polarity of the

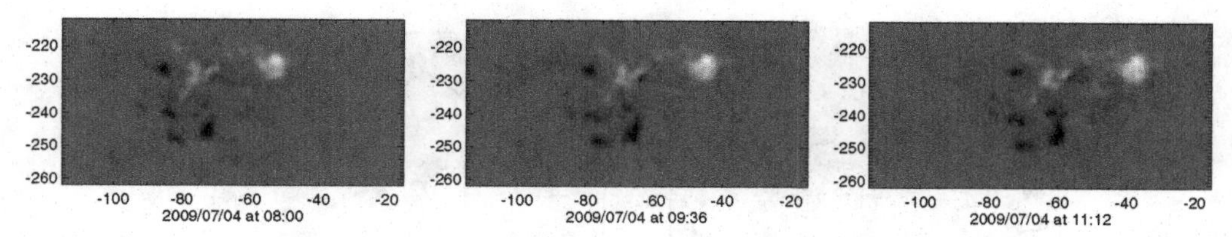

Figure 3. MDI magneto-grams at 08:00, 09:36, and 11:12 on July 4. The white polarity leading spot is followed by some more diverged black polarity. Yet, the white polarity is not fully contained in the spot as there is also some white polarity flux in the emerging site between the main bulk of the polarities.

active region. The spot further evolved, exhibiting a light bridge again two days later, and rotated off the solar disk after July 10.

3.1. *Formation of proto-spot*

Taking advantage of the MDI data base (Scherrer *et al.* 1995), we can track the formation of the proto-spot with MDI full disk continuum images. A series of maps is shown in Fig. 2 spanning the time between 19:11 on July 3 until 14:23 on July 4. While no pores are visible on the first image, some dark patches are seen at 20:47 on July 3. At 22:23, there is a pronounced pore leading the active region (upper right), being located at $(-100'', -225'')$. In the first image of the second row (23:59) a second leading pore has appeared. Now the pores are at $(-92'', -230'')$, and the magneto-grams in Fig. 3 indicate that they are of the same polarity. These pores are already very close to each other since their first appearance. Rather than migrating towards each other, it appears that they increase in size, and the granulation between them transforms into a light bridge. At about 07:59 (middle image of third row) they reach a state that we call proto-spot. Our observations at the VTT started at 08:30. At this stage the proto-spot started to develop penumbral segments.

The formation of the penumbra is also visible in the MDI images (bottom two rows of Fig. 2). It is seen that the penumbra forms on the side facing away from the active region. Zwaan (1992) – refering to Bumba & Suda (1984) and McIntosh (1981) – ascribes the formation of a sunspot to the coalescence of the existing pores. The case that we witness here is consistent with this picture, although the pores do not merge by approaching each other, but form very close to each other and further increase in size.

3.2. *Penumbra formation: the observational findings*

Morphology: Our G-band imaging and GFPI scans start around 08:30 and end at 13:05. The imaging data reveals that the penumbra forms in segments (paper I). Each segment forms in about 1 h. At the end of our time series, the penumbra does not fully spans around the umbra, but only 220 degrees (instead of 360). The segment where no stable penumbra forms is directed towards the emerging site. A snapshot of the forming sunspot at 11:39 is shown in the right panel of Fig. 1.

Elongated granules and emergent bipoles: The emerging site is characterized by elongated granulation and intense proper motions of such granules towards and away the spot. These elongated granules play a crucial role for the penumbra formation: Such granules are associated with emerging bipoles (cf. paper II). The bipole axes, as well as the axes of their elongated granules, are oriented radially away from the center of the spot. All bipoles that we measure pop up in the photosphere such that the bipole footpoint that has the polarity of the spot is directed towards the spot. As the elongated

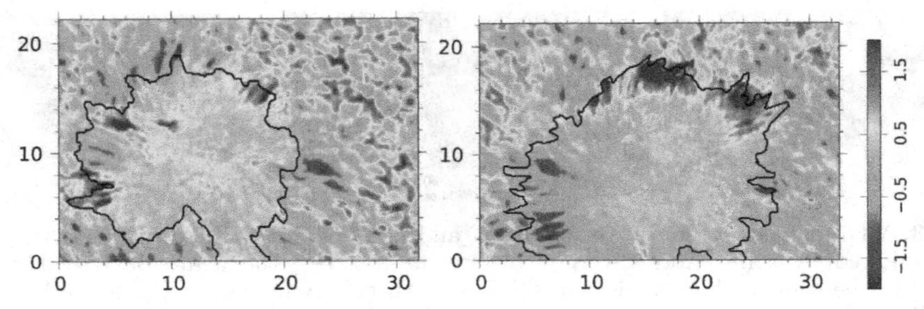

Figure 4. GFPI velocity maps inferred from inversions at 08:50 and 11:51 UT. Counter Evershed flows exist in the phase (08:50) prior to penumbra formation. Once the penumbra has formed, the flow pattern exhibits the typical Evershed flow pattern. The contours mark the white-light boundary of the spot. The color legend give the line-of-sight Doppler shift in km/s.

granule increases in size, that footpoint migrates towards the spot, while the footpoint of the other polarity migrates away from the forming spot.

Area: The area of the spot increases from 230 arcsec^2 to 360 arcsec^2 in 4.5 hours. This increase is exclusively taken up by the formation of the penumbra, the combined area of umbra and light bridges stays constant during that time (cf., paper I).

Magnetic flux and field strength: From the magnetic field strength and its inclination as inferred from the inversion, we can compute the magnetic flux in the plane perpendicular to the line-of-sight, commonly referred to as the longitudinal magnetic flux. We find that this flux increases from 1.6×10^{21} Mx to 2.4×10^{21} Mx, i.e. the spot increases by 8×10^{20} Mx in 4 hours (08:40 until 12:38). At this rate a sunspot of 10^{22} Mx would be formed in 2 days. Since no other pore merges with the spot while the penumbra forms, we assume that all the additional flux is supplied by granular scale bipoles. We find that a typical magnetic flux value for an individual bipole element amounts to some 2 – 3×10^{18} Mx. That means that about 1-2 emerging bipoles (of which the spot-polarity footpoints subsequently merge with the spot) are needed per minute to account for the increase of magnetic flux of the spot. As the penumbra forms, we track the magnetic field strength of the spot. We find that the mean umbral and penumbral field strength stays constant at 2.2 kG and at 1.5 kG, respectively. Averaging 100 pixels with the largest field strength of the umbra, we also find a constant value in time amounting to 2.7 kG (c.f., Rezaei, Bello González, & Schlichenmaier, in preparation).

Velocity field and inclination of magnetic field in penumbra: At a heliocentric angle of $\theta = 28°$, the 'classical' Evershed outflow should then appear as a redshift on the limb side of the spot and as a blueshift on the center side of the spot. This is of course also true for our sunspot penumbra, but only after the penumbra has formed (cf., right panel of Fig. 4). The flow pattern during the formation process is peculiar. Close inspection of flow, intensity, and inclination maps reveal that the areas of stable penumbral filaments are co-spatial with a classical Evershed flow pattern, i.e. radial outflow and large inclination. However, in areas where the penumbra has not yet formed, we observe Doppler shifts of opposite sign. These shifts can be seen in the left panel of Fig. 4: Red shifts are seen in the lower left portion of the spot which is the direction toward disk center, and blue shifts are seen in the upper part which points towards the limb (i.e. should be red shifted). Most interestingly these 'counter' shifts also show a filamentary shape and are co-spatial with field inclinations of up to 80° (w.r.t. vertical). Hence, if these flows are along magnetic field lines, they would be radially inward, i.e., could be described as counter Evershed flows. These flows reverse their sign as the penumbra forms (c.f., Bello González, Rezaei, & Schlichenmaier, in preparation).

Presently, we can not envisage a physical scenario that could explain these counter Evershed flows. But these properties of the flow and magnetic field just before the formation of penumbral filaments certainly are most interesting and may present the key to understand how a penumbra forms.

Acknowledgements

The German VTT is operated by the Kiepenheuer-Institut für Sonnenphysik at the Observatorio del Teide in Tenerife. We acknowledge the support by the VTT and KAOS group. SOHO is a project of international cooperation between ESA and NASA. NBG acknowleges the Pakt für Forschung, and RR the DFG grant Schm 1168/8-2. RS is grateful for travel support from the DAAD to attend the meeting and to ISSI for supporting the working group 'Filamentary Structure and Dynamics of Solar Magnetic Fields'.

References

Bello González, N. & Kneer, F. 2008, *Astron. Astrophys*, 480, 265

Bellot Rubio, L. R. 2003, in *Astronomical Society of the Pacific Conference Series*, ed. J. Trujillo-Bueno & J. Sanchez Almeida, 301

Bellot Rubio, L. R. 2004, *Rev. of Mod. Astron.*, 17, 21

Bellot Rubio, L. R. 2010, in *Magnetic Coupling between the Interior and Atmosphere of the Sun*, ed. S. S. Hasan & R. J. Rutten, 193

Bellot Rubio, L. R., Balthasar, H., & Collados, M. 2004, textit*Astron. Astrophys*, 427, 319

Bellot Rubio, L. R., Balthasar, H., Collados, M., & Schlichenmaier, R. 2003, *Astron. Astrophys*, 403, L47

Borrero, J. M. 2009, *Science in China G: Physics and Astronomy*, 52, 1670

Bumba, V. & Suda, J. 1984, *Bulletin of the Astronomical Institutes of Czechoslovakia*, 35, 28

Cheung, M. C. M., Rempel, M., Title, A. M., & Schüssler, M. 2010, *Astrophys. J.*, 720, 233

Cheung, M. C. M., Schüssler, M., Tarbell, T. D., & Title, A. M. 2008, *Astrophys. J.*, 687, 1373

Collados, M., Lagg, A., Díaz García, J. J., *et al.* 2007, in Astronomical Society of the Pacific Conference Series, Vol. 368, The Physics of Chromospheric Plasmas, ed. P. Heinzel, I. Dorotovič, & R. J. Rutten, 611

Franz, M. & Schlichenmaier, R. 2009, *Astron. Astrophys*, 508, 1453

Heinemann, T., Nordlund, Å., Scharmer, G. B., & Spruit, H. C. 2007, *Astrophys. J.*, 669, 1390

Leka, K. D. & Skumanich, A. 1998, *Astrophys. J.*, 507, 454

Lites, B. W., Skumanich, A., & Martinez Pillet, V. 1998, *Astron. Astrophys*, 333, 1053

McIntosh, P. S. 1981, in The Physics of Sunspots, ed. L. E. Cram & J. H. Thomas, 7–54

Puschmann, K. G., Kneer, F., Seelemann, T., & Wittmann, A. D. 2006, *Astron. Astrophys*, 451, 1151

Rempel, M., Schüssler, M., Cameron, R. H., & Knölker, M. 2009, *Science*, 325, 171

Ruiz Cobo, B. & del Toro Iniesta, J. C. 1992, *Astrophys. J.*, 398, 375

Ryutova, M., Berger, T., & Title, A. 2008, *Astrophys. J.*, 676, 1356

Scharmer, G. B. & Spruit, H. C. 2006, *Astron. Astrophys*, 460, 605

Scherrer, P. H., Bogart, R. S., Bush, R. I., *et al.* 1995, *Solar Phys.*, 162, 129

Schlichenmaier, R. 2002, *Astron. Nachr.*, 323, 303

Schlichenmaier, R. 2009, *Space Science Reviews*, 144, 213

Schlichenmaier, R., Bello González, N., Rezaei, R., & Waldmann, T. A. 2010a, *Astron. Nachr.*, 331, 563

Schlichenmaier, R., Jahn, K., & Schmidt, H. U. 1998, *Astron. Astrophys*, 337, 897

Schlichenmaier, R., Rezaei, R., Bello González, N., & Waldmann, T. A. 2010b, *Astron. Astrophys*, 512, L1

Solanki, S. K. 2003, *Astron. Astrophys. Rev.*, 11, 153

Tortosa-Andreu, A. & Moreno-Insertis, F. 2009, *Astron. Astrophys*, 507, 949 von der Lühe, O., Soltau, D., Berkefeld, T., & Schelenz, T. 2003, in *Society of Photo-Optical Instrumentation Engineers (SPIE) Conference Series*, ed. S. L. Keil & S. V. Avakyan, Vol. 4853, 187–193

Wöger, F., von der Lühe, O., & Reardon, K. 2008, *Astron. Astrophys*, 488, 375

Yang, G., Xu, Y., Wang, H., & Denker, C. 2003, *Astrophys. J.*, 597, 1190

Zwaan, C. 1992, in *NATO ASIC Proc. 375: Sunspots. Theory and Observations*, ed. J. H. Thomas & N. O. Weiss, 75–100

The physics of sun and star spots
Proceedings IAU Symposium No. 273, 2010
P.D. Choudhary & K.G. Strassmeier, eds.

doi:10.1017/S1743921311015146

Global MHD phenomena and their importance for stellar surfaces

Rainer Arlt

Astrophysikalisches Institut Potsdam, An der Sternwarte 16
D-14482 Potsdam, Germany
email: rarlt@aip.de

Abstract. This review is an attempt to elucidate MHD phenomena relevant for stellar magnetic fields. The full MHD treatment of a star is a problem which is numerically too demanding. Mean-field dynamo models use an approximation of the dynamo action from the small-scale motions and deliver global magnetic modes which can be cyclic, stationary, axisymmetric, and non-axisymmetric. Due to the lack of a momentum equation, MHD instabilities are not visible in this picture. However, magnetic instabilities must set in as a result of growing magnetic fields and/or buoyancy. Instabilities deliver new timescales, saturation limits and topologies to the system probably providing a key to the complex activity features observed on stars.

Keywords. MHD, turbulence, instabilities, stars: magnetic fields

1. Introduction

The imaging of the surfaces of stars other than the Sun provides us with a lot of observational facts about stellar activity. Temperature variations on the surfaces of rapidly rotating giants become visible, as shown, for example, for ζ And (Kővári *et al.* (2007) and II Peg (Carroll *et al.* 2007). Surface abundance maps of chemically peculiar stars have been published as well as maps of the surface magnetic field (e.g. Kochukhov *et al.* 2004 for 53 Cam). The magnetic fields are apparently fairly constant in time for half a century, and only a fraction of 10% or less of all A-type stars show magnetic fields. Dwarf stars are of course very interesting targets which are needed to place the Sun in the context of stellar activity. A compilation of the large-scale magnetic field geometries of, e.g., late M dwarfs is given by Morin *et al.* (2010). Since these stars are also supposed to have small-scale flares, the observations raise the question of which of the observed features represent the general topology generated by a dynamo, and which structures are local effects caused by secondary processes. The references given here are of course by no means complete. We refer the reader to other reviews for the status of the observations.

2. Dynamos from today's point of view

2.1. *Mean-field dynamos*

Motivated by convective motions of a conductive fluid, the net effect of the turbulence on the electromotive force was computed by Steenbeck *et al.* (1966). Up to a few years ago, next to all models of stellar cycles have been based on such mean-field models. The existence of this net effect, the α-effect, has been proven by numerical simulations.

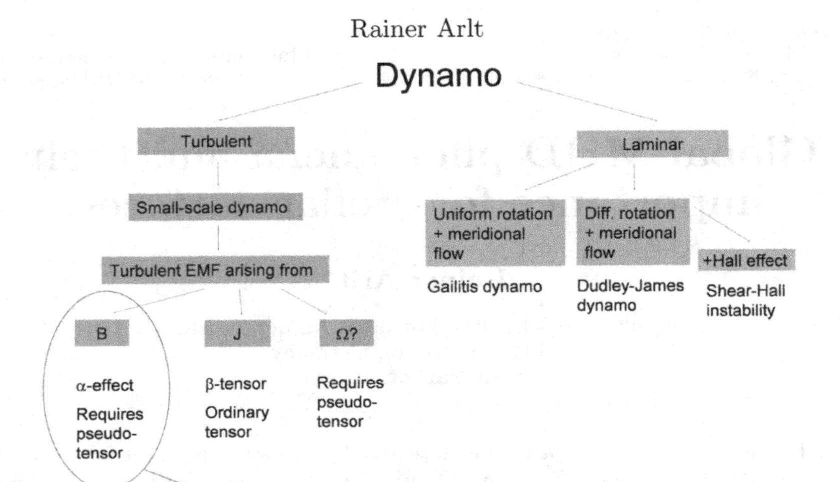

Figure 1. Possible dynamo action arising from various sources.

The presence of differential rotation leads to the generation of toroidal magnetic fields and is usually referred to as the Ω-effect. The α-effect, however, generates both components, poloidal fields from toroidal ones, and also toroidal fields from poloidal ones, just like the Ω-effect. The corresponding mean-field equation thus describes what is called an $\alpha^2\Omega$-dynamo. It is the neglect of the α-effect in the toroidal-field generation which makes the $\alpha\Omega$-dynamo which is often cited in stellar dynamos, but one should note that it is computationally inexpensive to use the full equations for any dynamo model of this kind, especially when additional subtleties play a major role in getting the cycle features right.

The other approximation is the α^2-dynamo in which the differential rotation is assumed to be zero. However, the effect of rotation is strong. Even if the differential rotation is only a 1%-effect, the nondimensional parameter for the Ω-effect in the mean-field equation is larger than the nondimensional α-parameter necessary to excite an α^2-dynamo. Differential rotation is never really negligible.

Figure 1 gives an overview of most of the possibilities to generate magnetic fields in stars continuously. The left-hand side shows the group of turbulent dynamos. Turbulence may deliver the small-scale dynamo which alone is not able to provide global magnetic fields. The generation of substantial large-scale magnetic fields through the turbulence requires the removal or cancellation of small-scale helicity. We are not going to review the issues of the problem of slow growth of the large-scale field here, but refer to Brandenburg & Subramanian (2005) for a review and assume that large-scale dynamo action can be expressed by a mean-field treatment. Numerous studies rely on the α-effect which is a pseudo-tensor acting on the axial vector $\boldsymbol{B}$. There is also the turbulent diffusivity tensor acting on gradients of $\boldsymbol{B}$ which does not only have destructive components. If rotation is present, generating terms are provided in this tensor and are called the $\boldsymbol{\Omega} \times \boldsymbol{J}$-effect. In principle, also a magnetic-field dependent pseudo-tensor acting on the angular velocity $\boldsymbol{\Omega}$ can lead to large-scale field generation (Yoshizawa 1993), but the applicability of this approach has not yet been fully explored.

While the models of kinematic dynamos based on the α-effect have become more and more sophisticated to show many of the features of the solar cycles, it is now time to

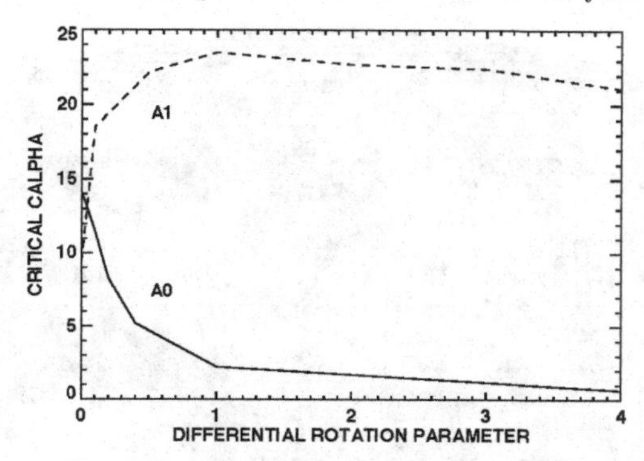

Figure 2. Critical dynamo-α in dimensionless units required for the onset of the axisymmetric dipole A0 and the oblique quadrupole A1 versus the differential rotation which mimics the solar rotation profile. In our units of the differential rotation the Sun has 6.87.

tackle the drawbacks and short-comings of the α-dynamos with a full MHD treatment that includes instabilities and the evolution of helicities in global spherical models.

2.2. *Laminar dynamos*

Only a brief mention of the group of laminar dynamos on the right-hand side of Fig. 1 is due here. They are also kinematic dynamos as there are known flow patterns – axisymmetric or non-axisymmetric – which lead to exponential growth of the magnetic field, without the addition of a turbulence source term like the α-effect. The field is always non-axisymmetric (Cowling 1934). Such dynamos would be interesting for radiative zones of Ap stars to explain their apparently strong magnetic fields, but no numerical confirmation with realistic flows from first principles has been found yet. Other problems are their slow growth and that the diversity of magnetic field topologies observed cannot be reproduced by the very regular, smooth solutions from a laminar dynamo.

2.3. *Active longitudes*

Several stars appear to exhibit persistent activity in a certain range in longitude for fractions of a stellar cycle. This long-lived non-axisymmetry raised the question of possible non-axisymmetric solutions of an $\alpha^2\Omega$-dynamo. Such solutions are in principle possible if the differential rotation of the stars is negligible. We have seen above that this is rarely the case, even if the differential rotation is only say 1%.

The observations even show changes of the active longitude by roughly 180°, mostly after a fraction of the stellar cycle length. The non-axisymmetric solutions of the $\alpha^2\Omega$-dynamo typically show, however, a stationary magnetic field pattern drifting continuously in longitude. Elstner & Korhonen (2005) found a solution in which both the axisymmetric "normal cycle" mode and the stationary non-axisymmetric mode are excited. The combination of the two resembles the flip-flop phenomenon, especially that of stars with spots at relatively high latitudes. The coexistence of the modes was basically achieved by a reduced differential rotation down to about 10% of the solar value. A small increase of the differential rotation leads to an immediate separation of the excitation thresholds of the non-axisymmetric mode and the axisymmetric one. This is demonstrated with an $\alpha^2\Omega$-dynamo with an α-effect distributed in the convection zone and a solar-like rotation profile of varying amplitude in Fig. 2.

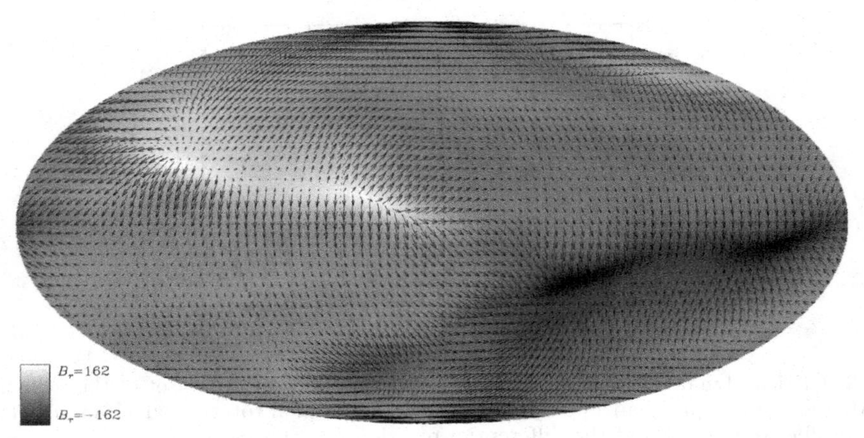

Figure 3. Surface plot of magnetic field from the Tayler instability as it appears after rising from the unstable region in the radiative envelope. The grey scale represents the radial field, the vectors the horizontal components.

It is very likely that other mechanisms than kinematic dynamo modes are at play here. Global magnetic instabilities seem to be an option, and we will elaborate on them in the following Section.

3. Instabilities

The kinematic dynamo is only a very small piece of the entire possible solutions of the MHD equations. These are the momentum equation, the induction equation, the energy equation, and mass conservation plus the constraint that $\nabla \cdot \boldsymbol{B} = 0$.

The number of known instabilities is large and the boundaries are not always clearly cut. We may group them into shear-driven instabilities, current-driven instabilities, and thermally driven instabilities. In the stellar case, shear-driven instabilities are fed by differential rotation and include the latitudinal shear instability (Watson 1981, and e.g. Dziembowski & Kosovichev 1987, Arlt *et al.* 2005) and the magneto-rotational instability (Velikhov 1959, and e.g. Arlt *et al.* 2003, Menou & Le Mer 2006). Current-driven instabilities can emerge from a magnetic field alone if it is not current-free. There is a variety of names for various types of instabilities but they all fall actually into the general class of instabilities fed by the magnetic field as studied by Vandakurov (1972) and Tayler (1973) and several authors afterwards. We will henceforth use the term Tayler instability even if modifying conditions such as differential rotation or density stratification are present. The Tayler instability very often favours non-axisymmetric modes that become unstable first.

In the first place, the Tayler instability is interesting for radiative zones like the solar tachocline – the transition from differential rotation to uniform rotation below the convection zone – and the extensive radiative envelopes of intermediate-mass stars from spectral types B to early F. Calculations by Arlt *et al.* (2007) and Kitchatinov & Rüdiger (2008) showed an upper limit for stable toroidal magnetic fields in the tachocline of less than 1000 Gauss. This contradicts the usual picture of fields up to 10^5 G which are necessary to explain a rise of flux to low latitudes forming the sunspots seen at latitudes of below about 40°. This illustrates that kinematic dynamos on the one hand (induction equation only) and thin flux-tube simulations (one-dimensional MHD) may not capture the full physics due to their approximations.

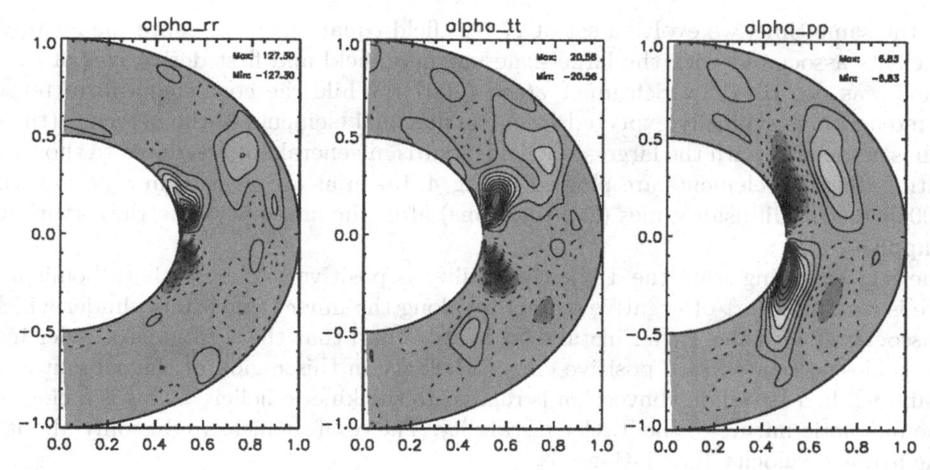

Figure 4. Diagonal elements of the α-tensor derived from the Tayler instability. Yellow areas denote positive values, blue areas represent negative values.

4. Possible dynamo action from the Tayler instability

Spruit (2002) suggested that the Tayler instability plus differential rotation may lead to dynamo action, since unstable toroidal magnetic fields will provide a source of poloidal field which in turn can be amplified by differential rotation. This idea, which is sometimes called the Tayler-Spruit dynamo, has been used by stellar evolution models. It is actually an enhanced, sustained diffusion from the continuously disrupting magnetic fields which enters these models, not the magnetic fields themselves. While simulations by Braithwaite (2006) show sustained magnetic fields when the differential rotation is maintained in the entire computational domain, other simulations, e.g. by Zahn *et al.* (2007) and Gellert *et al.* (2008), failed to obtain a dynamo. Their differential rotation was enforced only at the boundaries. Obtaining sustained dynamo action in a global, nonlinear simulation is a tricky thing because astrophysical parameters are not achievable numerically. It is thus elucidating to look for dynamo action through the presence of non-vanishing mean-field coefficients measured in nonlinear simulations.

The non-axisymmetric, unstable mode will develop other azimuthal wave numbers by nonlinear coupling and could be able to produce an average net effect of generating a large-scale axisymmetric magnetic field. If so, it may be expressed by mean-field coefficients. The following computations only employ an initial differential rotation and a poloidal magnetic field. When the generated toroidal magnetic field is supercritical, we impose a non-axisymmetric perturbation. The simulations are convectively stable and impose no rotation on the radial walls, so do not provide an energy source for sustained dynamo action. However, we can measure the mean-field coefficients arising from the Tayler instability.

We are using the MHD equations in an incompressible domain with a density constant in space and time, for the sake of simplicity. Details of the computational setup can be found in Arlt & Rüdiger (2011). The system quickly develops a toroidal magnetic field from the initial poloidal one. At the same time, Lorentz forces build up diminishing the differential rotation. The system thus evolves inevitably into a state which is Tayler unstable. When the configuration has reach the supercritical regime, we add a non-axisymmetric perturbation of azimuthal wavenumber $m = 1$ and allow the instability to grow (Fig. 3). Due to nonlinear coupling, a whole spectrum of spherical harmonics builds up including changes of the axisymmetric part of the solution.

At the same time, we evolve a set of 27 test field equations to measure the mean-field coefficients associated with the large-scale magnetic field and first derivatives of it. The method was described by Schrinner *et al.* (2007). While the coefficients form tensors, dynamo action is typically expected from the diagonal elements of the α-tensor (the one which is associated with the large-scale field), but is in general not restricted to those. The resulting diagonal elements are plotted in Fig. 4, taken as averages from a period which is 0.0005–0.0010 diffusion times (2–3 rotations) after the non-axisymmetric perturbation was applied.

The $\alpha_{\phi\phi}$ emerging from the Tayler instability is positive in the northern hemisphere. There is a counterpart of negative α roughly along the inner tangential cylinder which is not associated with the Tayler instability. We also find that the α from the Tayler instability is closely related to a positive current helicity in this region of the computational domain, while α based on convection is related to the kinetic helicity. This is a clear sign of the magnetic nature of the Tayler instability. The α-effect measured is only about 1% of the average velocity fluctuations.

5. Concluding remarks

Instabilities provide additional time-scales and magnetic-field topologies to the dynamo. Their consequences need to be incorporated in models of solar and stellar activity. Since full MHD simulations of entire stars are still not feasible, an incorporation of the characteristics of the instabilities in large-eddy simulations or mean-field models should be a suitable choice. With compressible spherical simulations coming into reach, the necessary quantities will become accessible for stellar contexts in the near future.

References

Arlt, R., Hollerbach, R., & Rüdiger, G. 2003, *Astron. Astrophys*, 401, 1087
Arlt, R., Sule, A., & Rüdiger, G. 2005, *Astron. Astrophys*, 441, 1171
Arlt, R., Sule, A., & Rüdiger, G. 2007, *Astron. Astrophys*, 461, 295
Arlt, R. & Rüdiger, G. 2011, *Mon. Not. Roy. Astron. Soc.*, 412, 107
Braithwaite, J. 2006, *Astron. Astrophys*, 449, 451
Brandenburg, A. & Subramanian, K. 2005, *Phys. Rep.*, 417, 1
Carroll, T. A., Kopf, M., Ilyin, I., & Strassmeier, K. G. 2007, *Astron. Nachr.*, 328, 1043
Cowling, T. G. 1934, *Mon. Not. Roy. Astron. Soc.*, 94, 39
Dziembowski, W. & Kosovichev, A. 1987, *Acta Astron.*, 37, 341
Elstner, D. & Korhonen, H. 2005, *Astron. Nachr.*, 326, 278
Gellert, M., Rüdiger, G., & Elstner, D. 2008, *Astron. Astrophys*, 479, L33
Kitchatinov, L. L. & Rüdiger, G. 2008, *Astron. Astrophys*, 478, 1
Kochukhov, O., Bagnulo, S., Wade, G. A. *et al.* 2004, *Astron. Astrophys*, 414,613
Kővári, Z. Bartus, J., Strassmeier, K. G. *et al.* 2007, *Astron. Astrophys*, 463, 1071
Menou, K. & Le Mer, J. 2006, *Astrophys. J.*, 650, 1208
Morin, J., Donati, J.-F., Petit, P. *et al.* 2010, *Mon. Not. Roy. Astron. Soc.*, 407, 2269
Schrinner, M., Rädler, K.-H., Schmitt, D. *et al.* 2007, *Geophys. Astrophys. Fluid Dyn.*, 101, 81
Spruit, H. 2002, *Astron. Astrophys*, 381, 923
Steenbeck, M., Krause, F., & Rädler, K.-H. 1966, *Z. Naturforsch.*, 21, 369
Tayler, R. J. 1973, *Mon. Not. Roy. Astron. Soc.*, 161, 365
Vandakurov, Yu.V. 1972, *Sov. Astron.*, 16, 265
Velikhov, E. P. 1959, *Sov. Phys. JETP*, 9, 995
Watson, M. 1981, *Geophys. Astrophys. Fluid Dyn.*, 16, 265
Yoshizawa, A. 1993, *Publ. Astron. Soc. Japan*, 45, 129
Zahn, J.-P., Brun, A. S., & Mathis, S. 2007, *Astron. Astrophys*, 474, 145

Discussion

KOSOVICHEV: Could you see the possible Tayler-Spruit dynamo from the flux-tube like magnetic field at the bottom of the convection zone?

ARLT: The example of magnetic flux rising through a convection zone, as shown in the presentation, was only a toy model demonstrating the convective processing and it was 2D, so cannot show any dynamo effect.

KOSOVICHEV: What is the magnetic field strength of the flux tube in the tachocline in your simulations of emerging magnetic flux?

ARLT: I chose the field strength such that it's balanced with the rotation simply that it stays in place, because otherwise a ring of magnetic field would just shrink. The field strength was somewhere with an Alfvén speed near the rotation speed, which is 10^5 Gauss or so.

The Physics of Sun and Star Spots
Proceedings IAU Symposium No. 273, 2010
D.P. Choudhary & K.G. Strassmeier, eds.

doi:10.1017/S1743921311015158

Solar subsurface flows of active regions: flux emergence and flare activity

Rudolf Komm, Rachel Howe, Frank Hill, and Kiran Jain

National Solar Observatory
950 N. Cherry Ave., Tucson, AZ 85719, USA
email: rkomm@nso.edu

Abstract. We study the temporal variation of subsurface flows associated with active regions within 16 Mm of the solar surface. We have analyzed the subsurface flows of nearly 1000 active and quiet regions applying ring-diagram analysis to Global Oscillation Network Group (GONG) Dopplergram data. We find that newly emerging active regions are characterized by enhanced upflows and fast zonal flows in the near-surface layers, as expected for a flux tube rising from deeper layers of the convection zone. The subsurface flows associated with strong active regions are highly twisted, as indicated by their large vorticity and helicity values. The dipolar pattern exhibited by the zonal and meridional vorticity component leads to the interpretation that these subsurface flows resemble vortex rings, when measured on the spatial scales of the standard ring-diagram analysis.

Keywords. Sun: helioseismology, Sun: magnetic fields, Sun: flares

---

1. Introduction

We study the temporal variation of subsurface flows of active regions from the solar surface to a depth of 16 Mm. We have measured the subsurface flows of nearly 1000 active and quiet regions by analyzing Global Oscillation Network Group (GONG) Dopplergram data with the ring-diagram technique. As part of an on-going study, we are focusing on (1) emerging active regions and (2) the twist of subsurface flows and its role in the flare activity of active regions. In previous studies (Komm *et al.* 2009), we have found that the vertical velocity component is a good indicator of temporal changes in magnetic flux. Here, we focus on emerging active regions and the zonal (east-west) velocity component in addition to the vertical one.

The other subject of interest here is the vorticity of subsurface flows. In previous studies (Komm *et al.* 2011; Reinard *et al.* 2010), we have shown that the flare activity of solar active regions is intrinsically linked with the vorticity of solar subsurface flows on spatial scales comparable to the size of active regions and on temporal scales from days to the lifetime of active regions. Here, we have a closer look at the vortical structure of subsurface flows of strong active regions.

2. Data and Method

We determine the horizontal components of solar subsurface flows with a ring-diagram analysis using the dense-pack technique (Haber *et al.* 2002). The full-disk Doppler images obtained with GONG are divided into 189 overlapping regions with centers spaced by 7.5° ranging from $\pm 52.5^{\circ}$ in latitude and central meridian distance (CMD). Each region is

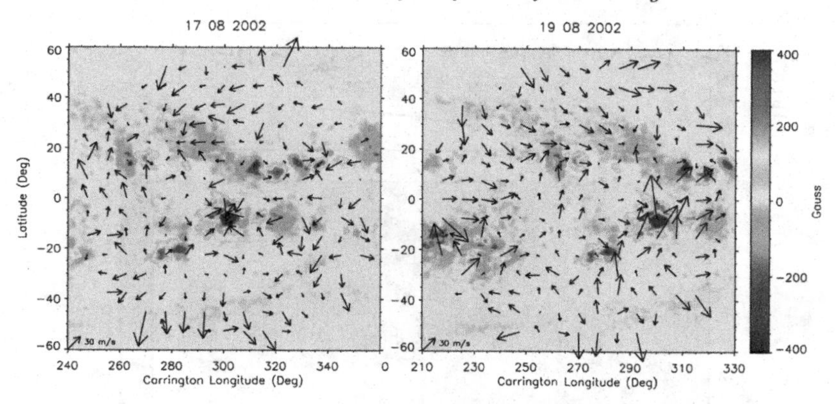

Figure 1. Daily horizontal-flow maps at a depth of 7 Mm with AR 10069 moving across the Sun. Arrows indicate flows; the background shows the magnetic activity (from a synoptic map).

apodized with a circular function reducing the effective diameter to 15° before calculating three-dimensional power spectra. The data are analyzed in "days" of 1664 minutes. Each of these regions is tracked throughout the sequence of images using the appropriate surface rate. For each dense-pack day, we derive maps of horizontal velocities at 189 locations in latitude and CMD for 16 depths from 0.6 to 16 Mm. Figure 1 shows two examples. The residual vertical velocity is estimated from the divergence of the horizontal flows assuming mass conservation (Komm 2007). To study the twist of the subsurface flows, we calculate the vorticity, defined as the curl of the velocity vector, and the kinetic helicity density, defined as the scalar product of the velocity and vorticity vector (Moffatt & Tsinober 1992).

As a measure of solar activity, we use Michelson Doppler Imager (MDI) magnetograms. We determine the change in unsigned magnetic flux during the disk passage of each active region by averaging 96-min MDI magnetograms in time over the length of a ring-diagram day and bin them to the dense-pack spatial grid. To study the emergence of active regions, we sort the data set by the average change in magnetic flux during the disk passage of each region. We then select 83 regions (10% of all regions) with the largest increase in activity starting from small activity values. For each emerging region, we determine the day of emergence as the first day with a flux increase of 10% and center each time series on this day.

3. Results

Figure 2 shows the average zonal and vertical velocity component associated with emerging active regions. For the zonal velocity component, a band of faster-than-average flow values moves from a depth of 16 Mm to the surface within two days of emergence. This agrees with the fact known from surface and helioseismic observations that active regions rotate faster than the ambient fluid. For the vertical velocity component, upflows increase with increasing flux at depths greater than 10 Mm. Between about 4 and 10 Mm, weak upflows change to downflows two days after flux emergence. This means that at the end of the time series, the average depth variation has been established with upflows in deeper layers and downflows in shallow ones. The time scale of two days coincides with the active phase of growth of newly emerging active regions.

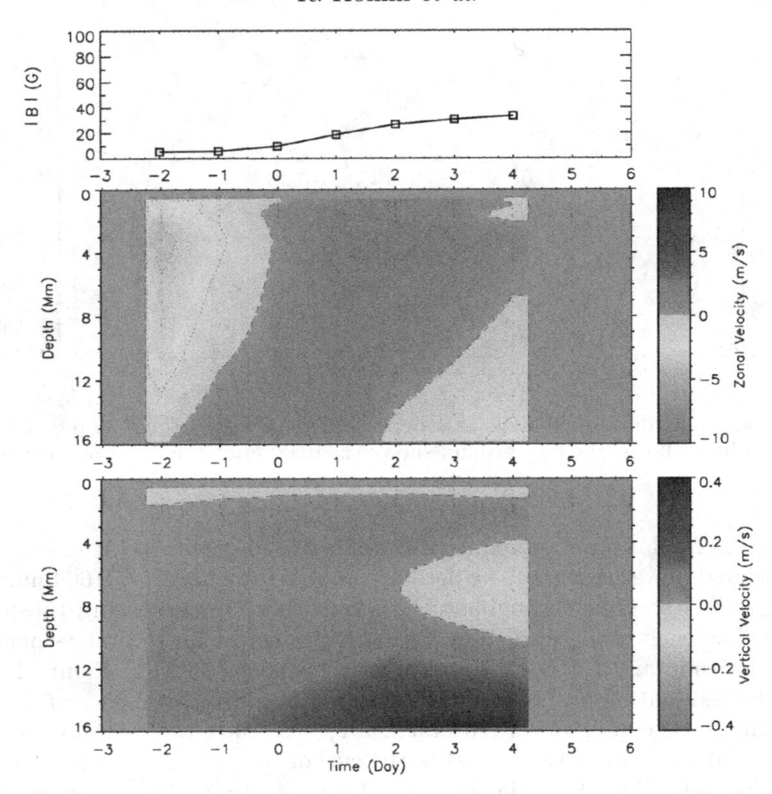

Figure 2. Emerging active regions and their subsurface flows. Top: The residual zonal velocity averaged over 83 emerging regions as a function of time and depth. Bottom: The corresponding residual vertical velocity. Positive (negative) values indicate faster (slower) zonal flows and upflows (downflows) in the vertical component. The mean value has been subtracted at every depth for the zonal flow. The line drawing indicates the unsigned magnetic flux.

Figure 3 shows, as example, the average kinetic helicity density of a strong flare-productive active region (10069). Strong active regions show a "dipolar" pattern in the kinetic helicity density that changes sign with depth. This pattern originates in the zonal and meridional vorticity component of the subsurface flows. The zonal vorticity shows this pattern in the north-south direction, while the meridional vorticity component shows a similar pattern aligned in the east-west direction, when measured on dense-pack scales. We interpret this pattern as representing two vortex rings stacked on top of each other, as shown in the schematic diagram. There might exist a third vortex ring in the shallow layers within 2 Mm of the solar surface.

4. Summary

We study subsurface flows of active regions within 16 Mm of the solar surface. We find that newly emerging active regions are characterized by enhanced upflows and fast zonal flows in the near-surface layers, as expected for a flux tube rising from deeper layers of the convection zone. After active regions are formed, downflows are established within two days of emergence in shallow layers between about 4 and 10 Mm. A detailed study will appear in the near future (Komm *et al.* 2011).

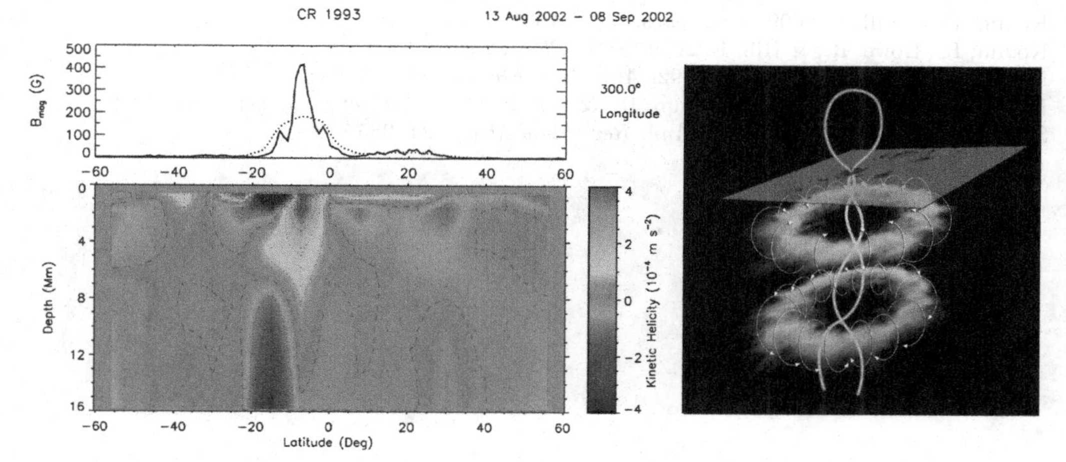

Figure 3. Left: The kinetic helicity density as a function of latitude and depth for AR 10069 (at 300° Carrington longitude). The peak of the unsigned magnetic flux in the line drawing indicates the location of the active region. Right: Schematic of the twisted subsurface flows associated with strong active regions. On dense-pack scales, these flows resemble stacked vortex rings centered on a flux tube (Courtesy: P. Marenfeld).

The subsurface flows associated with strong active regions are highly twisted, as indicated by the large vorticity and helicity values. The dipolar pattern exhibited by these quantities when measured on dense-pack scales leads to the interpretation that subsurface flows associated with strong active regions resemble vortex rings (Shariff & Leonard 1992). Previous studies have shown that flare-productive active regions have large values of magnetic activity and subsurface vorticity or helicity (Komm & Hill 2009). Subsurface vorticity and helicity thus provide information about the flare productivity of active regions. We plan to investigate the temporal variation of the vorticity pattern and its relation to the evolution of active regions in the near future.

Acknowledgements

This work utilizes data obtained by the Global Oscillation Network Group (GONG) program, managed by the National Solar Observatory, which is operated by the Association of Universities for Research in Astronomy (AURA), Inc. under a cooperative agreement with the National Science Foundation. The data were acquired by instruments operated by the Big Bear Solar Observatory, High Altitude Observatory, Learmonth Solar Observatory, Udaipur Solar Observatory, Instituto de Astrofísica de Canarias, and Cerro Tololo Interamerican Observatory. SOHO is a mission of international cooperation between ESA and NASA. This work was supported by NASA grant NNG08EI54I to the National Solar Observatory.

References

Haber D. A., Hindman, B. W., Toomre, J., Bogart, R. S., Larsen, R. M., & Hill, F. 2002, *ApJ*, 570, 885

Komm, R. 2007, *Astron. Nachr.*, 328, 269

Komm, R., Howe, R., & Hill, F. 2011, *Solar Phys.*, 268, 407

Komm, R., Ferguson, R., Hill, F., Barnes, G., & Leka K. D. 2011, *Solar Phys.*, 268, 389

Komm, R. & Hill, F. 2009, *J Geophys. Res.*, A06105
Komm, R., Howe, R., & Hill, F. 2009, *Solar Phys.*, 258, 13
Moffatt, H. K. & Tsinober, A. 1992, *Ann. Rev. Fluid Mech.*, 24, 281
Reinard, A. A., Henthorn, J., Komm, R., & Hill, F. 2010, *Astrophys. J. Lett.*, 710, L121
Shariff, K. & Leonard, A. 1992, *Ann. Rev. Fluid Mech.*, 24, 235

The Physics of Sun and Star Spots
Proceedings IAU Symposium No. 273, 2010
D.P. Choudhary & K.G. Strassmeier, eds.

doi:10.1017/S174392131101516X

Twist and writhe of δ-island active regions

M. C. López Fuentes[1], C. H. Mandrini[1] and P. Démoulin[2]

[1]Instituto de Astronomía y Física del Espacio (CONICET-UBA),
CC 67, Suc 28, 1428 Buenos Aires, Argentina
e-mail: lopezf@iafe.uba.ar

[2]Observatoire de Paris, LESIA, UMR 8109 (CNRS), F-92195, Meudon Principal Cedex, France

Abstract. We study the magnetic helicity properties of a set of peculiar active regions (ARs) including δ-islands and other high-tilt bipolar configurations. These ARs are usually identified as the most active in terms of flare and CME production. Due to their observed structure, they have been associated with the emergence of magnetic flux tubes that develop a kink instability. Our main goal is to determine the chirality of the twist and writhe components of the AR magnetic helicity in order to set constrains on the possible mechanisms producing the flux tube deformations. We determine the magnetic twist comparing observations of the AR coronal structure with force-free models of the magnetic field. We infer the flux-tube writhe from the rotation of the main magnetic bipole during the observed evolution. From the relation between the obtained twist and writhe signs we conclude that the development of the kink instability cannot be the single mechanism producing deformed flux-tubes.

Keywords. Sun: magnetic fields, sun: activity, sun: photosphere, sun: corona

1. Introduction

Since active regions (ARs) are produced by the emergence of subphotospheric toroidal magnetic flux tubes, they generally appear on the sun as bipolar configurations in the E-W direction (Hale's law, see Schrijver & Zwaan 2000). A small inclination (tilt angle) of the leading polarity towards the solar Equator (Joy's Law) is also observed due to the effect of the Coriolis force during the emergence (Fisher *et al.* 2000). There are, however, peculiar ARs with tilt angles that widely deviate from Hale's law. They are usually interpreted as the emergence of flux tubes that suffered some kind of abnormal deformation (López Fuentes *et al.* 2000). Among these ARs are the so-called δ-spots or δ-islands (an example in shown in Figure 1), which are of particular interest because they are the most active in terms of flare and CME production. Among the possible mechanisms to explain the flux tube deformations are the interaction with large scale vortices in the convection zone, the Coriolis force, and the development of a kink instability (Fan *et al.* 1999).

The study of the magnetic helicity properties of the flux tubes associated to peculiar regions can provide information about the mechanisms at the origin of the deformations. For instance, in the case of the kink instability there is an internal transfer of magnetic helicity from twist to writhe, so they must have the same sign. The writhe (W) of a magnetic flux tube is a measure of the deformation of its main axis as a whole, while the twist (T) corresponds to the winding of the magnetic field lines around the axis. Clearly, there are strong motivations to constrain the possible mechanisms at the origin of this kind of ARs. Although extensive statistical investigations have been done, they could not reach definitive conclusions (see e.g., López Fuentes *et al.* 2003, Holder *et al.* 2004, Tian *et al.* 2005). In the meantime, new mechanisms have been proposed to explain δ-spots (see e.g., Archontis & Hood 2010). We recently began to study in a "one by one" basis a small set of peculiar ARs to determine the signs of the twist and writhe helicities of their associated flux tubes (López Fuentes *et al.* 2009). Here, we extend the study to include a total of 10 ARs with these characteristics.

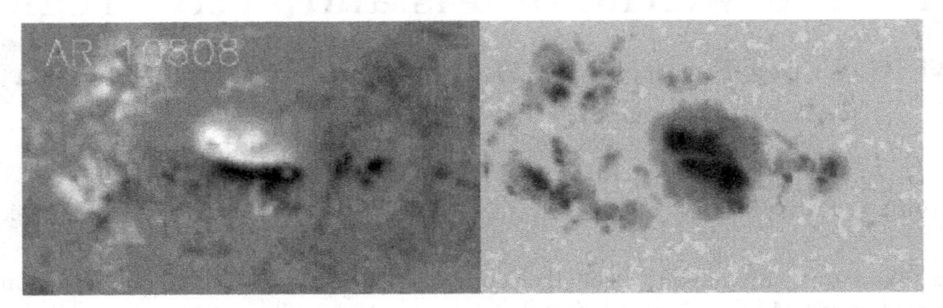

Figure 1. This is an example of δ-island AR: NOAA 10808, observed during Sept. 2005. The left panel corresponds to a SOHO/MDI magnetogram obtained in Sep. 14 2005, where the high-tilt and intense magnetic strength of the main polarities can be appreciated. The right panel illustrates how in δ-spots the umbrae corresponding to opposite polarities share the same penumbra.

2. Analysis

As we described in Section 1, we determine the signs of the twist and writhe components of the magnetic helicity for a set of peculiar ARs. In Table 1 we present a list of the studied cases, including the central meridian passage (CMP) date and the hemisphere in which the AR was observed. For the determination of the writhe we follow the evolution of the ARs main magnetic polarities using SOHO/MDI magnetograms in the way that is thoroughly described in López Fuentes & Mandrini (2008). As the deformed magnetic flux tubes emerge through the photosphere the magnetic bipoles that form the δ-spots rotate. We analyze this rotation during a total of 7 days around the CMP of the ARs. From the sense of rotation we can infer the positive or negative chirality of the flux tubes. In Figure 2 we show the rotation of the bipoles for two examples from our set. The centers of the plots correspond to the mean position of the positive polarity and the heads of the arrows indicate the relative location of the negative polarity. Initial and final dates of observation are indicated, so the sense of rotation can be appreciated and the writhe sign inferred.

For the twist sign determination we use a linear force-free field code to extrapolate SOHO/MDI magnetograms into the corona and we compare the model with coronal observations (EUV data from TRACE or soft X-ray data from Yohkoh/SXT). We use the sign of the α parameter from the force-free equation:

$$\nabla \times \mathbf{B} = \alpha \mathbf{B}, \tag{2.1}$$

as a proxy for the twist sign. Using different values of the α parameter we obtain the mean distance between model field lines and observed coronal loops. In Figure 3 we show the mean distance to the observed loops versus α for the same cases shown in Figure 2. The α that minimizes the distance to the observed loops provides the best sign of the AR twist.

For each of the studied ARs we compare the obtained signs of twist and writhe to determine if the development of a kink instability can be at the origin of the flux tube deformation (see Table 1). Cases with same sign of twist and writhe (60%) are consistent with the kink instability (Table 1, last column). Cases having different twist and writhe signs cannot be explained by that mechanism.

Table 1. Properties of the studied ARs.

AR	Date	Hemis.	Writhe	Twist	Kink Inst.
8108	1997 11 19	N	Neg	Pos	No
9026	2000 06 07	N	Neg	Neg	Yes
9165	2000 09 15	N	Neg	Neg	Yes
9415	2001 04 09	S	Pos	Pos	Yes
9632	2001 09 26	S	Neg	Pos	No
10314	2003 03 15	S	Pos	Pos	Yes
10386	2003 06 22	S	Neg	Pos	No
10484	2003 10 23	N	Pos	Neg	No
10696	2004 11 06	N	Neg	Neg	Yes
10808	2005 09 14	S	Pos	Pos	Yes

3. Conclusions

We study the magnetic helicity properties of a set of 10 δ-island or high-tilt ARs. Our main motivation is to constrain the models usually proposed to explain the deformation of emerging magnetic flux tubes. One of these mechanisms is the development of a kink instability, which implies that magnetic flux tubes must have the same sign of twist and writhe magnetic helicities. Although our set is still small for a conclusive statistical analysis, we find that only 6 of the 10 studied cases are consistent with this possibility. Therefore, other mechanisms should be considered to explain the remaining cases.

In future work we need to include more ARs in the analysis in order to obtain solid arguments for alternative mechanisms. To make a more complete inference on the magnetic structures it might be also necessary to study the photospheric evolution in more detail than just the main polarities rotation. We also plan to include vector magnetograms as an alternate source for determining the sign of the twist.

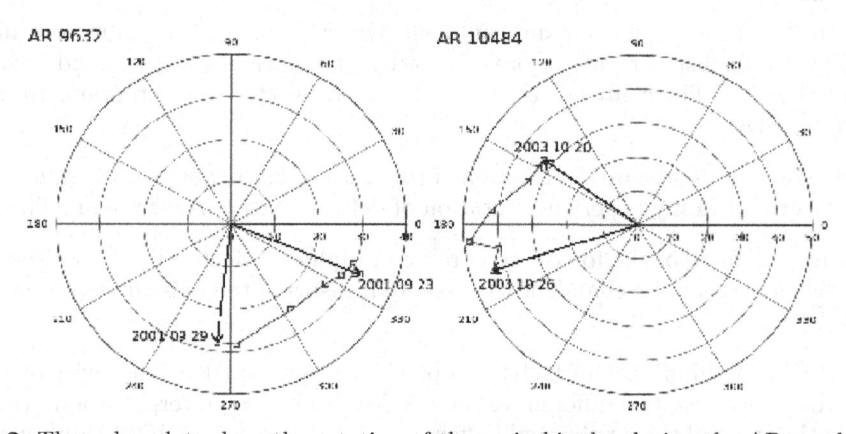

Figure 2. The polar plots show the rotation of the main bipoles during the ARs evolution for two of the studied cases. NOAA 9632 (left panel), observed in Sep. 2001, and NOAA 10484, observed during Oct. 2003. The centers of the plots correspond to the mean position of the positive polarities while the head of the arrows indicate the mean relative position of the negative polarities. From the observed senses of rotation it can be inferred that AR 9632 has negative writhe and AR 10484 has positive writhe.

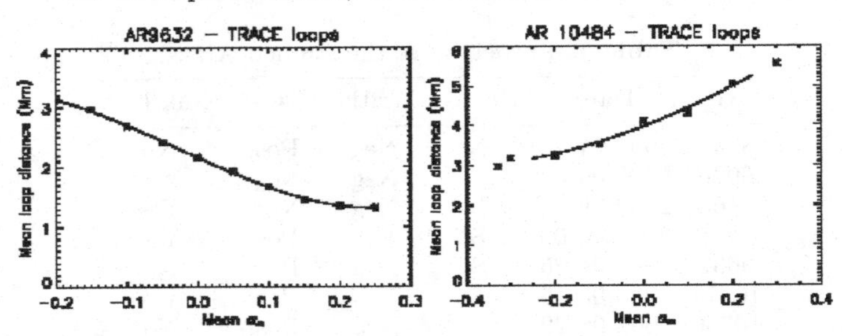

Figure 3. Mean distance from model magnetic field lines to observed coronal loops versus the force-free parameter α. The α value that minimizes the mean distance provides the sign of the AR twist.

References

Archontis, V. & Hood, A. W. 2010, *Astron. Astrophys*, 514, A56

Fan, Y., Zweibel, E. G., Linton, M. G., & Fisher, G. H. 1999, *Astrophys. J.*, 521, 460

Fisher, G. H., Fan, Y., Longcope, D. W., Linton, M. G., & Pevtsov, A. A. 2000, *Solar Phys.*, 192, 119

Holder, Z. A., Canfield, R. C., McMullen, R. A., Nandy, D., Howard, R. F., & Pevtsov, A. A. 2004, *Astrophys. J.*, 611, 1149

López Fuentes, M. C., Demoulin, P., Mandrini, C. H., & van Driel-Gesztelyi, L. 2000, *Astrophys. J.*, 544, 540

López Fuentes, M. C., Démoulin, P., Mandrini, C. H., Pevtsov, A. A., & van Driel-Gesztelyi, L. 2003, *Astron. Astrophys*, 397, 305

López Fuentes, M. C. & Mandrini, C. H. 2008, Boletín de la Asociación Argentina de Astronomía La Plata Argentina, 51, 31

López Fuentes, M. C., Mandrini, C. H., & Démoulin, P. 2009, Proceedings IAU Symposium No. 264, Solar and Stellar variability - Impact on Earth and Planets, A. G. Kosovichev Editor, 102

Schrijver, C. J. & Zwaan, C. 2000, Solar and stellar magnetic activity / Carolus J. Schrijver, Cornelius Zwaan. New York : Cambridge University Press, 2000. (Cambridge astrophysics series ; 34)

Tian, L., Alexander, D., Liu, Y., & Yang, J. 2005, *Solar Phys.*, 229, 63

Discussion

KOSOVICHEV: I have a question and comment. Question, is it seems from your model it follows that delta sunspot rotates more commonly than normal sunspots, and I wondered if this was the case. The comment is now that we have vector magnetograms from every sunspot from SDO.

FUENTES: We didn't compare the rotation of the studied delta-spots with regular sunspots. I can only say that in my experience rotation of delta-spot is fairly common. Thank you.

KLIMCHUK: So Marcelo, the force-free parameter alpha is affected by the writhe as well as the twist. So can you be confident that you're measuring the twist if you're measuring the alpha?

FUENTES: Well, I think that in fact the alpha is more affected by the twist of the flux tubes because these are two different spacial scales. This is my interpretation. You could be right in that part of the writhe could affect the coronal structure; but I think that when you consider the whole structure of the flux tube (the long term rotation of the magnetic polarities), that is the writhe. I think that the observed coronal structure inherits the sign of the twist.

The Physics of Sun and Star Spots
Proceedings IAU Symposium No. 273, 2010
D.P. Choudhary & K.G. Strassmeier, eds.

doi:10.1017/S1743921311015171

Magnetic field evolution of active regions and sunspots in connection with chromospheric and coronal activities

Toshifumi Shimizu[1]

[1]Institute of Space and Astronautical Science, Japan Aerospace Exploration Agency, 3-1-1 Yoshinodai, Chuo, Sagamihara, Kanagawa 252-5210, Japan
email: shimizu@solar.isas.jaxa.jp

Abstract. Ca II H imaging observations by the Hinode Solar Optical Telescope (SOT) have revealed that the chromosphere is extremely dynamic and that ejections and jets are well observed in moat region around sunspots. X-ray and EUV observations show frequent occurrence of microflaring activities around sunspots; small emerging flux or moving magnetic features approaching opposite pre-existing magnetic flux can be identified on the footpoints for half of microflares studied, while no encounters of opposite polarities are observed at footpoints for the others even with SOT high spatial magnetorams (Kano *et al.* 2010). Another observations tell the involvement of twisted magnetic fields in the microflares accompanied by no polarity encounters at the footpoints. Some type of sunspot light bridges shows recurrent occurrence of chromospheric ejections, and photospheric vector magnetic field data suggests that twsited magnetic flux tubes lying along light bridge play vital roles in producing such ejections (Shimizu *et al.* 2009). This presentation reviewed observational findings from these studies. We will need to understand the 3D configuration of magnetic fields for better understanding of activity triggers in the solar atmosphere.

Keywords. Sun: activity, sun: chromosphere, sun: corona, sun: evolution, sun: magnetic fields, sun: photosphere, sunspots

1. Introduction

Soft X-ray and EUV images of the Sun show that intense X-ray and EUV emissions are mostly concentrated in active regions. The active region corona consists of a lot of loop-like structures, which trace magnetic field bundles filled with hot plasma in the corona. Bright loops are mainly rooted around penumbra of sunspots and nearby areas. It is dark in X-rays above umbra of sunspots, although the umbra is filled with strong magnetic flux.

The time series of soft X-ray and EUV images reveal that weak transient brightenings of small coronal loops, i.e., microflares, are frequently produced in the active region corona. The released energy of each brightening is in order of 10^{24} to 10^{28} ergs, which is about 6 orders of magnitude smaller than that of solar flares (Shimizu 1995). The number of flares has a power-law distribution as a function of energy, which continues to much small energy range, at least down to 10^{24} erg (Aschwanden *et al.* 2000).

The *Hinode* Solar Optical telescope (SOT) has provided new views in the chromosphere, which is the interface layer between the photosphere and corona. The chromosphere observed with Ca II H line is much more dynamic than we have observed before. Plasma ejections and heating events are always observed around sunspots. This view can be seen in the Ca II H movies which capture sunspots and nearby areas located near the solar limb.

One of outstanding questions is what is origins of activities in the corona and chromosphere. The magnetic energy stored in the atmosphere is used for causing activities. Magnetic reconnection is believed as the converter from magnetic energy to kinetic and thermal energy, but details of magnetic reconnection is still poorly understood. We do not exactly understand magnetic configuration observed around the activities, which is essential to understand what triggers magnetic reconnection. How such magnetic configuration can be created in the solar atmosphere? Key information for the understanding is observations of magnetic fields well combined with observations of activities.

2. Magnetic Field Evolution in Active Regions

Magnetic field evolution observed in active region NOAA 11039 was discussed in this section. This active region was well observed with SOT. Ca II H images and Na D longitudinal magnetograms were continuously acquired every 3 minutes in a long period from 29 December 2009 until 2 January 2010, with very limited number of short interruptions during the period. It is almost the first time for SOT to continuously monitor an active region for such a long period and this unique observation was realized during the year-end and new-year holiday period. The field of view covered almost the entire area of the active region. The movie observation successfully captured time evolution of several sunspots, including the birth of sunspots, the evolution of well-developed sunspots, and the decay of sunspots.

Fig. 1 is the zoom up of the area where the successive emergences of magnetic flux form a pair of sunspots. In between the newly forming positive-polarity sunspot and

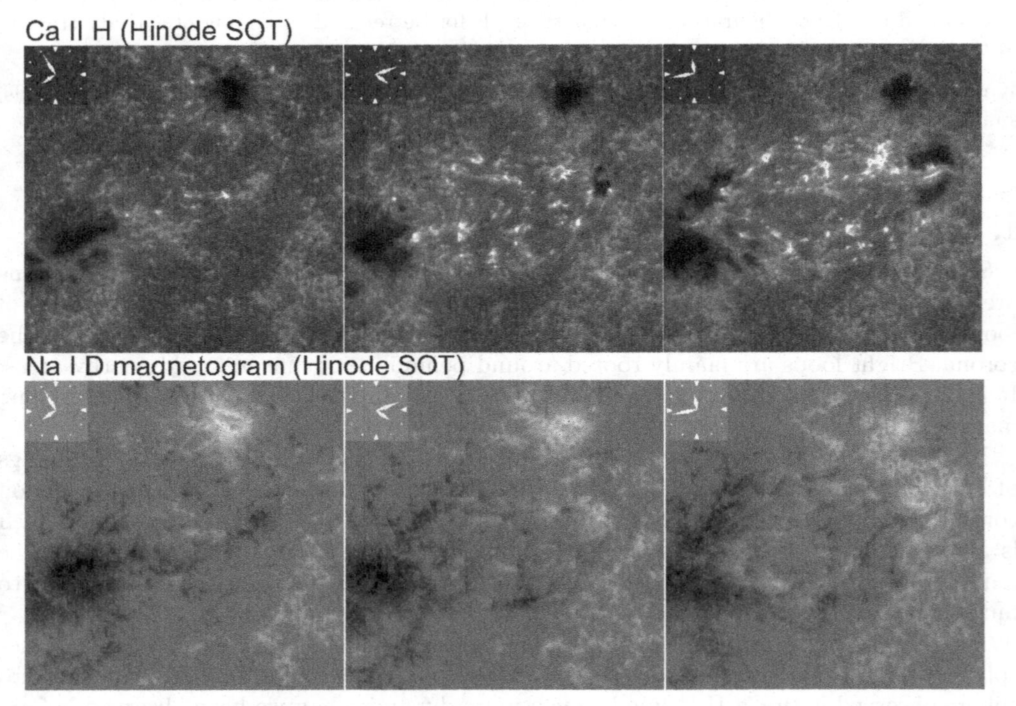

Figure 1. An emerging flux region appeared in NOAA 11039. Three snapshots were picked up from the continuous measurement of Ca II H (upper figures) and Na I D magnetogram (lower figures). Times are, from the left, 7:54, 16:12, and 23:44 UT on 30 December 2009. The field of view is 55.8 arcsec.

the following negative-polarity sunspot, a lot of small magnetic bipoles are emerged. Cancellation of opposite polarities may be observed when one polarity flux meets the opposite polarity flux of another emerging dipole. Transient brightenings (bright, white-colored patches in the figure) are observed in Ca II H above the area where the opposite polarities are in contact. They mean transient heating of chromospheric gas. Also, in many cases, microflaring activities of coronal loops may be observed in the corona. Bright transient brightenings in Ca II H mostly corresponds to the footprints of microflaring coronal loops. Merging of same polarities are also observed, which finally results in the development of sunspots at the both ends of the emerging flux region.

Fig. 2 shows the time evolution of a well-developed sunspot. Well-known, moving magnetic features, MMFs, are observed around the sunspot. They are small flux patches moving out from the outer edge of the penumbra in the radial direction. Some type of MMFs appears to be responsible for removing the magnetic flux from sunspots. The zone where MMFs are observed is called "moat." In the moat region, small flux emergences may be found, which is the major source for causing microflares around sunspots, as discussed in the next section. Cancelation of opposite polarities may be observed at the end of the moat.

Fig. 3 shows the disappearance of a sunspot. It is well observed that the concentrated sunspot flux is gradually fragmented. A sunspot light bridge may be observed when a sunspot is divided into two large segments. Ca II H movie shows less number of transient brightenings.

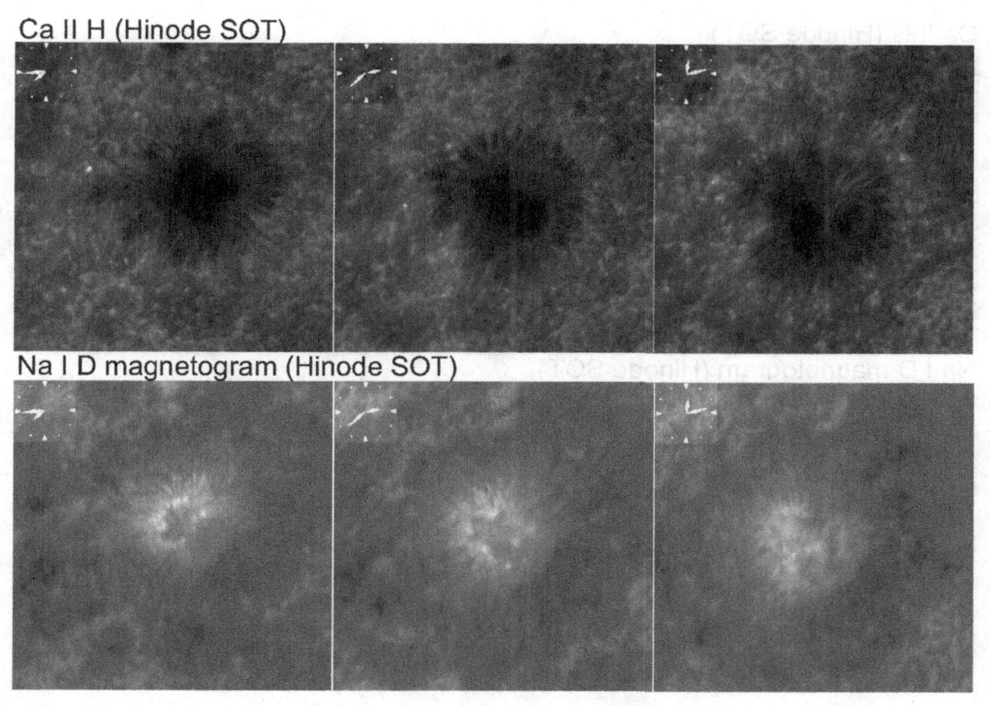

Figure 2. Evolution of a large sunspot developed in NOAA 11039. Three snapshots were picked up from the continuous measurement of Ca II H (upper figures) and Na I D magnetogram (lower figures). Times are, from the left, 7:44 and 14:38 UT on 1 January and 0:12 on 2 January 2010. The field of view is 55.8 arcsec.

3. Magnetic origins of microflares

Photospheric magnetic activities triggering X-ray microflares were studied by Kano *et al.* (2010). They identified 55 microflares around a well-developed sunspot surrounded by a moat with high-cadence X-ray images from the *Hinode* X-ray Telescope, and searched for their photospheric counterparts in high spatial resolution line-of-sight magnetograms taken by the SOT. They found opposite magnetic polarities encountering each other around the footpoints of 28 microflares (half of the samples), while such encounters could not be found around the footpoints in the rest of microlfares, even though we examined high quality and high resolution SOT magnetograms.

It is also identified that emerging magnetic fluxes in the moat were the dominant origin for causing the encounters of opposite polarities (21 of 28 microflares). Note that this result supports results from La Palma - Yohkoh coordinated observations (Shimizu *et al.* 2002). Additionally, uni-polar moving magnetic features with the polarity same as the sunspot definitely caused the encounters of opposite polarities for 5 cases. For these two types of microflares, the observational results lead to two magnetic configurations including magnetic reconnection, which can explain naturally most of observational signatures.

When a small-scale flux is emerged from below the photosphere, we may observe X-ray and Ca II H brightenings. X-ray brightening as well as Ca II H brightening are observed almost above the contact where a polarity patch of the emergence is approaching the pre-existing magnetic flux with the opposite polarity. At that time, we may observe that

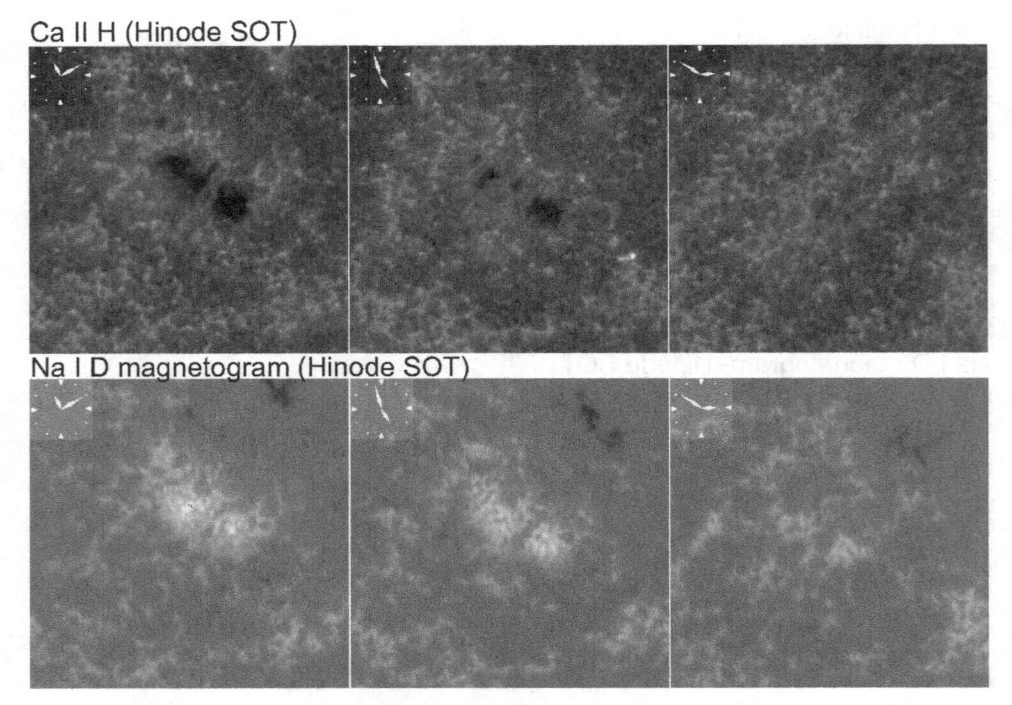

Figure 3. Disappearance of a sunspot observed in NOAA 11039. Three snapshots were picked up from the continuous measurement of Ca II H (upper figures) and Na I D magnetogram (lower figures). Times are, from the left, 23:09 on 29 December, 4:58 and 14:51 UT on 30 December 2009. The field of view is 55.8 arcsec.

the magnetic flux is decreased after the occurrence of the microflare. The observations suggest that the emergence of magnetic flux makes a field discontinuity (current sheet) with the pre-existing field, causing magnetic reconnection in the current sheet as a rapid converter of magnetic energy to thermal and kinetic energies. Submergence of fields as a result of reconnection is observed as the decrease of magnetic flux.

When a small magnetic flux patch is detached from the penumbral outer edge of sunspots and moves outward as a uni-polar moving magnetic feature, it may make an approach to the opposite polarity patches which are sometimes seen in some portions of the moat outer edge. A faint loop-like microflare may be observed with such approach, and one end of the brightening loop is located at the encounter of the polarities. With occurring a microflare, we may observe that the magnetic flux is reduced with time. The observations suggest that the approaching flux makes a field discontinuity with the opposite flux, which results in magnetic reconnection in the discontinuity, although the released energy may be small.

4. Twisted magnetic fields for creating activities

About half of microflares did not show any encounters of opposite polarities, even when high spatial resolution magnetograms were examined. According to a preliminary study of such microflares observed on 1 May 2007, the direction of magnetic field is same in multiple brightening loops. Note that the microflares discussed here are fairly large and we can easily identify more than two loop structures in the brightening. With the spectro-polarimeter (SP) data from the SOT, electric current density at the footpoints of brightening loops can be derived from the transverse field information. The electric current density map shows that the electric current is enhanced around the footpoints of brightening loops. Since one end of the brightening loops is located at the sunspot umbra, the current enhancement can be confirmed with high confidence. Assuming that the electric current is aligned along the vertical oriented umbral field, we interpret that magnetic field lines of the brightening loops are twisted. Field discontinuities may be easily formed in the corona, when the field lines are twisted at the base of loops. This suggests energy release among twisted magnetic fields in the corona.

Existence of twisted magnetic fields is one of important factors for causing various kinds of activities in the upper atmosphere. Shimizu *et al.* (2009) presented one of examples in which twisted fields play key roles in producing remarkable activities in the chromosphere. Long-lasting recurrent ejections of the chromospheric materials were found along a light bridge formed in the sunspot umbra. Ejections were recurrently and intermittently observed for almost 2 days. High-speed upward gas flow was observed in Ca II H images. At the same time, a downward flow signal was observed in the Stokes data measured in the photospheric level, meaning that a bi-directional flow is formed at the low chromosphere or photosphere. One remarkable feature revealed from vector magnetic field data is that the electric current density is significantly enhanced along the light bridge.

From the observations, we interpreted that twisted magnetic flux loop is trapped along the light bridge below the cusp-like magnetic field, as illustrated in Shimizu *et al.* (2009). Patchy and intermittent reconnections may be exited between the poloidal component of helical twisted loop and vertically oriented umbral fields. However, details of mechanism for producing reconnections intermittently for a long time are unknown and it is important to understand the 3D magnetic field configuration for further understanding.

5. Summary

Most of intense dynamical activities take place in corona and chromosphere above active regions, especially around sunspots. The presentation briefly discussed magnetic origins at the solar surface responsible for causing microflares in corona and transient plasma ejections in the chromosphere. Opposite magnetic polarities encountering each other, created by flux emergence and flux movement, are responsible for many microflares, while about half of microflares do not show such encounters near the microflares. For such cases, twisted magnetic fields, which can be observationally inferred from enhanced electric currents, may be formed in the corona. We need 3D configuration of magnetic fields for the better understanding. Twisted magnetic fields are also key for chromospheric activities, as demonstrated by long-lasting plasma ejections along light bridge.

Acknowledgements

Hinode is a Japanese mission developed by ISAS/JAXA with NAOJ as a domestic partner, and also with NASA (US) and STFC (UK) as international partners. It is operated by these agencies in corporation with ESA and NSC (Norway). The author would like to express his thanks to the SOC to invite me to present this paper in this symposium.

References

Aschwanden, M. J. *et al.* 2000, *Astrophys. J.*, 535, 1047
Kano, R. & Shimizu, T., Tarbell. T. D. 2010, *Astrophys. J.*, 720, 1136
Shimizu, T. 1995, *Pub. Astron. Soc. Jap.*, 47, 251
Shimizu, T., Shine, R. A., Title, A. M., Tarbell, T. D., & Frank, Z. 2002, *Astrophys. J.*, 574, 1074
Shimizu, T. *et al.* 2009, *ApJ*, 696, L66

Discussion

KOSOVICHEV: I'd like to comment that in the past this model is the connection of emerging magnetic flux developed by Heyvaerts and Priest. I also have a question. You had this event – the disappearance of a sunspot after a flare. How many events have you observed with Hinode where a sunspot disappears after a flare?

SHIMITZU: You mean the sunspot disappear after flare?

KOSOVICHEV: Yeah.

SHIMITZU: Actually no. I examined small micro-flares. Its energy is smaller than 10^{27} ergs. For such cases we see small magnetic flux as the photospheric counterpart, and such flux is not recognized as a sunspot. OKay? Large flares may be related to large sunspots. I'm not sure whether such complete disappearance of the sunspot can be caused by the occurrence of large flares.

PRIEST: It is very interesting that half of your micro flares are due to this process of either flux cancelation or flux emergence that Karen Harvey and Sarah Martin studied so well many years ago; but the other half, it would be great to study those in more detail because it looks like maybe there is some kind of reconnection process in the corona which is occurring spontaneously rather than being driven like the first class. So it would be very interesting to try to understand what kind of instabilities. There are many possibilities theoretically, but it would be interesting to try and study a few of those cases in very great detail.

SHIMITZU: Yes.

The Physics of Sun and Star Spots
Proceedings IAU Symposium No. 273, 2010
D.P. Choudhary & K.G. Strassmeier, eds.

doi:10.1017/S1743921311015183

Solar activity due to magnetic complexity of active regions

Brigitte Schmieder[1], Cristina Mandrini[2], Ramesh Chandra[1], Pascal Démoulin[1], Tibor Török[1], Etienne Pariat[1] and Wahab Uddin[3]

[1]Observatoire de Paris, LESIA, Meudon, 92195, France
email: brigitte.schmieder

[2]IAFE, CONICET-UBA (FCEN), Buenos Aires, Argentina
email: mandrini@iafe.uba.ar

[3]ARIES, Nainital, India

Abstract. Active regions (ARs), involved in the Halloween events during October- November 2003, were the source of unusual activity during the following solar rotation. The flares on 18-20 November 2003 that occur in the AR NOAA10501 were accompanied by coronal mass ejections associated to some particularly geoeffective magnetic clouds.

Our analysis of the magnetic flux and helicity injection revealed that a new emerging bipole and consequent shearing motions continuously energized the region during its disk passage. The stored energy was eventually released through the interaction of the various systems of magnetic loops by several magnetic reconnection events. Active events on November 18 (filament eruptions and CMEs) were originated by shearing motions along a section of the filament channel that injected magnetic helicity with sign opposite to that of the AR. Two homologous flares, that occurred on November 20, were apparently triggered by different mechanisms as inferred from the flare ribbons evolution (filament eruption and CMEs). We studied in detail the behaviour of two North-South oriented filaments on November 20 2003. They merged and split following a process suggestive of 'sling-shot' reconnection between two coronal flux ropes. We successfully tested this scenario in a 3D MHD simulation that is presented in this paper.

Keywords. Active region, magnetic helicity, filament reconnection

---

1. Introduction

It is now widely accepted that three phenomena, flare, CME, and filament eruption, are different observational manifestations of a sudden and violent disruption of the coronal magnetic field, often simply referred to as "solar eruption" (see e.g. Forbes T. G. 2000). To forecast solar activity we need to understand which are the pre-events and the mechanism that trigger and drive eruptions. Accordingly, a large variety of theoretical models for the initiation of eruptions has been proposed in the last decades: tether-cutting model, magnetic break-out model or loss of equilibrium. Multi-wavelength observations are needed to understand the role of flux emergence and shearing motions at the origin of solar eruptions. The large set of data that exist for the events on November 18 and 20, 2003 allow us to do a very detailed analysis.

2. Observations and Results

Active region NOAA 10501 was one of the most complex and eruptive regions during the decay phase of solar cycle 23. This AR is the successor of the also very flare productive active region NOAA 10484, as it was named in the previous solar rotation. AR 10501 produced 12 M flares, some of them accompanied by CMEs, from 18 to 20 November,

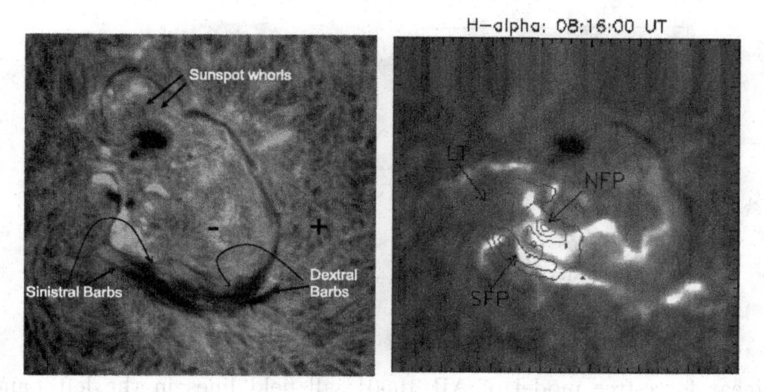

Figure 1. Chromospheric image of AR 10501 on November 18 with arrows pointing to several proxies for the helicity sign (left panel). Two-ribbon flare at 08:16 UT and contours representing RHESSI **(25-50 keV)** emission (right panel).

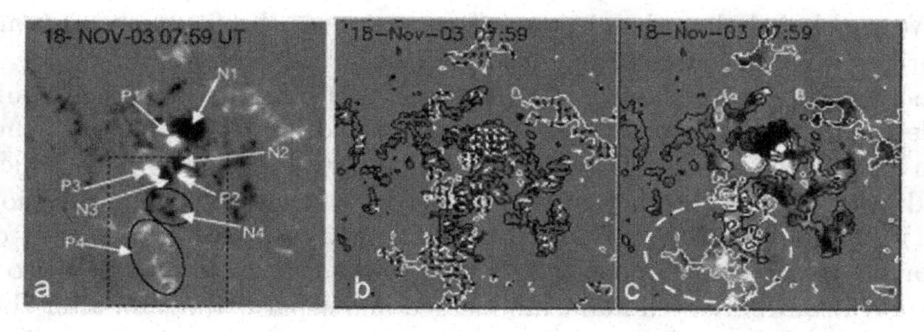

Figure 2. MDI magnetogram of AR 10501 illustrating the different bipoles present on November 18 (a). The same magnetogram with arrows representing the flux transport velocity field (b). Local helicity flux map (c) (white/black indicates positive/negative helicity injection). The oval surrounds a region of positive injection. A portion of a filament lying along this inversion line was ejected; this was probably the source of the magnetic cloud found later in the interplanetary medium.

2003. In particular, the early events on 18 November, 2003, have been extensively studied (Gopalswamy *et al.* 2005; Yurchyshyn, *et al.* 2005; Möstl *et al.* 2008; Srivastava *et al.* 2009; Chandra *et al.* 2010a). The peculiarity of this AR on that day was that it presented regions with different magnetic helicity signs, as was discussed by Chandra *et al.* (2010a).

Chandra *et al.* (2010a) discussed the multi-wavelength evolution of three solar flares (C3.8, M3.2, M3.9) on 18 November 2003 and concluded that the flares were associated with a filament eruption in the southwest direction which led to a fast CME (Fig 1). The CME was the source of a very geoeffective magnetic cloud (MC) with positive magnetic helicity on 20 November, 2003. The sign of the magnetic helicity of the active region is opposite to the MC sign.

Several bipoles were seen to emerge as the AR evolved (see Fig 2). Some of these bipoles showed strong rotational motions that injected helicity in the AR (i.e.P3-N3). We speculate that these motions were at the origin of the three flares (one of them shown in Fig 1) observed on November 18 2003. Analyzing the magnetic helicity injection with the method discussed in Pariat *et al.* (2006), we found that, though the injection had mainly a negative sign in most of the AR, there was a region of positive injection at the location of N4-P4. The topology of this complex active region confirms the existence of several sub-systems of different magnetic helicity sign, one of them is connecting N4-P4

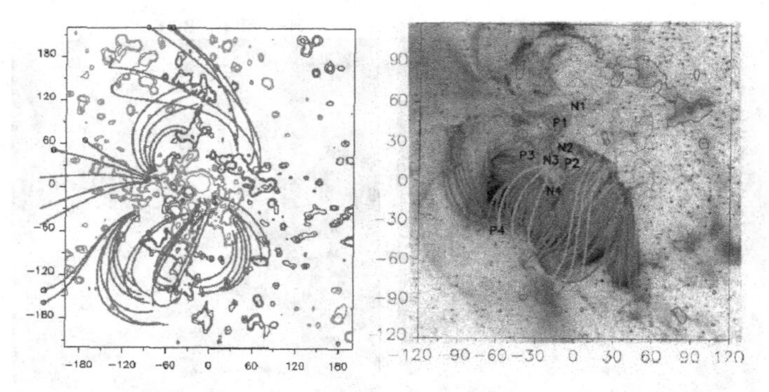

Figure 3. Linear force-free model of AR 10501, all field lines in the left panel have been computed using a negative value for alpha (except those leaving the figure towards the East for which alpha is zero). Field lines in the right panel have been computed using a positive (red and pink) or zero (light green) value for alpha.

(Fig 3). We concluded that the ejection of a portion of the filament lying in between these polarities was the source of the positive helicity MC observed *in-situ.*

Flux emergence, accompanied by the rotational motion of its polarities, could be also responsible for two homologous flares observed on November 20 2003 (Chandra *et al.* 2010b). In the first flare the central inner ribbons were observed first, which can be interpreted in terms of the tether-cutting model for the accompanying eruption (Moore *et al.* 2001). In the second flare the external ribbons appeared first, which is consistent with the break-out model (Antiochos S. K. 1998). Nearby the flare region two adjacent, elongated north-south oriented filaments merged and separated again; after this process two filaments with different footpoint connections were observed (Figure 4). This process was most probably forced mainly by the rotational motion of one polarity about the other in an emerging bipole at the center of the AR. This evolution was likely to decrease the remnant magnetic field between the two filaments, leading to their approach and interaction. The newly formed southern filament erupted later on, triggering a CME which was the source of an ICME on November 22, 2003.

3. Interaction of filaments

Using a 3D zero β magnetohydrodynamic (MHD) simulation, Török *et al.* (2010) showed that the interaction of the two filaments can be explained by "slingshot"

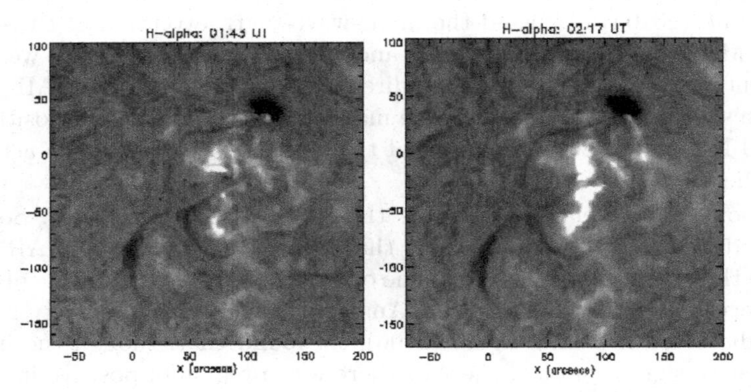

Figure 4. Observation of the interaction of filaments on November 20, 2003 and the inner ribbons of the first flare.

reconnection between two magnetic flux ropes (Linton *et al.* 2001). The simulation uses the coronal flux rope model by Titov & Démoulin (1999) for each filament. The magnetic field between the two flux ropes is first weakened by a diverging flow that moves most of the magnetic flux out of the region between the ropes (Figure 5, top). Then the flux ropes are driven towards each other by converging flows. As they meet, they reconnect and form two new flux ropes with exchanged footpoint connections (Figure 5, bottom). The magnetic dips of the flux ropes (where filament material is assumed to be located) follow this dynamic pattern, in very good agreement with the observed filaments.

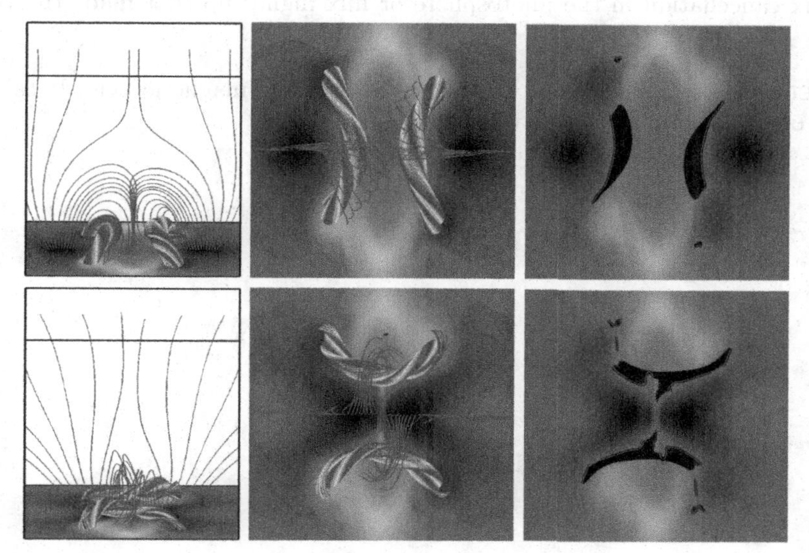

Figure 5. Numerical simulation of reconnection between two magnetic flux ropes by Török *et al.* (2010)

4. Conclusion

The roles of flux emergence and shearing motions are clearly put into evidence in this very complex AR. The stored energy was eventually released through the interaction of the various systems of magnetic loops by several magnetic reconnection events, leading to flares, filament eruption and CMEs.

References

Antiochos S. K., 1998, *Astrophys. J. Lett.*, 502, 181

Chandra R., Pariat E., Schmieder B., Mandrini C. H., & Uddin W. 2010a, *Solar Phys.*, 261, 127

Chandra R., Schmieder B., Mandrini C. H., Démoulin P., Pariat E., Török, T., & Uddin W. 2010b, *Solar Phys.* (Submitted)

Forbes T. G., 2000, *JGR*, 105, 23153

Gopalswamy N., Yashiro S., Michalek G., Xie H. Lepping H., & Howard R. A. 2005, *GRL*, 32, 12

Linton M. G., Dahlburg R. B., & Antiochos S. K. 2001, *Astrophys. J.*, 553, 905

Moore R. L., Sterling A. C., Hudson H. S., & Lemen J. R. 2001, *Astrophys. J.*, 552, 833

Möstl C., Miklenic C., Farrugia C. J., Temmer M., Veronig A. M., Galvin A., Vrsnak B., & Biernat H. K. 2008, *Ann. Geophys*, 26, 3139

Pariat E., Nindos A., Démoulin P., & Berger M. A. 2006, *Astron. Astrophys*, 452, 623

Srivastava N., Mathew S. K., Louis R. E., & Wiegelmann T. 2009, *JGR*, 114, 3107
Titov V. S. & Démoulin P. 1999, *Astron. Astrophys*, 351, 707
Török T., Chandra R., Démoulin P., Schmieder B., Aulanier G., Pariat E., & Mandrini C. H. 2010, *Astrophys. J.* (Submitted)
Yurchyshyn V., Hu Q., & Abramenko V. 2005, *Space Weather*, 3, 8

Discussion

GEORGOULIS: What was the instability mechanism that made the two flux ropes merge? Was it flux cancellation in the photosphere or flux higher up that made the two ropes merge?

SCHMIEDER: It's second solution, I think, it's decreasing magnetic field. It can be submergence or dispassion, or that is what we think.

The Physics of Sun and Star Spots
Proceedings IAU Symposium No. 273, 2010
D.P. Choudhary & K.G. Strassmeier, eds.

doi:10.1017/S1743921311015195

Nature of the unusually long solar cycles

Nadezhda Zolotova[1] **and Dmitri Ponyavin**[1]

[1]Institute of Physics, St. Petersburg State University, 198504 St. Petersburg, Russia
email: ned@geo.phys.spbu.ru

Abstract. The evolution of prolonged and unusually long solar cycles such as 23rd, 20th, and especially 4th is considered. Why the length of the 4th cycle was exceptionally large or really composed of two short cycles? Are the prolonged solar minima can be considered as precursor of low activity of the next cycles? Resolving these puzzles seems to be very important for dynamo theories trying to explain the solar long-term variations. We propose a possible model of the butterfly diagram during unusually long and prolonged cycles, based on (i) the Gnevyshev idea of sunspot distribution over the latitudes, and (ii) the phase differences of the northern and southern hemispheric activities.

Keywords. Sun: activity, sun: magnetic fields, sun: sunspots

1. Introduction

Analysing the latitudinal distribution of corona intensity, the appearance of prominences, and sunspots along the cycle 19, Gnevyshev found the existence of two maxima, spaced in time by 2–3 years (Gnevyshev 1963). He wrote: "The first maximum is characterized by an increase in the intensity at all latitudes between the equator and the poles with maximum at the latitude of 25°. During the second maximum there was an increase in the intensity only in the equatorial regions, with a maximum at latitudes of 10–15°". Late on, Gnevyshev shown that sunspots, faculae, filaments, prominences, and etc. are only the manifestations of a common process, which evolves as global pulses of activity (see Antalovà & Gnevyshev 1983). The times at which the pulses appear in both hemispheres do not coincide. Finally, the butterfly diagram is a superposition of two or more pulses, which peak at different times at different latitudes (Gnevyshev 1977). The activity distributes round centers of pulses according to the Gauss law (Gnevyshev 1977). Pulses are seen in the photosphere, chromosphere, and corona (Gnevyshev 1967, 1977). But for sunspots the double-peak structure of the cycle is less apparent, because the sunspots emerge only in the royal zones, not over the whole solar surface (Gnevyshev & Ohl 1966). He concluded, that there is not a gradual progression of a single maximum of activity from the high latitudes to equator, as it should be according to the Spörer's law (Gnevyshev 1966, 1977). But the Maunder butterfly diagram is the result of overlapping of two or more pulses. This intersection of the decreasing and increasing branches of pulses gives the artificial illusion of gradual shifting of activity equatorwards.

2. Gnevyshev scenario of solar cycle

In this section we visualize the Gnevyshev idea about solar cycle composed from pulses of strongly different features. At first (Fig. 1a), we reproduce the two-pulses scenario as a mixture of bivariate Gaussian distributions, separately in hemispheres. Parameters for this two-component model (listed in Table 1) are a location on coordinate grid, covariance (fits a shape and slope of distribution), and amount of random points for distribution.

Table 1. Parameters for the two-component model for the each hemisphere.

	Northern hemisphere		Southern hemisphere	
Components No.	1	2	1	2
Latitude	25° [1], time of maximum — 3 years after the cycle onset.	10°, time of maximum — 6 years after the cycle onset.	-25°, time of maximum — 4 years after the cycle onset.	-10°, time of maximum — 7 years after the cycle onset.
Covariance[2]	1/50	2.5/22	1.5/50	1.5/22
Amount of points[3]	3000	3000	2000	3500

Notes:
[1] The center of distribution is located at latitude 25°. The maximum is reached in 3 years after the cycle onset.
[2] Covariance is chosen as a scales ratio between x-y-axis. The slope was not used.
[3] Note, the total nominal number of sunspots, calculated for the northern hemisphere of the cycle 23, is about 10170 counts, for the southern hemisphere — about 11450 counts.

Parameters were specified in such a way to reproduce symmetrical ascending and descending branches of a cycle, and the Gnevyshev gap (Fig. 1b). We defined the constant phase lag (equal to year) between the north and south, and different magnitude of pulses (amount of points). Figure 1b shows only temporal representation. It is analog of sunspot area (or sunspot number) time-series. Blue line corresponds the northern hemisphere, red — the southern one, and black — their sum (total). The length of this simulated double-peak cycle is about 10–11 years (Fig. 1b).

Specifying the distributions parameters and mixture proportions, the model can be modified. Levels of the hemispheric phase and amplitude asymmetries, time interval

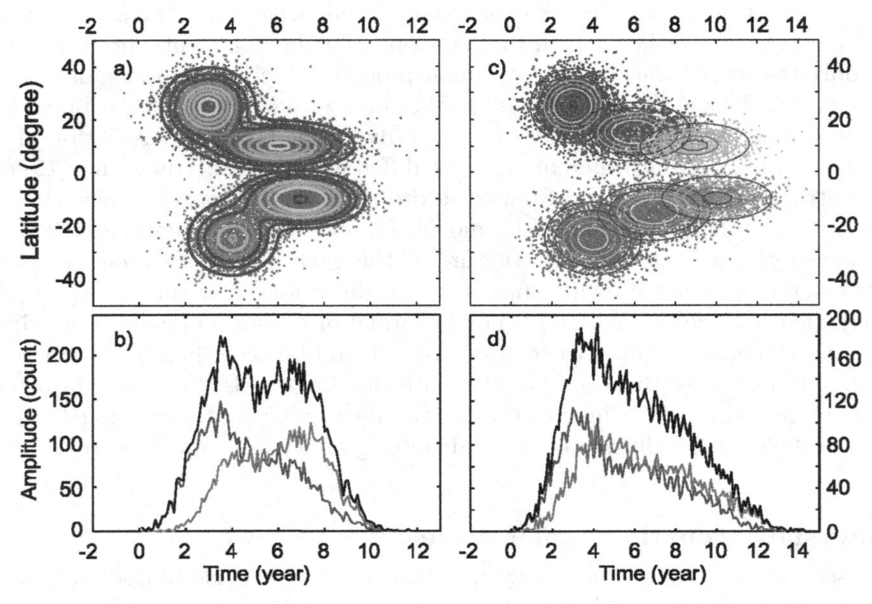

Figure 1. (a): Toy butterfly diagram composed of 4 pulses of activity. Solid lines represent levels of equal point density. (c): The same for 6 pulses. (b and d): Temporal evolutions of the same distributions summarized over latitudes. Analog of sunspots area time-series. Blue line corresponds to the northern hemisphere, red — the southern one, and black — their sum (total).

Table 2. Parameters for the three-component model for the each hemisphere.

Northern hemisphere			
Components No.	1	2	3
Latitude	25°, time of maximum — 3 years after the cycle onset.	15°, time of maximum — 6 years after the cycle onset.	10°, time of maximum — 9 years after the cycle onset.
Covariance	1/50	2/30	2/20
Amount of points	2500	2000	1000
Southern hemisphere			
Components No.	1	2	3
Latitude	-25°, time of maximum — 4 years after the cycle onset.	-15°, time of maximum — 7 years after the cycle onset.	-10°, time of maximum — 10 years after the cycle onset.
Covariance	1.5/50	2/30	2/20
Amount of points	2500	2000	1000

between the pulses, nonlinear stretch of distributions in time and latitudes, etc. will give a various butterfly wings geometry, different lengths and magnitudes of designed cycles.

To model prolonged cycles we added additional weak pulse in each hemisphere (Fig. 1c). Chosen parameters are listed in the Table 2. The length of this modeled double-peak cycle is about 13–14 years (Fig. 1d). Note that the length of the 4th cycle also is unusually long (13 year, from minimum to minimum). Parameters were specified to produce one-peak maximum of final black line (Fig. 1d). It is remarkable that the declining phase became longer than the ascending one. Only just from examining the Figure 1(d), one cannot recognize that the model consist of 3 pulses in each hemisphere. Thus, the overlapping hides internal structure of the cycle. If the cycle is composed of more additional weak impulses, then its shape became even more sophisticated.

Analysing the reconstructed butterfly diagram from the Staudacher drawings (Arlt 2008, 2009a,b) for the 4th cycle (1784–1798), Usoskin *et al.* (2009) found that the sunspot

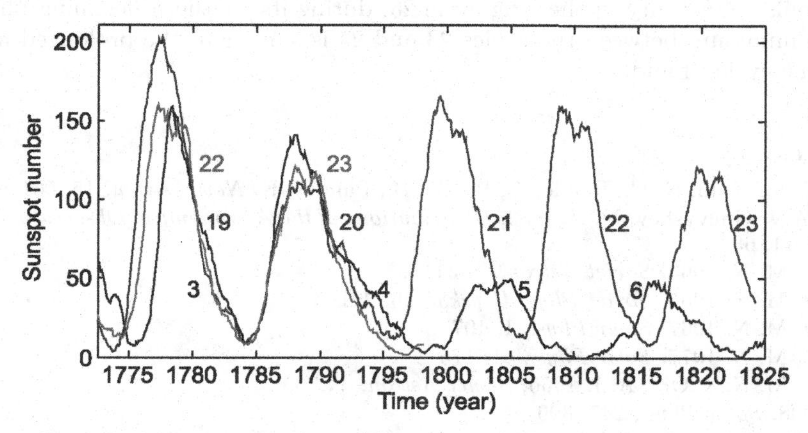

Figure 2. Smoothed Sunspot Numbers. The cycles from 3 to 6 are shown in black colour. X-axis is time for those cycles. The cycles 19–23 — blue colour and 22–23 — red colour.

occurrence rate in 1793–1796 is consistent with a typical ascending phase of the solar cycle.

Hoyt & Schatten (1998) stated that from 1795 to the present, the Group Sunspot Numbers (GSN) are well determined. From 1797 to the middle 1800 the sunspot activity was low. It was prolonged 3-year minimum. That is why the length of the cycle 4, determined from maximum to maximum, is 17 years. During the nowaday prolonged minimum between the cycles 23 and 24 there were already more then 800 spotless days. Hence, the 23rd cycle length from maximum to maximum should be also larger than usual (Fig. 2).

Figure 2 shows the smoothed monthly averages of Sunspot Number index. It is seen that the cycles 4, 20 and 23 are not strongly different in their shape, length and magnitude. Total activity lines (only temporal representation, convolute on latitudes) are the result of overlapping. However, a minor activity was happened in the southern hemisphere and strong north-south asymmetry was observed at the end of cycle 23 (Bankoti *et al.* 2010; Zolotova *et al.* 2009). Correspondingly, the unusually long 4th cycle can be explained by the pulse of activity in the northern hemisphere during the descending phase. We suggest that activity minimum in 1793 can be a gap between pulses. The another opposite suggestion is that this gap is illusion because of observations lack. Then, the descending phase of the 4th cycle is gradually declining, pulses are smeared by their overlapping (like in Fig. 1d). The idea of the "lost" tiny cycle also stays as a hypothesis. Thereby, lack, uncertainty, and scarcity of historical records can dramatically change our knowledge about the dynamics of solar cycle.

3. Conclusions

We realized the Gnevyshev idea about the composed structure of the solar cycles from the activity pulses with different features. Specifying the model parameters one- or two-peak 11-year cycle can be constructed. It is shown that by means of weak pulses the shape of the cycles can be changed. The declining phase became longer than the ascending one. Middle symmetrical and long asymmetrical cycles were presented. It was demonstrated that using only convolution of activity (like area or sunspot number data) it is difficult to recognize pulses. Even a monotonic decay may consist of them. The overlapping of pulses hides internal structure of cycle.

We suggested that the length of the 4th cycle can be explained by the pulse or impulse of activity in the northern declining phase of the cycle. Probably, it was somewhat similar to the impulse of activity in the long cycle 23, during its southern declining phase. The prolonged minimum between the cycles 23 and 24 is similar to the prolonged minimum between the cycles 4 and 5.

References

Bankoti, N. S., Joshi, N. C., Pande, S., Pande, B., Pandey, K. *New Astron.*, 15, 561

Antalová, A. & Gnevyshev, M. N. 1983, *Contributions of the Astronomical Observatory Skalnate Pleso*, 11, 63

Gnevyshev, M. N. 1963, *Soviet Astron.*, 7, 311

Gnevyshev, M. N. 1966, *Soviet Phys. Uspekhi*, 90, 291

Gnevyshev, M. N. 1967, *Solar Phys.*, 1, 107

Gnevyshev, M. N. 1977, *Solar Phys.*, 51, 175

Gnevyshev, M. N. & Ohl, A. I. 1966, *Soviet Astron.*, 9, 765

Arlt, R. 2008, *Solar Phys.*, 247, 399

Arlt, R. 2009, *Astron. Nachr.*, 330, 311

Arlt, R. 2009, *Solar Phys..*, 255, 143

Usoskin, I. G., Mursula, K., Arlt, R., & Kovaltsov, G. A. 2009, *Astrophys. J. Lett.*, 700, L154
Hoyt, D. V. & Schatten, K. H. 1998, *Solar Phys.*, 179, 189
Zolotova, N. V., Ponyavin, D. I., Marwan, N., & Kurths, J. 2009, *Astron. Astrophys*, 503, 197

Discussion

KITIASHVILI: Did you have explanation, a source of such pulses?

ZOLOTOVA: Unfortunately I don't know what can – which process can cause pulses. It was suggested it is kind of burst of magnetic flux near the tachocline.

The Physics of Sun and Star Spots
Proceedings IAU Symposium No. 273, 2010
D.P. Choudhary & K.G. Strassmeier, eds.

doi:10.1017/S1743921311015201

The zoo of starspots

Klaus G. Strassmeier

Astrophysical Institute Potsdam, An der Sternwarte 16, D-14482 Potsdam, Germany
email: kstrassmeier@aip.de

Abstract. Starspots are being observed with many different techniques but not always with coherent results. In particular not if model-dependent data analysis must be employed, e.g. through two-dimensional spot modelling of one-dimensional photometric light curves. I review the zoo of currently available physical spot parameters, i.e. their size, temperature and variability time scales, and also compare results from different techniques. Most of the current values come from Doppler imaging and multi-color photometry. I also list a few cases where starspot detections turned out to be very different to the solar analog.

Keywords. Magnetic fields, stars: spots, stars: magnetic fields, stars: rotation, methods: photometric, methods: spectroscopic

---

1. Introduction and context

Solar chromospheric emission-line flux is directly related to the underlying photospheric magnetic field (a description of this connection is given, e.g., in the book by Schrijver & Zwaan 2000). The majority of the photospheric magnetic field itself is thought to be linked again to the internal dynamo by myriads of magnetic field lines, possibly organized as flux tubes and, of course, as spots. For the majority of stars we can still make only qualitative statements for a connection between their surface magnetic tracers and the dynamo process, e.g., through detecting their activity cycles. This is partly because the evidence is based on disk-integrated measurements, most notably for Ca II S-index variability, X-ray luminosity or broad-band UV enhancement. Some surprises were encountered, e.g. that there is an activity "saturation" in the X-ray flux when the magnetic filling factor of the atmospheric volume comes close to 100%.

The most detailed stellar surface data come from optical (photospheric) Doppler imaging, a technique that currently can not be extended to the chromosphere, not to speak about the corona. But even for photospheric mapping, we are still biased towards stars rotating several times to tens of times faster than the Sun, where we may expect a qualitatively different link of the surface to the interior as compared to the Sun. Moreover, also the Sun bears still surprises. *Hinode* has told us recently that there is an ubiquitous horizontal photospheric magnetic field of super-equipartition kG strength even at high solar latitudes (e.g. Ishikawa & Tsuneta 2009). Whether this is truly evidence for a turbulent sub-photospheric dynamo or just describes an advected field through granular motion from deeper layers is not clear yet. If such fields are assumed to exist also on other solar-type stars, stellar observations of unresolved surfaces are then left with the ambiguity whether a particular magnetic tracer, e.g. Ca II H&K emission, is more related to the radial or the horizontal field, or both, as expected. Clearly, we need to spatially resolve the full magnetic-field vector also on other stars and directly observe the field geometry in order to compare with the Sun. The only technique to do so on other star is Zeeman-Doppler-Imaging (ZDI) (see the papers by Berdyugina, Carroll *et al.* and Kochukhov *et al.* in this proceedings).

Besides that we need spatially resolved stellar disk data, we need time resolution on very long scales to provide more solid evidence for the existence of a systematically changing pattern of the magnetic tracers, like the solar butterfly diagram. This endeavor is still in an infant state and involves the consecutive solution of three major obstacles for stellar observational techniques. Firstly, we need to spatially resolve and sample the surface of stars that are otherwise ideal point sources. Secondly, we need to distill the geometry of the magnetic surface field in three dimensions and, thirdly, we need to timely sample this over various time scales, from the spot lifetime to the period of an activity cycle. The latter may not appear to be an acceptable proposal for a telescope time allocation committee.

For the time being, one shall first review the collection of Doppler images and bring them into an appropriate physical order. This shall enable a meaningful comparison, despite of the different surface parameters of these maps, the different evolutionary state of the targets and their, e.g., vastly different angular velocities. This was partly already done in the recent review by Strassmeier (2009) and is not repeated here. In the course of the present zoological work, I stumbled over "starspot" observations that showed observations that were phenomenologically similar but very different in origin and physics, and sometimes completely unrelated to "Sunspots".

2. Some starspot numbers

2.1. *Stars with spots in our galaxy*

An educated guess of the expected number of spotted stars in the galaxy yields approximately 200 billion stars. This is basically the number of stars on the right side of the granulation boundary of the H-R diagram. So far, ≈500 spotted stars were observed with any kind of method. Of these, ≈80 were Doppler imaged, of which approximately half are close binaries and half are single stars. The latter are a mixture from nearly all stellar evolutionary stages. This makes clear that we are talking small-number statistics and that many more stars of various evolutionary state and physical parameters must be mapped before we can tie in the Sun (as a star).

2.2. *Spot sizes*

The spot sizes from the list of Doppler images collected recently in Strassmeier (2009) range between 0.1% - 11% of the total stellar surface. The record holders are the K0III component in the RS CVn binary XX Tri (Strassmeier 1999) with its huge high latitude to polar spot of size 10,000 times larger than the largest sunspot group ever observed, and the ultra rapidly-rotating K3 dwarf BO Mic with spots of size as small as 0.15% (Barnes 2005, Wolter *et al.* 2005). Such small spots are only resolved if the number of resolution elements across the stellar disk is sufficiently large, i.e. small spots are only detectable on very rapidly rotating stars. Most of the spotted stars that were Doppler imaged so far rotate with an average $v \sin i$ of ≈30 km s^{-1} and have an average spot size of "a few per cent". Clearly, Doppler imaging can recover just the large-scale spot distribution and is biased from the cross-talk with the unresolved spot component for the cases when fine surface detail is seen in the maps.

A special case are the W UMa systems which are very close binaries in a common outer envelope, usually convective. Their surfaces are distorted along the apsidal line by the respective gravitational pull and oblate due to their fast rotation. For example for

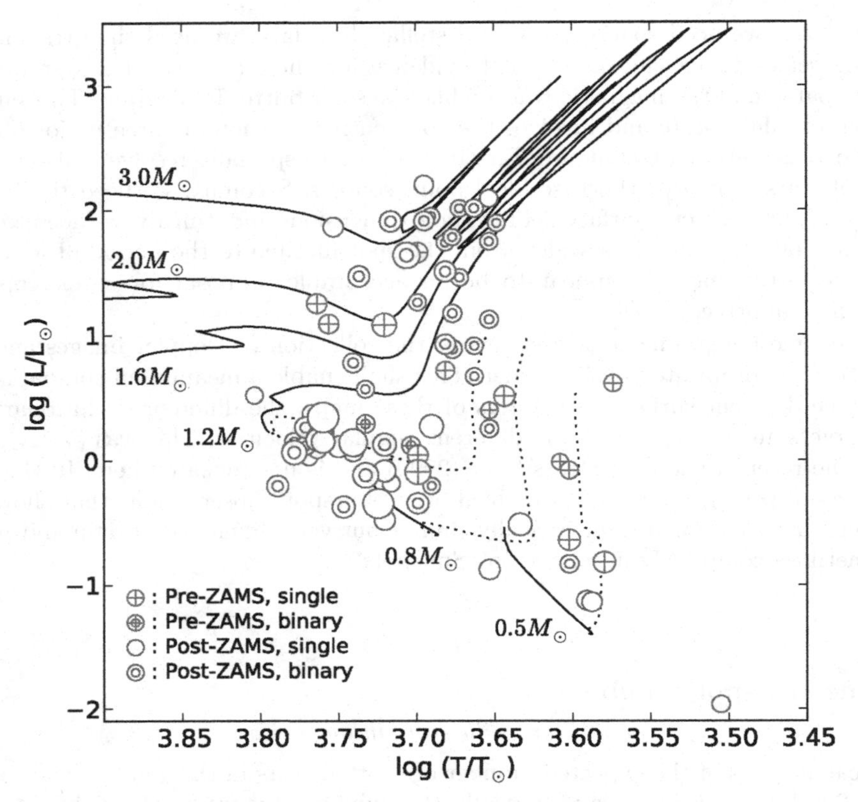

Figure 1. Cool stars with Doppler images in the H-R diagram. Tracks are shown for post-ZAMS evolution (solid lines) and for pre-ZAMS evolution (dotted lines) for masses between 0.5 and 3.0 M$_\odot$. The symbols are specified in the panel. The size of the symbol indicates the equatorial rotational velocity. The smallest size in the plot represents $v_{\rm equ}$ = 13 km s^{-1}, the largest size $v_{\rm equ}$ = 245 km s^{-1}.

VW Cep (Hendry & Mochnacki 2000), spots covered 66% of the surface of the primary and 55% of the secondary mostly at the rotational poles, much more than what has been detected for non-contact binary stars and over-active single stars. It remains to be determined whether all their spots are also magnetic.

On the smallest size scales, transit mapping of surface spots from the occultation by an orbiting exoplanet revealed two astounding spots of size just 0.025% for HD189733 from HST photometry (Pont *et al.* 2007). This is still larger than but similar to the size of a very large sunspot group. More data are now coming online from the Kepler spacecraft (e.g. Basri *et al.* 2010, Brown *et al.*, this proceedings) as well as from CoRoT (e.g. Silva-Valio *et al.* 2010).

2.3. *Spot temperatures*

The observational problem here is the cross talk between spot temperature and spot size and the ambiguity to separate these two quantities during the modelling approach. This problem is most evident for photometric spot modelling and does not exists in Doppler imaging. Photometric data are so prone to this ambiguity that the modelling results are fully model dependent. This was realized very early on by, e.g. Poe & Eaton (1979), and led to the common practice to use at least two well-defined photometric bandpasses to minimize the cross talk, usually one bandpass optimized for the unspotted photosphere and the other optimized for the spotted photosphere, e.g., Johnson V and I, respectively. All of the space-based exoplanet-transit photometry is done in "white light" and can not be used for classical spot modelling (or should not be used) unless one is very well aware that it just allows the reconstruction of spot longitudes, i.e. zonal rotational periods (e.g. Lanza *et al.* 2009, Fröhlich *et al.* 2009).

The spot temperature is much better constrained in Doppler imaging (DI; see the discussion in Strassmeier 2009 a.o. why this is so). Not all DI codes invert the spectral line profiles into a temperature map and thus our statistics are less convincing. However, from temperature-DI it appears that the cooler the star the smaller is the temperature difference between the spot and the unspotted photosphere. Berdyugina (2005) lists an average of 200 K temperature difference for M dwarfs and up to 2000 K for the F-stars.

From broad-band multi-color photometry alone, various authors obtained spot temperatures for a number of stars of a mixture of stellar parameters. The values range between 300–1700 K (see also Oláh *et al.*, this proceedings). The literature on this is just too extensive in order to comment here on individual results and I may refer to our "Thinkshop" proceedings on *Sunspots & Starspots* (Strassmeier *et al.* 2002) for further references. Multi-color photometry combined with modern light-curve inversion techniques (e.g. Savanov & Strassmeier 2008, Berdyugina 2005) is still a powerful tool to get good estimates of spot temperatures for a large number of stars.

A differential technique, based on two nearby spectral lines of different temperature sensitivity, was explored by David Gray and collaborators (e.g. Gray & Johanson 1991). From such line-ratios, plus the accompanying continuum variation from photometry, Catalano *et al.* (2002) and Frasca *et al.* (2008) obtained spot temperatures in the range of 700–1000 K.

Vogt (1979) was the first to realize the potential of molecular spectral lines as a temperature indicator for spots. Certain molecular absorption band heads can only be formed at temperatures significantly below the typical photospheric temperature of a cool star and thus a simple detection of its rotationally modulated strength can constrain the spot temperature. Employing the TiO band heads at 705 nm and 886 nm, O'Neal *et al.* (2004) obtained relative spot temperatures for five stars in the range 1000–2000 K. The 2000 K for the G1.5 dwarf EK Eri is the largest temperature difference obtained so far, even among all the techniques mentioned before. Previous spot-temperature determinations of this star ranged between 500–1200 K. Just recently, Rice *et al.* (2010) obtained separate Doppler images of the WTTS V410 Tauri from atomic lines as well as from the TiO 705.5-nm lines and found an average temperature difference between these images of 150 K in the sense that the TiO-based temperatures appeared to be generally cooler. Part of this difference likely stems from the difficulties to find the correct continuum at the TiO-band wavelengths but the basic conclusion is that atomic-line Doppler imaging may underestimate the true spot temperatures (on a 100 K level for a $T_{\rm eff} \approx 4500$-K star).

2.4. *Spot time scales*

Among the most important time scales associated with sunspots and starspots are their lifetimes and decay times. No single starspot or starspot group has been observed and followed from its formation to its death so far. Not to speak about a proper time sampling. Even for sunspots such observations are rare, hard to disentangle from intra-group evolution, and biased towards the visibility due to the one-month rotational period of the Sun. The new *SDO* data movies, now shown at various conferences, bear a great potential to settle also this question (tbd).

Idealized sunspots appear to follow a mean decay law of the form $dA/dt = C\, r/r_{\rm max}$, with $C = 32.0 \pm 0.26$ a constant and r the relative spot radius (Petrovay *et al.* 1999). The area decay rate is in units of one millionth solar hemisphere per day. It is directly related to the magnetic diffusivity, η, by $dA/dt = -4\pi\eta$. Numerical values for the magnetic diffusivity are not known exactly and adopted values range from 10^{10} cm^2/s to 10^{13} cm^2/s (see Hathaway & Coudhary 2008). This value may also appear to be important for dynamo models that are tailored to predict the next solar activity cycle. A recent determination from spot variations on CoRoT-2a inferred an η of 1.2×10^{13} cm^2/s (Fröhlich *et al.* 2009), which is a first step in the right direction but likely an overestimation given the spot model assumptions in this and comparable papers.

For a summary of starspot lifetimes I refer again to Strassmeier (2009) and the references therein. More recently, we have some first preliminary results from time-series Doppler images obtained with STELLA. STELLA is a new robotic observatory in Tenerife, and dedicated to the observation of spotted stars. It consists of two 1.2-m telescopes, one fiber feeding a high-resolution echelle spectrograph and the other feeding a wide-field imaging photometer (Strassmeier *et al.* 2010). Among the targets is the famous active binary XX Triangulum (HD 12545).

3. Other fierce creatures of the zoo

A "starspot" may not always be what we believe it is supposed to be, i.e. somehow an analog of a sunspot or a sunspot group. They can be impact regions from magnetospheric accretion, obscurations from circumstellar material, or planetary transits. Even if a magnetic field can be associated the field can be of rather different origin, e.g. a fossil field (31 Com, HR3162), a mixture of surface and core dynamo fields or due to some sort of magneto-gravitational coupling of the components in a close binary.

Betelgeuse (α Ori) is a neat example. It exhibits a well observed bright spot in ultraviolet observations, in particular at Mg II h&k (see Dupree, this proceedings, for a detailed account and further references). Is has been speculated that the spot is a granulation cell, a certain non-radial pulsation mode, an outflow structure, and a chromospheric magnetic faculae. Most likely it is a combination of the latter two. Another example is ϵ Aurigae, the eclipsing binary with the longest known period (27 years). It is during mid eclipse while we are meeting, and this time, the stellar disk has been resolved by phase-coherent interferometry with the CHARA array in the H band (Kloppenborg *et al.* 2010). What is seen is an obscuration of the stellar disk due to the accretion disk around the secondary star, itself possibly a neutron star. While the eclipse is progressing, the disk is "eating away" parts of the stellar disk. If one would not know about the eclipsing phenomenon one could interpret the obscuration as due to cool starspots, in particular if one has only snapshot images.

Another fierce case is the low-mass, single, pre-main sequence star MN Lupi (Strassmeier *et al.* 2005). Doppler imaging detected not only two high-latitude hot spots (2000 K warmer than the effective temperature) but also generally a warm polar cap (1200 K warmer than photospheric), some cool spots 400-500 K below the photospheric temperature, and a generally 500-K cooler "southern" than "northern" hemisphere. The hot spots are interpreted as the heating points of accretion shocks, the shock itself is evident in emission lines like He I and Balmer H 1-13 and not seen in optically thin lines. The warm polar cap is the trailed and redistributed impact energy. The isolated cool spots are likely of local magnetic origin as we know it from the Sun, but the cool "southern" hemisphere is an obscuration due to the inner rim of the accretion disk, or whatever is left of it (the star does not have an IR excess and is classified as a weak-line T Tauri star).

4. Outlook

Among future observational tasks is time-series Doppler imaging of a representative number of stars over the entire activity cycle, if one exists. The reward should be stellar butterfly diagrams as a function of relevant global parameters like rotation rate or age, i.e. internal stellar structure. If this could be done with polarized spectra would allow, in principle, to follow the magnetic field evolution as well. It has the potential to revise our common understanding of stellar evolution, in particular for the young Sun and its infant planetary system. The magnetic field possibly played a decisive role during the star and planetary formation and its broader impact on Earth may have been underestimated (e.g. Moore & Horwitz 2007). As a small nuisance, high spectral resolution combined with polarimetry requires a fairly large telescope for good S/N, and the need for dense phase cadence in order to do Doppler imaging in the first place requires most of its time. Not the best of all conditions for an observing proposal.

Acknowledgements

I would like to thank my AIP colleagues Thorsten Carroll, Michi Weber and Thomas Granzer for many discussions, and Carsten Denker and Rainer Arlt for sharing their overall subject wisdom with me. As always, it is a pleasure to thank John Rice for his continuous collaboration. I acknowledge grant STR645-1 from the Deutsche Forschungsgemeinschaft (DFG) and the support from the Bundeministerium für Bildung & Forschung (BMBF) through the Verbundforschung grant 05AL2BA1/3.

References

Barnes, J. R. 2005, *Mon. Not. Roy. Astron. Soc.* 364, 537
Basri, G., Walkowicz, L. M., Batalha, N. *et al.* 2010, *Astrophys. J.* 713, 155
Berdyugina, S. V. 2005, *Living Rev. Solar Phys.* 2
Catalano, S., Biazzo, K., Frasca, A., Marilli, E., Messina, S., & Rodonó, M. 2002, *Astron. Nachr.* 323, 260
Frasca, A., Biazzo, K., Tas, G., Evren, S., & Lanzafama, A. C. 2008, *Astron. Astrophys.* 479, 557
Fröhlich, H.-E., Küker, M., Hatzes, A. P., & Strassmeier, K. G. 2009, *Astron. Astrophys.* 506, 263
Gray, D. F. & Johanson, H. L. 1991, *Pub. Astron. Soc. Pac.* 103, 439
Hathaway, D. H. & Choudhary, D. P. 2008, *Solar Phys.* 250, 269
Hendry, P. A. & Mochnacki, S. W. 2000, *Astrophys. J.* 531, 467
Ishikawa, R. & Tsuneta, S. 2009, *Astron. Astrophys.* 495, 607
Kloppenborg, B., Stencel, R., Monnier, J. D. *et al.* 2010, Nature 464, 870
Lanza, A. F., Pagano, I., Leto, G., *et al.* 2009, *Astron. Astrophys.* 493, 193
Moore, T. & Horwitz, J. 2007, Rev. Geophys. 45, RG3002
O'Neal, D., Neff, J. E., Saar, S. H., & Cuntz, M. 2004, *AJ* 128, 1802
Petrovay, K., Martínez Pillet, V., & van Driel-Gesztelyi, L. 1999, SP 188, 315
Poe, C. H. & Eaton, J. A. 1985, *Astrophys. J.* 289, 644
Pont, F., Gilliland, R. L., Moutou, C. *et al.* 2007, *Astron. Astrophys.* 476, 1347
Rice, J. B., Strassmeier, K. G., & Kopf, M. 2011, *Astrophys. J.*, 728, 69
Savanov, I. S. & Strassmeier, K. G. 2008, *Astron. Nachr.* 329, 364
Schrijver, C. J. & Zwaan, C. 2000, Solar and stellar magnetic activity, New York, CUP

Silva-Valio, A., Lanza, A. F., Alonso, R., & Barge, P. 2010, *Astron. Astrophys.* 510, 25
Strassmeier, K. G. 1999, *Astron. Astrophys.* 347, 225
Strassmeier, K. G. 2009, *A&AR* 17, 251
Strassmeier, K. G., Granzer, T., Weber, M. *et al.*, 2010, *Adv. in Astr.* 2010, p.19
Strassmeier, K. G., Rice, J. B., Ritter, A. *et al.* 2005, *Astron. Astrophys.* 440, 1105
Strassmeier, K. G., Washuettl, A., & Schwope, A. 2002, *Astron. Nachr.* 323, 155
Vogt, S. S. 1979, *Pub. Astron. Soc. Pac.* 91, 616
Wolter, U., Schmitt, J. H. M. M., & vanWyk, F. 2005, *Astron. Astrophys.* 435, 261

The Physics of Sun and Star Spots
Proceedings IAU Symposium No. 273, 2010
D.P. Choudhary & K.G. Strassmeier, eds.

doi:10.1017/S1743921311015213

Exploring the magnetic topologies of cool stars

J. Morin[1,2], J.-F. Donati[2], P. Petit[2], L. Albert[3], M. Auriére[2], R. Cabanac[2], C. Catala[4], X. Delfosse[5], B. Dintrans[2], R. Fares[2], T. Forveille[5], T. Gastine[2], M. Jardine[7], R. Konstantinova-Antova[6], J. Lanoux[8], F. Lignires[2], A. Morgenthaler[2], F. Paletou[2], J.C. Ramirez Velez[4], S.K. Solanki[9], S. Thado[2] and V. Van Grootel[2]

[1]Dublin Institute for Advanced Studies, 31 Fitzwilliam Place, Dublin 2, Ireland
email: jmorin@cp.dias.ie

[2]LATT, Universit de Toulouse, CNRS, 14 Av. E. Belin, 31400 Toulouse, France

[3]CFHT, 65-1238 Mamalahoa Hwy, Kamuela HI 96743, USA

[4]LESIA, Observatoire de Paris-Meudon, 92195 Meudon, France

[5]LAOG, UMR5571 CNRS, Universit Joseph Fourier, BP 53, 38041 Grenoble, France

[6]Institute of Astronomy, Bulgarian Academy of Sciences, 72 Tsarigradsko shose, Sofia, Bulgaria

[7]School of Physics and Astronomy, University of St Andrews, St Andrews, Scotland KY16 9SS

[8]Centre dEtude Spatiale des Rayonnements, Universit de Toulouse, CNRS, France

[9]Max-Planck Institut fr Sonnensystemforschung, Katlenburg-Lindau, Germany

Abstract. Magnetic fields of cool stars can be directly investigated through the study of the Zeeman effect on photospheric spectral lines using several approaches. With spectroscopic measurement in unpolarised light, the total magnetic flux averaged over the stellar disc can be derived but very little information on the field geometry is available. Spectropolarimetry provides a complementary information on the large-scale magnetic topology. With Zeeman-Doppler Imaging (ZDI), this information can be retrieved to produce a map of the vector magnetic field at the surface of the star, and in particular to assess the relative importance of the poloidal and toroidal components as well as the degree of axisymmetry of the field distribution.

The development of high-performance spectropolarimeters associated with multi-lines techniques and ZDI allows us to explore magnetic topologies throughout the Hertzsprung-Russel diagram, on stars spanning a wide range of mass, age and rotation period. These observations bring novel constraints on magnetic field generation by dynamo effect in cool stars. In particular, the study of solar twins brings new insight on the impact of rotation on the solar dynamo, whereas the detection of strong and stable dipolar magnetic fields on fully convective stars questions the precise role of the tachocline in this process.

Keywords. Stars: low-mass, brown dwarfs, stars: magnetic fields, stars: activity, stars: rotation, techniques: spectroscopic, techniques: polarimetric

---

1. Context: stellar dynamos

Magnetic field is a key parameter to understand stellar formation and evolution. In cool stars, it powers activity phenomena that are observed across a large part of the electromagnetic spectrum and a wide range of timescales. Since the early XXth century, the cyclic solar magnetic field has been thought to be constantly regenerated against ohmic dissipation by a magnetohydrodynamical process: the dynamo. Although the solar dynamo is still far from being fully understood, the basic concepts, as exposed by Parker (1955) in his $\alpha\Omega$ dynamo model are rather simple: (i) an initially poloidal magnetic field is converted into a stronger toroidal one by differential rotation, (ii) a poloidal field component is regenerated from the toroidal field by a second mechanism, such as the α effect. Two decades ago, helioseismology revealed the existence of the tachocline (e.g., Spiegel & Zahn 1992), a thin layer of strong shear located at the interface between the inner radiative zone and the convective envelope. Since then, many theoretical and numerical studies have stressed the crucial role of the tachocline in the solar dynamo, being the place where large-scale toroidal fields can be stored and strongly amplified (e.g., Charbonneau & MacGregor 1997).

Partly convective cool stars possess an internal structure similar to that of the Sun, i.e. an inner radiative zone and an outer convective envelope supposedly separated by a tachocline. Hence, it is generally assumed that their magnetic fields — as revealed by activity or direct measurements — are generated by a solar-like dynamo. However, some cool partly-convective stars strongly differ from the Sun, either in depth of their convective zone or rotation rate, and the impact of these differences on their dynamo is mostly unknown. On the other hand, main sequence stars less massive than $\sim 0.35\ M_\odot$ ($\sim$M3) are fully convective (e.g., Chabrier & Baraffe 1997) and therefore do not possess a tachocline. If the tachocline is indeed an essential part of the solar dynamo, magnetic field generation in these fully convective objects must rely on different physical processes.

2. Magnetic field measurement and modelling

Direct measurements of stellar magnetic fields rely on the properties of the Zeeman effect. Two complementary methods are successfully applied to cool stars. By measuring Zeeman broadening of photospheric spectral lines it is possible to assess the magnetic flux averaged over the visible stellar disc (e.g., Saar 1988). This method is therefore able to probe magnetic fields regardless of their complexity but provides virtually no information about the field geometry. On its part, the analysis of Zeeman polarisation in spectral lines provides information on the vector properties (i.e. strength, orientation and polarity). However, as neighboring magnetic regions of opposite polarities result in polarised signatures of opposite sign that cancel each other when integrating over the stellar disc, spectropolarimetry can only detect the large-scale component of stellar magnetic fields.

Polarised signatures in spectral lines of cools active stars have a small amplitude, making their measurement a hard task. Two advances have brought a large number of cool stars within reach of spectropolarimetric measurements. First, multi-line techniques (such as LSD, Donati *et al.* 1997) extract the polarimetric information from a large number of photospheric lines resulting in a S/N multiplex gain that can reach as high as several tens when thousands of lines are used. Secondly, the new generation spectropolarimeters ESPaDOnS and NARVAL (see Donati 2003) feature a high overall efficiency and cover the full optical domain allowing to take full advantage of multi-line techniques.

Thanks to (*i*) the sensitivity of the Zeeman effect to field lines orientation (*ii*) rotational modulation and (*iii*) Doppler effect, the temporal evolution of polarised signatures in stellar lines strongly characterizes the parent magnetic topology. Thus, from a time-series of circularly polarised spectra sampling the stellar rotation cycle, Zeeman-Doppler Imaging can perform a maximum entropy reconstruction of the vector magnetic field distribution at the surface of the star (Semel 1989; Donati & Brown 1997). Although the resulting magnetic map has a better resolution for high $v \sin i$ values, this technique is also successfully applied to slow rotators (e.g., Morin *et al.* 2008b). For high S/N data

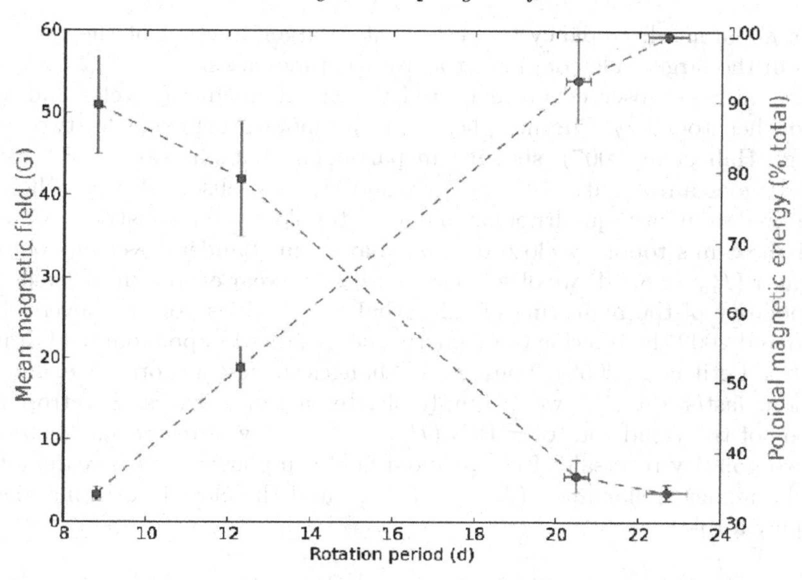

Figure 1. Rotational dependence of the mean reconstructed magnetic flux (green line), and of the fraction of magnetic energy stored in the poloidal field component (red line).

spanning several stellar rotations, differential rotation can also be constrained (e.g., Petit *et al.* 2002).

3. Solar twins

We selected a sample of 4 nearby dwarfs with stellar parameters as close as possible to the solar ones, except for the rotation period (see Petit *et al.* 2008). In particular, their internal structure is expected to be very similar to the Sun's. We aim at studying the effect of rotation alone on the solar dynamo. For each star, a set of $\sim$ 10 pairs of unpolarised and circularly polarised spectra was collected with NARVAL, from which we can map the surface magnetic field and to precisely determine the rotation period.

The first conclusion of this study is that the shorter the rotation period, the stronger is the large-scale magnetic field (see Fig. 1, green line). For the 2 slow rotators ($P_{\rm rot}$ = 22.7 and 20.5 d), the magnetic maps reconstructed by ZDI are dominated by low order multipole modes, this is reminiscent of the solar global magnetic field. The fast rotators ($P_{\rm rot}$ = 12.3 and 8.8 d) mainly feature a strong belt of toroidal field roughly encircling the pole, similar to the magnetic topologies of very active cool stars (e.g., Donati *et al.* 2003). The transition from an almost purely poloidal magnetic field to a strongly toroidal one, is apparently due to rotation, with a $P_{\rm rot}$ threshold located between 12 and 20 d (see Fig. 1). These observations are in agreement with recent MHD simulations of solar-type stars where dynamo action produces mostly toroidal magnetic topologies, featuring strong belts throughout the bulk of the convective envelope for $\Omega = 3\ \Omega_{\odot}$ (Brown *et al.* (2010), although the simulation domain does not encompass the stellar surface).

From our unpolarised spectra we measure the Ca II activity, and find that the $R'_{HK}(B_{mean})$ relation follows a power-law with an exponent of 0.32. This is significantly different from the solar relation (established from observations of the quiet Sun and active regions) which has an exponent of 0.6 (e.g., Schrijver *et al.* 1989). As the chromospheric flux is also sensitive to spatial scales smaller than those contributing to the polarimetric signal, this apparent discrepancy suggests that a larger fraction of the total magnetic energy lies in the large-scale component as rotation increases.

The stars we have observed are expected to exhibit magnetic cycles and associated evolution of their topology. Chromospheric activity monitoring exists for two stars of our sample (e.g., Hall *et al.* 2007), showing in particular that our slowest rotator ($P_{\rm rot}$ = 22.7 d) undergoes an activity cycle of 7 yr, and that we observed it in a high activity state. The predominantly quadrupolar magnetic topology we reconstruct is indeed reminiscent of the Sun's topology close to solar maximum (Sanderson *et al.* 2003). On the fastest rotator ($P_{\rm rot}$ = 8.8 d) we observe strong year-to-year evolution: between 2007 and 2008 the polarity of the main ring of azimuthal field had its polarity reversed, and between 2008 and 2009 the fraction of magnetic energy stored in poloidal field dramatically increased (see Petit *et al.* 2009). These rapid changes suggest a short cycle, and therefore a correlation: faster rotation would imply shorter activity cycles. Spectropolarimetric observations of the rapid rotator τ Boo ($P_{\rm rot}$ = 3.3 d) by Fares *et al.* (2009) have also revealed two polarity reversals of the poloidal field component in two years, although in this case the higher stellar mass ($M_\star$ = 1.3 $M_\odot$) and the close-in orbiting giant planet may also play a role.

4. Fully convective stars

Following the first detection in polarised light of a large-scale magnetic field on a fully convective star by Donati *et al.* (2006), we have carried out the first spectropolarimetric survey of a sample M dwarfs lying on both sides of the fully convective boundary ($0.08 < M_\star < 0.75$ $M_\odot$) and spanning a wide range of periods ($0.33 < P_{\rm rot} < 18.6$ d). A total of 23 stars were selected, all are active so that we can supposedly detect polarised signatures and map their surface magnetic field. For each star we collected one or more time-series of unpolarised and circularly polarised spectra. We could reconstruct a map of the large-scale surface magnetic field and derive an accurate period measurement of 18 stars. For the 5 remaining stars some constraints about the magnetic topology and an upper limit for the rotation period could still be derived. For more details see Donati *et al.* (2008) and Morin *et al.* (2008a,b, 2010).

The main results of this study are presented on Fig. 2, presenting the main properties of the reconstructed magnetic topologies as a function of stellar mass and rotation period. Our analysis reveals 3 distinct groups in this diagram.

(*a*) M dwarfs more massive than ~ 0.5 $M_\odot$ (partly convective) exhibit magnetic topologies with a strong toroidal component, even dominant in some cases; the poloidal component is strongly non-axisymmetric. For most of these stars, we can measure surface differential rotation, values are between 60 and 120 mrad d^{-1} (i.e. between once and twice the solar rate approximately), and the topologies evolve beyond recognition on a timescale of a few months. These properties are reminiscent of the observations of more massive (G and K) active stars (e.g., Donati *et al.* 2003).

(*b*) Stars with masses between ~0.2 and 0.5 $M_\odot$ (close the fully convective limit) host much stronger large-scale magnetic field with radically different topologies: almost purely poloidal, generally nearly axisymmetric, always close to a dipole more or less tilted with respect to the rotation axis. These topologies are observed to be stable on timescales of several years, and differential rotation (when measurable) is of the order or a tenth of the solar rate. Our findings are in partial agreement with the recent numerical study by Browning (2008). Similarly, we observe that fully convective stars can generate strong and long-lived large-scale magnetic fields featuring a strong axisymmetric component, that are able to quench differential rotation. But we observe almost purely poloidal surface

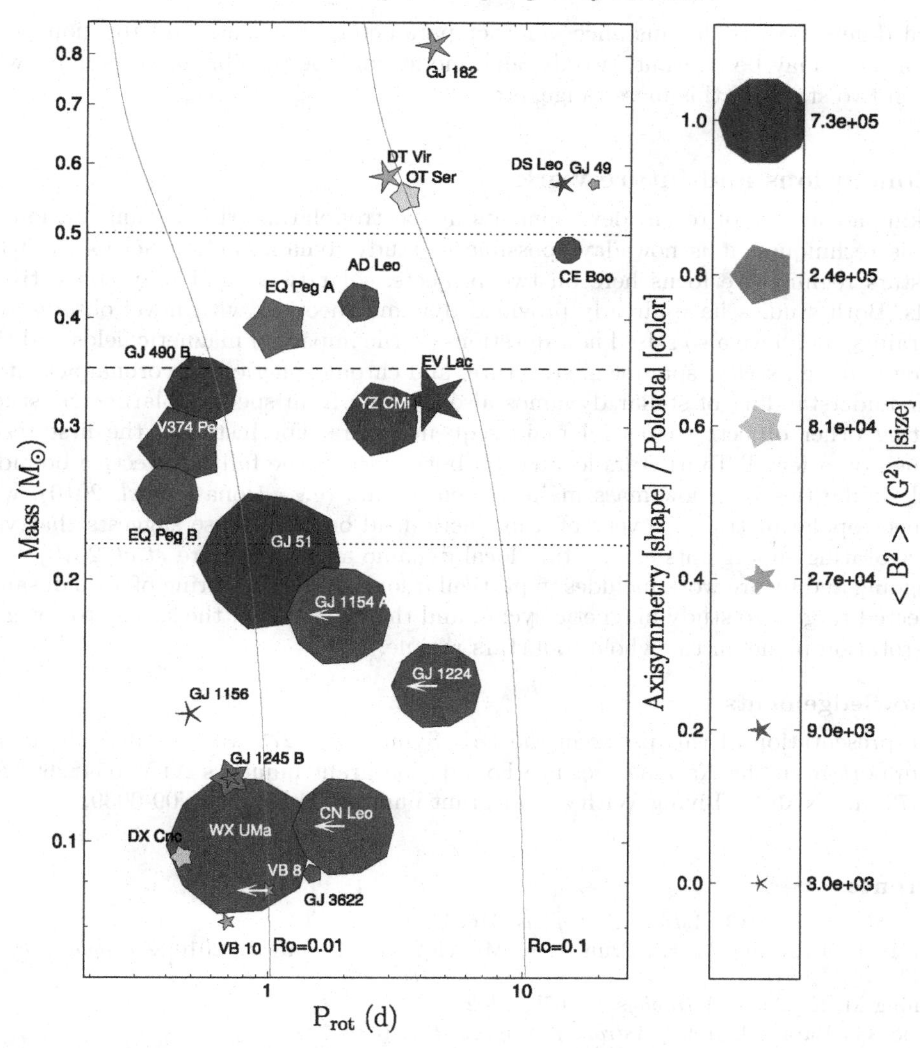

Figure 2. Properties of the magnetic topologies of our sample of M dwarfs (plus GJ 490 B, Phan-Bao *et al.* 2009) as a function of rotation period and mass. Larger symbols indicate stronger fields, symbol shapes depict the degree of axisymmetry of the reconstructed magnetic field (from decagons for purely axisymmetric to sharp stars for purely non axisymmetric), and colours the field configuration (from blue for purely toroidal to red for purely poloidal). Solid lines represent contours of constant Rossby number $Ro = 0.1$ (saturation threshold) and 0.01. The theoretical full-convection limit ($M_\star \sim 0.35$ M$_\odot$) is plotted as a horizontal dashed line, and the approximate limits of the three stellar groups discussed in the text are represented as horizontal dotted lines.

magnetic fields, whereas in the simulation the axisymmetric component of the field is mainly toroidal (although the simulation does not encompass the stellar surface).

(*c*) Below ~0.2 M$_\odot$, we observe 2 different categories of magnetic fields: either a very strong dipole (similar to group *b* above) or a much weaker field generally featuring a significant non-axisymmetric component, and in some cases a toroidal one. Strong temporal variability is also observed on some objects of this second category. However stars in both categories have similar stellar parameters and cannot be separated in the mass-rotation diagram. This unexpected observation is not yet understood, and may be explained in several different ways: for instance, another parameter than mass and rotation period (such as age) may be relevant, two dynamo modes may be possible or stars may switch between two states in this mass range, etc.

5. Conclusions and future work

Taking advantage of recent developments in spectropolarimetric instrumentation and analysis techniques, it is nowadays possible to study dynamo action across the whole cool stars regime. We focus here on two projects: solar twins and fully convective M dwarfs. Both studies have already provided dynamo theorists with novel observational constraints, and have also raised new questions on the impact of magnetic fields and their topology on e.g., stellar spindown, structure, and chromospheric and coronal activity.

Our understanding of stellar dynamos also benefits from spectropolarimetric studies targeting other objects than cool main sequence stars. For instance, the first results obtained on a few T Tauri stars located on both sides of the fully convective boundary reveal similarities with low-mass main sequence stars (e.g., Donati *et al.* 2010), while the spectropolarimetric discovery of a magnetic field on Betelgeuse suggests that very-slowly-rotating supergiants can sustain local dynamo action (Aurière *et al.* 2010).

Ongoing and future work includes in particular long-term monitoring of a small sample of selected targets to study magnetic cycles, and the extension of the survey to cover the mass-rotation plane on the whole cool stars regime.

Acknowledgements

The presentation of this paper in the IAU Symposium 273 was possible due to partial support from the National Science Foundation grant numbers ATM 0548260, AST 0968672 and NASA - Living With a Star grant number 09-LWSTRT09-0039.

References

Aurière M., *et al.*, 2010, *Astron. Astrophys*, 516, L2

Brown B. P., Browning M. K., Brun A. S., Miesch M. S., Toomre J., 2010, *Astrophys. J.*, 711, 424

Browning M. K., 2008, *Astrophys. J.*, 676, 1262

Chabrier G., Baraffe I., 1997, *Astron. Astrophys*, 327, 1039

Charbonneau P., MacGregor K. B., 1997, *Astrophys. J.*, 486, 502

Donati J.-F., Semel M., Carter B. D., Rees D. E., Cameron A. C., 1997, *Mon. Not. Roy. Astron. Soc.*, 291, 658

Donati J.-F. & Brown S. F., 1997, *Astron. Astrophys*, 326, 1135

Donati J.-F., *et al.*, 2003, *Mon. Not. Roy. Astron. Soc.*, 345, 1145

Donati J.-F., 2003, *ASPC*, 307, 41

Donati J.-F., *et al.*, 2006, *Science*, 311, 633

Donati J.-F., *et al.*, 2008, *Mon. Not. Roy. Astron. Soc.*, 390, 545

Donati J.-F., *et al.*, 2010, *Mon. Not. Roy. Astron. Soc.*, 402, 1426

Fares R., *et al.*, 2009, *Mon. Not. Roy. Astron. Soc.*, 398, 1383

Hall J. C., Lockwood G. W., & Skiff B. A., 2007, *AJ*, 133, 862

Morin J., *et al.*, 2008a, *Mon. Not. Roy. Astron. Soc.*, 384, 77

Morin J., *et al.*, 2008b, *Mon. Not. Roy. Astron. Soc.*, 390, 567

Morin J., Donati J.-F., Petit P., Delfosse X., Forveille T., & Jardine M. M., 2010, *Mon. Not. Roy. Astron. Soc.*, 407, 2269

Parker E. N., 1955, *Astrophys. J.*, 122, 293

Petit P., Donati J.-F., Collier Cameron A., 2002, *Mon. Not. Roy. Astron. Soc.*, 334, 374

Petit P., *et al.*, 2008, *Mon. Not. Roy. Astron. Soc.*, 388, 80

Petit P., *et al.*, 2009, *Astron. Astrophys*, 508, L9
Phan-Bao N., Lim J., Donati J.-F., Johns-Krull C. M., & Martín E. L., 2009, *Astrophys. J.*, 704, 1721
Saar S. H., 1988, *Astrophys. J.*, 324, 441
Sanderson T. R., Appourchaux T., Hoeksema J. T., & Harvey K. L., 2003, *JGRA*, 108, 1035
Schrijver C. J., Cote J., Zwaan C., & Saar S. H., 1989, *Astrophys. J.*, 337, 964
Semel M., 1989, *Astron. Astrophys*, 225, 456
Spiegel E. A. & Zahn J.-P., 1992, *Astron. Astrophys*, 265, 106

The Physics of Sun and Star Spots
Proceedings IAU Symposium No. 273, 2010
D.P. Choudhary & K.G. Strassmeier, eds.

doi:10.1017/S1743921311015225

Spots on Betelgeuse, what are they?

Andrea K. Dupree[1]

[1]Smithsonian Astrophysical Observatory/Harvard-Smithsonian Center for Astrophysics
60 Garden Street, Cambridge, MA 02138 USA
email: dupree@cfa.harvard.edu

Abstract. The supergiant star Alpha Orionis (*Betelgeuse*) is the only star other than the Sun to be spatially resolved either through direct imaging or through reconstruction of interferometric observations. Centimeter-radio wavelength, infrared and ultraviolet images reveal a few bright hot spots in the photosphere and chromosphere that possess characteristics different from sunspots. Large photospheric spots on *Betelgeuse* appear to result from convective motions, consistent with radiative hydrodynamic modeling; the chromospheric hot spots may be produced by shock waves in the chromosphere excited by the convective motions or pulsation in the photosphere. Bright chromospheric spots that cluster around the pole of *Betelgeuse* could be a natural result of shock breakout in a rotating star.

Keywords. stars: supergiants, stars: imaging

1. Introduction

The helium-burning red supergiant *Betelgeuse* (α Orionis; HD 38901) is a bright star in the sky, large in apparent size (~50 mas angular diameter in the optical), and hence a favorite target for numerous observational programs and numerical simulations. In fact, over 460 papers discussing *Betelgeuse* have appeared in the refereed literature over the past 20 years. The star is massive, currently $\sim 18 M_{\odot}$, large ($\sim 950 R_{\odot}$), with luminosity $\sim 10^5 L_{\odot}$, and cool, $T_{\rm eff} = 3650$K (Harper *et al.* 2008). It is classified as a semi-regular pulsating star of spectral type M2Iab. In contrast to the low T_{eff}, chromospheric emission in the form of Mg II is observed, but higher ion species such as C IV, O VI, and X-rays are absent (Dupree *et al.* 2005) suggesting that temperatures in excess of ~10000K do not exist. Centimeter-wave radio observations indicate the presence of much cool material (Lim *et al.* 1998; Harper & Brown 2003) coexisting with the warm chromospheric component. The star possesses a massive wind amounting to $\sim 10^{-5} M_{\odot} {\rm yr}^{-1}$. This rate exceeds the solar wind by a factor of 10^9 (!) at least, and it is not known definitively what drives the wind. Dust is formed at a great distance ($\sim 30 R_{\star}$) and has no effect on the atmospheric extension or acceleration of the warm chromosphere. Most recently, a weak longitudinal magnetic field of ~ 1 G has been detected through the measurement of the Stokes V parameter (Auriére *et al.* 2010). Ultraviolet monitoring of the total flux from *Betelgeuse* provided the first evidence that its light behavior and 'spottedness' are quite different from the signals of magnetic activity found in the Sun and active cool stars. In subsequent sections, the appearance of the photosphere and chromosphere from spatially resolved observations are discussed as well as our current understanding of the processes producing these phenomena.

2. Spatially unresolved ultraviolet observations

Quantitative studies of the *Betelgeuse* atmosphere started in earnest with synoptic spectra from the *International Ultraviolet Explorer* (*IUE*) satellite in conjunction with

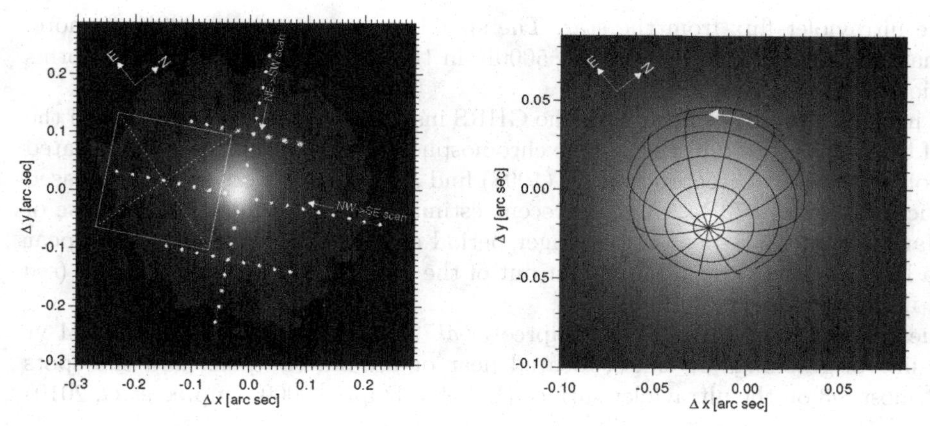

Figure 1. Alpha Ori observations with the FOC and GHRS: *Left panel:* Positions of the GHRS aperture (200 mas square) placed across the ultraviolet image (Gilliland & Dupree 1996) indicating position and direction of spectral scans. *Right panel:* Wire frame model of *Betelgeuse* overlaid on the ultraviolet image (JD.)The arrow indicates the direction of rotation. The bright spot appears at the rotation pole of the star. From Uitenbroek *et al.* (1998).

ground-based optical photometry. These measurements of course have no spatial resolution and monitor the radiation from the whole star. Several indications of the physical processes in the *Betelgeuse* atmosphere emerged from the *IUE* studies (Dupree *et al.* 1987). First, the continuum brightenings and chromospheric (Mg II) brightenings are correlated which is opposite to the behavior of magnetically active stars where dark spots cause diminution of continuum light coincident with chromospheric brightenings. The amplitude of this flux modulation (about a factor of two) exceeds that found in other low gravity stars. Secondly, a photometric period of 420 days lasted for 3 years (1984-1986), and was obvious in the optical (λ4530) and ultraviolet (λ3000) continua as well as the Mg II emission cores. If the period of 420 days results from rotation of the star, a rotational velocity is demanded which exceeds the observed photospheric line widths. The fundamental mode of pulsation for a star like *Betelgeuse* (Stothers & Leung 1971; Lovy *et al.* 1984) is approximately 400 days. Radial velocity variations with this period were also found (Smith *et al.* 1989). Thirdly, a lag occurs between initial variations in the optical and ultraviolet continua and the Mg II h emission, and subsequently the Mg II k line emission. These quantities are formed in successively higher atmospheric layers, and the sequential variation suggests the presence of a travelling disturbance. The 420-day variability disappeared in the years that followed (Dupree *et al.* 2010) confirming the 'semi-regular' classification of the supergiant. In sum, more than two decades ago, it was clear that while *Betelgeuse* varied in radiation output, the star did not present a classical case of magnetic sunspot/starspot variability found in the Sun or active cool stars. And, pulsation appeared to be a likely candidate for the variability.

3. Spatially resolved ultraviolet imaging

The advent of the *Hubble Space Telescope* (*HST*) and its Faint Object Camera (*FOC*) with spatial sampling of 14.35 mas per pixel made a direct image of *Betelgeuse* possible in the near-ultraviolet continuum λ2550 (Fig. 1). The ultraviolet diameter of the star turned out to be about 2.2 times larger at 108$\pm$4 mas than the optical diameter (Gilliland & Dupree 1996) directly indicating an extended chromosphere. Most dramatically, one unresolved bright spot appeared on the disk. It occupied 10% of the area and produced

20% of the ultraviolet flux from the star. The spot is hot and at least 200K or more, warmer than the surrounding material at 5000K in the low chromosphere which forms the ultraviolet continuum at λ2550.

Spectra in the UV were obtained with the GHRS instrument on HST and revealed the rotation of the star as shown in Fig. 1. The chromospheric hot spot appears to be located over the pole of the star. Uitenbroek *et al.* (1998) find a rotation period of 17 years based on an optical radius of 770$R_\odot$. A more recent estimate of the radius of *Betelgeuse* of 950$R_\odot$ (Harper *et al.* 2008) leads to a longer period of 21 yr. Yet both periods appear short for a large evolved star, although not out of the question for certain scenarios (see discussion in Uitenbroek *et al.* 1998).

Subsequent imaging with the *FOC* (Dupree *et al.* 2010) spanning a period of ~4 yr shows arc-like brightenings at times, located near or around the pole, and the spots number at most ~3 on the ultraviolet surface (Lobel & Dupree 2000; Dupree *et al.* 2010)

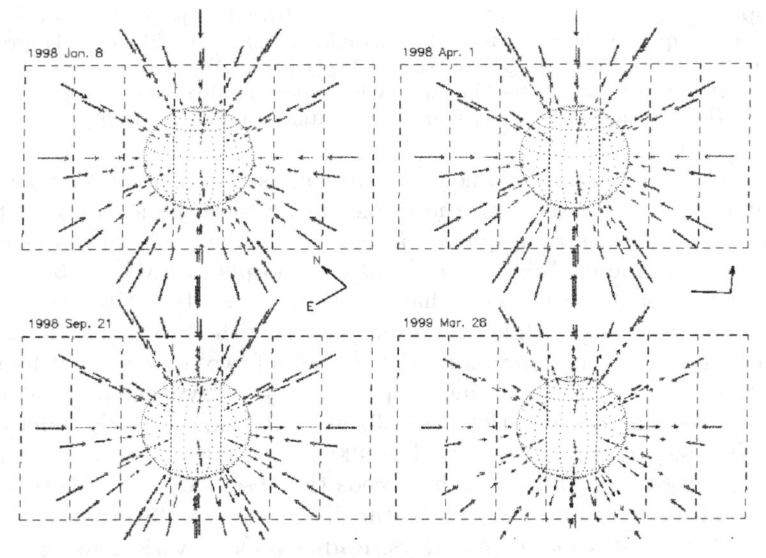

Figure 2. Velocity fields in the low chromosphere of *Betelgeuse* at 4 successive times. Arrows indicate the direction and magnitude of the material flow as inferred from measurements and semi-empirical non-LTE models of the Si I line profile (From: Lobel & Dupree 2001).

4. Spatially resolved UV spectroscopy: chromospheric motions

Spatially resolved UV spectroscopy with the STIS slit (25×100 mas) revealed the dynamics of the supergiant's atmosphere (Lobel & Dupree 2001). Ultraviolet spectra were obtained using a scanning motion in several directions across the extended stellar image of *Betelgeuse.* The raster pattern was executed 4 times over a time span of 15 months. The Si I λ2516 line was measured and found to display varying asymmetries with position and in time; these profiles indicate the presence of outflows and inflows of material. The Si I line was modeled in non-LTE formulation which included the presence of velocity fields. Fig. 2 shows a three-dimensional representation of the motions in the extended chromosphere. The global downflow observed over the first four months reverses into subsonic outflow and this outflow is enhanced in the last observation, 15 months after the first. Clearly nonradial movements of the chromosphere are present in confined regions as a part of the general chromospheric oscillations. The upper chromosphere, as

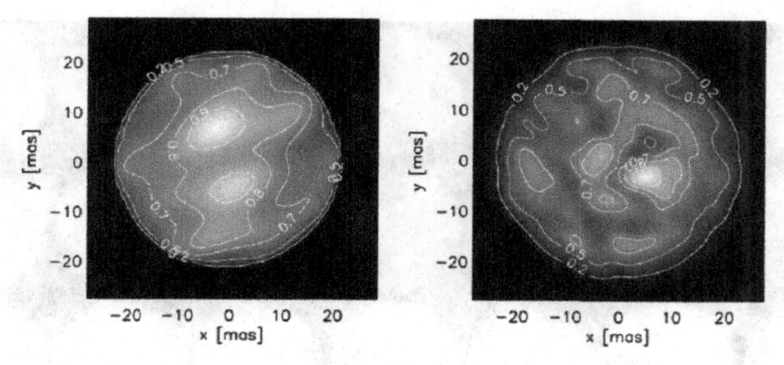

Figure 3. IR observations and simulation from Chiavassa *et al.* (2010). *Left panel:* H-band image of photosphere obtained with IOTA reconstructed from Haubois *et al.* (2009) showing the presence of two bright spots. *Right panel:* Snapshot from the 3D radiative hydrodynamic simulations of Chiavassa *et al.* (2010) which best matches the H-band image.

probed with the Mg II k line (formed higher in the atmosphere) shows a clear outflow when profiles sample the extended disk (Uitenbroek *et al.* 1998). Thus a warm (~8000K) wind emanates from the chromosphere.

5. Interferometric observations

Centimeter-wave observations of *Betelgeuse* with the VLA also detected one bright spot in the chromosphere (Lim *et al.* 1998; Harper & Brown 2003) which is believed to have enhanced temperature as well (Harper *et al.* 2001). These radio observations probe the cool component of the chromosphere (~3800K) but no velocity measures are possible from the continuous emission.

Interferometers have recently achieved phase closure in the infrared allowing spatially resolved imaging. Ohnaka *et al.* (2009) imaged *Betelgeuse* using AMBER with the VLTI in the K-band and also lines of CO in January 2008. At the time of the observations, the photosphere as measured in K-band appeared as a uniform limb-darkened disk, but the CO lines, formed higher up in the atmosphere show inhomogeneous structures across the surface, and differing velocity patterns of inflow and outflow with amplitudes of 10–15 km s^{-1}. AO narrow-band imaging with the VLT/NACO instrument between 1.04-2.17μm was carried out in January 2008 (Kervella *et al.* 2009). Selection of good images (the 'lucky' approach) revealed an asymmetric envelope and a bright plume extending out to 6 stellar radii, possibly originating from the pole of the star. These observations again document the asymmetric irregular outer atmosphere of *Betelgeuse* as found earlier in H-α speckle images (Hebden *et al.* 1986). H-band (1.86μm) interferometry with IOTA in October 2005 revealed two bright spots on the temperature-minimum surface of *Betelgeuse* (Haubois *et al.* 2009). These infrared measures probe the surface structures in the photosphere where convective motions are thought to occur.

6. Hydrodynamic modeling of betelgeuse: star in a box

Simulations of whole star convection ('Star in a Box') appear to confirm the presence and scale of the large convective cells in the photosphere of *Betelgeuse*. Simulations of time sequences of a supergiant are used to generate intensity maps from which the parameters measured by interferometry (visibility amplitudes and phases) are constructed (Chiavassa *et al.* 2009, 2010). These simulations give reasonable agreement with the interferometric results (Fig. 3) producing a few bright cells on the stellar surface. An attempt (Steffen & Freytag 2007) to evaluate the appearance of convection in a rotating

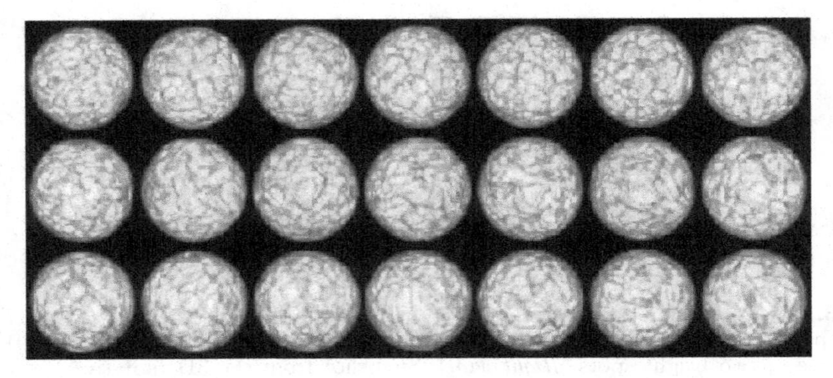

Figure 4. Simulations of global convection in a rotating reference frame from Steffen & Freytag (2007) showing a 3 hour time series of the appearance of the photosphere. The idealized model consists of a scaled-down Sun with parameters similar to those of a red supergiant, but here with a rigid rotating core. The top row is a non-rotating calculation. The middle row displays the star viewed at the pole. The lower row represents the star as viewed at the equator. Large scale structures develop in the photosphere in the rotating model.

star with a rigidly rotating core by assuming a ratio of convective scale to stellar radius similar to a supergiant, shows the presence of large scale features (Fig. 4). If the model is made fully convective (as in a real supergiant) the meridional flow on the photospheric surface moves poleward above 60° latitude and towards the equator at lower latitudes. All of these calculations assume a temperature structure that decreases outwards; there is no additional heating to produce a chromosphere. Thus while comparison to the infrared imaging observations is valid, it is not appropriate to use these models to interpret ultraviolet observations. High resolution spectroscopy of *Betelgeuse* in photospheric absorption lines (Gray 2008) suggests motions consistent with convection. The depths of neutral lines (V I, Fe I, and Ti I) show a pattern as a function of line core velocity that can be interpreted an increase in temperature of a volume of gas followed by an outflow (rising of hot cells), then followed by cooling and a falling back of the cooler material.

We need to be mindful that the oft-cited conjecture of Schwarzschild (1975) as confirmation of convective cells on the photosphere of a supergiant, really does not apply in detail as Uitenbroek *et al.* (1998) emphasized. The scalings used by Schwarzschild predicted 90 giant convective cells on a supergiant at one time, with a lifetime of 150 to 2000 days. In fact, photospheric observations (and the simulations) generally show 1-3 bright spots with lifetimes of 21-90 days. (Energy transport by convection stops at the photosphere, so this provides no insight into the upper layers in the chromosphere.) Numerical simulations currently provide the best support for convection as the source of photospheric spots.

7. So what *are* the spots on Betelgeuse?

In the photosphere (measured via the optical and infrared spectral regions), the large bright structures appear to be caused by convective motions, and these can be reproduced through a variety of radiative hydrodynamic simulations. Photospheric lines do not exhibit a predominant outflow, thus the convective cell motions do not initiate a mass outflow from the photosphere.

To understand the bright hot regions in the chromosphere, a calculation by Asida & Tuchman (1995) is relevant. They argued that convective photospheric motions and pulsations in the photosphere would generate shock waves in the chromosphere and act to extend the upper atmosphere, facilitating outflow. And a steady warm outflow occurs at the level of the chromospheric Mg II k line (Uitenbroek *et al.* 1998). Rotation causes preferential break-out of the shockwave near the stellar pole. This occurs because rotation of an extended atmosphere causes shock waves traveling towards the pole to encounter a shorter path length and a steeper density gradient than in the equatorial direction. The amplitude of the wave towards the pole is larger, making the break-out region appear brighter. The discovery of a longitudinal magnetic field on *Betelgeuse* (Auriére *et al.* 2010) of ~1 G (similar to the Sun-as-a-star value) could aid in the channeling of ionized gas, in addition to offering the possibility of Alfvén wave acceleration and heating. The warm (8000K) wind signaled by the outflowing Mg II k line may be driven by waves too.

References

Asida, S. M. & Tuchman, Y. 1995, *Astrophys. J.*, 455, 286

Auriére, M., Donati, J.-F., Konstantinova-Antova, R., Perrin, G., Petit, P., & Roudier, T. 2010, *Astron. Astrophys*, 516, L2

Chiavassa, A., *et al.* 2010, *Astron. Astrophys*, 515, A12

Chiavassa, A., Plez, B., Josselin, E., & Freytag, B. 2009, *Astron. Astrophys*, 506, 1351

Dupree, A. K., Baliunas, S. L., Guinan, E. F., Hartmann, L., Nassiopoulos, G., & Sonneborn, G. 1987, *Astrophys. J.*, 317, L85

Dupree, A. K., Lobel, A., & Stefanik, R. 2010, to be submitted.

Dupree, A. K., Lobel, A., Young, P. R., Ake, T. B., Linsky, J. L., & Redfield, S. 2005, *Astrophys. J.*, 622, 629

Gilliland, R. L. & Dupree, A. K. 1996, *Astrophys. J.*, 466, L29

Gray, D. F. 2008, *AJ*, 135, 1450

Harper, G. M. & Brown, A. 2003, in: N. E. Piskunov, W. W. Weiss, & D. F. Gray (eds.), *Modelling of Stellar Atmospheres*, Proc. IAU Symposium No. 210 (San Francisco: ASP), CDROM, F11

Harper, G. M., Brown, A., & Guinan, E. F. 2008, *AJ*, 135, 1430

Harper, G. M., Brown, A., & Lim, J. 2001, *Astrophys. J.*, 531, 1073

Haubois, X. *et al.* 2009, *Astron. Astrophys*, 508, 923

Hebden, J. C. *et al.* 1986, *Astrophys. J.*, 309, 745

Kervella, P. *et al.* 2009, *A&A*, 504, 115

Lim, J., Carilli, C. L., White, S. M., Beasley, A. J., & Marson, R. G. 1998, *Nature*, 392, 575

Lobel, A. & Dupree, A. K. 2000, *Astrophys. J.*, 545, 454

Lobel, A. & Dupree, A. K. 2001, *Astrophys. J.*, 558, 815

Lovy, D., Maeder, A., Noels, A., & Gabriel, M. 1984, *Astron. Astrophys*, 133, 307

Ohnaka, K., et al. 2009, *Astron. Astrophys*, 503, 183

Schwarzschild, M. 1975, *Astrophys. J.*, 195, 137

Smith, M. A., Patten, B. M., & Goldberg, L. 1989, *AJ*, 98, 2233

Steffen, M. & Freytag, B. 2007, *Astron. Nachr.*, 328, 1054

Stothers, R. & Leung, K. C. 1971, *Astron. Astrophys*, 10, 290

Uitenbroek, H., Dupree, A. K., & Gilliland, R. L. 1998, *AJ*, 116, 2501

Discussion

K. STRASSMEIER: I'm a little bit confused about a rotation period now. What is the rotation period?

DUPREE: Yes, 17 years. That is a high value if you consider that Betelgeuse was about 15 solar masses on the main sequence. Assuming that it spins down as it evolves and becomes larger, a period of 17 years is a fast rotation. However, there may be reasons for this fast rotation. Theorists are inventive, and there have been suggestions that Betelgeuse may have swallowed its planets, so that the angular momentum of the star increased.

Remember that Betelgeuse, as far as we know , is a single star. There is no confirmed companion. Calculations of stellar structure have been made, and it may be possible to have changes in the interior that would speed up the star. Another possibility is that we have an uncertainty of perhaps a factor of two in our measurement. So maybe the period is longer, but currently it appears a little fast.

MARK GIAMPAPA: Are there observations of bisectors or inverse C shapes and of the amplitude?

DUPREE: Yes. David Gray has a paper in 2008 where he did that, and he contends that with his bisectors he can identify rising expanding convective cells in the photosphere and so that he – he would confirm what I have been saying about infrared observations of the infrared imaging. It's Gray 2008, and he has done that, and he does it very carefully. He does confirm that and does think he sees the signatures of hot rising elements and cool falling elements in his C. This is measuring bisectors of photospheric lines. You need the signal to noise of 1,000, and you need resolution of a 100,000, and you can measure the profile very carefully and infer the velocity fields in the photosphere.

The Physics of Sun and Star Spots
Proceedings IAU Symposium No. 273, 2010
D.P. Choudhary & K.G. Strassmeier, eds.

doi:10.1017/S1743921311015237

The Butterfly diagram leopard skin pattern

Maurizio Ternullo[1]

[1]INAF – Osservatorio Astrofisico di Catania,
v. S. Sofia 78, 95123 Catania, Italia
email: maurizio.ternullo@oact.inaf.it

Abstract. A time-latitude diagram where spotgroups are given proportional relevance to their area is presented. The diagram reveals that the spotted area distribution is higly dishomogeneous, most of it being concentrated in few, small portions ("knots") of the Butterfly Diagram; because of this structure, the BD may be properly described as a cluster of knots. The description, assuming that spots scatter around the "spot mean latitude" steadily drifting equatorward, is challenged. Indeed, spots cluster around at as many latitudes as knots; a knot may appear at either lower or higher latitudes than previous ones, in a seemingly random way; accordingly, the spot mean latitude abruptly drifts equatorward or even poleward at any knot activation, in spite of any smoothing procedure. Preliminary analyses suggest that the activity splits, in any hemisphere, into two or more distinct "activity waves", drifting equatorward at a rate higher than the spot zone as a whole.

Keywords. Sun: activity, sun: magnetic fields, sun: sunspots

---

1. Introduction

Throughout a century, the Butterfly Diagram (BD) bidimensional character has been described as the scattering of spots around a mean latitude, which has been assumed to *steadily* drift equatorward, through the center of the butterfly wings. As a consequence, most theoretical works are aimed at predicting the location and evolution of such a latitude. On the other hand, Ternullo (2007a,b,c) found, from the examination of cycles 20 to 22, that the trace of the spot zone centroid results – in any hemisphere – from the quasi-biennial alternation of high-speed prograde phases with stationary or even retrograde phases, the average duration of the latter phases amounting to $\approx$35% of the cycle total duration. More recently, Ternullo (2008, 2010a,b) has shown that most of the spotted area is concentrated in small portions ("*knots*") of the BD, giving it a *"leopard skin"* aspect. Knots gather into two main streams per semicycle, drifting equatorward at a rate higher than the spot zone as a whole.

The present work is based on data sets compiled at the Royal Greenwich Observatory, integrated with data compiled by the US Air Force Solar Observing Optical Network and the National Oceanic and Atmospheric Administration. The valuable work of merging different archives was performed by Hathaway *et al.* (2003).

2. Results

For each one of the 1696 Carrington rotations from the 330th to the 2025th and for each one of the 84 1°−wide latitude strips in the interval −42, +42°, the spotted area average values have been computed. The resulting figures are the elements of an 84 × 1696 array representing the quantitative counterpart of the "short line" diagram drawn by Maunder. This array has been smoothed by a triangular running window covering 5 Carrington rotations and visualized by means of level curves. A portion of the resulting diagram, concerned with cycles 15 to 17, is shown in Figure 1. A glance at this Figure reveals that the spotted area is distributed in butterfly wings with remarkable dishomogeneity: small, overspotted portions (*"knots"*) of the BD give it a *"leopard skin"* aspect. Knots are the signature of *complexes of activity* (Bumba & Howard 1965; Gaizauskas *et al.* 1983) or *sunspot nests* (Castenmiller *et al.* 1986).

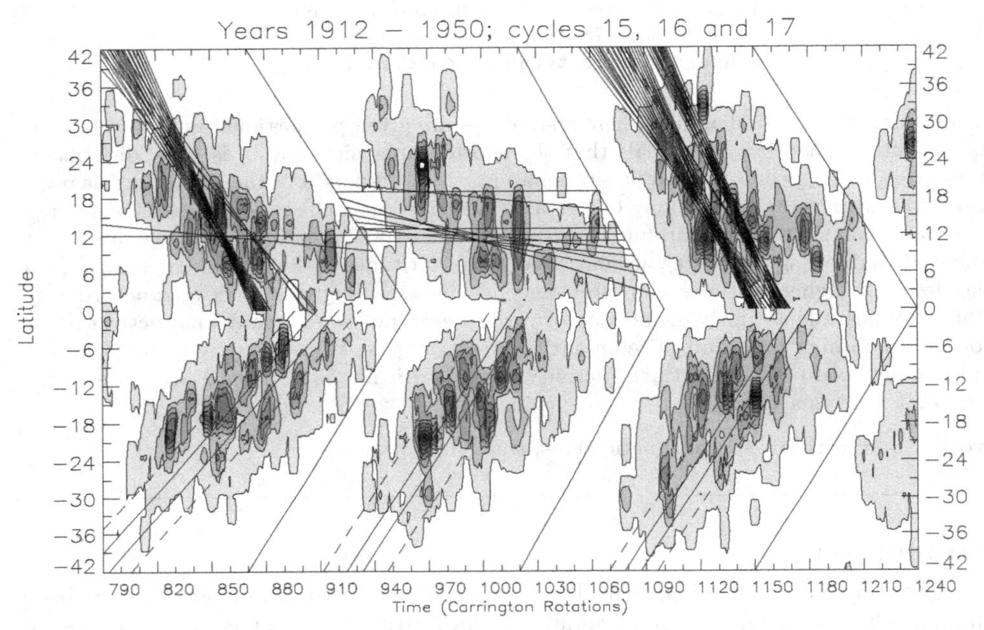

Figure 1. Butterfly Diagram for Carrington rotations 790–1300 (years 1912–1950). In any wing, the spotted area distribution is visualized by means of 10 level curves, dividing the range of spotted area values into equal intervals. The levels of grey qualitatively correspond to the spotted area density. The portion of the BD area lying between the two most external lines amounts to ≈70% of the butterfly total area, but hosts only ≈20% of the total spotted area. On the other hand, the 3rd level lines (circumscribing less than ≈15% of the butterfly total area) contain ≈50% of the total spotted area. Accordingly, the BD can be described as a cluster of small, highly concentrated spotted area aggregations (*knots*), distributed in a low-density spotted area population: this is the *leopard skin* pattern. In any southern-hemisphere wing, a couple of solid, oblique lines, 3 to 5° from each other, marks the *"depletion channel"*; this is a band where the spotted area density is lower than in both adjacent comparison channels (dashed lines), both differences being significant at a level of significance not lower than $7\,\sigma$; since many triplets of channels fulfil these statistical requirements, only the triplet associated with the most significant differences has been depicted for any wing. As regards the northern hemisphere wings, any triplet of channels fulfilling the same statistical requirements is schematically represented by the line passing through its center

Figure 1 shows that a knot may appear at latitudes either higher or lower than those of previous ones. Accordingly, because of the BD fragmentation into knots, and of their seemingly random occurrence, no continuous, steadily drifting equatorward line can represent the "mean latitude of spots": spots do not scatter on a line of this kind, but on *as many* latitudes as the knots.

Moreover, a glance at Figure 1 shows that, in cycles 15 s.h., 16 s.h. and 17 s.h., knots seem arranged into two oblique, discontinuous, roughly parallel streams, between which an underspotted band may be recognized. In order to get an objective definition

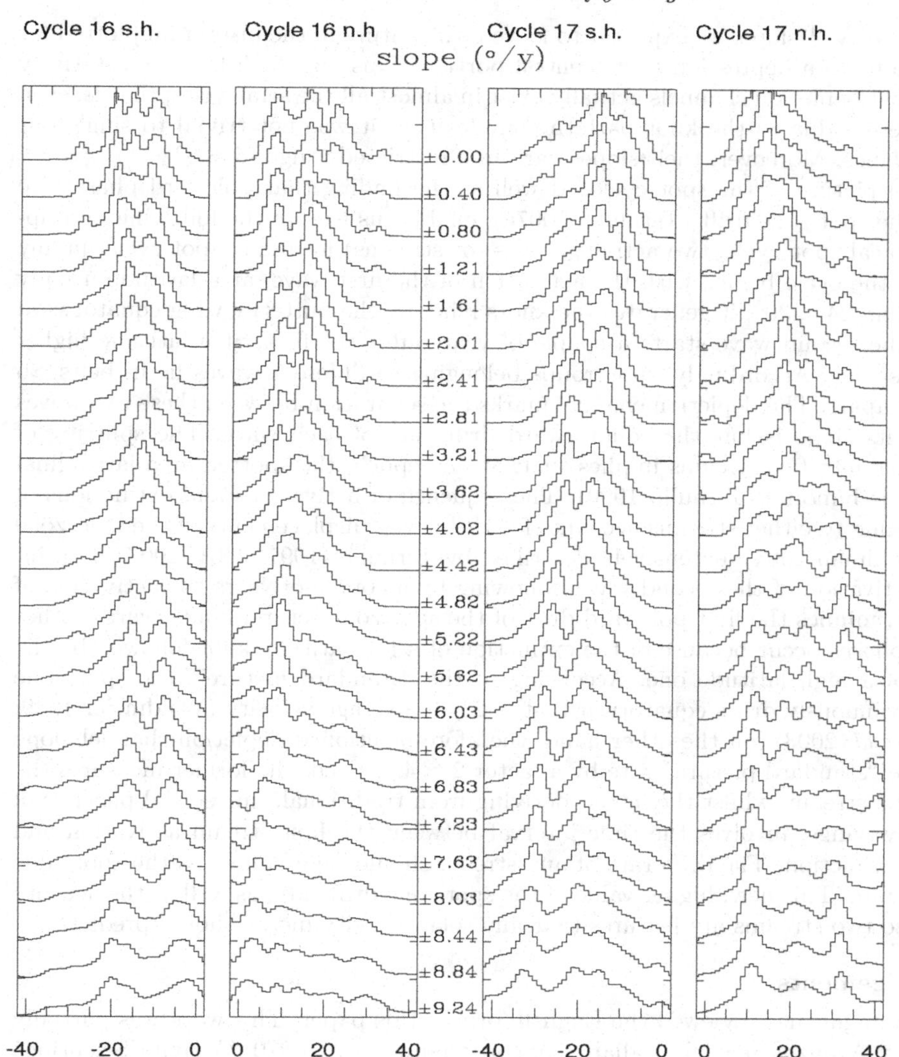

Figure 2. Histograms showing the spotted area distribution in sets of parallel, oblique, 1°-wide elementary bands covering the cycle 15 and 16 butterfly diagram. The slope scans the range [0, $\pm 9.24°\mathrm{y}^{-1}$] (slopes are positive for the s.h. and negative for the n.h.). Sequentially examining histograms related to a given semicycle, it is easy to find that a small dip becomes a sharper and sharper depression until a special slope value is attained; for further slopes, the inverse process occurs.

of this band, I have adopted the procedure described in the following. I have virtually superimposed on any wing a set of n parallel, oblique lines, 1° from each other, so as to completely cover the wing surface with $n-1$ *"elementary bands"*. Their slope is allowed to vary at small steps (0.03° per Carrington rotation (CR), equivalent to $0.402°\mathrm{y}^{-1}$), so as to finely scan the range [0; $9.24°\mathrm{y}^{-1}$] in the s.h. and [0; $-9.24°\mathrm{y}^{-1}$] in the n.e. For any slope, the spotted areas, contained inside each elementary band, have been summed and the resulting sequence of sums (a sum for any elementary band) has been visualized by a histogram. Accordingly, a collection of histograms (a histogram for any slope) has been obtained for any wing (Figure 2). If an underspotted band actually crosses a wing with the unknown $\sigma^\circ\,y^{-1}$ slope, we expect it to leave its signature in the histogram plotted for this slope value as a depression in its central portion. This approach has quantitatively confirmed that depletion channels actually exist in almost all the semicycles; that is, even in cases where – due to the knot pattern complexity – it was not trivial to single out them at a glance. Moreover, their slopes vary in a restricted range ($4 \approx 8^\circ\,y^{-1}$).

This novel picture of the spot cycle, as well as the finding that poleward phases are present in the spot zone drift (Ternullo, 2007a), enables us to give the following description of a typical spot cycle: two activity waves (or streams) form any spot cycle, in any hemisphere; the cycle begins with the activation of the first wave, at a latitude usually not larger than $24 \approx 30^\circ$; it generates the knots lying by the butterfly wing equatorward boundary; the second wave starts a couple of years after the first, at a latitude higher than the first one. Accordingly, spotgroups belonging to different waves lie in belts $\approx$6 through 10° apart. The depletion channel marks the separation between these two waves of activity; its slope is but the equatorward drift rate of each wave. The splitting of the spot cycle into two streams implies that, at any epoch, the spotted area latitudinal distribution is bimodal (Ternullo 1990). The sequence of activations and extinctions of knots belonging to either streams accounts for the zigzag displacements of the spot zone (Norton & Gilman 2004), extensively described by Ternullo (1997, 2001, 2007a,b,c): indeed, the activation of the second wave, following by a couple of years the activation of the first one, mimics the first poleward drift of the spot zone centroid; afterwards, other retrograde phases occur because of the extinction of a low latitude knot followed by the activation of a high latitude one. According to the "standard picture", the spot zone centroid continuously drifts equatorward, at $\approx 2^\circ y^{-1}$ average velocity (Waldmeier 1939; Hathaway *et al.* 2003); on the other hand, the afore-mentioned depletion-channel slope exceedes the "standard picture" rate by a factor $2 \approx 4$. Indeed, the former measures the drift rate of a stream, whilst the latter, deriving from traditional, unresolved pictures of the butterfly wings, involves the time interval between the first stream activation and the second extinction. The drift rate of any stream should be assumed as the spot zone actual drift rate. This new, higher value of the spot zone drift rate, as well as the activity splitting into two streams are features that any solar activity model should predict.

Acknowledgements

Ms. L. Santagati has reviewed the English form of this paper. This work was partially supported by Agenzia Spaziale Italiana (Accordo num. I/023/09/0 "Attività Scientifica per l'Analisi Dati Sole e Plasma - Fase E2" siglato fra INAF ed ASI)

References

Bumba, V. & Howard, R., 1965, *Astrophys. J.*, 141, 1502
Castenmiller, M. J.M. *et al.*,1986, *Solar Phys.*, 105, 237
Gaizauskas, V. *et al.*, 1983 *Astrophys. J.*, 265, 1056
Hathaway, D. H., *et al.*, 2003, *Astrophys. J.*, 589, 665
Ternullo, M., 1990, *Solar Phys.*, 127, 29
Ternullo, M., 1997 *Solar Phys.*, 172, 37
Ternullo, M., 2001a *Mem. S.A.It.* 72, 565
Ternullo, M., 2001b *Mem. S.A.It.* 72, 681
Ternullo, M., 2007a *Solar Phys.*, 240, 153
Ternullo, M., 2007b *Mem. S.A.It.*, 78, 596
Ternullo, M., 2007c *Astron. Nachr.*, 328, 1023
Ternullo, M., 2008 in *Electronic Proceedings - 12th European Solar Physics Meeting, 8-12 September 2008 - Freiburg, Germany: http://espm.kis.uni-freiburg.de*

Ternullo, M., 2010a, *Astrophys. Sp. Science*, 328, 301; DOI: 10.1007/s10509-010-0270-9
Ternullo, M., 2010b, *Mem. S.A.It. Suppl.* 14, 202
Ternullo, M. 2010c, *Advances in Plasma Astrophysics, Proceedings of the International Astronomical Union, IAU Symposium*, Volume 274, 195
Waldmeier, M., 1939, *Astron. Mitt Zürich*, 14, 470

The Physics of Sun and Star Spots
Proceedings IAU Symposium No. 273, 2010
D.P. Choudhary & K.G. Strassmeier, eds.

doi:10.1017/S1743921311015249

Turbulence and magnetic spots at the surface of hot massive stars

Matteo Cantiello[1], Jonathan Braithwaite[1], Axel Brandenburg[2,3], Fabio Del Sordo[2,3], Petri Käpylä[2,4] and Norbert Langer[1]

[1]Argelander-Institut für Astronomie der Universität Bonn, Auf dem Hügel 71, D–53121 Bonn, Germany email: cantiello@astro.uni-bonn.de; [2]NORDITA, AlbaNova University Center, Roslagstullsbacken 23, SE-10691 Stockholm, Sweden; [3]Department of Astronomy, AlbaNova University Center, Stockholm University, SE–10691 Stockholm, Sweden; [4]Department of Physics, Gustaf Hällströmin katu 2a (PO Box 64), FI-00014, University of Helsinki, Finland

Abstract. Hot luminous stars show a variety of phenomena in their photospheres and in their winds which still lack clear physical explanations at this time. Among these phenomena are non-thermal line broadening, line profile variability (LPVs), discrete absorption components (DACs), wind clumping and stochastically excited pulsations. Cantiello *et al.* (2009) argued that a convection zone close to the surface of hot, massive stars, could be responsible for some of these phenomena. This convective zone is caused by a peak in the opacity due to iron recombination and for this reason is referred to as the "iron convection zone" (FeCZ). 3D MHD simulations are used to explore the possible effects of such subsurface convection on the surface properties of hot, massive stars. We argue that turbulence and localized magnetic spots at the surface are the likely consequence of subsurface convection in early type stars.

Keywords. Convection, hydrodynamics, waves, stars: activity, stars: atmospheres, stars: evolution, stars: magnetic fields, stars: spots, stars: winds, outflows

---

1. Introduction

During their main sequence evolution, massive stars can develop convective regions very close to their surface. These regions are caused by an opacity peak associated with iron ionization. Cantiello *et al.* (2009) found a correlation between the occurrence and properties of subsurface convection, and microturbulence at the surface of hot massive stars. This correlation has been recently corroborated by new observations of microturbulence in massive stars (Fraser *et al.* 2010). Moreover there is growing evidence that the FeCZ is responsible for the observed solar-like oscillations at the surface of OB stars (Belkacem *et al.* 2009; Degroote *et al.* 2010). These observations seem to confirm the occurrence of such a convective region and its importance for the surface properties of early type stars.

2. 3D MHD Simulations of subsurface convection

The transport of energy by convection in the FeCZ is relatively inefficient. Radiation dominates and transports more than 95% of the total flux. The convective layer is very close to the photosphere, above which strong winds are accelerated. In rotating stars, the associated angular momentum loss might also drive strong differential rotation in these layers.

We perform 3D MHD simulations of the FeCZ. We use a setup similar to the one of Käpylä *et al.* (2008). This is described in more detail in Cantiello *et al.* (2010). As a preliminary study we perform simulations with modest resolution, where the density contrast between the bottom of the convective layer and the top of the domain is only ~20. This is about ten times smaller than in the case of the FeCZ. Moreover, the ratio of the convective to radiative flux is about 0.3, higher than in the FeCZ case. Therefore, at this stage, the velocities of convective motions cannot be directly compared with those of more realistic models, even though they still use mixing length theory. However, already in these preliminary runs we could follow the excitation and propagation of gravity waves above the convective region. Energy is transported up to the top layer by gravity waves, where the maximum of the energy is deposited at those wavelengths that are resonant with the scale of convective motions, as predicted, for example, by Goldreich & Kumar (1990). Käpylä *et al.* (2008) found excitation of a large scale dynamo in simulations of turbulent convection including rotation and shear. Our computational setup is very similar, so it is not surprising that we confirm this result. Dynamo action reaching equipartition is found in our simulations that include shear and rotation, with magnetic fields on scales larger than the scale of convection.

3. Surface turbulence

Microturbulence measures the amplitude of plasma motions that are of non-thermal origin and have a correlation length smaller than the region of line formation. In spectroscopy the microturbulence parameter needs to be estimated to derive consistent surface abundances for one element from different photospheric absorption lines through stellar model atmospheres (among many others Rolleston *et al.* 1996; Hibbins *et al.* 1998). It is unfortunate that so far microturbulence has always been used as a fudge-factor, as its physical origin is not understood. However Cantiello *et al.* (2009) suggested that the presence of convection below the surface of hot massive stars could explain turbulence in their photospheres, as measured by the microturbulence parameter. In fact a

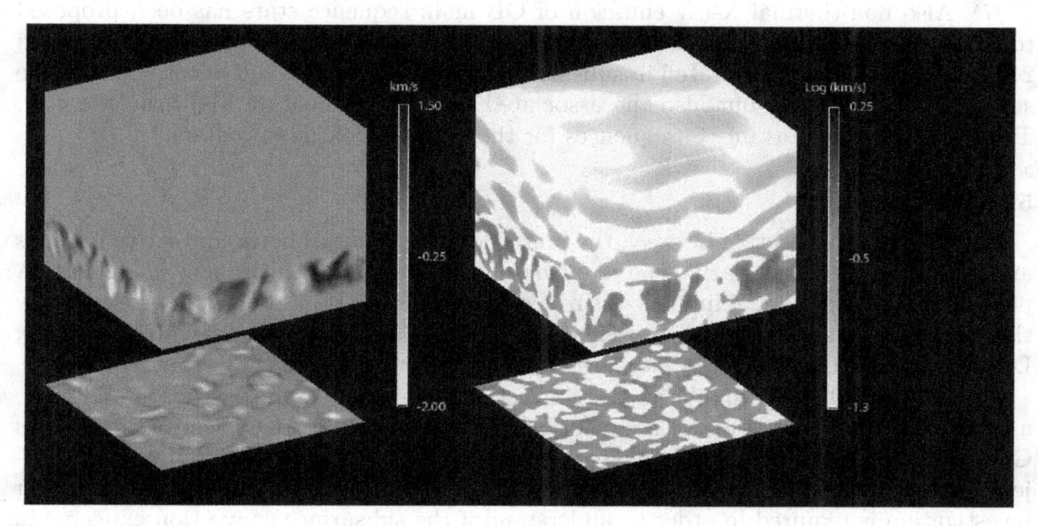

Figure 1. Simulation of subsurface convection. Starting from the top, the computational box is divided into three layers: a radiative layer with an upper cooling boundary, a convectively unstable layer and another stable layer at the bottom. The left panel shows a snapshot of the vertical velocity field in the simulation. In the right panel we show, for the same calculation, the logarithm of the vertical velocity field with $v_z > 0$. The plane below each box show the vertical velocity field at the lower boundary of the convective layer. The right box shows the gravity waves propagating in the radiative layer. Due to the low resolution of these preliminary runs, the amplitude of the convective velocities and of gravity waves can not be directly compared to the stellar case.

clear correlation has been found between the presence and properties of subsurface convection and the amplitude of microturbulent velocities in the photospheres of early type stars. Using our 3D MHD simulations we can study the excitation of gravity waves and their propagation. Such waves are excited in the convection zone and, once they reach the surface, they could produce the observed microturbulent velocity field. Further work still needs to be done, but ideally our simulations might lead to a situation in which the microturbulence is no longer a fudge-parameter, but a function of the stellar parameters.

4. Magnetic spots

The occurrence of convection zones close to the surface of hot massive stars opens a new scenario: If dynamo action is excited in the FeCZ, magnetic fields can be readily produced in the envelopes of OB stars. Such magnetic fields could reach the surface due to magnetic buoyancy. Following Cantiello *et al.* (2009) and supported by our 3D MHD simulations, we can assume that magnetic fields at equipartition level are generated in the FeCZ. This means that the magnetic energy density is equal to the kinetic energy density, giving amplitudes up to 3kG. Such fields may reach the stellar surface and result in localized magnetic spots. Details of how the magnetic fields produced in the subsurface convection may reach the stellar surface and with which amplitude, will be discussed in a forthcoming paper (Cantiello & Braithwaite, in prep.). Such magnetic spots, if they exist, could have remarkable effects on observable properties of early type stars. Surface magnetic fields have been linked to several observed phenomena in OB stars, e.g. discrete absorption components (DACs) in UV resonance lines (e.g., Prinja & Howarth 1988; Massa *et al.* 1995; Kaper *et al.* 1997; Prinja *et al.* 2002), which are thought to diagnose large scale coherent wind anisotropies (Cranmer & Owocki 1996; Lobel & Blomme 2008), or the less coherent line profile variability (Fullerton *et al.* 1996, 1997). Also non-thermal X-ray emission of OB main sequence stars has been proposed to relate to surface magnetic fields (e.g., Babel & Montmerle 1997; ud-Doula & Owocki 2002). Magnetic fields generated in subsurface convective zones could affect not only the stellar wind mass loss, but also the associated angular momentum loss from the star. This could have important consequences for the evolution of massive stars.

5. Discussion

An intriguing connection between the presence of sub-photospheric convective motions and microturbulence in early-type stars has been found by Cantiello *et al.* (2009). A picture in which the FeCZ influences surface properties of OB stars is supported also by the recent discovery of solar-like oscillations in early type stars (Belkacem *et al.* 2009; Degroote *et al.* 2010) and new measurements of microturbulence (Fraser *et al.* 2010).

We perform 3D MHD simulations of convection to investigate the excitation and propagation of gravity waves above a subsurface convection zone. Analytical predictions of Goldreich & Kumar (1990) on the spatial scale at which the maximum of energy is injected in gravity waves seem to be confirmed by our preliminary calculations. Further investigation is required in order to understand if the subsurface convection expected in OB stars excites gravity waves of the required amplitude to explain the observed microturbulence in massive stars. In particular we need higher resolution to increase the Reynolds number of our simulations and to be able to decrease the ratio of convective to radiative flux, which is an important parameter in determining the convective velocities (Brandenburg *et al.* 2005).

Simulations of turbulent convection in the presence of rotation and shear, show dynamo action with magnetic fields reaching equipartition (Käpylä *et al.* 2008). Since massive stars are usually fast rotators, perhaps the interplay between convection, rotation and shear is able to drive a dynamo in OB stars. Indeed our simulations of subsurface convection including rotation and shear show dynamo-generated magnetic fields with equipartition values. This means that fields of ~kG could be present in the FeCZ. These magnetic fields might experience buoyant rise and reach the surface of OB stars, where they could have important observational consequences. In particular it has already been suggested that the discrete absorption components observed in UV lines of massive stars could be produced by low amplitude, small scale magnetic fields at the stellar surface (Kaper & Henrichs 1994). We will discuss the emergence and appearance of localized magnetic spots at the surface of hot massive stars in a forthcoming paper (Cantiello & Braithwaite, in prep.).

References

Babel, J. & Montmerle, T. 1997, *Astrophys. J.*, 485, L29

Belkacem, K., Samadi, R., Goupil, M., *et al.* 2009, *Science*, 324, 1540

Brandenburg, A., Chan, K. L., Nordlund, Å., & Stein, R. F. 2005, *Astron. Nachr.*, 326, 681

Cantiello, M., Braithwaite, J., Brandenburg, A., *et al.* 2010, ArXiv e-prints

Cantiello, M., Langer, N., Brott, I., *et al.* 2009, *Astron. Astrophys.*, 499, 279

Cranmer, S. R. & Owocki, S. P. 1996, *Astrophys. J.*, 462, 469

Degroote, P., Briquet, M., Auvergne, M., *et al.* 2010, *Astron. Astrophys.*, 519, A38

Fraser, M., Dufton, P. L., Hunter, I., & Ryans, R. S. I. 2010, *Mon. Not. Roy. Astron. Soc.*, 404, 1306

Fullerton, A. W., Gies, D. R., & Bolton, C. T. 1996, *Astrophys. J. Suppl.*, 103, 475

Fullerton, A. W., Massa, D. L., Prinja, R. K., Owocki, S. P., & Cranmer, S. R. 1997, *Astron. Astrophys.*, 327, 699

Goldreich, P. & Kumar, P. 1990, *Astrophys. J.*, 363, 694

Hibbins, R. E., Dufton, P. L., Smartt, S. J., & Rolleston, W. R. J. 1998, *Astron. Astrophys.*, 332, 681

Kaper, L. & Henrichs, H. F. 1994, *Astrophys. Space Sci.*, 221, 115

Kaper, L., Henrichs, H. F., Fullerton, A. W., *et al.* 1997, *Astron. Astrophys.*, 327, 281

Käpylä, P. J., Korpi, M. J., & Brandenburg, A. 2008, *Astron. Astrophys.*, 491, 353

Lobel, A. & Blomme, R. 2008, *Astrophys. J.*, 678, 408

Massa, D., Fullerton, A. W., Nichols, J. S., *et al.* 1995, *Astrophys. J.*, 452, L53

Prinja, R. K. & Howarth, I. D. 1988, *Mon. Not. Roy. Astron. Soc.*, 233, 123

Prinja, R. K., Massa, D., & Fullerton, A. W. 2002, *Astron. Astrophys.*, 388, 587

Rolleston, W. R. J., Brown, P. J. F., Dufton, P. L., & Howarth, I. D. 1996, *Astron. Astrophys.*, 315, 95

ud-Doula, A. & Owocki, S. P. 2002, *Astrophys. J.*, 576, 413

The Physics of Sun and Star Spots
Proceedings IAU Symposium No. 273, 2010
D.P. Choudhary & K.G. Strassmeier, eds.

doi:10.1017/S1743921311015250

Velocity fields in and around sunspots at the highest resolution

Carsten Denker and Meetu Verma

Astrophysikalisches Institut Potsdam, An der Sternwarte 16, D-14482 Potsdam, Germany
email: cdenker@aip.de and mverma@aip.de

Abstract. The flows in and around sunspots are rich in detail. Starting with the Evershed flow along low-lying flow channels, which are cospatial with the horizontal penumbral magnetic fields, Evershed clouds may continue this motion at the periphery of the sunspot as moving magnetic features in the sunspot moat. Besides these well-ordered flows, peculiar motions are found in complex sunspots, where they contribute to the build-up or relaxation of magnetic shear. In principle, the three-dimensional structure of these velocity fields can be captured. The line-of-sight component of the velocity vector is accessible with spectroscopic measurements, whereas local correlation or feature tracking techniques provide the means to assess horizontal proper motions. The next generation of ground-based solar telescopes will provide spectropolarimetric data resolving solar fine structure with sizes below 50 km. Thus, these new telescopes with advanced post-focus instruments act as a 'zoom lens' to study the intricate surface flows associated with sunspots. Accompanied by 'wide-angle' observations from space, we have now the opportunity to describe sunspots as a system. This review reports recent findings related to flows in and around sunpots and highlights the role of advanced instrumentation in the discovery process.

Keywords. Sun: atmospheric motions, Sun: photosphere, Sun: sunspots, Sun: magnetic fields

1. Probing the velocity fields in and around sunspots

New instruments and observing capabilities have advanced our knowledge about plasma motions in and around sunspots and their interaction with the magnetic fields. Therefore, by introducing current and future instruments for high-resolution studies of the Sun, we will set the stage for this review of the intricate flow fields of sunspots. The line-of-sight velocity can be derived from a Doppler-shifted spectral line profile, and its height dependence can be inferred by carefully selecting lines, which originate at different layers in the solar atmosphere, or by measuring the bisectors of spectral lines. Local correlation or feature tracking are the methods of choice in determining horizontal proper motions (see Fig. 1). In principle, spectroscopic and imaging techniques provide access to the three-dimensional velocity field.

Instruments commonly used to measure solar velocity fields can be placed into five broad categories ordered according to their spectral resolving power: (1) imaging with interference filters, (2) imaging with Lyot filters, (3) line-of-sight velocity and magnetic field measurements using filtergraphs, (4) imaging spectropolarimeters, and (5) long-slit spectrographs. We introduce observations obtained with the instruments[1,2,3,5] (superscripts indicate the instrument category) of the *Hinode* Solar Optical Telescope (SOT, Tsuneta *et al.* 2008), the Interferometric Bidimensional Imaging Spectrometer (IBIS[4], Cavallini 2006), the Crisp Imaging Spectropolarimeter (CRISP[4], Scharmer 2006), the Göttingen/GREGOR Fabry-Pérot Interferometer (GFPI[4], Bello González & Kneer 2008;

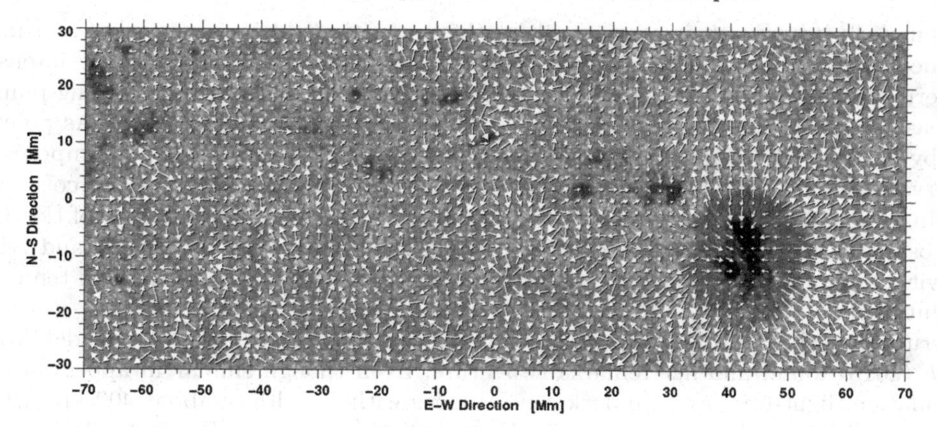

Figure 1. The Japanese *Hinode* mission placed for the first time a telescope into space, which can resolve the fine structure of sunspots. SOT feeds an instrument suite for high-resolution imaging and spectropolarimetry. These novel instruments provide access to the flow fields in and around sunspots. Displayed is a G-band image of active region NOAA 10921 observed near disk center on 2006 November 3. Direction and magnitude of horizontal flows are indicated by vectors. These high-resolution data set the stage for this review providing motivation for the next generation of ground-based telescopes and instruments.

Denker *et al.* 2010), the Tenerife Infrared Polarimeter (TIP[5], Collados *et al.* 2007), and the Polarimetric Littrow Spectrograph (POLIS[5], Beck *et al.* 2005).

The recent success of the Japanese *Hinode* mission (Kosugi *et al.* 2007) has certainly proven that high resolution observations pave the way for advancing solar physics. On the ground first observations have been obtained with the New Solar Telescope (NST, Denker *et al.* 2006; Goode *et al.* 2010), while the GREGOR solar telescope (Volkmer *et al.* 2010) awaits commissioning upon delivery of its primary mirror. With the Advanced Technology Solar Telescope (ATST, Rimmele *et al.* 2010) beginning construction and the European Solar Telescope (EST, Collados *et al.* 2010) finishing the design and development phase, further progress and new discoveries are expected when approaching the fundamental spatial scales for physical processes on the Sun.

A short review can never be complete but we at least strove to be up-to-date. Consequently, our approach had to be a very subjective one motivated by the imminent commissioning of the GREGOR solar telescope and the upcoming science demonstration time with the GFPI.

2. The intricate flow fields of simple sunspots

2.1. *Convective signatures in the umbra*

The existence of convective signatures in the umbra has been debated for many years. High-resolution observations with CRISP pinpointed strong upflows of up to 1.4 km s^{-1} in deep photospheric layers, which are associated with bright umbral dots (Ortiz *et al.* 2010). The height dependence of line-of-sight velocities was determined from the bisectors of coarsely sampled spectral lines. The umbral dots are identified with columns of hot, rising material with weaker and more inclined magnetic fields as compared to the umbral background. To establish overturning convective energy transport in the umbra, downflows have to exist, which were indeed observed (0.4–1.0 km s^{-1}) in confined areas at the periphery of umbral dots as predicted by theory. Some elongated umbral dots exhibit dark lanes in their centers, where the magnetic field is even weaker and more inclined. These umbral dots resemble 'coffee beans' because of this peculiar intensity pattern, which originates from the accumulation of material at the top of rising plumes increasing the density and elevating the opacity (Schüssler & Vögler 2006). As pointed out by Ortiz *et al.* (2010), the velocity gradient with height in umbral dots imposes an observational challenge, *i.e.*, the Doppler shifts are hard to observe in the line core – explaining the scarcity of reported umbral downflows, which are a phenomenon of the deep photosphere. In addition, small-scale features such as umbral dots change size and velocity within a few minutes, thus necessitating observations with high spatial and temporal resolution.

Scrutinizing light-bridges separating different umbral cores, Rouppe van der Voort *et al.* (2010) found another remarkable feature. Even though similar in appearance to granulation, light-bridges have dark central lanes with a width of about 400 km, where upflows reside, which are strongest in deep photospheric layers. Typical velocities are around 0.5 km s^{-1} but reach as high as 1.0 km s^{-1}. However, the direction of the flows is just opposite to the downflows in intergranular lanes. Similar to the dark lanes in umbral dots, the narrow, dark channels of light-bridges have weaker and more inclined magnetic fields.

2.2. *Penumbral fine structure and Evershed flow*

In the penumbra, the Evershed flow denotes a radial outflow of gas, which is aligned with more horizontally placed magnetic flux tubes in dark penumbral filaments. Theoretical models based on the thin flux tube approximation arrive at steady as well as time-dependent flows driven by pressure differences. The 'moving flux tube' model (*e.g*, Schlichenmaier 2002) explains much of the observed fine structure. However, 'steady siphon-flow' models still remain viable and provide strong arguments against super-Alfvénic, 'sea-serpent'-like flows (Thomas 2005), namely that they are gravitationally unstable.

The average vertical velocity field of the quiet Sun as observed by the *Hinode* spectropolarimeter (Franz & Schlichenmaier 2009) is always dominated by upflows, whereas the penumbra shows a different behavior. Here, upflows cover a larger area for Doppler velocities below 0.4 km s^{-1}, in particular in the inner penumbra. However, at velocities above 0.6 km s^{-1} downflows have a larger areal coverage. This leads to a net downflow of more than 0.1 km s^{-1} for the entire penumbra. The upflows of the inner penumbra are typically elongated an possess an aspect ratio of about five. In contrast, the largest downflows in the penumbra of up to 9 km s^{-1} at the outer penumbral boundary exceed even the largest quiet Sun values of about 3 km s^{-1} and have a shape closer to circular.

Penumbral grains migrate inwards in the inner penumbra. The Evershed flow begins at the leading edge of the penumbral grains (Ichimoto *et al.* 2007), which have been identified with the footpoints of hot upflows in strongly inclined flux tubes (Rimmele & Marino 2006). The Evershed flow then turns horizontal and follows preferentially the dark cores of the penumbral filaments (Scharmer *et al.* 2002), where the magnetic field is more horizontal. The striking dark-cored penumbral filaments are much easier to discern in polarized light than in continuum images (Bellot Rubio *et al.* 2007). Their spectral line profiles are very asymmetric, which hints at multiple magnetic field components along the line-of-sight or within the resolution element. The dark cores have a lateral extend of less than 200 km and exhibit magnetic fields, which are weaker by 100–150 G as compared to the lateral brightenings. The Evershed flow reaches velocities up to and even exceeding the photospheric sound speed of about 7 km s^{-1}. The Evershed flow shows temporal variations on scales of 10–15 min, which also corresponds to the intensity variations of Evershed clouds. Small patches of opposite polarity and strong downflows are observed throughout the outer penumbra indicating that some penumbral field lines already return to the interior well within the penumbra itself (Sainz Dalda & Bellot Rubio 2008). Selecting spectral lines with contribution functions covering the deep photosphere or bisector analysis are the means to determine the height dependence of the Evershed flow, which increases in strength with depth. The Evershed flow is not stationary. Coherent flow patches (Evershed clouds) can be traced from within the sunspot to move away from the spot (Cabrera Solana *et al.* 2006), where they can be associated with moving magnetic features (MMFs).

2.3. *Moat flow and moving magnetic features*

Martínez Pillet *et al.* (2009) report that the Evershed flow continues at least sporadically outside the penumbra into the sunspot moat, *i.e.*, not all field lines, which carry the Evershed flow, submerge below the photosphere at the penumbra's outer boundary. Analyzing spectropolarimetric *Hinode* data, Shimizu *et al.* (2008) find high-speed (supersonic) downflows by analyzing Stokes-V profiles. If such profiles become more complicated, *i.e.*, if they have multiple lobes, then using the zero-crossing of the Stokes-V profile is not a good indicator of flow velocities. This argues in favor of instruments capable of resolving spectral line profiles and not filtergraph systems, which deliver Dopplergrams and magnetograms prone to erroneous interpretation. The observed downflows occur in three distinct locations: (1) the outer penumbral boundary (MMFs), (2) the edge of the umbra in absence of penumbral structures, and (3) near small-scale field concentrations in the sunspot moat (convective collapse). These features have in common a pointlike appearance with diameters of about $1''$. They are transient features with lifetimes from a few minutes to about 30 min but on average the lifetimes tend towards the lower end of this range. On the other hand, in long-duration observations, filamentary magnetic features become visible in the sunspot moat (Sainz Dalda & Martínez Pillet 2005) revealing that MMFs preferentially move along certain pathways. Balthasar and Muglach (2010) present another interesting finding, namely that in the inner moat the flow velocities are higher in the ultra-violet (170 nm) as compared to visible (500 nm) continuum, while reversing this relationship in the outer moat.

Active region NOAA 10977 contained a bipolar group of pores, which never developed a penumbra, even though infrared Ca II λ854.2 nm observations revealed superpenumbral structures (a worthwhile research topic on its own). In a coordinated observing campaign with *Hinode* and IBIS, Zuccarello *et al.* (2009) detected short, radially aligned magnetic structures at the periphery of the pore, which however did not bear any resemblance to penumbral filaments. Consequently, in their absence no indications of the Evershed effect could be detected. Surprisingly, both moat flow and MMFs were surrounding the pore questioning their close ties to the Evershed flow. The presence of MMFs was interpreted as twisted magnetic field lines, which were pealed away from the vertical flux bundle of the pore by the (super)granular flow.

2.4. *Decay of sunspots*

A mechanism to remove magnetic flux from a sunspot is the interaction of penumbral filaments at the edge of the sunspot with the granulation, which erodes the magnetic field of the sunspot (Kubo *et al.* 2008). Flux detaches in the form of MMFs from the more vertical background fields of the uncombed penumbra, thus contributing to the decay of sunspots. Other types of MMFs, which are related to the magnetized Evershed flow, do not change the net magnetic flux. These dark penumbral filaments with strong horizontal magnetic fields often reach into the moat region. This goes along with a non-stationary penumbral boundary, which advances and retracts with respect to an average position.

In the last stage of sunspot decay, only a pore without a penumbra remains (Bellot Rubio *et al.* 2008), where small finger-like, weak, and almost horizontal magnetic features of opposite polarity can be recognized at the pore's boundary. They extend up to about 1.5 Mm and have blue-shifted Stokes profiles indicative of upflows with speeds of 1–2 km s^{-1}. This could be the remnants of magnetic field lines, which previously carried the Evershed flow. No longer held down by the mass provided by the Evershed flow, they become boyant and lift of to vanish in the chromosphere. In general, the question remains open, where penumbral flux tubes end. At least in the final stages of a decaying sunspot, alternatives might exist to the notion that they just bend below the surface at the periphery of the penumbra. This problem is also tied to the matter of mass continuity in the Evershed flow, since sources and sinks still need to be unambiguously identified and the balance between inflows and outflows has to be established. If the mass would be supplied by the flux rope rising through the convection zone, which initially led to the emergence of the sunspot, then decoupling from this flux rope would shut off the mass flow causing the sunspot to decay. It is noteworthy that the divergence line observed in the middle penumbra survives the decay process (Deng *et al.* 2007) and even the moat flow is still detectable long after the penumbra has vanished (Zuccarello *et al.* 2009).

3. Peculiar flows in the context of eruptive events

Flares occur near magnetic neutral lines, where strong magnetic field gradients exist, and where the horizontal component of the magnetic field is strongly sheared. The height dependence of horizontal proper motions can be derived by applying local correlation tracking techniques to images obtained in multiple spectral regions. Near-infrared images at the opacity minimum probe the deepest photospheric layers, whereas G-band images provide access to higher layers as compared to continuum images observed in the visible part of the solar spectrum. Deng *et al.* (2006) presented such a multi-wavelength study of active region NOAA 10486 – one of the most flare-prolific regions of solar cycle 23. Both horizontal and vertical shear flows with speeds of about 1 km s^{-1} exist in the vicinity of magnetic neutral lines. These flows are long-lived and persistent. Thus, the magnetic field might not be the only agent trigerring solar flares. The flow speed in shear regions is diminishing with height while the direction essentially the same. Therefore, shear flows are dominant in the deeper layers of the photosphere. In response to an X10 flare the shear flows significantly increased, which was interpreted as shear release in the overlying magnetic fields or as the emergence of twisted and sheared flux infusing energy from subphotospheric layers. Shear flows are not just limited to horizontal flow fields. In a spectropolarimetric study of a B7.8 flare in active region NOAA 10904, Hirzberger *et al.* (2009) detected (supersonic) downflows of up to 7 km s^{-1} in the penumbra. Twisting and interlaced penumbral filaments are no longer radially aligned with respect to the major sunspot and become almost tangential. However, the Evershed flow remains aligned with the magnetic field. Islands of opposite polarity and complex flows require a three-dimensional topology necessitating high spectral resolution to capture these features in multi-lobed Stokes-V profiles.

Twisted and sheared penumbral filaments, photospheric shuffling of footpoints, and rapid motion as well as rotation of sunspots can all destabilize the magnetic fields above an active region. In a multi-wavelength study of active region NOAA 10960, Kumar *et al.* (2010) find evidence that helical twist is accumulated before the flare and then activated to release the stored energy during the flare. The orientation of penumbral filaments can strongly deviate from the radial direction in satellite sunspots and δ-configurations. Photospheric signatures of the flare are the now well-established rapid penumbral decay and umbral enhancement (Liu *et al.* 2005), which are indicative of a rearranged magnetic field topology. In addition to these signatures, Gosain *et al.* (2009) noticed the lateral displacements of penumbral filaments during an X3.4 flare in active region NOAA 10930 on 2006 December 13. This occurs immediately before (4 min) the flare initiation, when penumbral filaments move laterally towards the magnetic neutral line, while changing direction and moving away from it for about 40 min after the flare. The energy involved in the lateral displacements is only a few percent of the total energy released in the flare.

4. Conclusions

Space observations are not only free from seeing but, depending on orbit, do not have to cope with the day night-cycle on Earth. Thus, in principle, high cadence observations with long coverage become possible to follow magnetic structures of active regions while they evolve. Such observations from space are only limited by on-board processing and telemetry. Despite the availability of long time-series with consitstent quality only few systematic and comprehensive studies of flow fields are available. The majority of investigations have been limited to case studies. This is an opportunity for the time yet to come to explore this enormous database with the aim to put the still fragmented puzzle of flows in and around sunspots together into a comprehensive picture. In this respect, high-cadence vector magnetograms of the recently launched Solar Dynamics Observatory (SDO) will certainly advance our knowledge regarding eruptive events on the Sun.

The new and upcoming instrumental capabilities on the ground will drastically improve angular resolution. Adaptive optics and future multi-conjugate adaptive optics will allow to capture time-series of spectropolarimetric data, which are longer than the lifetime of solar (small-scale) structures. The polarimetric sensitivity and spectral resolution of these instruments including infrared capabilities will produce multi-dimensional data sets suitable for advanced spectral inversion techniques. For the first time, physical quantities become accessible at the fundamental scales of physical processes on the Sun.

Acknowledgements

MV expresses her gratitude for the generous financial support by the German Academic Exchange Service (DAAD) in the form of a PhD scholarship. CD acknowledges a DAAD travel grant facilitating his attendance at the IAU Symposium.

References

Balthasar, H. & Muglach, K. 2010, *Astron. Astrophys* 511, A67
Beck, C., Schmidt, W., Kentischer, T., & Elmore, D. 2005, *Astron. Astrophys* 437, 1159
Bello González, N. & Kneer, F. 2008, *Astron. Astrophys* 480, 265
Bellot Rubio, L. R., Tritschler, A., & Martínez Pillet, V. 2008, *Astrophys. J.* 676, 698
Bellot Rubio, L. R., Tsuneta, S., Ichimoto, K., *et al.* 2007, *Astrophys. J. Lett.* 668, L91
Cabrera Solana, D., Bellot Rubio, L.R., Beck, C., & del Toro Iniesta, J.C. 2006, *Astrophys. J.* 649, L41
Cavallini, F. 2006, *Solar Phys.* 236, 415
Collados, M., Lagg, A., Díaz García, J.J., *et al.* 2007, *ASP Conf. Ser.* 368, 611
Collados, M., Bettonvil, F., Cavaller, L., *et al.* 2010, *Proc. SPIE* 7733, 77330H
Deng, N., Xu, Y., Yang, G., *et al.* 2006, *Astrophys. J.* 644, 1278
Deng, N., Choudhary, D.P., Tritschler, A., *et al.* 2007, *Astrophys. J.* 671, 1013.
Denker, C., Goode, P.R., Ren, D., *et al.* 2006, *Proc. SPIE* 6267, 62670A

Denker, C., Balthasar, H., Hofmann, A., *et al.* 2010, *Proc. SPIE* 7735, 77356M
Franz, M. & Schlichenmaier, R. 2009, *Astron. Astrophys* 508, 1453
Goode, P.R., Yurchyshyn, V., Cao, W., *et al.* 2010, *Astrophys. J. Lett.* 714, L31
Gosain, S., Venkatakrishnan, P., & Tiwari, S.K. 2009, *Astrophys. J. Lett.* 706, L240
Hirzberger, J., Riethmüller, T., Lagg, A., *et al.* 2009, *Astron. Astrophys* 505, 771
Ichimoto, K., Suematsu, Y., Tsuneta, S., *et al.* 2007, *Pub. Astron. Soc. Jap.* 318, 1597
Kosugi, T., Matsuzaki, K., Sakao, T., *et al.* 2007, *Solar Phys.* 243, 3
Kubo, M., Lites, B.W., Ichimoto, K., *et al.* 2008, *Astrophys. J.* 681, 1677
Kumar, P., Srivastava, A.K., Filippov, B., & Uddin, W. 2010, *Solar Phys.* 266, 39
Liu, C., Deng, N., Liu, Y., *et al.* 2005, *Astrophys. J.* 622, 722
Martínez Pillet, V., Katsukawa, Y., Puschmann, K.G., & Ruiz Cobo, B. 2009, *Astrophys. J. Lett.* 701, L79
Ortiz, A., Bellot Rubio, L.R., & Rouppe van der Voort, L. 2010, *Astrophys. J.* 713, 1282
Rimmele, T.R. & Marino, J. 2006, *Astrophys. J.* 646, 593
Rimmele, T.R., Wagner, J., Keil, S., *et al.* 2010, *Proc. SPIE* 7733, 77330G
Rouppe van der Voort, L., Bellot Rubio, L.R., & Ortiz, A. 2010, *Astrophys. J. Lett.* 718, L78
Sainz Dalda, A. & Bellot Rubio, L.R. 2008, *Astron. Astrophys* 481, L21
Sainz Dalda, A. & Martínez Pillet, V. 2005, *Astrophys. J.* 632, 1176
Scharmer, G.B. 2006, *Astron. Astrophys* 447, 1111
Scharmer, G.B., Gudiksen, B.V., Kiselman, D., *et al.* 2002, *Nature* 420, 151
Schlichenmaier, R. 2002, *Astron. Nachr.* 323, 303
Schüssler, M. & Vögler, A. 2006, *Astrophys. J. Lett.* 641, L73
Shimizu, T., Lites, B.W., Katsukawa, Y., *et al.* 2008, *Astrophys. J.* 680, 1467
Thomas, J.H. 2005, *Astron. Astrophys* 440, 29
Tsuneta, S., Ichimoto, K., Katsukawa, Y., *et al.* 2008, *Solar Phys.* 249, 167
Volkmer, R., von der Lühe, O., Denker, C., *et al.* 2010, *Proc. SPIE* 7733, 77330K
Zuccarello, F., Romano, P., Guglielmino, S.L., *et al.* 2009, *Astron. Astrophys* 500, L5

Discussion

AULANIER: Very nice observations. It's very interesting. I have a question which is maybe delicate. You talked at the end of your presentation about space weather forecasting. So what do you think these nice high resolution and small focused on one sunspot can teach us about space weather?

DENKER: Okay. What we can do is try to understand what's happening within the regions – say, along the neutral line what is really happening there. I mean, if you look at space weather forecasting, people look at the neutral line do some crazy averages over sheer flows, magnetic sheer angle, and so on.

But if you look at the quantities, they don't tell you much about whether there will be a flare or not. I think the reason for that is because you don't understand what happens, let's say, more at the microscopic level. If you can bring instruments like the GFPI to any of space weather forecasting system, I doubt it as well.

But it will tell you, if there are certain regions and active regions which are more important than others – and you might actually need different instruments to do that – you might not need the highest resolution once you know what feature you have to look at.

PRIEST: One question is to ask – what the main capabilities of GREGOR would be. And the second question was, when you're identifying the upflows and downflows in the penumbra, – can you roughly estimate there's mass balance between the two? In other words, whether you're seeing all of the upflows going down or whether some of them will continue into the moat.

DENKER: Let's answer the last one first. I think it's difficult because many of the you probably won't see or haven't seen yet because, if you only see at the top, you might upflows because they are maybe just scale deeper. It's not much. It's 40 kilometers deeper. Then how can you answer that question? But there are lines where you can go deeper. For example, the infrared at 1.56 microns, and that maybe ties in a little bit with the first part of your question. The nice thing about GREGOR, you have a big aperture. Can actually get three times higher resolution than with Hinode – not all the time, I mean, you have seen. But maybe for two hours three hours a day. Will get actually data sets where it can follow small scale features. The aperture is one and a half meters. Three times larger than the Hinode. Also bigger than what we see with the Swedish telescope, and already there we see that fine structure has substructure. Hope is you actually get to the substructure as well, but the other thing I wanted to mention have the GFPI for visible light observations – we have the instrument or which is the new name for it, where you have infrared observation and you can do it at the same time. Is another thing you want to get as much information as possible, and our problem is you have just very small field of view. Let's just say 60 by 60 arc seconds. Need a space based instrument to get the big picture.

I think that's where all the synergies will be in the future to basically have something like a magnifying glass looking at the small details and then a space data to get the big picture and bring it all together.

The Physics of Sun and Star Spots
Proceedings IAU Symposium No. 273, 2010
D.P. Choudhary & K.G. Strassmeier, eds.

doi:10.1017/S1743921311015262

Evolution of twist-shear and dip-shear in flaring active region NOAA 10930

Sanjay Gosain and P. Venkatakrishnan

Udaipur Solar Observatory, Physical Research Laboratory,
P. Box No. 198, Udaipur 313001, Rajasthan, India
email: sgosain@prl.res.in

Abstract. We study the evolution of magnetic shear angle in a flare productive active region NOAA 10930. The magnetic shear angle is defined as the deviation in the orientation of the observed magnetic field vector with respect to the potential field vector. The shear angle is measured in horizontal as well as vertical plane. The former is computed by taking the difference between the azimuth angles of the observed and potential field and is called the twist-shear, while the latter is computed by taking the difference between the inclination angles of the observed and potential field and is called the dip-shear. The evolution of the two shear angles is then tracked over a small region located over the sheared penumbra of the delta sunspot in NOAA 10930. We find that, while the twist-shear shows an increasing trend after the flare the dip-shear shows a significant drop after the flare.

Keywords. Sunspots, flares

1. Introduction

The non-potential magnetic field in solar active regions stores the free-energy which is needed to fuel the energetic events like solar flares. The conventional measure of non-potentiality has been the so-called magnetic shear angle (Hagyard *et al.* 1984). This angle measures the difference between the observed and potential field azimuths and has been studied in relation to the flares (Venkatakrishnan *et al.* 1988). However, this angle measures only the deviations of the observed field from potential field vector in the horizontal plane alone. Such deviations are also possible in the vertical plane i.e., in the magnetic field inclination angles of the observed and potential field. In order to distinguish between these two types of shear we call the shear in the horizontal plane as the twist-shear while the shear in the vertical plane as the dip-shear.

In this paper we show how the observed magnetic field deviates from the potential field in the vertical plane in flare productive active region NOAA 10930. Further, we show the evolution of these two shear parameters in the penumbral region located close to the polarity inversion line (PIL) of the delta sunspot before and after the flare. The high-resolution vector magnetograms were derived by using the spectropolarimetric observations from *Hinode* Solar Optical Telescope (SOT) (Tsuneta *et al.* 2008). We describe these results in the following sections.

2. Observations and Analysis Methods

The delta sunspot in NOAA 10930 was observed during 12–13 December 2006 by *Hinode* space mission (Kosugi *et al.* 2007). The spectropolarimetric data was obtained

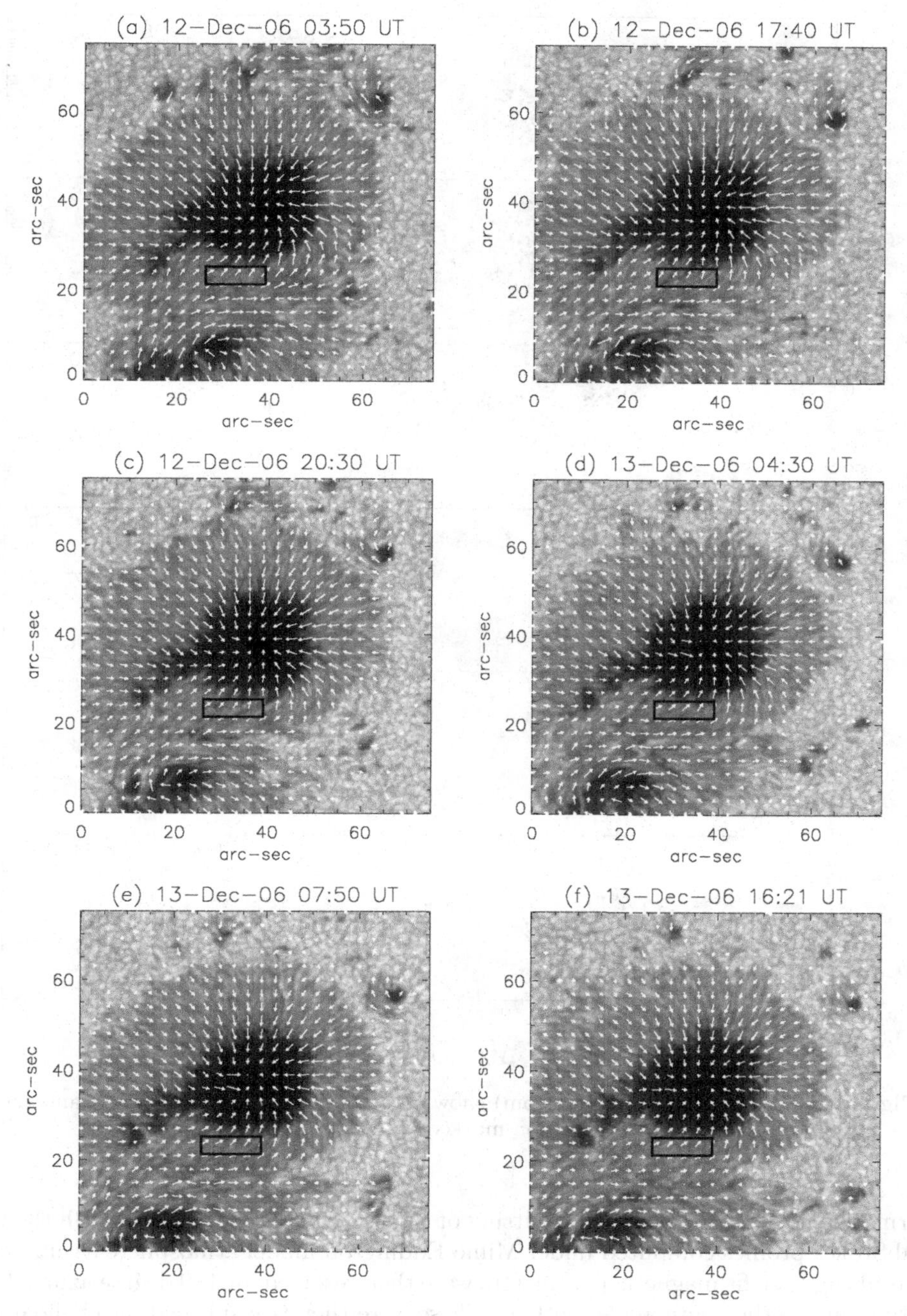

Figure 1. Panels (from top to bottom) show a continuum intensity map of the delta-sunspot in NOAA 10930 during the times mentioned at the top. The transverse magnetic field vectors are shown by arrows overlaid upon these maps. The black rectangle, as shown in all panels, is the region where we monitor the evolution of the twist-shear and dip-shear.

from *Hinode* SOT/SP instrument (Ichimoto *et al.* 2008) and was reduced and calibrated using SolarSoft package. The calibrated spectropolarimetric data was then inverted using MERLIN inversion code (Lites *et al.* 2007) at HAO, Boulder, USA. This inversion code

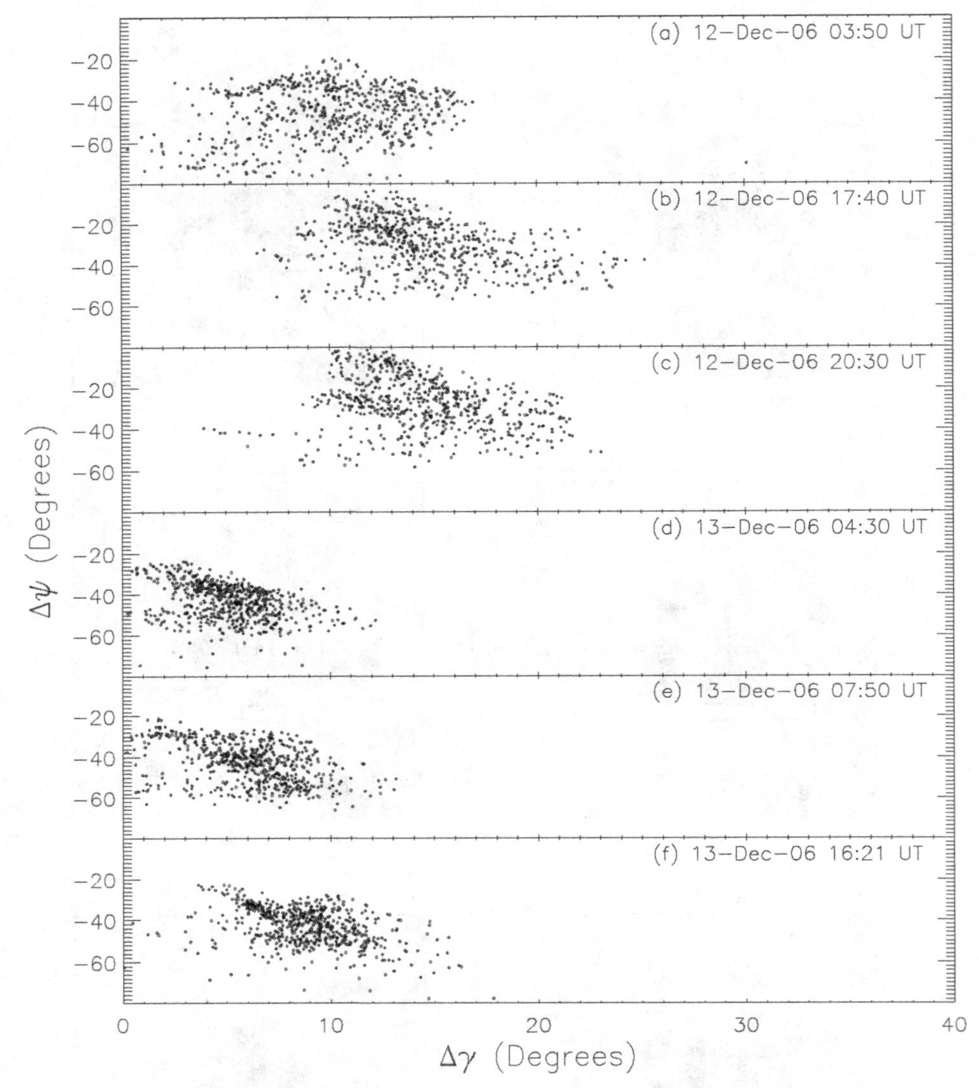

Figure 2. Panels (from top to bottom) show the evolution of the twist-shear and the dip-shear inside the region marked by the rectangle in figure 1.

performs the non-linear least squares fitting of the observed Stokes profiles with the theoretical Stokes profiles computed under Milne-Eddington model atmosphere assumptions. The resulting best-fit magnetic parameters were then resolved for 180 degree azimuth ambiguity by using the acute angle method. These were then transformed into heliographic coordinates using the method of Venkatakrishnan *et al.* (1988). The potential field was computed using the method of Alissandrakis (1981). The magnetograms were registered by applying the image cross-correlation method on the continuum intensity images. The figure 1 shows the continuum intensity maps of the six magnetograms obtained during 12-13 December 2006. The black rectangular box is the location where we monitor the dip-shear and the twist-shear. The figure 2 shows the evolution of the shear parameters inside this box during the observations.

3. Results and Discussions

It can be clearly noticed that: (i) the twist-shear and dip-shear are correlated i.e., the pixels with large twist shear also tend to have large value of dip-shear and vice-versa, with some spread in either parameter, (ii) the dip-shear shows an increasing trend before the flare, (iii) the dip-shear decreases significantly after the flare, (iv) the twist-shear increases after the flare which was also observed by Jing *et al.* (2008).

Any flare related change in the observed parameters of the active regions is useful in order to understand the nature of the energy build-up and its subsequent release in flares and CMEs. The changes in the line-of-sight magnetic field was been studied by Sudol & Harvey (2005) in large number of powerful flares and firmly established that there is abrupt and permanent flare related change in active regions. The present study tries to establish those results on more firm footing by detecting the changes in the magnetic field vector directly. However, the slow cadence of the *Hinode* SOT/SP observations present the biggest limitation in moving forward with such studies. We plan to conduct similar study in near future by using high-cadence vector magnetograms from the recently launched Helioseismic and Magnetic Imager (HMI) onboard Solar Dynamics Observatory (SDO).

Acknowledgements

The presentation of this paper in the IAU Symposium 273 was possible due to financial support from the National Science Foundation grant numbers ATM 0548260, AST 0968672 and NASA - Living With a Star grant number 09-LWSTRT09-0039. Hinode is a Japanese mission developed and launched by ISAS/JAXA, with NAOJ as domestic partner and NASA and STFC (UK) as international partners. It is operated by these agencies in co-operation with ESA and NSC (Norway).

References

Alissandrakis, C. E. 1981, *Astron. Astrophys*, 100, 197
Hagyard, M. J., Teuber, D., West, E. A., & Smith, J. B. 1984, *Solar Phys.*, 91, 115
Ichimoto, K., *et al.* 2008, *Solar Phys.*, 249, 233
Jing, J., Wiegelmann, T., Suematsu, Y., Kubo, M., & Wang, H. 2008, *Astrophys. J. Lett.*, 676, L81
Kosugi, T., *et al.* 2007, *Solar Phys.*, 243, 3
Lites, B., Casini, R., Garcia, J., & Socas-Navarro, H. 2007, *Memorie della Societ Astronomica Italiana*, 78, 148.
Schmieder, B., Demoulin, P., Aulanier, G., & Golub, L. 1996, *Astrophys. J.*, 467, 881
Sudol, J. J. & Harvey, J. W. 2005, *Astrophys. J.*, 635, 647
Tsuneta, S., *et al.* 2008, *Solar Phys.*, 249, 167
Venkatakrishnan, P., Hagyard, M. J., & Hathaway, D. H. 1988, *Solar Phys.*, 115, 125

The Physics of Sun and Star Spots
Proceedings IAU Symposium No. 273, 2010
D.P. Choudhary & K.G. Strassmeier, eds.

doi:10.1017/S1743921311015274

What determines the penumbral size and Evershed flow speed?

Na Deng[1,2], Toshifumi Shimizu[3], Debi Prasad Choudhary[1] and Haimin Wang[2]

[1]Physics and Astronomy Department, California State University Northridge, 18111 Nordhoff St., Northridge, CA 91330, United States
email: na.deng@csun.edu

[2]Space Weather Research Laboratory, New Jersey Institute of Technology, 323 Martin Luther King Blvd., Newark, NJ 07102, United States

[3]Institute of Space and Astronautical Science, Japan Aerospace Exploration Agency, 3-1-1 Yoshinodai, Sagamihara, Kanagawa 229-8510, Japan

Abstract. Using Hinode SP and G-band observations, we examined the relationship between magnetic field structure and penumbral length as well as Evershed flow speed. The latter two are positively correlated with magnetic inclination angle or horizontal field strength within 1.5 kilogauss, which is in agreement with recent magnetoconvective simulations of Evershed effect. This work thus provides direct observational evidence supporting the magnetoconvection nature of penumbral structure and Evershed flow in the presence of strong and inclined magnetic field.

Keywords. Sunspots, Sun: magnetic fields, Sun: atmospheric motions, Sun: photosphere.

1. Introduction and Motivation

The penumbra along with the coupled Evershed flow has been one of the most intriguing phenomena on the Sun that motivated many detailed observational and theoretical studies. High resolution imaging and spectro-polarimetric analyses have revealed that the penumbra consists of two distinct magnetic components. The more inclined magnetic component (60°–100° from inner to outer penumbra, with respect to the surface normal) with weaker field strength (∼1200 Gauss) is embedded in the less inclined magnetic background (30°– 60° from inner to outer penumbra) that has stronger field strength (∼1700 Gauss), which is frequently referred as "uncombed" structure (Solanki & Montavon 1993; Langhans *et al.* 2005; Beck 2008; Borrero 2009, and references therein). The Evershed flow is magnetized and mainly carried by the more inclined magnetic component (e.g., Title *et al.* 1993; Stanchfield *et al.* 1997; Bellot Rubio *et al.* 2004; Borrero *et al.* 2005; Ichimoto *et al.* 2008; Deng *et al.* 2010).

Several theoretical models have been proposed to understand the penumbral structure and the mechanism of the Evershed flow, such as siphon-flow with downward pumping of magnetic flux (Montesinos & Thomas 1997; Thomas *et al.* 2002), embedded moving flux tube model (Schlichenmaier *et al.* 1998), "gappy penumbra" model (Scharmer & Spruit 2006), and elongated convective granular cells by the presence of inclined strong magnetic field (Hurlburt *et al.* 1996, 2000). In particular, recent realistic three-dimensional numerical MHD simulations have successfully reproduced the penumbral structure and Evershed flow as a natural consequence of thermal magnetoconvection when the average inclination of kilogauss magnetic field is larger than 45° (Rempel *et al.* 2009a). Moreover, the filamentary patten and the speed of the simulated Evershed horizontal surface

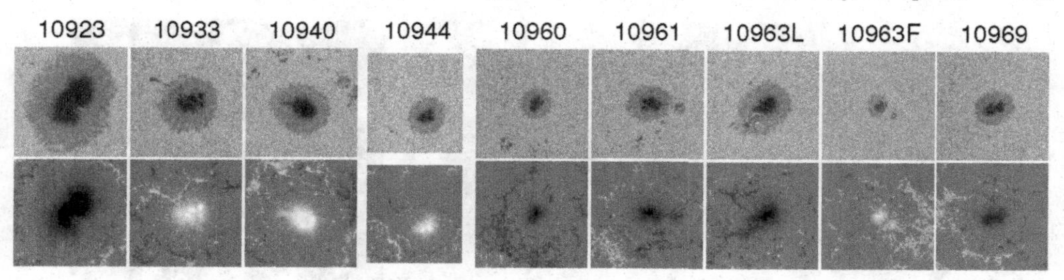

Figure 1. Hinode/SP Fe I 630 nm continuum images (top row) and B_z component magnetograms (bottom row) of the 9 sunspots with their NOAA AR numbers labeled on top. The FOV for each panel is $90'' \times 90''$.

flow are strongly controlled by both strength and inclination of the magnetic field with inclination playing most important role (Rempel *et al.* 2009b; Kitiashvili *et al.* 2009).

On the other hand, rapid penumbral decay (disappearance or length reduction) and change of associated Evershed flow were found right after major flares in the periphery of complex δ sunspots (Wang *et al.* 2004; Liu *et al.* 2005; Deng *et al.* 2005). The authors interpret the sudden change of penumbral white light structure to be a result of the change of overall magnetic field inclination down to the photosphere due to magnetic reconnection in the flare, which was then confirmed by other authors (Sudol & Harvey 2005; Li *et al.* 2009).

Both aforementioned magnetoconvective simulations and observations during flares hint a relationship between magnetic field structure (especially inclination) and penumbral length as well as Evershed flow speed. While a systematic examination of such relationship from real observation is still missing. Thanks to the high quality measurement of vector magnetic field by Hinode, we investigate how the magnetic parameters are related to the penumbral length and Evershed flow speed by analyzing sunspots of different sizes (i.e., in large scale) and the properties in different sectors within each sunspots (i.e., in small scale).

2. Observation and Data Reduction

We studied 9 simple α sunspots at the late phase of solar cycle 23. They are close to disk center (heliocentric angle $< 12°$) and exhibit different size. Fig. 1 shows their continuum images and B_z magnetograms obtained by Hinode Spectropolarimeter (SP). From the 13 parameters generated by a Milne-Eddington Stoke inversion of the SP data, we used or calculated the following most relevant parameters: continuum intensity, magnetic inclination angle, horizontal field strength, and field strength. The 180° azimuthal ambiguity of the inverted magnetic field was resolved using the "minimum energy" algorithm (Metcalf 1994). We transformed the measured magnetic field vectors to local Cartesian coordinates so that the inclination is with respect to the surface normal. To measure Evershed flow, we used Local Correlation Tracking (LCT) technique based on a 1hr series of Hinode G-band images (2min cadence) co-aligned and co-temporal with SP data. Same tracking window was used for all the sunspots. Fig. 2 shows the continuum intensity, LCT flow map and flow speed map of a sunspot.

The continuum images were smoothed and contoured to outline the penumbral areas, whose inner and outer boundaries are about $0.45I_0$ and $0.9I_0$ (I_0 is the quiet Sun continuum intensity), respectively. The penumbral lengthes were measured in each sector (i.e., the distance between the inner and outer boundaries, see panel a of Fig. 2). The

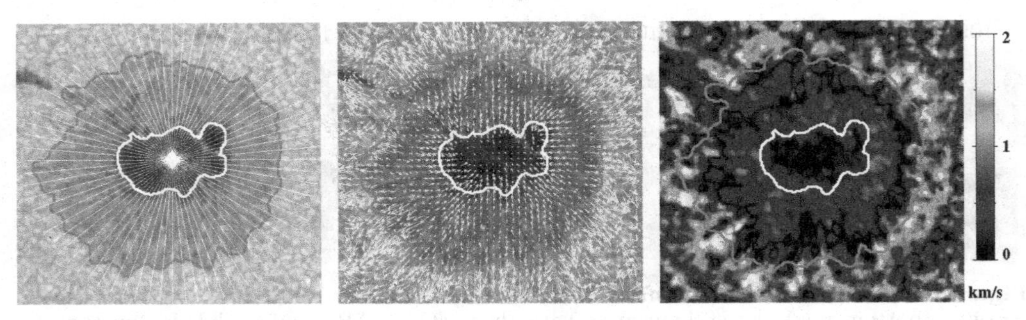

Figure 2. The SP continuum image, G-band LCT flow map and flow speed map of NOAA 10933. The sunspot is evenly divided into 360 sectors from the Center-Of-Mass of the umbra.

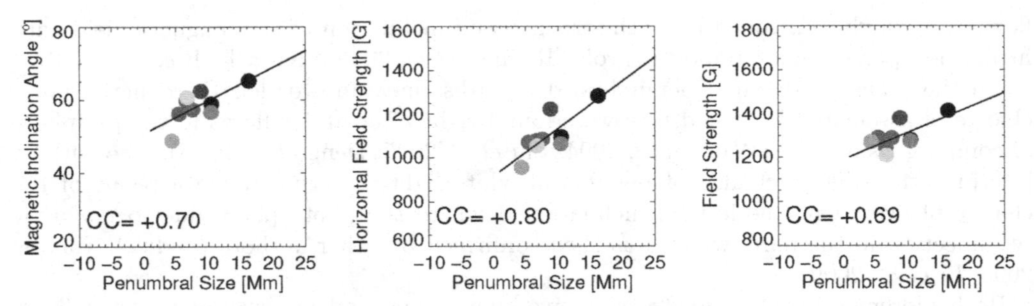

Figure 3. Scatter plots between penumbral length and magnetic parameters averaged over the whole penumbral areas of the 9 sunspots. The linear correlation coefficient (CC) are labeled.

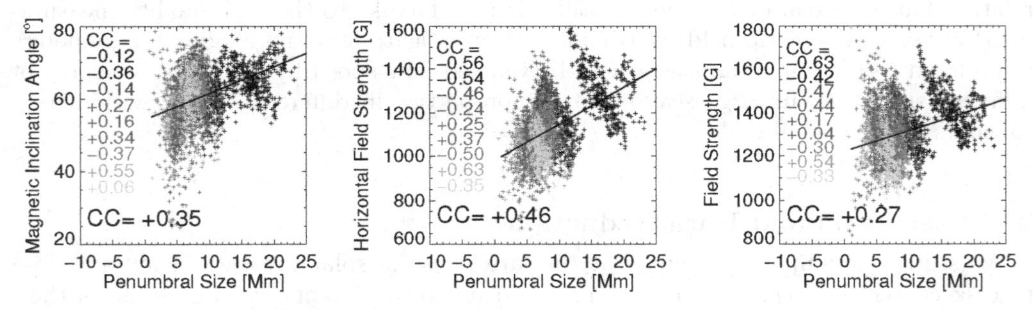

Figure 4. Scatter plots between penumbral length and magnetic parameters averaged in each sectors for the 9 sunspots. Each spot is represented by one color. The CC for each spot and for all the data points are labeled.

aforementioned magnetic parameters and LCT outward flow speed were also averaged over each sector's area. All these quantities were then averaged over the entire penumbral area for each sunspot. Sectors where the penumbral structure is complex or deviate from radial direction much were excluded from consideration.

3. Results and Conclusion

Fig. 3 shows that in large scale the penumbral size is well correlated with magnetic inclination angle and horizontal field strength. The mean inclination angles in penumbrae are all greater than 45° for the 9 sunspots. Fig. 4 plots the same but in small scale. While all the data points still follow the same trend as in Fig. 3, for each individual sunspot the

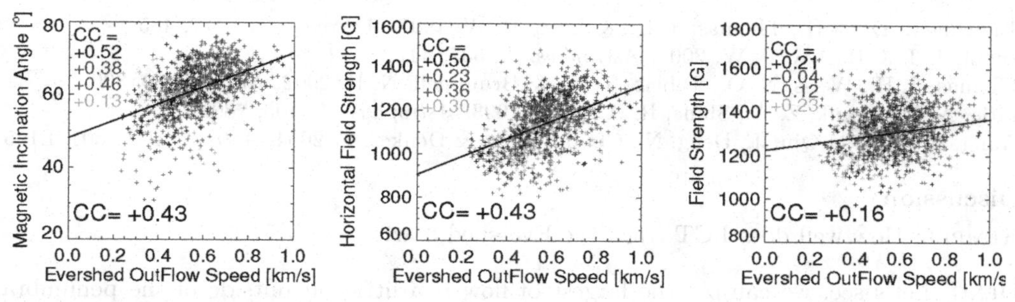

Figure 5. Scatter plots between outward Evershed flow speed and magnetic parameters averaged in each sectors for 4 sunspots that have co-temporal SP and G-band data.

trend is not consistent. This might be due to dynamic and chaotic fluctuations in small scale or over simplification of our method. Fig. 5 shows that the Evershed flow speed is always positively correlated with magnetic inclination angle and horizontal field strength in both small and large scales for the 4 sunspots studied. This result is consistent with magnetoconvective simulations of Evershed effect, where the horizontal flow speed increases with larger inclination angle within certain field strength range (Kitiashvili *et al.* 2009). This work thus provides direct observational evidence supporting the magnetoconvective nature of penumbral filamentary structure and Evershed flow under strong and inclined magnetic field.

Acknowledgements

N.D. and D.P.C. were supported by NASA grant NNX08AQ32G and NSF grant ATM 05-48260. N.D. thank valuable discussions with Drs. Ichimoto, Kitiashvili, Martinez Pillet, Sainz Dalda, and Schlichenmaier.

References

Beck, C. 2008, *Astron. Astrophys*, 480, 825
Bellot Rubio, L. R., Balthasar, H., & Collados, M. 2004, *Astron. Astrophys*, 427, 319
Borrero, J. M. 2009, Science in China G: Physics and Astronomy, 52, 1670
Borrero, J. M., Lagg, A., Solanki, S. K., & Collados, M. 2005, *Astron. Astrophys*, 436, 333
Deng, N., Liu, C., Yang, G., Wang, H., & Denker, C. 2005, *Astrophys. J.*, 623, 1195
Deng, N., Prasad Choudhary, D., & Balasubramaniam, K. S. 2010, *Astrophys. J.*, 719, 385
Hurlburt, N. E., Matthews, P. C., & Proctor, M. R. E. 1996, *Astrophys. J.*, 457, 933
Hurlburt, N. E., Matthews, P. C., & Rucklidge, A. M. 2000, *Solar Phys.*, 192, 109
Ichimoto, K., Tsuneta, S., Suematsu, Y., *et al.* 2008, *Astron. Astrophys*, 481, L9
Kitiashvili, I. N., Kosovichev, A. G., Wray, A. A., & Mansour, N. N. 2009, *Astrophys. J.*, 700, L178
Langhans, K., Scharmer, G., Kiselman, D., Löfdahl, M., & Berger, T. 2005, *Astron. Astrophys*, 436, 1087
Li, Y., Jing, J., Tan, C., & Wang, H. 2009, Science in China G: 52, 1702
Liu, C., Deng, N., Liu, Y., Falconer, D., Goode, P., Denker, C., & Wang, H. 2005, *Astrophys. J.*, 622, 722
Metcalf, T. R. 1994, *Solar Phys.*, 155, 235
Montesinos, B. & Thomas, J. H. 1997, *Nature*, 390, 485
Rempel, M., Schüssler, M., Cameron, R. H., & Knölker, M. 2009a, Science, 325, 171
Rempel, M., Schüssler, M., & Knölker, M. 2009b, *Astrophys. J.*, 691, 640
Scharmer, G. B. & Spruit, H. C. 2006, *Astron. Astrophys*, 460, 605
Schlichenmaier, R., Jahn, K., & Schmidt, H. U. 1998, *Astrophys. J.*, 493, L121
Solanki, S. K. & Montavon, C. A. P. 1993, *Astron. Astrophys*, 275, 283

Stanchfield, D. C. H., Thomas, J. H., & Lites, B. W. 1997, *Astrophys. J.*, 477, 485
Sudol, J. J. & Harvey, J. W. 2005, *Astrophys. J.*, 635, 647
Thomas, J. H., Weiss, N. O., Tobias, S. M., & Brummell, N. H. 2002, *Nature*, 420, 390
Title, A. M., Frank, Z. A., Shine, R. A., *et al.* 1993, *Astrophys. J.*, 403, 780
Wang, H., Liu, C., Qiu, J., Deng, N., Goode, P. R., & Denker, C. 2004, *Astrophys. J.*, 601, L195

Discussion

REMPEL: How well does LCT track the Evershed flow?

DENG: Let's see. we can see the largest of flow is a little bit outside of the penumbra area, I think maybe – because the Evershed flow extends outside to the moat region, as Dr. Denker showed, we can trace the feature outside of the penumbra. It's hard to tell how the LCT method can measure accurately the Evershed flow. Maybe in the future, we can combine the dopplergram and LCT together and see how we can get the Evershed flow more precisely.

The Physics of Sun and Star Spots
Proceedings IAU Symposium No. 273, 2010
D.P. Choudhary & K.G. Strassmeier, eds.

doi:10.1017/S1743921311015286

In-depth survey of sunspot and active region catalogs

Laure Lefèvre[1], Frédéric Clette[1] and Tunde Baranyi[2]

[1]Royal Observatory of Belgium, 3 Avenue Circulaire, 1180, Uccle, Belgium
email: laure.lefevre@oma.be and frederic.clette@oma.be

[2]Heliophysical Observatory of the Hungarian Academy of Sciences, H-4010 Debrecen, Hungary
email: baranyi@tigris.unideb.hu

Abstract. When consulting detailed photospheric catalogs for solar activity studies spanning long time intervals, solar physicists face multiple limitations in the existing catalogs: finite or fragmented time coverage, limited time overlap between catalogs and even more importantly, a mismatch in contents and conventions. In view of a study of new sunspot-based activity indices, we have conducted a comprehensive survey of existing catalogs.

In a first approach, we illustrate how the information from parallel catalogs can be merged to form a much more comprehensive record of sunspot groups. For this, we use the unique Debrecen Photoheliographic Data (DPD), which is already a composite of several ground observatories and SOHO data, and the USAF/Mount Wilson catalog from the Solar Optical Observing Network (SOON). We also describe our semi-interactive cross-identification method, which was needed to match the non-overlapping solar active region nomenclature, the most critical and subtle step when working with multiple catalogs. This effort, focused here first on the last two solar cycles, should lead to a better central database collecting all available sunspot group parameters to address future solar cycle studies beyond the traditional sunspot index time series R_i.

Keywords. Catalogs, surveys, sun: photosphere, sunspots, methods: data analysis, statistical

1. Introduction

So far, the main sunspot time series available for research has been the International Sunspot Index R_i obtained from a large number of observatories and visual observers. Until the last decade, the limitations in digitization equipment, in image processing techniques and computing capacity larglely prevented building more detailed sunspot series. However, the current advances in theoretical modeling require additonal long-term observational constraints and we now finally have appropriate means to create more complete long-duration catalogs.

Previous research based on catalog sets focused on single or just a few parameters for more detailed studies, often dropping the rest of the information (Balmaceda *et al.*, 2009). As an example, in an earlier comparative study of the discrepancies between different measurement of sunspot areas (Gyori & Baranyi 2006, Gyori *et al.*, 2005), the scope was limited to combining areas and sometimes also the corresponding latitudes and longitudes, thus leaving out other sunspot data.

The purpose of the work presented here is to extend the exploration to the complete parameters-space over extended periods of time. We considered a base of 20 catalogs. Most of them can be found on the NGDC website (see bibliography). To initiate this study, we use two particularly rich catalogs. The best choice for a multi-parameter study of sunspot activity is the Debrecen Photoheliographic Data catalog (Mezo & Baranyi, 2005, Gyori *et al.*, 2005). It is well complemented by the USAF multi-station catalog, which combines data from a worlwide network of solar observatories, mostly run by the US Air Force. We will now describe in more detail our comparative analysis of those catalogs.

2. The Debrecen and USAF data

The DPD catalog covers the last two solar cycles and is still in construction and the digitization work is still in progress in the framework of the *SoTerIA* project (Lapenta, 2007). Most of the DPD catalog is based on photographic and CCD images from the *Gyulia and Debrecen Observatories.* However, for the period from 2004 to 2006, in order to fill in promptly the catalog in support of this study, SoHO images were used instead: they do not offer the same level of details, but their preprocessing is faster (Baranyi T., private communication). *USAF* observations come from 9 stations between Greece and Australia. The sunspot group entries for each day are based on a maximum of 6 of those observatories.

Both datasets differ by 3 main aspects : *(1)* the time coverages are different: 86.3% for the USAF series versus 98.9% for the DPD series. *f (2)* The USAF catalog only lists groups while the DPD catalog lists individual sunspots, with grouping added only afterwards. (3) The USAF catalog is structured according to the evolution of sunspot groups, while the DPD adopts a chronological ordering.

In order to assess the consistency of these datasets, we computed the standard *Wolf Number (W_{OBS})* with the sunspot and sunspot group counts extracted from the catalogs. We compared W_{USAF} and W_{DPD} to the International Sunspot Index (R_i), and Wolf Numbers from several individual stations : Catania (CA), Locarno (LO), Kanzelhöhe (KZ) and Uccle (UC). *Figure 1* plots the resulting series.

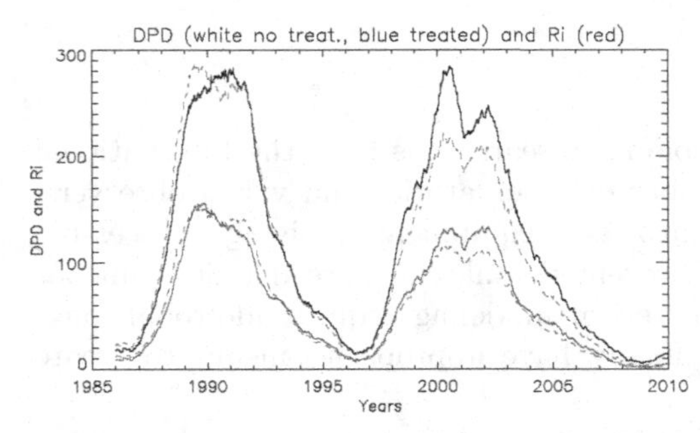

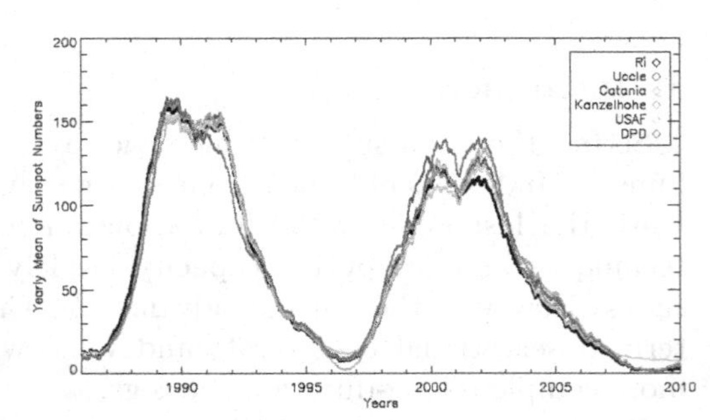

Figure 1. On the left, the Wolf number derived from the DPD catalog, W_{DPD}, from 1986 to 2010 (one-year running mean of daily values) is plotted for the full data set (black) and for the reduced set after filtering as described in the main text (blue). Each series is compared respectively with the R_i index (lower red dashed line)and with $R_i/0.6$ (upper red dashed line). On the right, W_{DPD*} after filtering (red curve) is compared with several independant sunspot number series: R_i (SIDC, black curve), Uccle, Catania, Kanzelhohe and the USAF catalog

While W_{USAF} agrees closely with other series, W_{DPD} shows significant deviations. The raw DPD values shown in Fig. 1 (left panel) give almost the same peak value for cycles 22 and 23 (max_{22}/max_{23} = 1.08), while all other series consistently show that #22 was significantly higher than #23, with ratios of 1.38, 1.36 , 1.33, 1.29 and 1.17 for W_{LO}, R_i, W_{USAF}, W_{UC} and W_{CA} respectively. Although there is an obvious scatter in the values, the systematic character of this discrepancy points at a significant bias in W_{DPD}. Moreover, the cycle #22 maximum does not show the caracteristic double peak present in all other series.

The cause of this difference resides in the details of the DPD catalog, which lists about 10% more sunspot groups than, e.g. the USAF catalog. After checking the characteristics of small sunspots listed in the DPD catalog, we see that approximately 50% of the spots are classified as penumbra without umbra and their status as spots should be carefully reconsidered. This seems to be true also for spots with a diameter smaller than 2 arcsec.

This R_i reconstruction indicates that the DPD catalog is highly comprehensive to the point of including some features that do not qualify as sunspots in long-term visual observations. After this filtering, W_{DPD*} matches better R_i and other individual series (Fig. 1, right panel). The ratio between maxima rises to 1.17, which is still on the low end but equivalent to the Catania ratio (Figure 1). However maximum #23 is still higher than in all other series and the maximum of cycle #22 still does not reproduce well the second peak.

3. Merging catalogs

After exploiting this sunspot-level selection only possible with the DPD catalog, as the DPD and USAF catalog contain different information about sunspot groups, we can take advantage of the time overlap between both catalogs to merge the contents, e.g. adding the USAF Zürich modified morphological classification (Zpc) to the DPD sunspot group catalog. In order to obtain a closer match in observing times, we first reorganized the USAF catalog entries by selecting the station observing at the time closest to the DPD time. We then matched each individual group listed in both catalogs for each date, using the Euclidian distance between groups with differing NOAA group numbers. In this process, the main difficulty came from the difference in the group splitting methods implemented in those two catalogs. This was further complicated by internal ambiguities in the respective group splitting and numbering schemes when the group evolution over successive days led to the insertion of extra "orphan" groups. By a lack of standardization, such groups are intercalated by adding suffixes to the official NOAA group number. This leads to a confusion between those sub-designations, e.g NN for NOAA, becoming NN, NN_a or NN_A in either the DPD or USAF catalog. The group identification program must then extend the search to all possible nomenclatures in both catalogs.

Overall, *about 80% of the DPD sunspot groups have a straight forward correspondence in the USAF group catalog. The remaining 20%, where the matching is problematic, are mainly related to groups from the DPD catalog that do not appear in the USAF catalog (11%), because of the higher level of details in the DPD catalog. They are also largely related to cases where the USAF and DPD groups are too distant to be matched (5%).*

As illustration, Figure 2 shows the sunspot regions for May 30th 1991 for DPD, USAF (black and red circles) and Kanzelhohe data. Three groups (6655, 6649, 6653 and 6653c) are listed only in the DPD catalog, suggesting that the latter is more comprehensive and includes spots that escaped detection in the other observations. Moreover, the size of the sunspot groups in the DPD data appears systematically larger than in the USAF data (see group 6644, fig 2. RIGHT panel), thus reflecting again the difference in standards and methods between catalogs of different origins and built at different epochs. Finally,

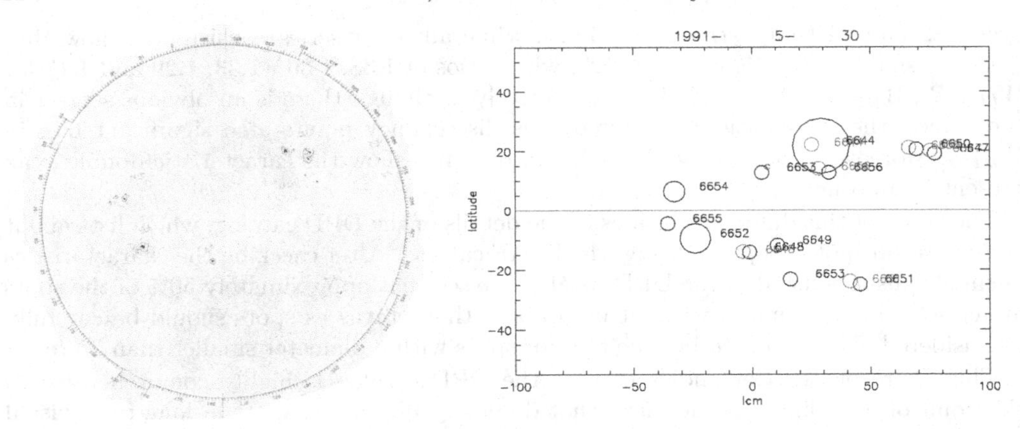

Figure 2. The left panel represents the drawing from Kanzelhöhe Observatory for May $30^t h$ 1991 while the right panel presents a synoptic map of the surface of the Sun with DPD (black) and USAF groups (red) for the same day. The size of the circles is proportional to the total area of the considered group.

groups 6652 and 6654 appear in the Kanzelhöhe and Debrecen data on May 30th but only on June 1st for the San Vito Observatory (USAF). Such mismatches can be caused either by the difference of sunspot grouping schemes or by a difference in observing times (date jump). Solving such cases often requires additional external information not necessarily included in other existing sunspot catalogs. This illustrates the work currently in progress and that will ultimately allow a full merging of those two catalogs, after removal of the unsolved ambiguities.

4. Conclusion

Presently, we have produced a first version of this extended composite catalog spanning the last two solar cycles, from 1986 until now. It contains groups, spots, areas and Zürich modified morphological types, as well as a basic computing of the extent of the groups. The catalog will also be enhanced by the addition of extra parameters derived from the base ones (e.g. dipole tilt angles) or extracted from other databases or catalogs, where they overlap the entries of this catalog. The resulting catalog will be made available through the SIDC website (www.sidc.be). Beyond the rich but time-limited DPD and USAF catalogs exploited here, we will then work backward in time by adding other older or sometimes longer catalogs. However, these catalogs contain less sunspot information and also pose more problems of standardisation, making the merging and group matching more complex. It will thus be necessary to implement advanced techniques like Multivariate Data Analysis (Murtagh F. & Heck, A., 1987). This effort will build on the knowledge acquired during this first test-bed offered by the DPD and USAF catalogs.

References

NGDC website: http://www.ngdc.noaa.gov/

Balmaceda, L. A., Solanki, S. K., Krivova, N. A., & Foster, S. 2009, *Journal of Geophysical Research (Space Physics)*, 114, 7104

Gyori, L., Baranyi, T., Muraközy, J., & Ludmány, A. 2005, *MmSAI*, 76, 985

Gyori, L., Baranyi, T., Muraközy, J., & Ludmány, A. 2005, *MmSAI*, 76, 981

Győri, L. & Baranyi, T. 2006, *SOHO-17. 10 Years of SOHO and Beyond*, 617,
Lapenta, G., SOTERIA Team 2007, *AGU Fall Meeting Abstracts*, A338
Mezo, G. & Baranyi, T. 2005, *MmSAI*, 76, 1004
Mezō, G., Baranyi, T., & Gyōri, L. 2005, *Solar Magnetic Phenomena*, 320, 247
Murtagh, F. & Heck, A. 1987, *Astrophysics and Space Science Library*, 131,

The Physics of Sun and Star Spots
Proceedings IAU Symposium No. 273, 2010
D.P. Choudhary & K.G. Strassmeier, eds.

doi:10.1017/S1743921311015298

The Sun at high resolution: first results from the SUNRISE mission

S. K. Solanki[1,8], P. Barthol[1], S. Danilovic[1], A. Feller[1], A. Gandorfer[1], J. Hirzberger[1], A. Lagg[1], T. L. Riethmüller[1], M. Schüssler[1], T. Wiegelmann[1], J. A. Bonet[2], V. Martínez Pillet[2], E. Khomenko[2], J. C. del Toro Iniesta[3], V. Domingo[4], J. Palacios[4], M. Knölker[5], N. Bello González[6], J. M. Borrero[6], T. Berkefeld[6], M. Franz[6], M. Roth[6], W. Schmidt[6], O. Steiner[6] and A. M. Title[7]

[1]Max-Planck-Institut für Sonnensystemforschung, Max-Planck-Str. 2, 37191 Katlenburg-Lindau, Germany; email: solanki@mps.mpg.de
[2]Instituto de Astrofísica de Canarias, C/Via Láctea s/n, 38200 La Laguna, Tenerife, Spain.
[3]Instituto de Astrofísica de Andalucía (CSIC), Apdo. de Correos 3004, E-18080, Granada, Spain
[4]Grupo de Astronomía y Ciencias del Espacio (Univ. de Valencia), E-46980, Paterna, Valencia, Spain
[5]High Altitude Observatory, National Center for Atmospheric Research, P.O. Box 3000, Boulder CO 80307-3000, USA. †
[6]Kiepenheuer-Institut für Sonnenphysik, Schöneckstr. 6, 79104 Freiburg, Germany.
[7]Lockheed-Martin Solar and Astrophysical Lab., Palo Alto, USA
[8]School of Space Research, Kyung Hee University, Yongin, Gyeonggi, 446-701, Korea

Abstract. The SUNRISE balloon-borne solar observatory consists of a 1m aperture Gregory telescope, a UV filter imager, an imaging vector polarimeter, an image stabilization system and further infrastructure. The first science flight of SUNRISE yielded high-quality data that reveal the structure, dynamics and evolution of solar convection, oscillations and magnetic fields at a resolution of around 100 km in the quiet Sun. Here we describe very briefly the mission and the first results obtained from the SUNRISE data, which include a number of discoveries.

Keywords. Sunrise, telescope, quiet sun, instrumentation

---

1. Introduction

SUNRISE is the latest in a long line of solar telescopes carried by stratospheric balloons. It combines high spatial resolution with sensitivity to ultraviolet radiation. At 1 m diameter, it is the largest solar telescope so far to leave the ground. It is equipped with state-of-the-art post-focus instruments, including a UV imager and a filter-based vector magnetograph.

The magnetic field in the solar photosphere shows a very complex and diverse structure. Concentrations of magnetic field with kilo-Gauss strength appear in a broad range of sizes reaching down to small flux concentrations on scales of 100 km or below. SUNRISE aims to determine the true size and brightness distribution of the concentrated magnetic features by spatially resolving them, as well as providing greatly improved properties of the internetwork fields. SUNRISE also aims to probe the convection in the solar photosphere, as well as the often complex effects of the interaction of the magnetic field with the turbulent convection.

† HAO/NCAR is sponsored by the NSF

2. Instrumentation and Mission

The SUNRISE stratospheric balloon-borne observatory is composed of a telescope, two post-focus science instruments (called SuFI and IMaX, see below), an Image Stabilization and Light Distribution (ISLiD) unit, and a Correlating Wave-front Sensor (CWS), all supported by a gondola, which possesses pointing capability.

The telescope is a Gregory-type reflector with 1 m clear aperture and an effective focal length of close to 25 m. A heat-rejection wedge at the prime focus reflects 99% of the light from the solar disk off to the side, reducing the heat load on the post-focus instruments to approximately 10 W. The post-focus instrumentation rests on top of the telescope. More details on the telescope, gondola and mission are given by Barthol *et al.* (2010), while the CWS is described in detail by Berkefeld *et al.* (2010).

The SUNRISE Filter Imager (SuFI) provides images in 5 narrow and medium bands at violet and near ultraviolet wavelengths between roughly 200 and 400 nm. The highest cadence that can be achieved is an image every 2 s. In order to overcome aberrations, a phase-diversity technique (e.g. Paxman *et al.* 1992) is employed, also by the other science instrument, IMaX. A description of SuFI can by found in Gandorfer *et al.* (2010).

The Imaging Magnetograph eXperiment (IMaX) operates in the Zeeman $g = 3$ Fe I 525.02 nm line. Images in polarized light covering 50×50 arcsec^2 are recorded at a spectral resolution of 85 mÅ, normally at 4 wavelengths within the spectral line and 1 in the nearby continuum. The full Stokes vector in these 5 wavelengths at a noise level of 10^{-3} is obtained in 30 sec, which is the typical cadence for most of the observations. A dual-beam approach is taken, with 2 synchronized 1k×1k CCD cameras. Detailed information on IMaX is provided by Martínez Pillet *et al.* (2010).

SUNRISE was flown on a zero-pressure stratospheric long-duration balloon launched on June 8, 2009 from ESRANGE near Kiruna in northern Sweden. It then floated westwards at a mean cruise altitude of 36 km and landed on Somerset island (northern Canada), suspended on a parachute, on June 13, 2009. At float altitudes, virtually seeing-free observations were possible all the time (since the payload was above more than 99% of the Earth's atmosphere).

The loss of high-speed telemetry relatively soon after reaching float altitude (due to the failure of a rented commercial telemetry system), meant that no full images could be downloaded during the entire mission. Consequently instrument commissioning and operations had to be carried out practically blindly. Nonetheless, the achieved spatial resolution was sufficiently high to resolve both, small-scale magnetic and convective features (Lagg *et al.* 2010, Khomenko *et al.* 2010).

3. First results

The Sun was extremely quiet during the entire flight of SUNRISE, so that almost all of the gathered data correspond to internetwork regions with occasional network elements. An overview of the data and first results is given by Solanki *et al.* (2010).

Images of the quiet Sun at disk centre in all 5 SuFI wavelengths are shown in Fig. 1a, whose grey scale is saturated at $\langle I \rangle \pm 3\sigma$ for each wavelength in order to allow a better intercomparison of the granulation, at the cost of overexposing the bright points. The brightness scale (see the gray-scale bars above the individual frames) already indicates the large rms contrasts of the imaged granulation, which reach up to 32% at 214 nm (cf. Hirzberger *et al.*, 2010a, for a complete study of the rms contrasts). At a number of the observed wavelengths the contrasts can be compared with those resulting from the 3D radiation-MHD simulations of Vögler *et al.* (2005) and are found to be in good

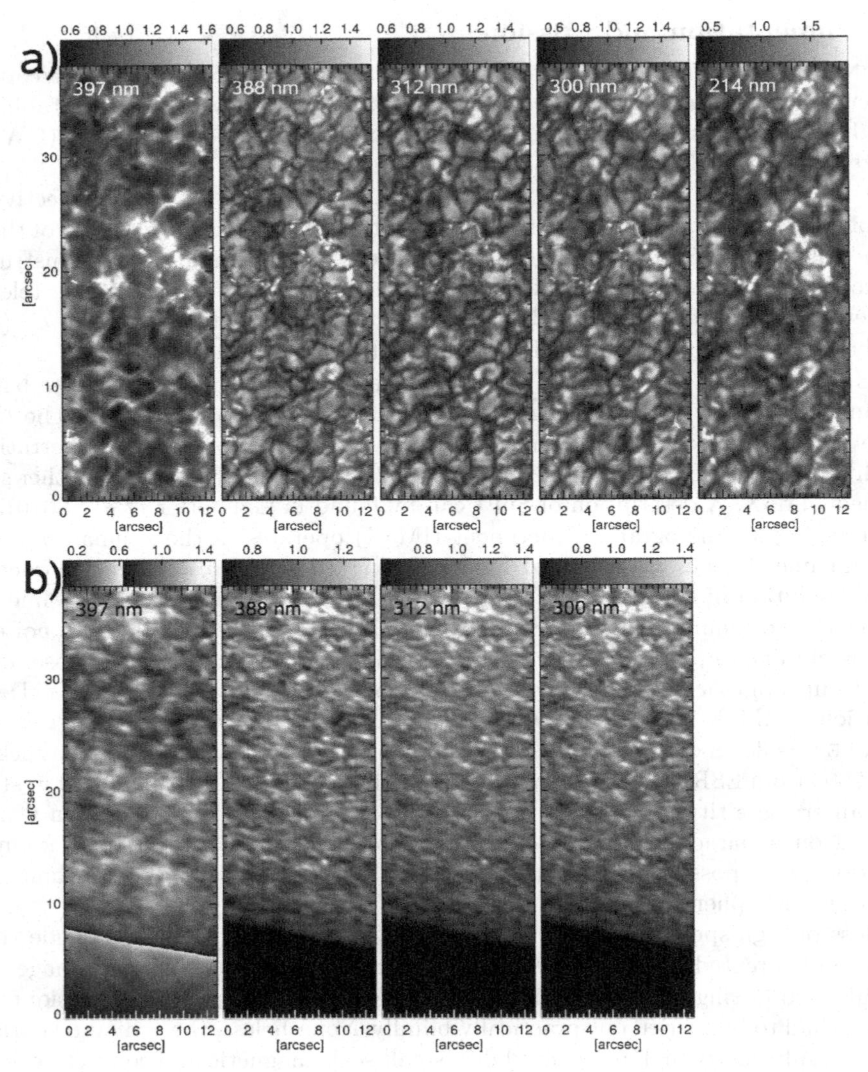

Figure 1. a) Images of a patch of quiet Sun near disk centre recorded by the SuFI instrument in wavelength bands centred on 397 nm, 388 nm, 312 nm, 300 nm and 214 nm (from left to right). The grey scale has been individually set to cover 3 times the RMS range of each image. b) Same as panel a, but at the solar limb. The Ca H image (397 nm) is plotted with an enhanced brightness scale for the off-limb parts. No 214 nm data are available at this position.

agreement. This supports both the high resolution and very low stray light of SUNRISE SuFI data.

Bright points are prominent at all wavelengths sampled by SuFI, but are particularly so at 214 nm (they are 2.3 times as bright as the background at this wavelength, see Riethmüller *et al.*, 2010), making them brighter than at any other wavelength studied so far.

Figure 2 shows a snapshot of IMaX data products. Stokes-V movies (see e.g. Solanki *et al.* 2010) reveal how dynamic the quiet Sun magnetic field is, with the weaker magnetic features, i.e., those in the internetwork, being particularly dynamic.

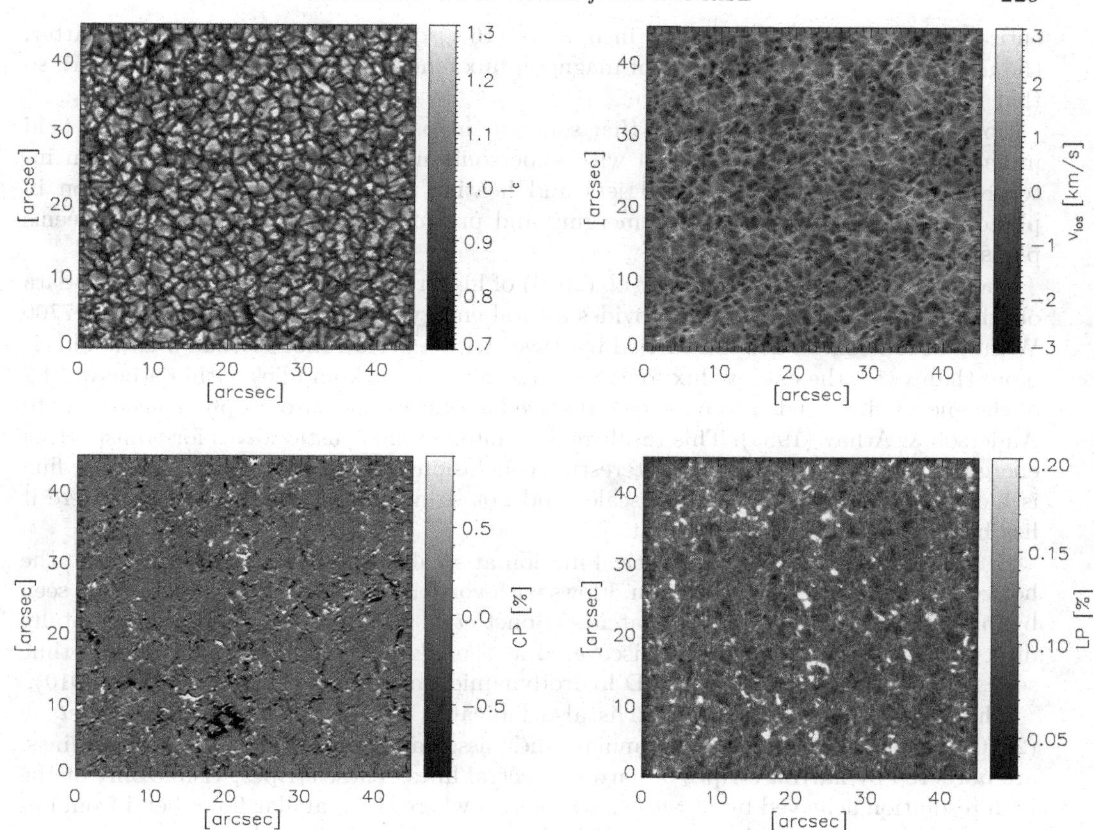

Figure 2. IMaX data. Clockwise from upper left: continuum intensity at 5250.4 Å, line-of-sight velocity, net linear polarization L_s averaged over the line, V_s, the line-averaged Stokes V obtained from Fe I 5250.2 Å. All images are based on phase-diversity reconstructed data except for the linear polarization image (reconstruction increases the noise, so that a number of significant L_s patches in the unreconstructed image are no longer sufficiently above the noise in the reconstructed data).

An investigation of the properties of small concentrations of strong magnetic fields in the quiet Sun showed that these have finally been resolved by IMaX on SUNRISE (Lagg *et al.* 2010). An inversion technique applied to retrieve the temperature stratification and the field strength could reproduce the observations well with a one-component, fully-magnetized atmosphere with a field strength exceeding 1 kG and a significantly enhanced temperature in the mid to upper photosphere compared to its surroundings. This is consistent with semi-empirical flux tube models describing magnetic elements. Consequently, it can be concluded that the SUNRISE measurements resolve the observed quiet Sun flux tubes. This result suggests that the SUNRISE data will allow further properties of these basic building blocks of the photospheric magnetic field to be determined rather directly.

SUNRISE observations show that the occurrence rate of patches of significant linear polarization signal (sensitive to the transverse component of photospheric magnetic field) is 1-2 orders of magnitude larger than values reported by previous studies (Danilovic *et al.* 2010). These features appear preferentially at granule boundaries with most of them being caught in downflow lanes at some point in their evolution. Only a small percentage are entirely and constantly embedded in upflows (16%) or downflows (8%). For the latter, the usual interpretation in terms of magnetic flux emerging from below cannot hold, so that they must have another source.

Borrero *et al.* (2010) discovered that some of the patches of horizontal magnetic field in the internetwork are associated with supersonic upflows of magnetized gas. An interpretation in terms of localized jets and heating due to magnetic reconnection in photopsheric layers between the emerging and previously present magnetic flux seems plausible.

An analysis by Bello Gonzalez *et al.* (2010) of high-resolution spectropolarimetric data obtained by IMaX on SUNRISE provides a total energy flux of approximately 6400–7700 W m^{-2} at a height of 250 km carried by waves with a period shorter than 3 min. This is more than twice the energy flux found in any previous work and lies within a factor of 2 of the energy flux needed to balance radiative losses from the chromosphere according to Anderson & Athay (1989). This result revives interest in acoustic waves for transporting energy to the chromosphere. An interesting conclusion is that the "missing" acoustic flux is hidden mainly at small spatial scales and not so much at high frequencies, where it has been assumed to lie in the past.

SUNRISE data have revealed vortical motion at small scales, with vorticity in both the horizontal and the vertical direction. Flows with vorticity in the vertical direction are seen by following small magnetic field patches (Bonet *et al.* 2010), while (smaller) horizontally directed vortex tubes have been discovered at the edges of granules by comparing time series of continuum images with 3D hydrodynamic simulations by Steiner *et al.* (2010).

The high resolution of the data is also indicated by the work of Khomenko *et al.* (2010), who show that the intergranular lanes, associated with broadened spectral lines, are bordered by narrow stripes of narrow spectral lines. These stripes, visible only at the high resolution achieved by SUNRISE, are located where the granular flows bend from up- to downflows, as comparisons with 3D hydrodynamic simulations show.

High degree p-modes are studied by Roth *et al.* (2010), who find that the power in waves with degree $l > 1000$ is enhanced over granules shortly before these start to split or explode, suggesting a connection between granule evolution and excitation of p-modes.

Finally, the magnetic field measured by SUNRISE in the photosphere has been extrapolated into the upper solar atmosphere by Wiegelmann *et al.* (2010). A statistical study of the connectivities of the extrapolated field shows that almost all of the magnetic field lines reaching the chromosphere or higher connect network with internetwork patches in the photosphere. Since the internetwork field is extremely dynamic and short-lived, this implies that the magnetic field in the upper atmosphere must also be changing very rapidly.

4. Conclusion

The SUNRISE observatory has provided high-quality, high-resolution images, Dopplergrams and vector magnetograms at different positions on the solar disk. The extremely low solar activity level at the time of the science flight of SUNRISE means that these data mainly enable new insights into the quiet Sun.

An initial analysis of these data has already led to new insights into the magnetism, convection, oscillations and waves in the quiet Sun. Given the richness and quality of the data and the fact that so far only a small fraction of them have been analyzed, we expect many more exciting results to follow. A flight of SUNRISE during a period of higher solar activity is greatly to be welcomed.

Acknowledgements

The German contribution to SUNRISE is funded by the Bundesministerium für Wirtschaft und Technologie through Deutsches Zentrum für Luft- und Raumfahrt e.V. (DLR), Grant No. 50 OU 0401, and by the Innovationsfond of the President of the Max Planck Society (MPG). The Spanish contribution has been funded by the Spanish MICINN under projects ESP2006-13030-C06 and AYA2009-14105-C06 (including European FEDER funds). The HAO contribution was partly funded through NASA grant number NNX08AH38G. This work has been partially supported by the WCU grant No. R31-10016 funded by the Korean Ministry of Education, Science & Technology.

References

Anderson, L. S. & Athay, R. G. 1989, *Astrophys. J.*, 336, 1089
Barthol, P., Gandorfer, A., Solanki, S. K., *et al.* 2010, *Solar Phys.*, 268, 1
Bello González, N., Franz, M., Martínez Pillet, V., *et al.* 2010, *Astrophys. J.*, 723, L134
Berkefeld, T., Schmidt, W., Soltau, D., *et al.* 2010, Solar Phys., 723, L139
Bonet, J. A., Márquez, I., Sánchez Almeida, J., *et al.* 2010, *Astrophys. J.*, 723, L134
Borrero, J. M., Martínez Pillet, V., Schlichenmaier, R., *et al. Astrophys. J.*, 723, L144
Danilovic, S., Beeck, B., Pietarila, A., *et al.* 2010a, *Astrophys. J.*, 723, L149
Gandorfer, A., Grauf, B., Barthol, P., *et al.* 2010, *Solar Phys.*, 268, 14
Hirzberger, J., Feller, A., Riethmüller, *et al.* 2010a, *Astrophys. J.*, 723, L254
Khomenko, E., Martínez Pillet, V., & Solanki, S. K., 2010, *Astrophys. J.*, 723, L159
Lagg, A., Solanki, S. K., Riethmüller, T. L., *et al.* 2010, *Astrophys. J.*, 723, L164
Martínez Pillet, V., del Toro Iniesta, J. C., Álvarez-Herrero, A., *et al.* 2010, *Solar Phys.*, in press
Paxman, R. G. and Schulz, T. J. , & Fienup, J. R. 1992, *J. Optical Soc. America A*, 9, 1072
Riethmüller, T. L., Solanki, S. K., Martínez Pillet, V., *et al.* 2010, *Astrophys. J.*, 723, L127
Roth, M., Franz, M., Bello González, N., *et al.* 2010, *Astrophys. J.*, 723, L175
Solanki, S. K., Barthol, P., Danilovic, S., *et al.* 2010, *Astrophys. J.*, 723, L127
Steiner, O., Franz, M., Bello González, N., *et al.* 2010, *Astrophys. J.*, 723, L164
Vögler, A., Shelyag, S., Schüssler, M., Cattaneo, F., Emonet, T., & Linde, T. 2005, *Astron. Astrophys*, 429, 335
Wiegelmann, T., Solanki, S. K., Borrero, J. M., *et al.* 2010, *Astrophys. J.*, 723, L185

Discussion

GEORGOULIS: What is the achieved spatial resolution, the magnetic filling factor, and the noise level of the data?

SOLANKI: The Noise level is a couple of times ten to the minus three if you take a single magnetogram. There are of course techniques of averaging magnetograms and reducing the noise level and so on, and that's something we're working on. These are just some of the results.

KLIMCHUK: So the Kilogauss really are existing in tubelike structures and not these thin stretched-out lane-type features?

SOLANKI: The noise level is very quiet region, yes; but that's because you have very little flux. So also in the simulations, if you take an image D simulation with a very low amount of magnetic flux, you will tend to get little point-like flux tubes. As you increase the amount of magnetic flux, it will not just stay there at the intersection of intergranules lanes, but it will start filling up these lanes, and then you will start getting these more elongated structures.

So we want refly again definitely and hopefully at a time when the sun is more active and you won't see just quiet features and hopefully also see these sheet-like structures.

KITIASHVILI: Do you want to repeat during the solar cycle maxima?

SOLANKI: Yes, we definitely plan to do that. So if all goes well, we would like to refly again in 2012 which at the moment is thought to be a time where the sun will be hopefully more active than in 2009. We still haven't got 100 percent of the funding yet, but we are relatively confident we will be able to do that.

The Physics of Sun and Star Spots
Proceedings IAU Symposium No. 273, 2010
D.P. Choudhary & K.G. Strassmeier, eds.

doi:10.1017/S1743921311015304

Coronal heating and flaring in QSLs

Guillaume Aulanier

Observatoire de Paris, LESIA, CNRS, UPMC, Universit Paris Diderot,
5 place Jules Janssen, 92190 Meudon, France
email: guillaume.aulanier@obspm.fr

Abstract. Quasi-Separatrix Layers (QSLs) are 3D geometrical objects that define narrow volumes across which magnetic field lines have strong, but finite, gradients of connectivity from one footpoint to another. QSLs extend the concept of separatrices, that are topological objects across which the connectivity is discontinuous. Based on analytical arguments, and on magnetic field extrapolations of the Sun's coronal force-free field above observed active regions, it has long since been conjectured that QSLs are favorable locations for current sheet (CS) formation, as well as for magnetic reconnection, and therefore are good predictors for the locations of magnetic energy release in flares and coronal heating. It is only up to recently that numerical MHD simulations and solar observations, as well as a laboratory experiment, have started to address the validity of these conjectures. When put all together, they suggest that QSL reconnection is involved in the displacement of EUV and SXR brightenings along chromospheric flare ribbons, that it is related with the heating of EUV coronal loops, and that the dissipation of QSL related CS may be the cause of coronal heating in initially homogeneous, braided and turbulent flux tubes, as well as in coronal arcades rooted in the slowly moving and numerous small-scale photospheric flux concentrations, both in active region faculae and in the quiet Sun. The apparent ubiquity of QSL-related CS in the Sun's corona, which will need to be quantified with new generation solar instruments, also suggests that QSLs play an important role in stellar's atmospheres, when their surface radial magnetic fields display complex patterns.

Keywords. Magnetic fields, magnetohydrosynamics, Sun: corona, stars: coronae

---

1. Introduction

Owing to the low plasma β of the solar corona, the energy which is needed to power flares and loop heating is believed to be primarily provided by the magnetic fields. Those are rooted in the photosphere, not only in large sunspots, but also in smaller-scale flux concentrations such as those in plage regions (e.g. the so called network elements) as well as in the quiet Sun (also involving granular and intergranular flux tubes). First, it is now well extablished that solar flares are caused by magnetic reconnection. Second, even though several alternative-current (AC) wave-based models have been proposed for coronal heating, observational arguments tend to favor direct-current (DC) mechanisms (Mandrini *et al.* 2000), the latter also being related to magnetic reconnection.

Understanding the global geometry of the magnetic field involved in reconnection is important, not only to infer potentially different fundamental reconnection regimes, but also when one wants to identify their potential occurence in observed events, when analyzed with magnetic field extrapolations (see Démoulin *et al.* 1997 and Wiegelmann 2011 for linear and non-linear force-free extrapolation methods). Two-dimensional (2D) models for magnetic reconnection at magnetic X-points have been long since studied. They have been extended to fully three-dimensional (3D) magnetic fiels configurations involving null points and separator field lines. Those magnetic configurations involve separatrix field lines or surfaces, across which the connectivity of field lines changes in a discontinuous way (see the review by Galsgaard 2011). General magnetic reconnection theory, however, does not require such topological singularities in 3D (Hesse & Schindler 1988). In this context, Priest & Forbes (1992) have first shown that the behavior of 2.5D magnetic field lines, reconnecting at a 2D X-point with a finite guide field perpendicular to the plane of the X-shaped separatrices, was different from standard separatrix reconnection: they argued in favor of a continuous flippage of the field lines across the remnants of the separatrices, instead of a classical cut-and-paste reconnection. Quasi-separatrix layers (QSLs), which are narrow volumes across which the magnetic field connectivity changes drastically, but in a continuous way, have then been developped as a likely geometrical extension of 2.5D flipping layers, and of 3D separatrices found in classical source models (Priest & Démoulin 1995; Démoulin *et al.* 1996a; Restante *et al.* 2009).

An impressive wealth of results has been obtained over the last two decades on the physics of QSL reconnection, on its application to solar observations, and on its relevance to several numerical and laboratory experiments. Instead of going through an exhaustive review of those, this author has chosen to briefly describe a personnal selection of only a few key results, which illustrate that many QSL-related studies actually show striking common features. To some extent, those highlight some unity which emerges from these disctinct approaches of coronal heating and flaring, around the concept of QSLs.

2. CS formation and magnetic reconnection in QSLs

The mathematically correct way to define a QSL is through the computation of the so-called squashing factor Q, which is constant along field lines, and which somehow measures the magnetic connectivity gradients around this field line (see Titov *et al.* (2002) for extensive details). A simpler way to identify the presence of QSLs, in a given magnetic field configuration, is to integrate field lines starting from small segments at the photospheric boundary: the spreading of their conjugate footpoints, along a curve that is much longer than the segment, indicates the presence of a QSL.

Démoulin *et al.* (1996a) were the first to give analytical arguments for QSLs being natural places for current sheet (CS) formation, driven by any line-tied photospheric motions, just like true separatrices are. Several groups have argued both in favor and against this idea (see e.g. Milano *et al.* 1999 and Galsgaard *et al.* 2003), until Aulanier *et al.* (2005) proved it right, with a series of numerical MHD of bipolar potential fields created by four flux concentrations, which initially possessed thin current-free QSLs. It has then been showed that QSLs are the 3D geometrical layers which support magnetic flipping, i.e. 3D finite-B reconnection (Aulanier *et al.* 2006; see Figure 2). Slip-running reconnection was there defined as the regime for which the slippage velocity of field lines (which does not correspond to a mass motion) is super-Alfvénic (which happens for high enough Q and reconnection rate). This regime cannot be misinterpreted as

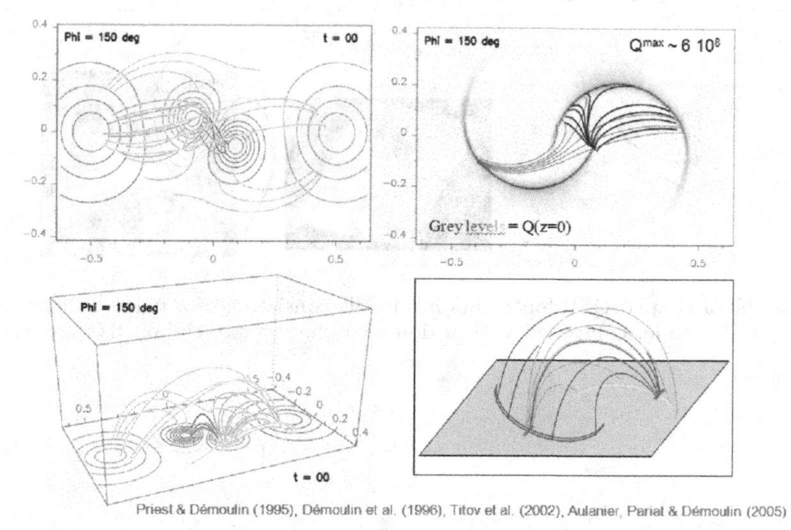

Figure 1. Example of QSLs formed in a 3D bipolar potential field, created by four flux concentrations. Note how field lines rooted in small segments, that cross high-Q photospheric regions (i.e. QSL footpoints), spread out and map the other QSL footpoint.

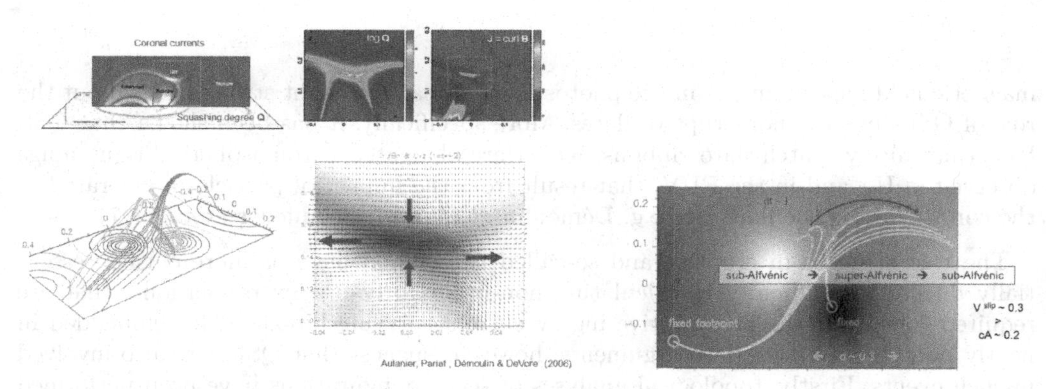

Figure 2. QSL related current sheet, that leads to finite-B magnetic reconnection. When analyzed in 2D, the standard X-type configuration and outflow jets are recovered, but the 3D picture shows a continuous slippage of field lines, which shows a slip-running regime which differs from mere diffusion.

magnetic diffusion, since it is local, and physically undistinguishable from mere null point or separator reconnection as far as MHD processes are concerned.

3. Observational evidences of QSLs in the Sun's corona

Before CS formation and reconnection was shown to occur in QSLs with MHD simulations, linear force-free field extrapolations of the coronal magnetic field, from line-of-sight

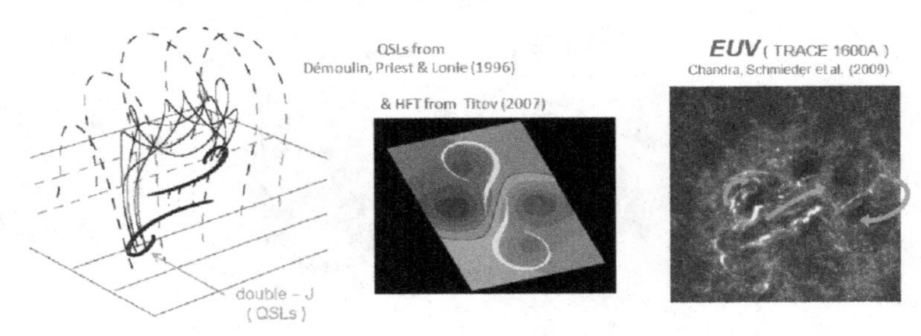

Figure 3. Double-J shaped QSL footprints in models consisting of a twisted flux tube embedded in a potential arcade, compared with a double-J shaped two-ribbon flare observed in the EUV.

magnetic field measurements in the photosphere, already brought strong hints about the role of QSLs in solar non-eruptive flares. More specifically, it has been shown that QSL footprints nicely match flare ribbons, which are elongated chromospheric brightenings observed in Hα and in the EUV, that result from the impact of particles accelerated in the corona during the flare (see e.g. Démoulin *et al.* 1997; Schmieder *et al.* 1997).

The role of QSLs in eruptive and so-called two-ribbon flares is more elusive, essentially due to the difficulty in calculating non-linear force-free extrapolations, that are required to recover the pre-eruptive highly current-carrying bipolar fields embedded in nearly potential arcades. Two arguments, however, suggest that QSLs are also involved in such events. Firstly, topological analyses of such configurations have been performed for purely theoretical models. Double-J shaped QSL footprints have been found to occur systematically around moderately (as well as strongly) twisted flux tubes (Démoulin *et al.* 1996b; Titov 2007; see Figure 3). Once knowing this, flare ribbons having such double-J shapes as recorded by various solar instruments can be identified in many observed eruptive events. One of the clearest cases, so far, has been reported by Chandra *et al.* (2009), and is shown in Figure 3.

Secondly, high-cadence HXR and EUV observations have revealed that strong brightenings are often seen to quickly propagate along eruptive flare ribbons (see Fletcher & Hudson. 2002 and Bogachev *et al.* 2005). While this can be interpreted in terms of a 2D moving reconnection site in the corona, or by sequences of 2D tether-cutting reconnections involving arcades with different shears from one another, it is also compatible with the gradual formation of elongated channels of propapations for accelerated particles, associated with slipping and slip-running field lines, hence with QSL reconnection. Such tether-cutting slip-running reconnection has indeed been found in MHD simulations of solar eruptions (Aulanier *et al.* 2010; Fan 2010).

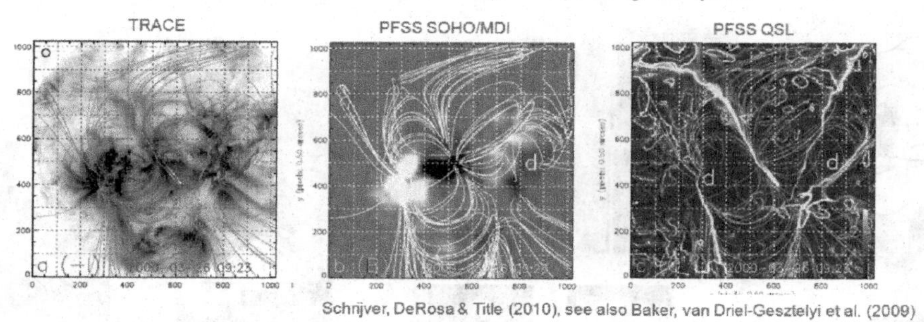

Figure 4. Warm EUV loops associated with QSL footpoints in a current-free magnetic field.

QSLs have also been associated with relatively quiet, non flaring, coronal loops in active regions, e.g. by combining magnetic field extrapolations and EUV or SXR observations.

Fletcher *et al.* (2001) associated Mg IX (resp. Si XII) emissions with QSL footprints (resp. loops), and concluded that transition region (resp. coronal) bright emissions could be associated with QSL heating. Schrijver *et al.* (2010) later identified that large-scale, current-free and warm coronal loops observed in Fe X, which interconnected several flux concentrations in active complexes, were typically rooted in QSL footprints (see Figure 4). They concluded that coronal heating should there be located at low heights (in accordance with current ideas of coronal heating in warm loops), since field line crossings (and therefore CS) produced by photospheric motions should remain confined at low altitudes, owing to the divergence with height of QSL flux tubes. Aulanier *et al.* (2007) reported on a rare observation of apparent bidirectional slipping of hot coronal loops observed in SXR, which arguably ressembles slipping and slip-running reconnection in QSL, and therefore concluded that QSLs may also play a role in the heating of hot active region loops. Baker *et al.* (2009) related the location of QSL footprints to the edge of large flow areas identified with Fe XII dopplershifts, and concluded that magnetic reconnection in QSLs could also be responsible for these hot coronal outflows, which are now typically observed on the edges of active regions.

4. Are there QSLs in braided homogeneous and complex flux tubes?

One of the main DC model for coronal heating is the Parker model, in which continuously and slowly braided field lines, in an initially homogeneous magnetic field, produce numerous CS whose dissipation produces heating. Over the years, this process was modeled with MHD simulations with larger and larger Reynolds numbers, first finding laminar/steady (van Ballegooijen 1986; Mikic 1989) and later turbulent/transient (Gómez *et al.* 1995; Hendrix & van Hoven 1996; Rappazzo *et al.* 2008) CS.

The question of the geometrical properties of the magnetic field in these models was addressed by Milano *et al.* (1999). They found that two simple rotational footpoint mo-

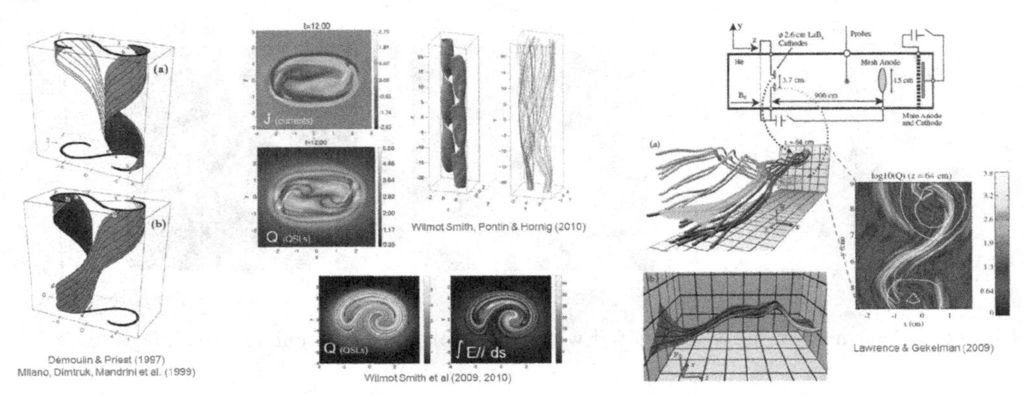

Figure 5. QSL located around and at the boundary between two twisted flux tubes (left), and QSL-related CS found with interacting twisted flux tubes, being produced numerically by boundary driving (middle) and experimentally by current injection (right).

tions, with a stagnation point in between, produce one single QSL-related S-shaped CS, at the boundary between the twisting flux tubes (in accordance with analytical predictions by Priest & Démoulin 1997). This CS eventually becomes turbulent as reconnection there proceeds. This association was somehow recovered in different numerical expermiments (Wilmot-Smith*et al.* 2009, 2010), which found a good match between the locations of large Q values and of strong currents, or integrated parallel electric fields, depending on the model. A QSL-related CS was also directly measured in a laboratory experiment of two interacting twisted flux tubes, there produced by a prescribed electric current instead of by footpoint motions (Lawrence & Gekelman 2009).

These results, which are illustrated in Figure 5, tend to argue in favor of an important role of QSLs in turbulent braided field with very large Reynolds numbers, which can exist in the Sun's and in star's coronae (see e.g. Rappazzo *et al.* 2008), hence in coronal heating at yet-unobservable subscales inside individual coronal loops.

At moderately larger scale, it is known that magnetoconvection typically produces numerous and slowly evolving flux concentrations in the photosphere. These not only lead to a departure from the homogeneous magnetic field models described above, but also to a complex magnetic field pattern (sometimes called the magnetic carpet, observed both in the quiet Sun and inside active region faculae). Such complex magnetic fields ought to result in numerous QSLs, hence in numerous CS when driven by any smooth photospheric motion. In this picture, Figures 1 and 2 only show a mild idea of the complex QSL and CS patterns that can exist in the real Sun's atmosphere. Such complex QSL patterns, here shown in Figure 6, were first modeled by Priest & Démoulin (1997), and later studied by Restante *et al.* (2009).

While the results of Schrijver *et al.* (2010) indicate that such complex QSL patterns can indeed be related to EUV coronal loops, quantifying their role in the real coronal heating process in the Sun, and in the one modeled in braiding simulations of action region magnetic fields obtained from force-free extrapolations (Gudiksen & Nordlund 2005; see Figure 6), will still require more investigation.

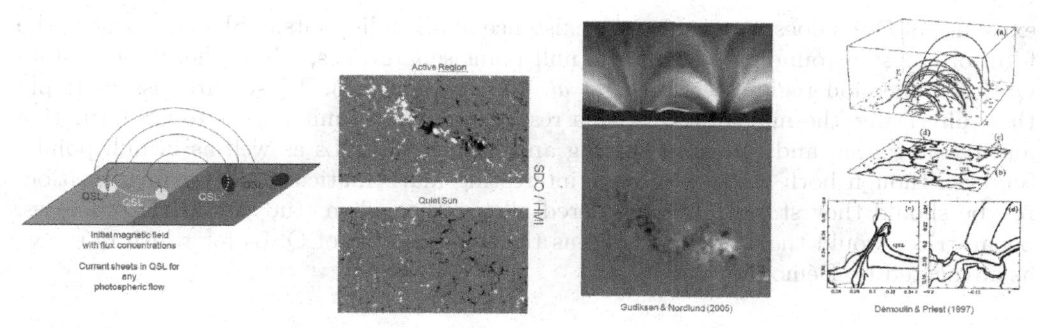

Figure 6. Observed complex flux concentrations (2nd column), resulting in QSLs, as hinted from geometrical arguments (1st column) indeed found in potential field models (4th column), and maybe occuring in MHD simulations of braided active regions (3rd column).

5. Discussion

QSLs are narrow layers of large spatial variations for the connectivity of magnetic field lines. QSLs naturally occur both in complex (but not necessarily multipolar) surface magnetic field distributions, and in the presence of large-scale volumetric field-aligned electric currents, that produce localized (but not necessarily large) magnetic twist.

The current highest-resolution photospheric magnetic field measurements and coronal images (such as those of the *Hinode*, *SDO* and *HiRISE*) show that both situations indeed exist at various scales in the Sun's atmosphere: close to sunspots, between or inside active regions, and also in the quiet Sun. Several studies illustrated in this paper, when put all together, argue in favor of the ubiquituous existence of QSLs, and of their associated slipping and slip-running reconnection regimes, in the heating of solar coronal loops and in driving reconnection in solar flares. By extension, it is also arguable that the dissipation regions which occur in line-tied MHD turbulence (Gudiksen & Nordlund 2005; Rappazzo *et al.* 2008) also correspond to QSLs. If so however, the causal relationship between QSL reconnections and the turbulent cascade will have to be analyzed.

Can these results be applied to star's coronae? The reliability of the Zeeman Doppler Imaging (ZDI) mapping of stellar surface magnetic fields is sill debated (see e.g. Carroll 2011). However, ZDI has recently provided impressive results, e.g. using the *ESPaDOnS* and *NARVAL* instruments (see e.g. Jardine *et al.* 2011). The obtained maps often show complex magnetic field distributions, which do fit the geometrical requirements for QSLs to exist. QSLs may therefore also be involved in the heating of some stellar's coronae. Testing this conjecture might be achieved through calculating QSLs and their localized enhanced EUV or X-ray emissions, using magnetic field extrapolations from ZDI maps, and comparing them with observed light curves as the stars rotate.

One may finally address the issue of the physical difference between QSLs and separatrices, as far as MHD processes only are concerned. First, CS naturally form in QSLs in the same way as they do in separatrices (Aulanier *et al.* 2005). Second, the slip-running reconnection regime in QSLs leads to super-Alfvénic field lines motions, which cannot be perceived by the magnetized plasma (Aulanier *et al.* 2006). Third, even though the existence of QSLs does not require the existence of 3D null points, QSLs also exist in the form of halos surrounding asymmetric null point separatrices, which there also sustain QSL reconnection regimes (Masson *et al.* 2009, submitted). These three issues imply that, physically, the magnetized plasma responds in very similar ways to CS formation and reconnection, and therefore heating and flaring, in QSLs as well as in null points. So, even though both features provide interesting mathematical tools for investigation, maybe should they start to be considered all together when studying MHD processes. Separatrices would then be considered as the limiting case of QSLs for which $Q \mapsto \infty$, as advocated by Démoulin (2006).

References

Aulanier, G., Pariat, E., & Démoulin, P. 2005, *Astron. Astrophys*, 444, 961
Aulanier, G., Pariat, E., Démoulin, P., & DeVore, C. R. 2006, *Solar Phys.*, 238, 347
Aulanier, G., Golub, L., DeLuca, E. E., Cirtain, J. W., & Kano, R., *et al.* 2007, *Science*, 318, 1588
Aulanier, G., Török, T., & Démoulin, P., DeLuca E. E. 2010, *Astrophys. J.*, 708, 314
Baker, D., van Driel-Gesztelyi, L., & Mandrini, C. H., *et al.* 2009, *Astrophys. J.*, 705, 926
Bogachev, S. A., Somov, B. V., & Kosugi, T. Sakao, T. 2005, *Astrophys. J.*, 630, 561
Carroll, T. A. 2011, *Proceedinds of the IAU Symposium 273*, this issue
Chandra, R., Schmieder, B., Aulanier, G., & Malherbe, J.-M. 2009, *Solar Phys.*, 258, 53
Démoulin, P., Hénoux, J.-C., Priest, E. R., Mandrini, C.H. 1996a *Astron. Astrophys*, 308, 643
Démoulin, P., Priest, E.R., Lonie, D. P. 1996b *JGR*, 101, A4, 7631
Démoulin, P., Bagala, L.G., Mandrini, C.H., Hénoux, J.C., Rovira, M.G. 1997 *Astron. Astrophys*, 325, 305
Démoulin, P. 2006 *Adv. Sp. Res.*, 37, 7, 1269
Fan, Y. 2010, *Astrophys. J.*, 719, 728
Fletcher, L., López Fuentes, M. C., & Mandrini, C. H., *et al.* 2001, *Solar Phys.*, 203, 255
Fletcher, L. & Hudson, H. S. 2002, *Solar Phys.*, 210, 307
Galsgaard, K., Titov, V. S., & Neukirch, T. 2003, *Astrophys. J.*, 595, 506
Galsgaard, K. 2011, *Proceedinds of the IAU Symposium 273*, this issue
Gómez, D. O., DeLuca, E. E., & McClymont, A. N. 1995, *Astrophys. J.*, 448, 954
Gudiksen, B. V. & Nordlund, A. 2005, *Astrophys. J.*, 618, 1031
Hendrix, D. L. & van Hoven, G. 1996, *Astrophys. J.*, 467, 887
Hesse, M., Schindler, K. 1988 *JGR*, 93, 5559
Jardine, M., Donati, J.-F. *et al.* 2011 *Proceedinds of the IAU Symposium 273*, this issue
Lawrence, E. E. & Gekelman, W. 2009, *PRL*, 103, 10, 105002
Mandrini, C.H., Démoulin, P., Klimchuk, J.A. 2000 *Astrophys. J.*, 530, 999
Masson, S., Pariat, E., Aulanier, G., & Schrijver, C. J. 2009, *Astrophys. J.*, 700, 559
Masson, S., Aulanier, G., Pariat, E., & Klein, K. L. 2010, *Astrophys. J.* (Letters), submitted
Mikic, Z., Schnack, D. D., & van Hoven, G., 1989, *Astrophys. J.*, 338, 1148
Milano, L. J., Dmitruk, P., Mandrini C. H., Gómez, D. O., & Dmoulin, P. 1999, *Astrophys. J.*, 521, 889
Priest, E. R. & Forbes, T. G. 1992, *JGR*, 97, A12, 1521
Priest, E. R. & Démoulin, P. 1995, *JGR*, 100, A12, 23443
Priest, E. R. & Démoulin, P. 1997, *Solar Phys.*, 175, 123
Rappazzo, A.F., Velli, M., Einaudi, G., Dahlburg, R.B. 2008 *Astrophys. J.*, 677, 1348
Restante, A. L., Aulanier, G., & Parnell, C. E. 2009, *Astron. Astrophys*, 508, 433
Schmieder, B., Aulanier, G., Démoulin, P., & van Driel-Gesztelyi, L., *et al.* 1997, *Astron. Astrophys*, 325, 1213
Schrijver, C. J., DeRosa, M. L., & Title, A. M. 2010, *Astrophys. J.*, 719, 1083
Titov, V. S. 2007, *Astrophys. J.*, 660, 863

Titov, V. S., Hornig, G., & Démoulin, P. 2002, *JGR*, 107, A8, SSH 3-1
van Ballegooijen, A. A. 1986, *Astrophys. J.*, 311, 1001
Wiegelmann, T. 2011, *Proceedinds of the IAU Symposium 273*, this issue
Wilmot-Smith, A. L., Hornig, G., & Pontin, D. I. 2009, *Astrophys. J.*, 696, 1339
Wilmot-Smith, A. L., Pontin, D. I., & Hornig, G. 2010, *Astron. Astrophys*, 516, 5

The Physics of Sun and Star Spots
Proceedings IAU Symposium No. 273, 2010
D.P. Choudhary & K.G. Strassmeier, eds.

doi:10.1017/S1743921311015316

Modelling stellar coronal magnetic fields

Moira Jardine[1], Jean-Francois Donati[2], Doris Arzoumanian[1] and Aline de Vidotto[1]

[1]SUPA, School of Physics and Astronomy, University of St Andrews, North Haugh, StAndrews, KY16 9SS, UK
email: *mmj@st-andrews.ac.uk, Aline.Vidotto@st-andrews.ac.uk, doris.arzoumanian@cea.fr*

[2]LATT, CNRS–UMR 5572, Obs. Midi-Pyrénées, 14 Av. E. Belin, F–31400 Toulouse, France
email: donati@ast.obs-mip.fr

Abstract. Our understanding of the structure and dynamics of stellar coronae has changed dramatically with the availability of surface maps of both star spots and also magnetic field vectors. Magnetic field extrapolations from these surface maps reveal surprising coronal structures for stars whose masses and hence internal structures and dynamo modes may be very different from that of the Sun. Crucial factors are the fraction of open magnetic flux (which determines the spin-down rate for the star as it ages) and the location and plasma density of closed-field regions, which determine the X-ray and radio emission properties. There has been recent progress in modelling stellar coronae, in particular the relative contributions of the field detected in the bright surface regions and the field that may be hidden in the dark star spots. For the Sun, the relationship between the field in the spots and the large scale field is well studied over the solar cycle. It appears, however, that other stars can show a very different relationship.

Keywords. Stars:magnetic fields, stars:coronae, stars:imaging, stars:spots

---

1. Introduction

Just as the surface distributions of spots on stars can vary considerably between different types of stars, so too can the structure of their coronae. For massive stars, the nature of the coronal magnetic field may have an influence on the structure of the wind (which can remove a significant fraction of the star's mass over its lifetime). We do not, however, know if the magnetic fields of these stars are generated by dynamo processes (Charbonneau & MacGregor 2001; Brun *et al.* 2005; Spruit 2002; Tout & Pringle 1995; MacDonald & Mullan 2004; Mullan & MacDonald 2005; Maeder & Meynet 2005) or if they are fossil fields (Moss 2001; Braithwaite & Spruit 2004; Braithwaite & Nordlund 2006). An equally challenging question is the source and generation mechanism of the observed X-ray emission (Ignace *et al.* 2010).

In the case of solar-mass stars, the nature of the magnetic cycles and the lifetimes of active regions are topics that will benefit enormously from the data that will soon be available from CoRoT and Kepler. As has been mention already in this symposium, the variation with mass of the differential rotation is another crucial question. At the bottom end of the main sequence, the low mass, fully convective stars appear to have much simpler, stronger fields than their higher-mass counterparts. While a decade or so ago, it was believed that these stars could only generate small-scale magnetic fields (Durney *et al.* 1993; Cattaneo 1999), more recent studies have suggested that large scale fields may be generated. These models differ, however, in their predictions for the form of this field and the associated latitudinal differential rotation. They predict that the fields should be either axisymmetric with pronounced differential rotation (Dobler *et al.* 2006), non-axisymmtric with minimal differential rotation (Küker & Rüdiger 1997, 1999; Chabrier

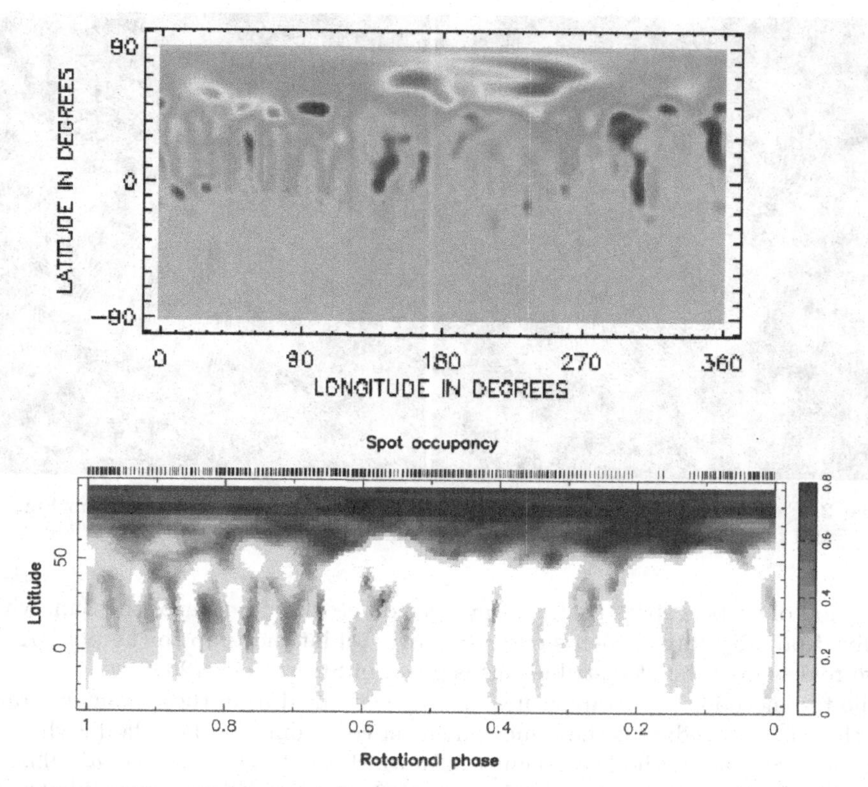

Figure 1. Zeeman-Doppler maps of the radial field component (top) and the corresponding spot occupancy map (bottom) for the AB Dor (observations carried out in 2004).

& Küker 2006). More recently, however, Browning (2008) has published a dynamo model that produces a highly symmetric field with little differential rotation. This field structure may have implications for the nature of the spin-down of these low mass stars, since they show puzzlingly-high rotation rates, even on the main sequence.

Even in the pre-main sequence phase, the structure of star's corona can be important since it affects the rate at which the star can accrete mass from its disk (Gregory *et al.* 2006b, 2010). Low mass stars remain fully convective throughout their evolution onto the main sequence, but higher mass stars will develop a radiative core at some stage - perhaps even while they are still accreting from their disks. This difference in internal structure may lead to different dynamo behaviours, and indeed there is a suggestion from observations of a handful of stars that the field structures of stars with a radiative core (and hence a tachocline where a solar-like interface dynamo might be active) are more complex than those of the fully-convective stars (Donati *et al.* 2007, 2008b; Jardine *et al.* 2008; Gregory *et al.* 2008; Hussain *et al.* 2009). The implications for the spin evolution of the stars have not yet been fully explored however.

2. Modelling stellar coronae

The first step in modelling the corona of a star is to determine the form of the surface field. This is most commonly done using the technique of Zeeman-Doppler imaging (Donati & Collier Cameron 1997; Donati *et al.* 1999). This typically shows a complex

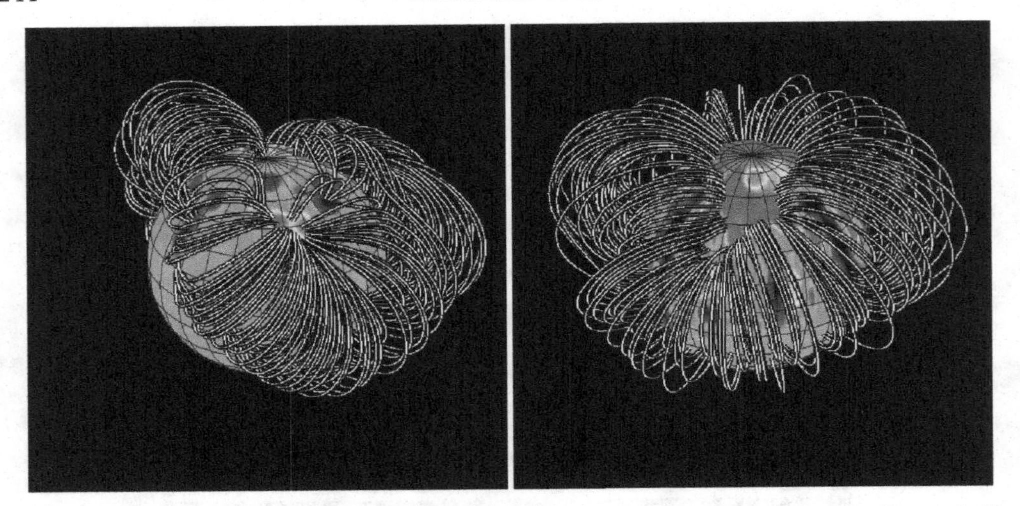

Figure 2. Field extrapolation from the original Zeeman-Doppler map shown in Fig. 1 (left) and from the maps of the combined Zeeman-Doppler and spot field (right).

distribution of surface spots that is often very different from that of the Sun, with spots and mixed polarity flux elements extending over all latitudes up to the pole. A comprehensive review of starspot distributions is given in Strassmeier (2009).

While the methods of extrapolating the magnetic field from these magnetograms may differ, the basic procedure is the same. The most commonly-used method is the *Potential Field Source Surface* method (Altschuler & Newkirk, Jr. 1969; Jardine *et al.* 1999; Jardine *et al.* 2001; Jardine *et al.* 2002a; McIvor *et al.* 2003), but it is also possible to use non-potential fields (Donati 2001; Hussain *et al.* 2002). Once the 3D structure of the magnetic field has been determined, the structure of the coronal gas can be determined by assuming that the gas trapped on these field lines is in isothermal, hydrostatic equilibrium. We can then determine the coronal gas pressure, subject to an assumption for the gas pressure at the base of the corona. We assume that it is proportional to the magnetic pressure, i.e. $p_0 \propto B_0^2$, where the constant of proportionality is determined by comparison with X-ray emission measures (Jardine *et al.* 2002b; Jardine *et al.* 2006; Gregory *et al.* 2006a). For an optically thin coronal plasma, this then allows us to produce images of the X-ray emission.

3. Effect of spot magnetic fields on coronal structure

The fraction of a star's surface that is covered in spots and the strength and degree of complexity of the coronal magnetic field all vary with stellar mass. While low mass, fully convective stars commonly have strong, simple surface fields with few spots (Donati *et al.* 2008a; Morin *et al.* 2008, 2010), solar mass stars such as AB Dor (shown in Fig 1) typically have complex fields and spots that can cover a significant fraction of their surfaces (Strassmeier 2009). Zeeman-Doppler imaging does not recover all of the magnetic field in these spotted regions - a factor that may be more important in stars with greater spot coverage (Johnstone *et al.* 2010). Although we do no know the polarity of the magnetic field in the spots, we can take a statistical approach. We randomly allocate a polarity to each spot (defined at a certain level of spot occupancy) and then add this field to the surface field reconstructed by Zeeman-Doppler imaging. By repeating this process many times and each time extrapolating the coronal field, we can create

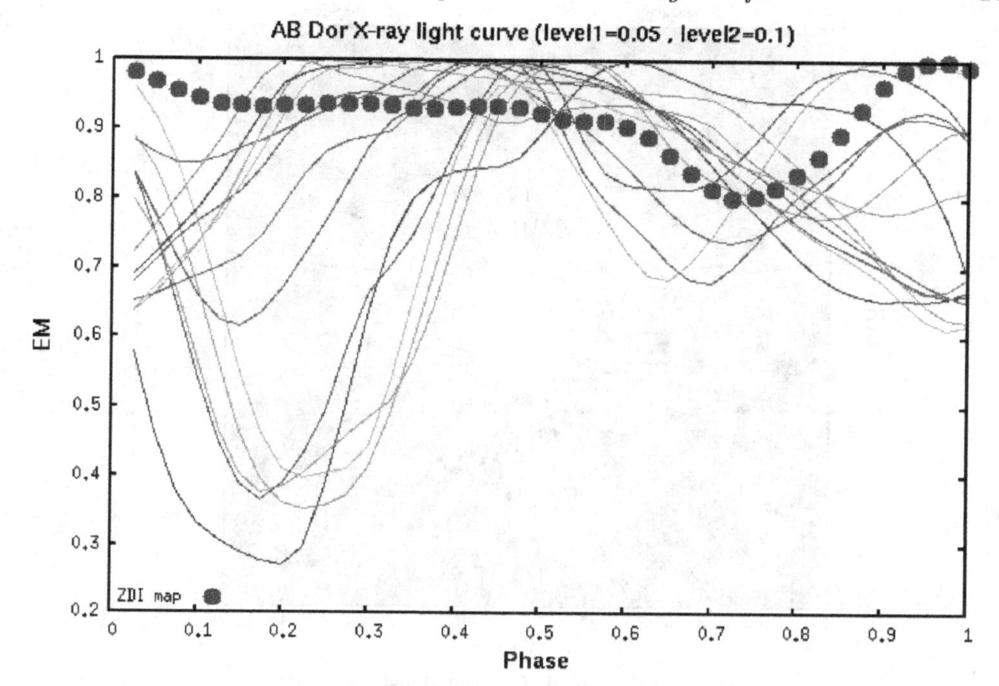

Figure 3. X-ray rotational modulation of AB Dor. The dots represent the X-ray rotational modulation calculated from the ZDI map. The lines represent the X-ray rotational modulation using the combined maps (ZDI + spot maps).

many realisations of the structure of the corona (see for example Figs 2). By doing this for stars of different masses, we can determine the possible effect that the flux hidden in the spots might have on the corona (Arzoumanian *et al.* 2010). Because the spot field typically has a greater degree of axial symmetry, this process often shifts the large-scale dipole axis closer to the rotation axis. The fraction of open flux and the magnitude of the X-ray emission measure also typically increase. Fig. 3 shows some examples of the X-ray light curves that might be produced.

4. Moving towards full MHD

While the potential field source surface models have the advantage of computational speed and simplicity, they only allow indirectly for the effect of the stellar wind. In order to improve our understanding of the interplay between coronal and wind processes in stars we need to incorporate the observed surface magnetograms into an MHD code (Cohen *et al.* 2010; Vidotto *et al.* 2010). First results from this for the fully-convective star V374 Peg have been extremely interesting (see Fig. 4). Like many low mass stars, V374 Peg rotates extremely rapidly and has a strong, almost dipolar field. This means that magnetic forces dominate over the other forces and results in a very non-solar, magneto-centrifugally driven wind. The general scalings that emerge from this study are that both the angular momentum loss rate and the mass loss rate scale as $n^{1/2}$ (where n is the density) and hence the spin-down time scales as $n^{-1/2}$. The fact that V374 Peg is a rapid rotator therefore suggests it has a coronal density that is low by stellar standards, if high relative to the Sun. The mass loss rate and speed of the wind are both also higher than

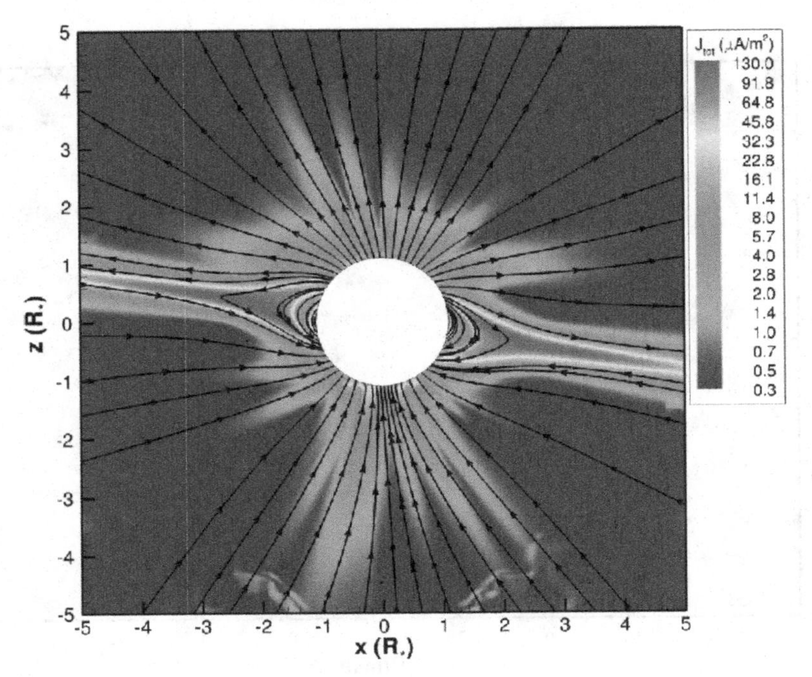

Figure 4. Current density.

solar values. Most interesting, however, is the ram pressure which may be five orders of magnitude higher than that of the solar wind. This has significant implications for planet habitability, since such a high ram pressure may crush planetary magnetospheres. If, for example, we place a planet in the habitable zone of an M-dwarf star, it needs a field of about 8G (around half that of Jupiter) to maintain a magnetosphere. The means by which such a field might be generated in the planetary interior are still a matter of debate.

5. Conclusions

The availability of spot maps and magnetograms for stars other than the Sun has revolutionised our view of stellar magnetic fields. By using these maps to model the coronal structure of a range of stars we can also gain insight into their winds, their rotational evolution and their coronal emission. Advances in MHD modelling that incorporate the three vector components of the surface magnetic field promise to bring about a similar revolution.

References

Altschuler, M. D. & Newkirk, Jr., G. 1969, *Solar Phys.*, 9, 131
Arzoumanian, D., Jardine, M., Donati, J., Morin, J., & Johnstone, C. 2010, ArXiv e-prints
Braithwaite, J. & Nordlund, A. 2006, *Astron. Astrophys*, 450, 1077
Braithwaite, J. & Spruit, H. C. 2004, *Nature*, 431, 819
Browning, M. K. 2008, *Astrophys. J.*, 676, 1262
Brun, A. S., Browning, M. K., & Toomre, J. 2005, *Astrophys. J.*, 629, 461
Cattaneo, F. 1999, *Astrophys. J.*, 515, L39

Chabrier, G. & Küker, M. 2006, *Astron. Astrophys*, 446, 1027
Charbonneau, P. & MacGregor, K. B. 2001, *Astrophys. J.*, 559, 1094
Cohen, O., Drake, J. J., Kashyap, V. L., Hussain, G. A. J., & Gombosi, T. I. 2010, ArXiv e-prints
Dobler, W., Stix, M., & Brandenburg, A. 2006, *Astrophys. J.*, 638, 336
Donati, J., Morin, J., Petit, P., *et al.* 2008a, *Mon. Not. Roy. Astron. Soc.*, 390, 545
Donati, J.-F. 2001, LNP Vol. 573: Astrotomography, *Indirect Imaging Methods in Observational Astronomy*, 573, 207
Donati, J.-F. & Collier Cameron, A. 1997, *Mon. Not. Roy. Astron. Soc.*, 291, 1
Donati, J.-F., Collier Cameron, A., Hussain, G., & Semel, M. 1999, *Mon. Not. Roy. Astron. Soc.*, 302, 437
Donati, J.-F., Jardine, M. M., Gregory, S. G., *et al.* 2007, *Mon. Not. Roy. Astron. Soc.*, 380, 1297
Donati, J.-F., Jardine, M. M., Gregory, S. G., *et al.* 2008b, *Mon. Not. Roy. Astron. Soc.*, 386, 1234
Durney, B. R., De Young, D. S., & Roxburgh, I. W. 1993, *Solar Phys.*, 145, 207
Gregory, S. G., Jardine, M., Cameron, A. C., & Donati, J.-F. 2006a, *Mon. Not. Roy. Astron. Soc.*, 373, 827
Gregory, S. G., Jardine, M., Gray, C. G., & Donati, J. 2010, ArXiv e-prints
Gregory, S. G., Jardine, M., Simpson, I., & Donati, J.-F. 2006b,*Mon. Not. Roy. Astron. Soc.*, 371, 999
Gregory, S. G., Matt, S. P., Donati, J., & Jardine, M. 2008, *Mon. Not. Roy. Astron. Soc.*, 389, 1839
Hussain, G. A. J., Collier Cameron, A., Jardine, M. M., *et al.* 2009, *Mon. Not. Roy. Astron. Soc.*, 398, 189
Hussain, G. A. J., van Ballegooijen, A. A., Jardine, M., & Collier Cameron, A. 2002, *Astrophys. J.*, 575, 1078
Ignace, R., Oskinova, L. M., Jardine, M., *et al.* 2010, ArXiv e-prints
Jardine, M., Barnes, J., Donati, J.-F., & Collier Cameron, A. 1999, *Mon. Not. Roy. Astron. Soc.*, 305, L35
Jardine, M., Collier Cameron, A., & Donati, J.-F. 2002a, *Mon. Not. Roy. Astron. Soc.*, 333, 339
Jardine, M., Collier Cameron, A., Donati, J.-F., Gregory, S. G., & Wood, K. 2006, *Mon. Not. Roy. Astron. Soc.*, 367, 917
Jardine, M., Collier Cameron, A., Donati, J.-F., & Pointer, G. 2001, MNRAS, 324, 201
Jardine, M., Wood, K., Collier Cameron, A., Donati, J.-F., & Mackay, D. H. 2002b, *Mon. Not. Roy. Astron. Soc.*, 336, 1364
Jardine, M. M., Gregory, S. G., & Donati, J. 2008, *Mon. Not. Roy. Astron. Soc.*, 386, 688
Johnstone, C., Jardine, M., & Mackay, D. H. 2010, *Mon. Not. Roy. Astron. Soc.*, 404, 101
Küker, M. & Rüdiger, G. 1997, *Astron. Astrophys*, 328, 253
Küker, M. & Rüdiger, G. 1999, in *ASP Conf. Ser. 178: Workshop on stellar dynamos*, Vol. 178, 87–96
MacDonald, J. & Mullan, D. J. 2004, *Mon. Not. Roy. Astron. Soc.*, 348, 702
Maeder, A. & Meynet, G. 2005, *Astron. Astrophys*, 440, 1041
McIvor, T., Jardine, M., Cameron, A. C., Wood, K., & Donati, J.-F. 2003, *Mon. Not. Roy. Astron. Soc.*, 345, 601
Morin, J., Donati, J., Petit, P., *et al.* 2008, *Mon. Not. Roy. Astron. Soc.*, 390, 567
Morin, J., Donati, J., Petit, P., *et al.* 2010, *Mon. Not. Roy. Astron. Soc.*, 1077
Moss, D. 2001, in ASP Conference Series, Vol. 248, *Magnetic fields across the Hertzsprung-Russell diagram*, ed. S. G. Mathys & D. Wickramasinghe (San Francisco), 305
Mullan, D. J. & MacDonald, J. 2005, *Mon. Not. Roy. Astron. Soc.*, 356, 1139
Spruit, H. 2002, *Astron. Astrophys*, 381, 923
Strassmeier, K. G. 2009, *A. Ast. Rev.*, 17, 251
Tout, C. A. & Pringle, J. E. 1995, *Mon. Not. Roy. Astron. Soc.*, 272, 528
Vidotto, A., Jardine, M., Opher, M., Donati, J.-F., & Gombosi, T. I. 2010, *Mon. Not. Roy. Astron. Soc.* (in press)

Discussion

STRASSMEIER: You are quite right with our inability to reconstruct magnetic field from flux deficient spots. However, we know that this is an artifact and recent results by Berdyugina and her group showed the existence of fields in starspots.

JARDINE: I think it would, and I think the solar community should solve the problem for the sun as well. It's – it's only true in its very [INAUDIBLE] limit where you have the very, very strong fields. It's not true for a solar type. So it's specific to these low mass stars in their incredibly strong field and rapid rotation. We tried several different types of assumption. We tried two types of star. We took Capella and V374Peg which has very small thin red pots, and we tried three different ways of allocating the field to the spot – so taking a different brightness contour to be a different field strength. The results you get depend on what you assume. I think it would be very nice actually to have some input perhaps from the solar community as to what might be a better way to do that. We recently tried three different parametration.

DUPREE: I have a wind question. If the sun was rapidly rotating and the centrifugally driven wind, how would the profile of the wind compare as to our standard solar wind?

JARDINE: It would be very much faster, and the RAM pressure would be very much greater. The critical point – the critical point essentially depends on the temperature. So it would depend on how hot the wind is. If the sun was still at its same temperature, the critical point would be the at the same point. So the sonic point would be the same; but the terminal velocity would be very different, much higher.

The Physics of Sun and Star Spots
Proceedings IAU Symposium No. 273, 2010
D.P. Choudhary & K.G. Strassmeier, eds.

doi:10.1017/S1743921311015328

The spots on Ap stars

Oleg Kochukhov

Department Physics and Astronomy, Uppsala University,
Box 516, SE-75120 Uppsala, Sweden
email: oleg.kochukhov@fysast.uu.se

Abstract. The upper main sequence magnetic chemically peculiar (Ap) stars exhibit a non-uniform distribution of chemical elements across their surfaces and with height in their atmospheres. These inhomogeneities, responsible for the conspicuous photometric and spectroscopic variation of Ap stars, are believed to be produced by atomic diffusion operating in the stellar atmospheres stabilized by multi-kG magnetic fields. Here I present an overview of the current state-of-the-art in understanding Ap-star spots and their relation to magnetic fields. In particular, I highlight recent 3-D chemical spot structure studies and summarize magnetic field mapping results based on the inversion of the full Stokes vector spectropolarimetric observations. I also discuss a puzzling new type of spotted stars, HgMn stars, in which the formation and evolution of heavy element spots is driven by a poorly understood mechanism, apparently unrelated to magnetic fields.

Keywords. Polarization, stars: chemically peculiar, stars: magnetic fields, stars: atmospheres

---

1. Introduction

The early-type magnetic chemically peculiar stars, often referred to as Ap/Bp stars, are main sequence objects found over the entire spectral type range from early B to early F, but most commonly around A0. These stars are not constrained to a particular evolutionary stage (Kochukhov & Bagnulo 2006) and are found both near the ZAMS in young open clusters (e.g., Folsom *et al.* 2008) and among the stars at the end of the hydrogen core burning phase (Kochukhov *et al.* 2006).

The most conspicuous and unusual property of Ap stars is the presence of strong, globally organized magnetic fields on their surfaces. The field intensity ranges from a well-defined lower threshold of ≈ 300 G (Aurière *et al.* 2007), probably representing the weakest global field able to withstand shearing by the differential rotation, to slightly over 30 kG for a few extreme objects (Babcock 1960; Elkin *et al.* 2010).

The slow rotation of Ap stars, which probably results from an enhanced angular momentum loss during the PMS evolutionary stage (Stępień 2000), the absence of surface convection zone, and the lack of intense mass loss provides ideal conditions for a build up of chemical anomalies by atomic diffusion (Michaud 1970). This slow separation of chemical elements under the influence of radiative pressure and gravity is believed to alter the surface chemistry in many types of stars with extended radiative zones. But it is manifested most strongly in Ap stars, which often exhibit enormous overabundances of rare-earth and other heavy elements as well as a noticeable enhancement of Cr and Fe. The richness of the sharp-lined Ap-star spectra makes these objects the prime astrophysical laboratories for studies of exotic chemical species and for astrophysical verification of the atomic line data (Ryabchikova *et al.* 2006; Quinet *et al.* 2007).

The presence of a super-equipartition magnetic field, which usually has a topology close to an oblique dipole, breaks the spherical symmetry of the diffusive segregation of chemical elements, leading to the formation of spots and rings of enhanced element abundance (Michaud *et al.* 1981). The surface magnetic and chemical structure modifies the local energy balance, resulting in a strong UV to optical flux redistribution for the regions of Ap star surface characterized by an overabundance of heavy elements. In combination with the stellar rotation, this yields a periodic variation of the average magnetic field characteristics, line profiles, spectral energy distribution, and brightness in different photometric bands. Both the field structure and the geometry of chemical spots remain stable on the time scale of many decades. This stability allows an accurate analysis of the surface properties of Ap stars through multi-epoch observations and a very precise measurements of rotation rates, even revealing rotational braking for some of these stars (Adelman *et al.* 2001; Mikulášek *et al.* 2008).

Rich phenomenological picture of Ap stars has been traditionally addressed with simple empirical models, based on the correlations between photometric variations and the phase curves of the equivalent widths and radial velocities of spectral lines (e.g. Polosukhina *et al.* 2000; Leone & Catanzaro 2001), while the magnetic field properties were inferred by fitting low-order multipolar models to measurement of the mean longitudinal magnetic field and the mean field modulus (Landstreet & Mathys 2000).

Here I review more recent advanced studies of the magnetic field and spots on Ap stars, which have the ambition of developing much more realistic and detailed 3-D models of the chemical element distributions, atmospheric structure, and magnetic field topology. The theoretical basis of these emerging modeling approaches is provided by the developments in the fields of model atmospheres, polarized radiative transfer, and numerical methods of the inverse problem solution, while the observational material – intensity and polarization spectra at high spectral and time resolution, and often with a coverage of the entire optical domain – is supplied by the new generation spectrometers and spectropolarimeters at the 4–8-m class telescopes.

2. Doppler imaging of chemical spots and magnetic fields

The Doppler imaging (DI) of chemical inhomogeneities on Ap stars was the first application of this powerful remote sensing method to spotted stars (Khokhlova *et al.* 1986). The basic principle of DI is to use a resolution of the stellar surface provided by the Doppler broadening of line profiles together with the modulation due to stellar rotation for the reconstruction of two-dimensional maps of the stellar surface. For a couple of decades since its invention DI was applied only to a small number of elements (e.g., Hatzes 1991), ignoring effects of magnetic field on line profiles. At the same, the resulting abundance maps were compared with primitive multipolar models of magnetic field inferred from the longitudinal magnetic field measurements.

The development of the magnetic DI technique by Piskunov & Kochukhov (2002) made it possible to obtain self-consistent maps of abundance distributions and magnetic field using high-resolution polarization spectra (Kochukhov *et al.* 2002; Lüftinger *et al.* 2010). Moreover, inclusion of the linear polarization spectra in the inversion enabled reconstruction of the vector magnetic field uniquely and without adopting an *a priori* multipolar field parameterization. Ap stars is the only class of spotted magnetically active stars permitting such a sophisticated full Stokes line profile analysis.

As demonstrated by Kochukhov *et al.* (2004a) for 53 Cam and recently by Kochukhov & Wade (2010) for the prototype magnetic Ap star α^2 CVn (Fig. 1), the inclusion of the Stokes Q and U spectra can radically change our view on the field structure of Ap stars. For both stars the overall field topology, determined by the radial field component, is roughly dipolar, but the field intensity distribution is much more complex due to small-scale patches of mainly horizontal magnetic field. These structures cannot be described

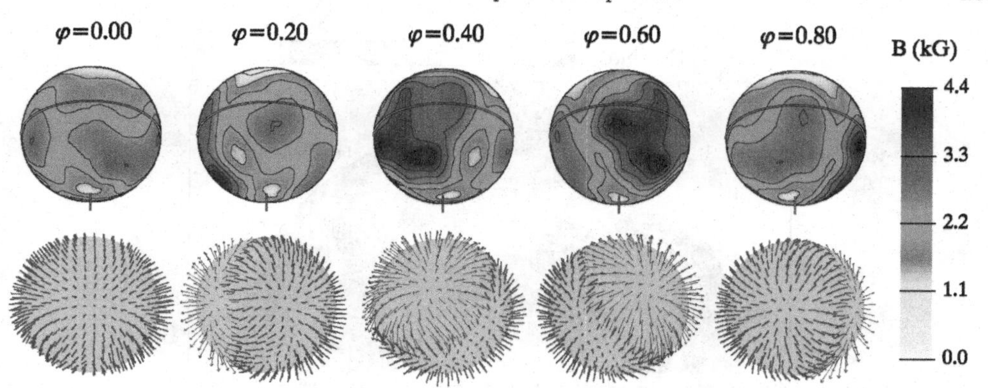

Figure 1. Representative results of the modern magnetic Doppler imaging analysis of the four Stokes parameter spectropolarimetric observations of a magnetic Ap star. The spherical plots show distribution of the magnetic field strength (*top row*) and field orientation (*bottom row*) over the surface of α^2 CVn. Adapted from Kochukhov & Wade (2010).

by any low-order multipolar field and are not detectable with the circular polarization spectra.

Meanwhile non-magnetic abundance mapping of Ap stars has reached maturity in its application to weak-field stars. Efficient numerical algorithms and the use of parallel supercomputers now allow reconstructing distributions for more than a dozen chemical elements (Kochukhov *et al.* 2004b). These comprehensive horizontal chemical structure studies generally reveal a great diversity of spot geometries, even for the chemical elements with similar spectra and position in the periodic table. A "typical" Ap-star chemical spot structure, with the iron peak elements concentrated at the magnetic equator and the rare-earth elements (REE) located at the poles of dipolar field is not confirmed. Only a few chemical elements (Li, O, Eu) show a well-defined spot or ring-like map correlating with the dipolar field component. Most other elements exhibit no such correspondence. It is possible that these elements are preferentially sensitive to the horizontal magnetic field component, which has a very complex topology according to the four Stokes parameter magnetic DI studies.

3. Vertical stratification of chemical abundances

Availability of very high quality spectroscopic observations of Ap stars and development of realistic magnetic spectrum synthesis codes (Wade *et al.* 2001) to interpret these observations opened an entirely new dimension of the Ap-spot research. It was recognized that previously dismissed spectral anomalies, such as ionization imbalance, large scatter of abundances derived from lines of different strength and excitation, etc., visible in the Ap-star spectra point to an inhomogeneous vertical distribution of chemical elements (Bagnulo *et al.* 2001; Ryabchikova *et al.* 2002). A separation of elements into chemically distinct layers was predicted by the theoretical diffusion studies (Babel 1992) but was not thoroughly studied observationally until now.

Many recent studies (e.g., Ryabchikova *et al.* 2006; Kochukhov *et al.* 2006, 2009; Shulyak *et al.* 2009) showed that all spectral indicators of a given element can be brought into a reasonable agreement with each other by introducing a sharp concentration gradient in the line forming region. Typically, the light and iron peak elements are found to have solar or subsolar abundance in the upper atmospheric layers, above $\log \tau_{5000} = 0$ to -2, and a 2–3 dex overabundance at the bottom of the atmosphere. The REEs show

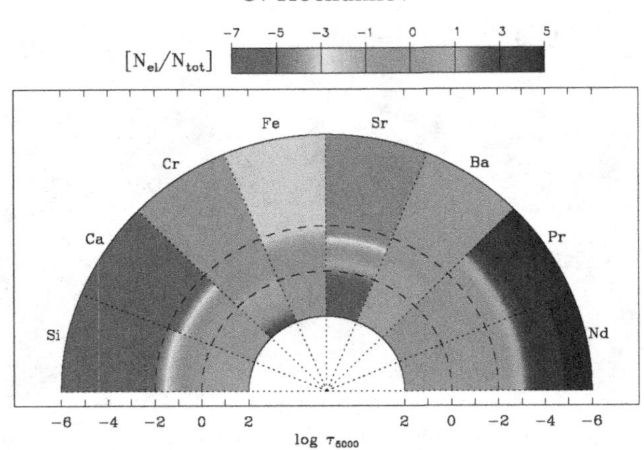

Figure 2. Chemical stratification in the atmosphere of the cool Ap star HR 1217 (Shulyak *et al.* 2009). Vertical abundance distributions are presented in a radial plot as a function of the continuum optical depth at λ = 5000 Å. Abundances relative to the Sun are given on a logarithmic scale.

an opposite chemical stratification profile, with an overabundance cloud located above $\log \tau_{5000} = -3$. Fig. 2 shows a typical chemical stratification in a cool Ap star, derived in this case for HR 1217 (Shulyak *et al.* 2009).

An independent confirmation of the reality of these enormous vertical abundance gradients came from the time-resolved spectroscopic studies of the pulsations in rapidly oscillating Ap (roAp) stars (Kochukhov 2006; Ryabchikova *et al.* 2007). All roAp stars show a large discrepancy of the pulsational characteristics between, on the one hand, the light and iron peak element spectral lines, which often show no detectable variability, and REE lines on the other hand, which pulsate with the radial velocity amplitudes of up to several km s^{-1}. Initially puzzling, this pulsational behaviour is now understood to be a natural consequence of the propagation of magneto-acoustic waves in a chemically stratified atmosphere (Khomenko & Kochukhov 2009). The wave amplitude increases rapidly with height, hence pulsations attain the largest amplitude in the uppermost atmospheric layers where REE lines are formed.

The presence of large vertical abundance gradients in many magnetic chemically peculiar stars calls for a revision of the very concept of Ap-star chemical "spot". The horizontal abundance maps derived with tradiational DI methods might represent an effective and simplified description of the vertical stratification variation across the surface due to its dependence on the local magnetic field, as anticipated by theoretical models (Alecian & Stift 2010). A study aiming to test this hypothesis with a simultaneous horizontal and vertical abundance mapping of chemical elements is currently underway.

4. Chemical weather in HgMn stars

According to the widely used classification scheme of the upper main sequence chemically peculiar stars (Preston 1974) these objects are separated in two distinct groups, depending on the presence of detectable magnetic field on the stellar surface. The objects in the first group, magnetic Ap/Bp stars, show large deviations from solar chemical composition, exhibit strong fossil fields, stable chemical spots, and show rotational variability of the photometric parameters and line profiles. The second group, non-magnetic CP stars, is represented by Am ($T_{eff} \leqslant$ 10000 K) and HgMn ($T_{eff} \geqslant$ 10000 K) stars, which are characterized by moderate chemical anomalies and were believed to show

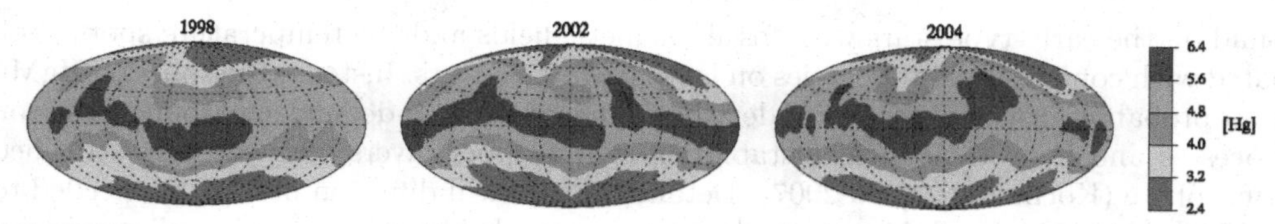

Figure 3. Evolution of mercury clouds on the surface of α And (Kochukhov *et al.* 2007). The DI mercury abundance maps in the Hammer-Aitoff projection show configuration of Hg spots for years 1998, 2002, and 2004. Abundances relative to the Sun are given on a logarithmic scale.

neither photometric nor spectroscopic variability. Spectropolarimetric surveys of Am and HgMn stars (Shorlin *et al.* 2002) and weak-field Ap stars (Aurière *et al.* 2007) reinforced the notion of this magnetic dichotomy by showing that the minimum magnetic field found in Ap stars significantly exceeds the upper magnetic field strength limits established for bright Am and HgMn stars. In this context, a strong magnetic field was thought to be a necessary prerequisite for the formation of chemical abundance spots on early-type stars.

This understanding of the relation between magnetism and spot formation on early-type stars has proved to be wrong. A series of recent spectroscopic studies demonstrated the presence of surface chemical inhomogeneities in several HgMn stars (Adelman *et al.* 2002; Kochukhov *et al.* 2005; Folsom *et al.* 2010; Briquet *et al.* 2010, Korhonen *et al.*, this meeting). These non-uniform abundance distributions typically have a lower contrast than those found on the magnetic Ap stars of similar T_{eff} and are observed only for a small number of chemical species, such as Ti, Y, Sr, Pt, and Hg, showing significant anomaly of the average abundance compared to the solar chemical composition.

The discovery of spots on HgMn stars revived the discussion of their magnetic status. Although no conclusive magnetic field detections were reported for these stars by studies using previous generation spectropolarimeters (Shorlin *et al.* 2002), the precision of these magnetic field measurements was not high enough to exclude the presence of ~ 100 G fields, which can still play an important role in the spot formation process. More recent sensitive searches for the magnetic field signatures using new generation spectropolarimeters targeted known HgMn stars with spots, α And (Wade *et al.* 2006) and AR Aur (Folsom *et al.* 2010), finding no longitudinal field stronger than ≈ 10 G for the former star and ≈ 30 G for the latter. Finally, Makaganiuk *et al.* (submitted) carried out a comprehensive magnetic survey of over 40 HgMn stars with the new HARPSpol instrument at the ESO 3.6-m telescope. None of their targets, which included several spotted HgMn stars, showed any evidence of magnetic field. For the majority of stars this survey achieved an upper limit of 10–20 G for the longitudinal magnetic field, and as low as 1–3 G for several sharp-lined stars. Analysis of the circular polarization profiles at the resolving power of $> 10^5$ also do not reveal complex magnetic fields similar to those seen in active late-type stars. These results confirm the non-magnetic status of HgMn stars and suggest that the spot formation on their surfaces is unrelated to magnetic field.

A new light on the puzzling nature of chemical inhomogeneities in HgMn stars was shed by the discovery of temporal evolution of the spot topology in α And by Kochukhov *et al.* (2007). In this study the Hg spots were followed over a time span of 7 years with the spectra of exquisite quality ($S/N \sim 1000$), revealing slow changes of the surface structure both in the original data and in the resulting Doppler images (Fig. 3). Later, a faster evolution of chemical spots was claimed for the HgMn star HD 11753 by Briquet *et al.* (2010) based on Doppler maps obtained for two epochs separated by only 65 days.

It is clear that HgMn stars represent an interesting new type of spotted stars, in which magnetic field is not instigating the chemical spot formation. In this respect the spots on HgMn stars are fundamentally different from both the stable abundance inhomogeneities found on the early-type stars with fossil magnetic fields and the temperature spots associated with complex field topologies on late-type active stars. Instead, the spots on HgMn stars probably form and evolve under the influence of time-dependent atomic diffusion processes and hydrodynamical instabilities in the upper layers of chemically stratified atmosphere (Kochukhov *et al.* 2007). Detailed theoretical diffusion models are needed to understand the origin of this newly discovered, remarkable spot formation phenomenon.

References

Adelman, S. J., Gulliver, A. F., Kochukhov, O. P., & Ryabchikova, T. A. 2002, *Astrophys. J.*, 575, 449

Adelman, S. J., Malanushenko, V., Ryabchikova, T. A., & Savanov, I. 2001, *Astron. Astrophys*, 375, 982

Alecian, G. & Stift, M. J. 2010, *Astron. Astrophys*, 516, A53

Aurière, M., Wade, G. A., & Silvester, J., *et al.* 2007, *Astron. Astrophys*, 475, 1053

Babcock, H. W. 1960, *Astrophys. J.*, 132, 521

Babel, J. 1992, *Astron. Astrophys*, 258, 449

Bagnulo, S., Wade, G. A., & Donati, J.-F., *et al.* 2001, *Astron. Astrophys*, 369, 889

Briquet, M., Korhonen, H., & González, J. F., *et al.* 2010, *Astron. Astrophys*, 511, A71

Elkin, V. G., Mathys, G., Kurtz, D. W., *et al. Mon. Not. Roy. Astron. Soc.*, 402, 1883

Folsom, C. P., Kochukhov, O., Wade, G. A., *et al.* 2010, *Mon. Not. Roy. Astron. Soc.*, 407, 2383

Folsom, C. P., Wade, G. A., & Kochukhov, O., *et al.* 2008, *Mon. Not. Roy. Astron. Soc.*, 391, 901

Hatzes, A. P. 1991, *Mon. Not. Roy. Astron. Soc.*, 248, 487

Khokhlova, V. L., Rice, J. B., & Wehlau, W. H. 1986, *Astrophys. J.*, 307, 768

Khomenko, E. & Kochukhov, O. 2009, *Astrophys. J.*, 704, 1218

Kochukhov, O. 2006, *Astron. Astrophys*, 446, 1051

Kochukhov, O., Adelman, S. J., Gulliver, A. F., & Piskunov, N. 2007, *Nature Physics*, 3, 526

Kochukhov, O. & Bagnulo, S. 2006, *Astron. Astrophys*, 450, 763

Kochukhov, O., Bagnulo, S., Wade, G. A., *et al.* 2004a, *Astron. Astrophys*, 414, 613

Kochukhov, O., Drake, N. A., Piskunov, N., & de la Reza, R. 2004b, *Astron. Astrophys*, 424, 935

Kochukhov, O., Piskunov, N., & Ilyin, I., *et al.* 2002, *Astron. Astrophys*, 389, 420

Kochukhov, O., Piskunov, N., Sachkov, M., & Kudryavtsev, D. 2005, *Astron. Astrophys*, 439, 1093

Kochukhov, O., Shulyak, D., & Ryabchikova, T. 2009, *Astron. Astrophys*, 499, 851

Kochukhov, O., Tsymbal, V., & Ryabchikova, T., *et al.* 2006, *Astron. Astrophys*, 460, 831

Kochukhov, O. & Wade, G. A. 2010, *Astron. Astrophys*, 513, A13

Landstreet, J. D. & Mathys, G. 2000, *Astron. Astrophys*, 359, 213

Leone, F. & Catanzaro, G. 2001, *Astron. Astrophys*, 365, 118

Lüftinger, T., Kochukhov, O., & Ryabchikova, T., *et al.* 2010, *Astron. Astrophys*, 509, A71

Michaud, G. 1970, *Astrophys. J.*, 160, 641

Michaud, G., Charland, Y., & Megessier, C. 1981, *Astron. Astrophys*, 103, 244

Mikulášek, Z., Krtička, J., & Henry, G. W., *et al.* 2008, *Astron. Astrophys*, 485, 585

Piskunov, N. & Kochukhov, O. 2002, *Astron. Astrophys*, 381, 736

Polosukhina, N. S., Shavrina, A. V., & Hack, M., *et al.* 2000, *Astron. Astrophys*, 357, 920

Preston, G. W. 1974, ARA&A, 12, 257

Quinet, P., Argante, C., & Fivet, V., *et al.* 2007, *Astron. Astrophys*, 474, 307

Ryabchikova, T., Piskunov, N., & Kochukhov, O., *et al.* 2002, *Astron. Astrophys*, 384, 545

Ryabchikova, T., Ryabtsev, A., Kochukhov, O., & Bagnulo, S. 2006, *Astron. Astrophys*, 456, 329

Ryabchikova, T., Sachkov, M., Kochukhov, O., & Lyashko, D. 2007, *Astron. Astrophys*, 473, 907

Shorlin, S. L. S., Wade, G. A., & Donati, J.-F., *et al.* 2002, *Astron. Astrophys*, 392, 637

Shulyak, D., Ryabchikova, T., Mashonkina, L., & Kochukhov, O. 2009, *Astron. Astrophys*, 499, 879

Stępień, K. 2000, *Astron. Astrophys*, 353, 227

Wade, G. A., Aurière, M., & Bagnulo, S., *et al.* 2006, *Astron. Astrophys*, 451, 293

Wade, G. A., Bagnulo, S., & Kochukhov, O., *et al.* 2001, *Astron. Astrophys*, 374, 265

The Physics of Sun and Star Spots
Proceedings IAU Symposium No. 273, 2010
D.P. Choudhary & K.G. Strassmeier, eds.

doi:10.1017/S174392131101533X

Dynamo generated field emergence through recurrent plasmoid ejections

Jörn Warnecke[1,2] and Axel Brandenburg[1,2]

[1]Nordita, AlbaNova University Center,
Roslagstullsbacken 23, SE-10691 Stockholm, Sweden
email: joern@nordita.org

[2]Department of Astronomy, AlbaNova University Center,
Stockholm University, SE 10691 Stockholm, Sweden

Abstract. Magnetic buoyancy is believed to drive the transport of magnetic flux tubes from the convection zone to the surface of the Sun. The magnetic fields form twisted loop-like structures in the solar atmosphere. In this paper we use helical forcing to produce a large-scale dynamo-generated magnetic field, which rises even without magnetic buoyancy. A two layer system is used as computational domain where the upper part represents the solar atmosphere. Here, the evolution of the magnetic field is solved with the stress–and–relax method. Below this region a magnetic field is produced by a helical forcing function in the momentum equation, which leads to dynamo action. We find twisted magnetic fields emerging frequently to the outer layer, forming arch-like structures. In addition, recurrent plasmoid ejections can be found by looking at space–time diagrams of the magnetic field. Recent simulations in spherical coordinates show similar results.

Keywords. MHD, Sun: magnetic fields, Sun: coronal mass ejections (CMEs), turbulence

1. Introduction

The solar magnetic field is broadly believed to be in the form of concentrated flux ropes in the bulk of the convection zone. At the solar surface they emerge to form bipolar regions and sunspots in the photosphere that appear as twisted loop-like structures in the higher atmosphere. However, there is no clear evidence that magnetic fields are generated in flux tubes that emerge from the tachocline all the way to the surface of the Sun. Numerical simulations have successfully shown that magnetic buoyancy, which has been thought to be the main driver of flux tube emergence, can be efficiently suppressed by downward pumping due to the stratification with concentrated downdraft in the solar convection zone (Nordlund *et al.* 1992; Tobias *et al.* 1998). Large-scale dynamo simulations suggest that flux tubes are primarily a feature of the kinematic dynamo regime, but tend to be less pronounced in the nonlinear stage (Käpylä *et al.* 2008). An alternative mechanism might simply be the relaxation of strongly twisted magnetic fields reaching the surface of the Sun. Twisted magnetic fields are produced by a large-scale dynamo mechanism which is generally believed to be the source of solar magnetic activity (Parker 1979). In order to study the emergence of helical magnetic fields from a dynamo, we consider a model that combines a direct simulation of a turbulent large-scale dynamo with a simple treatment for the evolution of nearly force-free magnetic fields above the surface of the dynamo. In the context of force-free magnetic field extrapolations this method is also known as the stress–and–relax method (Valori *et al.* 2005). Including a nearly force-free field in the upper part of the domain has the additional benefit of allowing a more realistic modeling of the dynamo itself. This is important, because it is known that the properties of the large-scale magnetic field depend strongly on the boundary conditions.

In the upper atmosphere, direct numerical simulation of the solar corona show force-free magnetic fields (Gudiksen & Nordlund 2005).

Above the solar surface, we expect helical magnetic fields to drive flares and coronal mass ejections through the Lorentz force. In the present paper we highlight some of the main results of our earlier work (Warnecke & Brandenburg 2010) and present recent applications and results using spherical coordinates.

2. The Model

A two layer system is used, where the upper part is modelled as a nearly force-free magnetic field by using the stress–and–relax method (Valori *et al.* 2005) and in the lower part a dynamo field is generated through helically forced turbulence. We combine these two layers by simply turning off terms that should not be included in the upper part of the domain. We do this with error function profiles of the form

$$\theta_w(z) = \tfrac{1}{2}\left(1 - \mathrm{erf}\frac{z}{w}\right), \tag{2.1}$$

where w is the width of the transition.

2.1. *Stress–and–relax method*

The equation for the velocity in the stress–and–relax method is similar to the usual momentum equation, except that there is neither pressure, nor gravity, nor other driving forces on the right-hand side, so we just have

$$\frac{\mathrm{d}\boldsymbol{U}}{\mathrm{d}t} = \boldsymbol{J} \times \boldsymbol{B}/\rho + \boldsymbol{F}_{\mathrm{visc}}, \tag{2.2}$$

where $\boldsymbol{J} \times \boldsymbol{B}$ is the Lorentz force, $\boldsymbol{J} = \boldsymbol{\nabla} \times \boldsymbol{B}/\mu_0$ is the current density, μ_0 is the vacuum permeability, $\boldsymbol{F}_{\mathrm{visc}}$ is the viscous force, and ρ is here treated as a constant that determines the strength of the velocity correction. Equation (2.2) is solved together with the uncurled induction equation,

$$\frac{\partial \boldsymbol{A}}{\partial t} = \boldsymbol{U} \times \boldsymbol{B} + \eta \nabla^2 \boldsymbol{A}, \tag{2.3}$$

with η being the magnetic diffusivity.

2.2. *Forced dynamo region*

In the lower part the velocity is driven by a forcing function and the density is evolved using the continuity equation,

$$\frac{\mathrm{d}\boldsymbol{U}}{\mathrm{d}t} = -\boldsymbol{\nabla} h + \boldsymbol{f} + \boldsymbol{J} \times \boldsymbol{B}/\rho + \boldsymbol{F}_{\mathrm{visc}}, \qquad \frac{\mathrm{d}h}{\mathrm{d}t} = -c_s^2 \boldsymbol{\nabla} \cdot \boldsymbol{U}, \tag{2.4}$$

where $\boldsymbol{F}_{\mathrm{visc}}$ is the viscous force, $h = c_{\mathrm{s}}^2 \ln \rho$ is the specific pseudo-enthalpy, $c_{\mathrm{s}} = \mathrm{const}$ is the isothermal sound speed, and $\boldsymbol{f}$ is a forcing function that drives turbulence in the interior and consists of random plane helical transverse waves with an average forcing wavenumber k_{f}. The pseudo-enthalpy h is given by $\rho^{-1}\boldsymbol{\nabla} p = c_{\mathrm{s}}^2 \boldsymbol{\nabla} \ln \rho = \boldsymbol{\nabla} h$. Equations (2.4) are solved together with the induction Eqs. (2.3).

The simulation box is horizontally periodic. For the magnetic field we adopt vertical-field and perfect-conductor conditions at the top and bottom boundaries, respectively. For the velocity we employ stress-free conditions at both boundaries. In this paper we present direct numerical simulations using the PENCIL CODE†, a modular high-order

† `http://pencil-code.googlecode.com`

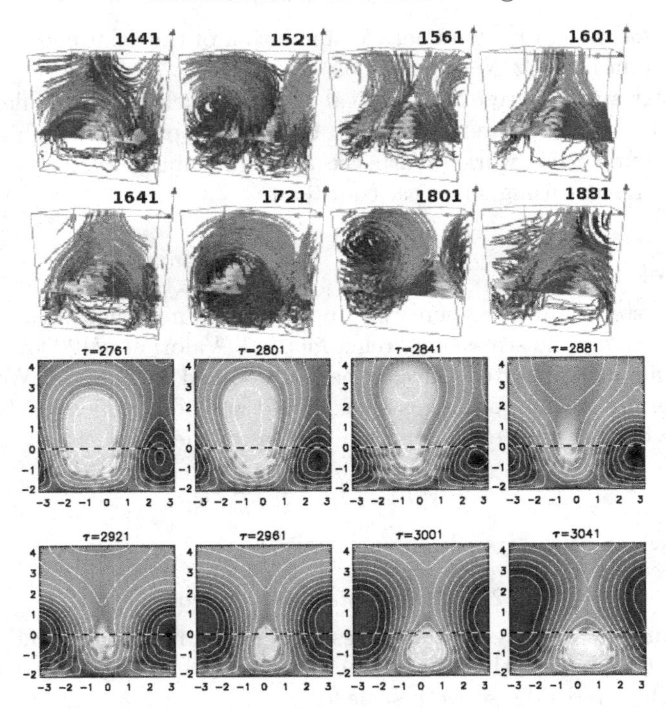

Figure 1. *Upper part*: Time series of arcade formation and decay. Field lines are colored by their local field strength which increases from purple to green. The inclined plane in the box shows B_z increasing from red (positive) to purple (negative). The normalized time τ is giving in each panel. *Lower part*: Time series of the formation of a plasmoid ejection. Contours of $\langle A_x \rangle_x$ are shown together with a color-scale representation of $\langle B_x \rangle_x$; dark/blue stands for negative and light/yellow for positive values. The contours of $\langle A_x \rangle_x$ correspond to field lines of $\langle \boldsymbol{B} \rangle_x$ in the yz plane. The dotted horizontal lines show the location of the surface at $z = 0$. Adapted from Warnecke & Brandenburg (2010).

code (sixth order in space and third-order in time) for solving a large range of partial differential equations.

3. Results

After a period of exponential growth, the magnetic field saturates at 78% of the equipartition field strength, $B_{\rm eq}$ in the turbulent layer. This behavior is typical for forced dynamo action. The structure of the magnetic field is a large-scale field in the turbulent zone. It always shows a systematic variation in one of the two horizontal directions. It is a matter of chance whether this variation is in the x or in the y direction. After the saturation phase the magnetic field extends well into the upper layer where it tends to produce an arcade-like structure, as seen in the upper panel of Fig. 1. The arcade opens up in the middle above the line where the vertical field component vanishes at the surface. This leads to the formation of anti-aligned field lines with a current sheet in the middle. The dynamical evolution is clearly seem in a sequence of field line images in the upper panel of Fig. 1, where anti-aligned vertical field lines reconnect above the neutral line and form a closed arch with plasmoid ejection. This arch then changes its connectivity at the foot points in one of the two horizontal directions (here the y direction), making the field lines bulge upward to produce a new reconnection site with anti-aligned field lines some distance above the surface. Field line reconnection is best seen for two-dimensional

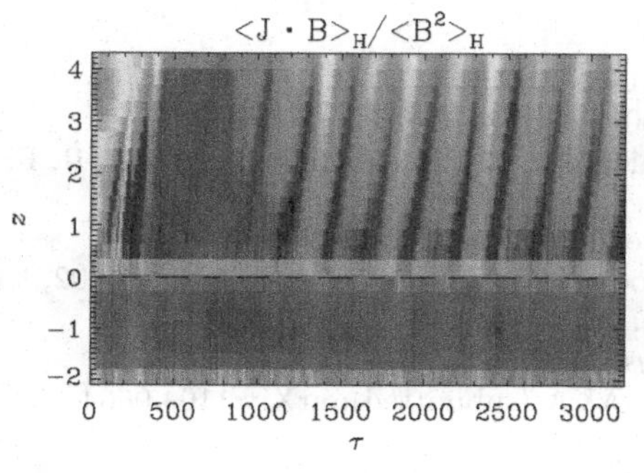

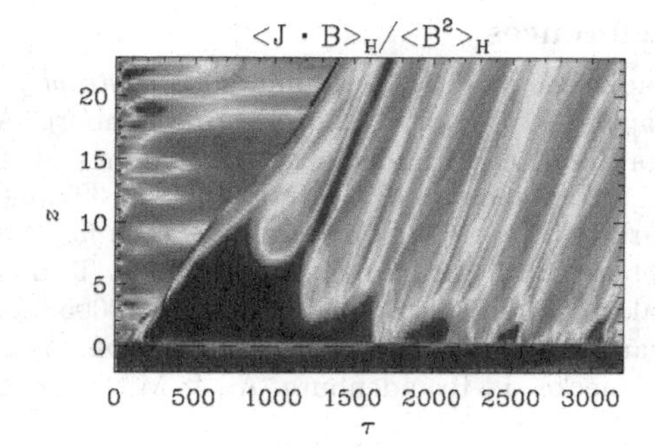

Figure 2. *Left panel*: Dependence of $\langle \boldsymbol{J}\cdot\boldsymbol{B}\rangle_{\rm H}/\langle \boldsymbol{B}^2\rangle_{\rm H}$ versus time τ and height z for $L_z = 6.4$ with $\mathrm{Re}_M = 3.4$. *Right panel*: Similar to the left panel, but for $L_z = 8\pi$ and $\mathrm{Re}_M = 6.7$. Adapted from Warnecke & Brandenburg (2010).

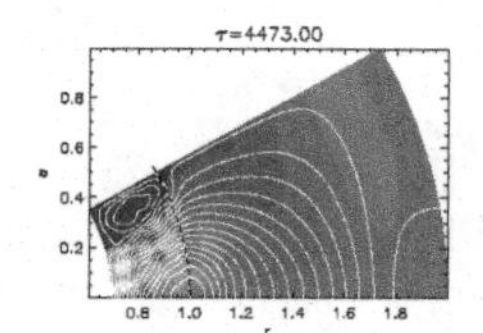

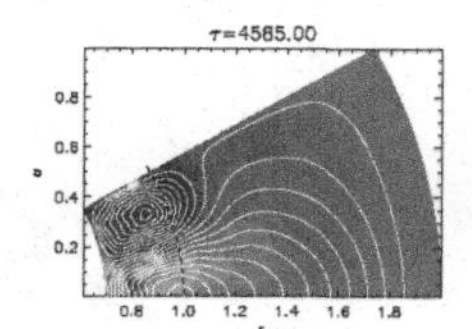

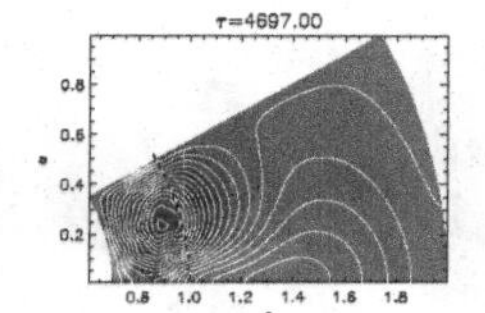

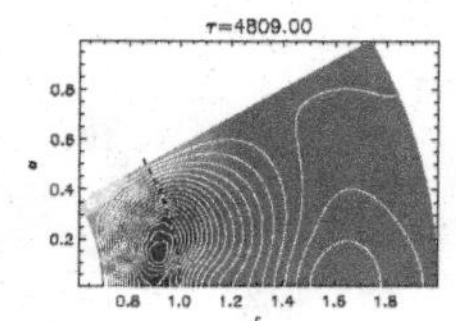

Figure 3. Time series of flux emergence in spherical coordinates. Contours of $r\sin\theta\langle A_\phi\rangle_\phi$ are shown together with a color-scale representation of $\langle B_\phi\rangle_\phi$; dark/blue stands for negative and light/yellow for positive values. The contours of $r\sin\theta\langle A_\phi\rangle_\phi$ correspond to field lines of $\langle \boldsymbol{B}\rangle_\phi$ in the $r\theta$ plane. The dotted horizontal lines show the location of the surface at $r = 1$ solar radii.

magnetic fields, because it is then possible to compute a flux function whose contours correspond to field lines in the corresponding plane. In the present case the large-scale component of the magnetic field varies little in the x direction, so it makes sense to visualize the field averaged over x (see lower panel of Fig. 1).

In order to demonstrate that plasmoid ejection is a recurrent phenomenon, it is convenient to look at the evolution of the ratio $\langle \boldsymbol{J}\cdot\boldsymbol{B}\rangle_{\rm H}/\langle \boldsymbol{B}^2\rangle_{\rm H}$ versus t and z. This is done in right panel of Fig. 2 for $L_z = 6.4$ and $\mathrm{Re}_M = 3.4$ and in the left panel of Fig. 2 for $L_z = 8\pi$ and $\mathrm{Re}_M = 6.7$. It turns out that in both cases the typical speed of plasmoid ejecta is about half the rms velocity of the turbulence in the interior region.

As an example of further work in this direction, we also present in this paper magnetic flux emergence in spherical coordinates. The dynamical evolution can be seen in Fig. 3, where a modulated slice covers the convection zone from 0.7 solar radii through the upper atmosphere to two solar radii. This meridional slice of a sphere consists, like the simulation box above, of two layers which contain the same physical properties. We solve the same equations as described in Section 2. Again, there is flux emerge through the surface above the turbulence zone. This can be seen as a recurrent event. Unfortunately, reconnection, current sheets and plasmoid ejections have not been seen in the present setup, although there are indications that they do occur in even more recent spherical models where gravity and density stratification are included (Warnecke *et al.* 2011).

Our first results are promising in that the dynamics of the magnetic field in the exterior is indeed found to mimic open boundary conditions at the interface between the turbulence zone and the exterior at $z = 0$. In particular, it turns out that a twisted magnetic field generated by the helical dynamo beneath the surface is able to produce flux emergence in ways that are reminiscent of that found in the Sun. The first results in spherical coordinates show recurrent flux emergence, but plasmoid ejections in a curved environment may only be possible if gravity and density stratification are included.

References

Gudiksen, B. V. & Nordlund. 2005, *Astrophys. J.*, 618, 1031

Käpylä, P. J., Korpi, M. J., & Brandenburg, A. 2008, *Astron. Astrophys*, 491, 353

Nordlund, Å., Brandenburg, A., Jennings, R. L., Rieutord, M., Ruokolainen, J., Stein, R. F., & Tuominen, I. 1992, *Astrophys. J.*, 392, 647

Parker, E. N. 1979, *Cosmical magnetic fields* (Clarendon Press, Oxford)

Tobias, S. M., Brummell, N. H., Clune, T. L., & Toomre, J.1998, *Astrophys. J. Lett.*, 502, L177

Valori, G., Kliem, B., & Keppens, R. 2005, *Astron. Astrophys*, 433, 335

Warnecke, J. & Brandenburg, A. *Astron. Astrophys*, 2010, 523, A19

Warnecke, J., Brandenburg, A., & Mitra, D., 2011, A&A (submitted), arXiv:1104.0664

The Physics of Sun and Star Spots
Proceedings IAU Symposium No. 273, 2010
D.P. Choudhary & K.G. Strassmeier, eds.

doi:10.1017/S1743921311015341

An "A star" on an M star during a flare within a flare

Adam F. Kowalski,[1] Suzanne L. Hawley,[1] Jon A. Holtzman,[2] John P. Wisniewski,[1,3] Eric J. Hilton[1]

[1]Astronomy Department, University of Washington
Box 351580, Seattle, WA 98195, USA
email: kowalski@astro.washington.edu

[2]Department of Astronomy, New Mexico State University
Box 30001, Las Cruces, NM 88003, USA

[3]NSF Astronomy & Astrophysics Postdoctoral Fellow

Abstract. M dwarfs produce explosive flare emission in the near-UV and optical continuum, and the mechanism responsible for this phenomenon is not well-understood. We present a near-UV/optical flare spectrum from the rise phase of a secondary flare, which occurred during the decay of a much larger flare. The newly formed flare emission resembles the spectrum of an early-type star, with the Balmer lines and continuum in *absorption*. We model this observation phenomenologically as a temperature bump (hot spot) near the photosphere of the M dwarf. The amount of heating implied by our model ($\Delta T_{phot} \sim 16,000$ K) is far more than predicted by chromospheric backwarming in current 1D RHD flare models ($\Delta T_{phot} \sim 1200$ K).

Keywords. Physical data and processes: radiative transfer, astronomical methods: numerical, atmospheric effects, techniques: spectroscopic, stars: atmospheres, stars: flare, stars: late-type

1. Introduction

Flares on M dwarfs are notorious for producing dramatic outbursts in the near-UV and optical (white light) continuum. The white light continuum has been observed in both the impulsive and decay phases of stellar flares, and the broadband shape of this emission resembles that of a hot blackbody with $T \sim 8500 - 10,000$ K (Hawley & Fisher 1992; Hawley *et al.* 2003). In contrast, radiative hydrodynamic (RHD) flare models predict a white light continuum with a prominent Balmer continuum in emission (Allred *et al.* 2006). However, when convolved with broadband filters the model spectrum *also* exhibits the shape of a hot blackbody (Allred *et al.* 2006). Spectra have been obtained during M dwarf flares (Hawley & Pettersen 1991; Eason E.L.E. *et al.* 1992; Garcia-Alvarez D. *et al.* 2002; Fuhrmeister B. *et al.* 2008), showing a clear rise into the near-UV without an abrupt discontinuity at the Balmer jump wavelength ($\lambda = 3646$ Å) or a prominent Balmer continuum in emission. To unravel this complexity in the white light continuum, we have begun a detailed investigation at wavelengths near the Balmer jump using new, time-resolved flare spectra and models.

On UT 2009 January 16, we observed an incredible flare on the dM4.5e star YZ CMi, obtaining high-cadence near-UV/optical (3350−9260 Å) spectra and simultaneous U-band photometry from the ARC 3.5-m and NMSU 1-m telescopes at APO (see Kowalski *et al.* 2010; hereafter K10). The spectra cover the section of the flare decay shown in Figure 1(a), which consists of several secondary flare peaks. The K10 analysis revealed two continuum components in the flare spectra: a Balmer continuum in emission as predicted by the RHD models and a hot (T ~10,000 K), compact blackbody as seen in previous flare observations.

2. The Secondary Flare Spectra

During the secondary flares, K10 found that the time evolution of the Balmer continuum and Balmer line fluxes are anti-correlated with the areal coverage of the $T \sim 10,000$ K blackbody component. For example, a secondary flare rise begins at $t \sim 124$ min and lasts ~5 min, resulting in an apparent decrease by 40% in the Balmer continuum flux and an increase by a factor of nearly 2 in the area of the blackbody-emitting region.

The total flare spectra at two times are shown in the inset in Figure 1(a). The times correspond to immediately before the secondary flare rise (F0; t = 123.4 min) and 2.2 min into the secondary flare rise (F1; t = 126.5 min). The spectra are clearly very complex, consisting of line and continuum emission from both previously heated and newly formed flare regions, in addition to molecular band absorption from the surrounding non-flaring

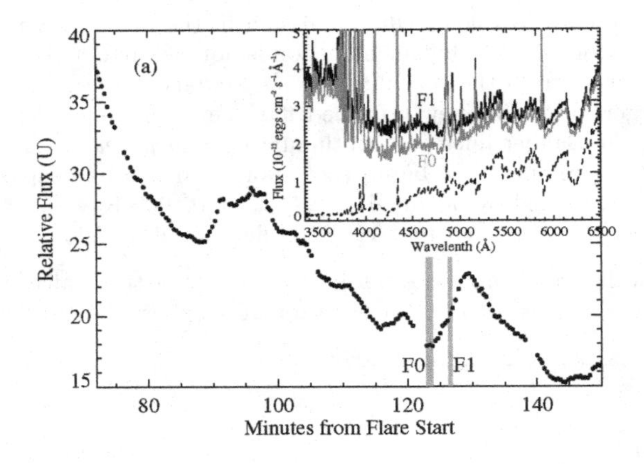

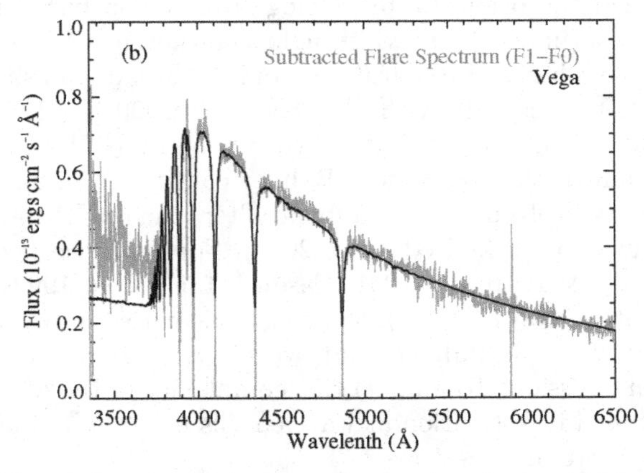

Figure 1. (a) The U-band photometry is shown for the time during which spectra were obtained (see K10 for the complete light curve of the flare). The inset displays averaged spectra immediately before a secondary flare (grey, F0), 2.2 min into the rise phase (black, F1), and during quiescence (dashed line). The times encompassed by each of the flare spectra are denoted by vertical grey bars in the U-band light curve. (b) Subtracting the F0 spectrum from the F1 spectrum isolates newly formed flare emission (grey) during the secondary flare rise. The resulting subtracted flare spectrum resembles the A star Vega with the Balmer features in absorption.

photosphere. We isolate the newly formed flare emission by subtracting the F0 spectrum from the F1 spectrum. The striking features of this subtracted flare spectrum (Figure 1b) include a Balmer continuum and lines in absorption†, in contrast to the total flare spectrum in which the Balmer continuum is in emission. Using continuum windows from 4000−6500 Å, the spectrum is fit by a blackbody with $T \sim 16{,}500$ K. We show the spectrum of the A0 V star Vega (from Bohlin, R.C. 2007) in Figure 1(b) to highlight the remarkably similar characteristics with the spectrum of an early-type star.

3. Phenomenological Modelling with RH

We model the flare spectrum by placing a hot spot, represented by a Gaussian temperature bump with peak temperature of 20,000 K, deep ($\log_{10}$ col mass $= 0.5$ g cm^{-2}) in the quiescent M dwarf atmosphere (Figure 2a). The emergent radiation is calculated with the static NLTE code RH (Uitenbroek 2001) with a 5 level (plus continuum) Hydrogen atom. We initially iterate to solve for the NLTE background opacities for Hydrogen, and we also consider some relevant molecular species that include Hydrogen (e.g., H_2). The emergent hot spot spectrum is shown in Figure 2(b). The model spectrum has a blackbody temperature of ∼18,000 K, and the Balmer continuum and lines are in absorption, similar to the subtracted flare spectrum. In future work, we will include Helium and metallic transitions, additional levels in the Hydrogen atom, and refined electron densities. We will also further constrain the parameters and time evolution of the hot spot.

4. Summary & Discussion

We find that newly formed emission during a secondary flare resembles the spectrum of an early-type star, such as Vega. Modelling the spectrum phenomenologically‡ by placing a hot spot near the quiescent M dwarf photosphere adequately reproduces the observed spectral shape and Hydrogen absorption features. In a future work, we will show how combining all flare continuum components reproduces the *total* flare spectrum, thereby providing an explanation for the anti-correlation in the time evolution between the Hydrogen emission and blackbody components.

Current, self-consistent 1D RHD models of stellar flares predict that the photosphere is heated by only ~ 1200 K, predominantly due to backwarming from the flare chromosphere (Allred *et al.* 2006). Heating of deep layers by the amount needed to produce a $T_{max} \sim 20,000$ K hot spot is clearly not achieved by a solar-type non-thermal electron beam as energetic as 10^{11} ergs cm^{-2} s^{-1}. More spectral observations of flares with time resolution much less than the rise time, and with spectral coverage farther into the near-UV, are needed to characterize the ubiquity of the phenomena presented in this work.

5. Acknowledgements

The presentation of this paper in the IAU Symposium 273 was possible due to partial support from the National Science Foundation grant numbers ATM 0548260, AST 0968672, AST 0807205 and NASA - Living With a Star grant number 09-LWSTRT09-0039. We thank H. Uitenbroek for his helpful discussions about RH.

† We use the term *absorption* throughout to refer to 'less emission than the neighboring spectral regions'.

‡ We note that our phenomenological flare models are very similar to the semi-empirical models for the absorption profiles seen during during Ellerman Bombs on the Sun (e.g., Fang C. *et al.* 2006).

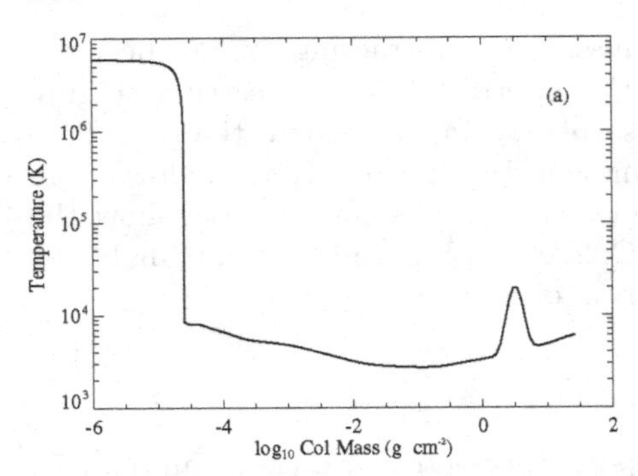

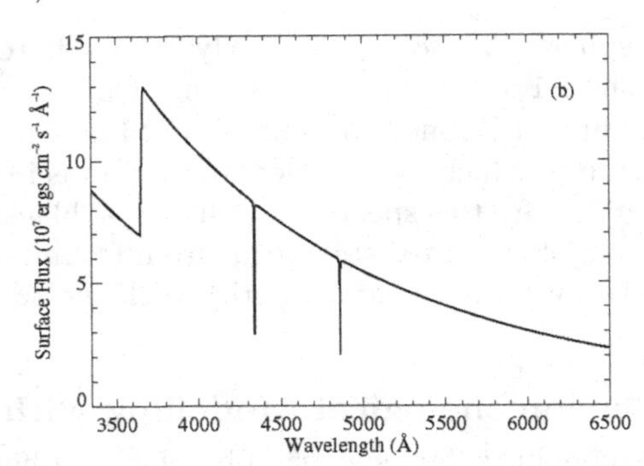

Figure 2. (a) The quiescent M dwarf atmosphere from Allred *et al.* (2006) with a phenomenological hot spot below the temperature minimum region. (b) The static code RH is used to calculate the emergent flux spectrum for a 6-level Hydrogen atom. The slope of this continuum closely resembles the continuum of the subtracted flare spectrum. Moreover, the temperature bump generates absorption features from Hydrogen.

References

Allred, J. C., Hawley, S. L., Abbett, W. P., & Carlsson, M. 2006, *Astrophys. J.*, 644, 484
Bohlin, R. C. 2007, *ASP-CS*, 364, 315
Eason, E. L. E.., Giampapa, M. S., Radick, R. R., Worden, S. P., & Hege, E. K. 1992, *AJ*, 104, 1161
Fang, C., Tang, Y. H., Xu, Z., Ding, M. D., & Chen, P. F. 2006, *Astrophys. J.*, 643, 1325
Fuhrmeister, B., Liefke, C., Schmitt, J. H. H.. M., & Reiners, A. 2008, *Astron. Astrophys*, 487, 293
Garcia-Alvarez, D., Jevremovic, D., Doyle, J. G., & Butler, C. J. 2002, *Astron. Astrophys*, 383, 548
Hawley, S. L. & Pettersen, B. R. 1991, *Astrophys. J.*, 378, 725
Hawley, S. L. & Fisher, G. H. 1992, *Astrophys. J. Suppl.*, 78, 565
Hawley, S. L., Allred, J. C., Johns-Krull, C. M., Fisher, G. H., Abbett, W. P., Alekseev, I., Avgoloupis, S. I., Deustua, S. E., Gunn, A., & Seiradakis, J. H. 2003, *apj*, 597, 535
Kowalski, A. F., Hawley, S. L., Holtzman, J. A., Wisniewski, J. P., & Hilton, E. J. 2010, *Astrophys. J.*, 714L, 98
Uitenbroek, H. 2001, *Astrophys. J.*, 557, 389

Discussion

KOSOVICHEV: 1) Have you been able to observe Doppler shifts? 2) Some theoretical models suggested that condensations behind the downward propagating shock may be responsible for white light emission. What is the satus of these models? Can this be ruled out by the new observations?

KOWALSKI: First, our spectra have a low spectral resolution, with R∼1000. We aren't concerned about getting very accurate wavelength calibration because we don't want to go off the slit very often to get an arc exposure, in case there is a flare. Flare rise times are very fast, typically 20-40s, so we don't want to miss it. We are more concerned with getting the flare and accurate flux calibration than with the wavlength calibration.
These RHD models predict a downward condensation wave with speeds of tens of km per second, but this doesn't reach the photosphere with sufficient energy. Most of the energy that reaches the photosphere is backwarming radiation from the Balmer continuum. I'd just like to add that X-ray backwarming was also once thought to be a candidate for the heating in deep layers; however, according to these models only ∼1% of the heating is caused by X-ray backwarming.

The Physics of the Sun and Star Spots
Proceedings IAU Symposium No. 273, 2010
D. P. Choudhary and K. G. Strassmeier, eds.

doi:10.1017/S1743921311015353

Sunspots at centimeter wavelengths

Mukul R. Kundu[1] **and Jeongwoo Lee**[2]

[1]Department of Astronomy, University of Maryland,
College Park, MD 20742, U.S.A.

[2]Physics Department, New Jersey Institute of Technology,
Newark, NJ 07102, U.S.A.
email: leej@njit.edu

Abstract. The early solar observations of Covington (1947) established a good relation between 10.7 cm solar flux and the presence of sunspots on solar disk. The first spatially resolved observation with a two-element interferometer at arc min resolution by Kundu (1959) found that the radio source at 3 cm has a core-halo structure; the core is highly polarized and corresponds to the umbra of a sunspot with magnetic fields of several hundred gauss, and the halo corresponds to the diffuse penumbra or plage region. The coronal temperature of the core was interpreted as due to gyroresonance opacity produced by acceleration of electrons gyrating in a magnetic field. Since the opacity is produced at resonant layers where the frequency matches harmonics of the gyrofrequency, the radio observation could be utilized to measure the coronal magnetic field. Since this simple interferometric observation, the next step for solar astronomers was to use arc second resolution offered by large arrays at cm wavelengths such as Westerbrock Synthesis Radio Telescope and the Very Large Array, which were primarily built for cosmic radio research. Currently, the Owens Valley Solar Array operating in the range 1-18 GHz and the Nobeyama Radio Heliograph at 17 and 34 GHz are the only solar dedicated radio telescopes. Using these telescopes at multiple wavelengths it is now possible to explore three dimensional structure of sunspot associated radio sources and therefore of coronal magnetic fields. We shall present these measurements at wavelengths ranging from 1.7 cm to 90 cm and associated theoretical developments.

Keywords. Sun: corona, Sun: magnetic fields, Sun: radio radiation, sunspots

1. Introduction

Sunspots appear very bright at cm wavelengths. This property has been known since Covington (1947) observed a sharp decrease in the 10.7 cm flux during a lunar occultation of a large sunspot group. It was then found in the first spatially resolved observation with a two-element interferometer by Kundu (1959) that solar radio source at 3 cm has a core-halo structure. The core emission is highly polarized and fully exhibits the coronal temperature and its radiation mechanism had been of a great curiosity while the halo emission was likely to be due to the well-known free-free emission. The most elegant interpretation of the core emission was then provided by Zheleznyakov (1962) in which hot electrons gyrating in strong magnetic fields produce the gyroresonance opacity strong enough to make the corona optically thick at cm wavelengths. This opacity is high only at the resonant layers where the frequency matches harmonics of the gyrofrequency: $\nu = \Omega_e, 2\Omega_e, 3\Omega_e$, and so on. Here $\Omega_e = 2.8B$ MHz is the electron gyrofrequency and B is magnetic field strength in gauss. This simple relation between ν and B suggests a powerful tool for measuring coronal magnetic fields without complications that arise at other wavelengths and thus opened a new science field of active region coronae. Major progress in this field is reviewed in this article. For more comprehensive reviews, we refer to White & Kundu (1997), Gary & Keller (2004) and Lee (2007).

2. Studies of Sunspot Emissions

2.1. *The WSRT and RATAN-600*

The early attempts to measure coronal magnetic field were dominated by the observations with the Westerbork Synthesis Radio Telescope (WSRT) and RATAN-600. The high spatial resolution observations with the WSRT showed that radio sunspots appear in a *ring* or *horseshoe* structure in polarization maps (Alissandrakis *et al.* 1980, Alissandrakis & Kundu 1982, 1984; Kundu & Alissandrakis, 1984) which could be due to either dependence of the gyroresonant opacity on the magnetic field orientation or the actual presence of cool plasma (~10^5 K) overlying the umbra. Alissandrakis & Kundu (1984) first applied the coronal field extrapolation technique to the interpretation of their center-to-limb observation of a sunspot-associated microwave source to discuss the 3-D magnetic field structure. This characteristic radio morphology found with the WSRT has continued to be discussed with the VLA observations (Vourlidas *et al.* 1997, Zlotnik *et al.* 1998). In contrast, observations with the RATAN-600 have emphasized the spectral information available at 5 frequencies with 1-D spatial resolution (Akhmedov *et al.* 1982; Bogod & Gelfreikh 1980, Krüger *et al.* 1986). In these works, extensive calculations of theoretical radio spectra and maps of intensity and polarization were compared with the corresponding observations for agreement. These two types of approaches had remained popular in later studies of sunspot emissions where observations are made at a limited number of frequencies.

2.2. *Imaging spectroscopy with the OVSA*

After the recognition of the importance of multifrequency solar observations, the Owens Valley solar interferometer was converted to frequency-agile operation capable of measuring the solar emission up to 86 frequencies in the range 1–18 GHz (Hurford *et al.* 1984). As a result spectral observations of a simple sunspot, AR 4741, with the Owens valley Solar Array (OVSA) confirmed that radio spectrum of a sunspot indeed takes the form of the gyroresonant spectrum as predicted by the theory (Lee *et al.* 1993*a*). The free-free spectrum was also found when the sunspot went over the limb and the gyroresonant layer was occulted by the limb (Lee *et al.* 1993*b*). As expected the radio source reduces in size with increasing frequency, which could then be converted to the spatial distribution of magnetic field strength using the above ν–B relation. The resulting magnetic field distribution has been compared with theoretical sunspot models to assess the force balance around sunspots (Lee *et al.* 1993*c*). Alternatively one can construct a data-cube of 2-D maps along the frequency axis, and then build a set of spectra at each common spatial point. Figure 1 shows the first demonstration of such imaging spectroscopy (Gary & Hurford 1994). When a spatially resolved spectrum (left panels) indicates the gyroresonant emission, they derived magnetic field strength from the spectral turnover frequency and electron temperature from the optically thick intensity at that location. In the regions where free-free emission dominates the electron density along with an upper limit on the magnetic field was derived. This approach was repeated by Komm *et al.* (1997) and Bong *et al.* (2003) for studies of other active regions. The advantage of this approach is that the information of the spatially resolved spectrum at each pixel is so self contained that one does not need to guess the dominant radiation mechanism. These works have established the microwave imaging spectroscopy as a standalone tool for determining the coronal temperature and magnetic field. One aspect overlooked in this procedure was, however, the possible presence of inhomogeneous temperature along the line of sight, which could, though, be appreciated by non-uniformity of the optically thick spectrum. This issue has been addressed in a recent spectral analysis of the joint OVSA-VLA ob-

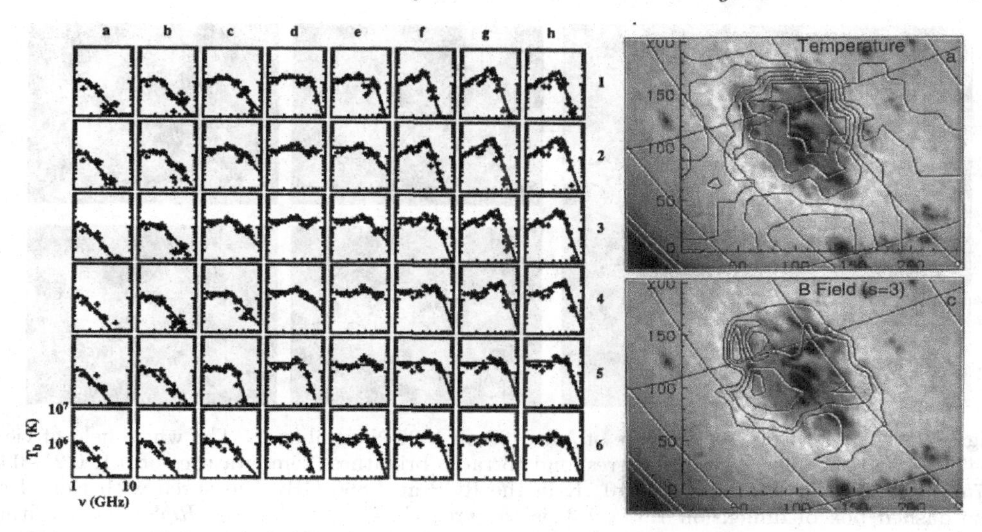

Figure 1. Physical maps from the OVSA imaging spectroscopy. *(Left)* Brightness temperature spectra obtained at 48 grid points in the data cube represented by the maps. The axis scales for the spectra are all the same, as shown on the spectrum at the lower left. *(Right)* Physical maps obtained by interpretation of these brightness temperature spectra: electron temperature with contour levels 0.9, 1.15, 1.4, 1.9 and 2.09 $\times 10^6$ K and total magnetic strength as measured from gyro-resonant spectra assuming the third harmonic everywhere. The contour levels are 200, 400, 600, 800 and 1000 gauss. (from Gary & Hurford 1994).

servations of AR 10923 where gyroresonant emissions from strong magnetic field region are found to be partly absorbed by over-lying cool materials via the free-free opacity (Tune 2010). As related, O mode polarization found above sunspots (Lee *et al.* 1993*a*) and the afore-mentioned ring structure could also be attributed to the inhomogeneous temperature structure over sunspot umbrae.

2.3. *Coronal Magnetic Structures Observing Campaign (CoMStOC)*

The Very Large Array (VLA) is capable of high quality imaging of sunspots but at a few discrete frequencies in which case information on coronal density and temperature obtained at other wavelengths is needed to assist interpretation of the radio data (Schmahl *et al.* 1982, Strong *et al.* 1984, Webb *et al.* 1987). This task has been systematically performed in a series of studies called the CoMStOC in which the plasma electron temperature and emission measures determined from the X-ray and EUV data are used to predict the radio emission expected at the observing frequencies, and these predictions are compared with the radio observations to ascertain the dominant radiation mechanism and derive information on the coronal conditions (Schmelz & Holman 1991). One example is shown in Figure 2, in which Brosius *et al.* (2002) used the EUV spectra and images of NOAA 8108 obtained with the Coronal Diagnostic Spectrometer (CDS) and the Extreme-ultraviolet Imaging Telescope (EIT) aboard the SOHO satellite to derive the differential emission measure (DEM) and the electron density for each spatial pixel within both regions. The VLA observations were used to constrain the magnetic scale height and the gross temperature structure of the atmosphere. These, along with the DEM, electron density, and observed radio brightness temperature maps, were then used to derive the height distribution of the coronal magnetic field strength that can reproduce the radio emissions at the observing frequencies for each spatial pixel in the images. Major findings of the CoMStOC include the presence of electric currents in the corona as

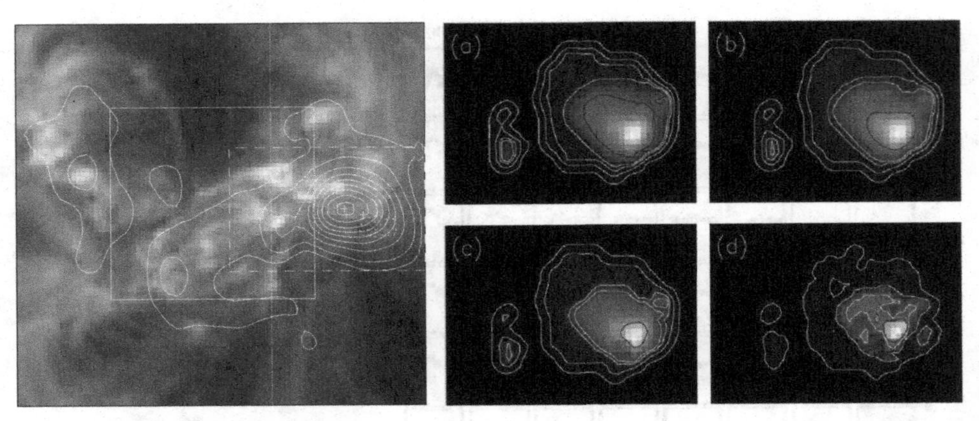

Figure 2. *Left*: an EIT 195 Å image of AR 8108 on 1997 November 18. The whole field of view is $4' \times 4'$ and the solid contours correspond to radio brightness temperatures of 0.1, 0.25, 0.5, 0.75, 1.0, 1.25, 1.5, 1.75, and 2.0×10^6 K in the RCP at 4.866 GHz, measured with the VLA. The dashed box of dimension $1.9' \times 1.3'$ is the area chosen for analysis. *Right*: the resulting coronal magnetic strengths are shown at heights (a) 5000, (b) 10000, (c) 15000 and (d) 25000 km above the photosphere. Contour levels are 200, 579, 868, 1005 and 1508 gauss where the last four values correspond, respectively, to $s = 3, 2$ at 4.866 GHz and $s = 3, 2$ at 8.450 GHz (from Brosius *et al.* 2002).

judged from the excess of field strength over the prediction made by the potential field extrapolation, and that of cool material, from the higher predicted brightness temperatures than those observed at 20 cm (Schmelz *et al.* 1994).

2.4. *Other Sciences Addressed with the VLA*

The importance of the magnetic topology in coronal temperature structure has been demonstrated by the modeling of high quality VLA maps of a complex active region AR6615 (Lee *et al.* 1999). The idea is based on the fact that radio maps at multifrequencies represent temperature on different isogauss surfaces and that temperatures measured at two different locations on a field line should be very well correlated. In their result, the correlation between the brightness temperatures measured at 5 and 8 GHz was indeed good for the nonlinear force-free extrapolation and much poorer for the other extrapolations. A more quantitative approach is to adopt a trial heating function and solve a heat transfer equation to determine the temperature along a given field line. Lee *et al.* (1998) applied several heating functions parameterized in terms of magnetic field, force-free parameter, and current density together with Rosner *et al.*'s (1978) scaling law to find that the most plausible heating function is proportional to the magnetic field strength at the footpoints, in favor of the hypothesis of coronal heating by nanoflares (Parker 1988).

Gopalswamy (1994, see also Zhang *et al.* 1998) found radio transient brightenings located near a sunspot umbra, perhaps at the footpoints of an X-ray transient loop observed with Yohkoh, which may indicate small-scale heating. Nindos *et al.* (2002) used VLA observations to detect spatially resolved oscillations in the total intensity and circular polarization emission of a sunspot-associated gyroresonant source and found that the oscillations are intermittent in time and patchy in space. They interpreted the radio oscillations as due to variations of the location of gyroresonant surfaces with respect to the base of the chromosphere-corona transition region.

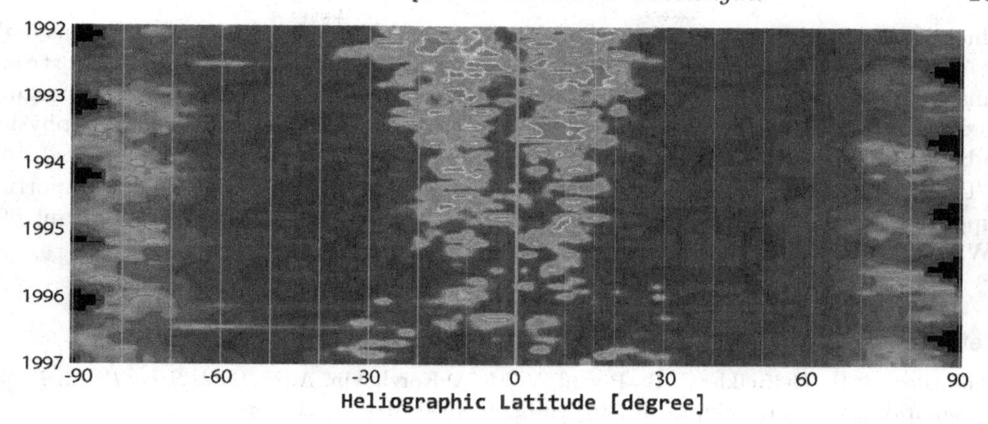

Figure 3. Microwave butterfly diagram constructed with the 17 GHz NoRH data. The bright features are active region belts and polar coronal holes. The dark features are radio counterparts of Hα dark filaments. (Courtesy: K. Shibasaki, 2010).

2.5. *Continuous solar observations with the NoRH*

As a solar dedicated heliograph at 17 GHz and 34 GHz, the NoRH has made many important contributions to sunspot emissions, especially, the temporally evolving nature of sunspot emissions as well as individual stationary structures. Shibasaki (1996) reported frequent occurrence of very weak radio bursts (lower than a few s.f.u.) above NOAA 7654 during its entire disk passage, and argued that it is due to short term heating in sunspots. Gelfreikh *et al.* (1999) measured the circular polarization of several 17 GHz sources as a function of time and found nearly harmonic oscillations with periods mostly between 120–220 s, which are identified with magnetohydrodynamic oscillations. Shibasaki (2001) detected 3-min oscillations in the 17 GHz emission of a sunspot for which he applied the values of density and temperature fluctuations deduced from Solar Ultraviolet Measurements of Emitted Radiation (SUMER) onboard SOHO to find a good agreement with the detected radio oscillation. Investigation of the long term variation of the whole sun emission turns out to be a unique advantage of the NoRH. Shibasaki (1998) produced a radio synoptic map based on the 5-year measurements of 17 GHz daily full disk images, and converted it to a radio butterfly diagram as shown in Figure 3. This radio butterfly diagram shows a similarity with the corresponding diagrams of photospheric magnetic fields and soft X-rays, but reveals that the polar cap brightening gradually expands toward low latitude during the declining phase of the solar cycle 22.

3. Concluding Remarks

Studies of the sunspot emissions at cm wavelengths have shown that these emissions can serve as an excellent indicator of the coronal magnetic field and temperature. The imaging spectroscopy of sunspots implemented with the OVSA would be one of the best procedures for unambiguous observational determination of these two quantities. The high quality imaging with the VLA and the solar dedicated full-sun imaging with NoRH have produced many important findings of the physical processes in sunspots that include the inhomogeneous structure of plasma, strong kilo gauss magnetic fields, electric currents and magnetohydrodynamic waves present in the sunspot coronae, and various transient activities of sunspot emissions. Sunspot radio emissions are thus particulary suitable for addressing topics related to coronal magnetic field strength and electron temperature. For the future, it is highly desirable to build a next generation solar-dedicated radio telescope which should feature as many frequencies as Owens Valley Solar Array offers and as high resolution imaging as the VLA performs. Observing targets may then extend from strong sunspot associated sources to fainter radio sources such as plages, filaments, emerging flux regions and coronal holes to allow an extremely broad range of solar coronal physics to be fully addressed.

The presentation of this paper in the IAU Symposium 273 was possible due to partial support from the NSF grants ATM 0548260, AST 0968672 and NASA LWS grant 09-LWSTRT09-0039. J.L. has been supported by NSF grant AST-0908344.

References

Akhmedov, S. B., Gelfreikh, G. B., Bogod, V. M., & Korzhavin, A. N. 1982, *Solar Phys.*, 79, 41
Alissandrakis, C. E., Kundu, M. R., & Lantos, P. 1980, *Astron. Astrophys.*, 82, 30
Alissandrakis, C. E. & Kundu, M. R. 1982, *ApJ Letters*, 253, L49
Alissandrakis, C. E. & Kundu, M. R. 1984, *Astron. Astrophys.*, 139, 271
Bong, S.-C., Lee, J., Gary, D. E., & Yun, H. S. 2003, *J. of the Korean Astron. Soc.*, 36, S29
Bogod, V. M. & Gelfreikh, G. B. 1980, *Solar Phys.*, 67, 29
Brosius, J. W., Landi, E., Cook, J. W., Newmark, J. S., Gopalswamy, N., & Lara, A. 2002, *ApJ*, 574, 453
Covington, A. E. 1947, *Nature*, 159, 405
Gary, D. E. & Hurford, G. J. 1994, *ApJ*, 420, 903
Gary, D. E. & Keller, C. U. 2004, *Solar and Space Weather Radiophysics - Current Status and Future Developments* (Dordrecht: Kluwer).
Gelfreikh, G. B., Grechnev, V., Kosugi, T., & Shibasaki, K. 1999, *Solar Phys.*, 185, 177
Gopalswamy, N., Payne, T. E. W., Schmahl, E. J., Kundu, M. R., Lemen, J. R., Strong, K. T., Canfield, R. C., & de La Beaujardiere, J. 1994, *ApJ*, 437, 522
Hurford, G. J., Read, R. B., & Zirin, H. 1984, *Solar Phys.*, 94, 413
Komm, R. W., Hurford, G. J., & Gary, D. E. 1997, *Astron. Astrophys. Suppl. Ser.*, 122, 181
Krüger, A., Hildebrandt, J., Bogod, V. M., Korzhavin, A. N., & Akhmedov, Sh. B. 1986, *Solar Phys.*, 105, 111
Kundu, M. R. 1959, *Paris Symposium on Radio Astronomy*, Ed. Ronald N. Bracewell, Stanford U. Press, Stanford, p. 222
Kundu, M. R. & Alissandrakis, C. E. 1984, *Solar Phys.*, 94, 249
Lee, J. 2007, *Space Science Reviews*, 133, 73
Lee, J., Hurford, G. J., & Gary, D. E. 1993a, *Solar Phys.*, 144, 45
Lee, J., Gary, D. E., & Hurford, G. J. 1993b, *Solar Phys.*, 144, 349
Lee, J., Gary, D. E., Hurford, G. J., & Zirin, H. 1993c, *ASPC*, 141, 287.
Lee, J., McClymont, A. N., Mikić, Z., White, S. M., & and Kundu, M. R. 1998, *ApJ* 501, 853
Lee, J., White, S. M., Kundu, M. R., Mikić, Z., & McClymont, A. N. 1999, *ApJ*, 510, 413
Nindos, A., Alissandrakis, C. E., Gelfreikh, G. B., Bogod, V. M., & Gontikakis, C. 2002, *ApJ*, 386, 658
Parker, E. N. 1988, *ApJ*, 330, 474
Rosner, R., Tucker, W. H., & Vaiana, G. S. 1978, *ApJ*, 220, 643
Schmahl, E. J., Kundu, M. R., Strong, K. T., Bentley, R. D., Smith, J. B., & Krall, K. R. 1982, *Solar Phys.*, 80, 233
Schmelz, J. T. & Holman, G. D. 1991, *Adv. Space Res.*, 11, 109
Schmelz, J. T., Holman, G. D., Brosius, J. W., & Willson, R. F. 1994, *ApJ*, 434, 786
Shibasaki, K. 1996, *Adv. Space Res.*, 17, 135
Shibasaki, K. 1998, *ASPC*, 140, 373
Shibasaki, K. 2001, *ApJ*, 550, 1113
Strong, K. T., Alissandrakis, C. E., & Kundu, M. R. 1984, *ApJ*, 277, 865
Tun, S. 2010, in preparation
Vourlidas, A., Bastian, T. S., & Aschwanden, M. J. 1997, *ApJ*, 489, 403

Webb, D. F., Holman, G. D., Davis, J. M., Kundu, M. R., & Shevgaonkar, R. K. 1987, *ApJ*, 315, 716

White, S. M. & Kundu, M. R. 1997, *Solar Phys.*, 174, 31

Zhang, J., Gopalswamy, N., Kundu, M. R., Schmahl, E. J., & Lemen, J. R. 1998, *Solar Phys.*, 180, 285

Zheleznyakov, V. V. 1962, *Soviet Astr.*, 6, 3

Zlotnik, E. Ya., White, S. M., & Kundu, M. R. 1998, *ASPC*, 155, 135

The Physics of the Sun and Star Spots
Proceedings IAU Symposium No. 273, 2010
D. P. Choudhary and K. Strassmeier, eds.

doi:10.1017/S1743921311015365

Global magnetic cycles in rapidly rotating younger suns

Nicholas J. Nelson[1], Benjamin P. Brown[2], Matthew K. Browning[3], Allan Sacha Brun[4], Mark S. Miesch[5] and Juri Toomre[1]

[1]JILA and Department of Astrophysical and Planetary Sciences, University of Colorado, Boulder, CO 80309-0440, USA, [2]Department of Astronomy, University of Wisconsin, 475 Charter St., Madison, WI 53706, USA, [3]Canadian Institute for Theoretical Astrophysics, University of Toronto, Toronto, ON M5S3H8, Canada, [4]DSM/IRFU/SAp, CEA-Saclay, 91191 Gif-sur-Yvette, France, [5]High Altitude Observatory, NCAR, Boulder, CO 80307-3000, USA, email: `nnelson@lcd.colorado.edu`

Abstract. Observations of sun-like stars rotating faster than our current sun tend to exhibit increased magnetic activity as well as magnetic cycles spanning multiple years. Using global simulations in spherical shells to study the coupling of large-scale convection, rotation, and magnetism in a younger sun, we have probed effects of rotation on stellar dynamos and the nature of magnetic cycles. Major 3-D MHD simulations carried out at three times the current solar rotation rate reveal hydromagnetic dynamo action that yields wreaths of strong toroidal magnetic field at low latitudes, often with opposite polarity in the two hemispheres. Our recent simulations have explored behavior in systems with considerably lower diffusivities, achieved with sub-grid scale models including a dynamic Smagorinsky treatment of unresolved turbulence. The lower diffusion promotes the generation of magnetic wreaths that undergo prominent temporal variations in field strength, exhibiting global magnetic cycles that involve polarity reversals. In our least diffusive simulation, we find that magnetic buoyancy coupled with advection by convective giant cells can lead to the rise of coherent loops of magnetic field toward the top of the simulated domain.

Keywords. Star: sun-like stars, star: magnetic activity

1. Coupling rotation, convection, and magnetism in younger suns

Global-scale magnetic fields and cycles of magnetic activity in sun-like stars are generated by the interplay of rotation and convection. At rotation rates greater than that of the current sun, such as when our sun was younger, observations tend to show increased magnetic activity indicating a strong global dynamo may be operating (Pizzolato *et al.* 2003). Here we explore large-scale dynamo action in sun-like stars rotating at three times the current solar rate, or $3\Omega_\odot$, with a rotational period of 9.32 days. As shown by helioseismology, the solar interior is in a state of prominent differential rotation in the convection zone (roughly the outer 30% by radius) whereas the radiative interior is in uniform rotation. A prominent shear layer, or tachocline, is evident at the interface between the convective and radiative regions. Motivated by these observations, a number of theoretical models have been proposed for the solar dynamo. The current paradigms for large-scale solar dynamo action favor a scenario in which the generation sites of toroidal and poloidal fields are spatially separated (e.g., Charbonneau 2005). Poloidal fields generated by cyclonic turbulence within the bulk of the convection zone, or by breakup of active regions, are pumped downward to the tachocline of rotational shear at its base. The differential rotation there stretches such poloidal fields into strong toroidal structures,

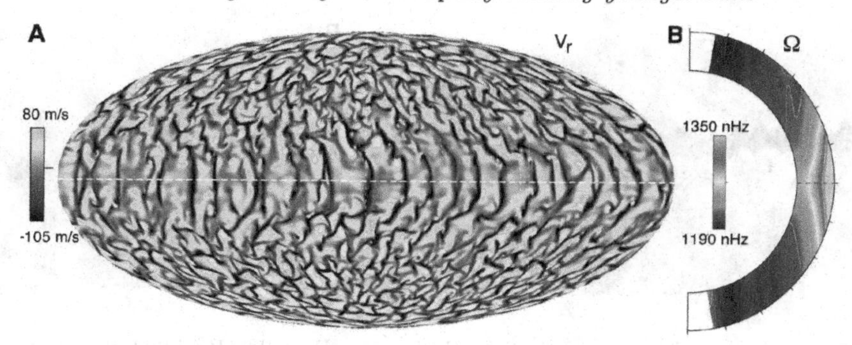

Figure 1. (*A*) Radial velocity in global Mollweide projection at $0.94R_\odot$ with fast, narrow downflows in dark tones and broad, slow upflows in light tones. (*B*) Differential rotation profile, with lines of constant angular velocity Ω largely along cylinders, as expected for rapidly rotating systems. Some deviation toward conical contours is seen at low latitudes. Magnetic wreaths tend to form in the regions of strong shear near the equator.

which may succumb to magnetic buoyancy instabilities and rise upward to pierce the photosphere as curved structures that form the observed active regions. Similiar dynamo processes are believed to be active in sun-like stars rotating several times faster than the current sun. Here we explore a variation to this paradigm by excluding the tachocline and the photosphere from our simulated domain, which extends from $0.72R_\odot$ to $0.96R_\odot$, in order to see if magnetic cycles can be realized in the bulk of the convection zone itself.

Using massively-parallel supercomputers, we solve the nonlinear anelastic MHD equations in rotating 3-D spherical shells using the anelastic spherical harmonic (ASH) code (Brun *et al.* 2004). The anelastic approximation filters out fast-moving sound and magneto-acoustic waves, allowing us to follow the decidedly subsonic flows in the solar convection zone with overturning times of days to months. In large-eddy simulation (LES) such as those using ASH, the effects of small, unresolved scales on larger scales must be parameterized using a turbulence closure model.

Previous ASH simulations of convective dynamos in sun-like stars rotating at $3\Omega_\odot$ have yielded large-scale wreaths of strong toroidal magnetic field in the bulk of their convection zones (Brown *et al.* 2010). These wreaths persist for decades of simulation time, remarkably coexisting with the strongly turbulent flows. Here we explore the effects of decreased levels of diffusion on these wreaths in two simulations, labeled case B and case S. Case B uses an eddy viscosity that varies with depth as the square root of the mean density. Case S uses the dynamic Smagorinsky model of Germano *et al.* (1991), which is based on the assumption of self-similarity in the inertial range of the velocity spectra. Case S has 50 times less diffusion on average than case B. Figure 1a shows the radial velocity field for case S near the top of the convection zone with columnar cells at low latitudes and smaller-scale helical convection at higher latitudes. Figure 1b shows the differential rotation profile for case S with roughly 20% (250 nHz) contrast in rotation rate between the equator and poles. The radial velocity patterns and differential rotation for case B are qualitatively similar to Figure 1.

2. Global magnetic cycles and buoyant magnetic loops

The most remarkable feature of case B is a cyclic variation in the toroidal wreaths of magnetic field. With significantly less diffusion than the simulation of Brown *et al.* (2010) that produced persistent wreaths with no reversals, case B creates strong toroidal bands of magnetic field as shown in Figure 2 with peak field strengths of about 38 kG.

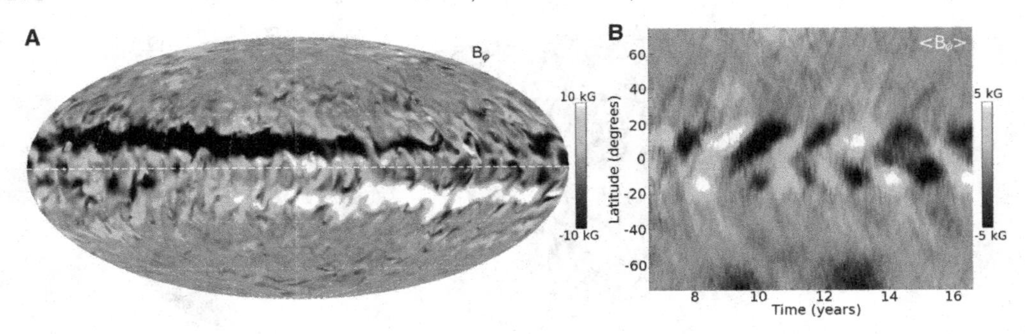

Figure 2. (A) Longitudinal magnetic field B_ϕ for case B at $0.84R_\odot$ in Mollweide projection, showing two strong but patchy magnetic wreaths of opposite polarity with peak field strengths of 38 kG. (B) Time-latitude plot of B_ϕ averaged over longitude $\langle B_\phi \rangle$ at the same depth over 15 years in case B, with strong negative-polarity wreaths shown in dark tones and strong positive-polarity wreaths shown in light tones, clearly indicating cyclic behavior and reversals in magnetic polarity.

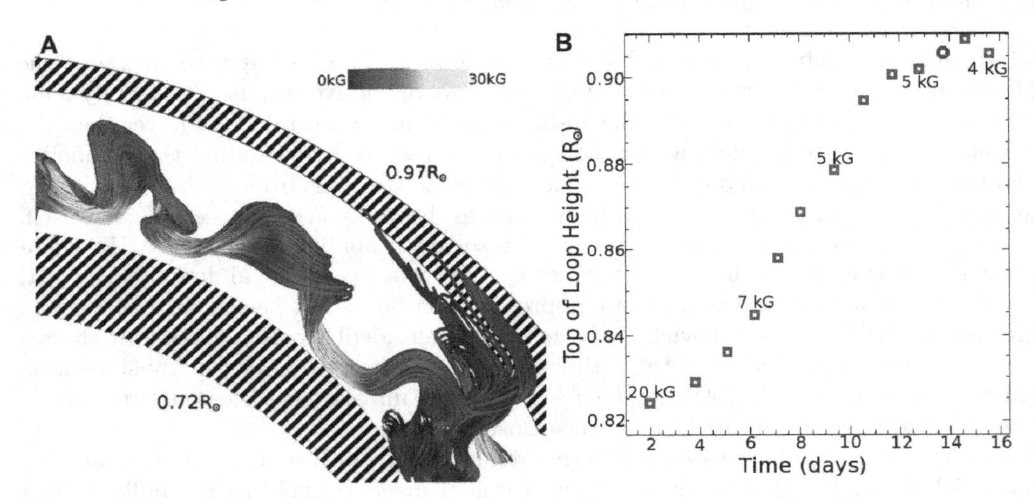

Figure 3. (A) From case S, 3-D volume visualization of magnetic field lines in the core of a wreath-segement with the inner and outer simulation boundaries shown as lined surfaces. View is looking at low latitudes along the rotation axis. (B) Radial location of the top of a buoyant loop as a function of time. Magnetic field strength at the top of the loop is indicated at representative times. Time corresponding to A is indicated by circular plotting symbol at day 13.7.

These wreaths of magnetic field vary strongly with time in both polarity and amplitude. Figure 2a shows B_ϕ in the lower convection zone when there are strong wreaths of opposite polarity in each hemisphere with significant longitudinal variation, which we term patchy wreaths. If we average over longitude, Figure 2b shows a time-latitude map of the $\langle B_\phi \rangle$ in the lower convection zone. The simulation clearly goes through reversals in the magnetic polarity of the wreaths in each hemisphere. At times the hemispheres are out of phase with each other, occasionally yielding wreaths of the same polarity in both hemispheres. Such behavior might be termed irregular magnetic activity cycles.

As we move to even less diffusive simulations, case S shows additional features in the strong toroidal wreaths, most notably buoyant loops of magnetic field. The wreaths are again patchy in longitude and roughly cyclic in time. The peak magnetic field strength rises to about 45 kG inside the wreaths. These strong fields combine with the very low levels of diffusion to allow regions of very strong field to coherently move upward without changing the magnetic topology via reconnection or simply diffusing away the strong fields. Such magnetic loops rise due to a combination of magnetic buoyancy and advection by convective giant cells that span the layer. Figure 3a shows a magnetic loop near its maximum size, extending from 0.765 to $0.908R_\odot$. Examination reveals that there is a significant amount of twist present in the loops and that there is a significant deflection poleward as they rise. The radial location of the top of one buoyant loop as a function of time is shown in Figure 3b. Initially the buoyancy of the wreath due to evacuation of fluid from magnetic pressure dominates over the advective force of the convective upflows, but within 6 days advection becomes dominant. After about 10 days the magnetic tension force begins to balance the advection, causing the top of the loop to stall near $0.905R_\odot$.

These simulations suggest that stars rotating slightly faster than the current sun may produce dynamos capable of cycles of magnetic activity and buoyant magnetic structures in the bulk of their convective envelopes despite the absence of a tachocline of shear. This both challenges and informs the interface dynamo paradigm for sun-like stars. The essential questions are what drives the magnetic reversals in these simulations and what are the conditions necessary to generate buoyant magnetic loops that can survive transit through the convection zone.

This work is supported by NASA Heliophysics Theory Program grants NNG05G124G and NNX08AI57G and major supercomputing support through NSF TeraGrid resources. The presentation of this paper in IAU Symposium 273 was aided by NSF grants ATM 0548260 and AST 0968672, and NASA grant 09-LWSTRT09-0039. Browning is supported by the Jeffrey L. Bishop fellowship at CITA.

References

Brown, B. P., Browning, M. K., Brun, A. S., Miesch, M. S., & Toomre, J., 2010, *Astrophys. J.*, 711 424

Brun, A. S., Miesch, M. S., & Toomre, J., 2004, *Astrophys. J.* **614** 1073

Charbonneau, P., 2005, *Living Rev. Sol. Phys.*, 2, 2

Germano, M., Piomelli, U., Moin, P., & Cabot, W., 1991, *Phys. Fluids A*, 3, 7

Pizzolato, N., Maggio, A., Micela, G., Sciortino, S., & Ventura, R., 2003, *Astron. Astrophys.*, 397, 147

The Physics of the Sun and Star Spots
Proceedings IAU Symposium No. 273, 2010
D. P. Choudhary and K. Strassmeier, eds.

doi:10.1017/S1743921311015377

Magnetohydrostatic equilibrium in starspots: dependences on color (T_{eff}) and surface gravity (g)

S. P. Rajaguru and S. S. Hasan

Indian Institute of Astrophysics, Bangalore - 560034, India,
email: rajaguru@iap.res.in

Abstract. Temperature contrasts and magnetic field strengths of sunspot umbrae broadly follow the thermal-magnetic relationship obtained from magnetohydrostatic equilibrium. Using a compilation of recent observations, especially in molecular bands, of temperature contrasts of starspots in cool stars, and a grid of Kurucz stellar model atmospheres constructed to cover layers of sub-surface convection zone, we examine how the above relationship scales with effective temperature (T_{eff}), surface gravity g and the associated changes in opacity of stellar photospheric gas. We calculate expected field strengths in starpots and find that a given relative reduction in temperatures (or the same darkness contrasts) yield increasing field strengths against decreasing T_{eff} due to a combination of pressure and opacity variations against T_{eff}.

Keywords. Sun: magnetic fields, sunspots, stars: magnetic fields, stars: spots, stars: activity

1. Introduction

Despite a lack of deductive magnetohydrodynamic explanation for the formation and the equilibrium of a sunspot, extensive observations in combination with magnetohydrostatic models have provided reasonable understanding of the thermal-magnetic structure of sunspots in the observable layers. Reduced temperature and gas pressure inside sunspots dictate dominantly the thermal-magnetic relationship derived from the magneto-hydrostatic balance. Such a relationship causes the Wilson depression – geometrical depression of the observable optical depth unity level within a sunspot –, which provides an explanation for the field strengths observed in sunspots and also relates the intensity (or the brightness) contrasts to field strengths. Here we examine how such a thermal-magnetic relationship scales with the stellar parameters, viz. the effective temperature T_{eff} and surface gravity g as well as the associated changes in the opacity of the stellar photospheric gas. We then discuss the implications of such scalings for the interpretations of observed field strengths. We also discuss how such scalings could be crucial players in the activity related photospheric brightness variations and their correlation with other activity measures.

2. Method of calculation

Thermal-magnetic relationship for starspots is obtained from a simplified magnetohydrostatic (MHS) condition, that relates the magnetic field strength to the temperature at the axis of a vertical column of magnetic field (Solanki *et al.* 1993)). The radial component of the magneto-hydrostatic force-balance equation (Maltby 1977, Solanki *et al.* 1993), after neglecting the magnetic curvature force and with the use of the equation of state $P = R\rho T/\mu$, yields the thermal-magnetic relation,

$$\frac{T(r,z)}{T_e(z)} = \frac{\mu(r,z)\rho_e(z)}{\mu_e(z)\rho(r,z)}\left[1 - \frac{B_z^2(r,z)}{8\pi P_e(z)}\right], \tag{2.1}$$

where P, T, ρ, μ and R denote the gas pressure, temperature, density, mean molecular weight and the gas constant respectively. The subscript e stands for the external atmosphere. B_z is the vertical component of magnetic field and r is the radial distance from the center of the spot and z is the depth measured from continuum optical depth $\tau_{c,e}$=1. We further neglect the r-dependence of quantities in the above equation, thus the calculated quantities refer to the axis of the spot, and we refer B_z simply as B hereafter. Eqn. 2.1 is valid for each level z. The variation of external atmospheric quantities with depth z are determined by g and T_{eff} of the parent star and are taken from the grid of Kurucz stellar models constructed to cover deeper regions of the convection zone using the ATLAS9 stellar atmosphere code (Kurucz 2001). We prescribe temperature contrasts for starspots under two cases, case (i): use the empirical relation $\Delta T = (590. * logg) - 680K$ (O'Neal *et al.* 1996) to determine the effective temperature of the spot $T_{eff,spt}$ in a parent star characterised by g and T_{eff}: $T_{spt}(\tau_c = 2/3) = T_{eff,spt} = T_{eff} - \Delta T$, case (ii): the temperature contrasts are independent of g and are of a constant ratio of T_{eff}: $\Delta T = 0.3T_{\mathit{eff}}$ (Berdyugina 2005).

Hydrostatic equilibrium inside the spot yields, to a very good approximation (Cox and Giuli 1968),

$$P(\tau_c = 2/3) = \frac{2}{3}\frac{g}{\kappa_R(\rho, T_{eff,spt})}. \tag{2.2}$$

The density inside the spot at $\tau_c = 2/3$ is determined by solving the above equation with the use of Rosseland mean opacities κ_R from the tables of Kurucz (1993) and Alexander and Ferguson (1994). Saha's equation is used to determine the mean molecular weight. The density difference between the $\tau_c = 2/3$ and $\tau_c = 1$ levels within the spot is assumed to be negligible, i.e., $\rho(\tau_c = 1) \approx \rho(\tau_c = 2/3)$. Using the assumption that $\rho(\tau_c = 1) = \rho_e(z = Z_w)$, i.e., the densities inside and outside the spot are equal at the

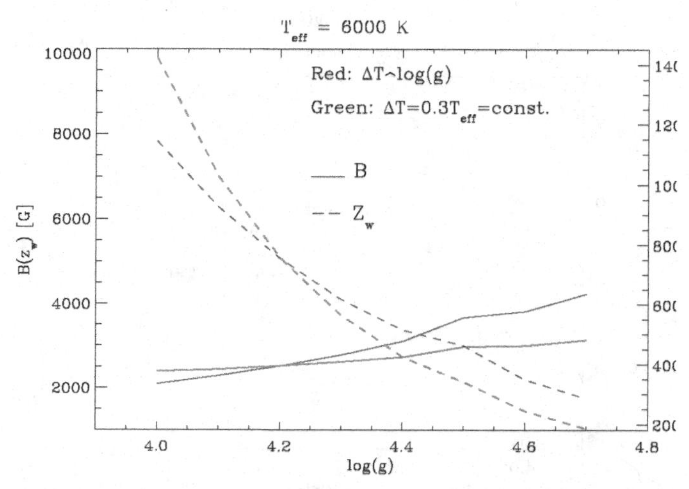

Figure 1. Variation of field strength B (solid lines) and the Wilson depression Z_w (dashed lines) against $log(g)$ for $T_{\mathit{eff}} = 6000K$

level of Wilson depression Z_w (Maltby 1977), we find the depth z at which the above equality is satisfied in the parent stellar model. This gives the value of Z_w, and now we determine the only unknown quantity B at this level using Eqn. 2.1.

3. Results and discussions

Figures 1 and 2 summarise the main results of our study. Each figure shows the variation of observable field strengths $B(Z_w)$ on the spot axis as a function of $log(g)$ for a particular T_{eff} for the two distinct kinds of variation of spot temperature given as case (i) (*red curves*) and (ii) (*green curves*) in the previous Section. The Wilson depressions, Z_w, are shown as dashed curves and the B values as solid curves. Z_w are scaled in terms of pressure scale heights H_p at the τ_c=1 level in the quiet photospheres. Fig. 1, for $T_{eff} = 6000K$, shows that the B values are not very different for the two cases above, and moreover the Wilson depressions Z_w are of the order of scale heights. In contrast, Figure 2, for T_{eff} = $5000K$, shows that the results for the above two cases are very

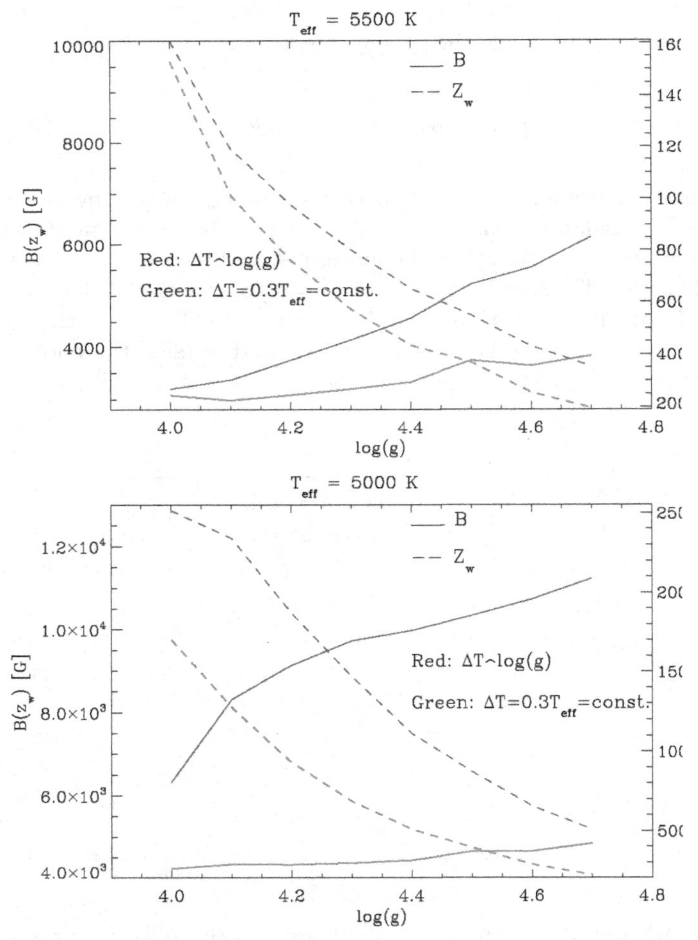

Figure 2. Same as Fig. 1 but for $T_{eff} = 5500K$ (top) and $5000K$ (bottom)

different: the gravity dependent temperature contrasts (as implied by the observed relation of O'Neal *et al.* (1996)) requires very strong magnetic fields for starspot equilibrium, which stems from the large values of Z_w. On the other hand, a constant value of temperature reduction (case ii above) over the range of surface gravity values used yields B values within the observed ranges (Saar 1990, Solanki 1992). Consequently, a relatively less temperature reduction and hence a less amount of gas evacuation is sufficient to attain a given value of field strength in cooler stars than that in stars hotter than about $5500K$. In other words, spots of a given field strength would appear much darker in a star of $T_{eff} = 6000K$ than those in $T_{eff} = 5000K$.

According to Lockwood *et al.* (1992), younger and faster rotating stars are 'spot dominated', i.e. more flux in spots than small-scale fields and faculae, and hence grow darker as activity increases. If such a 'spot domination' is purely dependent on age (rotation) but not on color (spectral type), then it could be thought of as a phenomenon not contradictory with the g dependent temperature contrasts for spots derived by O'Neal *et al.* (1996). However, our results in Figure 2 imply unrealistically large B and Z_w values for spots in such cases. On the other hand, consistency between Lockwood *et al.*'s results requiring 'spot domination' for younger stars and the situation of case (ii) of our results requires that there be a color T_{eff} dependence of spot properties in addition to the age dependence.

Alternatively, results of Lockwood *et al.* (1992), but without the requirement of 'spot domination' in younger stars, could be made consistent with our case (ii) of less gas evacuated spots if the small-scale fields forming the 'faculae' too are of such less evacuated state. This would imply that faculae in younger and cooler stars are less bright and therefore these stars grow darker as activity increases. Interestingly, the superadiabaticity that drives the convective collapse of small-scale flux tubes indeed linearly decrease with T_{eff} and is found not to intensify such fields to a high degree of evacuation as in stars hotter than about 5000 K (Rajaguru *et al.* 2002). Hence, it would appear that the near-surface thermal structure of cool stars crucially determine the key properties of magnetic structures small and large. We conclude that the scaling of thermal and magnetic properties of starspots with both the gravity g (age) and T_{eff} (color) are crucial for a consistent interpretation of observed correlations between activity measures that sample the different heights in the outer atmospheres.

Acknowledgements

The presentation of this paper in the IAU Symposium 273 was possible due to partial support from the National Science Foundation grant numbers ATM 0548260, AST 0968672 and NASA - Living With a Star grant number 09-LWSTRT09-0039.

References

Alexander, D. R. & Ferguson, J. W. 1994, *ApJ*, 437, 879

Berdyugina, S. 2005, *Living Rev. Solar Phys.*, 2, 8; *URL: www.livingreviews.org/lrsp-2008-8*

Cox, J. P. & Giuli, R. T. 1968, *Principles of Stellar Structure, Vol. 2*, New York: Gordon & Breach, p590

Kurucz, R. L. 1993, *ATLAS9 Stellar Atmosphere Programs and 2 km/s grid, Kurucz CD-ROM No.13. Cambridge, Mass.: Smithsonian Astrophysical Observatory.*

Kurucz, R. L. 2001, *private communication*

Lockwood, G. W., Skiff, B. A., Baliunas, S. L., & Radick, R. R. 1992, *Nature*, 360, 653

Maltby, P. 1977, *Solar Phys.*, 57, 335

O'Neal, D., Saar, S. H., & Neff, J. E. 1996, *ApJ*, 463, 766
Rajaguru, S. P., Kurucz, R. L., & Hasan, S. S. 2002, *ApJ*, 565, L101
Saar, S. H. 1990, in *IAU Symposium No. 138 Solar Photosphere: Structure, Convection, and Magnetic Fields*, ed. J. O. Stenflo (Kluwer: Dordrecht), p427
Solanki, S. K., Walther, U., & Livingston, W. 1993, *A&A*, 277, 639

The Physics of the Sun and Star Spots
Proceedings IAU Symposium No. 273, 2010
D. P. Choudhary and K. G. Strassmeier, eds.

doi:10.1017/S1743921311015389

Disentangling between stellar activity and planetary signals

Isabelle Boisse[1,5], François Bouchy[1,2], Guillaume Hébrard[1,2], Xavier Bonfils[3,4], Nuno Santos[5] and Sylvie Vauclair[6]

[1]Institut d'Astrophysique de Paris, Université Pierre et Marie Curie, UMR7095 CNRS, 98bis bd. Arago, 75014 Paris, France
email: iboisse@iap.fr

[2]Observatoire de Haute Provence, CNRS/OAMP, 04870 St Michel l'Observatoire, France
[3]Laboratoire d'Astrophysique de Grenoble, Observatoire de Grenoble, Université Joseph Fourier, CNRS, UMR 5571, 38041, Grenoble Cedex 09, France
[4]Observatoire de Genève, Université de Genève, 51 Ch. des Maillettes, 1290 Sauverny, Switzerland
[5]Centro de Astrofísica, Universidade do Porto, Rua das Estrelas, 4150-762 Porto, Portugal
[6]LATT-UMR 5572, CNRS & Université P. Sabatier, 14 Av. E. Belin, F-31400 Toulouse, France

Abstract. Photospheric stellar activity (i.e. dark spots or bright plages) might be an important source of noise and confusion in the radial-velocity (RV) measurements. Radial-velocimetry planet search surveys as well as follow-up of photometric transit surveys require a deeper understanding and characterization of the effects of stellar activities to disentangle it from planetary signals.

We simulate dark spots on a rotating stellar photosphere. The variations of the RV are characterized and analyzed according to the stellar inclination, the latitude and the number of spots. The Lomb-Scargle periodograms of the RV variations induced by activity present power at the rotational period P_{rot} of the star and its two-first harmonics $P_{rot}/2$ and $P_{rot}/3$. Three adjusted sinusoids fixed at the fundamental period and its two-first harmonics allow to remove about 90% of the RV jitter amplitude. We apply and validate our approach on four known active planet-host stars: HD 189733, GJ 674, CoRoT-7 and ι Hor. We succeed in fitting simultaneously activity and planetary signals on GJ674 and CoRoT-7. We excluded short-period low-mass exoplanets around ι Hor. Our approach is efficient to disentangle reflex-motion due to a planetary companion and stellar-activity induced-RV variations provided that 1) the planetary orbital period is not close to that of the stellar rotation or one of its two-first harmonics, 2) the rotational period of the star is accurately known, 3) the data cover more than one stellar rotational period.

Keywords. techniques: radial velocities - planetary systems - stars: activity - stars: individual: ι Hor, HD 189733, GJ 674, CoRoT-7

1. Introduction

High-precision radial-velocimetry is until now the more efficient way to discover planetary systems. However, an active star presents on its photosphere dark spots and/or bright plages rotating with the star. These inhomogeneities of the stellar surface can induce RV shifts due to changes in the spectral lines shape which may add confusions with the Doppler reflex-motion due to a planetary companion (e.g. Queloz *et al.* 2001; Huélamo *et al.* 2008). The amplitude of the RV shifts depend on the $v \sin i$ of the star, the spectrograph resolution, the size and temperature of spot (Saar & Donahue 1997; Hatzes

1999; Desort *et al.* 2007). For these reasons, active stars are then usually discarded from RV surveys using criteria based on activity index R'_{HK} and/or $v \sin i$. However, photometric transit search missions (like CoRoT and Kepler) require RV measurements to establish the planetary nature of the transiting candidates and to characterize their true masses. These candidates include active stars adding strong confusions and difficulties in the RV follow-up.

2. Dark spot simulations

SOAP is a program that calculates the photometric, RV and line shape modulations induced by one (or more) cool spots on a rotating stellar surface. SOAP computes the rotational broadening of a spectral line by sampling the stellar disk on a grid. For each grid cell, a Gaussian function represents the typical line of the emergent spectrum. The Gaussian is Doppler-shifted according to the projected rotational velocity ($v \sin i$) and weighted by a linear limb-darkening law. The stellar spectrum output by the program is the sum of all contributions from all grid cells. The spot is consider as a dark surface without emission of light, so we cannot compute different temperature for the spot. For a given spot (defined by its latitude, longitude and size), SOAP computes which of the grid cells are obscured and removes their contribution to the integrated stellar spectrum.

RV variations due to dark spots. Fig. 1 shows the RV modulations due to a spot as a function of time for different inclinations i of the star with the line of sight and different spot latitudes lat. These two parameters clearly modify the pattern of the RV modulation. Fig. 2 shows the Lomb-Scargle periodograms of the three cases showed in Fig. 1. Main peaks are clearly detected at the rotational period of the star P_{rot}, as well as the two-first harmonics $P_{rot}/2$ and $P_{rot}/3$. We noticed that the energy in each peak varies with the shape of the RV modulation. Multiples of the rotational period are never found.

Finally, the periods detected in the periodogram are the same for the following configurations : 1) a star with different inclinations, 2) spots at different latitudes, 3) spot size and/or temperature varying with time, 4) several spots on the stellar surface.

RV fit of dark spots. The purpose is to remove or at least to reduce the stellar activity signals in order to identify a planetary signal hidden in the RV jitter. The Lomb-Scargle periodogram corresponds to sinusoidal decompositions of the data. Three sinusoids with periods fixed at the rotational period P_{rot}, and its two-first harmonics $P_{rot}/2$ and $P_{rot}/3$ reduce the semi-amplitude of the RV jitter by more than 87%.

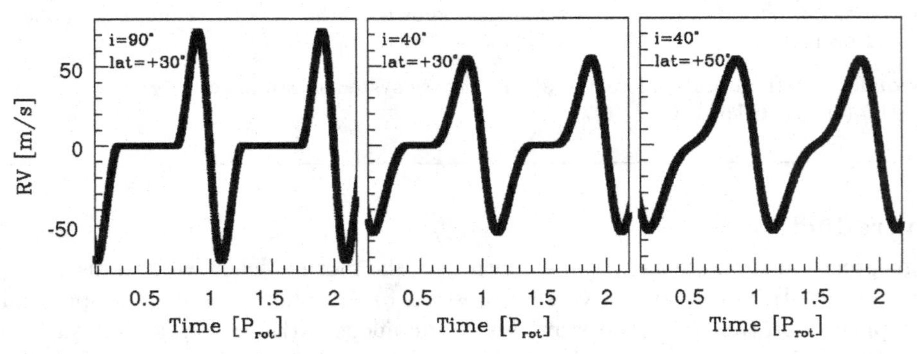

Figure 1. RV modulations due to one spot as a function of time (expressed in rotational period unit). At t=0, the dark spot of 1% of the visible stellar surface is in front of the line of sight. The shape of the signal changes with the inclination i of the star and the latitude lat of the spot, labelled in the top left of each panel.

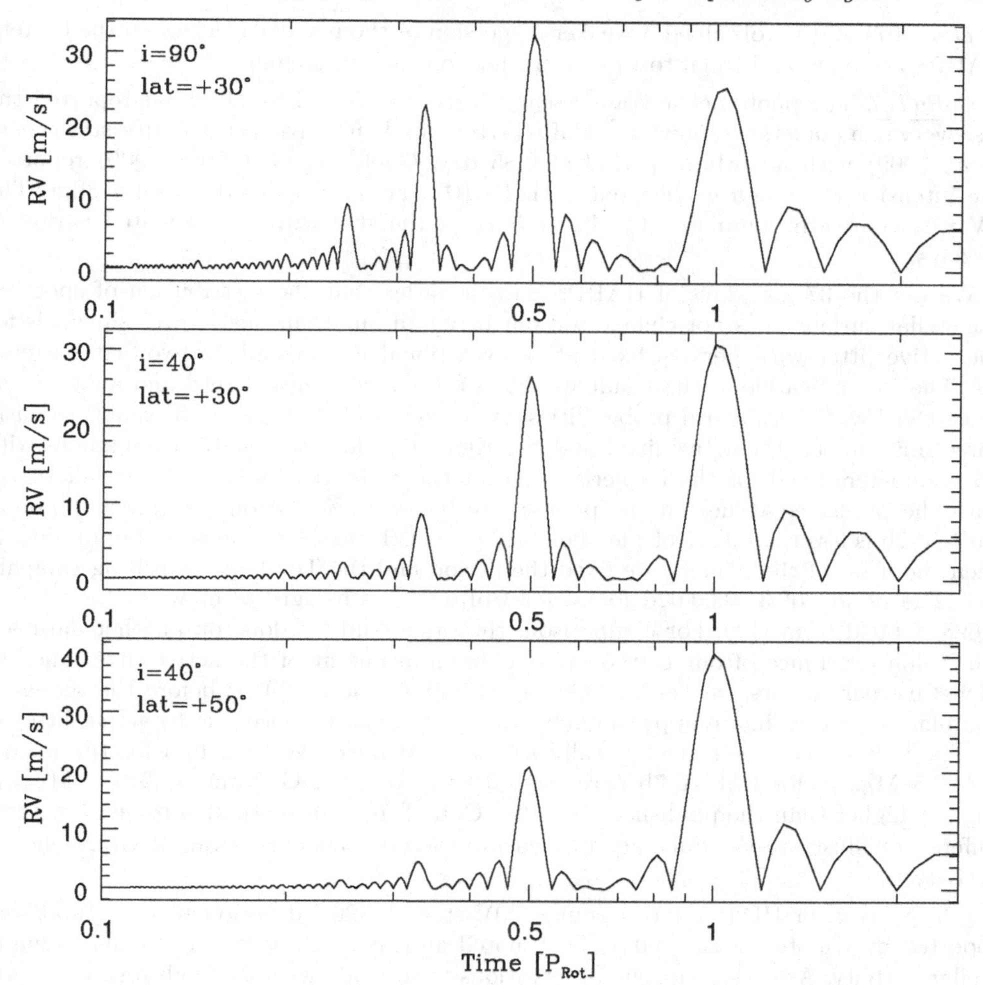

Figure 2. Lomb-Scargle periodograms of the three RV modulations showed in Fig. 1. The fundamental frequency, P_{rot}, and its first harmonics are detected.

3. Application to real data

HD 189733. The active K2V star HD 189733 and its transiting planetary companion was monitored by Boisse *et al.* (2009) with the high-resolution spectrograph *SOPHIE* mounted on the 1.93-m telescope at the Observatoire de Haute-Provence. The RV measurements subtracted from their fit of the planetary companion are variable due to the activity of the star. We computed the Lomb-Scargle periodogram of the residuals from the Keplerian fit. The fundamental period at P_{rot} ($\sim$12d) and its two-first harmonics are detected with false alarm probability lower than 10^{-1}.

GJ 674. GJ 674 is a moderately active M2.5V dwarf hosting a planet with 4.69-day period (Bonfils *et al.* 2007). A superimposed signal with a periodicity of 35 days due to stellar activity is also visible in the HARPS RV measurements. We fit the RV with three sinusoids with periods fixed at the rotational period and the two-first harmonics and one Keplerian that gives the planetary parameters. These are in agreement with Bonfils *et al.* (2007) and we obtained a weaker dispersion of the residuals closest to the current HARPS accuracy and equal to the uncertainty on each measurement.

CoRoT-7. The photometric transit search with the CoRoT satellite has reported the discovery of a planetary companion CoRoT-7b around an active V=11.7 G9V star (Léger *et al.* 2009) with an orbital period of 0.85 day. Queloz *et al.* (2009) (Q09) reported the intensive campaign carried out with HARPS at 3.6-m telescope at La Silla. The RV variations are dominated by the activity of the star with an estimated period of 23 days.

We use the 37 last days of HARPS data in order that the distribution of spots on the stellar surface does not change too much. We fit simultaneously three sinusoids for the active jitter with periods fixed at the rotational period and its two-first harmonics. The Lomb-Scargle of the residuals shows a clear peak near 3.69 d and another one near 0.85 d with false alarm probabilities lower than 5.10^{-4}. We then fit simultaneously three sinusoids for the active jitter and two Keplerians for the possible companions with no parameters fixed for the Keplerians except the eccentricities (e=0). The differences with the published values on the periods are below 0.5% and on the transit phase of CoRoT-7b is less than 0.2% of the planetary period. To measure the semi-amplitude and then the mass of the planets, we fixed the period and the T0 of the transiting companion. The period of 3.70±0.02d found for CoRoT-7c is in agreement with the value of 3.698±0.003d from Q09. For comparison, the same study is done on another data set. The main difference of our method is the simultaneous fit of the active jitter and the planetary parameters, instead of Q09 that fitted the active jitter before the search of the planets. We estimate approximately that a systematic noise due to active jitter of 1.5 ms^{-1} must be added quadratically to the error bars. We then find for the masses 5.7±2.5 M_{Earth} for CoRoT-7b agreeing with the value of Q09 and 13.2±4.1 M_{Earth}, slightly higher than the published value, for CoRoT-7c. Our method is robust but these differences illustrate the difficulty to measure the amplitudes accurately in presence of activity.

ι Hor. ι Hor, or HD 17051 is a young G0V star. A 320.1-d period planet ι Hor b was reported by Kürster *et al.* (2000). They noted an excess RV scatter of 27 ms^{-1} due to stellar activity. Asteroseismologic observations were made with the high-precision spectrograph HARPS on ι Hor (Vauclair *et al.* 2008). We studied these data to characterize the active jitter and to search for a possible hidden Doppler motion. Before studying the RV variations due to stellar activity, we subtract the long-period planet Doppler motion and we averaged the data by group of 20 measurements in order to average the p-modes signature. The mean RV photon noise uncertainty on averaged points is then about 26 cms^{-1}, but the actual precision is limited by the instrumental accuracy $\approx$ 80 cms^{-1}.

The ι Hor RV modulations are well explained by dark spots rotating with the photosphere with a rotational period in the range [7.9-8.4] days. The residuals are equal to $\sigma = 1.03\mathrm{ms}^{-1}$ reaching almost the instrumental accuracy. We do not detect in the ιHor data a short-period companion. Nevertheless, we would like to know if we have subtracted the RV shift due to a companion subtracting the effect of activity. We ran simulations and added RV due to fake planets to the ιHor data. We consider only the case of circular orbits. We fit the active jitter with three sinusoids with period fixed at the rotational period and its two-first harmonics and look afterwards at the Lomb-Scargle periodogram of the residuals. If a peak at the planetary period is detected, we fit simultaneously a Keplerian with null-eccentricity to obtain the planetary parameters. We excluded the presence of planet with a minimum mass between 6 and 10 M_{Earth} with periods respectively between 0.7 and 2.4 days.

References

Boisse, I., Moutou, C., & Vidal-Madjar, A. *et al.* 2009, *A&A*, 495, 959
Bonfils, X., Mayor, M., & Delfosse, X. *et al.* 2007, *A&A*, 474, 293
Desort, M., Lagrange, A.-M., & Galland, F. *et al.* 2007, *A&A*, 473, 983
Hatzes, A. P. 1999, *ASPC*, 185, 259
Huélamo, N., Figueira, P., & Bonfils, X. *et al.* 2008, *A&A* (Letters), 489, L9
Kürster, M., Endl, M., & Els, S. *et al.* 2000, *A&A* (Letters), 353, 33
Léger, A., Rouan, D., & Schneider, J. *et al.* 2009, *A&A* (Letters), 506, 287
Queloz, D., Henry, G. W., & Sivan, J. P. *et al.* 2001, *A&A*, 379, 279
Queloz, D., Bouchy, F., Moutou, C., Hatzes, A., & Hébrard, G. *et al.* 2009, *A&A*, 506, 303
Saar, S. H. & Donahue, R. 1997, *ApJ*, 485, 319
Vauclair, S., Laymand, M., & Bouchy, F. *et al.* 2008, *A&A*, 428, 5

The Physics of the Sun and Star Spots
Proceedings IAU Symposium No. 273, 2010
D. P. Choudhary and K. G. Strassmeier, eds.

doi:10.1017/S1743921311015390

First solar butterfly diagram from Schwabe's observations in 1825–1867

Rainer Arlt[1] and Anastasia Abdolvand[2]

[1] Astrophysikalisches Institut Potsdam,
An der Sternwarte 16, D-14482 Potsdam, Germany
email: rarlt@aip.de

[2] Lycée Fran çais de Berlin
Berlin, Germany
email: abdolvand_a@hotmail.de

Abstract. The original sunspot observations by Samuel Heinrich Schwabe of 1825–1867 were digitized and a first subset of spots was measured. In this initial project, we determined more than 14 000 sunspot positions and areas comprising about 11% of the total amount of spots available from that period. The resulting butterfly diagram has a typical appearance, but with evident north-south asymmetries.

Keywords. Sun: activity, sunspots, history and philosophy of astronomy

---

1. Introduction

A continuous set of sunspot positions was constructed from the observations at the Royal Greenwich Observatory starting in 1874 and the observations obtained by the USAF/NOAA starting in 1976 (cf. Hathaway *et al.* 2003). Numerous investigations regarding the butterfly diagram or sunspot area are based on this dataset. It is desirable to extend this information back in time to cover a larger part of the period for which sunspot *numbers* are known. Spörer (1874, 1878, 1880, 1886, 1894) gives sunspot positions for the period 1861–1894. A short set of observations is available from Richard Carrington (1863) covering the period 1853–1861. An even earlier set of sunspot positions was derived by Arlt (2009) from the observations by Staudacher in 1749–1799.

In this paper, we present a first set of sunspot positions and areas from the full-disk drawings of the sun by Samuel Heinrich Schwabe at his location in Dessau, Germany, in 1825–1867. The original observations are preserved by the Royal Astronomical Society, London, and have been kindly provided for digitization in 2009.

2. Determination of positions and areas

The total number of full-disk drawings in these records is 8468 consisting of circles of 5 cm in diameter. Within this set, 7299 drawings have a coordinate system which is found to be aligned with the celestial equator. Especially from mid-1830 onward, almost all drawings have this coordinate system. Special care was apparently taken by Schwabe, that the drawing represents the situation at 12^{h} local time. Schwabe gave descriptions of the spots several times a day, and in nearly all cases only the 12^{h} description matches the drawing. All observations were made looking through a Keplerian telescope equipped with one of a variety of solar filters. All images have thus been turned by 180° before being used for measurements.

We assume that the middle horizontal line of the drawings is parallel to the celestial equator and add the inclinations of the ecliptic as well as the tilt of the solar rotation axis against the ecliptic to the direction to the celestial pole. A preliminary heliographic coordinate system is drawn onto the sunspot drawing. The disks are enlarged to a size of 490 pixels radius to achieve a slight sub-pixel accuracy as compared to the original images. This corresponds to a scale of 0.05 mm/pixel which is below the probable plotting errors of Schwabe. If that was 1 mm (which is likely to be an upper limit), the error in heliographic position is $2.5° / \cos d$, where d is the distance from the disk center. The additional uncertainty in determining the position angle of the solar equator is included here with an approximate error of 0.3°; it is fairly well defined.

The cases without any coordinate system required special care. We noticed that the drawings were fairly well aligned with the celestial equator, simply by looking at the plausibility of the spot distribution from day to day. This alignment is remarkably consistent throughout the dataset, especially for the drawings explicitly marked to be made at 12^{h} local time. We utilized those by assuming that an imaginary horizontal line through the drawing is parallel to the celestial equator. Even if one of Schwabe's early telescopes was not mounted parallactically, an alignment with the horizon in an alt–azimuth system would deliver the same result for observations near noon.

For a few observations before 1830, we used a special matching algorithm for pairs of observations separated by no more than a few days. Two or more spots need to be on both drawings. Using the differential rotation derived by Balthasar *et al.* (1986), the individual position angles of the two drawings can be determined by a least-squares search. While the method may deliver additional positions, not measurable otherwise, we need to bear in mind that using the differential rotation as an input reduces the chances of an independent determination of the differential rotation of that Cycle 7.

The sunspot size was estimated by various cursor masks with radii from 1 to 11 pixels. As soon as Schwabe distinguished penumbra and umbra in his drawings, we used the umbral area for the size determinations, since we believe these are a more direct indication for the emerging magnetic flux than the total size of a sunspot. Note that this differs from the USAF/NOAA set which only reports the penumbral area after Dec 16, 1981. We also need to emphasize that we only employed the full-disk drawings by Schwabe. Numerous detailed drawings of individual groups are also available, showing more spots at greater detail, but placing them to the right scale and with correct orientation into the full-disk drawings is a delicate task.

3. Result

The tentative butterfly diagram of the period of 1825–1867 is shown in Fig. 1. The darkness of the individual spots is scaled with the area of the spots while the vertical (latitudinal) extent is scaled with the diameter of the spots. Note that the diagram contains only about 11% of the full set of observations available. The accuracy will improve significantly after measuring the remaining 89% of the drawings.

The solar activity cycle is clearly visible. The first cycle (Cycle 7) appears in darker colours, because Schwabe did not distinguish the umbra from the penumbra until 1830. The larger areas measured result in a "darker" cycle as compared to the following cycles. A recalibration of these data will be necessary for the final construction of the

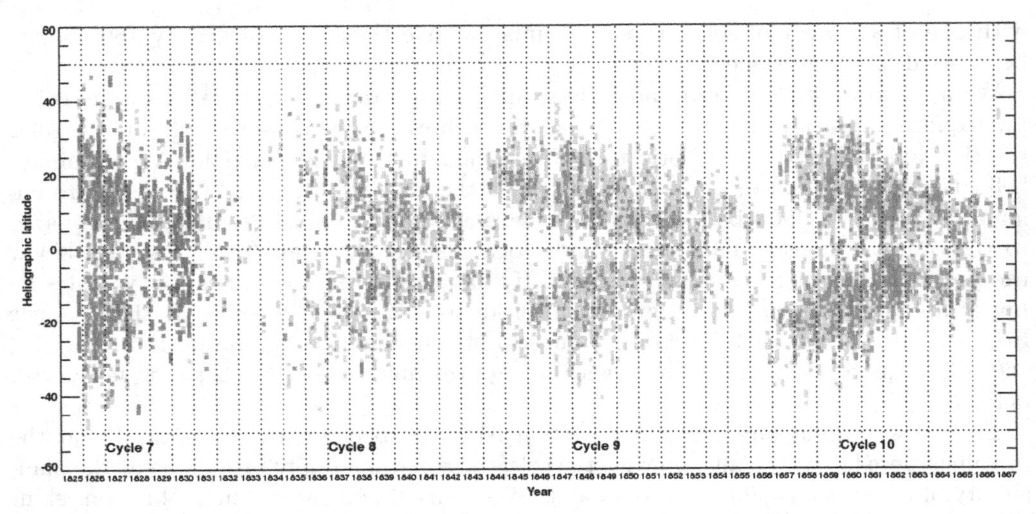

Figure 1. Butterfly diagram obtained from the observations by Heinrich Schwabe in 1825–1867. The sunspot areas up to 1830 are overestimated since Schwabe did not distinguish penumbrae from umbrae in the beginning.

butterfly diagram in the future. Also, there are some periods during Cycle 7 when we made measurements for each drawing, while other periods were covered by every 10th drawing to get a first picture of the entire observing period. These "daily" measurements are reflected as denser stripes in the butterfly diagram.

According to our results, Cycle 8 was a rather weak one, while the Wolf numbers indicate a stronger cycle. Cycle 9 shows a phase difference between the northern and southern hemispheres with the beginning of the northern cycle preceding the one of the southern hemisphere by at least half a year. The analysis as in Zolotova *et al.* (2010), once the full dataset is available, will show the entire development of the phase difference for the middle of the 19th century. Cycle 10 appears to be more symmetric, and the beginning of Cycle 11 is just barely visible in 1867.

The data will be available from the author once the full set of observations will be measured.

Acknowledgements

The authors are much obliged to the Royal Astronomical Society, London, and to Robert Massey and Peter Hingley in particular for the support with digitizing the original observations. We are grateful to Stela Frencheva, Jennifer Koch, and Christian Schmiel for their help with the utilization of the digital images.

References

Arlt, R. 2009, *Sol. Phys.*, 255, 143

Balthasar, H., Vázquez, M., & Wöhl, H. 1986, *A&A*, 155, 87

Carrington, R. C. 1863, *Observations of the Spots on the Sun from November 9, 1853 to March 24, 1861 made at Redhill.* Williams and Norgate, London, Edinburgh

Hathaway, D. H., Nandy, D., Wilson, R. M., & Reichmann, E. J. 2003, *ApJ*, 589, 665

Spörer, G. 1874, Publicationen der Astronomischen Gesellschaft, XIII

Spörer, G. 1878, Publicationen des Astrophysikalischen Observatoriums zu Potsdam Nr. 1, Vol. 1, part 1, Wilhelm Engelmann, Leipzig

Spörer, G. 1880, *ibid* Nr. 5, Vol. 2, part 1
Spörer, G. 1886, *ibid* Nr. 17, Vol. 4, part 4
Spörer, G. 1894, *ibid* Nr. 32, Vol. 10, part 1
Zolotova, N. V., Ponyavin, D. I., Arlt, R., & Tuominen, I. 2010, *AN*, 331, 765

The Physics of Sun and Star Spots
Proceedings IAU Symposium No. 273, 2010
D.P. Choudhary & K.G. Strassmeier, eds.

doi:10.1017/S1743921311015407

The structure and evolution of global solar magnetic fields

Duncan H. Mackay

School of Mathematics and Statistics, University of St Andrews, St Andrews, Fife, KY16 8HB
email: duncan@mcs.st-and.ac.uk

Abstract. This review will discuss both observational and theoretical aspects of the Sun's global magnetic field. First recent observations will be described, along with the main physical processes leading to the time evolution and structure of the global field. Following this, recent theoretical models of both the global surface and coronal magnetic field will be presented. The application of these models to the structure of the corona, formation of solar filaments, the onset of CMEs and finally the origin and variation of the Sun's open flux will be discussed.

Keywords. Sun: activity, sun: corona, sun: magnetic fields

1. Introduction

In this review, our present day understanding of the Sun's global magnetic field is considered from both observational and theoretical aspects. To date three solar cycles worth of continuous data has been collected on the distribution and evolution of the Sun's photospheric magnetic field. An illustration of this can be seen in Figure 1 which shows the magnetic butterfly diagram (from Hathaway 2010). The plot illustrates key features in the distribution and evolution of the magnetic field, such as: emergence latitudes, 11 year polar field reversal, Joys' Law and Hales polarity law. As the global field evolves through this cyclic evolution it produces a wide range of phenomena in the corona such as solar filaments (Tandberg-Hanssen 1995; Mackay *et al.* 2010), variations in the Open Flux (Balogh *et al.* 1995; Lockwood *et al.* 1999) and eruptive phenomena such as Coronal Mass Ejections (Forbes *et al.* 2006; Cremades & St. Cyr 2007). In the following we will describe models constructed to consider both the photospheric and coronal magnetic fields, along with the new generation of 3D MHD models.

2. Magnetic flux transport models

Magnetic flux transport simulations (Sheeley 2005) are designed to model the large-scale, long time evolution of the radial magnetic field, B_r, at the solar photosphere. The four key physical effects included in these models are flux emergence, differential rotation, meridional flow and supergranular diffusion. Since the early flux transport models were produced (Wang *et al.* 1989) new variations have emerged to include additional physical effects (Schrijver 2001; Baumann *et al.* 2006). Applications have shown that flux transport models may accurately reproduce the essential characteristics of the evolution of magnetic fields not only on the Sun (Sheeley 2005) but also in other stars (Schrijver & Title 2001), where in some cases different transport coefficients are required to match magnetic field observations (Mackay *et al.* 2004). Often, output from flux transport models are used as lower boundary conditions for coronal field modelling. Applications are now discussed.

3. PFSS models and corona null points

To date, the most common technique for modelling the global coronal magnetic field of the Sun is through Potential Field Source Surface Models (PFSS, Schatten *et al.* 1969). Such models are simple to construct, as they require only the radial magnetic field at the solar photosphere to be specified (either from observations or flux transport models) and assume zero electric current in the corona. Recently Cook *et al.* (2009) used magnetic flux transport models combined with PFSS models to consider the origin and variation of coronal null points over two solar cycles. Null points are a key topological feature in the Magnetic Breakout Model of CMEs (Antiochos *et al.* 1999). Results show that the number of coronal null points vary in phase with the solar cycle, with more at cycle maximum than minimum. The majority of coronal nulls form above active latitudes and lie in the low corona (below 175,000km). This shows that the complex active latitude field is more important for the existence of the nulls compared to the overlying global dipole. While a significant number of nulls (15-17) are present each day, due to their low height only 50% have more flux lying below the null than above. This is a necessary condition for breakout to occur. Therefore, while the Magnetic Breakout Model may be an important mechanism for CMEs, other mechanisms must also act to account for the observed numbers (Cremades & St. Cyr 2007; Barnes 2007). While potential field models are useful as a first approximation to the coronal field, it is known that the corona is non-potential due to eruptive phenomena that are present. Global non-potential models are now discussed. First, long-term reduced MHD models are considered, then steady state full MHD models.

4. Global, long term non-potential models

Recently van Ballegooijen *et al.* (2000) and Mackay & van Ballegooijen (2006) have developed a new technique to study the global, long-term evolution of coronal magnetic fields. The technique follows the build-up of free magnetic energy and electric currents in the corona. Such effects are required in order to explain many eruptive phenomena found on the Sun. The technique has two components. The first component, which represents the solar photosphere, is a data driven magnetic flux transport model. This reproduces a continuous time evolution of the observed magnetic flux seen on the Sun over periods of

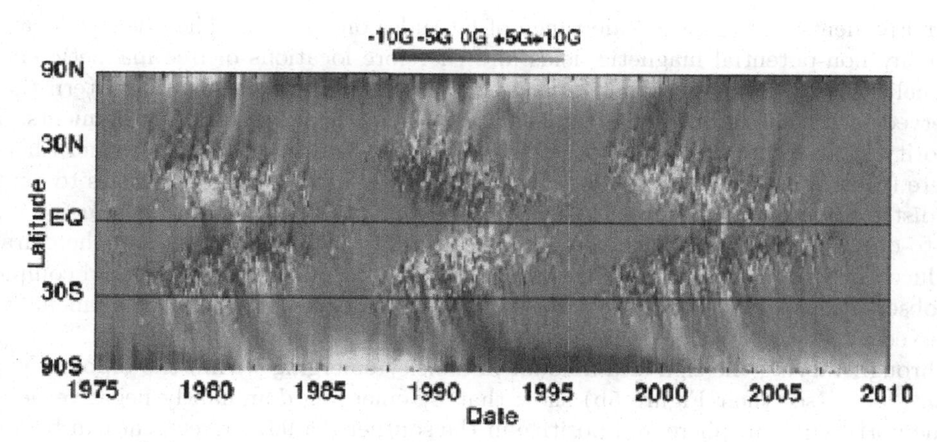

Figure 1. Magnetic butterfly Diagram (from Hathaway (2010))

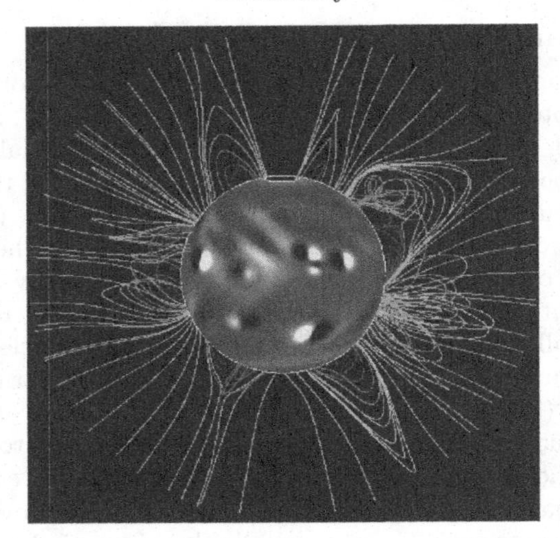

Figure 2. Example of the magnetic field distribution in the global simulation after 108 days of evolution, showing highly twisted flux ropes, weakly sheared arcades and near potential open fields. On the grey-scale image white/black represents positive/negative flux.

months to years (Yeates *et al.* 2007). Coupled to this is a quasi-static coronal evolution model, which uses a magnetofrictional relaxation technique (Yang *et al.* 1986) to evolve the coronal magnetic field through sequences of non-linear force-free fields in response to the observed photospheric evolution. In Figure 2 an example of the global field can be seen after 108 days of evolution where both the photospheric distribution (grey-scale image) and non-linear force-free coronal field are shown. The coronal field is made up of highly twisted flux ropes, slightly sheared coronal arcades and near potential open field lines. A key difference between this technique and previous studies, is that it can be run for extended periods without ever resetting the surface field back to that found in observations or the coronal field to potential. Therefore, the simulations are able to consider long-term helicity transport across the solar surface from low to high latitudes. Three applications of this model are now discussed.

4.1. *Hemispheric pattern of solar filaments*

Solar Filaments form over a wide range of latitudes on the Sun. They denote locations of highly non-potential magnetic fields and therefore locations of free magnetic energy and helicity storage. This helicity storage is seen through a hemispheric pattern that is observed in the direction of the axial magnetic field threading through filaments. The majority of filaments are found to be dextral/sinistral in the northern/southern hemipsphere (Martin *et al.* 1994). To test if the global code has the correct physics to explain the distribution and build-up of helicity in filaments, Yeates *et al.* (2008) carried out a one-to-one comparison between the observed chirality of 109 filaments and the chirality produced by the global model. The comparison covered a 6 month period and compared the observations and simulations at the exact time and location where the filaments were observed.

Through varying the sign and amount of helicity emerging within the bipoles, Yeates *et al.* (2008) (see their Figure 5b) show that by emerging dominantly negative helicity in the northern hemisphere and positive in the southern, a 96% agreement can be found between the observations and simulations, where the agreement is equally good for mi-

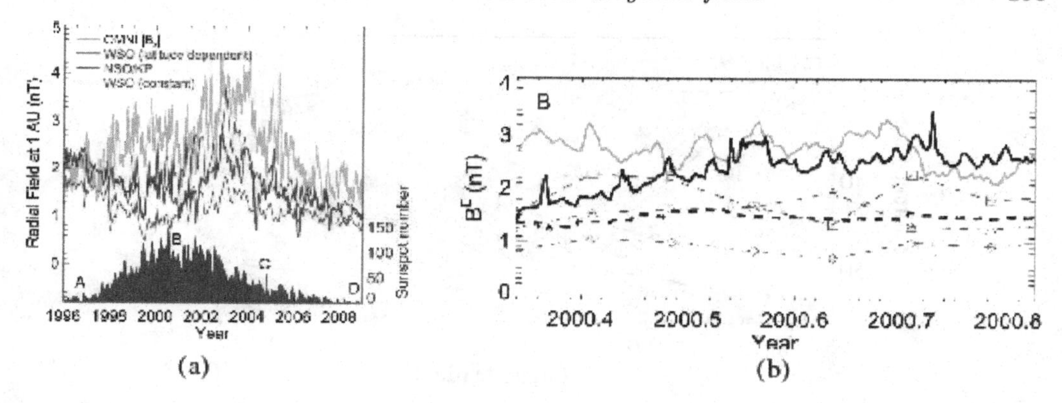

Figure 3. (a) Graph comparing open flux estimates from PFSS models to IMF field measurements (grey line). (b) Open flux estimate from non-potential simulations (black line) along with PFSS estimates (dashed) and IMF field measurements (grey line).

nority chirality filaments as well as for dominant chirality filaments. A key feature of the simulations is that a better agreement between the observations and simulations is found the longer the simulations are run. This indicates that the Sun has a long term memory of the transport of helicity from low to high latitudes. The reason for this high agreement is described in the paper of Yeates & Mackay (2009).

4.2. *Open flux*

The Sun's Open Magnetic Flux is part of its magnetic field that extents out into interplanetary space and forms the Interplanetary Magnetic Field (IMF). However, measurements of the IMF over several solar cycles do not agree with open flux values deduced from PFSS models. In general PFSS models underestimate the open flux, where this is particularly apparent around cycle maximum. This is clearly shown in Figure 3(a) which compares various PFSS extrapolations with the IMF field. Through simulating the global corona over 4 distinct, 6-month periods (labelled A-D in Figure 3(a)), Yeates *et al.* (2010), showed that this discrepancy may be resolved through allowing electric currents to form in the global corona. This is illustrated in Figure 3(b) which simulates period B. The dashed lines denote open flux from various PFSS extrapolations, the grey line the observed IMF field strength and the black solid line the open flux from the global simulation. This clearly gives a much better agreement in absolute magnitude terms. Yeates *et al.* (2010) deduced that to produce this agreement, the open flux has three main contributions. The first is a background level due to the location of the flux sources (this component is also present in potential field models). The second is an enhancement due to electric currents which results in an inflation of the magnetic field. This inflation can be seen as the steady increase of the curve over the first rotation to a higher base level. Finally, there is a sporadic component to the open flux as a result of flux rope ejections. While this is one explanation for the shortfall other explanations have been put forward by a variety of authors (see Schüssler & Baumann 2006; Riley 2007; Lockwood *et al.* 2009; Fisk & Zurbuchen 2006)

4.3. *Magnetic flux ropes and CMEs*

Over the years a wide variety of mechanisms have been proposed for the initiation of CMEs (Forbes *et al.* 2006). One such mechanism is the flux rope ejection model (Lin

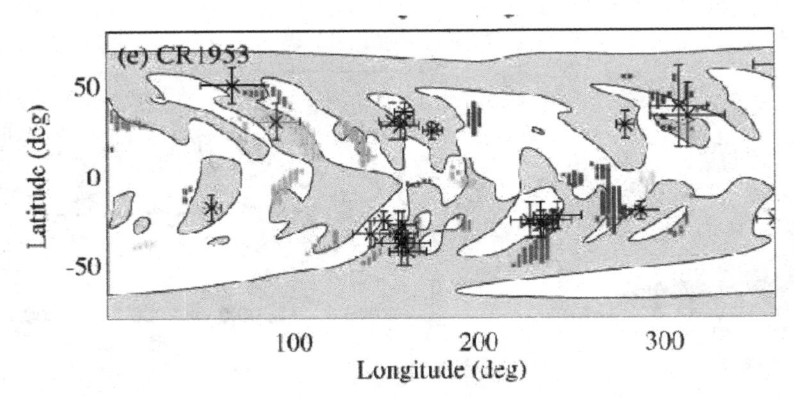

Figure 4. Comparison of CME source location from EUV images (crosses) with sites of flux rope ejections from the global non-potential simulations of Yeates *et al.* (2010) for CR1953.

et al. 2003) where surface shearing motions and flux cancellation produce a twisted structure which then loses equilibrium (van Ballegooijen & Martens 1989). Within the global simulations of Yeates & Mackay (2009) the formation of flux ropes is a natural consequence as flux cancels and is advected from low to high latitudes. As these flux ropes become larger, local losses of equilibrium may occur (Mackay & van Ballegooijen 2006). In the paper of Yeates *et al.* (2010) local losses of equilibrium were compared to the observed sites of CMEs. First, over a 4.5 month period (May-Sept 1999) 330 CMEs were identified from Lasco observations. However, from the corona-graphic images the initiation sites in the low corona could not be identified. To determine these, the CMEs were cross-correlated with EUV features in EIT 195 Åimages. Of the 330 CMEs, only 98 could be clearly associated with low coronal signatures in the EUV images. This illustrates a problem in identifying CME initiation sites.

When these 98 events where cross correlated with the sites of flux rope ejections from the global model (Figure 4), agreement could be found in some but not all cases. Overall the best correlation between the model and CMEs was 0.49. The authors however identified two separate classes of CME. The first are those which do identify with flux rope ejection locations in the simulation. These account for 1/2 and are the ones that produce the positive correlation. The other half were those located within the centres of active regions and frequently re-occur over short time scales. The global model was unsuccessful in re-producing these as it does not consider the internal structure or dynamics of active regions. While the comparison was only partially successful, it shows that flux rope formation and loss of equilibrium is an important model for CME initiation.

5. Global MHD models

In recent years significant advance has been made in the construction of realistic 3D global MHD models (Lionello *et al.* 2009; Downs *et al.* 2010). These models solve the resistive MHD equations, including a realistic energy equation (thermal conduction, radiative losses and coronal heating) along with modelling the upper chromosphere and transition region. Due to computational requirements, at present, these models are restricted to static photospheric boundary conditions and steady state coronal solutions. Through computing synthetic emission profiles and comparing these to the pass bands of 171, 195, 284 Åin EUV EIT images and X-ray images, the authors have shown that they may reproduce multi-spectral observations of the solar corona. The key element in doing this is the form of coronal heating used. In addition such models have reproduced the white light coronal emission seen during eclipse events (Rušin *et al.* 2010).

References

Antiochos, S. K., DeVore, C. R., & Klimchuk, J. A. 1999, *Astrophys. J.*, 510, 485

Balogh, A., Smith, E. J., Tsurutani, B. T., Southwood, D. J., Forsyth, R. J., & Horbury, T. S. 1995, *Science*, 268, 1007

Barnes, G. 2007, *Astrophys. J. Lett.*, 670, L53

Baumann, I., Schmitt, D., & Schüssler, M. 2006, *Astron. Astrophys*, 446, 307

Cook, G. R., Mackay, D. H., & Nandy, D. 2009, *Astrophys. J.*, 704, 1021

& Cremades, H., St. Cyr, O. C. 2007, *Advances in Space Research*, 40, 1042

Downs, C., Roussev, I. I., van der Holst, B., Lugaz, N., Sokolov, I. V., & and Gombosi, T. I. 2010, *Astrophys. J.*, 712, 1219

Fisk, L. A. & Zurbuchen, T. H. 2006, *Journal of Geophysical Research (Space Physics)*, 111, 9115

Forbes, T. G., *et al.* 2006, *Space Science Reviews*, 123, 251

Hathaway, D. H. 2010, *Living Reviews in Solar Physics*, 7, 1

Lin, J., Soon, W., & Baliunas, S. L., 2003, *New Astronomy Review*, 47, 53

Lionello, R., Linker, J. A., & Mikić, Z. 2009, *Astrophys. J.*, 690, 902

Lockwood, M., Stamper, R., & Wild, M. N. 1999, *Nature*, 399, 437

Lockwood, M., Rouillard, A. P., & Finch, I. D. 2009, *Astrophys. J.*, 700, 937

Mackay, D. H., Jardine, M., Cameron, A. C., Donati, J.-F., & Hussain, G. A. J. 2004, *Mon. Not. Roy. Astron. Soc.*, 354, 737

Mackay, D. H. & van Ballegooijen, A. A. 2006, *Astrophys. J.*, 641, 577

Mackay, D. H., Karpen, J. T., Ballester, J. L., Schmieder, B., & Aulanier, G. 2010, *Space Science Reviews*, 151, 333

Martin, S. F., Bilimoria, R., & Tracadas, P. W. 1994, *Solar Surface Magnetism*, 303

Riley, P. 2007, *Astrophys. J. Lett.*, 667, L97

Rušin, V., *et al.* 2010, *Astron. Astrophys*, 513, A45

Schatten, K. H., Wilcox, J. M., & Ness, N. F. 1969, *Solar Phys.*, 6, 442

Schrijver, C. J. 2001, *Astrophys. J.*, 547, 475

Schrijver, C. J. & Title, A. M. 2001, *Astrophys. J.*, 551, 1099

Schüssler, M. & Baumann, I. 2006, *Astron. Astrophys*, 459, 945

Sheeley, N. R., Jr. 2005, *Living Reviews in Solar Physics*, 2, 5

Tandberg-Hanssen, E. 1995, *Astrophysics and Space Science Library*, 199,

van Ballegooijen, A. A. & Martens, P. C. H. 1989, *Astrophys. J.*, 343, 971

van Ballegooijen, A. A., Priest, E. R., & Mackay, D. H. 2000, *Astrophys. J.*, 539, 983

Yang, W. H., Sturrock, P. A., & Antiochos, S. K. 1986, *Astrophys. J.*, 309, 383

Yeates, A. R., Mackay, D. H., & van Ballegooijen, A. A. 2007, *Solar Phys.*, 245, 87

Yeates, A. R., Mackay, D. H., & van Ballegooijen, A. A. 2008, *Solar Phys.*, 247, 103

Yeates, A. R. & Mackay, D. H. 2009, *Solar Phys.*, 254, 77

Yeates, A. R. & Mackay, D. H. 2009, *Astrophys. J.*, 699, 1024

Yeates, A. R., Attrill, G. D. R., Nandy, D., Mackay, D. H., Martens, P. C. H., & van Ballegooijen, A. A. 2010, *Astrophys. J.*, 709, 1238

Yeates, A. R., Mackay, D. H., van Ballegooijen, A. A., & Constable, J. A. 2010, arXiv:1006.4011

Wang, Y.-M., Nash, A. G., & Sheeley, N. R., Jr. 1989, *Science*, 245, 712

Discussion

MANALIS K. GEORGANLIS: How do you ensure that you have a unique solution for the global solar field?

MACKAY: With – yeah, yeah, well, first of all, the distinction – what to make here is the type of model you are talking about is an extrapolation model where you specify a photospheric boundary condition which is the normal field component and essentially the vertical component of electric current, and then you can't produce a unique connectivity. Here we preserve our connectivity throughout the simulation mostly within it. There are some regions where, of course, we don't do that where there is strong reconnection sites. But in general the connectivity is preserved.

So when we start off with the connectivity of a bipolar merging, that connectivity will be preserved as its transferred across the surface. So we are forming a unique solution in terms of the initial connectivity start off, the new flux emergence, and the transport process. So if you run it again and again and again, you will get the same result. So it will always be the same result. So we don't have this uniqueness problem where you connect one field light to the next.

KOSOVICHEV: A significant component of the Wang-Sheeley model is the process of diffusion of the leading magnitude flux across the equator. This is important for the polar field reversals. If the flux emerges only at high latitudes then it cannot diffuse to the equator, and then there will be no polarity reversals, which are important for the dynamo mechanism.

MACKEY: Well, we've – we've apply – we've only applied it to two stars, AB Duratus(spelled phonetically) and the other star, I can't remember. But Moira Jardine remembers the name, HD or something. What we find is, yes, we do for those two stars always need enhanced meridianal flow rates much larger than what occurs on the sun; but that does fit in with it being a rapid rotator.

Q.: So the only way you can get it faster is having a [INAUDIBLE]?

MACKEY: Yeah, well, yeah, I completely agree with you there. It's a tricky problem doing this. But remember the flux transport simulations are only representing the surface field. We don't take into account these other effects within them. So we do need to include more physics to really see what's going on. So this study was more initial speculative studies to see what was happening there.

But if we take the observed magnetograms to represent the surface field, that's what we find we need to include. Of course we have assumed many similar type solar parameters there. If the flux emergence is completely different, the profile is completely difference. Then of course you can get very different results, and we might need different surface transport parameters. So the more observations we get on the emergence latitudes of spots and the frequency and so on, the more we can be able to construe it.

Q.: How do you determine what type of helicity you inject?

MACKEY: That's as much as the question had previously there. At the moment we use statistical relationships, say, determined by Petsolve(spelled phonetically) *et al.* for a dominant sign of helicity in each hemisphere as the initial results because we can't constrain it any further than that. But with things like SDO and vector magnetic field measurements which will occur much more regularly, we can use them to constrain the helicity injection through new active regions. So that's the next stage in the models to make it much more realistic constraining it with observations.

The Physics of the Sun and Star Spots
Proceedings IAU Symposium No. 273, 2010
D. P. Choudhary and K. Strassmeier, eds.

doi:10.1017/S1743921311015419

Solar activity and differential rotation

Hari Om Vats[1] and Satish Chandra[2]

[1]Physical Research Laboratory,
Navrangpura, Ahmedabad, Gujarat, India
email: vats@prl.res.in

[2]Department of Physics, PPN College,
Parade, Kanpur, UP, India
email: satish0402@gmail.com

Abstract. The coronal sidereal rotation rate as a function of latitude for each year, extending from 1992 to 2001 for soft X-ray images and from 1998 - 2005 for radio images are obtained. The present analysis reveals that the equatorial rotation rate of the corona is comparable to the photosphere and the chromosphere, However, at the higher latitudes, the corona rotation quite differently than the photosphere and chromosphere. The latitude differential obtained by both radio and X-ray images is quite variable throughout the period of the study. The equatorial rotation period seems to vary almost systematically with sunspot numbers which indicates its dependence on the phases of the solar activity cycle.

Keywords. Sun: corona – Sun: rotation – Sun: radio radiation – Sun: X-rays, gamma rays

---

1. Introduction

The Sun rotates on its axis approximately once every 27 days as seen from the Earth. The rotation is not uniform, being substantially slower near the poles than at the equator. This superficial aspect of the solar differential rotation was well known from sunspot observations as early as the 17th century. However the details of its variability are known in this century only. It is only within the last 30 years that it has become possible to observe the rotation profile in the solar interior. The subtle temporal variations have become quite evident with the help of helioseismology. Heliosesmology is the study of the waves that propagate within the Sun. The inferences from the properties of these waves reveal the solar interior structure and dynamics. This is the most important tool we have to measure this internal rotation. Basu & Antia (2000) found that during the high activity period in 1998 the equatorial region was rotating slower than the smoothed average rotation rate. It was rotating faster at other times. On the other hand, Antia & Basu (2001) estimated using MDI data that the polar rotation rate decreases between 1995 and 1999 after which it has started increasing. Howe (2009) reviewed in detail the solar interior rotation and its variation.

The rotation of the solar atmosphere, the photosphere and the chromosphere has been investigated in quite some details; especially there are extensive reviews of these measurements and their temporal and latitudinal variations (Howard 1996 and Beck 2000) There are many references in these review papers. The rotation has been measured by three methods, namely, (1) Tracers: using quasi permanent features on the solar surface *e.g.* sunspots, filaments etc. (2) Spectroscopy: wherein one measures Doppler shift of emission lines and (3) Flux Modulation: this is the method that we have used extensively (Vats *et al.* 1998a and Vats *et al.* 1998b). The disk integrated radio flux at multi-frequency (11) radio emissions over 26 months. This showed that there is a systematic change in rotation period as a function of altitude in the solar corona (Vats *et al.* 2001). The disk

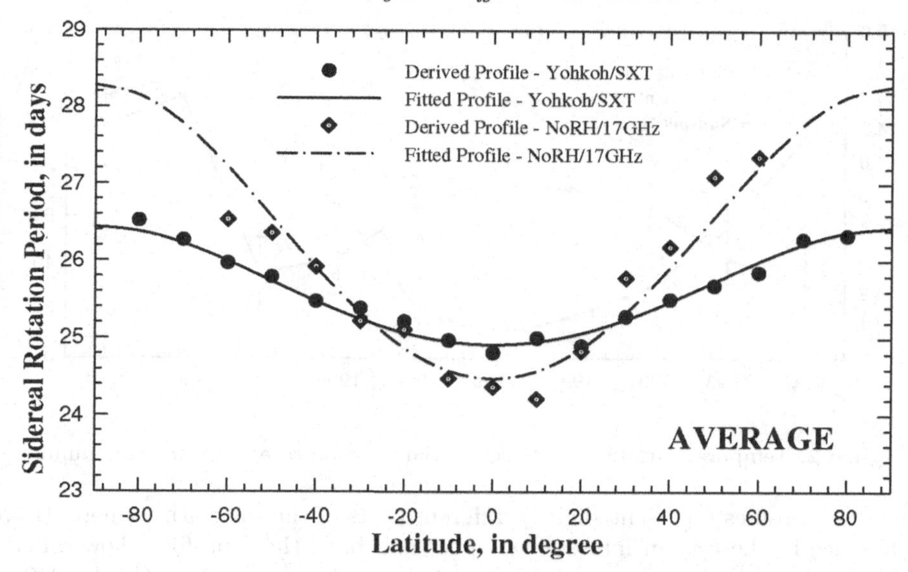

Figure 1. Average rotation profiles obtained by flux modulation method.

integrated measurements give no information on the latitude variation. Radio images at 17 GHz for the period 1999 - 2005 give very interesting information about the differential rotation of the solar corona at the height of this emission. The differential profile is shallower than those for photosphere and chromospheres. A comparison of radio and X-ray measurements is shown in Figure 1. This figure based on the recent work for estimating differential rotation in corona by flux modulation method using radio and X-ray images. The radio images are from Nobeyama Radioheliograph for the years 1999 - 2001 (Chandra, Vats & Iyer 2009). The X-ray images of Yohkoh SXT for the years 1991 - 2001 are used by Chandra, Vats & Iyer (2010). The aim of this paper is to discuss the temporal variation of the rotation as a function solar activity during the years 1998 - 2001 for radio observations and for the years 1991 - 2001 for X-ray measurements.

2. Results

Using radio images at 17 GHz of Nobeyama Radioheliograph and Yohkoh soft X-ray images (Chandra, Vats & Iyer 2009 and Chandra, Vats & Iyer 2010) have provided value of the coefficients A and B of the equation of differential rotation rate $\Omega(\psi) = A + B\sin^2\psi$. Here ψ is solar latitude. The mean rotation profiles of these extensive measurements in X-ray and at Radio wavelength are shown in Figure 1 by a dash-dot-dash and a continuous curve, respectively. The mean profiles show an interesting behaviour. The equatorial region radio measurements indicate a faster rotation rate, whereas at high latitude X-ray measurements indicate a faster rotation rate. Vats *et al.* (2001) used the coronal electron density model (Aschwanden & Benz 1995) and estimated that the radio emission at 17 GHz would be originating from a height $\sim 1.2 \times 10^4$ km above the photosphere. There is also as an assumption in this calculation that emission occurs at frequencies $\sim$ second harmonic of local plasma frequency in the corona. The Yohkoh images are known to represent an average X-ray emission over a height of ~ 0.5 solar radii above the photosphere.

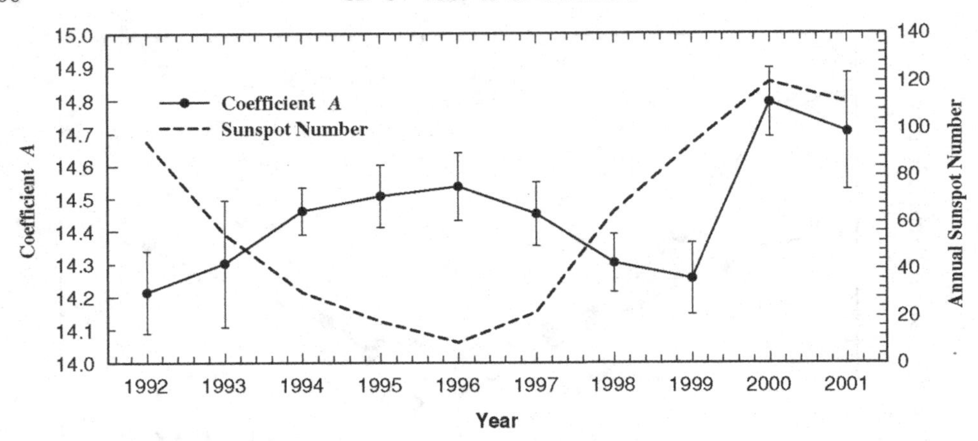

Figure 2. Temporal variation of A (equatorial rotation rate) and sunspots number.

Thus these profiles represent slightly different parts of the solar atmosphere. However, the difference in these is an interesting aspect and both these profiles show differential rotation. Review of many observations of surface rotation of the Sun (Beck 2000) show that the equatorial rotation period is $\sim$ 25 days and that near poles is $\sim$ 36 days. The profiles shown in Figure 1 are quite shallower than those in the photosphere and chromosphere. Thus the corona does have a differential rotation but it is shallower than in the lower parts of solar atmosphere. It is now believed that the sunspots and solar activity cycle are connected to the differential rotation of the Sun through a complex manner.

The coefficients A and B represent equatorial rotation rate and the mid-latitude differential rate, respectively. The variation of A from Yohkoh SXT measurements is shown in Figure 2. There are two curves in this figure as a function of time (for the entire period of Yohkoh observations for the years 1992 - 2001); the dashed curve is for the sunspot number and the continuous curve for coefficient A. The sunspot number (which is a measure of solar activity) decreases from 1992 to 1996, whereas the value of the coefficient A increases steadily during this period. The increasing in A is about 3%. After 1996, the behaviour of both (the sunspots and A) is almost opposite, except in the year 2000 when the value of coronal rotation rate is rather high. The coefficient A obtained from the observations of Nobeyama Radio Heliograph is shown in Figure 3. The curves of Figure 3 are for a slightly different epoch (1999 to 2005). Here, the variation of A and the sunspots are not opposite to each other. However, the peak of solar activity is in 2000, whereas peak of A is in 2003, so there is lag of 3 years. However, the change in the value of A is > 6 %. The change in the equatorial rotation rate as a function of time or the solar activity by X-ray measurements is only half of radio measurements.

The difference in the behavior of coefficient A obtained by Radio and X-ray could be either due to difference in the two epochs or could be due to a difference in the region of their emission in the solar atmosphere.

3. Summary and discussions

The presented study shows that corona has differential rotation. The equatorial part of the corona rotates slightly faster than that of the photosphere and chromospheres. The region of higher latitudes in the corona rotates much faster than the corresponding region of the lower solar atmosphere. Casas, Vaquero & Vazquez (2006) used sunspot

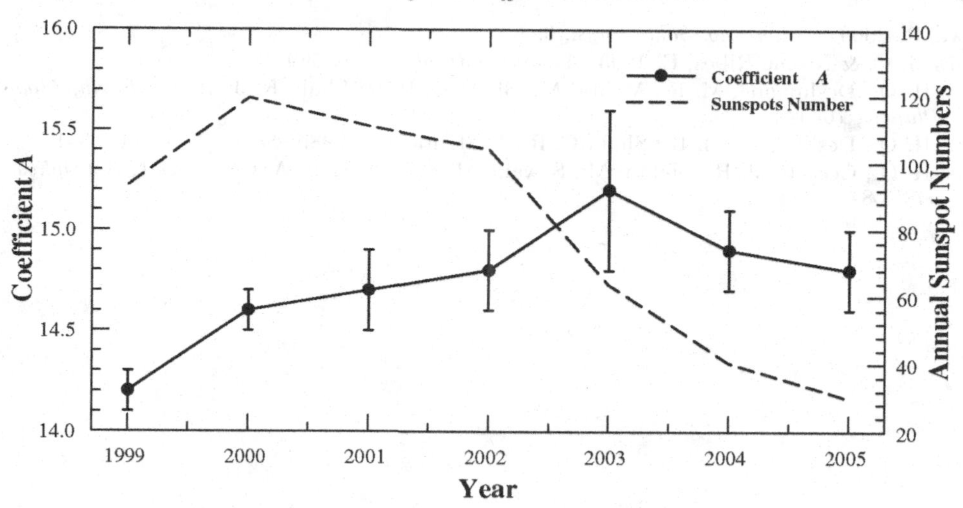

Figure 3. same as in Figure 2, this is for Radio measurements.

drawings to investigate the behavior of solar rotation during the episode of reduced solar activity, known as the Maunder Minimum (MM). This confirmed the conclusion of Ribes & Nesme-Ribes (1993), that during the MM the solar surface rotates slowly at the equator and it shows a higher level of differential rotation with latitude. Balthasar, Vázquez & Wöhl (1986) reported a similar, but smaller decrease in the equatorial rotation rate between cycles 13 and 14 ($\sim$ 1901), accompanied by a change in the decay rate of sunspots. This epoch did coincide with the minimum of the 80-year activity cycle. Present work using X-ray images suggest that the equatorial rotation rate is anti-correlated with the sunspot activity, except in the year 2000. The radio images at 17 GHz show that the equatorial rotation rate (A) and the sunspot number have a lag of 3 years. The B coefficient representing the differential part is almost in phase with the sunspot number. More studies are needed to ascertain the temporal and spatial variability of the rotation rate.

4. Acknowledgments

This was presented at IAU 273 symposium at Ventura. One of us (HOV) received travel support provided partially (1) IAU (2) NSF, USA, (3) INSA India and (4) Physical Research Laboratory, Ahmedabad India. This paper was prepared while HOV was visiting at LMSAL, Palo Alto USA. We thank Markus J. Aschwanden for proof reading the manuscript.

References

Antia, H. M. & Basu, S. 2001, *Astrophys. J.*, 559, L67
Aschwanden, M. J. & Benz, A. O. 1995, *Astrophys. J.*, 438, 997
Basu, S. & Antia, H. M. 2000, *Solar Phys.*, 192, 469
Balthasar, H., Vázquez, M., & Wöhl, H. 1986, *Astron. Astrophys.*, 155, 87
Beck J. G. 2000, *Solar Phys.*, 191, 47
Casas, R., Vaquero, J. M., & Vazquez, M. 2006, *Solar Phys.*, 234, 379
Chandra, S., Vats, H. O., & Iyer, K. N. 2009, *Mon. Not. Roy. Astron. Soc. Letters*, 400, L34
Chandra, S., Vats, H. O., & Iyer, K. N. 2009, *Mon. Not. Roy. Astron. Soc.*, 407, 1108
Howard, R. F. 1996, *ARAA*, 34, 75

Howe, R. 2009, *Living Rev. Solar Phys.*, 6, 1

Ribes, J. C. & Nesme-Ribes, E. 1993, *Astron. Astrophys*, 276, 549

Vats, H. O., Deshpande, M. R., Mehta, M., Shah, C. R., & Shah, K. J. 1998a, *Earth, Moon & Planets*, 76, 141

Vats, H. O., Deshpande, M. R., Shah, C. R., & Mehta, M. 1998b, *Solar Phys.*, 118, 351

Vats, H. O., Cecatto, J. R., Mehta, M., Sawant, H. S., & Neri, J. A. C.. F. 2001, *Astrophys. J.*, 548, L87

The Physics of the Sun and Star Spots
Proceedings IAU Symposium No. 273, 2010
D. P. Choudhary and K. Strassmeier, eds.

doi:10.1017/S1743921311015420

Flare induced penumbra formation in the sunspot of NOAA 10838

Sreejith Padinhatteeri[1,2] and Sankarasubramanian K.[1]

[1]Space Astronomy Group, ISRO Satellite Centre, Bangalore, India - 560017
email: sreejith.p@gmail.com, sankark@isac.gov.in

[2]Dept. of Physics, University of Calicut, Kerala, India.

Abstract. We have observed formation of penumbrae on a pore in the active region NOAA10838 using Dunn Solar Telescope at NSO,Sunpot,USA. Simultaneous observations using different instruments (DLSP,UBF,Gband and CaK) provide us with vector magnetic field at photosphere, intensity images and Doppler velocity at different heights from photosphere to chromosphere. Results from our analysis of this particular data-set suggests that penumbrae are formed as a result of relaxation of magnetic field due to a flare happening at the same time. Images in Hα show the flare (C 2.9 as per GOES) and vector magnetic fields show a re-orientation and reduction in the global α value (a measure of twist). We feel such relaxation of loop structures due to reconnections or flare could be one of the way by which field lines fall back to the photosphere to form penumbrae.

Keywords. Sunspot, penumbra, flare.

---

1. Introduction

Sunspots are the manifestation of strong magnetic fields that emerge in the solar photosphere (Bray & Loughhead (1964), Solanki (2003)). Although sunspots are stable configurations when compared to the dynamical time scales of other features on the Sun, the observed umbral and penumbral fine-structures are very dynamic and subjected to constant change and transformation on small spatial scales. Our understanding of these processes and the nature of the fine structures improved significantly in the last decade (Rimmele (2008), Brummell *et al.* (2008)), but we still lack detailed knowledge about the key process of penumbral formation and decay. Observations of this process are very rare, most prominent among them are by Leka & Skumanich (1998), Yang,*et al.* (2003) and Schlichenmaier *et al.* (2010). All of them observe the formation of a penumbra happening in few hours time. In all the cases they suggest the onset of penumbral formation is due the flux emergence. Leka & Skumanich (1998) suggests critical flux limit of $1 - 1.5 \times 10^{20}$ Mx for initiation of penumbra. Here in this paper we discuss multi-wavelength observations of the formation of penumbrae in the active region NOAA 10838 which was carried out on 2005 December 22. Coincidentally, a C-2.9 class flare was also observed in the same active region. These observations corresponding to different heights in the solar atmosphere suggest that the penumbral formation in this particular case is related to the flare at chromospheric and higher layers.

2. Observations and Data Analysis

The formation of the penumbra was observed on the leading sunspot of the active region NOAA 10838 on 2005 December 22 at the Dunn Solar Telescope in Sunspot,NM, USA.

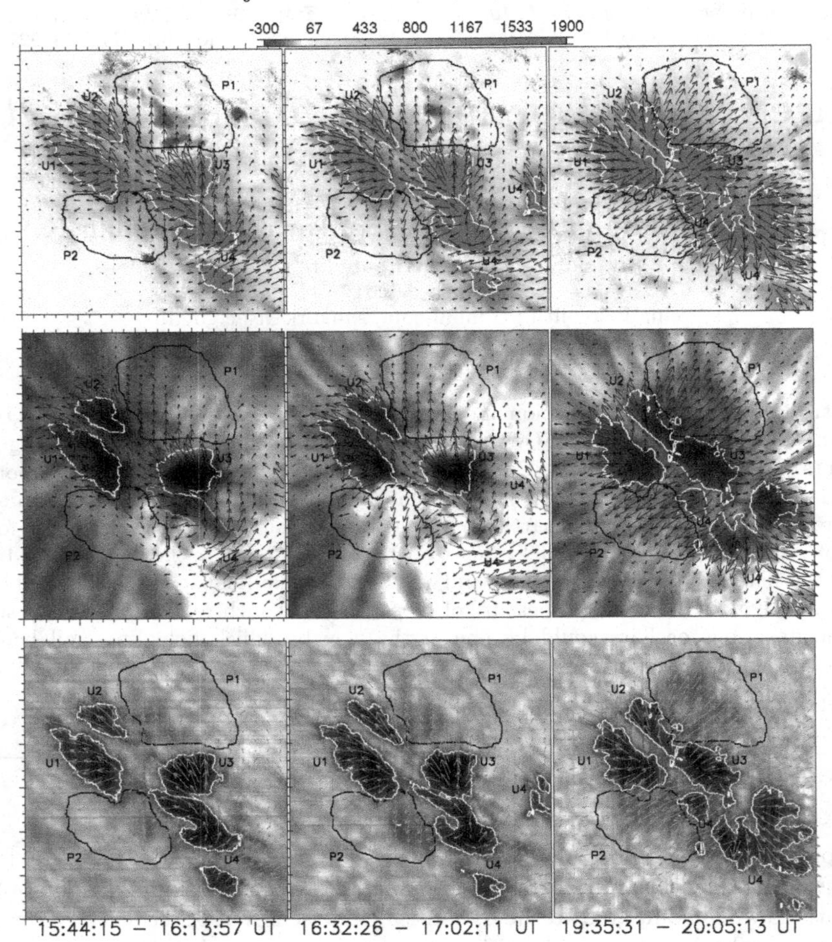

Figure 1. Figures showing Vector magnetic field evolution during the penumbral formation. Horizontal field vectors (arrows) are overplotted on vertical magnetic field (top row; with colour bar on the top showing the B_z values), Hα image (middle row) and continuum image (bottom row). Raster scan time is mentioned at the bottom of each column. Dark contours mark the area where penumbra forms, and white contours mark different umbrae.

The observed sunspot was situated at $\mu = cos\theta = 0.77$ (N15.5 E35.4) on the solar disk. The atmospheric seeing was moderate during the observation. The adaptive optics system at the DST (Rimmele *et al.* (2004)) was operated and the atmospheric seeing corrected beam was fed to a set of back-end instruments. The main back-end instruments used were: Diffraction Limited Spectro-Polarimeter (DLSP; Sankarasubramanian *et al.* (2006)) and Universal Birefringent Filter (UBF; Beckers *et al.* (1975)). DLSP was used to obtain stokes profiles of two magnetically sensitive lines, Fe I λ 6301.5 Å and 6302.5 Å. The stokes profiles were inverted with an assumption of Milne-Eddington atmosphere using the HAO inversion code. The magnetic field vector components were transformed from observer's to heliographic co-ordinates using simple spherical trigonometric transformations (see Smart & Green (1977)). The UBF is a tunable Lyot filter with a pass-band that varies between 120 to 250 mÅ in the visible wavelength of 5000 to 7000 Å. The UBF was tuned

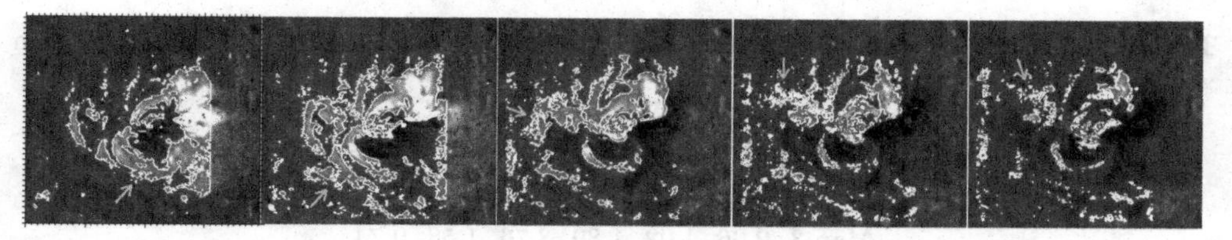

Figure 2. EIT images with white contours highlighting the bright loops. Red and blue contours represent positive and negative magnetic field, as derived from MDI, respectively. The arrow in each image show the loop that untwist as the flare happens.

in to two different spectral lines namely, Fe I λ 5434 and Hα λ 6562. We have also used the archived data from EIT and MDI onboard SOHO satellite, during the same observation time, to study the loop dynamics during the flare. EIT and MDI data were calibrated and matched each other using standard codes available in the SSW software package. Magnetic field contours derived from MDI were overplotted on EIT images. To bring up the loop structure, a time averaged EIT image was subtracted from each frame and the intensity level contours were overplotted.

3. Results and conclusion

We have observed the formation of penumbrae, and a flare happening simultaneously. Preliminary results suggests a relation between the two. Top row of the Figure 1 shows the magnetic filed orientations at three raster scan maps. The colour code is used for vertical field strength and the arrows represent horizontal field. Dark Solid contours mark the ROI, selected manually, where penumbral formation happens, and are marked as **P1** and **P2**. Four umbral areas are selected using the intensity values (less than 0.7 I_{QS} where I_{QS} is the quite sun mean intensity), and are marked in white contours. U1, U2 and U3 are the bottom left, top left, and top right umbrae respectively. The bottom right umbra undergoes lot of changes, in shape and area, and many other small pores come and join. Hence all the umbral area other than U1, U2 and U3 are marked as U4. Apart from the bottom right umbra seen in first column, small pores which later join this spot are all grouped under U4. In the second row the horizontal field vector is overplotted on Hα line core images obtained using UBF at similar time of the respective raster scan. The flare can be seen as high intensity values in Hα images. In the third row horizontal field vector is overplotted on continuum raster scan image. The formation of penumbra is clearly visible in the continuum images. It is clear from the middle and bottom row images that the horizontal vectors match quite well with the magnetic structures like penumbra seen in continuum images and the super penumbral structures in Hα line core images. This suggest that the performed analyses, like inversions, azimuth ambiguity corrections and coordinate transformation from observers to heliographic co-ordinate were proper.

The arrows, representing the horizontal magnetic field, show a twisted field orientation in first two maps (first and second column of Figure 1), and as the flare happens, the field lines untwist to a more potential like nature as seen in third column. The global alpha value, $\boldsymbol{\alpha_g}$, which is a measure of magnetic field twist was estimated (see for e.g. Tiwari *et al.* (2009)). The $\boldsymbol{\alpha_g}$ of this sunspot reduces from $1.77 \times 10^{-7}/m$ to $-0.43 \times 10^{-7}/m$ as the penumbrae forms. The reduction in the $\boldsymbol{\alpha_g}$ shows the relaxation of magnetic

Table 1. Magnetic Flux Φ in units of 10^{20}Mx. Errors to Flux values in umbra are less than 0.05% and in penumbra are less than 0.12%

	U1	U2	U3	U4	P1	P2
Map-1	1.67	0.50	1.45	2.44	0.37	0.89
Map-2	1.96	0.61	1.55	2.65	0.45	0.99
Map-3	2.06	1.02	1.89	3.76	1.58	0.74

field configuration at the photosphere, due to the flare. The flare was of C-2.9 class as measured by GOES. Flare started at around 16:32 UT and peaked at around 16:50 UT.

The total longitudinal magnetic flux ($\sum fB_z A$, where f is the fill fraction, A is the area and B_z is the longitudinal magnetic field intensity) of this spot is of the order 10^{21} Mx. It's little higher than the critical limit of 10^{20}Mx needed for penumbral formation as suggested by Leka & Skumanich (1998). There is an increase in the flux majorly due to similar polarity pores joining the main spot in U4. The flux values of four umbral region and the penumbral forming region are tabulated in Table 1. Apart from the flux values one also have to consider the twisted configuration of the magnetic fields. We believe that, in this case, the flare which transformed the twisted field structure to a more potential nature might have triggered the formation of penumbra.

Figure 2 gives the five images taken using EIT, with $\approx$ 12 minutes cadence. The white contours shows the loops, and the blue and red contours represent the negative and positive magnetic field strength, derived from MDI, respectively. The images were taken close to the peak of the flare. The images show untwisting of large loops, in the same direction as seen in the photosphere vector magnetic field configuration. Similar untwisting is also seen in Hα time sequence movie. It suggest that the untwisting of magnetic field and the subsequent relaxation happens all the way from corona to photosphere.

We conclude that, observation of this particular event on this sunspot suggest a relation between flare and penumbral formation. Apart from the coincidence of both events, the observation also confirms the magnetic field re-orientation, all the way from corona to photosphere. Hence, we speculate that the penumbral formation may not be simply a photospheric phenomena but a result of global reconfiguration of magnetic field due to activities like a flare.

Acknowledgements: SOHO is a project of international cooperation between NASA and ESA. The NSO is operated by the AURA under a cooperative agreement with NSF.

References

Beckers, J. M., Dickson, L., & Joyce, R. S., 1975, AFCRL Report No. AFCRL-TR-75-0090, A ir Force Cambridge Research Laboratory, Massachusetts.

Bray, R. J. & Loughhead, R. E. 1964, Sunspots (The International Astrophysics Series, London: Chapman Hall, 1964)

Brummell, N. H., Tobias, S. M., Thomas, J. H., & Weiss, N. O. 2008, *ApJ*, 686, 1454

Leka, K. D. & Skumanich, A. 1998, *ApJ*, 507, 454

Rimmele, T. 2008, *ApJ*, 672, 684

Rimmele, T. R., Richards, K., Hegwer, S., *et al.* 2004, in SPIE Conf. Ser., eds. S. Fineschi & M. A. Gummin, *SPIE Conf. Ser.*, 5171, 179

Sankarasubramanian, K., Lites, B., Gullixson, C., *et al.* 2006, in ASP Conf. Ser., eds. R. Casini & B. W. Lites, *ASP Conf. Ser.*, 358, 201

Schlichenmaier, R.,Rezaei, R., González, N. B., & Waldmann, T. A. 2010, *A&A*, 512, L1

Smart, W. M. & Green, R. M. 1977, Textbook on Spherical Astronomy, (Cambidge University Press, Cambidge, 1977)
Solanki, S. K. 2003, *A&AR*, 11, 153
Tiwari, S. K., Venkatakrishnan, P., Gosain, S., & Joshi, J. 2009, *ApJ*, 700, 199
Yang, G., Xu, Y., Wang, H., & Denker, C. 2003, *ApJ*, 597,1190

The Physics of the Sun and Star Spots
Proceedings IAU Symposium No. 273, 2010
D. P. Choudhary and K. G. Strassmeier, eds.

doi:10.1017/S1743921311015432

Dynamic responses of sunspots to their ambient magnetic configuration

S. P. Bagare

Indian Institute of Astrophysics, Bangalore, India
email: bagare@iiap.res.in

Abstract. In our earlier study of a revisit of the classic Wilson Effect, it was found that a large proportion of sunspots do not display the geometric effect which is ascribed to a depression of the umbra. It was shown that the presence or absence of the effect, observed close to the limb, depends upon the ambient magnetic configuration of the sunspot. In this follow up study, we look for the impact of changes in ambient magnetic configuration on the measurable properties of sunspots during their disk passage, using observations obtained at the Kodaikanal observatory during 1978-80. Digitized photoheliogram data were used to examine and measure areas of spots and their umbrae for 101 cases. Magnetic field measures published by the Academy of sciences, Leningrad were used to evaluate the ambient magnetic configuration. The results indicate that the extent of magnetic bipolarity is associated with changes in the proportion of the area of the penumbra to that of the umbra. In regular spots, the relative area of penumbra reduces with reduction in the strength of ambient opposite polarity spots and pores. However, in the presence of pore sized blob(s) in penumbra, and with associated emerging fluxes, the penumbra is significantly enlarged. But in the presence of a light bridge or a split umbra, the relative area of penumbra is considerably reduced.

Keywords. photosphere, sunspots, magnetic fields

---

1. Introduction

One of the early observations of the physical properties of sunspots was that single isolated sunspots, as they approach the solar limb, display a reduction of the width of the penumbra on the disk-center side more rapidly than the width on the limb-ward side. This geometric effect, known as the Wilson effect, is understood to be due to a depression of the umbra with respect to surrounding photosphere, in the range of 400-800 km (Gokhale & Zwaan 1972), and the increased transparency of the sunspot atmosphere owing to its low temperature and gas pressure (Thomas & Weiss 2008). Many extensive studies have, however, shown that a large proportion of sunspots either do not display the effect or show an inverse effect, but they were generally ignored as being attributable to nonuniformity in the shape of penumbrae or various difficulties in the actual measurement of the effect (McIntosh, P. S. 1981).

Bagare & Gupta (1998) concluded from a study of 20 sunspots that the ambient magnetic configuration appears to play a key role in the occurrence or absence of the classic Wilson effect. Bagare (2010) studied 580 epochs of 253 sunspots located within a longitude zone of 40^0 to 80^0 on either side of the central meridian and concluded that all the unipolar sunspots of magnetic class α, αp, or αf invariably display the classic Wilson effect while spots of magnetic class β, βp or βf do not display the Wilson effect, and spots of class $\beta\gamma$ and γ display an inverse effect. It was also shown that the occurrence of inverse effect was invariably associated with the presence of mixed polarity pores in the neighbourhood of the main spot.

Our case studies in the meantime suggested that various properties of sunspots probably depend upon the magnetic class of the sunspot. The properties noted were the changing location in penumbra of the peak velocity in Evershed flows, the presence or absence of a weak bright ring around the spot, the domination of three or five minute intensity oscillations, and various properties of umbral dots. Since these results indicated that the ambient magnetic configuration appears to have an observable impact on the sunspots, the study of a significant number of sunspots during their disk passage was taken up to evaluate their ambient magnetic configurations and to measure the areas of their umbrae and penumbrae, and thereby to look for possible correlations.

Further, the results reported by many authors, especially those using X-ray images of the corona and the extreme ultraviolet observations with TRACE spacecraft, show that the field lines that are less steeply inclined in penumbrae rise up to form loops that extend for great distances across the solar surface, connecting either to other sunspots or to distant footpoints (apparently those of opposite magnetic polarity). These results have been summarized by Thomas & Weiss (2008).

2. Obsrevation and Measurements

The Kodaikanal photoheliogram and spectroheliogram digitization program is going on for the past many years. CCD cameras (4k x 4k) are being used with high precision lab spheres and light tables for the digitization. Photoheliograms digitized for the years 1978-80 were used for the present study. The raw images in Fits format were processed for dark subtraction and flat fielding. Measurements of the full spot and umbral areas were carried out on IDL platform using the XROI program. The boundaries marked manually agreed well with the intensity values of 59% and 85% of the quiet photosphere for umbra-penumbra and penumbra-photosphere boundaries respectively, as adopted by Brandt, Schmidt, & Steinegger (1990). Only the spots with clearly measurable umbral and penumbral boundaries were selected for measurement. Spots of all magnetic categories were included. The measured areas, converted to millionths of solar hemisphere, are accurate to within a few percent.

The values of longitudinal component of magnetic field were obtained from the monthly Solar Data Bulletins (SDB) published by the Academy of Sciences, Leningrad, which provide daily sketches and magnetic field strengths of all sunspots and strong pores. Our earlier measurements of Zeeman splittings and the longitudinal field strengths at Kodaikanal agreed closely with the values reported in SDB. The extent of Zeeman splitting provides estimates accurate only to a few hundred Gauss and measurements are not made for the weak pores. There is also a time lag of about three hours in the observations used in SDB for evaluation but the ambient magnetic configuration, as well as the reported field strengths, remain comparable. These accuracies may be considered sufficient for the purpose of this study.

In order to represent the ambient magnetic configuration of sunspots, we define a magnetic bipolarity index as follows:

$$Q_s = \sum(B_i/d_i) + \sum 4(N_j/d_j) \tag{2.1}$$

where B_i are the magnetic field strengths of the opposite polarity sunspots and strong pores, in hundred Gauss measures, at heliographic distances of d_i from the sunspot, taken

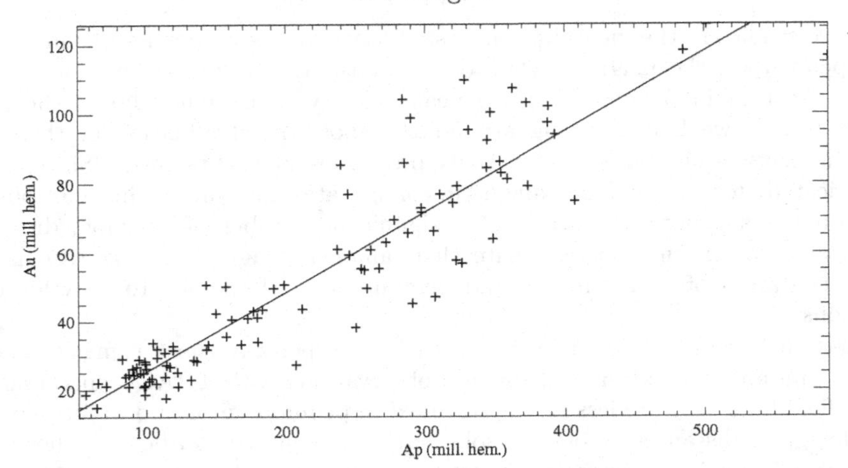

Figure 1. Area of umbra vs the area of penumbra for 101 spots, in millionths of visible hemisphere. A spread in the area of penumbra for a given area of umbra is cognizable.

up to a maximum distance of about 40 degrees. The heliographic distance of 40 degrees was chosen since it was observed in our earlier study that spots and pores at this distance appear to have an impact on the Wilson effect in sunspots. N_j is the number of pores at a group distance of d_j from the sunspot. N_j accounts for the pores for which the field strengths are not given in SDB. A mean field strength of 800 Gauss (8 in hundred Gauss measures) is assumed. It is also assumed that the pores emerge in pairs and hence the factor $8/2 = 4$ per pair is used in representing the opposite polarity pores. Pores within a heliographic distance of about 15^0 only are included. These measurements were carried out for 101 cases observed during the period 1978-80, corresponding to the maximum phase of solar cycle 21, providing a large number of active sunspot groups for the study.

3. Results and discussion

The areas of the entire spot (A_s), the umbra alone (A_u), and that of the penumbra alone (A_p), in millionths of the visible hemisphere, were computed for all the spots studied. A plot of A_u vs A_p is shown in Figure 1. A linear relationship between the parameters is evident, though with a significant spread in the values of A_p for any given value of A_u. A histogram of the ratio of A_u to that of A_s is shown in Figure 2. It is readily seen that majority of the sunspots have the area ratio in the range of 0.17 to 0.23, in general agreement with the earlier studies by Brandt, Schmidt, & Steinegger (1990) and Ringnes, T. S. (1964). However, we do find a significant number of spots in the lower range of 0.13 to 0.17 as well as in the higher range of 0.23 to 0.27.

Further, in Figure 3 the ratio of A_p to A_u is plotted against the magnetic bipolarity index Q_s, estimated using Equation 2.1. It is evident that the area of penumbra varies largely over a wide range of 2.7 to about 6.5 times that of the umbra. We classifed the spots into three major morphological categories and studied the properties of individual sunspots in each category, looking for possible patterns in the dynamic behaviour. The results of this analysis are discussed below.

3.1. *Penumbral area in regular spots*

A study of the distribution of the spots in Figure 3, in the light of their magnetic type, reveals the following characteristics. The spots represented with asterisks are regular

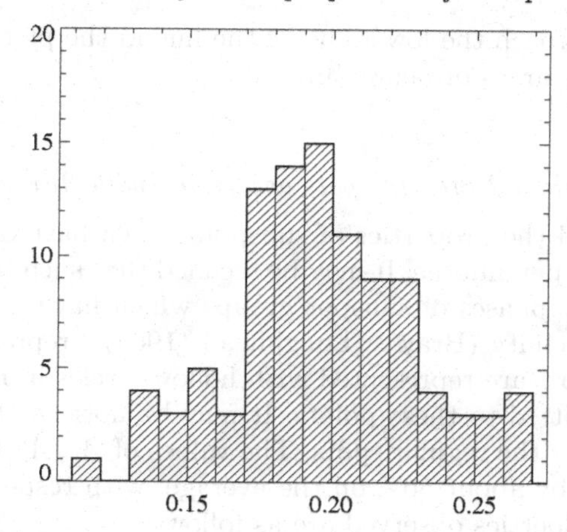

Figure 2. Distribution of the ratio of the area of umbra to the area of spot.

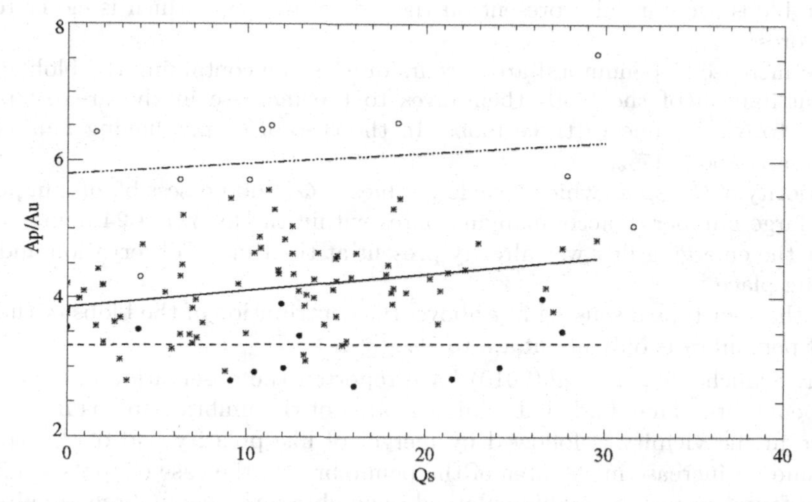

Figure 3. Ratio of area of penumbra to that of the umbra vs the magnetic bipolarity index defined by equation 2.1. Asterisks represent the regular spots, hollow circles represent spots having pore sized blob(s) in their penumbrae, and solid circles represent spots having light bridges or split umbrae. Regression fits are plotted for these using solid line, dash and dots line, and dashed line respectively.

spots having neither pore-sized blobs in their penumbrae nor having any evident light bridges or split umbrae. These spots show a clear increase in the value of A_p/A_u with increasing value of Q_s, the range of A_p/A_u being about 2.7 to 5.6. In general, single isolated spots of α type, which also have low values of Q_s say below 10, have low values for A_p/A_u. On the other hand, spots of class $\beta\gamma$ and γ, which are generally associated with large number of pores, apparently of mixed polarity, have high values of Q_s and relatively higher values of A_p/A_u. The spots of β type have intermediate values for both Q_s and A_p/A_u. A regression fit for these regular spots is shown with a solid line in the Figure. In the light of known properties of birth, evolution, and decay of sunspots, it may be inferred from this result that spots in the early stages of formation of the group begin with their position close to higher end of the solid line and as they transform into single isolated spots, they approach the lower end of the line in the plot, implying a significant reduction in the relative areas of penumbrae.

3.2. *Penumbral area in spots associated with flux emergence*

We separately examined the properties of sunspots which have one or more (upto four) pore sized blobs in their penumbrae. It may be recalled that such blobs are found predominantly during the early phases of sunspot groups which have a large number of mixed polarity pores in the vicinity (Bray & Loughhead (1964)), representing flux emergence in the region. These spots are represented with hollow circles in Figure 3. A line of dash and dots is regression fitted to these points. It may be noticed that the slope of this fit is comparable to that for the regular spots. The values of A_p/A_u for these spots are very significantly increased, by about 50% on the average, with respect to the regular spots. Some of the related properties observed are as follows;

(*a*) The blobs have intensities which are between those of the umbra and the penumbra.

(*b*) The blobs are normally present on the side of the spot which is closer to accompanying pores.

(*c*) The increase in penumbral area occurs on the side containing the blob(s).

(*d*) Contribution of the blobs themselves to the increase in the area of penumbra is about 5 to 6% for one to three blobs. In the case of a spot having four blobs, the contribution is about 17%.

(*e*) Majority of the spots which have low values of Q_s and possess blob(s) in penumbra, display a large number of accompanying pores within the following 24 hours. This indicates that the emerging flux was already present at the time of observation and mergers were taking place.

(*f*) For the spot type discussed in *e* above, the contribution of the blobs to the increase in area of penumbra is only about 1.5 to 5%.

Further, Schlichenmaier *et al.*(2010) have reported the observation of formation of a sunspot penumbra. They find that while the area of the umbra remains intact, the flux emergence in the vicinity is followed by merger of like polarity end of the bipole with the spot and an increase in the area of the penumbra. In the case of spots studied by us above and found to possess highly enlarged penumbrae, the penumbrae are already well developed but it appears that the process of flux merger is still in progress. Once the flux mergers are complete, that is when pore sized blobs are no more present in penumbra, the spot apparently settles down to its regular ratio of A_p/A_u.

3.3. *Penumbral area in spots having light bridges or split umbrae*

Another category we studied is that of spots having light bridges or split umbrae. These spots are represented with filled circles in Figure 3. The relative areas of penumbrae in these spots is less by about 12 to 27% with respect to those of the regular spots and they do not change practically over the entire range of Q_s values. It turns out that out of the 12 cases studied, 11 are spots in old active regions, in their second or third rotations and are fast approaching the phase of being a unipolar spot, followed by further decay. The lower its Q_s value, the closer is the spot to its transition to unipolar phase, 2 to 3 days for Q_s below 10 and 5 to 6 days for the rest (with the exception of one spot which decays faster). In the remaining one case out of 12, the spot has a light bridge in its young and complex phase. However, this too displays a low value of A_p/A_u. It may be summarized that in sunspots having light bridges or having split or nearly split umbrae, the relative area of penumbra is significantly decreased, as if the penumbra has shrunk in size. Curiously, this property appears to be independent of the ambient magnetic configuration.

4. Conclusion

It is evident from the study that the sunspot penumbra responds dynamically to ambient magnetic configuration. In regular sunspots, the ratio of the area of penumbra to that of the umbra shows an increase with increasing collective strength of opposite polarity spots and pores in the vicinity. In the presence of mixed polarity pores, implying flux emergence in the region, the increase in relative area of penumbra is the highest. This is exactly the phase when the spot shows an inverse Wilson effect if it is close to the limb. Therefore, if we assume that the extended penumbra is also elevated higher than in its normal phase, the inverse Wilson effect observed under these conditoins can perhaps be understood better.

In the presence of flux emergence in the form of pores and with pore sized blob(s) present in it, the penumbra is very significantly enlarged, by upto 50% on average, particularly in the direction of the blob and nearby pores. In many cases where only the blob is present, the accompanying mixed polarity pores show up within the following day implying that the emerging flux was present at the time of observation and the penumbra was responding to it. Therefore, there is a clear indication that spots have highly enlarged penumbrae in their formative and flux merger phase, and as they evolve to the status of a single isolated spot, the relative areas of their penumbrae decrease steadily and considerably.

Another aspect noted in this study is that in the presence of a light bridge and also when the umbra is in the process of splitting or is just split, the relative area of penumbra is significantly reduced implying that the penumbra has shrunk from its regular size. The relative area of penumbra in this phase is independent of the ambient magnetic configuration. It therefore appears that the penumbra is attempting to retain the spot together against attempts by convective processes to split it into fragments.

It emerges from this study that the relative area of penumbra plays a fundamental role in the stability of sunspots. It may be hypothesised that the spot responds to changes in ambient magnetic configuration and to the attempts to fragment it, through a balancing of magnetic and gas pressures. These responses are apparently readily reflected by changes in the relative area of the penumbra.

Acknowledgement

The data used in this study was taken from the Kodaikanal Observatory archives and digitized under the digitization program of IIA. The author's participation in the symposium was made possible with support from the IAU, IIA, and NASA-NSF funds.

References

Bagare, S. P. 2010, in: Hasan, S. S., & Rutten, R. J.(eds.) *Magnetic Coupling between the Interior and Atmosphere of the Sun*, p 398

Bagare, S. P. & Gupta, S. S. 1998, *Bull. Astron. Soc. India*, 26, 197

Brandt, P. N., Schimdt, W., & Steinegger, M. 1990, *Solar Phys.*, 129, 191

Bray, R. J. & Loughhead, R. E. 1964, in The International Astrophysics Series, Chapman & Hall, London *Sunspots*

Gokhale, M. H. & Zwaan, C. 1972, *Solar Phys.*, 26, 52

McIntosh, P. S. 1981, in: Cram, L. E., & Thomas, J. H. (eds.), Sacramento Peak Observatory *The Physics of Sunspots*

Ringnes, T. S. 1964, *Astrophys. Norvegica*, 8, 303

Schlichenmaier, R., Bello Gonzalez, N., Rezaei, R., & Waldmann, T. A. 2010, *Astron. Nachr.* 331, 563

Thomas, J. H. & Weiss, N. O. 2008, in: Cambridge Astrophysics Series 46 *Sunspots and Starspots* (Cambridge University Press)

The Physics of the Sun and Star Spots
Proceedings IAU Symposium No. 273, 2010
D. P. Choudhary and K. G. Strassmeier, eds.

doi:10.1017/S1743921311015444

Numerical simulations of magnetic structures

I. N. Kitiashvili[1,2], A. G. Kosovichev[1], A. A. Wray[1] and N. N. Mansour[3]

[1]W.W. Hansen Experimental Physics Laboratory, Stanford University,
Stanford, CA 94305, USA
email: sasha@sun.stanford.edu

[2]Center for Turbulence Research, Stanford University,
Stanford, CA 94305, USA
email: irinasun@stanford.edu

[3]NASA Ames Research Center,
Moffett Field, Mountain View, CA 94040, USA
email: Alan.A.Wray@nasa.gov, Nagi.N.Mansour@nasa.gov

Abstract. We use 3D radiative MHD simulations of the upper turbulent convection layer for investigation of physical mechanisms of formation of magnetic structures on the Sun. The simulations include all essential physical processes, and are based of the LES (Large-Eddy Simulations) approach for describing the sub-grid scale turbulence. The simulation domain covers the top layer of the convection zone and the lower atmosphere. The results reveal a process of spontaneous formation of stable magnetic structures from an initially weak vertical magnetic field, uniformly distributed in the simulation domain. The process starts concentration of magnetic patches at the boundaries of granular cells, which are subsequently merged together into a stable large-scale structure by converging downdrafts below the surface. The resulting structure represents a compact concentration of strong magnetic field, reaching 6 kG in the interior. It has a cluster-like internal structurization, and is maintained by strong downdrafts extending into the deep layers.

Keywords. Sun: magnetic fields, sunspots; methods: numerical

1. Introduction

The recent progress in the numerical modeling has made it possible to reproduce in simulations many observational effects in the quiet Sun region, active regions, magnetic flux emerging and whole magnetic structures (e.g., Stein & Nordlund 2001, Jacoutot *et al.* 2008, Kitiashvili *et al.* 2009, Cheung *et al.* 2008, Stein *et al.* 2010, Rempel *et al.* 2009). However, most of the modeling has been done by setting up the initial conditions with already existing magnetic structures, e.g. a horizontal flux tube for the modeling of magnetic flux emerging, or a vertical flux tube with strong field for the sunspot/pore structures simulations. It seems that so far only one study succeeded in reproducing a spontaneous formation of a micropore-like magnetic structure from an initially uniform field in the turbulent convection of the Sun (Stein *et al.* 2003), and the lifetime of this structure was very short.

Here, we present results of the realistic MHD simulations that show a process of spontaneous formation of a stable pore-like magnetic structure in fully developed convection from an uniform magnetic field. For the simulations we used a 3D radiative MHD code, "SolarBox" (Jacoutot *et al.* 2008). The code is built for 3D simulations of

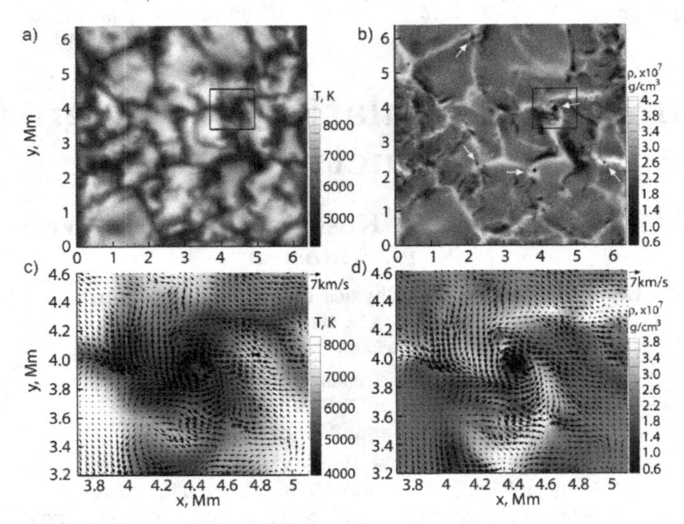

Figure 1. Snapshots of granular convection at the surface for the simulations without magnetic field, and the horizontal resolution of 12.5 km: temperature (left column) and density (right). The black square indicates a large whirlpool shown in more detail in panels c) – d). Black arrows show the flow velocity. White arrows in panel b) point to the centers of some vortices (dark low-density points).

compressible fluid flows in a magnetized and highly stratified medium of top layers of the convective zone and the low atmosphere, in the rectangular geometry. The simulation results are obtained for the computational domain of $6.4^2 \times 5.5$ Mm with the grid sizes: $50^2 \times 43$ km, $25^2 \times 21.7$ km and $12.5^2 \times 11$ km ($128^2 \times 127$, $256^2 \times 253$ and $512^2 \times 505$ mesh points). The domain includes a top, 5 Mm-deep, layer of the convective zone and the low atmosphere.

2. Simulations of granulation and vortex tubes

Initially we simulate the conditions of the quiet Sun in absence of magnetic field. Figure 1 shows snapshots for temperature (left column) and density (right) at the surface for the case without magnetic field. An interesting feature of the convective flows is the formation of whirlpool-like motions of different sizes ($\sim 0.2 - 1$ Mm) and lifetimes ($\sim 15 - 20$ min at the vertexes of the intergranular lanes (Kitiashvili *et al.* 2010). The vortical motions are particularly well seen in the density variations. The centers of the whirlpools are seen as dark dots (indicated by white arrows in Fig. 1b) in the intergranular space. The evolution of these vortices (swirls) is ultimately related to the dynamics of convective motions in the domain. The convective flows sometime may collect the swirls in a local area, merge together, and then destroy them. Stronger vortices usually correlate with downflows, and this is also found in our results.

From time to time, convection creates pretty big whirlpools, as the one indicated by square in Fig. 1, which can swallow up other smaller swirls around them. The big swirls are usually easy to see also in the surface temperature and intensity variations. The detailed structure of a large whirlpool around a compact vortex tube with a sharp boundary is shown in Figs 1c-d. The whirlpool structure is characterized by (Kitiashvili *et al.* 2010): 1) formation of downdraft lanes (visible as "arms" in the Figs 1c-d) of higher density that correlates with lower temperature; 2) a pronounced vortical

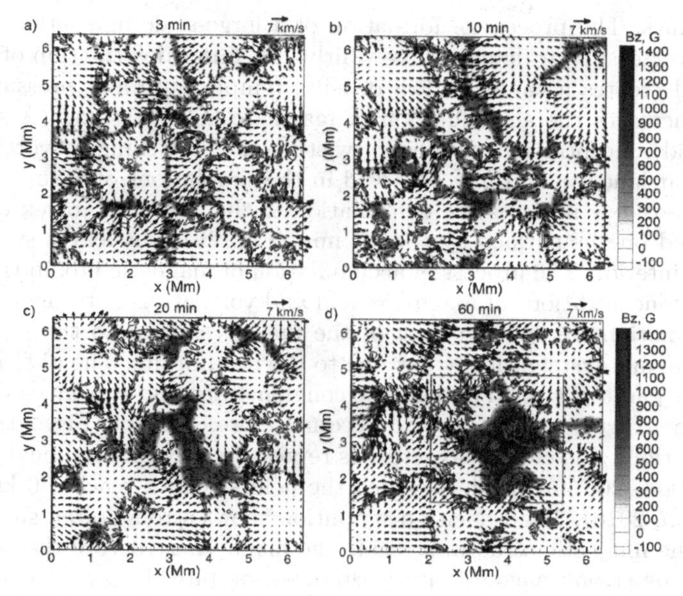

Figure 2. Snapshots of the surface distribution of vertical magnetic field (background), horizontal flows (arrows) and vorticity magnitude (contour lines) for four moments of time: 3, 10, 20 and 60 min, from the moment of initiation of a uniform magnetic field ($Bz_0 = 100$ G).

structure of the velocity flow; 3) increased magnitude of the horizontal velocity up to 7 – 9 km/s; 4) a sharply decreased density in the central core of the vortex, and a slightly higher temperature than in the surrounding. The typical depth of large swirls is about 100 – 200 km. The vortical motions in the solar granulation have been detected in high-resolution observations (e.g., Pötzi & Brandt 2005, Bonet *et al.* 2008, 2010), and the observational results generally agree with the simulations.

3. Spontaneous magnetic structure formation

We made a series of simulations for the initial vertical uniform magnetic field, $Bz_0 =$ 100 G, different computational grids and domain sizes. Qualitatively the simulation results are very similar in all these cases, and show the formation of stable magnetic pore-like structures. In Figures 2 we present the results for the horizontal resolution of 25 km, and the domain size of $6.4 \times 6.4 \times 5.5$ Mm, for which we have made the longest run ($\gtrsim$ 9hours). The periodic lateral boundary conditions allow us, for the illustration purpose, to shift the horizontal frame so that the structure is located close to the center. As we see in the simulations, the structure can be formed in any place of our computational domain, but usually the process starts at one of the strongest vortices (Kitiashvili *et al.* 2010).

Figure 2 shows snapshots of the vertical magnetic field (background image), horizontal flow field (arrows), and the vorticity magnitude (contour lines) for four moments of time: 3, 10, 20 and 60 min after the moment $t = 0$, when the 100 G vertical field was uniformly distributed in the computational domain. During few minutes the magnetic field is swept into the intergranular lines, and is significantly amplified up to $\sim 500 - 1000$ G. The vortices and magnetic field get concentrated at some locations in the intergranular lanes, where they are deformed and became elongated (or elliptically shaped) along the intergranular lines. The process of formation of a large-scale magnetic structure starts at a strongest vortex in our domain. The whirlwind causes deformation of the intergranular space, and creates a cavity of low density, temperature and pressure. The cavity expands and increases the accumulation of magnetic field (Fig. 2a). A similar process of magnetic field concentration, sweeping, twisting and stretching by vortical motions in the intergranular lane was initially observed in the simulations of Stein *et al.* (2003).

During the next few minutes the deformation of the "parent" vortex continues; then it gets destroyed on the surface by $t = 10$ min (Fig. 2b), but leaves strong downdraft motions in the interior. The process of accumulation of magnetic flux in this area continues. The local concentrations of magnetic field and vorticity get stronger and are moved by convective motions in the direction of the initial cavity, into the region where the gas pressure remains systematically low due to the downdrafts. The different small magnetic structures join together in a magnetic conglomerate that continues to attract other magnetic micro-structures (Fig. 2c) and becomes more compact (Fig. 2d).

In our simulations, the cluster structure is represented by internal field concentrations (flux tubes), $100 - 200$ km thick, in which the field strength reaches 6 kG after 1 hour (Kitiashvili *et al.* 2010). The velocity distribution shows strong, often supersonic, downflows around the magnetic structure. Inside the magnetic structure the convective flows are suppressed by strong magnetic field. However, despite the weak velocities ($\sim 0.1 - 0.2$ km/s) there are very small elongated convective cells resembling the umbral dots observed at the surface. We have followed the evolution of the magnetic pore-like structure for more than 8 solar hours, and did not see any indication of its decay. The structure continues to evolve, changes the shape and internal structurization, but remains compact.

4. Conclusion

Our simulations show that small-scale vortex tubes representing whirlpool-type motions at intersections of intergranular lanes may play important roles in dynamics of the quiet-Sun and magnetic regions. The results indicate that the process of spontaneous formation of small-scale magnetic structures and their accumulation into a large-scale magnetic structure is associated with strong vortical downdrafts developed around these structures. The resulting stable pore-like magnetic structure has the highest field strength of ~ 6 kG at the depth of $1 - 4$ Mm and ~ 1.5 kG at the surface (Kitiashvili *et al.* 2010). It has a cluster-like internal structurization, and seems to be maintained by strong downdrafts converging around this structure and extending into the deep layers. The simulations show that this internal dynamics plays a critical role in the magnetic self-organization of solar magnetic fields and formation of large-scale magnetic structures.

Acknowledgement

The presentation of this paper at the IAU Symposium 273 was possible due to the support from the National Science Foundation: grant numbers ATM 0548260, AST 0968672, and the NASA - Living With a Star grant, number 09-LWSTRT09-0039. Also, we thank the International Space Science Institute (Bern) for the opportunity to discuss these results at the international team meeting on solar magnetism.

References

Bonet, J. A., Márquez, I., Sánchez Almeida, J., Cabello, I. & Domingo, V., 2008, *Astrophys. J.*, 687, L131

Bonet, J. A., Márquez, I., Sánchez Almeida, J., Palacios, J. *et al.*, 2010, arXiv:1009.1992
Cheung, M. C. M., Schüssler, M., Tarbell, T. D. & Title, A. M., 2008, *Astrophys. J.*, 687, 1373
Jacoutot, L., Kosovichev, A. G., Wray, A. A., & Mansour, N. N., 2008, *Astrophys. J.*, 684, L51
Kitiashvili, I. N., Kosovichev, A. G., Wray, A. A., & Mansour, N. N., 2009, *Astrophys. J.*, 700, L178
Kitiashvili, I. N., Kosovichev, A. G., Wray, A. A., & Mansour, N. N., 2010, *Astrophys. J.*, 719, 307.
Pötzi, W. & Brandt, P. N., 2005, *Hvar Obs. Bull.*, 29, 61.
Rempel, M., Schüssler, M., & Knölker, M., 2009, *Astrophys. J.*, 691, 640
Stein, R. F. & Nordlund, Å., 2001, *Astrophys. J.*, 546, 585.
Stein, R. F., Bercik, D. & Nordlund, Å., 2003, *ASP Conf. Ser.*, 286, 121
Stein, R. F., Lagerfjärd, A., Nordlund, Å., & Georgobiani, D., 2010, *Solar Phys.* (in press)

The Physics of the Sun and Star Spots
Proceedings IAU Symposium No. 273, 2010
D. P. Choudhary and K. G. Strassmeier, eds.

doi:10.1017/S1743921311015456

Investigation of a sunspot complex by time-distance helioseismology

A. G. Kosovichev[1] and T. L. Duvall Jr[2]

[1]Stanford University, Stanford, CA 94305, USA
[2]Solar Physics Laboratory, Goddard Space Fight Center, NASA, Greenbelt, MD 20771, USA

Abstract. Sunspot regions often form complexes of activity that may live for several solar rotations, and represent a major component of the Sun's magnetic activity. It had been suggested that the close appearance of active regions in space and time might be related to common subsurface roots, or "nests" of activity. EUV images show that the active regions are magnetically connected in the corona, but subsurface connections have not been established. We investigate the subsurface structure and dynamics of a large complex of activity, NOAA 10987-10989, observed during the SOHO/MDI Dynamics run in March-April 2008, which was a part of the Whole Heliospheric Interval (WHI) campaign. The active regions in this complex appeared in a narrow latitudinal range, probably representing a subsurface toroidal flux tube. We use the MDI full-disk Dopplergrams to measure perturbations of travel times of acoustic waves traveling to various depths by using time-distance helioseismology, and obtain sound-speed and flow maps by inversion of the travel times. The subsurface flow maps show an interesting dynamics of decaying active regions with persistent shearing flows, which may be important for driving the flaring and CME activity, observed during the WHI campaign. Our analyses, including the seismic sound-speed inversion results and the distribution of deep-focus travel-time anomalies, gave indications of diverging roots of the magnetic structures, as could be expected from Ω-loop structures. However, no clear connection in the depth range of 0-48 Mm among the three active regions in this complex of activity was detected.

Keywords. Sun: helioseismology, sunspots, Sun: interior, Sun: magnetic field

1. Introduction

Local helioseismology provides important insight into the subsurface structure and dynamics of emerging magnetic flux, formation, evolution and decay of sunspots and active regions. Methods of local helioseismology and acoustic tomography, based on measurements and inversion of acoustic travel times, are intensively developed and tested via numerical simulations (for a recent review see, Kosovichev 2010). These methods provide important insights in the subsurface structure and dynamics of sunspots, which are important for understanding the origins of solar magnetism. In particular, our previous results obtained by a time-distance helioseismology techniques (Duvall *et al.* 1993) revealed significant changes in the subsurface flow patterns during the life cycle of active regions: strong diverging flows during the flux emergence, formation of localized converging flows around stable sunspots, and dominant outflows during the decay (Kosovichev & Duvall 2006; Kosovichev 2009). The sound-speed images indicate that the magnetic flux gets concentrated in strong field structures just below the surface during the sunspot formation, but the seismic perturbations of large sunspots extend at least up to 20-30 Mm, indicating that sunspots have deep roots.

A complex of decaying sunspots, NOAA 10987-10989, was observed by SOHO/MDI in March 2008 during Whole Heliospheric Interval (WHI) campaign. Three active regions

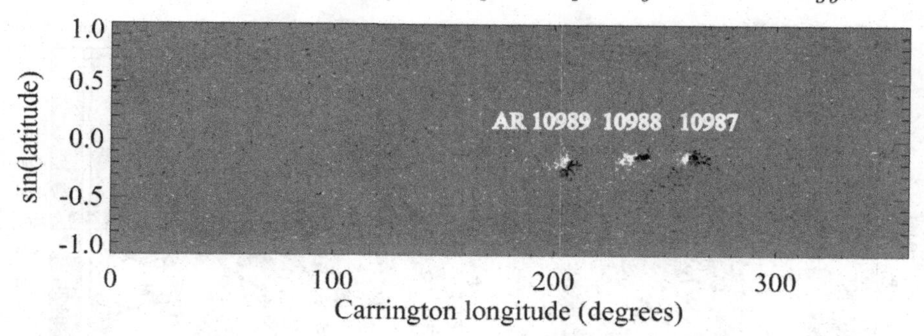

Figure 1. Synoptic magnetic field map from SOHO/MDI for Carrington rotation 2068. The grayscale shows the line-of-sight magnetic field saturated at ± 100 G.

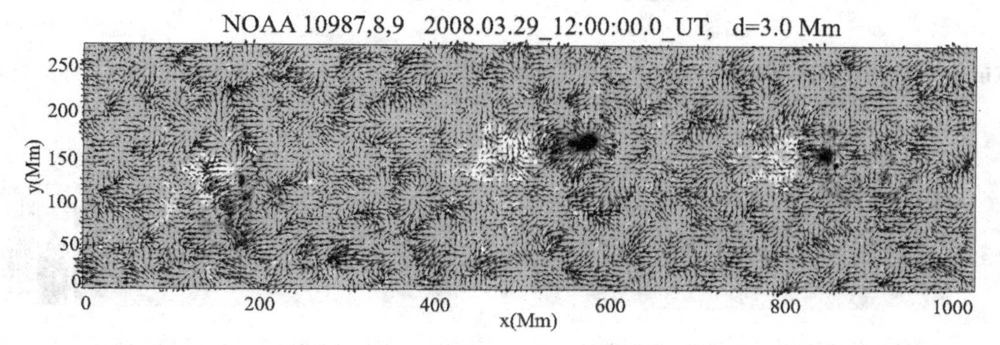

Figure 2. A map of subsurface flows at depth 3Mm (arrows) over the corresponding photospheric magnetogram for all three active regions, observed on March 29, 2008, 12:00 UT. The longest arrows correspond to 0.5 km/s. The magnetic field scale is from -100 to 500 G.

appeared in a narrow latitudinal and longitudinal range (Fig. 1). This suggests that they probably originated from a common subsurface nest of activity.

2. Data Analysis and Results

For the analysis we used full-disk Dopplergrams from SOHO/MDI (Scherrer *et al.* 1995) obtained with 1-min cadence and 2 arcsec/pixel resolution. The time-distance data analysis and inversion procedure is described by Duvall *et al.* (1997); Kosovichev (1996); Kosovichev & Duvall (1997). The acoustic travel-time maps were calculated for 12 distance ranges (annuli), from 0.78 to 11.76 heliographic degrees (or from 9.48 to 142.85 Mm) for fifteen 8-hour intervals during the period of March 27 - April 1, 2008. We used both the surface-focus and deep-focus measurement schemes (Duvall 1998). This coverage allowed us to investigate evolution of this decaying complex of activity.

Figure 2 shows a subsurface flow map of all three active regions at the depth of 3 Mm. The flow pattern shows supergranulation flows and strong diverging flows around sunspots of these active regions. The diverging flows are observed in the whole range of depth, 0 – 12 Mm obtained in our inversions. At greater depths the inversion results become too noisy. We did not attempt to recover the flow patterns at greater depths in this work. The dominance of diverging flows for the decaying sunspots is consistent with our previous results (Kosovichev & Duvall 2006; Kosovichev 2009).

The structures of the subsurface flows in AR 10989 and 10988 are shown in more detail in Figure 3. An interesting feature is strong shearing flows associated with the diverging

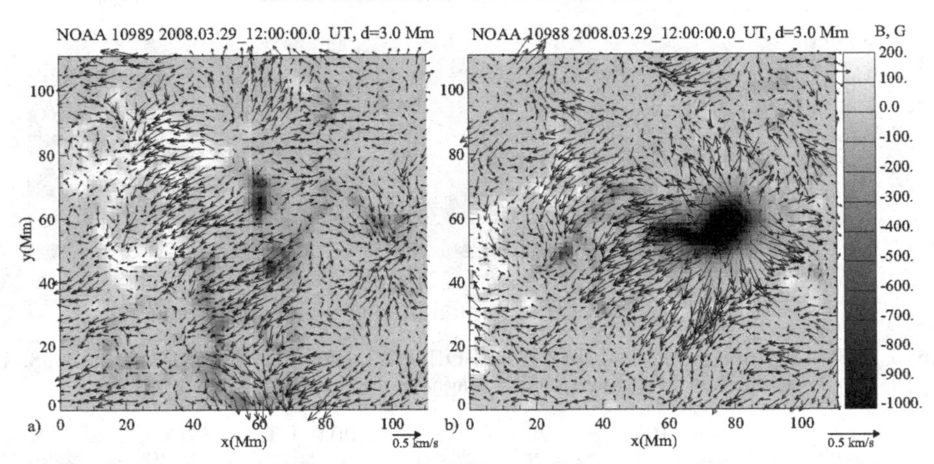

Figure 3. Subsurface flow maps at depth 3 Mm in active regions a) 10989, and b) 10988. The background images are the corresponding photospheric magnetograms.

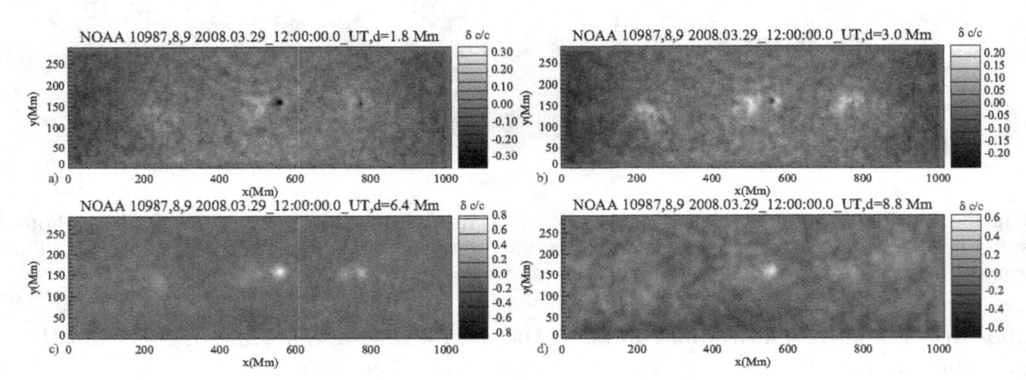

Figure 4. Maps of subsurface sound-speed variations beneath the active regions, NOAA 10987 – 9, at depth: a) 1.8 Mm, b) 3 Mm, c) 6.4 Mm, and d) 8.8 Mm.

flows of the decaying sunspots. These flows may be important for understanding the flaring and CME activity of decaying regions. In this case, during the WHI campaign several strong CMEs were detected from AR 10989, which was decaying most rapidly.

Our inspection of the flow maps did not reveal any large-scale pattern, which can be common for these active regions and indicate on their subsurface connections. Therefore, we investigated also helioseismic sound-speed maps, obtained by inversion of the surface-focus travel-time variations, using the ray-path approximation (Kosovichev 1999). Samples of the seismic sound-speed maps are shown in Fig. 4. Interestingly enough, a shallow layer of negative sound-speed variation, previously reported for isolated sunspots (e.g. Couvidat *et al.* 2006a,b; Kosovichev *et al.* 2000; Zhao *et al.* 2010), is observed only for the leading sunspots of AR 10988 and 10987. The seismic structure of the sunspot in AR 10989 showed only positive variations, probably because it was mostly decayed.

These maps, in which the seismic structure is clearly seen up to 12 Mm depth, also did not reveal potential subsurface links among the active regions. Therefore, we studied averaged properties in order to obtain a better signal-to-noise ratio at greater depths. Figure 5*a* shows the depth structure of the sound-speed inversions averaged over a narrow range of latitude. The seismic structures can be traced up to the depth of 24 Mm. It

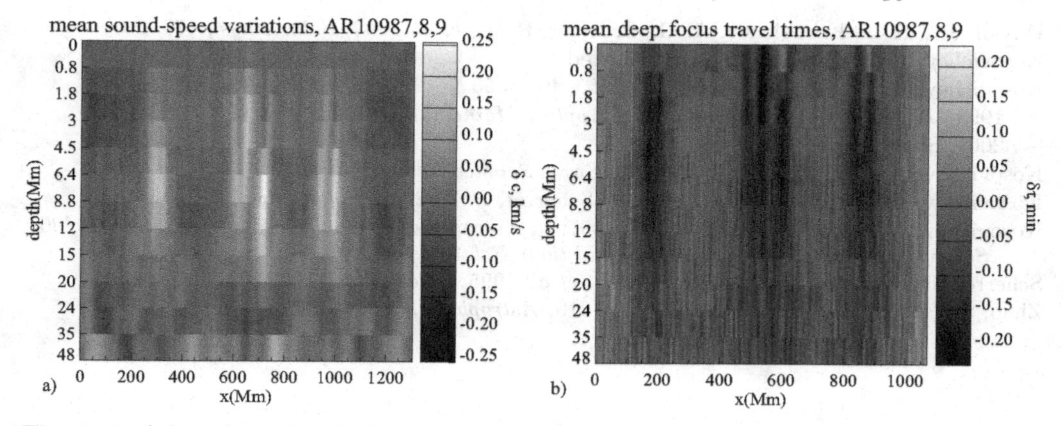

Figure 5. a) Sound-speed variations, averaged over a narrow range of latitude including all three active regions, as a function of longitude (x-coordinate) and depth. b) Travel-time variations calculated by using the deep-focus scheme and averaged over a narrow range of latitude, as a function of longitude and the focus depth.

seems that the tube-like structures become more inclined and diverge from each other with depth, like this can be expected from Ω-loop structures, but this evidence is weak. From these inversion data we cannot make a conclusion about the subsurface connections.

In addition, we attempted to detect the links directly from the deep-focus travel time anomalies. Figure 5*b* shows the deep-focus travel time perturbations (relatively to the quiet Sun) averaged over a range of latitudes including the active regions, as a function of longitude and the focus-point depth. The depth distribution of these travel-times is qualitatively similar to the sound-speed inversion results from the different (surface-focus) type of measurements. The deep-focus data allows us to trace the subsurface structures in deeper layers, up to 48 Mm. They also indicate divergence of the active region structures, which may be an evidence that these are parts of a Ω-loop toroidal structure. But this remains uncertain and requires further investigation.

3. Conclusions

We have used data from SOHO/MDI to investigate seismic sound-speed structures and mass flows in the solar interior, associated with a complex of three active regions. We found that all sunspots in this complex were decaying and were surrounded by strong deep outflows and shearing flows, which could be important for initiation of CMEs observed during this period. Our time-distance helioseismology analyses, including seismic sound-speed inversions of the surface-focus travel-time measurements and deep-focus travel-time anomalies, provided indications of diverging roots of the magnetic structures as this could be expected from Ω-loop structures, but were unable to detect connections in the depth range of 0-48 Mm among the three active regions in this complex.

References

Couvidat, S., Birch, A. C., & Kosovichev, A. G. 2006a, *Astrophys. J.*, 640, 516

Couvidat, S., Birch, A. C., Rajaguru, S. P., & Kosovichev, A. G. 2006b, in IAU Symposium, Vol. 233, *Solar Activity and its Magnetic Origin*, ed. V. Bothmer & A. A. Hady, 75–76

Duvall, Jr., T. L. 1998, in ESA Special Publication, Vol. 418, *Structure and Dynamics of the Interior of the Sun and Sun-like Stars*, ed. S. Korzennik, 581

Duvall, Jr., T. L., Jefferies, S. M., Harvey, J. W., & Pomerantz, M. A. 1993, *Nature*, 362, 430

Duvall, Jr., T. L., Kosovichev, A. G., Scherrer, P. H., *et al.* 1997, *Solar Phys.*, 170, 63
Kosovichev, A. 2010, *Solar Phys.*, submitted
Kosovichev, A. G. 1996, *Astrophys. J. Lett.*, 461, L55
—. 1999, *Journal of Computational and Applied Mathematics*, 109, 1
—. 2009, Space Science Reviews, 144, 175
Kosovichev, A. G. & Duvall, T. L. 2006, *Space Science Reviews*, 124, 1
Kosovichev, A. G., Duvall, Jr., T. L. ., & Scherrer, P. H. 2000, *Solar Phys.*, 192, 159
Kosovichev, A. G. & Duvall, Jr., T. L. 1997, in Astrophys. Space Sci. Lib., v.225, SCORe'96 : *Solar Convection and Oscillations and their Relationship*, 241–260
Scherrer, P. H., Bogart, R. S., Bush, R. I., *et al.* 1995, *Solar Phys.*, 162, 129
Zhao, J., Kosovichev, A. G., & Sekii, T. 2010, *Astrophys. J.*, 708, 304

The Physics of the Sun and Star Spots
Proceedings IAU Symposium No. 273, 2010
D. P. Choudhary and K. G. Strassmeier, eds.

doi:10.1017/S1743921311015468

Sunspot temperatures from red and blue photometry

G. A. Chapman, A. M. Cookson, & D. G. Preminger

[1]San Fernando Observatory, CSU Northridge, Northridge, CA 91330, USA

Abstract. Photometric images are used to measure the temperature of sunspots at different wavelengths. Images at 672.3 nm and 472.3 nm are obtained at the San Fernando Observatory using the CFDT2 (2.5" x 2.5" pixels). Images at 607.1 nm and 409.4 nm are obtained by the PSPT at Mauna Loa Observatory. Monochromatic intensities are converted to temperatures as in Steinegger *et al* (1990). The pixel by pixel temperature for a sunspot is converted into a bolometric contrast for that sunspot according to Chapman *et al* (1994). Sunspot temperatures, i.e., their bolometric contrasts, are calculated from both red (672.3 nm) and blue wavelengths (472.3 nm) and compared.

Keywords. Sun: irradiance, sun:temperature

---

1. Introduction

Sunspots create the largest short term fluctuations in the Total Solar Irradiance (TSI). Knowledge of their mean temperature compared to the photosphere can help in the understanding of their influence on TSI. In this paper, photometric images are used to measure the temperature of sunspots at different wavelengths. Images at 673.2 nm, 472.3 nm, 780 nm and 997 nm are obtained at the San Fernando Observatory using the CFDT2 on a daily basis. Figure 1 is a blue (472.3 nm) image from CFDT2.

Monochromatic intensities are converted to temperature using Eq. 1.1 (Steinegger *et al.* 1990)

$$T_i \;=\; \frac{c_2}{\lambda}\left[ln(e^{c_2/\lambda T_o} - 1)\frac{I_o(\lambda)}{I_i(\lambda)} + 1\right]^{-1} \tag{1.1}$$

where $c_2 = \frac{hc}{k}$, T_o is the temperature of the disk center, $I_o(\lambda)$ is the intensity of the disk center and $I_i(\lambda)$ is the intensity of pixel i.

The pixel by pixel temperature for a sunspot is converted into a bolometric contrast for that sunspot by Eq. 1.2

$$\alpha_{thermal} \;=\; \frac{1}{N}\sum_i [1 - (T_i/T_{ph})^4], \tag{1.2}$$

where the sum is over umbral and penumbral pixels. T_{ph} is the photospheric temperature at the same limb position as T_i.

Images are from the SFO Cartesian Full Disk Telescope no. 2 (CFDT2) which has 2.5" x 2.5" pixels. Data processing is described in Walton *et al.* 1998. Sunspot temperatures, ie. their bolometric contrasts, are calculated from red ($\lambda = 673.2\,\mathrm{nm}$), blue ($\lambda = 472.3\,\mathrm{nm}$), near infrared [NIR] ($\lambda = 780\,\mathrm{nm}$) and infrared [IR] ($\lambda = 997\,\mathrm{nm}$) wavelengths and compared in Tables 1 and 2.

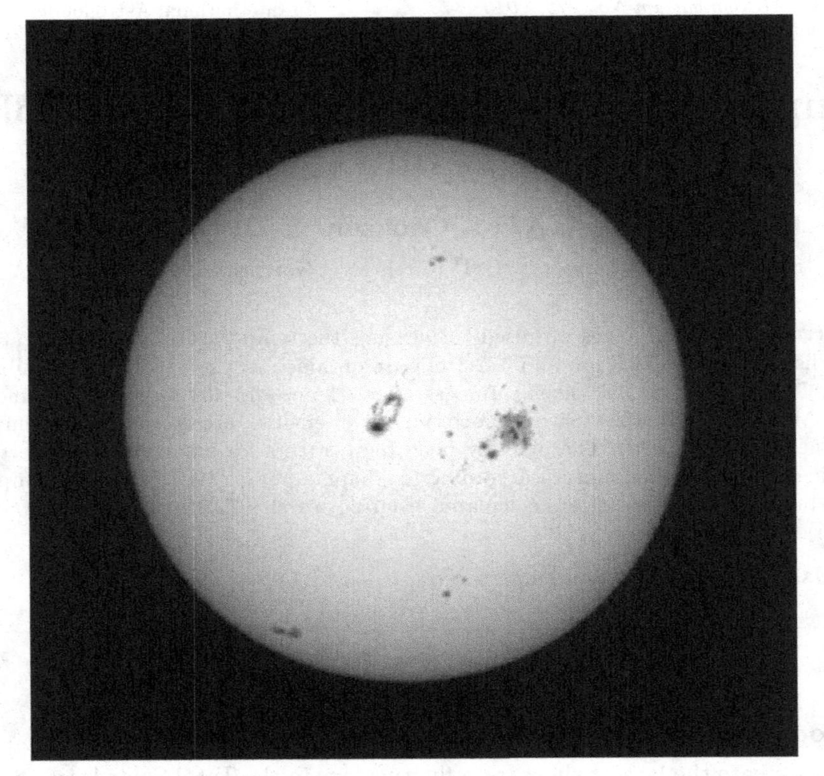

Figure 1. A CFDT2 blue image from 28 October 2003. Spot 1 is on the right (south) and spot 2 to the left of disk center. Geocentric north is at the left.

Table 1. α_x vs. color for spot 1

filter	α_{therm}	α_{eff}
blue	0.252	0.499
red	.255	.362
NIR	.255	.316
IR	.257	.263

Alpha effective (α_{eff}) was defined by Chapman *et al.* (1994)

$$\alpha_{eff} = \frac{DEF}{2PSI} \tag{1.3}$$

where DEF is the photometric sunspot deficit and PSI is the photometric sunspot index.

2. Results

For sunspot no. 1 on 28 October 2003, the following values were determined for α_x (α_{therm} or α_{eff}). These values are given in Table 1.

α_{eff} (Eq. 1.3) is strongly wavelength dependent, varying roughly as λ^{-1} due to the sensitivity of sunspot contrast to wavelength. Spot 2, with a large umbra gave similar α_{therm} values (see Table 2). It was shown in Chapman *et al.* (1994) that α_{eff} was strongly correlated with the ratio of umbral to penumbral area.

Table 2. α_x vs. color for spot 2

filter	α_{therm}	α_{eff}
blue	.247	.481
red	.256	.356
NIR	.253	.308
IR	.277	.281

Using restored red images (Walton and Preminger 1999) gave $\alpha_{therm} = 0.292$ for spot 1 and $\alpha_{therm} = 0.312$ for spot 2. Restoration removes stray light from sunspot images.

The bolometric contrast can also be calculated from mean umbral and penumbral temperatures (Eq. 2.1),

$$\alpha = \frac{A_u}{A_s}\left[1 - \left(\frac{T_u}{T_{ph}}\right)^4\right] + \frac{A_p}{A_s}\left[1 - \left(\frac{T_p}{T_{ph}}\right)^4\right] \tag{2.1}$$

where A respresents area and T represents temperature with subscripts u, p, ph and s representing umbra, penumbra, photosphere and total sunspot.

Using sunspot temperatures in the literature (Lites 1984) we find a mean penumbral intensity ratio of 0.83. Using an umbral temperature of 3700 K (Foukal 2004) we calculate $\alpha = 0.302$. This is only slightly greater than the values of α_{therm} in Tables 1 and 2. Hudson *et al.* (1982) obtained $\alpha = 0.315$ using mean sunspot values from Allen (1963)

3. Conclusion

The bolometric contrasts (related to the spot mean temperatures) from four different wavelengths agree closely for the two sunspots studied here. The bolometric contrasts from restored images agree closely with values of α found in the literature.

Acknowledgment

Observations have been obtained by CSUN students and SFO staff. This research was partially supported by NSF Grant ATM-0848518.

References

Allen, C. W. 1963 "Astrophysical Quantities," Athlone Press, London.
Chapman, G. A., Cookson, A. M., & Dobias, J. J. 1994, *Astrophys. J.* 432, 403.
Foukal, P. V. 2004 "Solar Astrophysics," Wiley-VCH, Weinheim.
Hudson, H. S., Silva, S., Woodard, M., & Willson, R. C. 1982, *Solar Phys.*. 76, 211.
Lites, B. 1984, *Astrophys. J.* 90, 1.
Steinegger, M., Brandt, P. N., Schmidt, W., & Pap, J. 1990, *Ap & SS* 170, 127.
Walton, S. R., Chapman, G. A., Cookson, A. M., Dobias, J. J., & Premiinger, D. G.. 1998, *Solar Phys.* 179, 31.
Walton, S. R. & Preminger, D. G. 1999, *Astrophys. J.* 514, 959.

The Physics of the Sun and Star Spots
Proceedings IAU Symposium No. 273, 2010
D. P. Choudhary and K. G. Strassmeier, eds.

doi:10.1017/S174392131101547X

A filament supported by different magnetic field configurations

Y. Guo[1,2], B. Schmieder[2], P. Démoulin[2], T. Wiegelmann[3], G. Aulanier[2], T. Török[2], and V. Bommier[4]

[1]Department of Astronomy, Nanjing University, Nanjing 210093, China
email: guoyang@nju.edu.cn

[2]LESIA, Observatoire de Paris, CNRS, UPMC, Université Paris Diderot, 5 place Jules Janssen, 92190 Meudon, France

[3]Max-Planck-Institut für Sonnensystemforschung, Max-Planck-Strasse 2, 37191 Katlenburg-Lindau, Germany

[4]LERMA, Observatoire de Paris, CNRS, UPMC, Université Paris Diderot, 5 place Jules Janssen, 92190 Meudon, France

Abstract. A nonlinear force-free magnetic field extrapolation of vector magnetogram data obtained by THEMIS/MTR on 2005 May 27 suggests the simultaneous existence of different magnetic configurations within one active region filament: one part of the filament is supported by field line dips within a flux rope, while the other part is located in dips within an arcade structure. Although the axial field chirality (dextral) and the magnetic helicity (negative) are the same along the whole filament, the chiralities of the filament barbs at different sections are opposite, i.e., right-bearing in the flux rope part and left-bearing in the arcade part. This argues against past suggestions that different barb chiralities imply different signs of helicity of the underlying magnetic field. This new finding about the chirality of filaments will be useful to associate eruptive filaments and magnetic cloud using the helicity parameter in the Space Weather Science.

Keywords. Sun: corona, Sun: filaments, Sun: magnetic fields

1. Introduction

A filament or prominence is constructed by plasma that is typically a hundred times cooler and denser than the coronal surroundings. The dense plasma is usually thought to be supported by Lorentz force provided by dipped magnetic field lines. Magnetic dips could be present in quadrupolar fields (Kippenhahn & Schlüter 1957), or in bipolar fields with a flux rope (Kuperus & Raadu 1974; Aulanier & Démoulin 1998), or highly sheared field lines below a strong arcade field (Antiochos *et al.* 1994). The flux rope model corresponds to the so-called inverse configuration ,i.e. the orientation of the magnetic field in the filament is inverse compared to the photospheric magnetic field, the later one is pointing from positive to negative polarity. The other two models correspond to the so-called normal configuration.

Magnetic field configurations play a key role in many aspects of a filament, such as its pattern, suspension, and formation. However, vector magnetic fields are currently not available for the chromosphere and the corona in practice. Therefore, these fields are usually extrapolated from the magnetic fields observed in the photosphere by the assumption of the force-free field model, i.e., $\nabla \times \mathbf{B} = \alpha \mathbf{B}$. The force-free parameter α usually changes in space in the corona with complicated magnetic field structures (Régnier *et al.* 2002; Schrijver *et al.* 2005), which yields a nonlinear force-free field (NLFFF). The gravity of the relatively denser plasma in a filament could affect the associated magnetic field

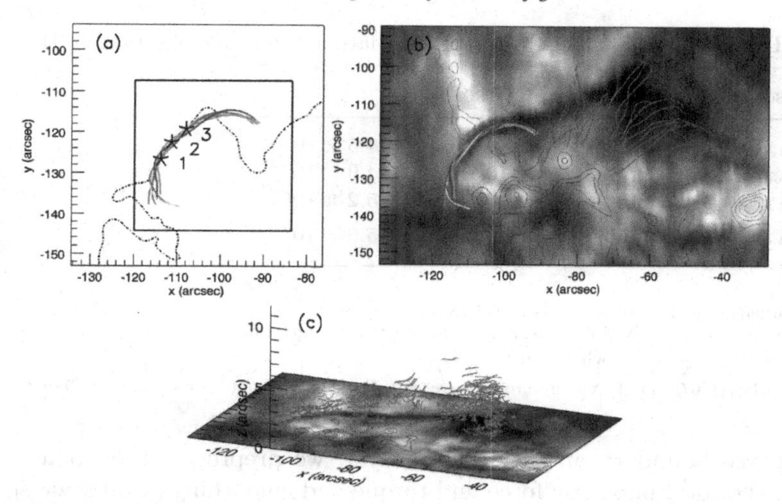

Figure 1. (a) Twisted field lines represent the flux rope. Dash-dotted line denotes the polarity inversion line. A third order polynomial is adopted to fit the polarity inversion line between points 1 and 3 to approximate the horizontal flux rope axis. The square is the projection of a rectangular box with $80 \times 80 \times 5$ grid points just above the bottom layer. (b) Hα filament overlaid with the flux rope and some selected field lines. (c) Dips ("+" signs) and field line sections in the extrapolated magnetic field.

configuration. However, the force-free field model may be still adequate in active regions, since the magnetic field strength in active region filaments could be very large, which is up to 600–700 G for instance (Kuckein *et al.* 2009).

Guo *et al.* (2010) performed a NLFFF extrapolation for the active region NOAA 10767 using Wiegelmann's method (2004). An Hα filament was observed in the active region. A magnetic flux rope was found by the NLFFF extrapolation corresponding to the eastern part of the filament, while magnetic dips of arcade field lines coincided with the other part. In this paper, we study how different grid resolutions in the optimization algorithm affect the extrapolation results in the definition of the flux rope. Four cases, where the spatial resolution along the height are 2, 1, $\frac{1}{2}$ and $\frac{1}{4}$ times that along the horizontal plane, have been carefully tested with Low and Lou's analytical NLFFF solution and applied to the active region NOAA 10767.

2. Analysis and Results

Active region NOAA 10767 was observed by *Télescope Héliographique pour l'Etude du Magnétisme et des Instabilités Solaires/Multi-Raies* (THEMIS/MTR) from 9:54 to 10:41 UT on 2005 May 27. The Stokes profiles obtained by the Fe 6302.5 Å line are fitted by the inversion code UNNOFIT to infer the vector magnetic field (Landolfi & Landi Degl'Innocenti 1982; Bommier et al. 2007). Due to the fact that opposite orientations of the transverse magnetic field generate indistinguishable linear polarizations, a 180° ambiguity in the direction of the transverse field is present. We remove the 180° ambiguity in two steps. First, it is removed preliminarily by an automatic algorithm, the non-potential magnetic field calculation method (Georgoulis 2005). Then, some regions with larger current borders, where transverse fields change abruptly, are reconsidered by an interactive code by removing the ambiguity manually. In order to simulate a force-free

Table 1. Metrics for NLFFF Extrapolations from the Observation Data

Grid Resolution[1]	$E/E_{\rm pot}$[2]	$< {\rm CW} \sin\theta >$[3]	$< \vert f_i \vert >$[4]
$\Delta z = 2\Delta x = 2\Delta y$	1.41	0.28	4.5×10^{-3}
$\Delta z = \Delta x = \Delta y$	1.36	0.28	4.9×10^{-3}
$\Delta z = \frac{1}{2}\Delta x = \frac{1}{2}\Delta y$	1.30	0.28	5.2×10^{-3}
$\Delta z = \frac{1}{4}\Delta x = \frac{1}{4}\Delta y$	1.25	0.26	5.6×10^{-3}

Notes:
[1] The optimization codes with different grid resolutions.
[2] Energy contained in the NLFFF referred to that in the potential field.
[3] $< {\rm CW} \sin\theta > = \frac{\sum_i J_i \sin\theta_i}{\sum_i J_i}$, where $\sin\theta_i = \frac{|\mathbf{J}_i \times \mathbf{B}_i|}{J_i B_i}$.
[4] $|f_i| = |(\nabla \cdot \mathbf{B})_i|/(6B_i/\Delta x)$. All the metrics are calculated in the selected volume in Figure 1a.

and torque-free boundary on the bottom layer, we preprocess the boundary data by minimizing the total magnetic force and torque and smoothing small scale structures.

The preprocessed bottom boundary are used by the optimization code to extrapolate the NLFFF (Wheatland *et al.* 2000; Wiegelmann 2004). Selected field lines are plotted in Figure 1. The magnetic field configurations associated with the Hα filament are partly flux rope and partly arcade as shown in Figure 1b. The magnetic dips of these two field configurations are thought to support the filament material (Figure 1c). The local magnetic helicity sign in this region is negative, as suggested by the left-handed flux rope and the left-skewed arcades. However, the filament barb chiralities are different, which is right-bearing in the flux rope part and left-bearing in the arcade part (refer to Guo *et al.* (2010) for analysis in detail).

We further test the extrapolation results with codes of different spatial resolutions in z, which is along the height. The horizontal resolutions keep unchanged and equal to each other. To test the performance of these codes ($\Delta z = 2\Delta x = 2\Delta y$, $\Delta z = \Delta x = \Delta y$, $\Delta z = \frac{1}{2}\Delta x = \frac{1}{2}\Delta y$, and $\Delta z = \frac{1}{4}\Delta x = \frac{1}{4}\Delta y$), we prepare four analytical solutions (Low & Lou 1990) with the four different spatial resolutions. The initial condition is specified by the potential field extrapolated from the bottom boundary. If all the six boundaries are specified, the analytical NLFFF can be recovered to a very good degree. If only the bottom boundary is specified, and the lateral and top boundaries are given by the potential field extrapolated from the bottom boundary, all the codes can still recover the NLFFF in the center of the computing volume to an acceptable degree.

We apply the four versions of the optimization procedure to the preprocessed vector magnetic field observed by THEMIS/MTR. Table 1 shows that the metrics of the nonlinear force-free fields extrapolated by all the codes do not differ from each other too much. We decompose the magnetic field components in a local reference frame of the flux rope and study their distributions in Figure 2. We find that B_v always changes its sign when it crosses a defined axis, which indicates the existence of twisted field lines of a flux rope. Two cases with grid resolution of $\Delta z = 2\Delta x = 2\Delta y$ and $\Delta z = \Delta x = \Delta y$ give similar heights of the flux rope axis. However, the other two codes ($\Delta z = \frac{1}{2}\Delta x = \frac{1}{2}\Delta y$, and $\Delta z = \frac{1}{4}\Delta x = \frac{1}{4}\Delta y$) cannot recover the flux rope to the same physical height, but only the same grid points as shown in Figure 2.

3. Conclusions

The chirality of a filament barb depends on both the magnetic helicity (positive/negative) and the magnetic field configuration (flux rope/arcade). Based on the NLFFF extrapo-

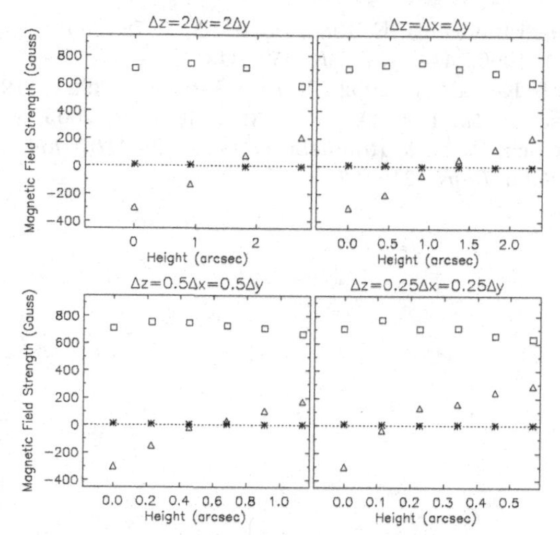

Figure 2. Distributions of B_p (square), B_v (triangle), and B_z (asterisk) along the height at the point 2 (Figure 1a). B_p and B_v are the horizontal components, parallel and orthogonal to the axis of the flux rope, respectively. Different panels show the results obtained with the codes of different grid resolutions.

lation of the active region NOAA 10767 magnetic field observed by THEMIS/MTR on 2005 May 27, we find that the chirality of filament barbs in different sections are different, i.e., right-bearing in the flux rope part and left-bearing in the arcade part, although the magnetic helicity (negative, which is equivalent to dextral axial field chirality) is the same along the whole filament.

The existence of the flux rope is further confirmed by extrapolations with four codes of different grid resolutions along the height. However, we cannot get more grid points to resolve the flux rope along height when the resolution is increased. We think this problem is caused by the boundary conditions. First, we do not have data on the full boundaries, such as the lateral and top surfaces for a box, which are specified by a potential field extrapolation instead. Secondly, As a prior requirement for the preprocessing, the bottom boundary should be well isolated and flux balanced, which is not fulfilled due to the limitation of the field of view in the observation. Appropriate preprocessing of boundary conditions is crucial for a credible extrapolation.

References

Antiochos, S. K., Dahlburg, R. B., & Klimchuk, J. A. 1994, *Astrophys. J. Lett.*, 420, L41

Aulanier, G. & Démoulin, P. 1998, *Astron. Astrophys*, 329, 1125

Bommier, V., Landi Degl'Innocenti, E., Landolfi, M., & Molodij, G. 2007, *Astron. Astrophys*, 464, 323

Georgoulis, M. K. 2005, *Astrophys. J. Lett.*, 629, L69

Guo, Y., Schmieder, B., Démoulin, P., Wiegelmann, T., Aulanier, G., Török, T., & Bommier, V. 2010, *Astrophys. J.*, 714, 343

Kippenhahn, R. & Schlüter, A. 1957, *Zeitschrift fur Astrophysik*, 43, 36

Kuckein, C., Centeno, R., Martínez Pillet, V., Casini, R., Manso Sainz, R., & Shimizu, T. 2009, *Astron. Astrophys*, 501, 1113

Kuperus, M. & Raadu, M. A. 1974, *Astron. Astrophys*, 31, 189

Landolfi, M. & Landi Degl'Innocenti, E. 1982, *Solar Phys.*, 78, 355
Low, B. C. & Lou, Y. Q. 1990, *Astrophys. J.*, 352, 343
Régnier, S., Amari, T., & Kersalé, E. 2002, *Astron. Astrophys*, 392, 1119
Schrijver, C. J., De Rosa, M. L., Title, A. M., & Metcalf, T. R. 2005, *Astrophys. J.*, 628, 501
Wheatland, M. S., Sturrock, P. A., & Roumeliotis, G. 2000, *Astrophys. J.*, 540, 1150
Wiegelmann, T. 2004, *Solar Phys.*, 219, 87

The Physics of the Sun and Star Spots
Proceedings IAU Symposium No. 273, 2010
D. P. Choudhary and K. G. Strassmeier, eds.

doi:10.1017/S1743921311015481

Are the photospheric sunspots magnetically force-free in nature?

Sanjiv Kumar Tiwari

Udaipur Solar Observatory, Physical Research Laboratory, Dewali, Bari Road,
Udaipur - 313 001, India.
email: stiwari@prl.res.in

Abstract. In a force-free magnetic field, there is no interaction of field and the plasma in the surrounding atmosphere i.e., electric currents are aligned with the magnetic field, giving rise to zero Lorentz force. The computation of many magnetic parameters like magnetic energy, gradient of twist of sunspot magnetic fields (computed from the force-free parameter α), including any kind of extrapolations heavily hinge on the force-free approximation of the photospheric magnetic fields. The force-free magnetic behaviour of the photospheric sunspot fields has been examined by Metcalf *et al.* (1995) and Moon *et al.* (2002) ending with inconsistent results. Metcalf *et al.* (1995) concluded that the photospheric magnetic fields are far from the force-free nature whereas Moon *et al.* (2002) found the that the photospheric magnetic fields are not so far from the force-free nature as conventionally regarded. The accurate photospheric vector field measurements with high resolution are needed to examine the force-free nature of sunspots. We use high resolution vector magnetograms obtained from the Solar Optical Telescope/Spectro-Polarimeter (SOT/SP) aboard Hinode to inspect the force-free behaviour of the photospheric sunspot magnetic fields. Both the necessary and sufficient conditions for force-freeness are examined by checking global as well as as local nature of sunspot magnetic fields. We find that the sunspot magnetic fields are very close to the force-free approximation, although they are not completely force-free on the photosphere.

Keywords. Sun: atmosphere, Sun: force-free fields, Sun: magnetic fields, Sun: sunspots

1. Introduction

A force-free magnetic field does physically mean a zero Lorentz force (Chandrasekhar 1961; Parker 1979; Low 1982a) i.e., $(\nabla \times \mathbf{B}) \times \mathbf{B} = \mathbf{0}$. This equation can be rewritten as

$$\nabla \times \mathbf{B} = \alpha \mathbf{B} \tag{1.1}$$

The z component of above condition allows us to compute the distribution of α on the photosphere (z = 0)

$$\alpha = \left[\frac{\partial B_y}{\partial x} - \frac{\partial B_x}{\partial y} \right] / B_z \tag{1.2}$$

Three cases may arise: (i) $\alpha = 0$ everywhere, i.e., no electric current in the atmosphere resulting in a potential field (Schmidt (1964); Semel (1967); Sakurai (1989); Régnier & Priest (2007) (ii) α = constant everywhere, i.e., linear force-free state (Nakagawa & Raadu 1972; Gary 1989; van Ballegooijen & Cranmer 2010, etc) which is not always valid and (iii) α varies spatially, i.e., nonlinear force-free magnetic field (Sakurai 1979; Low 1982b; Amari *et al.* 2006; Wiegelmann 2004; Schrijver *et al.* 2008; De Rosa *et al.* 2009; Mackay & van Ballegooijen 2009), this is the most common state expected (Low 1985). However, high resolution vector magnetograms are required to confirm this. In earlier works, perhaps the poor resolution of data obscured the conclusions about the validity of linear/non-linear force-free approximations. In the present work, we check the validity of linear/nonlinear assumptions along with examining the force-freeness over sunspot magnetic fields using high spatial resolution photospheric vector magnetograms obtained from Solar Optical Telescope/Spectro-Polarimeter onboard Hinode. The effect of polarimetric noise present in the data obtained from SOT/SP does not affect much in derivation of the magnetic field parameters (Tiwari *et al.* 2009a; Gosain *et al.* 2010).

2. Necessary and sufficient conditions

Necessary condition. Under the assumption that the magnetic field above the plane z = 0 (photosphere) falls off enough as z goes to infinity, the net Lorentz force in the volume z > 0 is just the Maxwell stress integrated over the plane z = 0 (Aly, 1984; Low, 1985). Thus the components of the net Lorentz force at the plane z = 0 can be expressed by the surface integrals as follows:

$$F_x = -\frac{1}{4\pi}\int B_x B_y dxdy; F_y = -\frac{1}{4\pi}\int B_y B_z dxdy; F_z = -\frac{1}{8\pi}\int B_z^2 - B_x^2 - B_y^2 dxdy \quad (2.1)$$

where F_x, F_y and F_z represent the components of the net Lorentz force. According to Low (1985) the necessary conditions for any magnetic field to be force-free are that

$$|F_x| << |F_p|; \quad |F_y| << |F_p|; \quad |F_z| << |F_p| \quad (2.2)$$

where F_p is force due to the distribution of magnetic pressure on z = 0, as given by,

$$F_p = -\frac{1}{8\pi}\int B_z^2 + B_x^2 + B_y^2 dxdy \quad (2.3)$$

It was discussed by Metcalf *et al.* (1995) that the magnetic field is force-free if the aforementioned ratios are less or equal to 0.1. It is to be noted that the conditions 2.2 are only necessary conditions for the fields to be force-free. The reason for this is that some information is lost in the surface integration in Equations 2.1.

Sufficient condition. In a force-free case the tension force will balance the gradient of magnetic pressure demanding for zero Lorentz force. We can split up the Lorentz force $(F = (1/c)J \times B)$ in two terms as,

$$\mathbf{F} = \frac{(\mathbf{B}\cdot\nabla)\mathbf{B}}{4\pi} - \frac{\nabla(\mathbf{B}\cdot\mathbf{B})}{8\pi} \quad (2.4)$$

The first term in the right hand side in the above equation is the tension force (**T**). The second term represents the gradient of the magnetic pressure i.e., the force due to magnetic pressure ($\mathbf{F_p}$). The vertical component of the tension force term can be simplified to,

$$T_z = \frac{1}{4\pi}\left[B_x\frac{\partial B_z}{\partial x} + B_y\frac{\partial B_z}{\partial y} - B_z\left(\frac{\partial B_x}{\partial x} + \frac{\partial B_y}{\partial y}\right)\right] \quad (2.5)$$

where, the last component has been drawn from the condition $\nabla\cdot\mathbf{B} = 0$. The usefulness of the tension force has not found much attention earlier in the literature but for few studies (Venkatakrishnan 1990; Venkatakrishnan *et al.* 1993; Venkatakrishnan & Tiwari 2010). Recently Venkatakrishnan & Tiwari (2010) pointed out the utility of tension force as a diagnostic of dynamical equilibrium of sunspots. It was found (Venkatakrishnan & Tiwari 2010) that the magnitude of vertical tension force attains values comparable to the force of gravity at several places over the sunspots meaning that the non-magnetic forces will not be able to balance this tension force. Only gradient of the magnetic pressure can match this force resulting into the force-free configurations. This serves as a sufficient condition for verifying the force-freeness of the sunspot magnetic fields. A detailed work is under preparation.

3. Data and analysis

We have used the high resolution vector magnetograms obtained from the Solar Optical Telescope/Spectro-polarimeter (SOT/SP: Tsuneta *et al.* (2008); Suematsu *et al.* (2008); Ichimoto *et al.* (2008); Shimizu *et al.* (2008)) onboard Hinode (Kosugi *et al.* (2007)). The data has been prepared as done successfully in Tiwari *et al.* (2009b, 2010); Venkatakrishnan & Tiwari (2009, 2010); Gosain *et al.* (2009, 2010)).

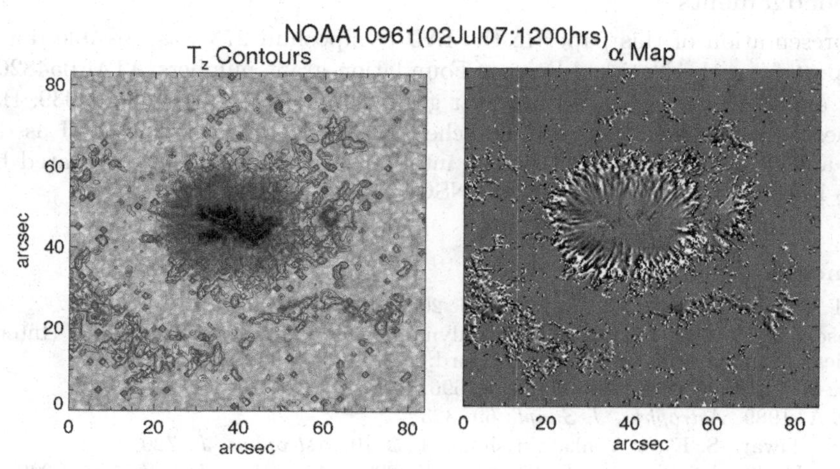

Figure 1. Left panel: Vertical tension force distribution of a sunspot NOAA AR 10961 over its continuum map. Blue (red) colors show negative (positive) contours of $\pm0.4, \pm1.2, \pm4, \pm12$ millidynes/cm^3. Right panel: Alpha map of the same active region.

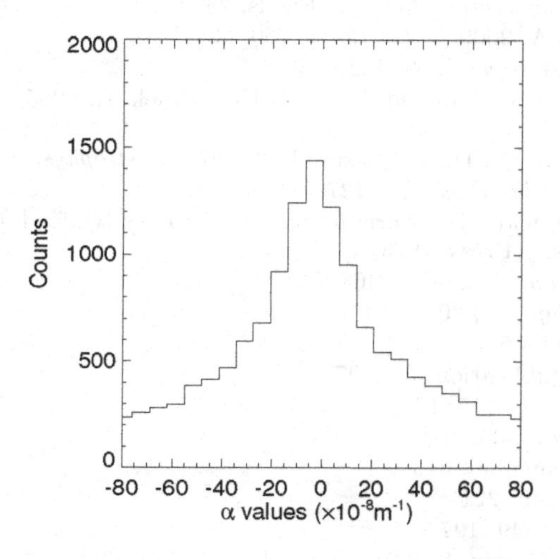

Figure 2. Histogram of the α values of NOAA AR 10961 observed on July 12, 2007 at 1200UT, as an example. We can see even in this simple sunspot, the alpha has a wide range of its distribution.

4. Results and Discussion

We conclude the following: 1. the sunspot magnetic fields are not far from the force-free nature as has been suspected for long time. 2. The non-linear force-free approximation is closer to validity in all sunspots, either it is a simple active region or a complex one.

It is well known that all extrapolation techniques rely on the photospheric vector field measurements and also on its force-free approximation. Coronal magnetic field reconstruction by extrapolations of photospheric magnetic fields under non-linear modeling have shown satisfactory results by roughly matching with the coronal observations McClymont & Mikic (1994); Wiegelmann *et al.* (2005); Schrijver *et al.* (2006). These results then also support our conclusion that the sunspot magnetic fields are close to non-linear force-free approximation. Greater details will be given in a forthcoming regular paper.

Acknowledgements

The presentation of this paper in the IAU Symposium 273 was possible due to partial support from the National Science Foundation grant numbers ATM 0548260, AST 0968672 and NASA - Living With a Star grant number 09-LWSTRT09-0039. Hinode is a Japanese mission developed and launched by ISAS/JAXA, with NAOJ as domestic partner and NASA and STFC (UK) as international partners. It is operated by these agencies in co-operation with ESA and NSC (Norway).

References

Amari, T., Boulmezaoud, T. Z., & Aly, J. J. 2006, *Astron. Astrophys*, 446, 691
Chandrasekhar, S. 1961, Chapter-2 : Hydrodynamic and hydromagnetic stability (International Series of Monographs on Physics, Oxford: Clarendon, 1961)
De Rosa, M. L., *et al.* 2009, *Astrophys. J.*, 696, 1780
Gary, G. A. 1989, *Astrophys. J. Suppl.*, 69, 323
Gosain, S., Tiwari, S. K., & Venkatakrishnan, P. 2010, *Astrophys. J.*, 720, 1281
Gosain, S., Venkatakrishnan, P., & Tiwari, S. K. 2009, *Astrophys. J. Lett.*, 706, L240
Ichimoto, K., *et al.* 2008, *Solar Phys.*, 249, 233
Kosugi, T., *et al.* 2007, *Solar Phys.*, 243, 3
Low, B. C. 1982a, *Solar Phys.*, 77, 43
Low, B. C. 1982b, *Reviews of Geophysics and Space Physics*, 20, 145
Low, B. C. 1985, Measurements of Solar Vector Magnetic Fields, 2374, 49
Mackay, D. H. & van Ballegooijen, A. A. 2009, *Solar Phys.*, 260, 321
McClymont, A. N. & Mikic, Z. 1994, *Astrophys. J.*, 422, 899
Metcalf, T. R., Jiao, L., McClymont, A. N., Canfield, R. C., & Uitenbroek, H. 1995, *Astrophys. J.*, 439, 474
Moon, Y., Choe, G. S., Yun, H. S., Park, Y. D., & Mickey, D. L. 2002, *Astrophys. J.*, 568, 422
Nakagawa, Y. & Raadu, M. A. 1972, *Solar Phys.*, 25, 127
Parker, E. N. 1979, Cosmical magnetic fields: Their origin and their activity (Oxford, Clarendon Press; New York, Oxford University Press, 1979)
Régnier, S. & Priest, E. R. 2007, *Astron. Astrophys*, 468, 701
Sakurai, T. 1979, *Pub. Astron. Soc. Jap.*, 31, 209
Sakurai, T. 1989, *Space Science Reviews*, 51, 11
Schmidt, H. U. 1964, NASA Special Publication, 50, 107
Schrijver, C. J., *et al.* 2008, *Astrophys. J.*, 675, 1637
Schrijver, C. J., *et al.* 2006, *Solar Phys.*, 235, 161
Semel, M. 1967, Annales d'Astrophysique, 30, 513
Shimizu, T., *et al.* 2008, *Solar Phys.*, 249, 221
Suematsu, Y., *et al.* 2008, *Solar Phys.*, 249, 197
Tiwari, S. K., Venkatakrishnan, P., & Gosain, S. 2010, *Astrophys. J.*, 721, 622

Tiwari, S. K., Venkatakrishnan, P., Gosain, S., & Joshi, J. 2009a, *Astrophys. J.*, 700, 199
Tiwari, S. K., Venkatakrishnan, P., & Sankarasubramanian, K. 2009b, *Astrophys. J. Lett.*, 702, L133
Tsuneta, S., *et al.* 2008, *Solar Phys.*, 249, 167
van Ballegooijen, A. A. & Cranmer, S. R. 2010, *Astrophys. J.*, 711, 164
Venkatakrishnan, P. 1990, *Solar Phys.*, 128, 371
Venkatakrishnan, P., Narayanan, R. S., & Prasad, N. D. N. 1993, *Solar Phys.*, 144, 315
Venkatakrishnan, P. & Tiwari, S. K. 2009, *Astrophys. J. Lett.*, 706, L114
Venkatakrishnan, P. & Tiwari, S. K. 2010, *Astron. Astrophys*, 516, L5
Wiegelmann, T. 2004, *Solar Phys.*, 219, 87
Wiegelmann, T., Lagg, A., Solanki, S. K., Inhester, B., & Woch, J. 2005, *Astron. Astrophys*, 433, 701

The Physics of Sun and Star Spots
Proceedings IAU Symposium No. 273, 2010
D. P. Choudhary & K. G. Strassmeier, eds.

doi:10.1017/S1743921311015493

Vector magnetic field and vector current density in and around the δ-spot NOAA 10808†

Véronique Bommier[1], Egidio Landi Degl'Innocenti[2], Brigitte Schmieder[3], and Bernard Gelly[4]

[1]LERMA, Observatoire de Paris, CNRS, ENS, UPMC, UCP; Place Jules Janssen, F-92190 Meudon, France, V.Bommier@obspm.fr

[2]Dipartimento di Fisica e Astronomia, SASS, Università degli Studi di Firenze, Largo E. Fermi 2, I-50125 Firenze, Italy

[3]LESIA, Observatoire de Paris, CNRS, UPMC, Université Paris-Diderot; Place Jules Janssen, F-92190 Meudon, France

[4]Télescope Héliographique pour l'Étude du Magnétisme et des Instabilités Solaires, CNRS UPS 853 – THEMIS, Vía Láctea s/n, E-38205 La Laguna, Tenerife, Spain

Abstract. The context is that of the so-called "fundamental ambiguity" (also azimuth ambiguity, or 180° ambiguity) in magnetic field vector measurements: two field vectors symmetrical with respect to the line-of-sight have the same polarimetric signature, so that they cannot be discriminated. We propose a method to solve this ambiguity by applying the "simulated annealing" algorithm to the minimization of the field divergence, added to the longitudinal current absolute value, the line-of-sight derivative of the magnetic field being inferred by the interpretation of the Zeeman effect observed by spectropolarimetry in two lines formed at different depths. We find that the line pair Fe I λ 6301.5 and Fe I λ 6302.5 is appropriate for this purpose. We treat the example case of the δ-spot of NOAA 10808 observed on 13 September 2005 between 14:25 and 15:25 UT with the THEMIS telescope. Besides the magnetic field resolved map, the electric current density vector map is also obtained. A strong horizontal current density flow is found surrounding each spot inside its penumbra, associated to a non-zero Lorentz force centripetal with respect to the spot center (*i.e.*, oriented towards the spot center). The current wrapping direction is found to depend on the spot polarity: clockwise for the positive polarity, counterclockwise for the negative one. This analysis is made possible thanks to the UNNOFIT2 Milne-Eddington inversion code, where the usual theory is generalized to the case of a line (Fe I λ 6301.5) that is not a normal Zeeman triplet line (like Fe I λ 6302.5).

Keywords. Magnetic fields, polarization, radiative transfer, sunspots, Sun: magnetic fields, techniques: polarimetric, techniques: spectroscopic

---

This paper is now submitted (revision submitted) under the title "Azimuth ambiguity solution in photospheric vector magnetic field measurements from multiline observations" by the same authors in the same order to the Journal *Astronomy and Astrophysics.*

† observed with THEMIS based on observations made with the French-Italian telescope THEMIS operated by the CNRS and CNR on the island of Tenerife in the Spanish Observatorio del Teide of the Instituto de Astrofísica de Canarias.

The Physics of Sun and Star Spots
Proceedings IAU Symposium No. 273, 2010
D. P. Choudhary & K. G. Strassmeier, eds.

doi:10.1017/S174392131101550X

Substructure of quiet sun bright points

Aleksandra Andic[1], Jongchul Chae[2] and Phillip R. Goode[1]

[1]BBSO, 40386 N. Shore Ln.
Big Bear City, CA-92314, USA
email: aandic@bbso.njit.edu

[2]Astronomy Program, Department of Physics and Astronomy, Seoul National University
Seoul 151-741, Korea
email: jcchae@snu.ac.kr

Abstract. Since photospheric bright points (BPs) were first observed, there has been a question as to how are they structured. Are they just single flux tubes or a bundle of the flux-tubes? Surface photometry of the quiet Sun (QS) has achieved resolution close to 0.1″ with the New Solar Telescope at Big Bear Solar Observatory. This resolution allowed us to detect a richer spectrum of BPs in the QS. The smallest BPs we observed with TiO 705.68 nm were 0.13″, and we were able to resolve individual components in some of the BPs clusters and ribbons observed in the QS, showing that they are composed of the individual BPs. Average size of observed BPs was 0.22″.

Keywords. Photosphere, bright points

1. Introduction

Observations of the solar photosphere, so far, revealed a plethora of tiny bright features, usually concentrated in active regions or bordering the supergranules in the quiet Sun (QS). Near disk center, they appear as "Bright Points" (BPs), roundish or elongated bright features located in the intergranular dark lanes Dunn & Zirker (1973), Mehltretter (1974), and Title *et al.* (1987).

So far, the areas chosen for the study of BPs were active regions or plage because with prior resolution, the plethora of BPs was visible only there. The achieved resolution of the New Solar Telescope (NST) at Big Bear Solar Observatory (BBSO) revealed to us large number of BPs structures in the QS. The smallest observed BPs were around 0.1″ leaving their substructure still a puzzling subject. Considering the recent result that BPs are possible source of the acoustic oscillations (Andic *et al.* 2010) the importance of their study increases.This work presents a details that might contribute to the answer of the possible substructure of the BP.

2. Data

The data set used here was obtained on 29 July 2009 using the 1.6 m clear aperture New Solar Telescope (NST) at Big Bear Solar Observatory (Goode *et al.* 2010). The optical setup at the Nasmyth focus of the NST contained as its main components a TiO broadband filter centered at 705.68 nm and a PCO.2000 camera. Both are described in detail in Andic *et al.* (2010). We observed the quietest Sun at the center of the solar disk.

The data sequence consists of 120 bursts with 100 images in each burst. Individual frames have an exposure time of 10 ms and the cadence, between bursts, is 15 s. The images have a sampling of 0.037″ per pixel.

The images were speckle reconstructed based on the speckle masking method of von der Lühe (1993) using the procedure and code described in Wöger *et al.* (2008). The cadence of our data provided us with a Nyquist frequency of 67 mHz. After speckle reconstruction, the images were aligned using a Fourier routine. This routine uses cross-correlation techniques and squared mean absolute deviations to provide sub-pixel alignment accuracy. However, we did not implement sub-pixel image shifting to avoid the substantial interpolation errors that sometimes accompany the use of this technique.

To accurately measure the size of the structures, we first located the maximum intensity of the structure in time and space within our time series and used full-width at half maximum (FWHM) of a spatial profile that contained that point. Due to the irregularity in the shape of the structures, we usually chose the profile that encompasses the longest dimension of the structure.

3. Results

We observed the quiet Sun area during the deep solar minimum. In this area we found the small scale structures that are typical for the active regions (Fig. 1).

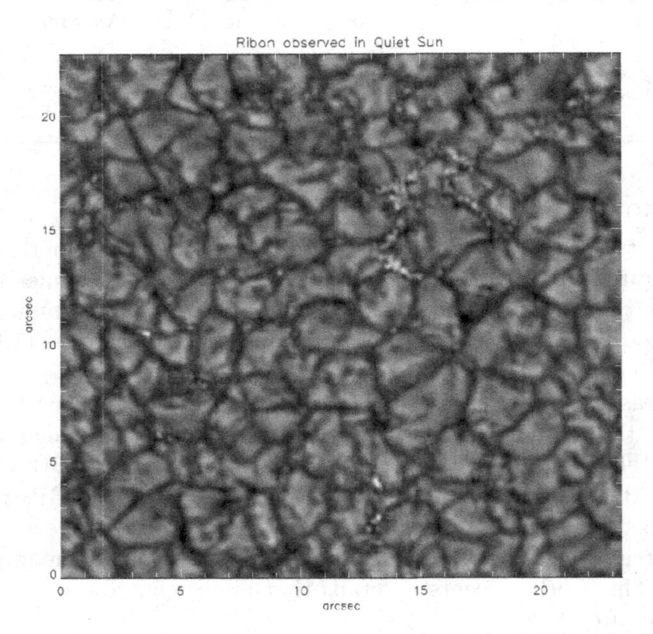

Figure 1. The image shows a formation consisting of individual BPs which were previously seen in plague regions. This formation is called a ribbon. This ribbon is observed in the quietest sun, that is in the quiet sun area during the deep solar minimum. Our resolution also made it possible to see the ribbon is composed of the individual BPs, which agrees with previous results.

The BPs analyzed formed the structures called clusters by Viticchie *et al.* (2009) terminology. The BPs tended to form persistent groups that stayed on the same location for duration of our time series. BPs in those groups tend to be attracted to each other until they reach a proximity that we cannot resolve, so they form a single structure, a cluster. An example of the process is visible at Fig. 2 where we can see in the left panel one large elongated BP, while right panel shows the same structure 90 sec later after structure split into 2 separate BPs.

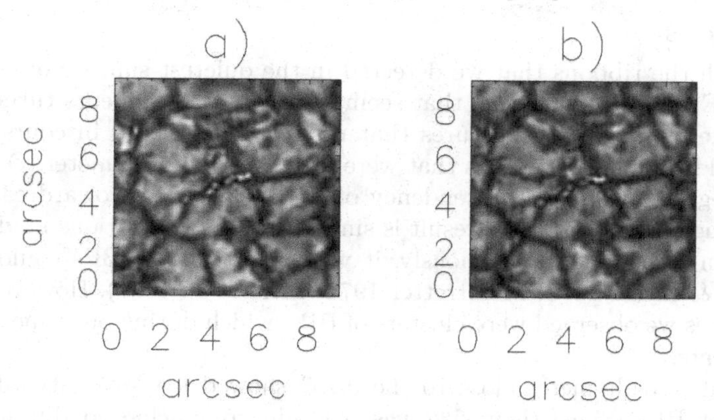

Figure 2. The image shows two different stages in the BPs evolution. Left panel shows 6 BPs in one lane, all individually resolved. The right panel shows the situation around 1 min later where 3 of BP s are merged in one structure. This is typical behavior when we have, at least, 2 BPs close together.

The larger clusters consisted of 3 BPs that were individually resolved at some point during our observations. Those clusters had the elongated shapes with the longest dimensions over 1″ in length. Most of the observed elongated shapes were clusters of BPs that during our time series unveiled their components, consisting of 2 or 3 separate BPs. Size of those clusters is varying from 0.48″ to 1.2″. The lifetime of the individual clusters varied from 0.7 min to 30 min.

The average size of the observed BPs was 0.22″, within a range from 0.13″ for the smallest observed structure to 0.48″ for the largest individual structure (Fig. 3). Right panel of Fig. 3 shows one of the smallest observed structure. Another typicality is that those structures tended to have rather low intensity. Considering the facts that their size was at the limit of our achieved diffraction limit and their low intensity, we can speculate that those structures are smaller than the resolution power of our telescope.

Their average lifetime was 4.28 min, in the range of our imposed minimum of 45 s to the longest lifetime of 29.75 min. There is no clear relationship between the size and the lifetime of the analyzed BPs.

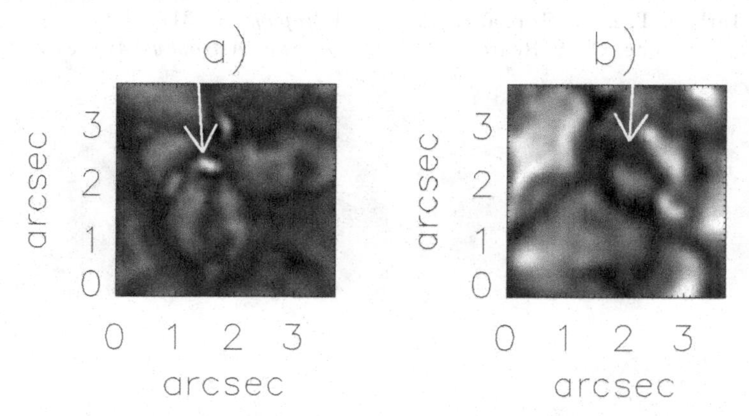

Figure 3. The largest and the smallest registered bright point are shown. Panel a) shows one of the large BP-like structures (arrow points to it) across the length of it measures 0.9″. Panel b) shows the smallest BP we detected. It's diameter measures 0.13″, at the limit of our achieved resolution indicating the possibility that the true structure might be even smaller.

4. Conclusions

Starting with the ribbons that we detected in the quietest sun, we observed constant substructure. Even individual BPs that seem as a compact elements turned to be composed from one or more substructures that revealed themselves in course of the time. This happened for most of the BPs that were at least 0.5" in diameter. Our observation of BPs in the groups showed the tendency of the BPs to drift toward each other until they form a single structure. This result is similar to the observations made by Viticchie *et al.* (2009) in G-band line. Previously, it was reported that BPs could have various shapes (Dunn & Zirker 1973, Mehltretter 1974, Title *et al.* 1987). However, most of the elongated shapes we observed were clusters of BPs, which during our time series unveiled their components.

All BPs that had diameter close to the 0.13" showed low intensity when compared with the larger BPs. Since their size was matching our achieved diffraction limit we might speculate that in reality they are much smaller. And that their intensity is low because it fills our resolution element.

To complement this result a signature of the flux-tubes collision inside a single bright point that did not separate itself has to be mentioned. Details concerning this finding are presented at the poster by A. Andic presented at this same conference. All this results encourage us to speculate that BPs, as we observe them now, are composed of the multiple flux tubes.

P.R.Goode and NST are supported by following grands: NSF (AGS-0745744), NASA (NNX08BA22G) and AFOSR (FA9550-09-1-0655).

References

Andic, A., Goode, P. R., Chae, J., Cao, W., Ahn, K., Yurchyshyn, V., & Abramenko, V. 2010, *Astrophys. J.*, 717L, 79

Andic, A. 2010, *IAU 273 proceedings*, poster 0036.

Dunn, R. B., & Zirker, J. B. 1973 *Solar Phys.*, 33, 281

Goode, P. R., Yurchyshyn, V., Cao, W., Abramenko. V., Andic, A., Ahn, K., & Chae, J. 2010, *Astrophys. J.*, 714, L31

von der Lühe, O. 1993 *Astron. Astrophys*, 268, 347

Mehltretter, J. P. 1974 *Solar Phys.*, 38, 43

Title, A. M., Tarbell, T. D., & Topka, K. P. 1987 *Astrophys. J.*, 317, 892

Wöger, F., von der Lühe, O., & Reardon, K. 2008, *Astron. Astrophys*, 488, 375

The Physics of Sun and Star Spots
Proceedings IAU Symposium No. 273, 2010
D.P. Choudhary & K.G. Strassmeier, eds.

doi:10.1017/S1743921311015511

Two types of coronal bright points their characteristics, and evolution

Isroil Sattarov[1], Nina V. Karachik[2], Chori T. Sherdanov[3], Azlarxon M. Tillaboev[1], and Alexei A. Pevtsov[2]

[1]Tashkent State Pedagogical University, 103 Yusif Kxos Kxojib str.,
Tashkent 100070, Uzbekistan
email: isattar@astrin.uz

[2]National Solar Observatory, Sunspot, NM 88349, USA
email: apevtsov@nso.edu

[3]Astronomical Institute of the Academy of Sciences of Uzbekistan
email: chori@astrin.uz

Abstract. Using maximum brightness of coronal bright point's (CBP) as a criterion, we separate them on two categories: dim CBPs, associated with areas of a quiet Sun, and bright CBPs, associated with an active Sun. This study reports on characteristics of two types of CBPs and their evolution.

Keywords. Solar corona, coronal bright points

1. Introduction and Data

Coronal Bright Points (CBPs, also referred to as X-ray bright points, XBPs) are compact coronal brightenings observed in X-ray, EUV, and radiowaves. Several authors had reported on existence of two types of XBPs: XBPs of quiet Sun, uniformly distributed over the solar surface (both temporally and spatially) and the XBPs with non-statistical variations in the longitudinal distributions e.g. Golub, Krieger, Vaiana (1975); Sattarov, *et al.* (2002, 2005a,b); McIntosh & Gurman (2005). Sattarov (2007) have found that the number of high latitude CBPs shows negative correlation with sunspot cycle, whereas number of CBPs in active region belts correlates positively with the cycle. This finding suggests a more complicated relationship between the CBP number and the solar activity cycle than had been previously thought. In this report we separate CBPs on two types using their brightness: "dim" CBPs are connected to quiet Sun regions and "bright" CBPs related to the active Sun. We investigate the characteristics of the two types of CBPs and their evolution, including the latitudinal distribution, temporal variation of CBP's maximal intensity, area, and the lifetime. We use full disk images observed by the Extreme-ultraviolet Imaging Telescope (EIT) on board of SOHO. We utilize EIT full disk synoptic data with spatial resolution of 2.64 arcsec per pixel and six hours cadence observed (full disk mode) in 195 Å from 1996–2008. For study the evolution of CBPs, we use EIT data obtained in "CME watch" mode with much higher time cadence. The data are calibrated following the standard EIT data reduction routine. To identify the coronal bright points, we employ the automatic procedure developed by us see (Sattarov, *et al.* 2010).

2. Temporal and latitudinal distribution of CBPs

If the CBPs are related to magnetic fields in the photosphere, one can make a reasonable assumption that the intensity of CBP must correlate with the magnetic flux. As is

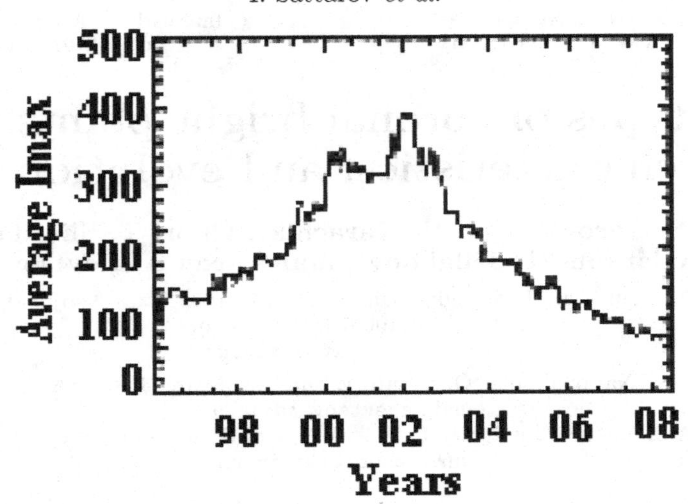

Figure 1. Cycle variation of average maximal intensity (Imax) of CBP for 1996-2008.

well-known, the magnetic flux on solar surface changes with solar activity, and therefore the increased level of solar activity would stimulate rise of maximal intensity of CBP and its decrease would followed by a decline of CBP brightness. Figure 1 presents cycle variation of average maximal intensity of CPBs during solar Cycle 23, which indicates that CBPs are (on average) brighter during maximum of solar activity cycle as compared with periods of solar minimum. At solar minimum in 1996, the average Imax intensity was about 150 DNs.Thus, we take it as a threshold between the "dim" and "bright" CBPs. In accordance with this criteria at solar minimum (1996) a majority of CBPs is "dim" ones. In 1997, number of bright CBPs starts rise quickly, while number of "dim" CBPs decreases. Latitudinal distribution of "dim" CBPs for 1996 (see Fig.6 in Sattarov, *et al.* (2010)) follows cosine function . The latitudinal distribution of bright CBPs (See Fig. 7 in Sattarov, *et al.* (2010)) for 1998 and 2002 exhibits two humps similar to latitudinal distribution of sunspots. However, in these years of high solar activity, both types of CBPs are present; bright CBP being predominant in active region belts, and dim CBPs are more abundant near the equator.

3. Evolution of coronal bright points

To study temporal evolution of CBPs we use EIT data obtained in "CME Watch" mode on 26 September 1997 (start of cycle 23) and 21 December 2002 (second maximum of the cycle). We find that CBPs have different area at maximum of its brightness, from 10 up to 200 pixels, and we study evolution of CBPs of different area (Figure 2).

On Fig.2 one can see the evolutions of two types CBP have small difference by area, but amplitude variation of intensity differs. The difference is small for bright CBP and large for "dim" ones. Fig.3 presents the temporal variations in intensity and area of small CBP of quiet Sun type. A small flare-like brightening had occurred at the beginning of CBP development (see peak in intensity in Figure 3, right). After the intensity spike, CBP area remains glaring during 5-10 hour. CBPs of larger size also show flare-like changes in intensity (see examples in Figure 2). Area of CBP shown in Figure 2 is 8 pixels or circle with radius 4 arc sec or 16 MHS. It represents an example of small, quiet Sun CBP during solar minimum of activity. Based on the examples shown in Figure 2, we speculate that

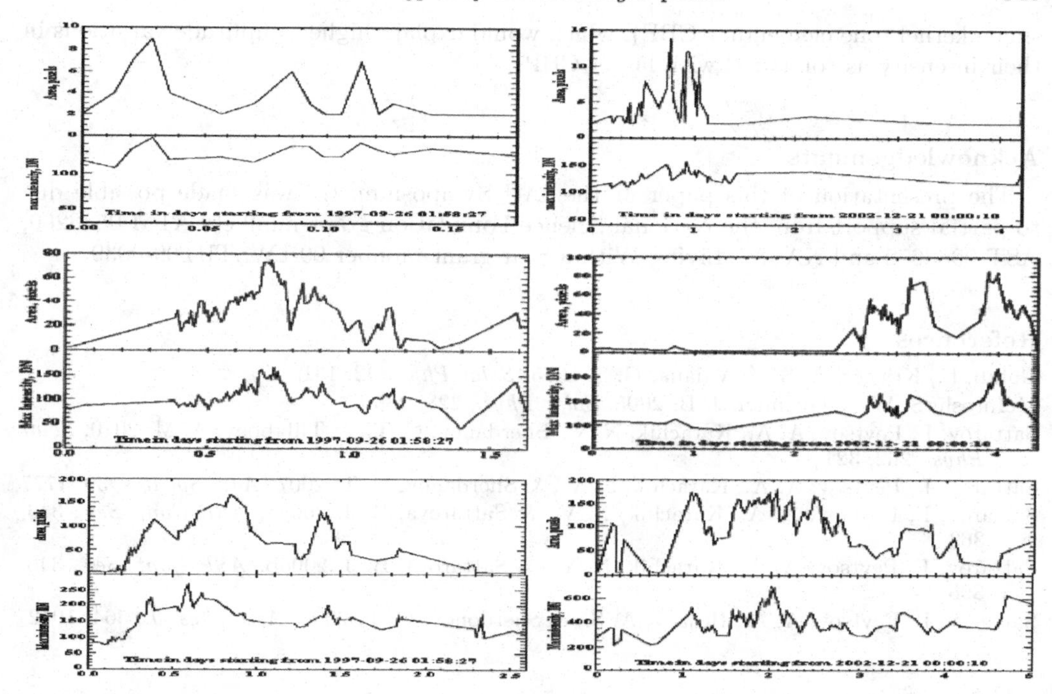

Figure 2. Examples of temporal variation an area (upper panel for each example) and intensity (lower panel for each example) of CBPs of different size and maximum intensity for periods of low (left column) and high (right) levels of solar activity.

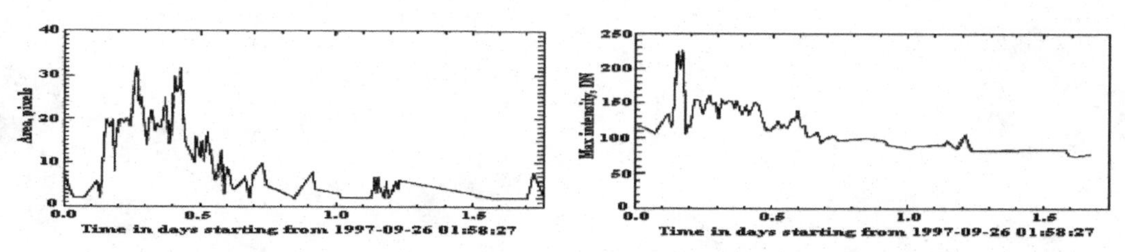

Figure 3. Temporal variation an area and intensity for CBP on September 26 1997.

large CBPs may be comprised of smaller CBPs that exhibit flare-like activity similar to example shown in Figure 3.

4. Discussion and Conclusions

We find that the active and quiet Sun coronal bright points exhibit flare-like variation in their maximum intensity. The amplitude of intensity variations is higher for small CBPs and lower for larger ones. Small quiet Sun CBPs may show repeated flare-like spikes with amplitude gradually decreasing during half a day period. This flaring slowly increases background around the CBP. Large CBPs also show such repeated flaring up albeit with smaller amplitude as compared with small CBPs. The repeat flaring in large CBPs may be related to elementary CBPs that collectively form large CBP. This our finding is in agreement with early studies, which found that X-ray bright point may consist of several kernels that flare-up one after the other. Small CBPs may consist of a single kernel (one elementary CBP), which would explain higher amplitude variations in their intensity as compared with larger CBPs.

Acknowledgements

The presentation of this paper at the IAU Symposium 273 was made possible due to partial support from the National Science Foundation grant numbers ATM 0548260, AST 0968672 and NASA - Living With a Star grant number 09-LWSTRT09-0039.

References

Golub, L., Krieger, A. S., & Vaiana, G. S. 1975, *Solar Phys.*, 42, 131

McIntosh, S. W. & Gurman, J. B. 2005, *Solar Phys.*, 228, 285

Sattarov, I., Pevtsov, A. A., Karachik, N. V., Sherdanov, C. T., & Tillaboev, A. M. 2010, *Solar Phys.*, 262, 321

Sattarov, I., Pevtsov, A. A., Karachik, N. V., & Sherdanov, C. T. 2007, *Adv. Sp. Res.* 39, 1777

Sattarov, I., Pevtsov, A. A., Karachik, N. V., & Sattarova, B. J. 2005a, *ASP Conf. Ser.*, 346, 363

Sattarov, I., Pevtsov, A. A., Karachik, N. V., & Sattarova, B. J. 2005b, *ASP Conf. Ser.*, 346, 395

Sattarov, I., Pevtsov, A. A., Hojaev, A. S., & Sherdonov, C. T. 2002, *Astrophys. J.*, 564, 1042

The Physics of Sun and Star Spots
Proceedings IAU Symposium No. 273, 2010
D.P. Choudhary & K.G. Strassmeier, eds.

doi:10.1017/S1743921311015523

Distribution of magnetic shear angle in an emerging flux region

Sanjay Gosain

Udaipur Solar Observatory, Physical Research Laboratory,
P. Box No. 198, Udaipur 313001, Rajasthan, India
email: sgosain@prl.res.in

Abstract. We study the distribution of magnetic shear in an emerging flux region using the high-resolution Hinode/SOT SP observations. The distribution of mean magnetic shear angle across the active region shows large values near region of flux emergence i.e., in the middle of existing bipolar region and decreases while approaching the periphery of the active region.

Keywords. Sun: active regions, sunspot: flux emergence

1. Introduction

The non-potentiality of the magnetic field in solar active regions provides the free-energy needed to fuel the energetic events like solar flares and CMEs. The non-potentiality may result either due to the shearing motions of the footpoint or due to the emergence of pre-stressed magnetic flux from the convection zone into the photosphere. The study of magnetic flux emergence in active regions is therefore very important for understanding the flaring activity. The high-quality vector magnetograms derived from spectro-polarimetric (SP) observations by space-based Hinode Solar Optical Telescope (SOT) (Tsuneta *et al.* 2008) are very useful for such studies. Although, the cadence of such SP observations is not fast enough to capture the dynamic evolution of magnetic field. One can, however, study the spatial distribution of magnetic non-potentiality in such active regions.

One of the important parameters used to characterize the non-potentiality of the magnetic field is the magnetic shear angle (Hagyard *et al.* 1984). The magnetic shear angle $\Delta\psi$ is defined as the angle between the azimuths of the observed and the potential field i.e., $\Delta\psi = \psi_o - \psi_p$. The larger is the magnetic shear angle (henceforth MSA) more non-potential is the magnetic field.

The distribution of MSA in an active region, which is going through flux emergence process, is useful in order to understand the buildup of non-potentiality in active regions. In the present paper we present results from such a study.

2. Observations and analysis methods

The magnetic flux was emerging in the middle of an existing bipolar region NOAA 10978, during 13 December 2007. Several thin elongated fibril structures were seen in the region of flux emergence. These emerging flux tubes displace the photospheric plasma, causing an appearance of diverging flow pattern. As the flux emergence progresses the

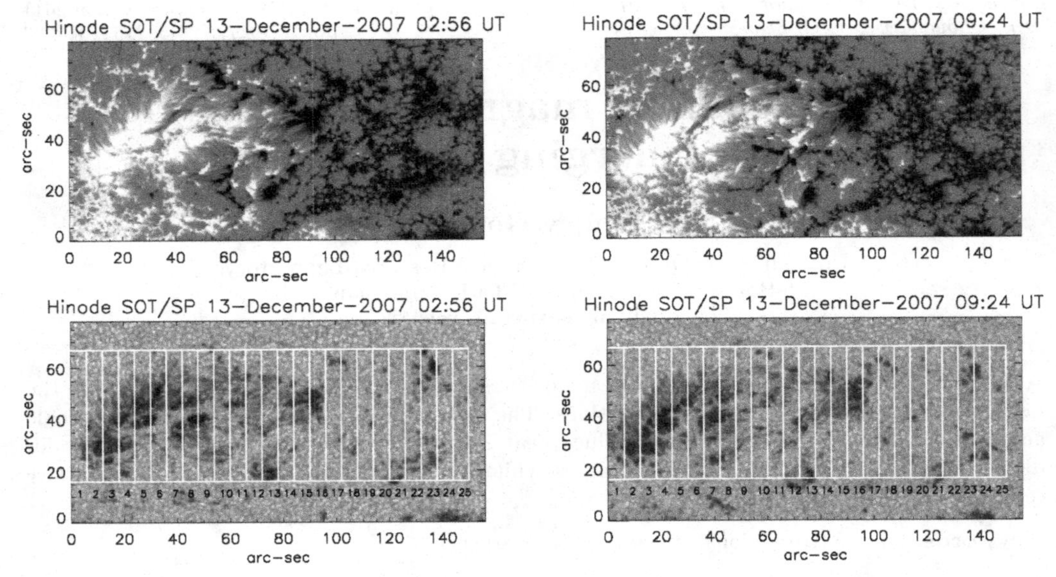

Figure 1. The top panels show the longitudinal field map of the emerging flux region in NOAA 10978 during 13 December 2007. The bottom panels show the continuum intensity maps of this region. The distribution of magnetic shear across the region is estimated by calculating the mean shear angle inside rectangular boxes labeled 1 through 25. The variation of magnetic shear angle across the region from boxes 1 to 25 is shown in figure 2.

foot points of the fibrils move apart and become more and more vertical and bundle together to form sunspots.

The spectropolarimetric observations of this active region were obtained by the spectropolarimeter instrument (Ichimoto *et al.* 2008) onboard Hinode space mission (Kosugi *et al.* 2007). The magnetic field vector was derived by using a Stokes inversion code called MERLIN (Lites *et al.* 2007). The code performs a non-linear least squares fit of the observed Stokes profiles with theoretical Stokes profiles computed under the Milne-Eddington model atmosphere assumptions.

The top panel of Figure 1 shows the longitudinal magnetic field map of this active region at two different stages during its evolution. The small bipoles near the polarity inversion line (PIL), with opposite polarity orientation as compared to the overall active region polarity suggests the presence of undulatory U-loops. The bottom panel of Figure 1 shows the continuum intensity maps of the active region. The typical features of an emerging flux region, like elongated thin fibril like structures, can be clearly noticed. We evaluate the variation of shear angle across the active region by computing the mean shear angle within each of the rectangular box labeled 1–25. The boxes are distributed along the bipolar axis of the active region so as to get spatial profile of shear across the active region. We remove the tilt of the active region by rotating the map so that the bipolar axis is horizontal. The bipolar axis is computed by joining the centroid of the longitudinal flux in either polarity of the active region. The shear angle is computed for each pixel of the map and is averaged inside each of the box 1–25. The plot in Figure 2 shows the distribution of mean magnetic shear angle across the active region.

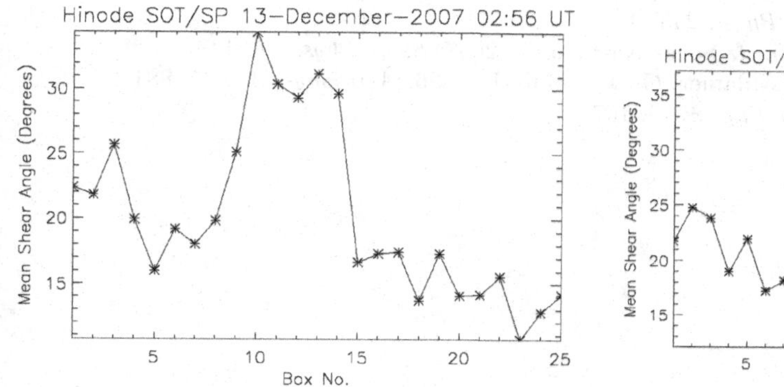

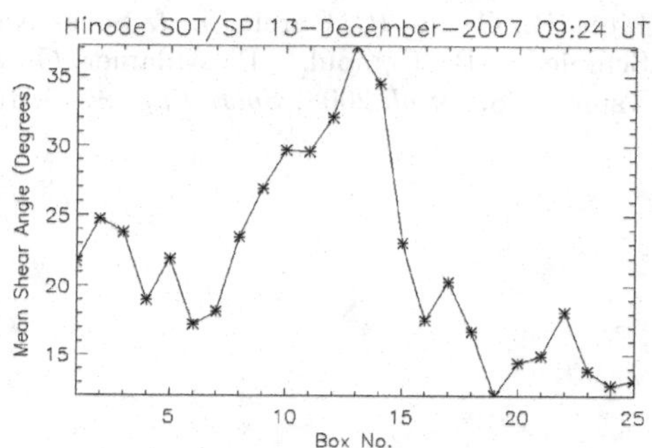

Figure 2. The left and right panels show the distribution of magnetic shear angle across the active region at two different times during 13 December 2007. The abscissa shows the box number and the ordinate shows the mean shear angle inside the corresponding box.

3. Results and discussions

It can be seen from Figure 2 that the mean shear angle is not uniformly distributed across the active region. The mean shear angle is higher in the central portion of the active region where the flux tubes emerge. The peak value of the shear angle in the central portion is about 30–35 degrees, while at the periphery it is about 25 degrees. The pattern persists during the two time intervals which are separated by about six hours as shown in Figure 1.

These observations show that the magnetic field in the middle portion of the emerging flux region consists of the largest amount of non-potentiality. The present results obtained with the high-resolution and high-sensitive polarimetry put shear distribution studies on a firm footing. Similar results were obtained by Schmieder *et al.* (1996) for NOAA 6718 where the comparison of coronal field structure with linear force-free field computation yielded a differential magnetic field shear model. The present observations support their argument that the decreased shear in the outer portions of the active region is probably due to continual relaxation of the magnetic field to lower energy state in the older portions of the active region. In future we plan to carry out a similar analysis of magnetic shear in decaying or dormant active regions to see how different the pattern appears as compared to the emerging flux regions.

Acknowledgements

The presentation of this paper in the IAU Symposium 273 was possible due to financial support from the National Science Foundation grant numbers ATM 0548260, AST 0968672 and NASA - Living With a Star grant number 09-LWSTRT09-0039. Hinode is a Japanese mission developed and launched by ISAS/JAXA, with NAOJ as domestic partner and NASA and STFC (UK) as international partners. It is operated by these agencies in co-operation with ESA and NSC (Norway).

References

Hagyard, M. J., Teuber, D., West, E. A., & Smith, J. B. 1984, *Solar Phys.*, 91, 115
Ichimoto, K., *et al.* 2008, *Solar Phys.*, 249, 233

Kosugi, T., *et al.* 2007, *Solar Phys.*, 243, 3
Lites, B., Casini, R., Garcia, J., & Socas-Navarro, H. 2007, *Solar Phys.*, 78, 148.
Schmieder, B., Demoulin, P., Aulanier, G., & Golub, L. 1996, *Astrophys. J.*, 467, 881
Tsuneta, S., *et al.* 2008, *Solar Phys.*, 249, 167

The Physics of Sun and Star Spots
Proceedings IAU Symposium No. 273, 2010
D.P. Choudhary & K.G. Strassmeier, eds.

doi:10.1017/S1743921311015535

Damping rates of p-modes by an ensemble of randomly distributed thin magnetic flux tubes

Andrew Gascoyne and Rekha Jain

Department of Applied Mathematics, University of Sheffield, UK
email: app07adg@sheffield.ac.uk, R.Jain@sheffield.ac.uk

Abstract. The magnetohydrodynamic (MHD) sausage tube waves are excited in the magnetic flux tubes by p-mode forcing. These tube waves thus carry energy away from the p-mode cavity which results in the deficit of incident p-mode energy. We calculate the loss of incident p-mode energy as a damping rate of f- and p-modes. We calculate the damping rates of f- and p-modes by a model Sun consisting of an ensemble of many thin magnetic flux tubes with varying plasma properties and distributions. Each magnetic flux tube is modelled as axisymmetric, vertically oriented and untwisted. We find that the magnitude and the form of the damping rates are sensitive to the plasma-β of the tubes and the upper boundary condition used.

Keywords. Sun: magnetic fields, Sun: helioseismology, Sun: oscillations, MHD, Scattering

1. Introduction

Solar acoustic modes (p-modes) can get absorbed and scattered when travelling through active regions such as sunspots or plages. This scattering of p-modes has been well observed using various helioseismic techniques (see Braun *et al.* 1988; Couvidat *et al.* 2006). Braun & Birch (2008) found that sunspots can absorb over half of incident p-mode power and plage regions can absorb around 20%. Also, it is found that the absorption is frequency dependent. Thus, the importance of sub-surface field structure in modifying the properties of f- and p-modes has been realised and many theoretical investigations have concentrated on understanding the physical mechanism responsible for this absorption (e.g., Spruit 1991; Bogdan & Cally 1995; Bogdan *et al.* 1996; Crouch & Cally 2005; Hindman & Jain 2008; Jain *et al.* 2009).

In this paper we wish to study the combined effects of the upper boundary condition and the tube parameter plasma-β on the damping rates of p-modes. We construct a model consistent with Hindman & Jain (2008) in order to calculate damping rates produced by the excitation of sausage tube waves across the whole Sun. In Hindman & Jain (2008) they used an idealised model where by each tube of the ensemble was identical; this need not be the case. Thus we use a random ensemble of thin, axisymmteric, vertical magnetic flux tubes with varying plasma-β.

2. Equilibrium configuration

We model the nonmagnetic medium as a polytropic atmosphere with gravity acting downwards $\mathbf{g} = -g\hat{\mathbf{z}}$; z is increasing upwards. The pressure, density and sound speed vary as power laws (see Bogdan *et al.* 1996 for details). Following Bogdan *et al.* (1996), Hindman & Jain (2008) and Jain *et al.* (2009) we set the truncation depth at $z = -z_0$

which is our model photosphere where the quantities ρ_0 and P_0 are the values of the mass density and gas pressure at this depth. Above the truncation depth $z > -z_0$ we assume the existence of a hot vacuum ($\rho_{ext} \to 0$ with temperature $T_{ext} \to \infty$). The external pressure, density and sound speed increases with depth and as our atmosphere is in convective equilibrium, the polytropic index a is related to the ratio of specific heats γ via $a = 1/(\gamma - 1)$; which we set to $a = 1.5$. We set the following characteristic physical parameters for our model photosphere, $\rho_0 = 2.78 \times 10^{-7}$ g cm^{-3}, $P_0 = 1.21 \times 10^5$ g cm^{-1} s^{-1}, and $g = 2.775 \times 10^4$ cm s^{-1}, which coincide with the photospheric reference model of Maltby *et al.* (1986). The choice of polytropic index $a = 1.5$ yields the truncation depth $z_0 = 392$ km and thus the sound speed at the surface is 8.52 km s^{-1}.

We also assume that the magnetic fibrils threading the atmosphere are untwisted, straight, vertically aligned, thin tubes with a circular cross-section. For these thin magnetic flux tubes, the lateral variation of the magnetic field across the tube is ignored and the plasma β is constant with depth (see Bogdan *et al.* 1996 and Hindman & Jain 2008 for details).

3. The governing wave equations

The incident acoustic wavefield can be expressed as a single partial differential equation for the displacement potential Φ,

$$\frac{\partial^2 \Phi}{\partial t^2} = c^2 \nabla^2 \Phi - g \frac{\partial \Phi}{\partial z}. \tag{3.1}$$

This equation supports plane wave solutions of the form,

$$\Phi(\boldsymbol{x}, t) = \mathcal{A} e^{i(kx - \omega t)} Q(z), \quad Q(z) = w^{-(\mu + 1/2)} W_{\kappa, \mu}(w), \tag{3.2}$$

where $\mathcal{A}$ is the complex wave amplitude, ω the temporal frequency, and k the wavenumber. $Q(z)$ is the vertical eigenfunction proportional to Whittaker's W function (Abramowitz & Stegun 1964), with $\mu = (a+1)/2$, $\nu^2 = a\omega^2 z_0/g$ and $\kappa = \nu^2/(2kz_0)$. We calculate the eigenvalues and eigenfunctions for the truncated polytrope by requiring that the Lagrangian pressure perturbation vanishes at the truncation depth, i.e., $\nabla \cdot \xi = 0$. Mathematically this takes the form $W_{\kappa, \mu+1}(2kz_0) = 0$.

We are seeking waves on thin magnetic tubes that lack internal structure, so only three types of MHD waves can possibly satisfy this criterion: torsional Alfvén waves, longitudinal (sausage) waves and transverse (kink) waves. We ignore torsional Alfvén waves as our driving force, the p-mode oscillations, are irrotational in this model and therefore, will not generate these types of waves. In this paper we will also ignore the kink waves and specifically concentrate on the $m = 0$ mode (m being the azimuthal order). Using the formulation of Jain *et al.* (2009) (see also Hindman & Jain 2008 and Bogdan *et al.* 1996), the fluid displacement due to the excitation of sausage waves within the tube can be described by the following equation,

$$\left(\frac{\partial^2}{\partial t^2} + \frac{2gz}{2a + \beta(1+a)} \frac{\partial^2}{\partial z^2} + \frac{g(1+a)}{2a + \beta(1+a)} \frac{\partial}{\partial z}\right) \xi_\parallel = \frac{(1+a)(\beta+1)}{2a + \beta(1+a)} \frac{\partial^3 \Phi}{\partial z \partial t^2}. \tag{3.3}$$

These sausage waves $\xi_\parallel(z, t)$ can be described as axisymmetric pressure pulses that produce displacements that are primarily parallel to the magnetic field. The p-mode driver term appears on the right hand side of equation (3.3) represented by the displacement potential Φ.

4. Damping rates for a random ensemble of tubes

In order to calculate the damping rates Γ for the excitation of sausage waves we follow Hindman & Jain (2008). Thus we derive for a single flux tube, the energy flux of sausage tube waves that escape the p-mode cavity as

$$E = -\frac{\gamma\beta}{2+\gamma\beta}\frac{\pi g\rho_0\omega A_0}{4(a+1)(\beta+1)}\frac{|\mathcal{A}|^2}{z_0^2}(|\Omega+\mathcal{I}^*|^2+|\mathcal{I}|^2-|\Omega|^2+\mathcal{S}). \quad (4.1)$$

Since we are interested in calculating the energy flux produced over the entire Sun, the energy flux produced by a single flux tube is not enough. Instead we need to calculate the total energy flux E_{tot} produced by an ensemble of flux tubes. Thus E_{tot} is simply the energy flux E for a single tube multiplied by the number of tubes N on the solar surface. The number of flux tubes depends only on the surface cross-sectional area of the tubes A_0 and the filling factor f, thus $N = 4\pi R_\odot^2 f/A_0$, where $R_\odot$ is the radius of the Sun. We then define the damping rate for the excitation of sausage waves as,

$$\begin{aligned}\Gamma &= \frac{1}{2\pi}\frac{E_{tot}}{E_n}\\ &= \frac{\beta}{4(\beta+1)\epsilon}\frac{\omega A_0}{4\pi R_\odot^2}\frac{\lambda^a}{\nu^4}\frac{(|\Omega+\mathcal{I}^*|^2+|\mathcal{I}|^2-|\Omega|^2+\mathcal{S})}{\mathcal{N}_n}N, \end{aligned} \quad (4.2)$$

where $\lambda = 2kz_0$, $E_n = 4\pi R_\odot^2 \frac{g\rho_0\nu^4|\mathcal{A}|^2}{4a\lambda^a z_0^2}\mathcal{N}_n$ is the energy contained in f- and p-mode oscillations, ($\mathcal{N}_n$ is an integral which can be found in Hindman & Jain 2008 eqn. 3.14).

The next step is to simulate the aggregate behavior of the Sun by considering a random distribution of flux tubes for a range of β values between 0 and 10. We shall use three different random distributions, a uniform distribution so all β values have equal probability. We shall also consider a normal (Gaussian) distribution of flux tubes and a exponential distribution such that the probability decreases as β increases. We are given the number of tubes required by the parameter N, which requires us to keep the cross-sectional A_0 of each tube the same. Thus β has the effect of changing the magnetic flux Θ of the tube.

5. Results and discussion

We now present the results for theoretically calculated damping rates of p-modes by the generation of magnetic tube waves in the convection zone through external p-mode buffeting. We calculated damping rates for two different upper boundary conditions (BC), the stress-free condition which essentially means that all propagating tube waves are reflected at the surface and the maximal-flux condition which means all upward propagating tube waves escape into the atmosphere; this BC gives an upper bound on the amount of energy escaping the p-mode cavity.

We consider a random distribution of magnetic fibrils with β values ranging from 0 to 10 as discussed in §4. We use three different random distributions, uniform, normal (Gaussian), and an exponential distribution and plot our results in figure 1. For comparison we have also over plotted the case $\beta = 5$ with dashed lines for each mode order. For

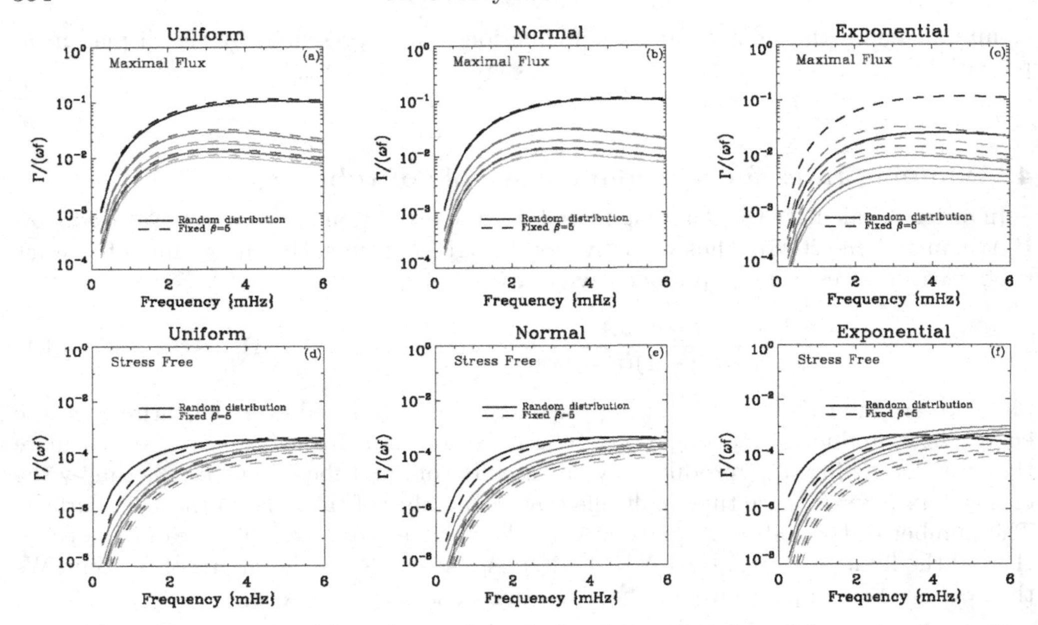

Figure 1. Damping rate of f- and p-modes calculated as a function of frequency for various random distributions in β, between 0-10. Over-plotted with dashed lines, is the case for $\beta = 5$. Panels (a)–(c) are for the maximal flux BC and panels (d)-(f) are for the stress free BC. Each mode order is denoted by a different colour: black (f), red (p1), green (p2), blue (p3) etc.

the maximal-flux case, the solid and dashed curves are similar for the uniform distribution and increasingly so for the normal distribution. For the exponential case the solid curve drops well below the dashed, this is due to the fact that this distribution dominates for lower β values. For the stress free condition there is less correlation with the fixed $\beta = 5$ case this is due to the fact that the damping rates depend nonlinearly on β.

6. Conclusion

In this paper we have presented a calculation for the damping rates for a random ensemble of vertical, axisymmetric, thin magnetic flux tubes. The mechanism used here is the generation of longitudinal (sausage) tube waves through buffeting of the magnetic fibril by external solar p-modes. The generation of these tube waves carry energy out of the p-mode cavity and thus result in the damping of p-modes. We have shown that there are a number of factors to consider when investigating the damping rates of p-modes by magnetic regions. In particular, the effect of the upper boundary condition and the distribution used to construct the random ensemble.

References

Abramowitz, M. & Stegun, I. A. 1964, *Handbook of Mathematical Functions (New York: Dover)*

Bogdan, T. J. & Cally, P. S. 1995, *Astrophys. J.*, 453, 919

Bogdan, T. J., Hindman, B. W., Cally, P. S., & Charbonneau, P. 1996, *Astrophys. J.*, 465, 406

Braun, D. C., Duvall, T. L., & LaBonte, B. J. 1988, *Astrophys. J.*, 335, 1015

Braun D. C., Birch, A. C. 2008, *Solar Phys.*, 251, 267

Couvidat, S., Birch, A. C., & Kosovichev, A. G. 2006, *Astrophys. J.*, 640, 516

Crouch, A. D. & Cally, P. S. 2005, *Solar Phys.*, 227, 1

Hindman, B. W. & Jain, R. 2008, *Astrophys. J.*, 677, 769

Jain R., Hindman, B., Braun, D. C., & Birch, A. C. 2009, *Astrophys. J.*, April 10 issue.

Maltby, P., Avrett, E. H., Carlsson, M., Kjelsdeth-Moe, O., Kurucz, R. L., & Loeser, R. 1986, *Astrophys. J.*, 306, 284

Spruit, H. C. 1991, in Toomre J., Gough D. O., eds, *Lecture Notes in Physics, Vol. 388, Challenges to Theories of the Structure of Moderate Mass Stars.* Springer-Verlag, Berlin, p. 121

The Physics of Sun and Star Spots
Proceedings IAU Symposium No. 273, 2010
D.P. Choudhary & K.G. Strassmeier, eds.

doi:10.1017/S1743921311015547

Subsurface flows associated with rotating sunspots

Kiran Jain, Rudolf Komm, Irene González Hernández, Sushant C. Tripathy, and Frank Hill

National Solar Observatory
950 N. Cherry Av., Tucson, AZ 85719,USA
email: kjain@noao.edu

Abstract. In this paper, we compare components of the horizontal flow below the solar surface in and around regions consisting of rotating and non-rotating sunspots. Our analysis suggests that there is a significant variation in both components of the horizontal flow at the beginning of sunspot rotation as compared to the non-rotating sunspot. The flows in surrounding areas are in most cases relatively small. However, there is a significant influence of the motion on flows in an area closest to the sunspot rotation.

Keywords. Sun : helioseismology, sun : interior, sun: sunspots

1. Introduction

Sunspots that rotate around their umbral centers are termed as rotating sunspots. In some cases, these are identified by the rotation around another sunspot within the same active region (AR). The identification of these sunspots has been made easier by advances in the spatial and temporal resolution of recent satellite-borne telescopes (e.g. Brown *et al.* 2003). The origin of rotational motion is believed to be due to the shear and twist in magnetic field lines or vice-versa. It is also suggested that the magnetic twist may result from large-scale flows in the solar convection zone and the photosphere or in sub-photospheric layers. In an earlier study, Zhao & Kosovichev (2003) found evidence for two opposite sub-photospheric vortical flows in the depth range of 0–12 Mm around a fast rotating sunspot in AR 9114. In this paper, we extend our earlier study of the flows beneath a rotating sunspot (Jain *et al.* 2010a) to the surrounding regions, and investigate how flow fields change with depth compared to a non-rotating case.

2. Technique

We apply the technique of ring diagrams to obtain the depth dependence beneath the solar surface (Hill 1988). In this technique, the high-degree waves in localized areas over the solar surface are used to infer the characteristics of these propagating waves. We use high-cadence continuous Dopplergrams obtained by the Global Oscillation Network Group (GONG). These images are processed through the GONG ring-diagram pipeline (Corbard *et al.* 2003) where we track and remap each region for 1680 min, and apply a three-dimensional FFT on both spatial and temporal direction to obtain a 3D power spectrum. The corresponding power spectrum is fitted using a Lorentzian profile model (Haber *et al.* 2000) that includes the perturbation term due to horizontal flow fields in the region. Finally, the obtained velocities are inverted using regularized least square method to obtain the depth dependence of the horizontal velocity flows.

3. Data and analysis

We study flows in photospheric layers beneath regions in and around AR 10930 and AR 10953. AR 10930 was observed during the declining phase of solar cycle 23 and located at E19S04 on 2006 December 10. It had two major sunspots; the big sunspot did not show any visible change during the period of disk passage while the small sunspot in the southern part exhibited rapid counterclockwise rotation about its umbral center. Since we do not have the resolution to discriminate between two sunspots, for our purpose, we will consider the group as a "rotating sunspot". The sunspot started to rotate after mid day on December 10 and continued to rotate until December 13. Since the ring-diagram inferences are affected by the location and size of the region,we compare flows in the region of rotating sunspot with those for an active region with non-rotating sunspots located at around the same location as AR 10930. For this purpose, we analyze AR 10953 that was located at S10E09 on 2007 May 1 with non-rotating sunspots. Further, both active regions are approximately of the same size.We use three consecutive time series to study x- and y-components of the horizontal flow; these are chosen in such a way that they represent the period of before, during and after the beginning of sunspot rotation in AR 10930. For comparison, we also use three time series of equal length for AR 10953.

We calculate flows in a mosaic of 9 regions where each region is $\sim 11^{\circ}\times 11^{\circ}$ in size. The central region in this mosaic has the active region at its center and is surrounded by eight regions spaced by 5° in each direction. The depth variations of x- and y-components of horizontal velocity in the mosaic of AR 10930 for all three time samples are shown in Figures 1 and 2. Note that the active region is located at the center of Panel (e). Since AR 10930 was a medium sized active region, the central patch covers an area larger than the actual sunspot group. It is clearly seen that the relative values of u_x and u_y in Panel (e) of both figures are significantly larger than in the neighboring regions which are relatively quiet. This is in agreement with earlier findings where flows in active regions are larger than their quieter surroundings (Komm *et al.* 2005).

It is evident from Figure 1(e) that the maximum variation in u_x with depth is obtained at the beginning of sunspot rotation while no significant variation is seen on the other two days. Here we focus on the depth range between 3 Mm and 12 Mm, where the errors are smaller due to the set of modes contained in the analysis combined with the limitations of the model used near the surface. Below 3 Mm, we find that u_x, at the beginning of sunspot rotation, increases rapidly in deeper layers. Although u_x has significantly decreased in surrounding regions, there is a significant influence of rotation on the profiles of u_x in the regions adjacent to the rotating part of the sunspot group. The values of u_x in Panels (d) and (h) are again larger at the beginning of sunspot rotation while profiles for all three cases did not change significantly in other neighboring regions.

On the other hand, even if profiles of u_y in Panel (e) of Figure 2 are similar for all three time samples, these exhibit variations in magnitude. It starts increasing below 6 Mm and the maximum value is again achieved at the beginning of sunspot rotation. Since the region is located in the southern hemisphere and the u_y shows a strong poleward trend, this significantly higher positive value of u_y during the initial stages of sunspot rotation indicates an equator-ward meridional flow. However, neighboring regions in this case are less influenced by the sunspot rotation. In contrast to the rotating sunspot region, the flows in a non-rotating sunspot region show less variation in both components. As an

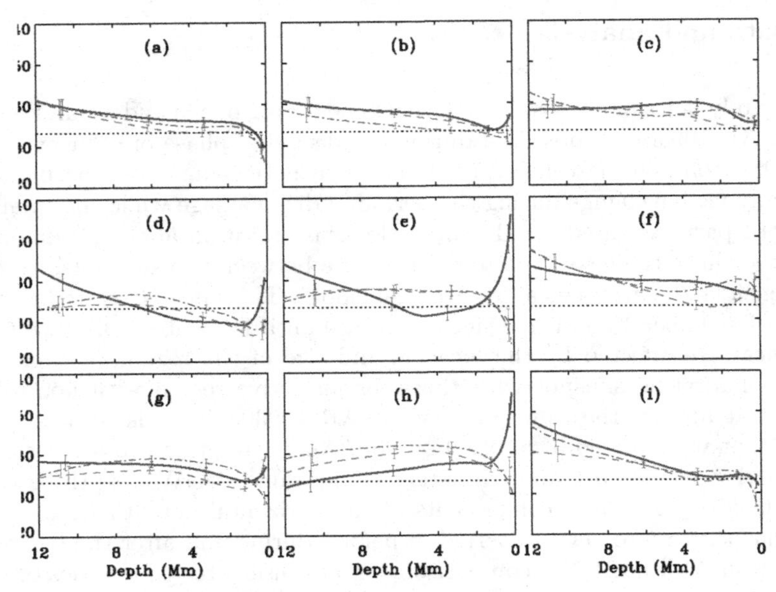

Figure 1. Depth variation of x-component of horizontal flows in and around AR10930 for three overlapping time series; green/dashed line represents the period before the beginning of rotation, blue/solid line includes the period when the rotation started, and red/dash-dot line is for after the rotation began. Error bars are shown for a few points. The active region is located at the center of Panel (e).

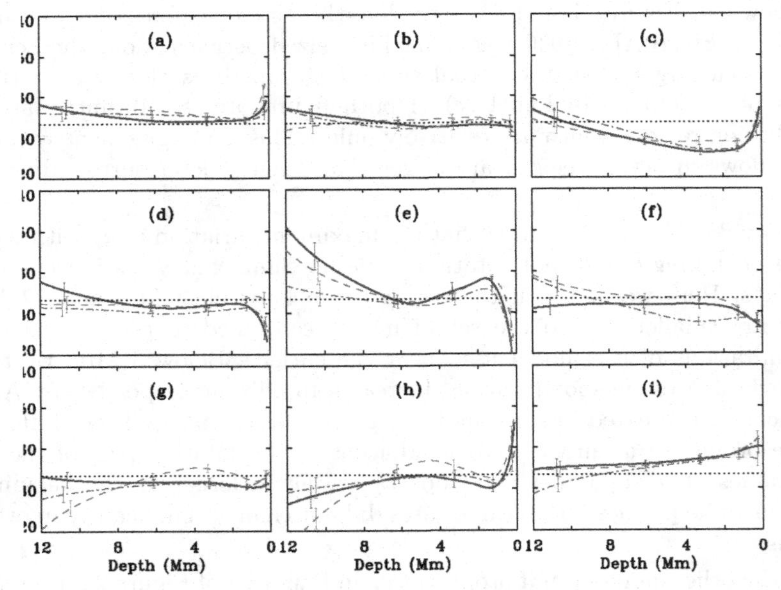

Figure 2. Same as Figure 1 but for the y-component of the horizontal flow.

example, we exhibit the depth variation of u_x and u_y beneath AR 10953 in Figure 3. The profiles for both components are similar for all three time samples. We also find that flows in surrounding area are comparable for all three time samples but with smaller magnitude.

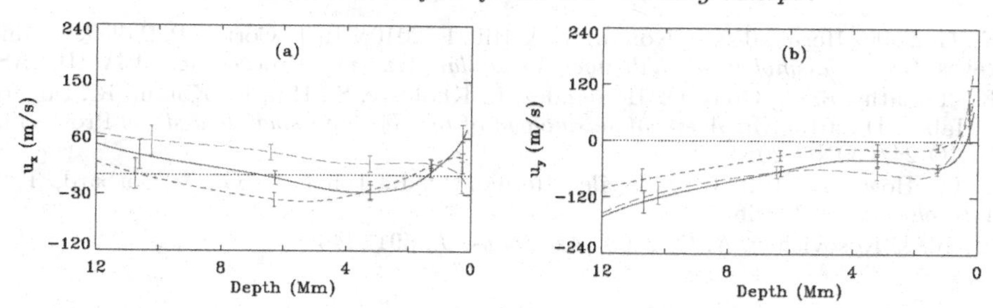

Figure 3. Depth variation of (*a*) x- and, (*b*) y-components of the horizontal velocity flow for AR10953 for three overlapping time series; green/long-dash, blue/solid, and purple/small-dash are for time samples 1, 2 and 3 respectively. The errors are shown only at selected depths.

4. Summary

We have compared the horizontal flow components in and around two sample cases of rotating and non-rotating sunspots. Our analysis suggests that both u_x and u_y show significant variation with depth during the course of sunspot rotation as compared to the non-rotating sunspot. The maximum change in u_x is found at the beginning of sunspot rotation. The flows in surrounding regions in both cases are relatively small, however there is a significant variation in regions neighboring to the rotating sunspot. In the present study, we have considered a region that includes both rotating and non-rotating sunspots, and the quiet area around them. This is mainly due to the limitation imposed by the resolution of images on the technique. However, the high-resolution images from Helioseismic Magnetic Imager (HMI) onboard *Solar Dynamics Observatory (SDO)* will allow us to study smaller regions within an active region that will provide a deeper insight on the dynamics of subsurface layers beneath individual sunspots, e.g. using HMI test Dopplergrams, Jain *et al.* (2010b) have shown that the ring-diagram technique can be reliably applied to regions as small as 5°.

Acknowledgements

This work utilizes data obtained by the Global Oscillation Network Group (GONG) project, managed by the National Solar Observatory, which is operated by AURA, Inc. under a cooperative agreement with the National Science Foundation. The data were acquired by instruments operated by the Big Bear Solar Observatory, High Altitude Observatory, Learmonth Solar Observatory, Udaipur Solar Observatory, Instituto de Astrofísica de Canarias, and Cerro Tololo Interamerican Observatory. This work was supported by NASA-GI grant NNG 08EI54I.

References

Brown, D. S., Nightingale, R. W., Alexander, D., Schrijver, C. J., Metcalf, T. R., Shine, R. A., Title, A. M., & Wolfson, C. J. 2003, *Solar Phys.*, 216, 79

Corbard, T., Toner, C., Hill, F., Hanna, K. D., Haber, D. A., Hindman, B. W., & Bogart, R. S. 2003, In H. Sawaya-Lacoste (ed.) *Local and global helioseismology: the present and future*, Proceedings of SOHO 12/GONG+ 2002 (ESA SP 517), p 255

Haber, D. A., Hindman, B. W., Toomre, J., Bogart, R. S., Thompson, M. J., & Hill, F. 2000, *Solar Phys.*, 192, 335

Hill, F. 1988, *Astrophys. J.*, 333, 996

Jain, K., González Hernández, I., Komm, R., & Hill, F. 2010a, In T. Cortés, P. Pallé & S. Jiménez Reyes. (eds.) *Seismological challenges for stellar structure*, Proceedings of IV HELAS

Jain, K., Tripathy, S. C., González Hernández, I., Kholikov, S., Hill, F., Komm, R., Bogart, R., & Haber, D. 2010b, In *A era of seismology of the Sun and solar-like stars*, Proceedings of SOHO 24/GONG 2010

Komm, R., Howe, R., Hill, F., González Hernández, I., Toner, C. G., & Corbard, T. 2005. *Astrophys. J.*, 631, 636

Zhao, J. W. & Kosovichev, A. G. 2003, *Astrophys. J.*, 591, 446

The Physics of the Sun and Star Spots
Proceedings IAU Symposium No. 273, 2010
D. P. Choudhary and K. Strassmeier

doi:10.1017/S1743921311015559

Studies of waves in sunspots using spectropolarimetric observations

Gordon A. MacDonald[1] and S. P. Rajaguru[2]

[1]California State University, Northridge,
18111 Nordhoff St., Northridge, CA 91330, USA
email: gordon.macdonald.31@my.csun.edu

[2]Indian Institute of Astrophysics,
II Block, Koramangala, Bangalore 560 034, INDIA
email: rajaguru@iiap.res.in

Abstract. We observe the acoustic velocity oscillations in and near active region NOAA 10960 on 8 June, 2007 using observations from the IBIS instrument at the Dunn Solar Telescope at NSO/Sacramento Peak and simultaneous Hinode BFI/SP data. Inversions were performed on the spectropolarimetric datasets in order to get magnetic field information for the AR. A time series of Doppler maps from line bisectors and Stokes V zero-crossing was constructed and allowed us to construct power maps for the AR. Past works by various authors have shown that acoustic power in the solar atmosphere is strongly influenced by magnetic field strength and inclination. Our study also explores this, but in addition, we also discuss the role of oscillations due to purely magnetized gas in the photosphere. Our preliminary results for this study are presented.

Keywords. Local helieseismology, active regions

---

1. Introduction

It was shown by Hindman & Brown (1998) that acoustic oscillations within sunspots are strongly influenced by magnetic field strength. Later, Nicolas *et al.* (2005) demonstrated that the amount of influence cannot be solely determined by the photospheric magnetic field strength and Schunker & Braun (2010) showed that the acoustic power is strongly influenced by magnetic field inclination. We attempt to demonstrate how the magnetic field of AR's change the oscillatory behavior of the solar atmosphere.

Our datum are observations of NOAA 10960 on 8 June, 2007 at S07W17 from the IBIS instrument located at the Dunn Solar Telescope at Sacramento Peak. The IBIS data observes simultaneous Stokes I, Q, U, V parameters in the magnetic line FeI 6173Å and also continuum intensity in the non-magnetic line FeI 7090Å. Our set has a cadence of 47.5 seconds and a total observation time of 7.5 hours.

Using these data, we find the time series of the line bisector and also the Stokes V zero crossing and extract the oscillation frequencies by performing a FFT. It will also be possible to obtain a time series of intensity variations using the CaII H images. The zero crossing Doppler shift will be of particular interest since it gives the Doppler speed of only the magnetized portion of the gas. Inversions were performed on the IBIS 6173Å line to extract information concerning the local magnetic field.

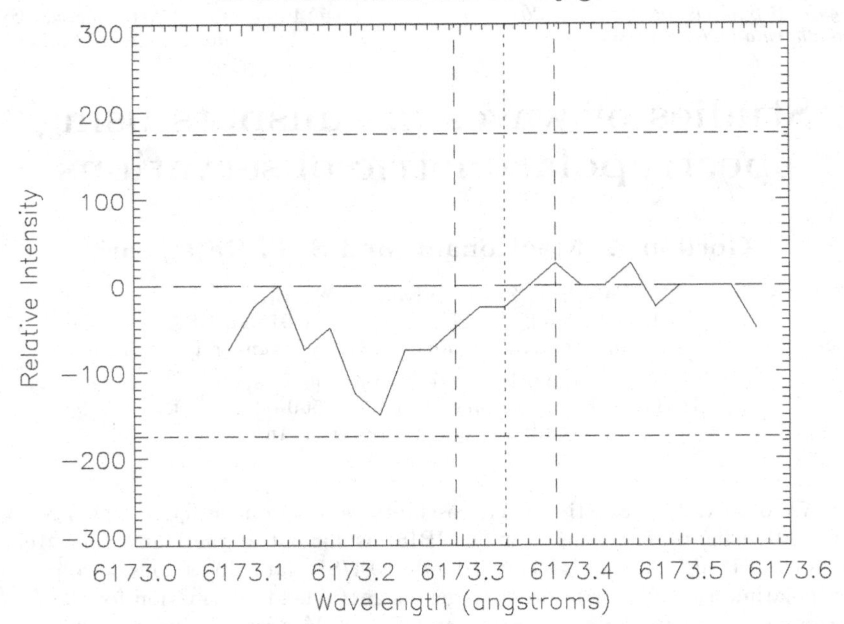

Figure 1. Seen here is a single-lobed profile which is created by the weak fields normally found in the quiet sun Sigwarth (2001). The two horizontal dashed lines indicate the cutoff intensity of ± 175 and the vertical dashed lines indicate the ± 45mÅ interval where the ZC is expected to be found. This is an example of a profile that would not be analyzed, as the amplitude of each lobe is far too small to be considered for analysis. In addition, the fact that there exists only one lobe excludes it from analysis. All profiles which do not meet this criteria are assigned a Doppler speed of 0.

2. Data reduction and analysis

Our study begins with the creation of Doppler maps obtained from FeI 6173Å line core, wings and Stokes V zero-crossing (ZC). Stokes V were analyzed based on their lobe amplitude. A 582 element-long time series of these maps is thus obtained, and by performing a FFT on these maps, we are able to get the power spectrum of each pixel in the FOV. By binning these spectra, we may construct power maps for the AR. The intervals used in this study are 2–3 mHz, 3–4 mHz and 4–6 mHz.

3. Findings

The obvious continuation of quiet region oscillations in the penumbra as seen in the line core may be explained by the field lines being nearly transverse to the LOS. Indeed, this is reflected in the power correlations in Figure 3 where we see a rise in acoustic power in line wings between approximately 50° and 65° and a corresponding drop in ZC acoustic power. Horizontal fields seem to suppress these ZC oscillations and enhance those seen in the line wing. Conversely, we see a node-like structure in the Doppler power maps but not in the ZC power maps. In fact, the umbral acoustic power in the ZC is more than twice as large as the acoustic power viewed in the wings. This is the site of very strong, nearly-vertical magnetic field. Sites such as these appear to play host to much higher power in ZC oscillations. The contributions of these magnetic os-

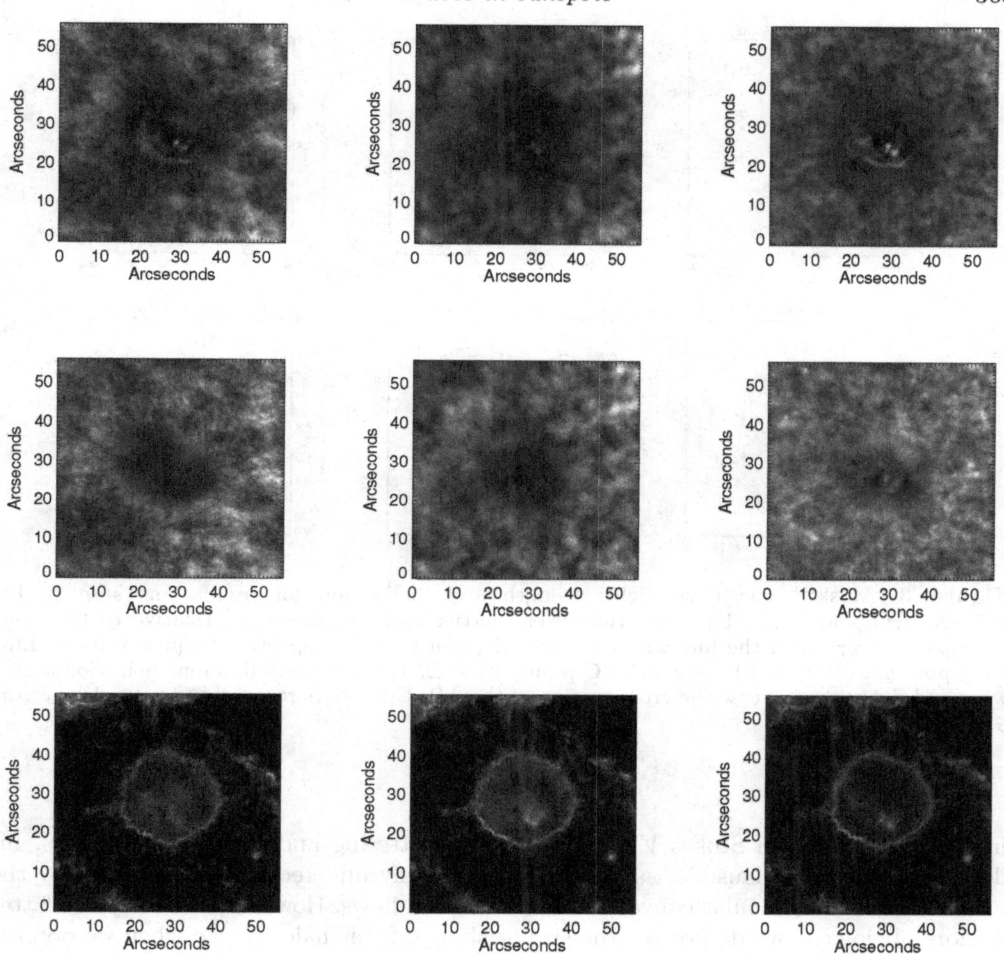

Figure 2. *From top to bottom row:* power maps for FeI 6173Å line wing, core and ZC. *From left to right column:* power spectra for frequency bands 2-3mHz, 3-4mHz and 4-6mHz. We observe umbral node-like structures in the power maps constructed from line wing and core. A thin halo of enhanced power is observed around the outer penumbra in power maps from Stokes V ZC. In each set of three power maps, the units are not scaled and are purely arbitrary.

cillations may simply be greater when strong fields are nearly parallel or anti-parallel to the LOS while those seen in the line core and wings are suppressed by these same fields.

We also observe the localized regions of ZC acoustic power in the umbra and also an acoustic halo surrounding the outer penumbra. We believe the regions of enhanced oscillations in the umbra to be caused by strong fields in those regions, which give rise to high concentrations of dynamic Stokes V profiles. Since these profiles have 3 ZC's, obtaining an accurate time series of ZC velocities is not trivial and we see that if they are not treated carefully, one may end up with false regions of high-power oscillations. In the case of the halo surrounding the outer penumbra, this may be caused by one of two things: *i*) high power oscillations in the magnetic field itself in sites where the field lines dive back into the surface or *ii*) the disappearance and reappearance of

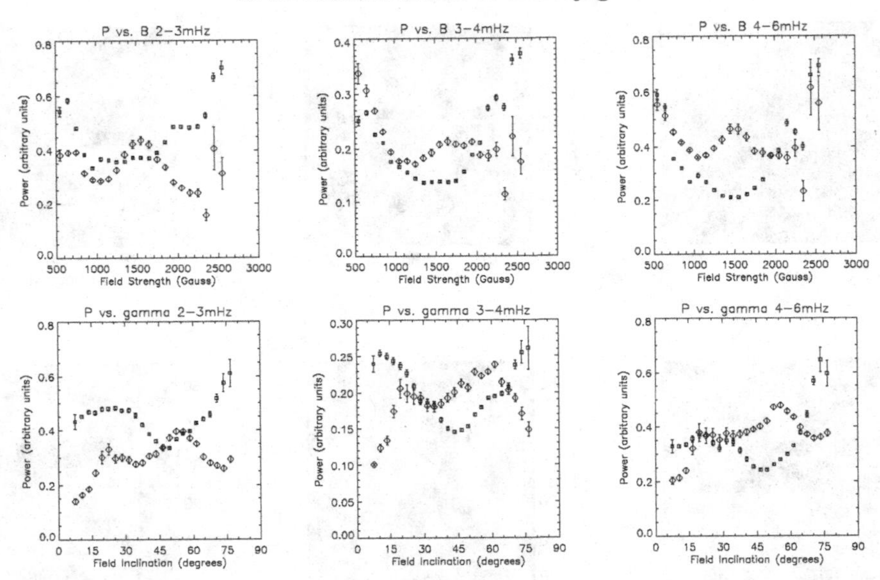

Figure 3. Masked power vs. field strength and inclination for oscillations seen in FeI 6173 Å wings and ZC. The acoustic power (vertical axis) is measured relative to the average power observed in the line wings in the quiet sun in each respective frequency band. Line wing points are denoted by a ◇ and ZC points by a □. We see vastly different behavior of line wing and ZC as we traverse the cross-section of the AR. Error bars represent the standard error, $\sigma/\sqrt{N}$.

intermediate strength Stokes V profiles due to scattering and other seeing effects. Indeed, the former is plausible, as there may previously undetected effects caused by the interface between granular convection and Evershed flows. However, currently the latter is more obvious as we do not see the size or shape of this halo change when we observe in different bands. Furthermore, power in this halo does not change with the frequency band chosen. Thus, the halo may be a consequence of our selection criteria for the ZC analysis.

Acknowledgements

This work was funded by the National Science Foundation's Office of International Science and Education, Grant Number 0854436: International Research Experience for Students, and managed by the National Solar Observatory's Global Oscillation Network. Our sincere thanks also goes to the academic faculty and support staff at the Indian Institute of Astrophysics. We would also like to thank K. Sankarasubramanian for his counsel on case difficulties encountered during this study.

References

B. W. Hindman & T. M. Brown 1998, *Astrophys. J.*, 504, 1029

K. Nagashima, T. Sekii, A. G. Kosovichev, H. Shibahashi, S. Tsuneta, K. Ichimoto, Y. Katsukawa, B. Lites, S. Nagata, T. Shimizu, R. A. Shine, Y. Suematsu, T. D. Tarbell, & A. M. Title, 2007, *Pub. Astron. Soc. Jap.*, 59, 631

C. J. Nicholas, M. J. Thompson & S. P. Rajaguru, 2005, *Solar Phys.*, 225, 213

A. A. Norton, J. P. Graham, R. K. Ulrich, J. Schou, S. Tomczyk, Y. Liu, B. W. Lites, A. López Ariste, R. I. Bush, H. Socas-Navarro, & P. H. Scherrer, 2006, *Solar Phys.*, 239, 69

S. P. Rajaguru, R. Wachter, K. Sankarsubramanian, & S. Couvidat, 2010, *Astrophys. J.*, 721, L86

H. Schunker & D. .C. Braun, 2010, *Solar Phys.*, DOI 10.1007/s11207-010-9550-3

M. Sigwarth, 2001, *Astrophys. J.*, 563, 1031

The Physics of Sun and Star Spots
Proceedings IAU Symposium No. 273, 2010
D.P. Choudhary & K.G. Strassmeier, eds.

doi:10.1017/S1743921311015560

A theoretical model of torsional oscillations from a flux transport dynamo model

Piyali Chatterjee[1], Sagar Chakraborty[2] and Arnab Rai Choudhuri[3]

[1]NORDITA, AlbaNova University Center, Roslagstullsbacken 23,
SE 10691 Stockholm, Sweden
email: piyalic@nordita.org

[2]NBIA, Niels Bohr Institute, Blegdamsvej 17,
2100 Copenhagen, Denmark
email: sagar@nbi.dk

[3]Department of Physics, Indian Institute of Science,
Bangalore – 560012, India
email: arnab@physics.iisc.ernet.in

Abstract. Assuming that the torsional oscillation is driven by the Lorentz force of the magnetic field associated with the sunspot cycle, we use a flux transport dynamo to model it and explain its initiation at a high latitude before the beginning of the sunspot cycle.

Keywords. Flux transport dynamo, magnetic field, sunspot cycle

---

1. Introduction

The small periodic variation in the Sun's rotation with the sunspot cycle, first discovered on the solar surface by Howard & LaBonte (1980), is called torsional oscillations. Helioseismology has now established its existence throughout the convection zone (see Howe *et al.* 2005 and references therein). Its amplitude near the surface is of order 5 m s^{-1} or about 1% of the angular velocity. Apart from the equatorward-propagating branch which moves with the sunspot belt, there is also a poleward-propagating branch at high latitudes. One intriguing aspect of the equatorward-propagating branch is that it begins a couple of years before the sunspots of a particular cycle appear and at a latitude higher than where the first sunspots are seen. The top panel of Fig. 1 shows the torsional oscillations at the solar surface with the butterfly diagram of sunspots. If the torsional oscillation is caused by the Lorentz force of the dynamo-generated magnetic field as generally believed, then the early initiation of this oscillation at a higher latitude does look like a violation of causality! Our main aim is to explain this which could not be explained by the earlier theoretical models (Durney 1980; Covas *et al.* 2000; Bushby 2006; Rempel 2006). The details of our work can be found in a recent paper (Chakraborty, Choudhuri & Chatterjee 2009a, hereafter CCC). Please note that this paper has an erratum (Chakraborty, Choudhuri & Chatterjee 2009b).

2. Theoretical model

The flux transport dynamo model first developed by Choudhuri, Schüssler & Dikapti (1995) appears to be the most promising model for explaining the sunspot cycle. We use the model presented by Chatterjee, Nandy & Choudhuri (2004). Some details of the model with the basic equations can be found in Choudhuri (2011). In order to model

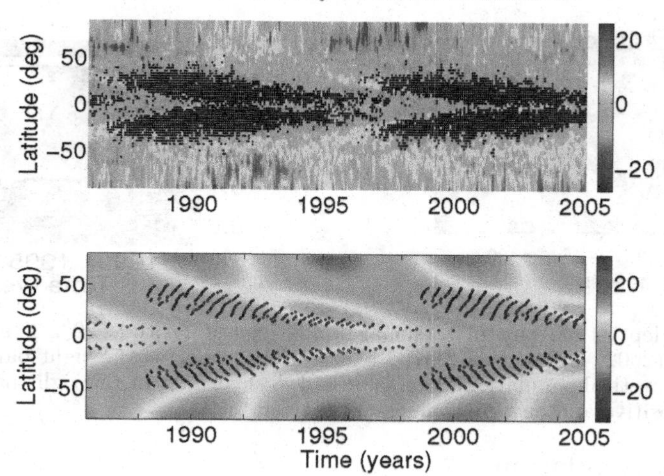

Figure 1. The time-latitude plot of torsional oscillation on the solar surface with the butterfly diagram of sunspots superposed on it. The upper panel is based on observational data of surface velocity v_ϕ measured at Mount Wilson Observatory (courtesy: Roger Ulrich). The bottom panel is from our theoretical simulation.

torsional oscillations, in addition to the basic equations of the dynamo, we simultaneously have to solve the Navier–Stokes equation in the form

$$\rho\left\{\frac{\partial v_\phi}{\partial t} + D_v[v_\phi]\right\} = D_\nu[v_\phi] + (\mathbf{F}_L)_\phi, \tag{1}$$

where $D_v[v_\phi]$ is the term corresponding to advection by the meridional circulation, $D_\nu[v_\phi]$ is the diffusion term, and $(\mathbf{F}_L)_\phi$ is the ϕ component of the Lorentz force. If the magnetic field is assumed to have the standard form

$$\mathbf{B} = B(r,\theta,t)\mathbf{e}_\phi + \nabla \times [A(r,\theta,t)\mathbf{e}_\phi], \tag{2}$$

then the Lorentz force is given by the Jacobian

$$4\pi(\mathbf{F}_L)_\phi = \frac{1}{s^3} J\left(\frac{sB_\phi, sA}{r,\theta}\right), \tag{3}$$

where $s = r\sin\theta$. On the basis of flux tube simulations suggesting that the magnetic field in the tachocline should be of order 10^5 G (Choudhuri & Gilman 1987; Choudhuri 1989; D'Silva & Choudhuri 1993), it is argued by Choudhuri (2003) that the magnetic field has to be intermittent in the tachocline. Hence the full expression of Lorentz force involves a filling factor as explained by CCC.

Our theoretical model incorporates a hypothesis proposed by Nandy & Choudhuri (2002), which is essential for explaining the early initiation of the torsional oscillation at high latitudes. According to this Nandy–Choudhuri (NC) hypothesis, the meridional flow penetrating in stable layers below convection zone causes formation of toroidal field in high latitude tachocline. Sunspots form a few years later when this field is advected to lower latitudes and brought inside convection zone. We also assume that the stress of the magnetic field formed in the tachocline is carried upward by Alfven waves propagating along vertical flux concentrations conjectured by Choudhuri (2003).

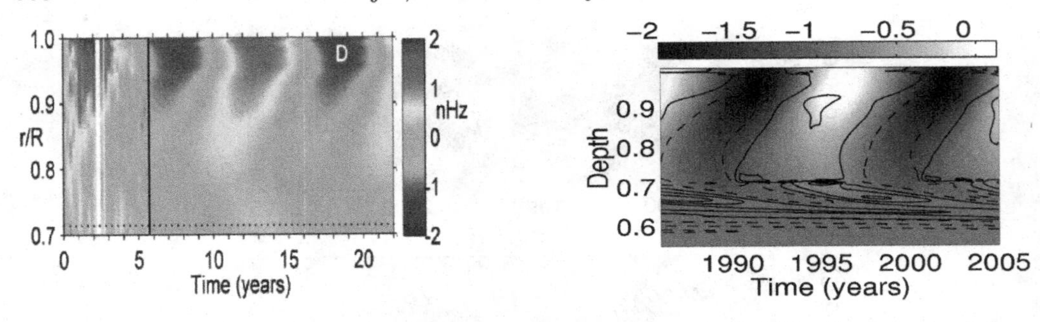

Figure 2. The depth-time plot of torsional oscillations at latitude 20°. The left panel from Vorontsov *et al.* (2002) is based on SOHO observations, whereas the right panel from CCC is based on our theoretical simulation. The solid and dashed lines in the right panel indicate the Lorentz force (positive and negative values respectively).

3. Results of simulation

The incorporation of the NC hypothesis in our theoretical model causes magnetic stresses to build up at higher latitudes before sunspots of the cycle appear, leading to the early initiation of torsional oscillations. The bottom panel of Fig. 1 shows theoretical results of torsional oscillations at the surface with the theoretical butterfly diagram. This bottom panel can be compared with the observational upper panel in Fig. 1. Our theoretical model also gives a satisfactory account of the evolution of torsional oscillations within the convection zone. The depth-time plot of torsional oscillations at a certain latitude given in Fig. 3 of CCC compares favourably with the observational plot given in Fig. 3(D) of Vorontsov *et al.* (2002). This is reproduced in Fig. 2 for completeness.

References

Bushby, P. J. 2006, *MNRAS* , 371, 772
Chakraborty, S., Choudhuri, A. R., & Chatterjee, P. 2009a, *Phys. Rev. Lett.*, 102, 041102
Chakraborty, S., Choudhuri, A. R., & Chatterjee, P. 2009b, *Phys. Rev. Lett.*, 103, 099902
Chatterjee, P., Nandy, D., & Choudhuri, A. R. 2004, *Astron. Astrophys*, 427, 1019
Choudhuri, A. R. 1989, *Solar Phys.*, 123, 217
Choudhuri, A. R. 2003, *Solar Phys.*, 215, 31
Choudhuri, A. R. 2011, In IAU Symp. 273: Physics of Sun and Star Spots (eds. D. P. Choudhary & K. G. Strassmeier), p. 28
Choudhuri, A. R. & Gilman, P. A. 1987, *Astrophys. J.*, 316, 788
Choudhuri, A. R., Schüssler, M., & Dikpati, M. 1995, *Astron. Astrophys* , 303, L29
Covas, E., Tavakol, R., Moss, D., & Tworkowski, A. 2000, *Astron. Astrophys*, 360, L21
D'Silva, S. & Choudhuri, A. R. 1993, *Astron. Astrophys*, 272, 621
Durney, B. R. 2000, *Solar Phys.*, 196, 1
Howard, R. & LaBonte, B. J. 1980, *Astrophys. J.*, 239, L33
Howe, R. et al. 2005, *Astrophys. J.*, 634, 1405
Nandy, D. & Choudhuri, A. R. 2002, *Science*, 296, 1671
Rempel, M. 2006, *Astrophys. J.*, 647, 662
Vorontsov, S. V. et al. 2002, *Science*, 296, 101

The Physics of Sun and Star Spots
Proceedings IAU Symposium No. 273, 2010
D.P. Choudhary & K.G. Strassmeier, eds.

doi:10.1017/S1743921311015572

The solar active region magnetic field and energetics

Qiang Hu[1], Na Deng[2], Debi P. Choudhary[2], B. Dasgupta[1], & Jiangtao Su[3]

[1]CSPAR, University of Alabama in Huntsville, Huntsville, AL, United States
email: qh0001@uah.edu

[2]California State University Northridge, Northridge, CA, United States

[3]National Astronomical Observatories, Beijing, China

Abstract. Motivated by increasingly more advanced solar observations, we recently develop a method of coronal magnetic field extrapolation, especially for an active region (sunspot region). Based on a more complex variational principle, the principle of minimum (energy) dissipation rate (MDR), we adopt and solve a more complex equation governing the coronal magnetic field that is non-force-free in general. We employ the vector magnetograms from multiple instruments, including Hinode, NSO, and HSOS, and particularly observations at both photospheric and chromospheric levels for one active region. We discuss our results in the context of quantitative characterization of active region magnetic energy and magnetic topology. These quantitative analyses aid in better understanding and developing prediction capability of the solar activity that is largely driven by the solar magnetic field.

Keywords. MHD, methods: data analysis, Sun: corona

---

1. Motivation and Approach

The magnetograph measurements of the solar magnetic field are probably the most direct and quantitative solar observations. The solar magnetic field plays a critical role in controlling solar activity that ultimately affects human life on Earth. There has been a long history of observing solar magnetic field, especially that in a solar active region (AR; sunspot region). Recent effort in ground-based and space-borne observations has yielded full vector magnetograms of solar ARs with modern instrumentations from, for instance, Huairou Solar Observing Station (HSOS), Big Bear Solar Observatory (BBSO), National Solar Observatory (NSO), Hinode, and Solar Dynamics Observatory (SDO). These observations provide magnetic field measurements on the solar surface at a limited number of heights, mostly on the photosphere only. Therefore the quantitative characterization of the coronal magnetic field largely depends on the numerical modeling, such as extrapolation (e.g., Wiegelmann 2010, this volume).

Solar coronal magnetic field extrapolation is to extrapolate the unknown coronal magnetic field in a finite volume from bottom boundary where the magnetogram provides the necessary boundary conditions. A common model of AR magnetic field is the force-free assumption that can be derived from the variational principle of minimum energy (Freidberg, 1987). We adopted an alternative variational principle, the principle of minimum (energy) dissipation rate (MDR), initially developed for fusion plasmas in the regime of magnetohydrodynamics (Montgomery & Phillips, 1988; Dasgupta *et al.* 1998). Later it was extended to an open and driven system, like the solar corona, to establish a more complex governing equation of the coronal magnetic field (Bhattacharya *et al.* 2007).

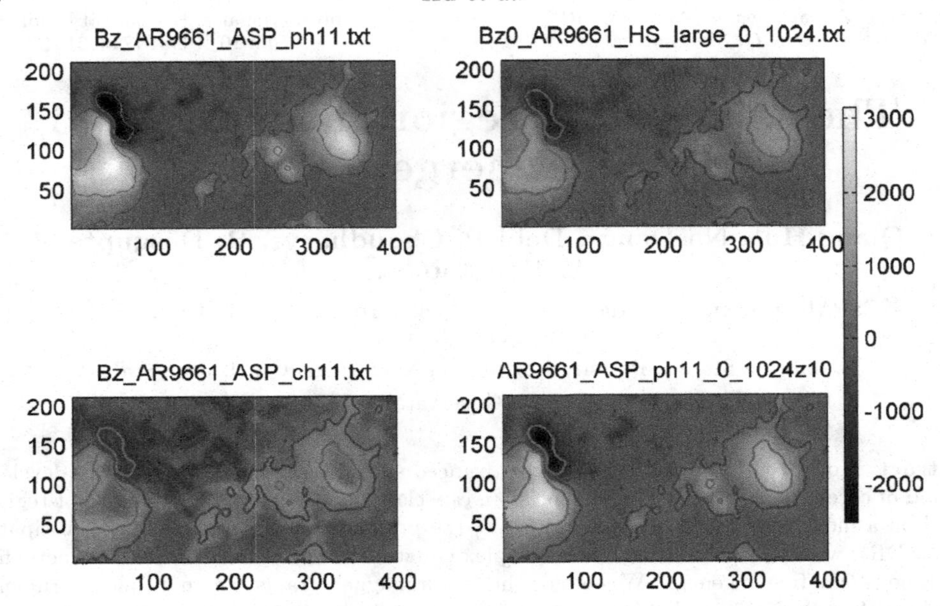

Figure 1. The B_z component within the same field-of-view on the same gray scale in Gauss for AR 9661 on 17 Oct. 2001. Clock wise from the top left panel: ASP photospheric magnetogram, HSOS photospheric magnetogram, extrapolation result at $z \approx 611$km, and ASP chromospheric magnetogram. The contours are $|B_z^{ASP}| = [300, 1500]$ G.

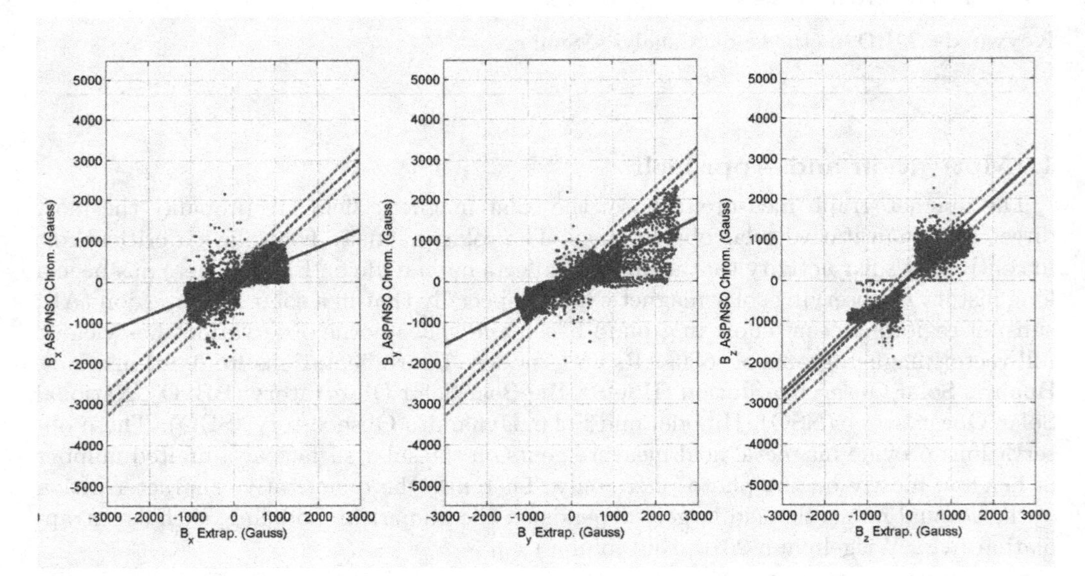

Figure 2. Component-wise comparison (from left to right: x, y, and z component) of extrapolated (horizontal axis) and measured (vertical axis) magnetic field (in Gauss) at chromosphere.

Its solutions include force-free fields (e.g., Hu & Dasgupta, 2008), but are more general, supporting non-vanishing plasma pressure gradient.

We devised an algorithm to extrapolate the coronal magnetic field into a Cartesian volume, utilizing measured vector magnetogram on the bottom boundary, based on the MDR theorem (Hu & Dasgupta, 2006, 2008; Hu *et al.* 2008, 2010a,b). Ideally two layers of vector magnetograms at two heights are needed to construct the bottom boundary conditions (Hu *et al.* 2008). An iterative procedure was recently developed to obtain the solution with one layer vector magnetogram (Hu *et al.* 2010a). The agreement between the measured and computed transverse magnetic field components on the bottom boundary is quantitatively assessed to judge in part the goodness of the solution.

2. A case study: AR 9661

We select AR 9661 to examine its magnetic field topology by the MDR-based extrapolation because a series of vector magnetograms and related observations are available. These data include both photospheric and chromospheric vector magnetograms from NSO Advanced Stokes Polarimeter (ASP) (Deng *et al.* 2010a), and vector magnetogram from HSOS, both on 17 Oct. 2001, ∼5 hours apart. Fig. 1 shows the multiple vertical component B_z distribution within the same field-of-view on the same gray scale. There exists some variations for $300 \leqslant |B_z| \leqslant 1500$ G within which the ASP measurements are believed to be most reliable for both photosphere and chromosphere (Deng *et al.* 2010b). The HSOS measurements are generally smaller in magnitude than those of ASP. The chromospheric data (at a height ∼600 km) are noisier and their magnitudes decrease by up to 1/3 from photospheric values. The extrapolated magnetic field at a height ∼611 km is generally greater in magnitude than observed, especially the transverse component, as further demonstrated in Fig. 2. The comparison is slightly improved upon a corresponding potential field extrapolation (not shown).

The extrapolated 3D field-line configurations are shown in Fig. 3 on grid $400\times212\times200$ for ASP photospheric data and $782\times715\times400$ for HSOS data, respectively. All have a grid size 0.42" in each dimension, while the latter has a larger field-of-view. All configurations deviate significantly from the corresponding potential fields (not shown). Because of different computation domain and measurements on the bottom boundary, the magnetic field topology and connectivity are different. But significantly sheared low-lying arcades exist in the region of polarity inversion to the left in both plots. The total magnetic energy is 2.0×10^{33} erg, and 2.2×10^{33} erg, respectively.

Fig. 4, left panel, shows the MDI magnetogram with consecutive flare loop footpoints superimposed on 19 Oct. 2001 during a flare eruption (Qiu *et al.* 2009). The right panel shows the calculated squashing factor Q (Titov *et al.* 2002) of the extrapolated field from ASP photopheric magnetogram (Fig. 3 left panel). Although the flare occurred two days later, the flare loop footpoints seem to correspond to the large Q values near the polarity inversion region to the left where the flare initiated (Aulanier 2010, this volume). And they seem to match the root points of the corresponding sheared arcades in Fig. 3, especially those in the right panel.

3. Discussion and Outlook

All the extrapolations are carried out on 1024×1024 transverse grid. Better chromospheric data are needed to calculate the vertical gradient of the magnetic field over the whole domain. Clearly the finite computation domain and the boundary conditions greatly affect the extrapolation result and larger field-of-view is often more desirable (see also, DeRosa *et al.* 2009). More validation and applications combined with data-driven MHD simulation utilizing vector magnetograms are underway (Hu *et al.* 2010b). In anticipation of the SDO full-disk vector magnetograms, it is desirable to extend the extrapolation to global scale.

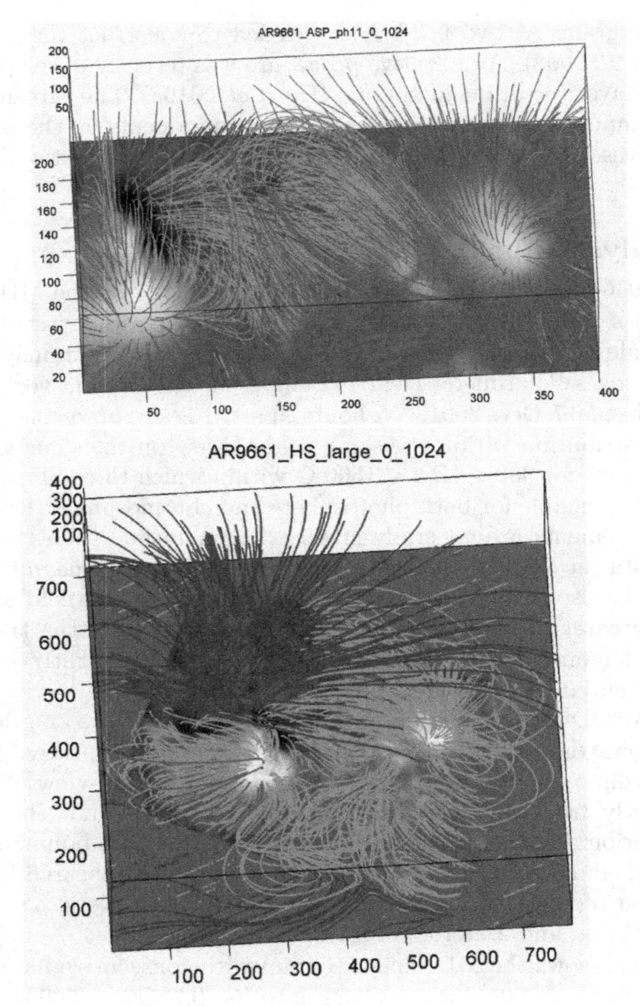

Figure 3. Extrapolated field lines from ASP photospheric vector magnetogram (left panel), and HSOS vector magnetogram (right panel). The bottom images are the B_z distributions.

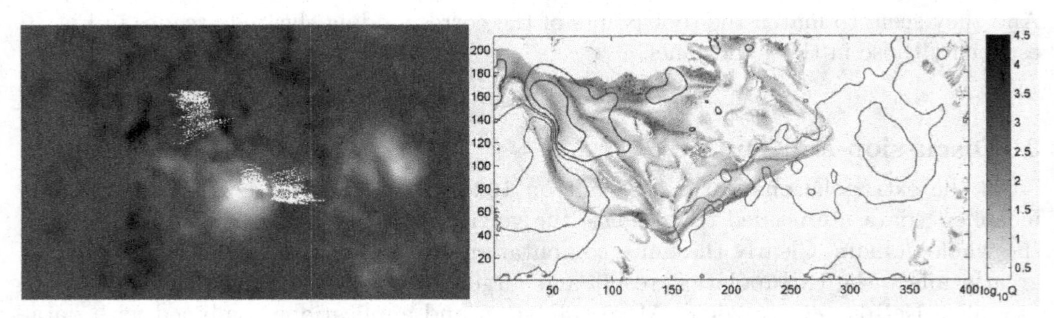

Figure 4. Left panel: the MDI magnetogram and flare loop footpoints (white dots) derived from TRACE images (courtesy of J. Qiu) on 19 Oct. 2001; Right panel: the squashing factor Q derived from Fig. 3, left panel. Contours are $B_z = \pm[300, 1500]$ G.

Acknowledgements

HQ and BD acknowledge NASA grants NNX07AO73G and NNX10AG03G for partial support. ND and DPC were supported by NASA grant NNX08AQ32G and NSF grant ATM 05-48260. HQ thanks Prof. J. Qiu for providing part of the plots.

References

Aulanier, G. 2010, this volume
Bhattacharyya, R., Janaki, M. S., Dasgupta, B., & Zank, G. 2007, *Solar Phys.*, 240, 63
Dasgupta, B., Dasgupta, P., Janaki, M. S., Watanabe T., & Sato, T. 1998, *Phys. Rev. Lett.* 81, 3144
Deng, N., Choudhary, D. P., & Balasubramaniam, K. S. 2010a, *Astrophys. J.*, 719, 385
Deng, N., Choudhary, D. P., & Balasubramaniam, K. S. 2010b, *ASP Proceedings*, submitted
DeRosa, M. L., *et al.* 2009, *Astrophys. J.*, 696, 1780
Freidberg, J. P. 1987, *Ideal Magnetohydrodynamics*, Published by Plenum Press
Hu, Q., Wang, A., Wu, S. T., & Gary, G. A. 2010b, *Astrophys. J.*, submitted
Hu, Q., Dasgupta, B., DeRosa, M. L., Büchner, J., & Gary, G. A. 2010a, *J. Atmos. Sol. Terres. Phys.* 72, 219
Hu, Q. & Dasgupta, B. 2008, *Solar Phys.*, 247, 87
Hu, Q., Dasgupta, B., Choudhary, D. P., & Büchner, J. 2008, *Astrophys. J.*, 679, 848
Hu, Q. & Dasgupta, B. 2006, *Geophysical Research Letters* 33, L15106
Montgomery, D. & Phillips, L. 1988, *Phys. Rev. A* 38, 2953
Qiu, J., Gary, D. E., & Fleishman, G. D. 2009, *Solar Phys.*, 255, 107
Titov, V. S., Hornig, G., & Démoulin, P. 2002, *J. Geophys. Res.* 107, 1164
Wiegelmann, T. 2010, this volume

The Physics of Sun and Star Spots
Proceedings IAU Symposium No. 273, 2010
D.P. Choudhary & K.G. Strassmeier, eds.

doi:10.1017/S1743921311015584

How reliable are observations of solar magnetic fields? Comparison of full-disk measurements in different spectral lines and calibration issues of space missions SOHO, Hinode, and SDO

Mikhail Demidov

Institute of Solar-Terresrial Physics, P.O.Box 291, 664033 Irkutsk, Russia
email: demid@iszf.irk.ru

Abstract. An urgent problem in modern solar physics, which is not completely solved up to now, is to obtain realistic magnetic field strength values from parameters measured magnetographs or Stokes-meter instruments. One of the important tools on this way is a comparison of observations made in different spectral lines with the same or with the different telescopes. This issue is an actual task in the analysis of the new data sets provided by the space missions SOHO and Hinode, which measurements are available for several years already, and SDO, which data appeared recently. The main aim of this study is a cross-comparison of magnetic field observations made in different spectral lines used on the above mentioned space observatories: Ni I λ676.77 nm (SOHO/MDI), Fe I λ630.152 nm and Fe I λ630.25 nm (Hinode/SP), and Fe I λ617.33 nm (SDO/HMI). Full-disk high-precision Stokes-meter measurements with the STOP telescope at the Sayan observatory in these lines are used basically, as well as some observations in other spectral lines having a great diagnostic impact, such as Fe I λ525.02 nm, Fe I λ523.29 nm and Fe I λ532.42 nm. The difference between one-instrument (STOP) simultaneous or quasi-simultaneous observations in different spectral lines do not exceed the factor of 2-3 depending on the combination of spectral lines and the position on the solar disk. This is significantly less than in some other studies devoted to cross-comparison of different data sets. Importance and consequences of the obtained results are discussed.

Keywords. Sun: photosphere, Sun: magnetic fields

---

1. Introduction and Motivation

A significant progress in many solar physics problems has been achieved during the recent years, but some questions are still waiting of their solution. The problem of the determination of the true magnetic fields on the Sun is one of them. Indeed, due to an extremely complicated spatial structure of solar atmosphere, it is not a simple task to connect the parameters, measured with instruments, with the magnetic field strength in the point of observation. It is possible only in the frameworks of some assumptions, simple or rather complicated ones. A powerful tool to test which of the assumptions is better (closer to a reality) is a comparison of observations made in different spectral lines. Of course, to avoid many instrumental problems, it is better to use measurements from the same instruments. However, a comparison of observations made with different instruments is very important as well.

During the first decades of magnetographic measurements such comparisons were made, naturally, only for the ground-based solar observatories. But with the launch of the space missions, starting with SOHO in 1996, then Hinode (2006) and Solar

Dynamics Observatory (SDO) (2010), it became possible to include in the analysis the new, space-borne, data sets. Some of the most important references on the papers devoted to comparisons of different magnetic field observations, including SOHO/MDI, are listed in Demidov *et al.* (2008). The main scientific result of this paper is the discovered complicated spatial distribution of magnetic field strength ratios across the solar disk in some combinations of data sets. A comparison of SOHO/MDI full-disk magnetograms (remind, made in spectral line Ni I λ676.77 nm) with Sayan Solar observatory (SSO) measurements in spectral line Fe I λ525.02 nm was made. The mean value of the magnetic strength ratio $R = B(SOHO/MDI)/B(SSO)$ is 2.75. This number is important in the context of the following discussion.

The SOHO/MDI magnetograms are widely used in many studies, thus it is quite obvious that the question of the reliability of such data is very important. When the papers by Tran *et al.* (2005) and Ulrich *et al.* (2009) have appeared, where the necessity of an essential re-calibration (increase of strengths by factor of about 2) of the SOHO/MDI magnetograms was suggested, they attracted a significant attention of the solar physics community. But some of the results (connected with SOHO/MDI calibration issue) of these two papers from Mount Wilson observatory (MWO) were doubted by Demidov and Balthasar (2009).

The conclusions of Tran *et al.* (2005) and Ulrich *et al.* (2009) are based mainly on the comparisons of magnetic field observations in the Fe I λ525.02 nm and Fe I λ523.29 nm spectral lines. The same true (but with different conclusions) for the paper by Demidov and Balthasar (2009), where simultaneous high-precision Stokes-meter measurements in this pair of lines (both of them are registered on the same linear CCD detector) are analysed. For the further progress in the SOHO/MDI calibration problem it is important, of course, to compare observations made at the same instrument in these two lines Ni I λ676.77 nm and Fe I λ525.02 nm. Such experiments were made by the author with the STOP telescope (Solar Telescope for Operative Predictions) at SSO in the beginning of 2010. Additionally, results of the analogous measurements made in the spectral lines used for magnetic field measurements on Hinode (Fe I λ630.152 nm and Fe I λ630.25 nm) and SDO/HMI (Fe I λ617.33 nm) are presented here as well. Besides, observations in the line Fe I λ532.42 nm are added in the analysis because of two reasons: (1) this line is used at the new Chinese solar telescope SMAT (Solar Magnetism and Activity Telescope) (Zhang *et al.*, 2007) which provides the full-disk magnetograms, (2) atomic parameters of this line are very similar to those of the line Fe I λ523.29 nm, and therefore it is extremely important in the context of the SOHO/MDI calibration issue.

Table 1 presents the basic information about spectral lines involved in the following analysis. Mutual comparisons of observations in different combinations of these spectral lines are useful in the context of calibration problems of corresponding instruments.

2. Results

Figure 1 shows an example of the Stokes I and Stokes V spectra in the vicinity of the spectral line Ni I λ676.77 nm for one of the points (with magnetic field strength of about 11 G) of the magnetogram, The scatter plot for combination of observations in the lines Fe I λ617.33 nm and Fe I λ525.02 nm is shown in the left panel, and in the lines Fe I λ532.42 nm and Fe I λ523.29 nm in the right one of Figure 2. The results of the statistical correlation and regression analysis for different combinations of spectral lines are summarized in the Table 1.

In contrast to Demidov and Balthasar (2009), where the wavelength difference between the explored spectral lines (Fe I λ523.29 nm and Fe I λ525.02 nm) is small enough to allow

Table 1. Basic parameters of spectral lines, used in this study

Spectral line [nm]	Landé factor, g_{eff}	EP [eV]	W [mÅ]	Instrument or observatory	Reference
Fe I λ523.29	1.3	2.94	346	MWO	Ulrich *et al.* (2009)
Fe I λ525.02	3.0	0.12	62	SSO	Demidov *et al.* (2008)
Fe I λ532.42	1.5	3.21	334	SMAT	Zhang *et al.* (2007)
Fe I λ617.33	2.5	2.22	50	SDO/HMI	Norton *et al.* (2006)
Fe I λ630.15	1.66	3.65	127	Hinode/SP	Tsuneta *et al.* (2008)
Fe I λ630.25	2.5	3.69	83	Hinode/SP	Tsuneta *et al.* (2008)
Ni I λ676.78	1.43	1.83	83	SOHO/MDI	Scherrer *et al.* (1995)

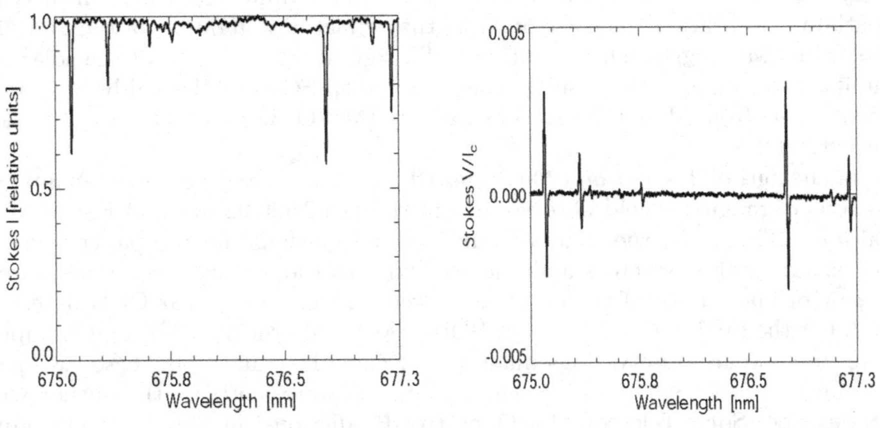

Figure 1. Spectra of Stokes I (left panel) and V/I_c (right panel) for the vicinity of the Ni I λ676.77 nm spectral lines, used at SOHO/MDI. The magnetic field strength in the point of observation is 11 G.

simultaneous observations in both lines on the same CCD detector, it was necessary in this study to change the setting of the spectrograph. The time difference between magnetograms in the corresponding different spectral lines was no more than 2 hours. To diminish the influence of the time differences on the results, which is, at least partly, responsible for the scatter of the points in the corresponding scatter-plots, observations were made with the low (typical for regular observation on STOP) spatial resolution of 100″, instead of 10″ in Demidov and Balthasar (2009). Comparisons of observations in the spectral lines Fe I λ630.152 nm and Fe I λ630.25 nm were made for solar mean magnetic field (SMMF) measurements (Demidov *et al.*, 2002).

From the consideration of Table 1 and Figure 2 it is obvious, that the result of the present study concerning the $B(523.29)/B(525.02)$ strength ratio is in excellent agreement with Demidov and Balthasar (2009), and observations in the lines Fe I λ532.42 nm and Fe I λ523.29 nm confirm this statement.

3. Summary

Considering Table 1 and Figure 2, it is concluded that the regression coefficient between quasi simultaneous observations with the same instrument in the lines Ni I λ676.77 nm and Fe I λ525.02 nm is only 1.65. That means that, probably, even the old, before the

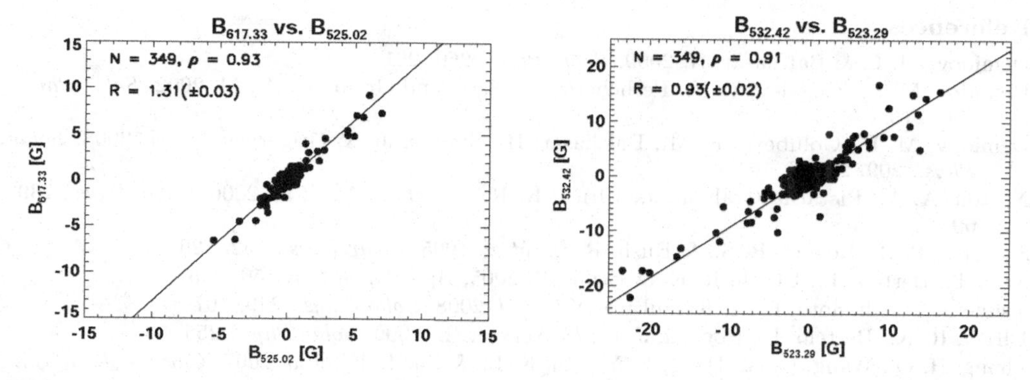

Figure 2. Correlation and regression analysis of solar magnetic field observations in Fe I λ617.33 nm and Fe I λ525.02 nm (left panel), and Fe I λ532.42 nm and Fe I λ523.29 nm (right panel). N is the number of points, ρ is the correlation coefficient, and R is the coefficient of linear regression (the slope of the line through the scatter plot).

Table 2. Results of correletion and regeression analysis of solar magnetic field measurements in different spectral lines. $A(\pm\Delta A)$, $R(\pm\Delta R)$ are parameters of the linear regression equation $B_{lineY} = A(\pm\Delta A) + R(\pm\Delta R)B_{lineR}$, ρ - is correletion coefficient.

Line X, nm	Line Y, nm	R	ΔR	A	ΔA	ρ
Fe I λ525.02	Ni I λ676.77	1.65	0.06	0.00	0.17	0.76
Fe I λ525.02	Fe I λ617.33	1.31	0.03	0.00	0.05	0.93
Fe I λ525.02	Fe I λ630.15	2.52	0.09	-0.03	0.13	0.99
Fe I λ525.02	Fe I λ630.23	1.53	0.10	0.01	0.14	0.98
Fe I λ525.02	Fe I λ523.29	1.92	0.04	-0.01	0.12	0.92
Fe I λ523.29	Fe I λ532.42	0.93	0.02	0.01	0.17	0.91

re-calibrations (two new calibrations were suggested and realized) SOHO/MDI data yield too high magnetic field strengths (by a factor of 2.75/1.65 ≈1.7). The difference between Hinode data and observations in the line Fe I λ525.02 nm should be of the order of 2.5 (Fe I λ630.152 nm) or of 1.5 (Fe I λ630.25 nm). At last, the strength ratio $B(617.33)/B(5250.02)$ is 1.3, what allows us to judge about possible differences between SDO/HMI magnetograms and traditional observations in the spectral line Fe I λ525.02 nm

The next near future natural step in the investigations in this direction will be a comparison of the SOHO/MDI, Hinode/SP, SDO/HMI and SMAT observations with SSO measurements in the corresponding spectral lines.

Acknowledgments

The presentation of this paper in the IAU Symposium 273 was possible due to partial support from the National Science Foundation grant numbers ATM 0548260, AST 0968672 and NASA - Living With a Star grant number 09-LWSTRT09-0039, and from the Russian Foundation for Basic Researches travel grant 10-02-08457-3. The author acknowledges Dr. H. Balthasar for reading the manuscript and discussions. This work has been supported by grant BA 1875/6-1 of the Deutsche Forschungsgemeinschaft (DFG).

References

Demidov, M. L. & Balthasar, H. 2009, *Solar Phys.*, 260, 261

Demidov, M. L., Zhigalov, V. V., Peshcherov, V. S., & and Grigoryev, V. M. 2002, *Solar Phys.*, 250, 279

Demidov, M. L., Golubeva, E. M., Balthasar, H., Staude, J., & Grigoryev, V. M. 2008, *Solar Phys.*, 209, 217

Norton, A. A., Pietarila Graham, J., Ulrich, R. K., & Schou, J., *et al.* 2006, *Solar Phys.*, 239, 69

Scherrer, P. H., Bogart, R. S., & Bush, R. L., *et al.* 1995, *Solar Phys.*, 162, 129

Tran T., Bertelle L., Ulrich, R. K. & Evans, S. 2005, *ApJ Suppl. Ser.* 156, 295

Tsuneta, S., Ichimoto, K., & Katsukava, Y., *et al.* 2008, *Solar Phys.*, 249, 167

Ulrich, R. K., Bertelo, L., Boyden, J. E., & Webster, L. 2009, *Solar Phys.*, 255, 53

Zhang, H. Q., Wang, D. G., Deng, Y. Y., Hu, K. L., & Su, J. T., *et al.* 2007, *Chinese J. Astron. Astroph.* 7, 281

The Physics of Sun and Star Spots
Proceedings IAU Symposium No. 273, 2010
D. P. Choudhary & K. G. Strassmeier, eds.

doi:10.1017/S1743921311015596

Towards physics-based helioseismic inversions of subsurface sunspot structure

D. C. Braun[1], A. C. Birch[1], A. D. Crouch[1] and M. Rempel[2]

[1]NorthWest Research Assoc, CoRA Div,
3380 Mitchell Ln, Boulder, CO, USA
email: `dbraun@cora.nwra.com` `aaronb@cora.nwra.com` `ash@cora.nwra.com`

[2]NCAR, HAO Div,
3080 Center Green Dr,
Boulder, CO, USA
email: `rempel@ucar.edu`

Abstract. Numerical computations of wave propagation through sunspot-like magnetic field structures are critical to developing and testing methods to deduce the subsurface structure of sunspots and active regions. We show that helioseismic analysis applied to the MHD sunspot simulations of Rempel and collaborators, as well as to translation-invariant models of umbral-like fields, yield wave travel-time measurements in qualitative agreement with those obtained in real sunspots. However, standard inversion methods applied to these data fail to reproduce the true wave-speed structure beneath the surface of the model. Inversion methods which incorporate direct effects of the magnetic field, including mode conversion, may be required.

Keywords. Sun: helioseismology, Sun: magnetic fields

1. Introduction

Current controversy exists in the interpretation and modeling of helioseismic measurements of sunspots (see the review by Gizon, Birch & Spruit 2010). A major issue is the discrepancy between the relatively deep two-layer wave-speed models derived from standard time-distance helioseismic inversions (Kosovichev *et al.* 2000; Couvidat, *et al.* 2005) and shallow, positive wave-speed perturbations inferred from forward models which explicitly include magnetic fields (e.g. Crouch *et al.* 2005; Cameron *et al.* 2010).

Structural (i.e. wave-speed) models of sunspots are inferred from p-mode travel-time perturbations relative to travel times through quiet Sun. These travel-time perturbations typically show a variation with phase-speed w (the temporal frequency divided by the horizontal wavenumber), ranging from positive (longer times) at small phase-speeds, to negative (shorter times) at larger phase-speeds. Deeper penetrating modes have increasing phase-speed, so the travel-time variation with w provides the basis for the two-layer wave-speed models (e.g. Kosovichev *et al.* 2000) which extend downward to approximately 10 Mm below the surface. However, recently observed variations of travel-time perturbations with frequency (at fixed phase-speed) have been suggested as evidence of strong near-surface perturbations (Couvidat & Rajaguru 2007; Braun & Birch 2006; 2008). In addition, the nature and interpretation of the positive (slower) travel-time perturbations and the sensitivity of the travel-time perturbations to the analysis methodology have been questioned (Braun & Birch 2008; Birch *et al.* 2009; Gizon *et al.* 2009; Moradi *et al.* 2010). Positive travel-time shifts (slower waves) have also been measured in artificial datasets where the wave-speed perturbations are positive (Birch *et al.* 2009; Moradi *et al.* 2009).

2. Simulated sunspot models for helioseismology

We discuss here the use of two types of models for developing and testing helioseismic methods. One of these consists of a realistic magnetoconvection simulation using the MURaM (Max Planck Institute for Solar System Research/University of Chicago Radiative MHD) code described by Vögler *et al.* (2005) and modified to simulate realistic sunspot structures (Rempel *et al.* 2009). The simulation spans a 48 by 48 Mm box extending 8 Mm into a solar-like stratification. The data saved for helioseismic use consists of slices of vertical velocity sampled every minute at a constant optical depth near the photosphere.

A second type of model propagates waves through (horizontally) translation-invariant background models (Crouch *et al.* 2010). The power spectra from the translation-invariant models (TIMs) are then converted into a time series of synthetic helioseismic data using the algorithm outlined by Gizon & Birch (2004).

Using helioseismic holography, we produce sets of maps of travel-time perturbations, using an analysis analogous to those involving center-annulus time-distance correlations (Braun & Birch 2008). Our analysis is applied to MDI observations of two sunspots (AR 9787 & AR 10615), 27 hrs of the Rempel simulation, and a TIM consisting of a vertical 3 kG field embedded into a model S background stratification. Our analysis is similar to previous time-distance methods in that we employ standard phase-speed filters and their corresponding annuli (see Table 1 of Couvidat *et al.* 2006), but differs in the additional use of box-car filters to isolate narrow ranges in temporal frequency.

Comparing the spatial average of the travel-time perturbations over the umbrae (Figure 1) and penumbrae (Figure 2), we see at least a qualitative similarity between results for the real and artificial sunspots. In many cases (especially in the penumbral measurements) the agreement is remarkably quantitative as well. Of particular note is the presence, in both real and artificial spots, of positive travel-time perturbations at smaller phase-speeds, which tend to decrease to negative values with increasing frequency, and the predominantly negative values at higher phase-speeds.

3. Inversions of the Rempel simulation

We carried out three-dimensional ray-approximation based wave-speed inversions for a 12 hour time-series of the MURaM simulation. We used the ray approximation for the travel-time shifts caused by wave-speed perturbations as described by Kosovichev & Duvall (1997) using a RLS MCD approach. Figure 3 shows the results of applying the inversion procedure to travel-time maps measured from the simulation. The maps were made using a wide frequency bandpass (i.e. 2.5–5.5 mHz) for each phase-speed filter. The wave-speed structure that is recovered from the inversion (left panel) is reminiscent of the two-layer wave-speed structure seen from inversions of solar data (e.g. Kosovichev *et al.* 2000; Couvidat, *et al.* 2005). While the results are consistent with the mean-travel time measurements (right panels of Figure 3) it is clear that the inversion procedure does not recover either the sound-speed or fast-mode speed perturbations present in the model. This shows that one or more of the assumptions of the inversion is not met.

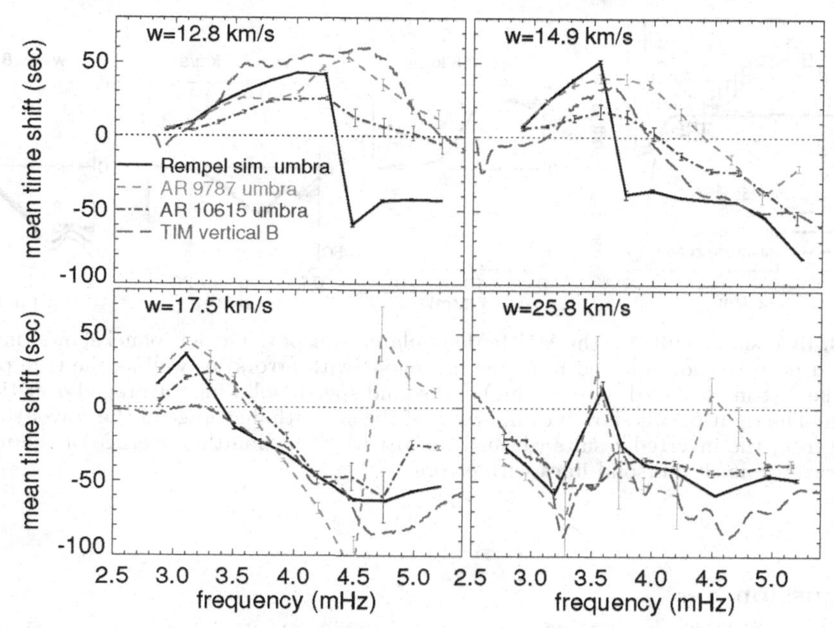

Figure 1. Comparison of mean travel-time shifts, averaged over the umbrae, determined from both MDI observations of two sunspots as well as the umbra of the MURaM simulated spot of Rempel and collaborators and a translation-invariant model (TIM) with a vertical magnetic field embedded within a solar model. The phase-speed filters used are indicated in each panel. Box-car frequency filters of width 0.5 mHz were used in all cases except for the TIM results which use a considerably narrower frequency range.

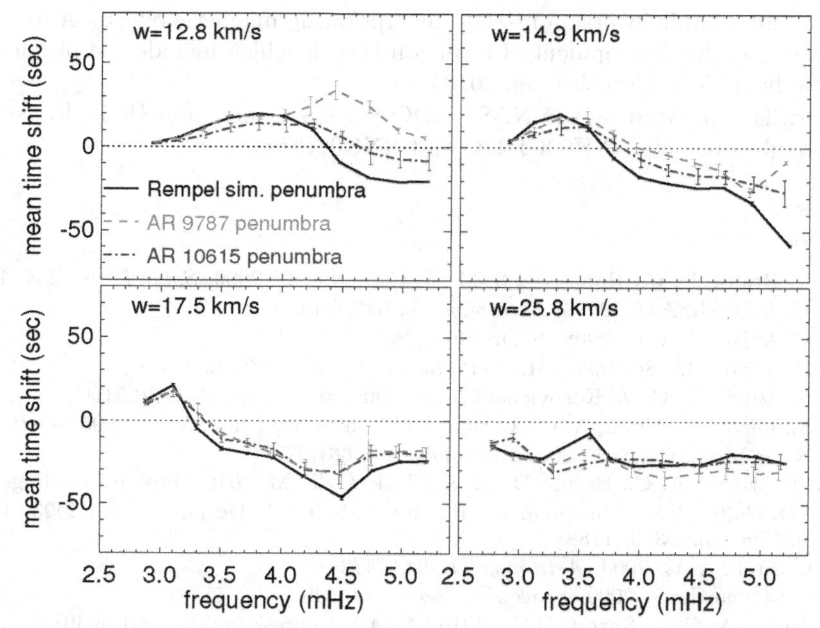

Figure 2. Comparison of mean travel-time shifts, averaged over the penumbrae, determined from both MDI observations of two sunspots as well as the MURaM simulated sunspot.

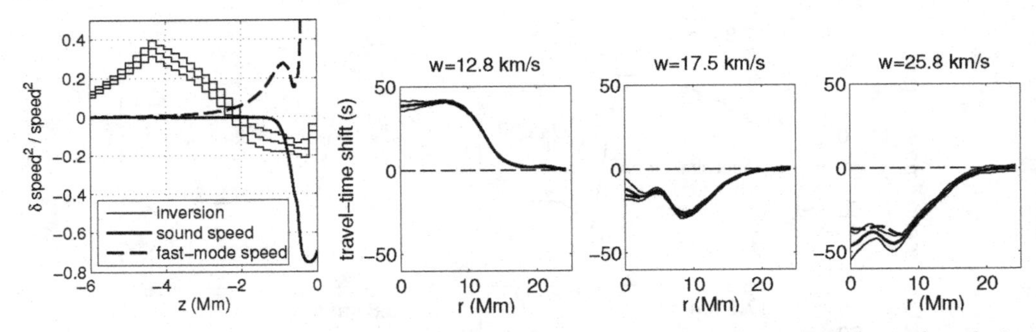

Figure 3. Inversion results for the MURaM simulated sunspot. The left panel shows the relative wave-speed perturbation inferred from the inversion (with errors) as well as the true perturbations to the fast-mode speed (dashed line) and sound speed (solid line) averaged over the region $r < 6$ Mm. The right panels show a comparison of the azimuthal average of the travel-time shifts predicted from the inverted results (dashed lines) with the azimuthal average of the measured (input) travel-time shifts (solid lines with errors).

4. Discussion

It is clear that standard helioseismic inversion methods fail to recover the subsurface wave-speed structure within the Rempel sunspot simulation. Consequently, the similarities between the helioseismic travel-time measurements made using the artificial sunspot and actual observations of sunspots adds to the uncertainty in our inferences of subsurface structure below strong magnetic fields. Numerical and semi-analytic MHD models provide considerable insight into the physics of sunspots and the propagation of waves in magnetic regions. Making full use of the year-round, almost limb-to-limb, coverage provided by HMI onboard the Solar Dynamics Observatory will likely require an efficient and reliable inversion method incorporating magnetic effects. A first step towards this end is the development of inversion kernels which include the physical effects of magnetic fields (e.g. Crouch *et al.* 2010).

This work is supported by the NASA SDO Science Center and Heliophysics GI programs through contracts NNH09CE41C and NNG07EI51C.

References

Birch, A. C., Braun, D. C., Hanasoge, S. M., & Cameron, R. 2009, *Solar Phys.*, 254, 17
Braun, D. C. & Birch, A. C. 2006, *Astrophys. J.*, 647, L187
Braun, D. C. & Birch, A. C. 2008, *Solar Phys.*, 251, 267
Cameron, R., Gizon, L., Schunker, H., & Pietarila, A. 2010, 268, 293–308
Couvidat, S., Birch, A. C., & Kosovichev, A. G. 2006, *Astrophys. J.*, 640, 516
Couvidat, S., Gizon, L., Birch, A. C., Larsen, R. M., & Kosovichev, A. G. 2005, *ApJS*, 158, 217
Couvidat, S., & Rajaguru, S. P. 2007, *Astrophys. J.*, 661, 558
Crouch, A. D., Birch, A. C., Braun, D. C., & Clack, C. T. M. 2010, these proceedings
Crouch, A. D, Cally, P. S., Charbonneau, P., Braun, D. C., & Desjardins, M. 2005, *Mon. Not. Roy. Astron. Soc.*, 363, 1188
Gizon, L. & Birch, A. C. 2004, *Astrophys. J.*, 614, 472
Gizon, L. & 14 coauthors 2009, *Space Sci. Revs*, 144, 249
Gizon, L., Birch, A. C., & Spruit, H. C. 2010, *ARAA*, in press (arXiv:1001.0930)
Kosovichev, A. G., & Duvall, T. L., Jr. 1997, *SCORe'96: Solar Convection and Oscillations and their Relationship*, Astrophysics and Space Science Library 225, 241

Kosovichev, A. G., Duvall, T. L., Jr., & Scherrer, P. H. 2000, *Solar Phys.*, 192, 159
Moradi, H., Hanasoge, S. M., & Cally, P. S. 2009, *Astrophys. J.*, 690, L72
Moradi, H. & 21 coauthors 2010, *Solar Phys.*, in press (arXiv:0912.4982)
Rempel, M., Schüssler, M., & Knölker, M. 2009, *Astrophys. J.*, 691, 640
Vögler, A., Shelyag, S., Schüssler, M., Cattaneo, F., Emonet, T., & Linde, T. 2005, *Astron. Astrophys*, 429, 335

The Physics of Sun and Star Spots
Proceedings IAU Symposium No. 273, 2010
D. P. Choudhary & K. G. Strassmeier, eds.

doi:10.1017/S1743921311015602

Helioseismic probing of the subsurface structure of sunspots

A.D. Crouch, A.C. Birch, D.C. Braun, and C.T.M. Clack

NorthWest Research Associates, Colorado Research Associates Division
3380 Mitchell Lane, Boulder, CO

Abstract. We discuss recent progress in the helioseismic probing of the subsurface structure of solar magnetic regions. To simulate the interaction of helioseismic waves with magnetic fields and thermal perturbations we use a simple model that is translation invariant in the horizontal directions, has a realistic stratification in the vertical direction, and has physically consistent boundary conditions for the waves at the upper and lower boundaries of the computational domain. Using this model we generate synthetic helioseismic data and subsequently measure time-distance travel times. We evaluate a model for the wave-speed perturbation below sunspots that replaces the sound speed in a non-magnetic model by the fast-mode speed from a magnetic model; our results indicate that this approach is unlikely to be useful in modeling wave-speed perturbations in sunspots. We develop and test an inversion algorithm for inferring the sound-speed perturbation in magnetic regions. We show that this algorithm retrieves the correct sound-speed perturbation only when the sensitivity kernels employed account for the effects of the magnetic field on the waves and the subsurface structure.

Keywords. Sun: helioseismology, Sun: magnetic fields, sunspots

1. Introduction

One goal of local helioseismology is to infer the subsurface structure of sunspots and magnetic regions. At present there are two different general classes of helioseismic models for the wave-speed perturbation in sunspots (e.g., Gizon, Birch, & Spruit, 2010). On the one hand, a shallow, positive wave-speed perturbation is favored by Fan, Braun, & Chou (1995), Crouch *et al.*(2005), and Cameron *et al.*(2010). On the other hand, standard inversions from time-distance measurements favor a two-layer model with a negative wave-speed perturbation near the surface and a positive wave-speed perturbation in deeper layers (e.g., Kosovichev, Duvall, & Scherrer, 2000, Couvidat *et al.*, 2005, Hughes, Rajaguru, & Thompson, 2005). One possible source of uncertainty is the effect of the magnetic field on both solar oscillations and structure. The aim of this investigation is to evaluate how magnetic fields may affect helioseismic inversion methods.

2. Translation-invariant model for wave propagation (TIM)

The translation-invariant model for wave propagation (TIM) solves the linearized MHD equations in a horizontally uniform, three-layer background model that may be permeated by a uniform, inclined magnetic field. The upper layer is an isothermal slab that simulates the influence of the solar atmosphere. The bottom layer is a polytrope that represents the deep solar interior. In the polytrope and separately in the isothermal slab semi-analytic solutions for the linearized MHD equations are developed (e.g., Crouch & Cally, 2003, Crouch *et al.* 2005, Cally & Goossens, 2008). Physical boundary conditions

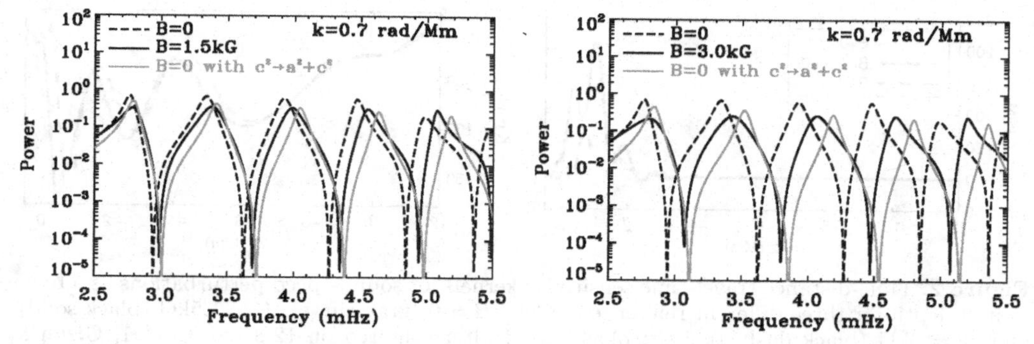

Figure 1. Power as a function of frequency at fixed horizontal wavenumber. In each panel the dashed curve is for a non-magnetic model and the solid black curve is for a model with a vertical magnetic field (left: $B = 1.5$kG, right: $B = 3$kG). A vertical field affects the power spectra in two ways: the ridges are shifted to higher frequency / horizontal phase speed and they have a larger linewidth (due to the additional damping from mode conversion to waves that propagate along the field). The gray curve corresponds to the case where the sound speed in the non-magnetic model is replaced by the fast-mode speed from the magnetic model (i.e., $c^2 \rightarrow c_f^2 = c^2 + a^2$).

are then imposed at large height and large depth that ensure wave-like disturbances propagate out of the domain and evanescent modes decay. The central layer of the background model can have a general (tabulated) stratification. The models used in this investigation are based on model S (Christensen-Dalsgaard *et al.*, 1996), with the gas pressure reduced to account for the magnetic pressure below the photosphere as in Crouch *et al.* (2005). The momentum equation,

$$\rho_0 \frac{\partial^2 \xi}{\partial t^2} = -\nabla p_1 + \frac{1}{\mu_B} (\nabla \times \mathbf{B}_1) \times \mathbf{B}_0 - \rho_1 g \hat{\mathbf{e}}_z + \rho_0 g F_z (x, y, z, t) \hat{\mathbf{e}}_z - \Gamma (\mathbf{k}, \omega, z) \rho_0 \frac{\partial \xi}{\partial t},$$

where the subscript 0 refers to background quantities, subscript 1 is for Eulerian perturbations, includes finite-sized forcing, F_z, and damping, Γ, operators that peak near the photosphere (to simulate the influence of granulation on the waves). Examples of the power spectra produced by these models are shown in Figure 1. Although these models are simple in many respects, travel-time measurements taken from synthetic data produced by these models show qualitative similarities to measurements from sunspot-like magnetoconvection simulations and observational data of sunspots (Braun *et al.*, 2010).

3. A simple model for the wave-speed perturbation in sunspots

We have tested the possibility that a simple model for the wave interactions with a sunspot can be constructed by replacing the sound speed in a non-magnetic model with the fast-mode speed, $c_f^2 = a^2 + c^2$, from a magnetic model. If this worked it would greatly simplify helioseismic inversions for subsurface structure below sunspots. In order to determine whether this is a viable approach we performed a series of tests using the TIM. Power spectra produced by this simple model are also shown in Figure 1. These preliminary results indicate that this simplified approach is not useful for modeling the wave-speed perturbation caused by a sunspot because the location of the peaks in the power and the corresponding linewidths do not agree with those for the magnetic model.

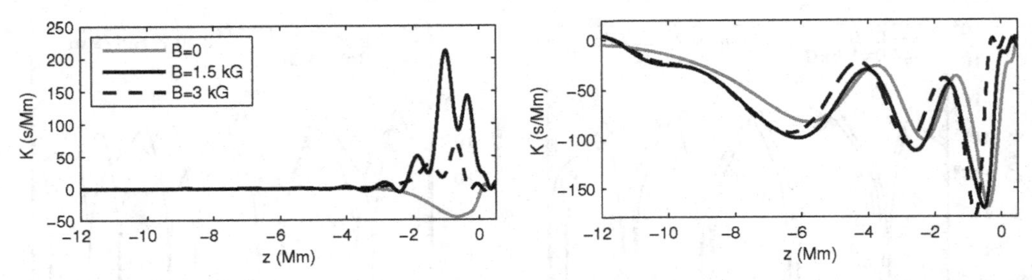

Figure 2. Time-distance travel-time sensitivity kernels for sound-speed perturbations as a function of depth for three different reference models: $B = 0$ (gray curve), $B = 1.5$kG (black solid), and $B = 3$kG (black dashed). Left: phase-speed filter centered on 12.8 km/s (TD1, Gizon & Birch, 2005). Right: phase-speed filter centered on 35.5 km/s (TD5). A vertical magnetic field clearly alters the kernels. In some cases (generally larger phase speed, e.g., TD5) the depth-variation is similar for the non-magnetic and magnetic cases, although the magnitude of the sensitivity is different. In other cases the effect of the field is drastic – both the sign and the depth-variation of the kernels is completely different in the non-magnetic and strong magnetic cases

4. Inversions for sound speed in the presence of a magnetic field

We now present an approach for inferring the sound speed in magnetic regions that includes the effects of the magnetic field on both the background structure and the wave mechanics. To construct sensitivity kernels for sound-speed perturbations we use a set of known, localized sound-speed perturbations based on B-splines. To produce a perturbation in c^2 we perturb the adiabatic index, Γ_1, keeping the pressure and density fixed. For each localized perturbation we generate power spectra with the TIM and then compute the time-distance correlation (filtering in phase speed and selecting different frequency ranges). From the correlation we compute travel times (using the method of Gizon & Birch, 2002) and the travel-time shifts between each perturbed model and the (unperturbed) reference model. With the set of known sound-speed perturbations and corresponding travel-time shifts we can then compute kernels (see e.g., Fig. 2).

To infer the subsurface sound-speed profile we use regularized least-squares, which involves the minimization of

$$\chi^2 = \sum_i \frac{1}{\sigma_i^2}\left[\delta\tau_i - \int K_i(z)\frac{\delta c^2}{c^2}(z)dz\right]^2 + \lambda \int |\frac{\delta c^2}{c^2}(z)|^2 dz \,,$$

where $\delta\tau_i$ are the measured travel-time shifts, σ_i are the measurement errors, K_i are the sensitivity kernels for sound-speed perturbations $\delta c^2/c^2$, and λ is the regularization parameter. We perform several "hare and hounds" tests using the translation-invariant model both to compute sensitivity kernels K_i (for the "hounds") and to generate synthetic measurements (for the "hares"); results for sound-speed inversions in a background model with 3kG vertical field are presented in Figure 3. For this approach to produce reliable results we find that the travel-time shifts must be measured relative to a magnetic reference model (also with 3kG vertical field); travel-time shifts induced by a magnetic model measured relative to unperturbed quiet-Sun tend to be very large and hence outside the range of validity for linear inversions.

In the left-hand panel of Figure 3 we test the inversion algorithm using kernels for a non-magnetic background model; we emphasize that this approach is not formally consistent, however, if it worked it would be advantageous because the construction of kernels for a three-dimensional, magnetic sunspot model is computationally expensive.

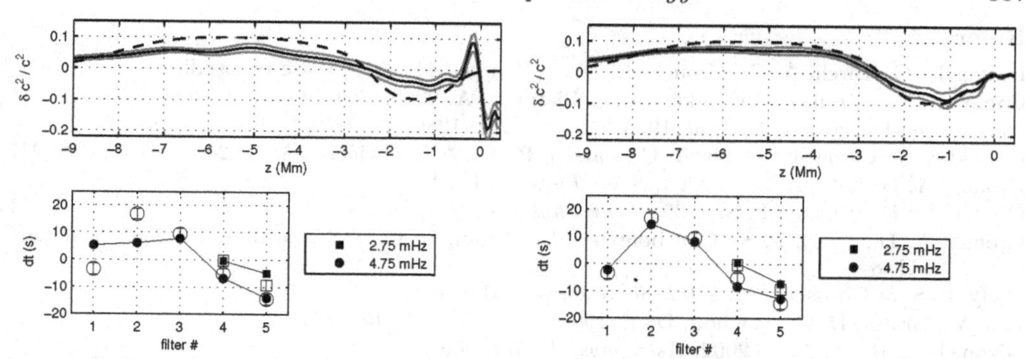

Figure 3. A "hare and hounds" test of inversions for a sound-speed perturbation in a background model with 3 kG vertical magnetic field. Upper panels: the dashed curve is the sound-speed perturbation that was used to generate the synthetic measurements. The solid black curve is the perturbation retrieved from the inversion; the gray curves are the formal 1-σ error estimates. Lower panels: travel-time shifts as a function of filter number for two representative frequency bands as labeled. The open symbols with error bars are the synthetic travel-time shifts and corresponding noise estimates. The closed symbols connected by lines are the travel-time shifts produced by the retrieved perturbation. Left: Inversion results using kernels for a non-magnetic background model. Right: Same as the left but using kernels for a background model with a 3 kG vertical magnetic field.

We find that the results are dependent on the magnitude of the regularization parameter λ. After exploring a range of values we conclude that it is not possible to retrieve a solution using non-magnetic kernels that is both consistent with the travel-time measurements and has the expected sound-speed profile. In the right-hand panel of Figure 3 we test the self-consistent approach, using kernels constructed from simulations with a background model with a 3 kG vertical field (matching the "hare"). These results show that when the correct kernels are employed this algorithm can retrieve the correct sound-speed profile.

5. Conclusions

We are developing inversion techniques for probing the subsurface structure and dynamics of sunspots and magnetic regions. This is challenging because the magnetic field in sunspots may affect both the background model (sound speed, density, etc.) and the wave mechanics (e.g., through mode conversion). We have evaluated a simple model for the wave-speed perturbation below sunspots that replaces the sound speed in a non-magnetic model by the fast-mode speed, $c_f^2 = a^2 + c^2$, from a magnetic model; our results indicate that this approach is unlikely to be useful in modeling the wave-speed perturbations in sunspots. We develop and test an inversion algorithm for inferring the sound-speed perturbation in magnetic regions. Our preliminary results indicate that this algorithm is successful only when travel-time shifts are measured relative to a magnetic reference model (with a magnetic field that is close to that of the target of the inversion) and the models used to construct the sensitivity kernels properly account for the effects of the magnetic field on both the background model and the wave mechanics.

Acknowledgement

This work was supported by NASA contracts NNH09CE41C and NNG07EI51C.

References

Braun, D. C., Birch, A. C., Crouch, A. D., & Rempel, M. 2010, these proceedings

Cameron, R., Gizon, L., Schunker, H., & Pietarila, A. 2010, *Solar Phys.*, in press

Christensen-Dalsgaard, J., *et al.* 1996, *Science*, 272, 1286

Couvidat, S., Gizon, L., Birch, A. C., Larsen, R. M., & Kosovichev, A. G. 2005, *ApJS*, 158, 217

Crouch, A. D. & Cally, P. S. 2003, *Solar Phys.*, 214, 201

Crouch, A. D. & Cally, P. S. 2005, *Solar Phys.*, 227, 1

Crouch, A. D., Cally, P. S., Charbonneau, P., Braun, D. C., & Desjardins, M. 2005, *MNRAS*, 363, 1188

Cally, P. S. & Goossens, M. 2008, *Solar Phys.*, 251, 251

Fan, Y., Braun, D. C., & Chou, D.-Y. 1995, *Astrophys. J.*, 451, 877

Gizon, L. & Birch, A. C. 2002, *Astrophys. J.*, 571, 966

Gizon, L. & Birch, A. C. 2005, *Living Reviews in Solar Physics*, 2, 6

Gizon, L., Birch, A. C., & Spruit, H. C. 2010, *ARA&A*, 48, 289

Hughes, S. J., Rajaguru, S. P., & Thompson, M. J. 2005, *Astrophys. J.*, 627, 1040

Kosovichev, A. G., Duvall, T. L., Jr., & Scherrer, P. H. 2000, *Solar Phys.*, 192, 159

The Physics of Sun and Star Spots
Proceedings IAU Symposium No. 273, 2010
D. P. Choudhary & K. G. Strassmeier, eds.

doi:10.1017/S1743921311015614

Temporal changes in the frequencies of the solar p-mode oscillations during solar cycle 23

E J Rhodes, Jr[1,2], J Reiter[3], J Schou[4], T Larson[4], P Scherrer[4], J Brooks[1], P McFaddin[1], B Miller[1], J Rodriguez[1], J Yoo[1]

[1]Department of Physics & Astronomy, University of Southern California, Los Angeles, CA 90089-1342, U.S.A.

[2]Space Physics Research Element, Jet Propulsion Laboratory, Pasadena, CA 91101, U.S.A.

[3]Zentrum Mathematik, Technische Universität München, D-80333 Munich, Germany

[4]HEPL 4085, Department of Physics, Stanford University, Stanford, CA 94305-4085, U.S.A.

Abstract. We present a study of the temporal changes in the sensitivities of the frequencies of the solar p-mode oscillations to corresponding changes in the levels of solar activity during Solar Cycle 23. From MDI and GONG++ full-disk Dopplergram three-day time series obtained between 1996 and 2008 we have computed a total of 221 sets of m-averaged power spectra for spherical harmonic degrees ranging up to 1000. We have then fit these 284 sets of m-averaged power spectra using our WMLTP fitting code and both symmetric Lorentzian profiles for the peaks as well as the asymmetric profile of Nigam and Kosovichev to obtain 568 tables of p-mode parameters. We then inter-compared these 568 tables, and we performed linear regression analyses of the differences in p-mode frequencies, widths, amplitudes, and asymmetries as functions of the differences in as many as ten different solar activity indices. From the linear regression analyses that we performed on the frequency difference data sets, we have discovered a new signature of the frequency shifts of the p-modes. Specifically, we have discovered that the temporal shifts of the solar oscillation frequencies are positively correlated with the changes in solar activity below a limiting frequency. They then become anti-correlated with the changes in activity for a range of frequencies before once again becoming positively-correlated with the activity changes at very high frequencies. We have also discovered that the two frequencies where the sensitivities of the temporal frequency shifts change sign also change in phase with the average level of solar activity.

Keywords. Sun:P-mode, sun: magnetic field, sun: solar activity

1. Introduction

At the present time, there is still disagreement over the nature of the physical mechanism which causes the frequencies of the solar *p*-mode oscillations to change during each solar cycle. Our main motivation for this study was our desire to provide more-detailed evidence of the frequency dependence of the responses of the oscillation frequencies to changes in solar activity than has been available in the past. Our hope has been that we could provide additional constraints on the possible mechanism that causes these frequency shifts.

2. Results

From MDI and GONG++ full-disk Dopplergram three-day time series obtained between 1996 and 2008 we have computed a total of 284 sets of m-averaged power spectra

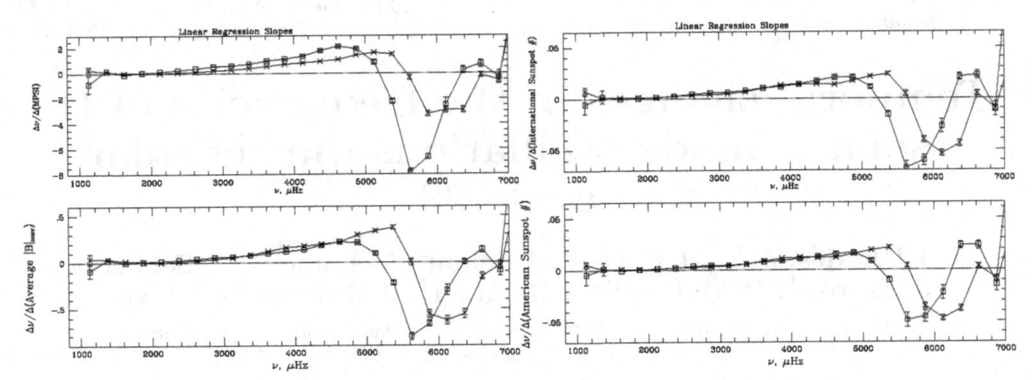

Figure 1. a.(top-left) The frequency dependence of the slopes of 25 separate pairs of linear regression analyses of binned symmetric-fit p-mode frequency shifts upon differences in Magnetic Plage Strength Index, MPSI. The 2001 GONG+ (solar maximum) results are shown as the Xs, while the 1996 MDI (solar minimum) results are shown as the squares. The $\pm$ zero-crossing frequency increased systematically with increasing solar activity. Similarly, the $\mp$ zero-crossing frequency also increased systematically between 1996 and late 2001, as can be seen at the far right side of the panel. b.(lower-left) The frequency dependence of the 25 pairs of linear regression slopes for which changes in the average of the unsigned magnitude of the corrected photospheric magnetic field, $|B|_{\mathrm{corr}}$, served as the independent variable. c.(top-right) Same as in panels a and b except that here differences in the three-day averages of the International Sunspot Number were used as the independent variable in the regression analyses. d.(bottom-right) In this case, differences in the three-day averages of the American Sunspot Number were used as the independent variable.

for spherical harmonic degrees ranging up to 1000. We have then fit these 284 sets of m-averaged power spectra using our WMLTP fitting code and both symmetric Lorentzian profiles for the peaks as well as the asymmetric profile of Nigam and Kosovichev to obtain 568 tables of p-mode parameters. We then inter-compared these 568 tables, and we performed linear regression analyses of the differences in p-mode frequencies, widths, amplitudes, and asymmetries as functions of the differences in as many as ten different solar activity indices. Examples of the frequency dependence of some of the slopes of these linear regression comparisons are shown here in Figure 1. In Figure 1a, we show the frequency dependence of the 25 different slopes which we obtained from the linear regression analyses of the GONG++ frequency shifts upon the differences in the Magnetic Plage Strength Indicator (MPSI) as the Xs, while the 25 corresponding MDI regression slopes are shown as the squares. In this panel we can see that the regression slopes did in fact change sign at two different frequencies in the case of the GONG++ and MDI analyses. Based upon these results, we have defined the frequency at which the sign of the regression slopes change from being positive to being negative as the $\pm$ zero-crossing frequency. Figure 1a also shows that both sets of regression slopes later changed sign again from negative to positive at much higher frequencies. We refer to the locations of these changes in sign as the $\mp$ zero-crossing frequencies. In fact, both the $\pm$ and the $\mp$ zero-crossing frequencies were shifted toward higher values for the GONG++ observations from solar maximum in comparison with the zero-crossing frequencies of the MDI observations from solar minimum conditions. We believe that the most intriguing new feature that can be seen in Figure 1a is the obvious displacement of the solar maximum curve (i.e., the Xs) toward the right side of the panel. The location where the frequency shifts change sign from positive to negative was moved to a higher frequency when the level of solar activity increased from the minimum levels of 1996 to the maximum levels of late-2001. To our knowledge, this frequency shift in the zero-crossing frequency of the temporal frequency shifts has never been seen before by any other helioseismology studies. Not only is the zero-crossing frequency shifted toward higher frequencies at the time of solar maximum, but the entire high-frequency portion of the GONG++ curve is shifted toward the right side of the plot. The same behavior can be seen in Figure 1b where the independent variable used in the regression analyses was the difference in average value of the unsigned magnitude of the corrected photospheric magnetic field, $|B|_{\rm corr}$. The same behavior can also be seen in Figures 1c and 1d, where we have plotted the regression slopes which we obtained when we employed the differences in the International and American Sunspot Numbers as the independent variables, respectively.

The dependence of the 19 mean $\pm$ zero-crossing frequencies upon the mean levels of the International Sunspot Number (ISN) are shown as the set of Xs in Figure 2a. There is a clear tendency for the zero-crossing frequencies to increase as the mean value of the ISN increases. In fact, the lower solid line shown in Figure 2a is the linear regression fit to all 19 of the $\pm$ zero-crossing frequencies. We also show the dependence of the 18 mean $\mp$ zero-crossing frequencies upon the mean levels of the ISN as the set of circles in Figure 2a. We have included the linear regression fit to those 18 $\mp$ zero-crossing frequencies as the upper solid line. The similarity of the two linear regression fits show that both the $\pm$ and the $\mp$ zero-crossing frequencies were positively correlated with the changes in the mean level of solar activity during each time interval. In Figure 2b we present similar results to those shown in Figure 2a except that the asymmetric fitting profiles were used instead of the symmetric Lorentzian profiles. The $\mp$ zero-crossing frequencies and regression line are very similar to the results obtained using the symmetric profiles. However, the 19 $\pm$ zero-crossing frequencies computed using the asymmetric profiles are systematically lower than the corresponding points in Figure 2a.

3. Conclusion

After discovering some of the basic behaviors of zero-crossing frequencies in recent years, we have now been able to track the way that these frequencies change on a yearly basis from 1996 to 2008. We have been pleasantly surprised to find that the newest points on our graph exhibit a linear relationship between the zero-crossing frequency and average solar activity. We believe that this change in the location of the zero-crossing frequency can be used as a new tool for probing changes in the sub-surface shear layer since some of the previous theoretical models which have been developed to explain the temporal frequency shifts have found that various combinations of temperature and magnetic field changes can result in a change in the sign of the frequency shifts at high frequencies (Jain & Roberts, 1993; Jain, 1995; Johnston *et al.*, 1995). Hence, we believe that this new observational signature will allow estimates to be made of the changes in both temperature and field strength which occurred between 1996 and 2001. To summarize this important set of results, we have demonstrated

– that the p-mode oscillations demonstrate a complicated pattern in which their frequencies are correlated with changes in solar activity at low frequencies and then become anti-correlated with those changes in activity at intermediate frequencies before once again being positively correlated with the changes in activity at the highest frequencies;

– that the frequencies where the frequency shifts change from being correlated to being anti-correlated with solar activity change with the mean level of activity;

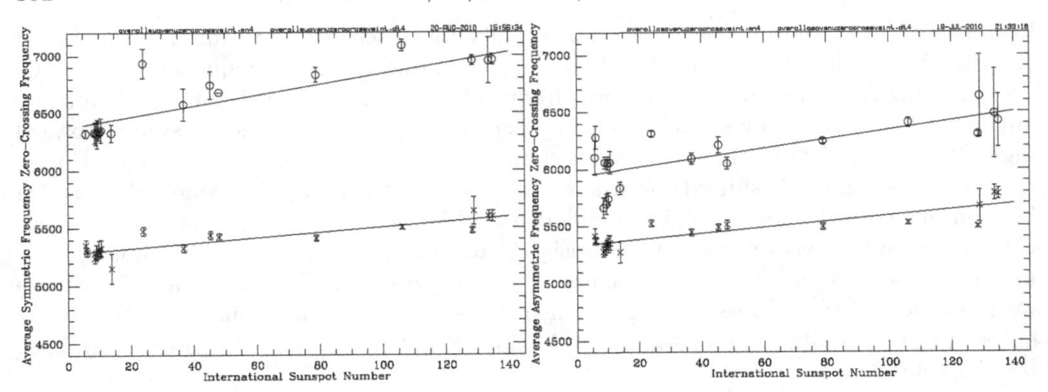

Figure 2. a.(left) Dependence of the mean ± zero-crossing frequencies (the Xs) for all 19 different time intervals upon the mean value of the ISN during each of those 19 time periods. The straight line is the linear regression fit to the 19 data points. It shows that the differences in the zero-crossing frequencies which were evident in Figure 1 were the extreme cases of a solar cycle dependence of those zero-crossing frequencies. The 18 corresponding mean ∓ zero-crossing frequencies are shown as the 18 open circles in the upper portion of the panel. The linear regression fit to the 18 mean ∓ zero-crossing frequencies is shown as the upper solid line. b.(right) Similar results to those shown in the left-hand panel except that the asymmetric fitting profiles were used instead of the symmetric Lorentzian profiles. The ± zero-crossing frequencies and regression line were very similar to the results obtained using the symmetric profiles. However, the 19 ∓ zero-crossing frequencies computed using the asymmetric profiles were systematically lower than the corresponding points in panel a.

– that the frequencies at which the frequency shifts go back to being positively correlated with the activity changes also changes with the mean level of activity;
Taken together, all of these new signatures of the temporal shifts in the p-mode frequencies should give the theorists new information to include in their refined models of the causes of these temporal frequency shifts.

Acknowledgements

In this work we utilized data from the Michelson Doppler Imager (MDI) on the Solar and Heliospheric Observatory (SOHO). SOHO is a project of international cooperation between ESA and NASA. The Stanford component of this work was supported by NASA Awards NNX07AK36G and predecessors and NAS5-02139. The portion of this research which was conducted at USC and at the Technical University of Munich was supported by the following grants to USC: NASA Grants NNX08AJ24G, NNX06AC24G, NAG5-13510, NAG5-11582, and NAG5-11001, NSF Grant AST-0307934, and by the following sub-awards from Stanford University to USC: Number 1503169-33789-A and Number 14405890-26967.Some of the computations for this work were carried out at USC using the HPCC and some of them were carried out with the use of the JPL Origin 2000 supercomputers. This work utilizes data obtained by the Global Oscillation Network Group (GONG) project, managed by the National Solar Observatory, which is operated by AURA, Inc. under a cooperative agreement with the National Science Foundation. The data were acquired by instruments operated by the Big Bear Solar Observatory, High Altitude Observatory, Learmonth Solar Observatory, Udaipur Solar Observatory, Instituto de Astrofisico de Canarias, and Cerro Tololo Interamerican Observatory.

References

Jain R., 1995, *ESA SP-376*, 69
Jain R. & Roberts B., 1993, *Astrophys. J.* 414, 898
Johnston A., Roberts B., & Wright A. N., 1995, *ASP Conf. Ser.*, 76, 264

The Physics of Sun and Star Spots
Proceedings IAU Symposium No. 273, 2010
D. P. Choudhary & K. G. Strassmeier, eds.

doi:10.1017/S1743921311015626

Correlations of magnetic features and the torsional pattern

Judit Muraközy and András Ludmány

Heliophysical Observatory, H-4010 Debrecen P.O.Box 30, Hungary

Abstract. The striking similarity between the cyclic equatorward migration of the torsional oscillation belts and the sunspot activity latitudes inspired several attempts to seek an explanation of the torsional phenomenon in terms of interactions between flux ropes and plasma motions. The aim of the present work is to examine the spatial and temporal coincidences of the torsional waves and the emergence of sunspot magnetic fields. The locations of the shearing latitudes have been compared with the distributions of certain sunspot parameters by using sunspot data of more than two cycles. The bulges of the sunspot number and area distributions tend to be located within the 'forward' belts close to their poleward shearing borders. A possible geometry of the magnetic and velocity field interacion is proposed.

Keywords. Torsional wave, sunspots

1. Introduction

The torsional oscillation was firstly reported by Howard and LaBonte 1980 as a travelling wave pattern superposed on the differential rotation. At the beginning the similarity of the equatorward migration of these belts and the butterfly diagram was very remarkable and inspiring, Labonte and Howard (1982) already provided a comparison of the belts and the latitudinal distribution of the emerging magnetic field by using magnetograms. Snodgrass (1987a, 1987b), Wilson (1987), Snodgrass and Wilson (1987) presented arguments that the torsional pattern is related to large convection cells. Subsurface measurements were reported by Kosovichev and Schou (1997) and Schou *et al.* (1998), Howe *et al.*, (2000a) published results of GONG and MDI subsurface measurements and reported torsional pattern down to about 0.92 $R_{\odot}$, this was confirmed by Komm *et al.* (2001) and Antia and Basu (2001). The torsional oscillation can be regarded to be a persistent feature extending down to about 0.92 $R_{\odot}$.

The early models followed global approach, the works of Yoshimura (1981) and Schüssler (1981) were based on the Lorentz force. Later models followed local treatment. Küker *et al.* (1996) considered magnetic quenching of Reynolds stresses modifying the differential rotation profile. In the model of Petrovay and Forgács-Dajka (2002) the sunspots modify the turbulent viscosity in the convective zone which leads to the modulation of the differential rotation.

The present paper aims at finding any spatial correlation between the torsional wave and any sunspot feature. Earlier works also indicated some spatial connections but only based on magnetograms. LaBonte and Howard (1982) averaged the latitudinal magnetic activity distribution for a longer time, whereas Zhao and Kosovichev (2004) only indicated the location of the activity belt with no distribution information. By using sunspot data, one can study the role of the most intensive magnetic fluxes hopefully leading to an answer for the question: which properties of the magnetic flux ropes could be able to modify the ambient flow resulting in the observed zonal velocity pattern.

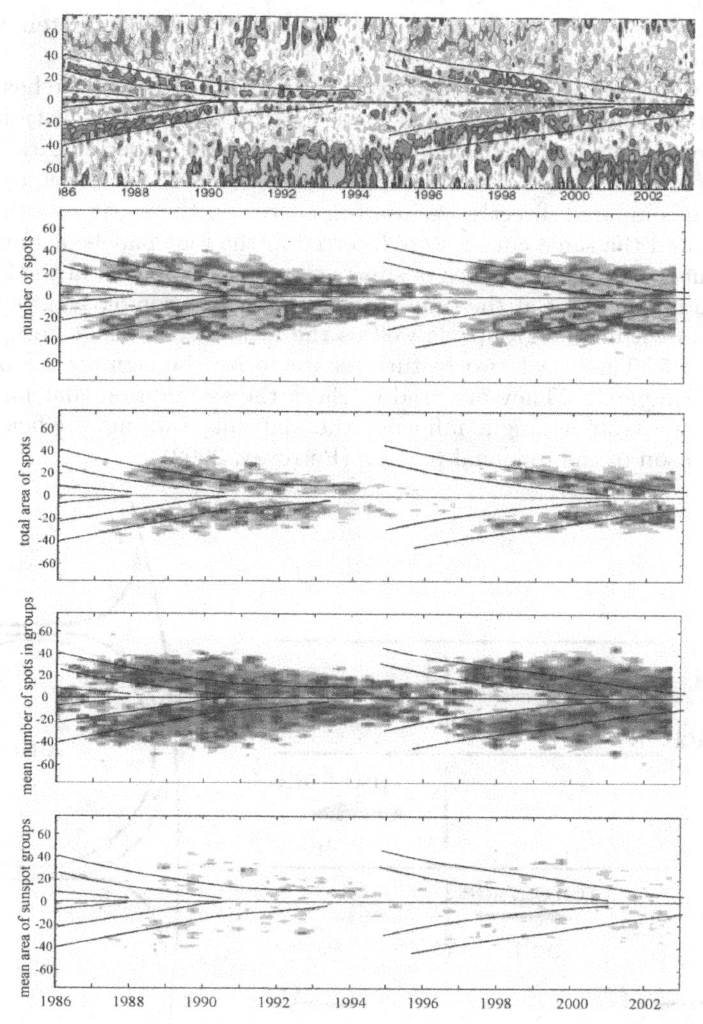

Figure 1. The upper panel shows the azimuthal velocity data of Ulrich and Boyden (2005) in black-and-white reproduction along with additionally inserted borderlines of the prograde belts. These borderlines are also inserted into the rest of the panels. The second and third panels show the distributions of the number and the area of spot groups. The fourth and fifth panels show the mean number of spots in the groups and the mean area of groups.

2. Observational material and method of analysing

The aim is to follow the temporal variation of the latitudinal distribution of sunspot parameters in comparison with the migration of the torsional waves. Four parameters were considered: the number and the total area of sunspots as well as the mean number of spots in groups and the mean area of groups. These latter two parameters can characterise the complexity of the sunspot groups to check the assumption that larger clusters of flux ropes can modify more efficiently the ambient velocity fields.

The sunspot data were taken from the most detailed sunspot database, the Debrecen Photoheliographic Data (DPD, Győri *et al.* 2010). The present study analyses the years 1986-2002. The latitudinal distributions were determined in such a way that latitudinal stripes of one degree were considered, and all applied sunspot parameters were summed up for all stripes on a three-monthly basis.

To compare the obtained distributions with the torsional pattern one has to determine the latitudinal location of the torsional wave i.e. the latitudes of fast/slow belts and the shearing zones. The torsional data were taken from the paper of Ulrich and Boyden (2005), Figure 1. shows these data for the period 1986-2002. Since fractal-like distributions cannot be compared directly we inserted curves on the shear zones indicating the prograde belts and the same curves were inserted in the four panels of sunspot distributions. The number and the total area of sunspot groups are plotted in the 2-3 panels, the areas are taken into account at the time of the largest extension of the groups. The mean number of spots within the groups as well as the mean area of spot groups are plotted in panels 4 and 5. These last two features characterize the significance of the groups, their size and complexity. They are used to check the assumption that larger and more complex flux rope systems might influence the ambient flows more efficiently and this could be the reason of the torsional pattern (Petrovay, 2002).

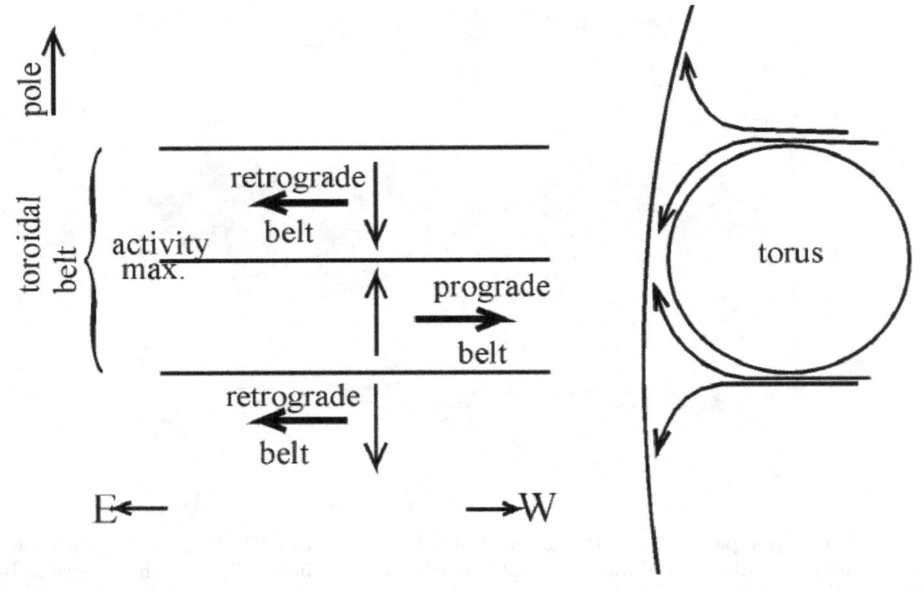

Figure 2. Schematic representation of the proposed mechanism in the northern hemisphere. Left: view from outside, right: view in the meridional plane from west.

3. Possible scenario of a flow system

The panels do not support the assumption that the torsional belts are caused by the sunspots themselves because the onset of the belts precedes the appearance of the first spot in the new cycles and the complexity of spot groups does not seem to play any role, at least in the present representation of spot group complexity. The following properties of the distributions are conspicuous:

1. The line of weight of the area occupied by sunspots is located on the borderline between the prograde and the poleward retrograde belts.

2. The equatorward borderline of the prograde belt is a definite borderline of the sunspot occurrence (see the second panel of Fig. 1).

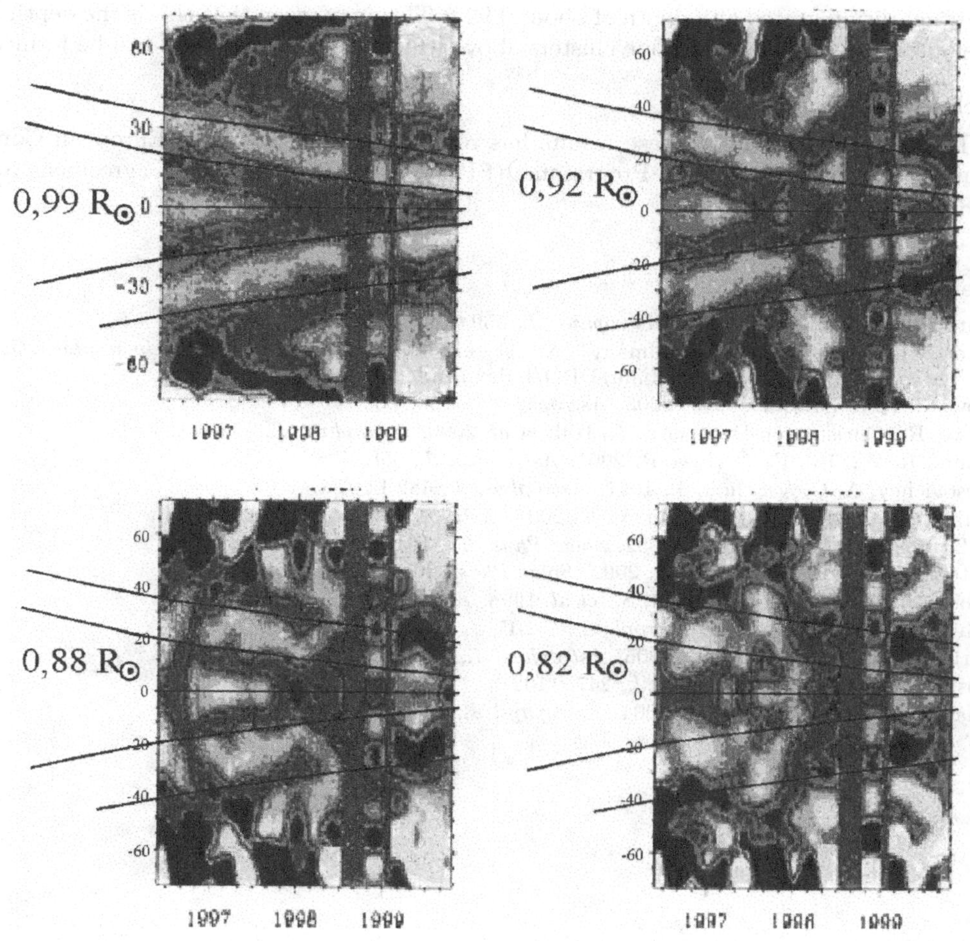

Figure 3. Internal torsional belts between 1996-1999 reported by Howe *et al.* (2000) in black-and-white reproduction. The borderlines of Fig. 1 are also inserted.

The second feature may be a signature of certain internal streams, it seems to allow the following interpretation. The cluster of toroidal magnetic field ropes may be present in the convective zone up to a certain height and may act as an obstacle for the emerging material. Out of this region (poleward and equatorward from the toroidal belt) the uprising streams may be stronger than above the torus and they turn around the obstacle, the torus. Above the torus these turning around streams flow equatorward in the poleward half of the toroidal belt and they flow poleward in the equatorward half. The former flow is then oriented eastward and the latter one westward due to the Coriolis force, these produce the retrograde/prograde belts, see Fig. 2. The inward motion at active regions has been observed by Zhao and Kosovichev (2004). This mechanism avoids the problem of the apparent violation of causality arising in models based on sunspots. In this model the source region of the sunspots, i.e. the toroidal field is the cause of the torsional pattern which is present prior to the emergence of the first sunspot in a cycle. This restricts the onset of the belts to middle latitudes.

If this scenario is correct it may have implications for the depth of the toroidal field rope clusters. Howe *et al.* (2000) reported internal torsional patterns and found that the belts get disintegrated at a depth of about $0.92_{\odot}$. This may mean that this is the depth of the outermost toroidal field rope clusters above which the torsional belts can be formed.

Acknowledgements

The research leading to these results has received funding from the European Community's Seventh Framework Programme (FP7/2007-2013) under grant agreement No. 218816.

References

Antia, H. M. & Basu, S. 2001, *Astrophys. J.*, 559, L67

Győri, L., Baranyi, T., Ludmány, A. *et al.* 2010 *Debrecen Photoheliographic Data* http://fenyi.solarobs.unideb.hu/DPD/index.html

Howard, R. & LaBonte, B. J. 2005, *Astrophys. J.*, 239, L33

Howe, R., Christensen-Dalsgaard, J., Hill, *et al.* 2000, *Astrophys. J.*, 533, L163

Komm R. W., Hill F., & Howe R. 2001, *Astrophys. J.*, 558, 428

Kosovichev, A. G. & Schou, J., 1997, *Astrophys. J.*, 482 L207

Küker, M., Rüdiger, G., & Pipin, V. V. 2005, *Astron. Astrophys*, 312, 615

LaBonte, B. J. & Howard, R. 1982, *Solar Phys.*, 75, 161

Petrovay, K., Forgács - Dajka E. 2002, *Solar Phys.*, 205, 39

Schou, J., Antia, H. M., & Basu, S., *et al.* 1998, *Astrophys. J.*, 505, 390

Schüssler, M. 1981, *Astron. Astrophys*, 94, L17

Ulrich R. K. & Boyden, J. E. 2005, *Astrophys. J.*, 620, L123

Yoshimura, H. 1981, *Astrophys. J.*, 247, 1102

Zhao, J. & Kosovichev, A. G. 2004, *Astrophys. J.*, 603, 776

The Physics Sun and Star Spots
Proceedings IAU Symposium No. 273, 2010
D. P. Choudhary & K. G. Strassmeier, eds.

doi:10.1017/S1743921311015638

Signature of collision of magnetic flux tubes in the quiet solar photosphere

Aleksandra Andic
BBSO, 40386 N. Shore Ln.
Big Bear City, CA-92314, USA
email: aandic@bbso.njit.edu

Abstract. Collision of the magnetic flux tubes in the Quiet Sun was proposed as one of the possible sources for the heating of the solar atmosphere (Furusawa and Sakai, 2000). The solar photosphere was observed using the New Solar Telescope ad Big Bear Solar Observatory. In TiO spectral line at 705.68 nm we approached resolution of 0.1″. The horizontal plasma wave was observed spreading from the larger bright point. Shorty after this wave an increase in the oscillatory power appeared at the same location as the observed bright point. This behavior matches some of the results from the simulation of the collision of the two flux tubes with a weak current.

Keywords. Bright points, oscillations

1. Introduction

The heating of the solar atmosphere is not well understood. Through the years of the research various solutions for the heating of the atmosphere were suggested and tested. One of the suggestions for the possible energy supply is collision of the magnetic flux tubes in the Quiet Sun (Furusawa & Sakai, 2000). Furusawa and Sakai model present these collisions as the source of the fast magneto acoustic waves. Those waves should form shocks higher up in the atmosphere and deposit the energy trough the dissipation of the shock front.

The solar photosphere has ubiquitous magnetic field. This fact is confirmed with the observations (Orzoco Suárez *et al.*, 2007; Lites *et al.*, 2008) and simulations (Schüssler and Vögler, 2008; Steiner *et al.* 2008). Because the magnetic field is ubiquitous the importance of its study increases in the the quiet Sun, as well. The New Solar Telescope (NST) at Big Bear Solar Observatory (BBSO) revealed previously unresolved small structures in the quiet Sun photosphere. Moreover, previously unresolved bright points turned out to be a source for the part of the oscillations that were previously exclusively contributed to the intergranular lanes (Andic *et al.* 2010). We are now able to resolve those small structures and study their dynamic in detail.

With NST at BBSO we were able to observe bright points in such detail that we're able to detect some of the signature proposed for collision of the magnetic flux tubes.

2. Data and analysis

The data-set used in this analysis was obtained on 29 July 2009. We used NST at BBSO (Goode *et al.* 2010) and Nasymut optical setup that consists from TiO broadband filter and PCO.2000 camera. A filter has central wavelength at 705.68 nm with the bandpass of 1 nm (Andic *et al.* 2010). The target of the observations was the center of the quiet Sun disk. The data sequence consist of 120 speckle reconstructed images with the temporal cadence of 15 s. The performed image reconstruction is based on the speckle masking method of von der Lühe (1993). To reconstruct images we used the procedure and code described in Wöger *et al.* (2008).

The intensity oscillations were detected using the wavelet analysis (Torrence & Compo, 1998) with the approach to the automated analysis described in detail in Andic *et al.* (2010). The speed was calculated using the nonlinear affine velocity estimator (NAVE) method which applies an affine velocity profile instead of a uniform velocity profile commonly used in the local correlation tracking (LCT) method (Chae & Sakurai, 2008).

3. Results

The Doppler velocity maps, calculated from the intensity images of the quietest Sun region, showed that some of the BP have a ring-like speed distribution. Around 11% of the total number of analyzed bright points had similar speed distribution. We speculate that this kind of distribution is connected with the collision of the magnetic flux tubes with the weak current (Furusawa & Sakai, 2000). No circular ring-like distributions were observed; the speed distribution was distorted by the presence of the surrounding granules or nearby bright points.

The analysis of the intensity oscillations of the BPs showed a spike in the oscillatory power appearing shortly after the ring-like speed distribution. This finding indicate the increased oscillatory emission. However, since we had only intensity images from one formation height in the photosphere, there was no reliable method to confirm what kind of oscillations we observed. Furusawa & Sakai (2000) expected strong fast magnetosonic waves.

Fig. 1 shows an example of the speed distribution. In this figure we have an example of two neighboring bright points from which only one shows this signature. The peak of the oscillatory power occurred few minutes after we observed the ring like distribution as visible at the Fig. 3.

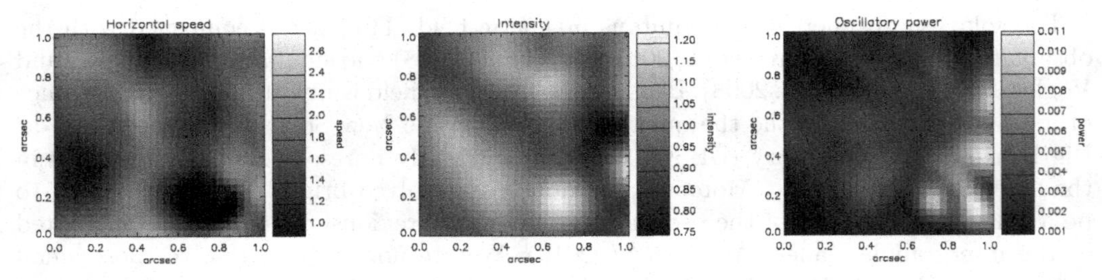

Figure 1. The image shows an example of the signature of the collision of two parallel flux tubes with weak current. The panel marked 'Intensity map' shows the analyzed bright point. The two close bright points are visible with the analyzed one centered. Panel marked with 'Horizontal speed' shows the plasma flow we detected around these bright points. It is visible that there is ring-like increase in the speed of the plasma flow all around the analyzed larger bright point. Panel marked with 'Oscillatory distribution' shows the spatial distribution of the oscillatory power around and on top of these bright points. The largest power in the area is detected at the same location as the largest bright point.

Fig. 2 shows example of the horizontal plasma flow when the analyzed bright point is surrounded only by granules, while Fig. 4 shows corresponding change in the maximum power for the same bright point. This lone bright point also demonstrate the plasma flow that has ring-like speed distribution. The deviation of the circular shape is caused by plasma flow influences of the nearby granules.

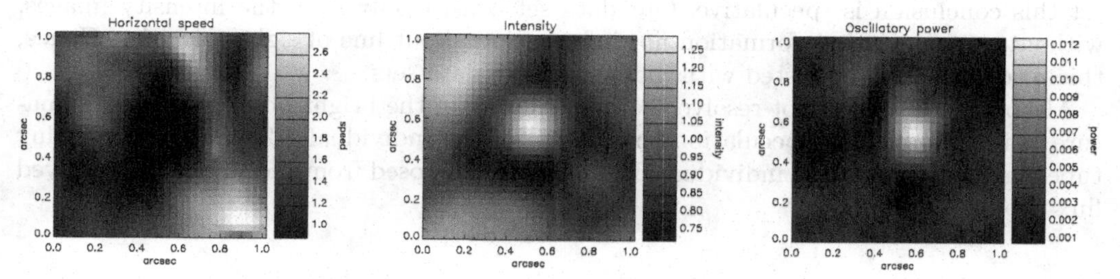

Figure 2. Another example of the slightly deformed signature appearing around the single bright point. Panels illustrate the same quantities as in image above.

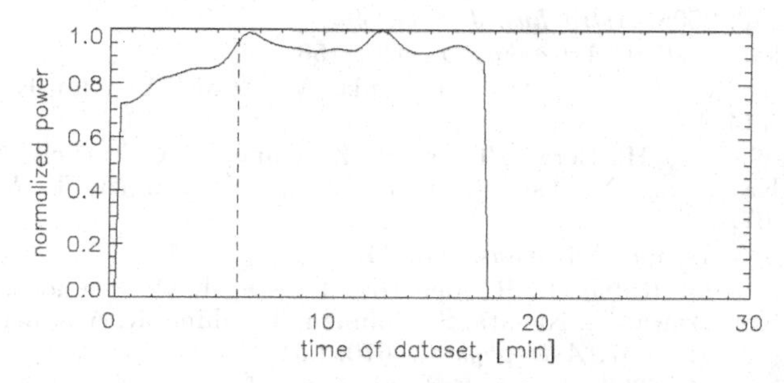

Figure 3. The image shows a temporal change in integrated oscillatory power above the brightest bright point of the couple of the bright points shown at Fig. 1. Vertical dotted line marks a moment when the ring like signature was observed. It is noteworthy that after appearance of the ring-like speed distribution we have a significant increase in the oscillatory power emitted by the bright point. The graph illustrates only power registered during the lifetime of the BP.

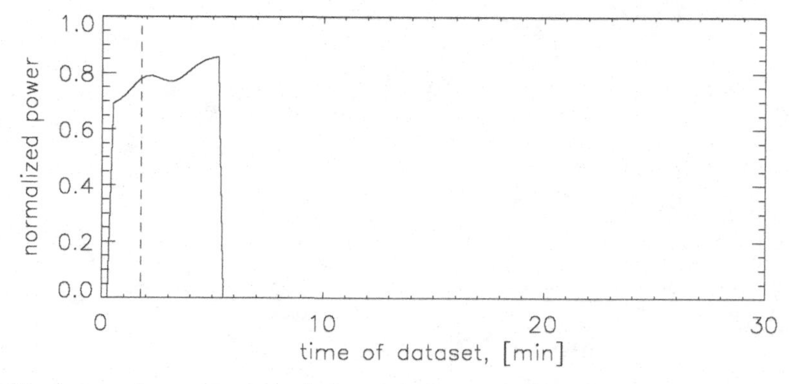

Figure 4. The image shows a temporal change in integrated oscillatory power above the bright point from the second example shown at Fig. 2. Vertical dotted line marks a moment when the ring-like signature was observed.

4. Conclusions

The ring-like speed distribution around the bright point was followed closely in time by rise in oscillatory power detected at the same location at which was the bright point in question. This sequence of the events might indicate that there was a collision of the magnetic flux tubes happening inside that bright point. Due to the limitation of our data set this conclusion is speculative. Our data set consist only from the intensity images, without any magnetic information nor information about line of sight velocities. Hence, this result needs to be tested with more complete data-sets.

Although we still cannot resolve the substructure of the bright point itself, this signature points to another speculation that inside of the individual BP are individual flux tubes, indicating that the individual BP might be composed from the smaller, unresolved flux tubes.

References

Andic, A., Goode, P. R., Chae, J., Cao, W., Ahn, K., Yurchyshyn, V., & Abramenko, V. 2010, *Astrophys. J.*,717L, 79

Chae, J. & Sakurai, T. 2008, *Astrophys. J.*, 689, 593

Furusawa, K. & Sakai, J. 2000, *Astrophys. J.*, 540,1156

Goode, P. R., Yurchyshyn, V., Cao, W., Abramenko. V., Andic, A., Ahn, K., & Chae, J. 2010, *Astrophys. J.*, 714, L31

Lites, B. W., Socas-Navarro, H., Berger, T., Frank, Z., Shine, R. A., Tarbell, T. D., Title, A. M., Ichimoto, K., Katsukawa, Y., Tsuneta, S., Suematsu, Y., Shimizu, T., & Nagata, S. 2008, *Astrophys. J.*, 672, 1237

von der Lühe, O. 1993 *Astron. Astrophys*, 268, 347

Orozco Suárez, D., Bellot Rubio, L. R., del Toro Iniesta, J. C., Tsuneta, S., Lites, B. W., Ichimoto, K., Katsukawa, Y., Nagata, S., Shimizu, T., Shine, R. A., Suematsu, Y., Tarbell, T. D., & Title, A. M. 2007, *Astrophys. J.*, 670, L61

Schüssler, M. & Vögler, A. 2008, *Astron. Astrophys*, 481, L5

Steiner, O., Rezaei, R., & Schaffenberger, W., Wedemeyer-Böhm 2008, *Astrophys. J.*, 680, L85

Torrence, C. & Compo, G. P. 1998, *Bull. Amer. Meteor. Soc.* 79, 61

Wöger, F., von der Lühe, O., & Reardon, K. 2008, *Astron. Astrophys*, 488, 375

The Physics of Sun and Star Spots
Proceedings IAU Symposium No. 273, 2010
D. P. Choudhary & K. G. Strassmeier, eds.

doi:10.1017/S174392131101564X

Photospheric data programs at the Debrecen Observatory

L. Győri, T. Baranyi, and A. Ludmány
Heliophysical Observatory of the Hungarian Academy of Sciences,
4010 Debrecen, P.O. Box 30, Hungary
email: gylajos@tigris.unideb.hu, baranyi@tigris.unideb.hu,
ludmany@tigris.unideb.hu

Abstract. The primary task of the Debrecen Observatory is the most detailed, reliable and precise documentation of the solar photospheric activity. This long-term effort started with the continuation of the Greenwich photoheliograph program, this is the Debrecen Photoheliographic Data (DPD) sunspot catalogue based on ground-based observations. The profile of the work has later been extended to space-borne observations (SOHO/MDI and SDO), to magnetic fields and faculae as well as to higher temporal resolution (one hour) and nearly real-time data supply. The database also includes historical observations. The web-presentation developed for the material is easy to search and browse. We describe the main characteristics of these catalogs, and their advantages. We summarize the recent advances in the procedure of their compilation, and the available sets of the data and images.

Keywords. Sun: sunspots, faculae, activity, photosphere, magnetic fields

1. Debrecen Photoheliographic Data

Position and area data of sunspot groups for every day were published at Greenwich (Greenwich Photoheliographic Results, GPR) until 1976. After that, Debrecen Heliophysical Observatory took over this responsibility. The Debrecen Photoheliographic Data (DPD) is mainly compiled by using white light full disk observations taken at Debrecen Observatory and its Gyula Observing Station with an archive containing more than 150,000 photoheliograms observed since 1958. Observations of several other observatories help in making the catalogue complete.

The final version of DPD completely covers the whole year with one observation/day time resolution similar to the GPR, and it contains the same data for the sunspot groups. In addition, it gives account of all sunspots, even the smallest observable ones, both umbrae and penumbrae. The estimated mean precision of position data is ~ 0.1 heliographic degrees, and that of area data is ~ 10 percents.

The instrumental background for DPD catalogue compilation has changed over the years. Recently a professional scanner is used to obtain full disk scans of photoheliograms on films or glass plates with 8kx8k spatial resolution to avoid any loss of information. Recently a 4kx4k CCD camera is being used for the observations in the Gyula Observing Station. In addition, several 2kx2k or 1kx1k CCD observations of different observatories are measured for the DPD. The rate of space-borne observations applied for DPD has also increased recently. The quality of the available films is decreasing with the increasing usage of CCD cameras, but the spatial resolution of CCD cameras has not yet reached the resolution of the former high quality films. Thus, a large amount of SOHO/MDI images were measured for DPD in the recent years to provide data of the best available quality. It is probable that this trend will continue in the next years by using the SDO images of the highest quality. However, the ground-based observations remain essential to contribute to the completeness of the material, data validation, comparison, and the security of data flow.

The software package called SAM (Sunspot Automatic Measurement) (Győri 1998, 2003, 2005) was developed to handle the scans of the ground-based photographic observations. After that, it was further developed to handle the ground-based and space-borne CCD FITS images. Now it can be applied to any full disk digitized solar image if the orientation of the image is known or can be derived from the image itself. The automatic sunspot-measuring program determines the borders of umbrae and penumbrae, finds the centers of spots. The package automates (as far as possible) the compilation of the sunspot catalogue which contains all relevant pieces of information about the spots.

The DPD is available for 1986-2010 at present. For some years, the final version is published but for other years only a preliminary version is available. However, the DPD is continuously improved, revised, and extended in its content. The efforts to achieve the full coverage of the post GPR-era also continue. In addition, we have started compiling catalogs of higher-level data (e.g. catalogue of sunspot group tilt angles, catalogue of recurrent active regions).

2. SOHO/MDI-Debrecen Data

The SOHO/MDI images (1024x1024 pixel) obtained as proxies for the continuum intensity near the NiI absorption line at 676.8 nm by combining the standard five filtergrams (Scherrer *at al.* 1995). Recently the compilation of SOHO/MDI-Debrecen Data (SDD) hourly sunspot catalogue has been started by using the SOHO/MDI continuum intensity images and magnetograms. This catalogue is similar to that of DPD in its data format and image products but the time cadence is 1 hour when MDI observations allow. The other difference is that the SDD contains magnetic information because the SAM is suitably modified to determine the mean line-of-site magnetic field of umbral and penumbral parts of spots from the quasi-simultaneous magnetogram.

We use the Full Disk Continuum images from the Hourly Data Sets Level 1.8 and the recently recalibrated magnetograms. In this data set, there are observations with different time resolution (from one/day to one/min) in different time intervals. In the latest years, usually there is one intensity image per hour but not in every hour. To get a data set with a more or less regular time cadence we decided to use roughly one image/hour time resolution. The best available intensity image is chosen within the time interval (hour - 30 min, hour + 30 min) which is the closest to the center of the time interval and the time-difference between the next images is at least 30 minutes. The best available magnetograms closest in time to the intensity images are chosen from the merged data sets of Full Disk Magnetograms of the Hourly Data Sets and Daily Data Sets. The set of images contains the intensity images with the nearest magnetograms selected after using quality-filters and time sequence criteria. If an hourly data is missing from the catalogue it means that there is no intensity observation within the given time interval, or the available intensity observation is not measurable (perfect), or there are no measurable intensity images and magnetograms available with less than 50 min time difference. The full disk SOHO/MDI intensity images are enlarged (3x), corrected for limb darkening, flat fielded and transformed to negative. If the position angle of the image is 180 degree, the processed image is rotated to the normal position; otherwise, the image preserves its original orientation.

The full-disk version of the SDD catalogue (fdSDD) is based on the same hourly time resolution MDI observations, but the spots are numbered by the computer program on the basis of their longitude and they is not assigned to sunspot groups. In the fdSDD catalogue, all sunspots have their area, position and magnetic field values similar to SDD sunspot group catalog.

3. Presentation of sunspot data

The electronic versions of the DPD and SDD are freely downloadable and usable. Their formats and contents are developed continuously to provide an easily usable set of sunspot data and solar images. Starting from this year the dataset has been refreshed with the data for recent days on a daily basis.

Several kinds of numerical data are presented in ASCII files: 1. yearly tables for daily (DPD) and hourly (SDD) sums of area data 2. time series of daily data 3. tables containing the whole area of sunspot groups and their mean position data, 4. tables for the sunspot area and position data, 5. combined datasets containing all data.

The appended images show the detailed morphology. Scans of sunspot groups are published in JPG and FITS format with numbered spots. Full disk white light images and magnetic observations are appended to provide the most information available concerning the sunspots. The processed white-light or continuum images are published on-line in about 4kx4k or 3kx3k image size. The magnetograms are available in its original format and converted to JPG images too.

These data and images are accessible by ftp to provide an easy bulk download. The whole material is also provided in a graphical presentation containing daily full-disk drawings of the observed sunspot groups, and the full-disk white-light and magnetic observations. The days can be surveyed by turning the pages, and all the information on a sunspot group can be reached by clicking on the group number.

A new facility has been developed recently. There is an on-line MySQL query at the website of the catalogues which makes possible a quick and easy selection of the numerical data.

4. Catalogue of white-light faculae

A further novelty at the Debrecen Observatory is the publication of a catalogue of continuum faculae. It is based on the SOHO/MDI full disc images for the whole SOHO era. The full-disk facular data have the same format as full-disk sunspot data, but umbral areas are obviously not measured for faculae, and the first column of magnetic data contains the line-of-site magnetic field value at the brightest pixel within the facular contour in the intensity image.

5. Historical Solar Image Database (HSID)

A long-term effort is in progress to reveal, gather, digitize and evaluate all existing historical full-disc solar observations. The Debrecen Observatory preserves historical drawings between 1872 and 1919 gathered by two observatories: the material of Ógyalla Observatory (founded by Miklós Konkoly-Thege, 1842-1916) covers the period 1872-1891, the observations of Haynald Observatory in Kalocsa are made by Julius Fényi S.J. between 1880 and 1919. The old observations are indispensable in the long-term studies of solar activity, and they contain an invaluable amount of information, which can and must be reconstructed with the recent technology.

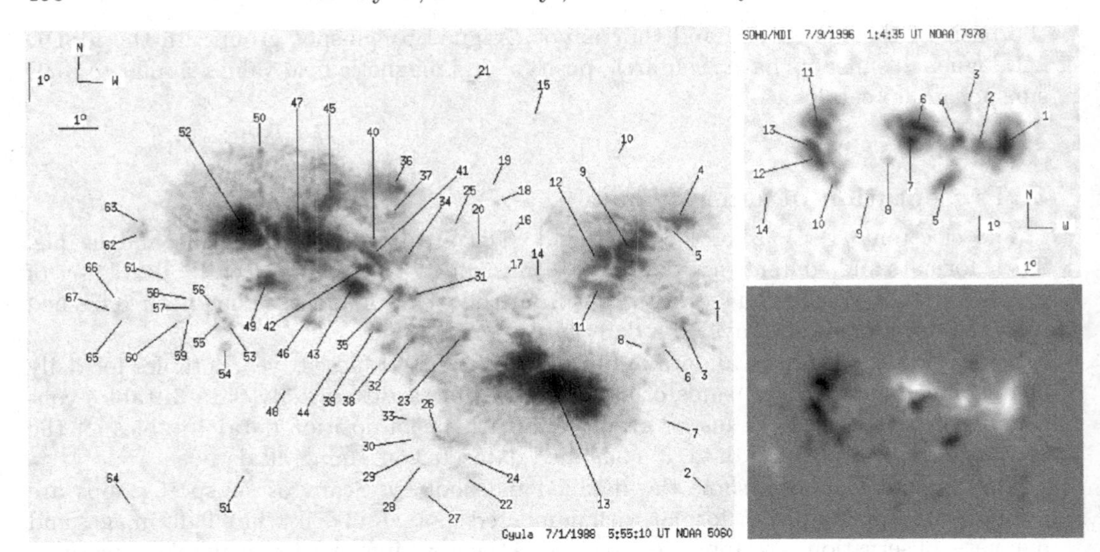

Figure 1. Example for an image of a sunspot group in DPD, and an image of a sunspot group from SDD with its magnetogram.

6. Summary

Both DPD and SDD contain data of all sunspot groups and sunspots, even those of the smallest observable ones. The estimated mean precision of position data is $\sim$ 0.1 heliographic degrees, and that of area data is $\sim$ 10 percents.

The DPD is mostly suitable for long-term studies because of the daily sampling, its spatial scale extends from full disc to about 1 arcsec. Additional data are also provided, e.g. catalogues of sunspot group tilts and recurrent active regions.

An additional advantage of the SDD catalogue is the roughly hourly time resolution, furthermore, it also contains magnetic information for the sunspots.

The content of the online database and its presentation: numerical datasets; appended images of active regions with numeration of spots (jpg, fits); full disc white-light or continuum images appended; magnetic observations appended; graphical HTML-presentation with full-disk and sunspot group images; on-line MySQL query for the numerical database; quick-look catalogue refreshed on a daily basis.

All the data and images mentioned in this paper are available at the site of Debrecen Observatory at http://fenyi.solarobs.unideb.hu/. Some further information can also be found here concerning the data, the observations, and the contributing observatories. The data sets are available without restrictions, but the users should refer to this paper in the References when the data or images are used in a scientific publication.

Acknowledgements

This work was supported by the European Community's Seventh Framework Program (FP7 SP1-Cooperation) under grant agreement No. 218816.

SOHO is a mission of international cooperation between ESA and NASA. The authors gratefully acknowledge the past and ongoing effort of the MDI team.

We express our deepest gratitude to the colleagues at the 16 collaborating observatories for putting the necessary material at our disposal.

References

Győri, L. 1998, *Solar Phys.*, 180, 109

Győri, L., 2003, *Compiling sunspot catalogue: the principles, Proc. Solar Image Recognition Workshop (SIRW)*, Brussels, 2003

Győri, L., 2005, Automated determination of the alignment of solar images, *Hvar Obs. Bull.*, 29, 299

Scherrer, P. H., Bogart, R. S., Bush, R. I., *et al.* 1995, *Solar Phys.*, 162, 129

The Physics of Sun and Star Spots
Proceedings IAU Symposium No. 273, 2010
D. P. Choudhary & K. G. Strassmeier, eds.

doi:10.1017/S1743921311015651

Chromosphere above sunspots as seen at millimeter wavelengths

Maria A. Loukitcheva[1,2], Sami K. Solanki[2] and Stephen M. White[3]

[1]Astronomical Institute, St.Petersburg University, Universitetskii pr. 28,
198504 St.Petersburg,Russia
email: lukicheva@mps.mpg.de

[2]Max-Planck-Institut für Sonnensystemforschung, D-37191 Katlenburg-Lindau, Germany
email: solanki@mps.mpg.de

[3]Air Force Research Lab, Albuquerque, NM, 87117 USA
email: AFRL.RVB.PA@Hanscom.af.mil

Abstract. Millimeter emission is known to be a sensitive diagnostic of temperature and density in the solar chromosphere. In this work we use millimeter wave data to distinguish between various atmospheric models of sunspots, whose temperature structure in the upper photosphere and chromosphere has been the source of some controversy. From mm brightness simulations we expect a radio umbra to change its appearance from dark to bright (compared to the Quiet Sun) at a given wavelength in the millimeter spectrum (depending on the exact temperature in the model used). Thereby the millimeter brightness observed above an umbra at several wavelengths imposes strong constraints on temperature and density stratification of the sunspot atmosphere, in particular on the location and depth of the temperature minimum and the location of the transition region. Current mm/submm observational data suggest that brightness observed at short wavelengths is unexpectedly low compared to the most widely used sunspot models such as of Maltby *et al.* (1986). A successful model that is in agreement with millimeter umbral brightness should have an extended and deep temperature minimum (below 3000 K), such as in the models of Severino *et al.* (1994). However, we are not able to resolve the umbra cleanly with the presently available observations and better resolution as well as better wavelength coverage are needed for accurate diagnostics of umbral brightness at millimeter wavelengths. This adds one more scientific objective for the Atacama Large Millimeter/Submillimeter Array (ALMA).

Keywords. Sun: atmosphere, sun: chromosphere, sun: radio radiation, sun: sunspots

1. Introduction

Since the 1980s there have been several attempts to build a comprehensive semiempirical model of the atmosphere above a sunspot umbra. Deriving such models is a complicated process that tries to balance observations of a range of optical/UV lines (mostly formed in non-LTE conditions, see, e.g., Solanki (2003) and, when available, radio measurements of brightness spectra, together with ionization equilibrium and radiative transfer calculations that include heat transfer down from the corona as well as other factors (e.g. Fontenla *et al.* 1993). The radio data are particularly valuable because the measurements are in the Rayleigh–Jeans limit, meaning that measured brightness temperatures actually represent thermal electron temperatures in the optically thick atmosphere. The temperature at which a given frequency is optically thick (via thermal bremsstrahlung and, at very high frequencies, $H-$ opacity) is sensitive to density and temperature, and hence the radio data provide important constraints for modeling (e.g., Loukitcheva *et al.* 2004). However, while there are numerous radio temperature measurements of the quiet–Sun atmosphere, there are very few for sunspots, particularly at millimeter wavelengths, since good spatial resolution is required to isolate the sunspot temperature from its surroundings, and single–dish measurements generally do not have enough resolution. The highest resolution single–dish data, at 10–20 arcsec, are from JCMT (Lindsey & Kopp 1995) and CSO (Bastian *et al.* 1993) but at submm, not mm, wavelengths. The sunspot umbra appears darker than the quiet Sun at these wavelengths. At millimeter wavelengths the instrument best suited for high resolution observations of sunspots is the 10-element Berkeley-Illinois-Maryland Array (BIMA) operating at 3.5 mm (Welch *et al.* (1996)). In this work we used BIMA observations of the temperature decrement above a sunspot umbra at 3.5 mm, where model differences are large, to distinguish between various atmospheric models of sunspots.

2. BIMA 3.5 mm observations

On August 31, 2003 BIMA observed a small active region NOAA 10448 north of the solar equator. The images were reconstructed using the maximum entropy method (MEM) and restored with a Gaussian beam of 12 arcsec (White *et al.* (2006)). The 3.5 mm image from BIMA is shown in Fig. 1 together with contemporaneous MDI photospheric magnetogram and BBSO CaII K images. The largest sunspot in the region is at the trailing (eastern) end of the region (see corresponding umbral and penumbral contours in Fig. 1). Its photospheric magnetic field does not exceed 2000 G. Clear depressions over this sunspot are seen both in radio and in calcium. The loop of bright emission encircling the sunspot in the CaII K image is essentially reproduced in the 3.5 mm image. From the BIMA maps we evaluate the "3.5 mm umbra" to be approx. 400 K cooler than the Quiet Sun at this wavelength. Note that due to the use of MEM for image restoration we probably underestimate the umbral brightness decrement and 400 K is likely the lower limit for it. For comparison with the models we completed 3.5 mm brightness with the mm data at 0.35 mm, 0.85 mm, and 1.2 mm from Lindsey & Kopp (1995), plotted with circles in Fig. 2 (see next Section).

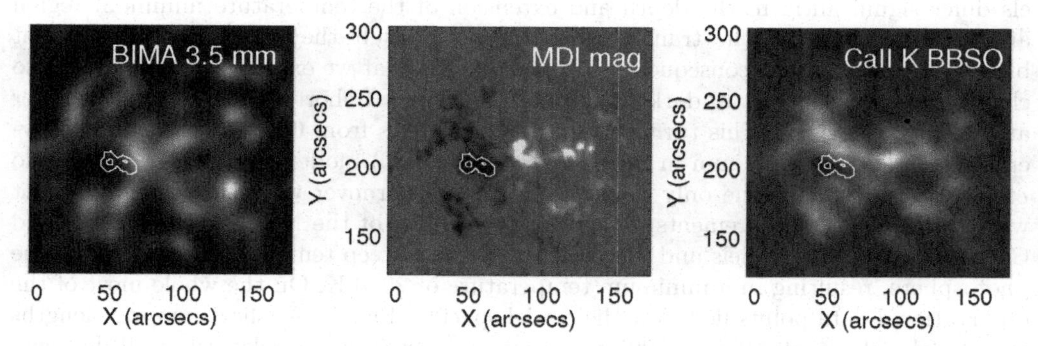

Figure 1. Active region NOAA 10448 observed on August 31, 2003. From left to right: BIMA image at 3.5 mm, MDI photospheric magnetogram and BBSO CaII K image. The images were rotated to the same time and coaligned. White contours mark umbra and penumbra of the analyzed sunspot.

3. Millimeter brightness spectra expected from sunspot models

In order to distinguish the sunspot models, we have calculated the expected sub/mm brightness temperatures at 24 selected wavelengths in the range 0.1-20 mm. We analyzed classical sunspot models as well as their recent updates. These include the sunspot model of Avrett (1981), model M of Maltby *et al.* (1986), the sunspot model of Severino *et al.* (1994), model B of Socas–Navarro (2007), and model S of Fontenla *et al.* (2009). The temperature versus height dependence of each of these models is shown in Fig. 2, together with the standard reference quiet–Sun atmosphere (model C of Fontenla *et al.* (1993), commonly referred to as $FALC$). The calculations were done under the assumption

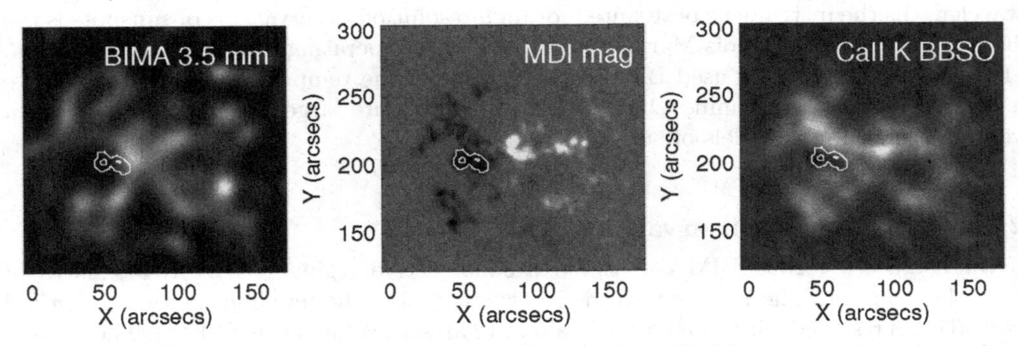

Figure 2. *Left panel:* The electron temperature as function of height in a number of standard models of the solar chromosphere above a sunspot, as labeled (see text). The solid black line is the reference quiet-Sun atmosphere known as $FALC$. *Right panel:* The brightness temperature as a function of wavelength for each model. Circles mark reported brightness temperature measurements

that bremsstrahlung opacity is responsible for the mm continuum radiation. We assume that gyroresonance opacity is negligible at these high frequencies. The expression for bremsstrahlung opacity in magnetic media was taken from Zheleznyakov (1996). The magnetic field was modelled by a vertical dipole buried under the photosphere according to Zlotnik (1968).

Figure 2 also shows the predicted brightness temperature spectra from each model, together with measurements culled from Lindsey & Kopp (1995). The sunspot models differ significantly in the depth and extension of the temperature minimum region and in the location of the transition region. As a result they produce very different brightness spectra. One consequence of the models is that we expect the radio umbra to change its appearance from darker than the quiet Sun at short wavelengths to brighter at longer wavelengths. This turnover wavelength ranges from 6 mm in the case of Severino *et al.* (1994) to 0.3 mm in the Maltby *et al.* (1986) model. The model of Severino *et al.* (1994), which is the only model that has the turnover wavelength in agreement with the 3.5 mm measurements, is a modified version of the Maltby *et al.* (1986) and Caccin *et al.* (1993) models and is characterized by a steep temperature gradient in the photosphere, resulting in a minimum temperature of 2900 K. On the whole most of the observational data points lie below the model spectra (Fig. 2). At short mm wavelengths the model of Fontenla *et al.* (2009) represents the observations relatively well, however longward of 3 mm the discrepancy between the observed and calculated brightness is substantial.

4. Conclusions

Millimeter brightness observations impose strong constraints on temperature and density stratifications of the sunspot atmosphere, in particular on the location and depth of the temperature minimum and the location of the transition region. Current mm/submm observational data suggest that sunspot models, including the most widely used sunspot models of Maltby *et al.* (1986), the classical model of Avrett (1981), and recent models of Socas–Navarro (2007) and Fontenla *et al.* (2009), are too hot at chromospheric heights. A successful model that is in agreement with millimeter umbral brightness should have an extended and deep temperature minimum (below 3000 K) as in the model of Severino *et al.* (1994). A new sunspot model by Avrett (this Proceedings) also possesses a very deep temperature minimum and therefore might result in reasonable agreement with the mm observations.

However, due to the spatial resolution limit of 12 arcsec we are not able to resolve the umbra cleanly in the present observations and the results obtained in this work are preliminary. A detailed study of the appearance of sunspot umbrae at mm waves requires substantially better resolution. Furthermore, good wavelength coverage is needed for accurate diagnostics of the turnover wavelength and hence successful modeling of the sunspot atmospheric temperature structure. We place high expectations on the Atacama Large Millimeter/Submillimeter Array (ALMA), which will start Early Science observations in 2011. With the instrument wavelength range of 0.3 to 10 mm and up to 64 antennas, ALMA will be an extraordinarily powerful instrument for studying the three-dimensional thermal structure of sunspots at chromospheric heights.

References

Avrett, E. H. 1981, in: L. E. Cram, J. H. Thomas (eds.), *The Physics of Sunspots* (National Solar Obs., Sunspot, NM), p. 235

Bastian, T. S., Ewell, M. W., Jr., & Zirin, H. 1993, *Astrophys. J.*, 415, 364

Caccin, B., Gomez, M. T., & Severino, G. 1993, *Astron. Astrophys*, 276, 219

Fontenla, J. M., Avrett, E. H., & Loeser, R. 1993, *Astrophys. J.*, 406, 319

Fontenla, J. M., Curdt, W., Haberreiter, M., Harder, J., & Tian, H. 2009, *Astrophys. J.*, 707, 482

Lindsey, C. & Kopp, G. 1995, *Astrophys. J.*, 453, 517

Loukitcheva, M., Solanki, S. K., Carlsson, M., & Stein, R. F. 2004, *Astron. Astrophys*, 419, 747

Maltby, P., Avrett, E. H., Carlsson, M., Kjeldseth-Moe, O., Kurucz, R. L., & Loeser, R. 1986, *Astrophys. J.*, 306, 284

Severino, G., Gomez, M. T., & Caccin, B. 1994, in: R. J. Rutten, C. J. Schrijver (eds.), *Solar Surface Magnetism* (Dordrecht: Kluwer), p. 169

Socas-Navarro, H. 2007, *Astrophys. J. Sup.*, 169, 439

Solanki, S. K. 2003, *Astron. Astrophys. Rev.*, 11, 153

Welch, W. J., *et al.* 1996, *Pub. Astr. Soc. Paci.*, 108, 93

White, S., Loukitcheva, M., & Solanki, S. K. 2006, *Astron. Astrophys*, 456, 697

Zheleznyakov, V. V. 1996, *Radiation in Astrophysical Plasmas*

Zlotnik, E. Y.a. 1968, *Soviet Astron.*, 12, 245

The Physics of Sun and Star Spots
Proceedings IAU Symposium No. 273, 2010
D. P. Choudhary & K. G. Strassmeier, eds.

doi:10.1017/S1743921311015663

Study of sunspot motion and flow fields associated with solar flares

Shuo Wang[1], Chang Liu[1] and Haimin Wang[1]

[1]Space Weather Research Laboratory, New Jersey Institute of Technology,
University Heights, Newark, NJ 07102-1982, USA
email: sw84@njit.edu

Abstract. Evolution of sunspot structure and photospheric magnetic fields are important to understand how the flare energy is built up and released. With high-resolution optical data, it is possible to examine in details the optical flows of the photosphere and their relationship to the flaring process. Using G-band and Stokes-V data obtained with Hinode Solar Optical Telescope (SOT), we study the sunspot motion and flow fields associated with the 2006 December 13 X3.4 flare in NOAA AR 10930. We calculate the centroids of the delta spot umbrae lying in opposite magnetic polarities, and use two different methods to derive the photospheric flow fields of the AR. We find that the shearing motion before the flare changes to unshearing motion associated with the eruption. A decrease of average velocity of shear flow is found to be associated with the flare, with a magnitude of 0.2 km s^{-1}.

As a related study, we also test implementing the recently developed differential affine velocity estimator for vector magnetograms (DAVE4VM; Schuck, P. W 2008) technique for the magnetic field observations obtained by the Big Bear Solar Observatory (BBSO) and Helioseismic Magnetic Imager (HMI) on board the Solar Dynamic Observatory (SDO). Using this method to analyze changes of active region magnetic fields associated with flares may shed new light on the cause and effect of flaring process.

Keywords. Sun: magnetic fields, sun: flares, sunspots

1. Introduction

Evolution of sunspot structure and photospheric magnetic field associated with solar flares have recently drawn increasing attention. A sudden change of center-of-mass (CoM) separation of the two opposite polarities of a δ spot structure was found to be associated with large flares (Wang 2006). In the direction parallel to the magnetic polarity inversion line (PIL), the CoM separation always decreases, while in the direction perpendicular to the PIL, the CoM separation increases and decreases when the active region magnetic fields have a divergence and convergence motions, respectively. To further understand the variation of CoM separation of sunspots related to the flaring process, we investigated the high-resolution Hinode observations (Kosugi *et al.* 2007) to obtain the spot motion and flow fields associated with the 2006 December 13 X3.4 flare. Such a study can be greatly advanced with the aid of temporal evolution of the three-dimensional (3D) velocity fields of the source active region, which can now be achieved by the DAVE4VM technique using vector magnetic field observations.

The DAVE4VM models motion of a vector of images with normal component of the ideal magnetic induction equation: $\partial_t B_z + \nabla_h \cdot (B_z V_h - V_z B_h) = 0$, where the plasma velocity V and the magnetic fields B are decomposed into a local Cartesian coordinate system with vertical direction along the z-axis and the horizontal plane, denoted generically by the subscript h, containing the x- and y-axes. Here we applied this method to

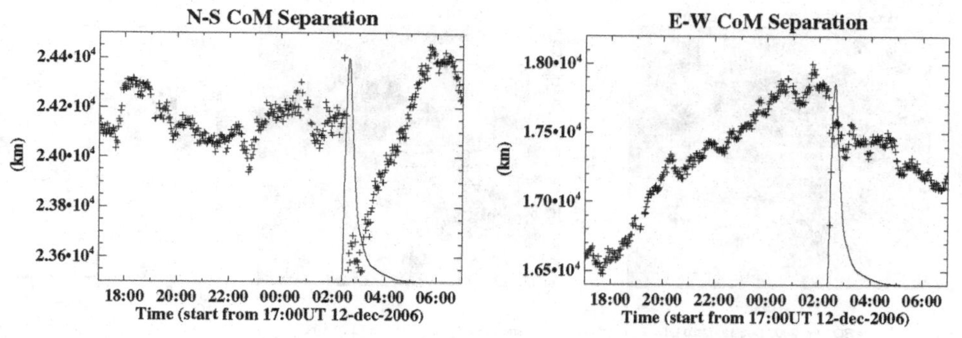

Figure 1. Time profile of CoM separation (pluses) in G-band intensity of NOAA AR 10930 between the northern (positive) and southern (negative) umbrae in the north-south (left) and east-west (right) direction, overplotted with GOES 1–8 Å soft X-ray light curve.

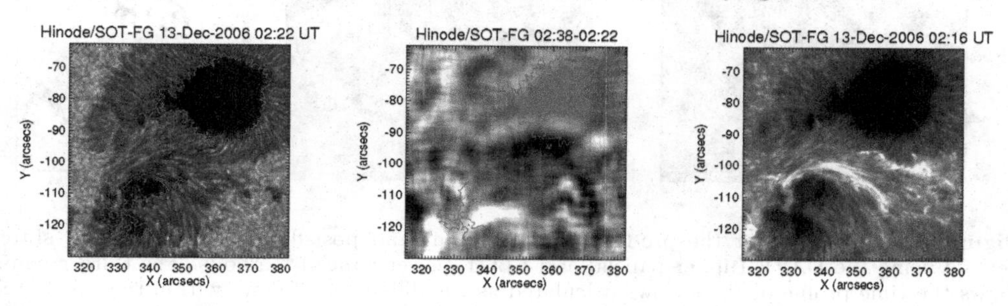

Figure 2. From left to right: Pre-flare G-band image, G-band difference image between post- and pre-flare states, and a Ca II H image of the flare. Red contour shows position of umbra boundary.

NOAA AR 10365 on 2003 May 28 and NOAA AR 11057 on 2010 March 29, the latter of which belongs to the new solar cycle 24.

2. Data Sets and analysis methods

On 2006 December 13, an X3.4 two-ribbon flare occurred in NOAA AR 10930. Hinode fully covered this event, and we used G-band (430.5 nm) and Stokes-V (Fe I 630.2 nm) images obtained by its onboard SOT (Tsuneta *et al.* 2008). Active region flow fields from before to after the flare were derived and compared using both the DAVE (Schuck, P. W 2006) and Fourier LCT (Welsch *et al.* 2004) techniques.

Vector magnetograms of NOAA AR 10365 associated with an X3.6 flare on 2003 May 28 were taken by BBSO. For NOAA AR 11057 on 2010 March 29, magnetic field observation was made by the state-of-the-art SDO/HMI. The 3D active region velocity fields of both active regions were derived using DAVE4VM. The initial test shows the promise that we will be able to carry out systematic studies of flare-related temporal evolution of photospheric flow field in the near future.

3. Results

Figure 1 shows that the shearing motion of the two sunspots with opposite polarities of the δ configuration of NOAA AR 10930 changes to unshearing motion in the direction parallel to the PIL (east-west), which seems to be cotemporal with the GOES-class X3.4 flare. In Figure 2, the difference images in G-band intensity are taken immediately

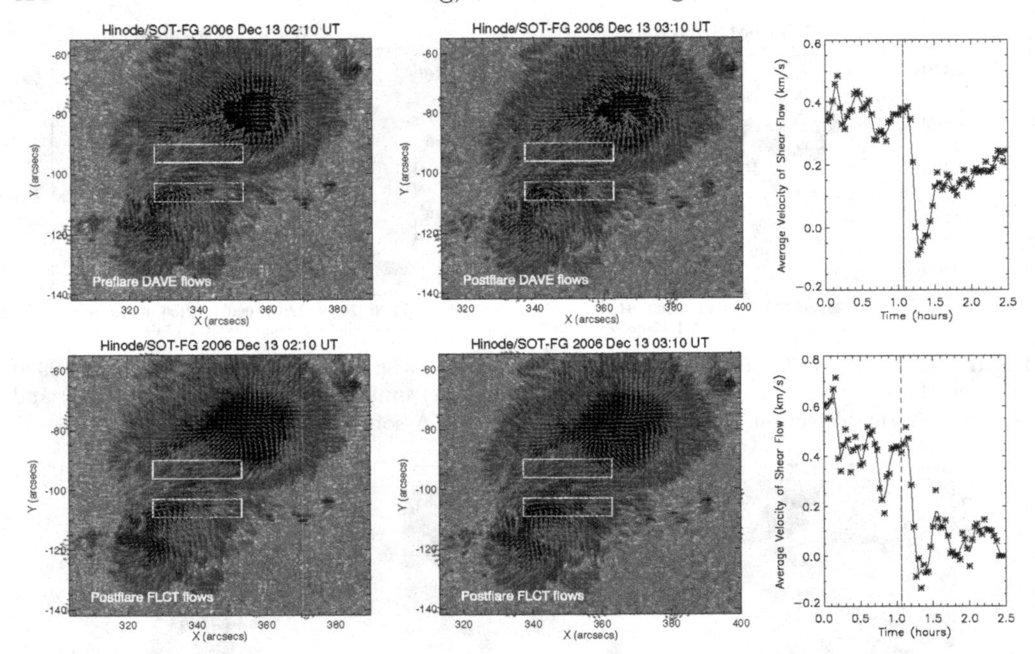

Figure 3. Flow fields for the pre-flare (first column) and post-flare (second column) states derived using the DAVE (upper panels) and FLCT (lower panels) methods. The third column shows the time profile of shear flows calculated as the difference of flows within the two boxed regions, and the start time is 1:10:40 UT 2006 December 13.

before and after the flare (post-flare 02:38:36 UT minus pre-flare 02:22:35 UT state), which reveals the enhancement (black) and decay (white) of sunspot structure that mainly occurred at penumbral regions. The enhancement and decay patterns appear to be consistent with the sudden change in the CoM separation. A Ca II image at 02:16 UT shows the position of initial flare ribbons. The locations of penumbral decay and enhancement are related to flare emissions similar to previous studies (e.g., Liu *et al.* 2005). In Figure 3, the white boxes denote the regions where the temporal evolution of the mean shear flow (curves with asterisks) is calculated. Overall, the DAVE method seems to produce more consistent results of flow field with the sunspot morphology. The decrease of average velocity of shear flow around the flaring PIL with a magnitude of 0.2 km s^{-1} is comparable to that obtained by Tan *et al.* 2009 using LCT method (0.3 km s^{-1}).

In Figure 4 (left), we present the longitudinal magnetic field B_z of NOAA AR 10365 overplotted with the 3D DAVE4VM velocities associated with the occurrence of 2003 May 28 X3.6 flare. The horizontal flows are up to 6 km s^{-1}, and the contour levels for vertical flows are 0.1, 0.2, and 0.3 km s^{-1}. Enhancement of the 3D flow fields (i.e., diverging and vertical flows) seem to be spatially correlated with the flare around the same segment of the PIL. Note that these are mainly to test the implementation of the DAVE4VM technique on the ground-based magnetograph data. The observed flow signal during the flare may not be true as the magnetic field measurement can be seriously affected by flare emissions. Moreover, seeing variation can also contribute to about 1 km s^{-1} noise in the velocity measurement.

Figure 4 (right) shows B_z of NOAA AR 11057 on 2010 March 29 overplotted with the 3D DAVE4VM velocities (same as above but the contour levels are 0.01 and

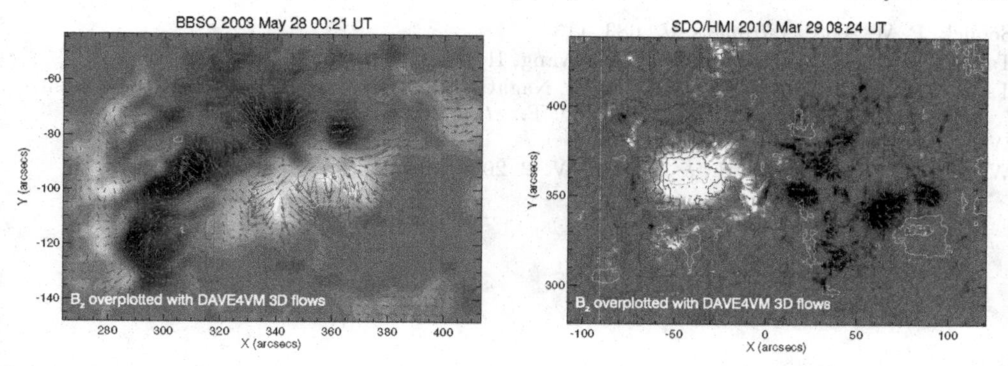

Figure 4. 3D flow fields of NOAA AR 10365 (left) and NOAA AR 11057 (right). Red arrows denote horizontal flows, and blue and yellow contours represent upflows and downflows, respectively.

0.03 km s^{-1}). The horizontal flow field exhibits super-penumbral flows corresponding to moving magnetic features, the magnitude of which is about 0.3 km s^{-1}.

4. Summary

Using Hinode G-band data spanning the X3.4 flare occurred in NOAA AR 10930 on 2006 December 13, we found a sudden change of shearing motion to unshearing motion between the two spots with opposite polarities after the flare. We obtained the flow fields with both DAVE and FLCT methods, and computed the evolution of shear flow around the flaring PIL. A gradual decrease of average velocity of the shear flow is found to be associated with the flare, with a magnitude of 0.2 km s^{-1}.

We obtained the 3D velocity field of NOAA AR 10365 on May 28, 2003 using DAVE4VM. Enhancement of 3D flow fields seem to be associated with the occurrence of an X3.6 flare. Note that these are mainly to test the implementation of the DAVE4VM technique on the ground-based magnetograph data. The observed flow signal during the flare may be affected by flare emissions as well as seeing variation.

DAVE4VM 3D velocity field of NOAA AR 11057 on 2010 March 29 was also derived using HMI vector magnetogram. This active region shows a horizontal super-penumbral flows corresponding to moving magnetic features, the magnitude of which is about 0.3 km s^{-1}. The vertical flow is smaller than 0.1 km s^{-1}.

Acknowledgement

We thank Dr. Peter Schuck for providing the DAVE and DAVE4VM codes. We thank Dr. Yang Liu and the HMI team for processing and providing the excellent HMI magnetograms. The presentation of this paper in the IAU Symposium 273 was made possible due to partial support from the National Science Foundation grants ATM 0548260 and AST 0968672, and NASA Living With a Star grant 09-LWSTRT09-0039.

References

Kosugi, T., Matsuzaki, K., Sakao, T., Shimizu, T., & Sone, Y., *et al.* 2007, *Solar Phys.*, 243, 3

Liu, C., Deng, N., Liu, Y., Falconer, D., Goode, P. R., Denker, C., & Wang, H. 2005, *Astrophys. J.*, 622, 722

Schuck, P. W 2006, *Astrophys. J.*, 646, 1358

Schuck, P. W 2008, *Astrophys. J.*, 683, 1134

Tan, C., Chen, P. F., Abramenko,V., & Wang, H. 2009, *Astrophys. J.*, 690, 1820

Tsuneta, S., Ichimoto, K., Katsukawa, Y., Nagata, S., Otsubo, M., Shimizu, T., Suematsu, Y., Nakagiri, M., Noguchi, M., & Tarbell, T., *et al.* 2008, *Solar Phys.*, 249, 167

Wang, H. 2006, *Astrophys. J.*, 649, 490

Welsch, B. T., Fisher, G. H., & Abbett, W. P. 2004, *Astrophys. J.*, 620, 1148

The Physics of Sun and Star Spots
Proceedings IAU Symposium No. 273, 2010
D. P. Choudhary & K. G. Strassmeier, eds.

doi:10.1017/S1743921311015675

Study of the change of surface magnetic field associated with flares

Yixuan Li[1], Ju Jing[1], Yuhong Fan[2] and Haimin Wang[1]

[1]Space Weather Research Lab, New Jersey Institute of Technology, Newark, USA
email: yl89@njit.edu, jj4@njit.edu, haimin@flare.njit.edu

[2]High Altitude Observatory, National Center for Atmospheric Research†, Boulder, USA
email: yfan@ucar.edu

Abstract. How magnetic field structure changes with eruptive events (e.g., flares and CMEs) has been a long-standing problem in solar physics. Here we present the analysis of eruption-associated changes in the magnetic inclination angle, the transverse component of magnetic field and the Lorentz force. The analysis is based on an observation of the X3.4 flare on Dec.13 2006 and a numerical simulation of a solar eruption made by Yuhong Fan. Both observation and simulation show that (1) the magnetic inclination angle in the decayed peripheral penumbra increases, while that in the central area close to flaring polarity inversion line (PIL) deceases after the flare; (2) the transverse component of magnetic field increases at the lower altitude near flaring PIL after the flare. The result suggests that the field lines at flaring neutral line turn to more horizontal near the surface, that is in agreement with the prediction of Hudson, Fisher & Welsch (2008).

Keywords. Sun: flares; sun: magnetic fields

1. Introduction

The 3-D magnetic topology of active regions from the photosphere to corona is of fundamental importance to our understanding of the energy storage and release process that accounts for flares and coronal mass ejections (CMEs). Recently, it has been widely reported that photosphere magnetic fields can experience some rapid, significant and permanent changes during X-and M-class flares. In the reconnection picture proposed by Liu *et al.* (2005): part of the penumbral segments in the outer δ spot vanish rapidly after flares, and meanwhile, the umbral cores and/or inner penumbral regions are darkened. There are two separate magnetic systems in a δ sunspot before the flare; after the flare these two become strongly connected. The outer field lines stretched out by the eruption turn the field of the outside penumbrae to become more vertical, explaining the penumbral decay. Meanwhile, the lower field lines near the neutral line (NL) become more horizontal, and thus explain the darkening of the central umbral regions.

Although the flare-associated magnetic changes have been reported, the flare models which could reveal the physical mechanism for these changes have not been achieved. Part of the reason for this is that we can only make observations of the solar magnetic fields directly and precisely at the photosphere level. A good understanding of the role that the magnetic fields play in powering the flares is essentially based on study the 3-D structure of magnetic fields and their evolution associated with flares. We will compare

† The National Center for Atmospheric Research is Sponsored by the National Science Foundation

our observational findings with numerical simulations of eruptive flares to examine how the observational signatures compare with the predictions of the specific models.

2. Data sets and model description

The data base for this paper consists of the X3.4 flare occurred in NOAA active region 10930 on 2006 December 13 and a numerical simulation of an eruption in the solar corona in a spherical domain (Fan 2010).

For this X3.4 event, we used the *Hinode* vector magnetograms taken before and after the 4B/X3.4 flare to study the changes of magnetic field. The vector magnetograms were obtained in two time bins, 20:30-21:33 UT on 2006 December 12 (before the flare) and 4:30-5:36 UT on 2006 December 13 (after the flare). The pixel resolution of the magnetograms is 0.16" pixel^{-1}. Combined with the corresponding G-band (430nm) data, we identified the decayed and the enhanced regions of the spot. Since the active region is not located at the center of solar disk, the projection effect was corrected with coordinate transformation. The 180° ambiguity in vector magnetograms was resolved with the "minimum energy" method (Metcalf 1994).

For the numerical simulation shown in this letter, the detailed magnetic field structure was obtained by solving the isothermal MHD equations in a spherical domain representing the solar corona, given by $r \in [R_{\odot}, 5.496R_{\odot}], \theta \in [5\pi/12, 7\pi/12], \phi \in [-\pi/9.6, \pi/9.6]$(Fan 2010). The simulated domain has been resolved in $432 \times 192 \times 240$ grids, which is uniform in θ and ϕ direction, and non-uniform along the solar radius, and the step size in the range from $r = R_s$ to $r = 1.788R_s$ is $dr = 0.0027271R_s$. The initial state of this simulation is assumed to contain a pre-existing potential arcade field in a hydrostatic isothermal atmosphere domain. The emergence of a twisted and arched flux tube is imposed at the lower boundary with an upward advection velocity $\mathbf{v}_0$ (small compared to the Alfvén speed) in the area where the emerging tube intersects the lower boundary. After the emergence is stopped, a quasi-equilibrium of coronal flux rope with an underlying sigmoid shaped current sheet is established immediately. Subsequently, the flux rope continued to build up due to reconnections in the current sheet and the ejective eruption is triggered when flux rope exceeds a critical height. In this study, we concentrate on the evolution of the magnetic field immediately above the line-tying lower boundary, at the grid level 2.

3. Calculation of lorentz force change

Given a change $\delta\mathbf{B}$ in the photospheric magnetic field, Hudson *et al.* (2008) predicted the back reaction in the photosphere by estimating the change in Lorentz force per unit area as $\delta f_z = (B_z\delta B_z - B_x\delta B_x - B_y\delta B_y)/4\pi$. We calculate the change of Lorentz force by integrating the δf_z over an area at flaring PIL.

4. Results

For this X3.4 event, we compare the sunspot structures before and after the flare in Figure 1. The brightened/darkened areas in the difference image represent the penumbral regions with decayed/enhanced G-band intensity. The right two panels show the time evolution of the mean G-band intensity in the brightened area (*top*) and the darkened area (*bottom*) during a 3.5 hr time period around the flare.

After aligning the G-band images with the corresponding SP vector magnetograms, we are able to study the changes of the photospheric magnetic parameters in the decayed and

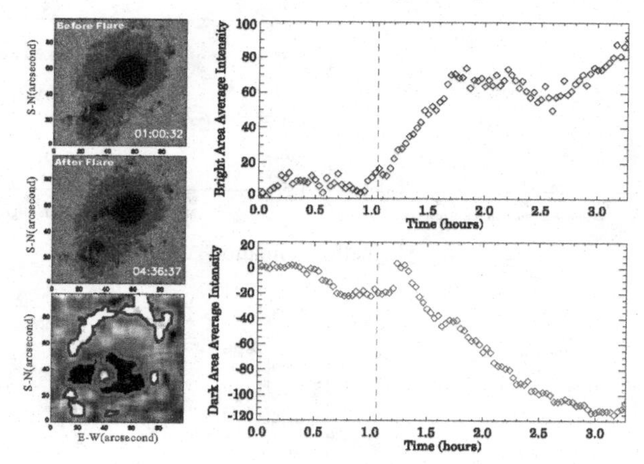

Figure 1. The left column shows *Hinode* G-band images taken before (*top*) and after the flare (*middle*), and their difference image (the post-flare image subtracted by the pre-flare image; *bottom*). The field-of-view (FOV) is 100"×100". In the difference image, the brightened area and the darkened area are outlined with blue and red contours, respectively. In the right column, *Top Panel:* Evolution of the mean G-band intensity of the decayed region, starting from 01:10 December 13, 2006;*Bottom Panel:* Evolution of the mean G-band intensity of the enhanced region. The dashed lines indicate the flaring peak time of hard X-rays.

enhanced penumbral areas. Figure 2 compares the pre-flare and the post-flare distributions of magnetic inclination angle, transverse field strength in the decayed (*left column*) and the enhanced (*right column*) areas. The mean values of parameters before and after the flare are indicated by the vertical green and orange lines, respectively. Evidently we can see that the inclination angle increases by $\sim 3.3°$ after the flare in the decayed region, and decreases by 5° in the enhanced region. The transverse magnetic field strength within the centered enhanced region increased by 20% after the flare, meanwhile which falls the amount of 16% in the peripheral decayed region. In this case, it demonstrates that the magnetic fields change from a more inclined to a more vertical configuration in the peripheral penumbral region.

As a comparison, we have analyzed a sequence of the MHD simulations. In Figure 3, the left two columns compare the magnetic transverse field and Light-of-sight(LOS) field structure taken before and after the eruption in the second layer of grid along r, which is about 4.745Mm above the solar surface. The decayed and enhanced areas are represented by the red and blue boxes. The blue boxes contour the enhanced areas after eruption, which correspond to the darkening inner penumbra, whereas the red boxes correspond to the outer decaying penumbra. The right panels (from top to bottom) show respectively the time profile of magnetic inclination angle, the transverse field strength and the change of Lorentz force comparing to $t = 150$ (the unit for time is $\tau = 356.8s$). The dashed lines indicate the flaring time. We can see that the inclination angle increases and the B_t decreases after the flare in the outer decayed region, while the inclination angle decreases and the B_t increases at flaring neutral line. The sudden enhancement of the downward Lorentz force after the flare, which may indicate the back reaction on the photosphere and solar interior due to releasing of flare energy, is consistent with the prediction by Hudson, Fisher & Welsch.

We note that Moore *et al.* (2001) proposed the Tether-cutting model to explain the onset of flares and subsequent eruptions. In this reconnection model, a short and flat loop forms near the photosphere after the eruption, which is consistent with our observation

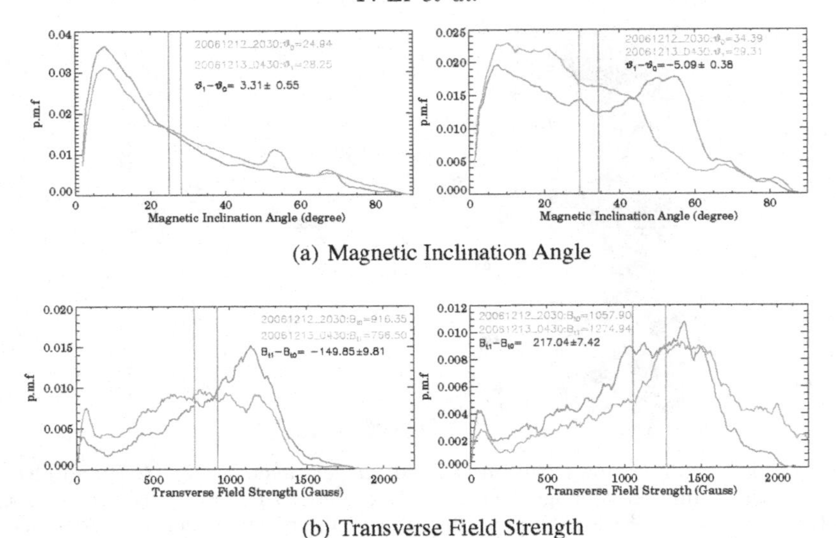

(a) Magnetic Inclination Angle

(b) Transverse Field Strength

Figure 2. The pre-flare (*green lines*) and the post-flare (*orange lines*) distributions of magnetic inclination angle, transverse field strength in the decayed (*left column*) and the enhanced (*right column*) areas. The vertical green lines indicate the mean value of parameters before the flare, while the vertical orange lines indicate that after the flare. The 95% confidence intervals of each parameter were shown in corresponding panel.

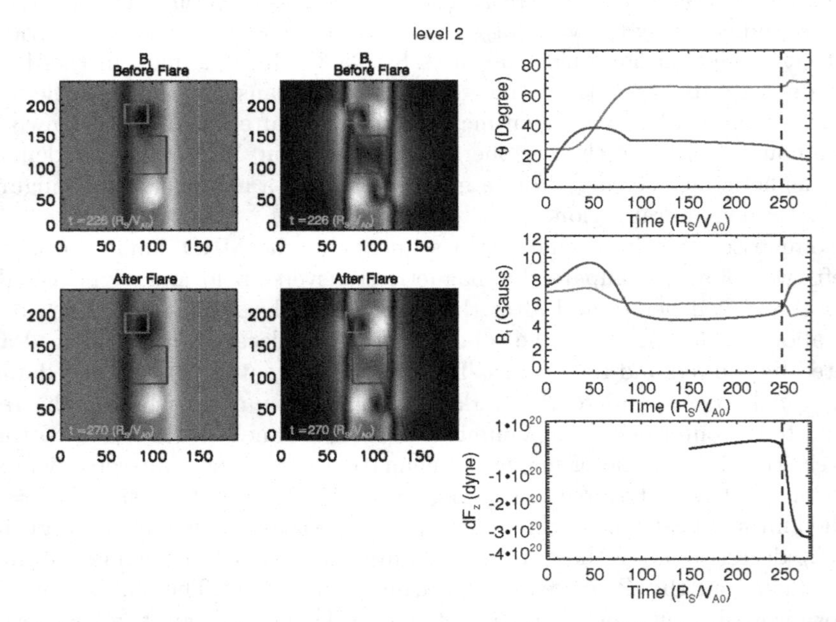

Figure 3. *Left two columns:* The magnetic transverse fields and LOS fields before and after the eruption with the red and blue boxes indicated the decayed and enhanced regions, respectively. *Right column:* The time profile, from top to bottom, of magnetic inclination angle, the transverse field strength and the change of Lorentz force comparing that at $t = 150\tau$.

of the enhanced transverse field near flaring NLs. This is an example of predicting permanent field changes associated with flares, also consistent with our observational and simulated results.

This work was supported by NSF grant AGS-0936665 and NASA grants NNX08-BA22G and NNX08-AQ90G to New Jersey Institute of Technology.

References

Hudson, H. S., Fisher, G. H., & Welsch, B. T. 2008, *Subsurface and Atmospheric Influences on Solar Activity*, 383, 221

Liu, C., Deng, N., Liu, Y., Falconer, D., Goode, P. R., Denker, C., & Wang, H. 2005, *Astrophys. J.*, 622, 722

Fan, Y. 2010, *Astrophys. J.*, 719, 728

Metcalf, T. R. 1994, *Solar Phys.*, 155, 235

Moore, R. L., Sterling, A. C., Hudson, H. S., & Lemen, J. R. 2001, *Astrophys. J.*, 552, 833

The Physics of Sun and Star Spots
Proceedings IAU Symposium No. 273, 2010
D. P. Choudhary & K. G. Strassmeier, eds.

doi:10.1017/S1743921311015687

Comparison of numerical simulations and observations of helioseismic MHD waves in sunspots

K. V. Parchevsky[1], J. Zhao[1], A. G. Kosovichev[1] and M. Rempel[2]

[1]Hansen Experimental Physics Laboratory, Stanford University,
Stanford, CA 94305, USA
email: kparchevsky@solar.stanford.edu

[2]HAO/NCAR, Boulder,CO 80307, USA
email: rempel@ucar.edu

Abstract. Numerical 3D simulations of MHD waves in magnetized regions with background flows are very important for the understanding of propagation and transformation of waves in sunspots. Such simulations provide artificial data for testing and calibration of helioseismic techniques used for analysis of data from space missions SOHO/MDI, SDO/HMI, and HINODE. We compare with helioseismic observations results of numerical simulations of MHD waves in different models of sunspots. The simulations of waves excited by a localized source provide a detailed picture of the interaction of the MHD waves with the magnetic field and background flows (deformation of the waveform, wave transformation, amplitude variations and anisotropy). The observed cross-covariance function represents an effective Green's function of helioseismic waves. As an initial step, we compare it with simulations of waves generated by a localized source. More thorough analysis implies using multiple sources and comparison of the observed and simulated cross-covariance functions. We plan to do such calculations in the nearest future. Both, the simulations and observations show that the wavefront inside the sunspot travels ahead of a reference "quiet Sun" wavefront, when the wave enters the sunspot. However, when the wave passes the sunspot, the time lag between the wavefronts becomes unnoticeable.

Keywords. Helioseismology, MHD, sunspots

1. Introduction

Understanding of acoustics wave generation, propagation, and scattering in the convective zone of the Sun is one of the high-priority tasks of helioseismology. The commonly used acoustic ray theory is not applicable near the top turning point (near the photosphere) where the background model changes quickly due to large gradients of pressure and density, and wave effects must be taken into account. Different types of MHD waves (fast, slow, Alfven) and magneto-gravity wave exist in magnetized regions on the Sun. Their interference, transformation, and reflection from the top boundary creates very complicated picture. In this situation numerical simulations of MHD waves in sunspots are important for understanding the behavior of the acoustic waves in the complicated solar environment. They also provide artificial data for testing and calibrating the helioseismic measurement and inversion techniques used for processing data from space missions SOHO, Hinode and SDO.

2. MHD wave modeling

Propagation of waves inside the Sun, in the presence of magnetic fields and velocities in the background state, is described by linearized MHD equations (see Parchevsky &

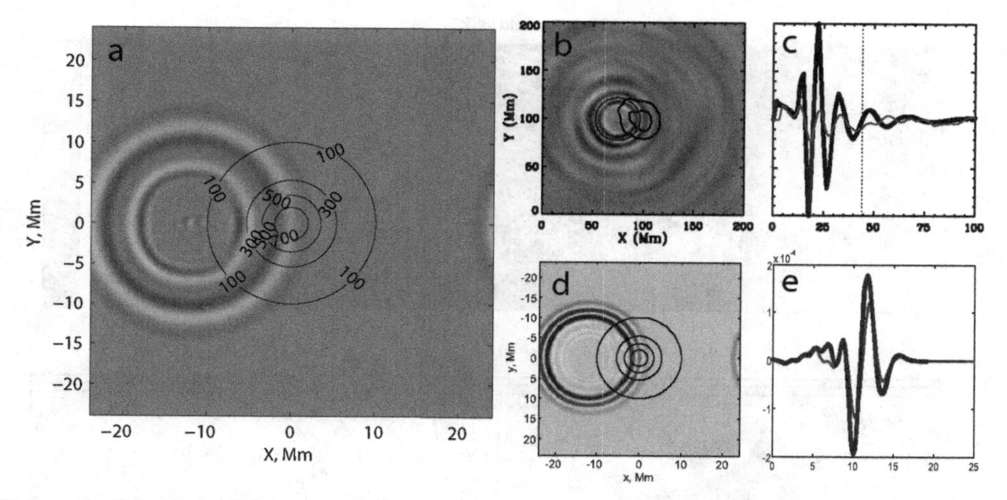

Figure 1. Snapshot of the z-component of the velocity at the level of the photosphere (panel a). Cross-correlation obtained from SOHO/MDI observations by J. Zhao for active region AR9787 (panel b) and a snapshot of the p-mode front (panel d) from simulations in the magnetostatic sunspot model. The light and heavy curves (panels c and e) represent cuts through the sunspot and the quiet Sun region respectively.

Kosovichev (2009) for details):

$$
\begin{aligned}
\frac{\partial \rho}{\partial t} &= -\nabla \cdot \boldsymbol{m}' - \nabla \cdot (\rho' \boldsymbol{V}_0) \\
\frac{\partial \boldsymbol{m}'}{\partial t} &= -\nabla p' - (\boldsymbol{m}' \cdot \nabla)\boldsymbol{V}_0 - (\boldsymbol{V}_0 \cdot \nabla)\boldsymbol{m}' - \boldsymbol{m}'\nabla \cdot \boldsymbol{V}_0 - \rho'(\boldsymbol{V}_0 \cdot \nabla)\boldsymbol{V}_0 \\
&\quad -\frac{1}{4\pi}\nabla(\boldsymbol{B}_0 \cdot \boldsymbol{B}') + \frac{1}{4\pi}[(\boldsymbol{B}_0 \cdot \nabla)\boldsymbol{B}' + (\boldsymbol{B}' \cdot \nabla)\boldsymbol{B}_0] + \rho' \boldsymbol{g}_0 + \boldsymbol{S}(\boldsymbol{r}, t) \\
\frac{\partial \boldsymbol{B}'}{\partial t} &= \nabla \times \left(\frac{\boldsymbol{m}'}{\rho_0} \times \boldsymbol{B}_0\right) + \nabla \times (\boldsymbol{V}_0 \times \boldsymbol{B}') \\
\frac{\partial p'}{\partial t} &= c_s^2 \frac{\partial \rho}{\partial t} - c_s^2 \boldsymbol{m}' \cdot \left(\frac{\nabla p_0}{\Gamma_1 p_0} - \frac{\nabla \rho_0}{\rho_0}\right) - c_s^2 \rho_0 \left(\frac{p'}{p_0} - \frac{\rho'}{\rho_0}\right) \nabla \cdot \boldsymbol{V}_0 \\
&\quad -\boldsymbol{V}_0 \cdot \nabla p' + c_s^2 \boldsymbol{V}_0 \cdot \nabla \rho'
\end{aligned}
\tag{2.1}
$$

where ρ', p', $\boldsymbol{m}' = \rho_0 \boldsymbol{v}'$, and $\boldsymbol{B}'$ are the perturbations of the density, pressure, momentum, and magnetic field respectively. Quantities p_0, ρ_0, c_s, $\boldsymbol{g}_0$, $\boldsymbol{V}_0$, and $\boldsymbol{B}_0$ are the background pressure, density, sound speed, gravitational acceleration, flow velocity, and the magnetic field respectively. The source term $\boldsymbol{S} = (0, 0, S_z(\boldsymbol{r}, t))$ represents the acoustic source, localized in space and explicitly depending on time as Rickers wavelet. This source model provides the wave spectrum, which closely resembles the solar spectrum (see details in Parchevsky & Kosovichev (2007)).

A semi-discrete finite-difference numerical scheme is used. It easily permits to combine different versions of spatial discretization and time advancing schemes. A dispersion-relation-preserving scheme developed by Tam & Webb (1993) is used for the spatial discretization, and a strong stability preserving Runge-Kutta scheme is used for time advancing.

The waves with frequencies lower than the acoustic cut-off frequency are reflected below the photosphere. The waves with higher frequencies pass through and penetrate

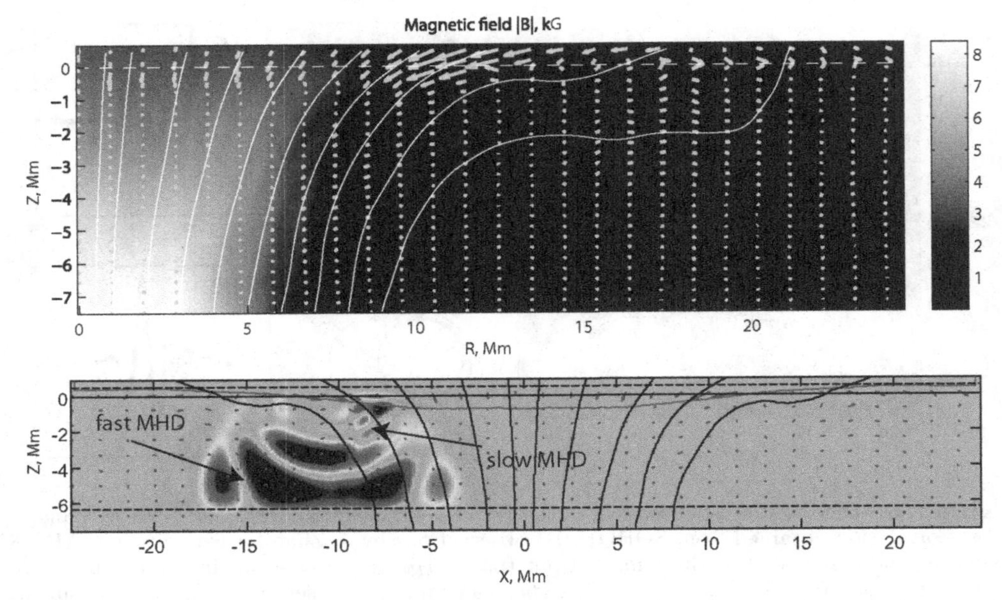

Figure 2. Magnetic field strength of the axially symmetric (angular averaged) realistic model of the sunspot with the background flows (top) obtained from numerical non-linear MHD simulations (M. Rempel). The white curves show the magnetic field lines, arrows represent the background flows. The snapshot of the model is used as the background state for the wave simulations. The bottom panel shows perturbations of the vertical momentum of the wave generated by a single source. A mixture of slow MHD and Alfven waves at the source location is observed. The rigth part of the wave front is skewed by the background flows. This skewness is not observed in simulations with magnetostatic models.

into the chromosphere. To simulate such frequency-dependent reflecting property we put the Perfectly Matched Layer (PML) at 500 km above the photosphere. On the Sun all waves which reach this layer propagate further to chromosphere. In simulations they will be absorbed by PML and do not pollute the computational domain. To prevent spurious reflection from the bottom of the domain we put the PML at the bottom boundary. The lateral boundary conditions are chosen to be periodic, which mimics incoming waves.

3. Results and discussion

Observational cross-correlation function. In this section we compare the behavior of the observational cross-correlation function inside a sunspot observed by SOHO/MDI and simulations of the wave front from a single source inside the magnetostatic sunspot model (Parchevsky *et al.* (2010)). The backgrouns model was provided to us by Khomenko & Collados (2008). For both simulations and observations the f-mode signal has been filtered out. Qualitatively, the top and bottom panels show similar behavior. In both figures, the waves propagating through the sunspot have smaller amplitude than the waves propagating in the quiet regions. Both, simulations and observations, show that the wave front inside the sunspot travels ahead of the "quiet" wave front when the wave enters the sunspot. Both, simulations and observations, show that when the wave passes the sunspot, the time lag between wave fronts becomes unnoticeable.

Realistic sunspot model with the background flows In Figure 2 we present snapshot of the vertical momentum of MHD waves (bottom panel) in a realistic sunspot model with background flows (top panel). The background model was provided us by Rempel *et al.*

(2009). In deep layers the right part of the wave front is noticeably skewed due to the background flows (the wave front in the same region of the magnetostatic model is almost symmetric). It seems that in deep layers the shape of the wave front of the fast MHD wave is affected stronger by flows than by magnetic fields.

4. Conclusion

Comparison of simulations for magnetostatic models of sunspots with observations shows that the observational wave forms cross-correlation functions show similarities to the behavior of the numerically simulated wave front of *p*-modes: (i) the amplitude is reduced inside the sunspot; (ii) when the wave enters the sunspot, the wave front inside the sunspot travels ahead of the wave front in the quiet region. We developed a 3D code for numerical simulation of MHD waves in models of sunspots in presence of the background flows. In addition to the primary fast MHD wave, used for helioseismic studies, simulations show mixture of slow MHD and Alfven waves excited by the source, and traveling along field lines into the interior (as in the magnetostatic model). The simulations reveal also a strong distortion of the wave front due to the background velocities (not present in the magnetostatic models). This code is used for simulations of stochastic oscillations excited by multiple random sources for testing local helioseismology and calibration of helioseismic inferences of the subsurface structure of sunspots.

Acknowledgements

We are grateful to E. Khomenko and M. Collados for providing us magnetostatic models of sunspots. This research was supported by NASA LWS grant NNG05GM85G to Stanford University.

References

Khomenko, E. & Collados, M. 2008, *Astrophys. J.*, 689, 1379
Low, B. C. 1975, *Astrophys. J.*, 197, 251
Parchevsky, K. V. & Kosovichev, A. G. 2007, *Astrophys. J.*, 666, 547
Parchevsky, K. V. & Kosovichev, A. G. 2009, *Astrophys. J.*, 694, 573
Parchevsky, K. V., Kosovichev, A. G., Khomenko, E., Olshevsky, V., & Collados, M. 2010, *arXiv:1002.1117v1*
Pizzo, V. J. 1986, *Astrophys. J.*, 302, 785
Rempel, M., Schüssler, M., Cameron, R. H., & Knölker, M. 2009, *Science*, 325, 171
Tam, C. & Webb, J. 1993, *J. Comput. Phys.* 107, 262

The Physics of Sun and Star Spots
Proceedings IAU Symposium No. 273, 2010
D.P. Choudhary & K.G. Strassmeier, eds.

doi:10.1017/S1743921311015699

Are the umbral dots, penumbral grains, and G band bright points formed by the same type of magnetic flux tubes?

Isroil Sattarov

Tashkent State Pedagogical University, 103 Yusif Kxos Kxojib str.,Tashkent 100070, Uzbekistan
email: isattar@astrin.uz

Abstract. Today's Solar Physics comes across of different type of fine structures in solar atmosphere including umbral dots and penumbral grains in sunspots, and G-band bright points in quiet Sun. In this report, we present evidence that umbral dots, penumbral grains, and, possibly, G band bright points are related to a common type of features in solar atmosphere magnetic flux tubes.

Keywords. Umbral dots, penumbral grains, sunspots, G-band bright points

1. Introduction

Stratospheric balloon-borne observations of sunspots in USA under leadership of M. Schwarzchild in 1959 (Danielson 1964), and balloon experiments in USSR under leadership of V. Krat (Krat *et al.* 1972) excited great interest and put forward many problems in physics of sunspot fine structure. The mechanism of emergence of so-called umbral dots, their relation with penumbral filaments are examples of questions raised by these observations. These earlier stratospheric observations had turned out to be insufficient for study the dots evolution, which require long-term observations; and the author was among the first observers, who begun studying the sunspots' fine structure using ground-based telescopes. Today's Solar Physics comes across different types of fine structures in solar atmosphere. In this report we present evidence to support that umbral dots, penumbral grains, and, possibly, G-band bright points are related to the same type of features magnetic flux tubes in solar atmosphere.

2. The telescope and observations

Horizontal solar telescope ATSU-5 used by us is one of the serial instruments (D=44 cm, F=1700 cm, where D is diameter of main mirror, and F is its focal length) developed by Russian optician V.N. Ponamarev 55 years ago. Ten years later, the majority of Soviet Union observatories were equipped with such telescopes. In 1964, ATSU-5 telescope was mounted in Tashkent Astronomical Observatory in the city of Tashkent. The territory of Uzbekistan is very sunny. In Tashkent, the average sunshine time is near 2900 hours per a year with a consistently very good seeing from June 20 to the beginning of October. The very good seeing takes place during windless and hot weather, frequently for several days before a cyclone. White light observations of sunspots, discussed in this article, were carried out in Cassegrain focus (diameter of solar disk is 60 cm) using high-speed 35 mm photographic camera (exposure time 0.01 sec). Bursts of images were taken during the periods of best seeing, based on visual control by the observer (the author of this article). Spectral observations were conducted in Newton (prime) focus of the telescope

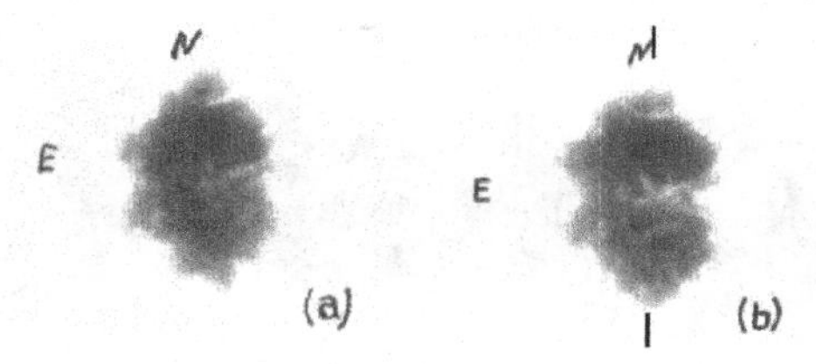

Figure 1. Two prints (a 0650 UT, b 0828 UT) of umbra of the unipolar sunspot observed 8 August 1981 at the ATSU-5 in Tashkent. Vertical touch on Fig. 1b shows position of spectrograph's slit at receiving spectrum presented on Fig. 3.

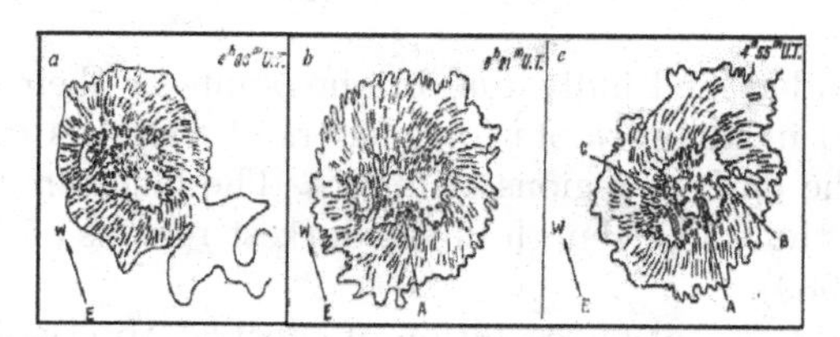

Figure 2. Drawing of penumbral grains and umbral dots (short filigree) of the large sunspot exhibiting counter-clockwise rotation for 14(a), 15(b) and 16(c) September 1977.

using ASP-20 horizontal spectrograph with linear dispersion of 0.6 Å / mm at optical wavelength. The spectra were recorded using photographic plates and film with exposure time of 1-3 seconds. The observations were used to study the magnetic field, fine structure and evolution penumbral filaments and umbral dots.

3. The results of sunspots fine structure studies

It was found that in most studied cases the umbral dots persist for more than 30 minute: 43 % of dots persist more than 60 minutes and 10% - more than 90 minutes. In their evolution, bright dots increase the brightness reaching the maximum during a relativity short time, and after that, gradually become weaker and expand. Finally, the dots break up and disappear (Sattarov 1981a). Most of bright dots are observed at the disk center side of penumbra umbra boundary (as in right side of umbra on Fig. 1 b).

In September 1977, a rotation of an unipolar sunspot with two umbrae (small northern and large southern umbra) was observed. During 13-16 September the spot rotated by 700 around its northern small umbra in counter-clockwise direction. The rotation affected the orientation of penumbral filaments (grains) and umbral fine structure, twisting them in clockwise direction. In the course of rotation, the main (large) umbra was deformed; its east boundary was "attacked" by brushes of penumbral grains, which appear as "peninsula" intruding into the umbra. The dots in front of "peninsula" were transformed into the bright penumbral grains. On the west boundary of main umbra, the opposite affect took place, i.e. the penumbral west boundary joined the umbra and the bright penumbral grains turned into the umbral dots (Sattarov 1981b). It appears that the magnetic flux tube that forms the large unipolar sunspot rotates, indeed. The rotation may be caused by interaction of the magnetic tube forming unipolar sunspot and the magnetic tubes of new sunspot group developed near the southern boundary of the penumbra of unipolar sunspot. The magnetic field of sunspot had southern (S) polarity, and the sunspot of leading (N) polarity of this new group had developed near it later (Sattarov 1981b). Between the white light observations we took spectrograms of sunspots by scanning their umbrae in selected (narrow or wide) spectral bands. In the

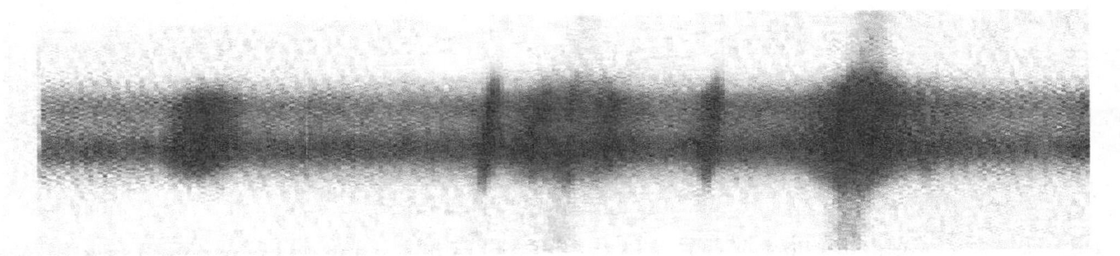

Figure 3. A part of the central section (noted in Fig. 1b by vertical stroke) spectrum with the line Fe I 6302.5 of the umbra printed on Fig. 1.

spectrograms we have identified individual bright points or their clusters. It was found that the magnetic field in umbral dots (or clusters of them) is more longitudinal than the field strength in the darkest regions of umbra. The field strength in umbral dots is about 20% lower than the field strength in the darkest regions of umbra (Sattarov 1982, Litvinov & Sattarov 1989).

As one can see on Fig. 3 spatial resolution of spectral observations is lower than the resolution of the white light images (compare Fig. 1 with Fig. 3). It is the result of blurring effect, which is larger near the boundary of umbra. In our estimate, the contribution of scatter light in brightness of sunspot is not significant. Fig. 3 shows the spectral line Fe I 6302.5 with telluric lines of both sides. On Fig. 3, one can see the simple doublet Zeeman splitting (middle white band on Fig. 3) on bright dot's cluster (Fig. 1b). If the scatted light was strong, we would see strong central component, which is not present on the middle white band. The observations of clusters of dots do not show measurable Doppler displacement. This implies that if there were plasma flows in dots the velocity of these flows should not exceed 100 m/s (Litvinov & Sattarov 1989).

4. Discussion and conclusions

Bright dots were observed near the disk side of penumbra-umbra boundary in the "intrusion" of penumbral material into the umbra. Rotation of the sunspot had transformed umbral dots into penumbral grains and penumbral grains into umbral dots. Umbral dots are associated with vertical magnetic tubes and, it seems, the rotation deflected umbral dots flux tubes from its vertical direction to more horizontal orientation prior to formation of penumbral peninsula in umbra. Magnetic field in penumbral grain flux tubes have more horizontal direction and sunspot rotation had deflected penumbral flux tubes on the opposite (limb) side of the umbra from its horizontal direction to more vertical position. Umbral dots and penumbral grains are the very small structural elements of solar atmosphere. The movie of a newly developing sunspot (Fig. 4 presents one of images of this movie) shows how the penumbral grains are formed. They develop in intergranular spaces from low contrast small photosphere features (it seems, from photosphere bright points), but not from granules themselves. Plasma flow velocities in umbral dots are as small as in intergranular spaces in the photosphere(Litvinov & Sattarov 1989, Lites *et al.* 1991). The umbral dots live longer than the photospheric granules, and it seems they may have a different mechanism of flaring up than granulation As it appears from our study, the granules cannot transform to the penumbral grains (as was suggested by some models of sunspot penumbra). The granules show plasma flows of sufficiently larger velocity directed vertically upward, whereas the dots have not show such strong plasma flows. Intergranular lanes exhibit bright points - a very small structural element of solar atmosphere. A sunspot movie (Nemirooff & Bonnel 2000) shows how the formation of

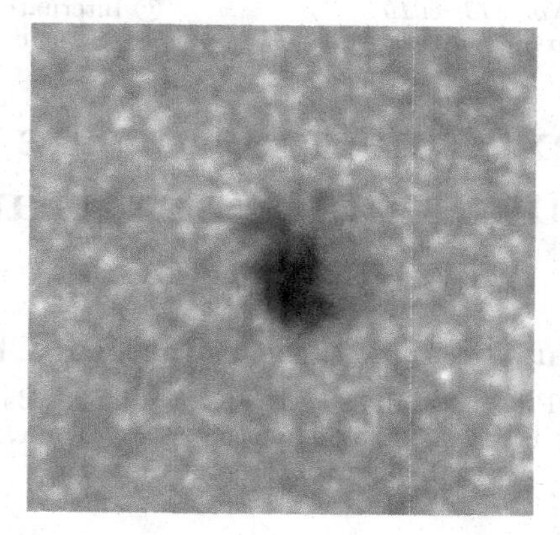

Figure 4. Small new sunspot observed at 8 August 1981 in ATSU-5 in Tashkent.

grains and photosphere bright points takes place near outside boundary of penumbra; both features go away: the grains move to umbra, while the bright points move to the surrounding photosphere. The grains and the bright points have opposite magnetic polarity. On sunspot images of low resolution the grains present inner and the photosphere bright points outer rings of Secchi. Therefore, we propose that the umbral dots, penumbral grains, and photosphere bright points are associated with the feature - magnetic flux tubes - which float to the solar surface in outside parts of sunspot penumbra.

Acknowledgments

The presentation of this paper in the IAU Symposium 273 was possible due to partial support from the National Science Foundation grant numbers ATM 0548260, AST 0968672 and NASA - Living With a Star grant number 09-LWSTRT09-0039.

References

Danielson, R. E. 1964, *Astrophys. J.* 139, 45
Krat, V. A., Karpinsky, V. N., & Pravdjuk, L. M. 1972, *Solar Phys.* 26, 305
Sattarov, I. 1981a, *Morphology and Cyclicity of Solar Activity*, 122
Sattarov, I. 1981b, *Morphology and Cyclicity of Solar Activity*, 107
Sattarov, I. 1982, *Sun and Planetary System*, 96, 129
Sattarov, I. 1980, *Soviet Astron.* 24, 352
Litvinov, O.V., & Sattarov, I. 1989,Magnetic Fields and Corona, (Moscow, Nauka), 210
Lites, B. W., Bida, T. A., Johannesson, A., & Scharmer, G. B. 1991, *Astrophys. J.* 373, 683
Nemirooff, R.& Bonnel, J. 2000, Astronomy Picture Of the Day, 23/02/2000.

The Physics of Sun and Star Spots
Proceedings IAU Symposium No. 273, 2010
D. P. Choudhary & K. G. Strassmeier, eds.

doi:10.1017/S1743921311015705

Possible explanations of the Maunder minimum from a flux transport dynamo model

Bidya Binay Karak and Arnab Rai Choudhuri

Department of Physics, Indian Institute of Science, Bangalore-560012
email: bidya_karak@physics.iisc.ernet.in,
arnab@physics.iisc.ernet.in

Abstract. We propose that at the beginning of the Maunder minimum the poloidal field or amplitude of meridional circulation or both fell abruptly to low values. With this proposition, a flux transport dynamo model is able to reproduce various important aspects of the historical records of the Maunder minimum remarkably well.

Keywords. Sun: activity, sun: magnetic field, meridional circulation

1. Introduction

One important aspect of the solar cycle is the Maunder minimum during 1645–1715 when the solar activity was strongly reduced (Ribes & Nesme-Ribes 1993). It was not an artifact of few observations, but a real phenomenon (Hoyt & Schatten 1996). From the study of historical data (Ribes & Nesme-Ribes 1993), it has been confirmed that the sunspot numbers in both the hemisphere fell abruptly to nearly zero value at the beginning of the Maunder minimum, whereas a few sunspots appeared in the southern hemisphere during the last phase. It is also established from the cosmogenic isotopes data (Beer *et al.* 1998; Miyahara *et al.* 2004) that the cyclic oscillations of solar activity continued in the heliosphere at a weaker level during the Maunder minimum, but with a period of 13–15 years instead of the regular 11-year period.

The most promising model of studying solar cycle at present is the flux transport dynamo model (Choudhuri *et al.* 1995; Durney 1995; Dikpati & Charbonneau 1999; Chatterjee *et al.* 2004). The main sources of irregularities in this model are the stochastic fluctuations in the Babcock–Leighton process of poloidal field generation (Choudhuri 1992; Choudhuri *et al.* 2007) and the stochastic fluctuations of meridional circulation (hereafter MC) (Hathaway 1996). Therefore we propose that the polar field or amplitude of MC or both decreased at the beginning of Maunder minimum. With this proposition, we use a flux transport dynamo model to reproduce a Maunder minimum. The details of this work can be found in Choudhuri & Karak (2009) and Karak (2010).

2. Methodology

We cary out all the analyses with the flux transport dynamo model described in Chatterjee *et al.* (2004). To reproduce the Maunder minimum, we perform the following three separate sets of experiments. Similar to Choudhuri *et al.* (2007), first, we decrease the polar field above $0.8R_{\odot}$ by a factor γ after stopping the code at a solar minimum. We change the polar field by different amount in two hemispheres. In northern hemisphere, we take $\gamma = 0.0$, whereas in southern hemisphere, it is 0.4. In addition, in this calculation,

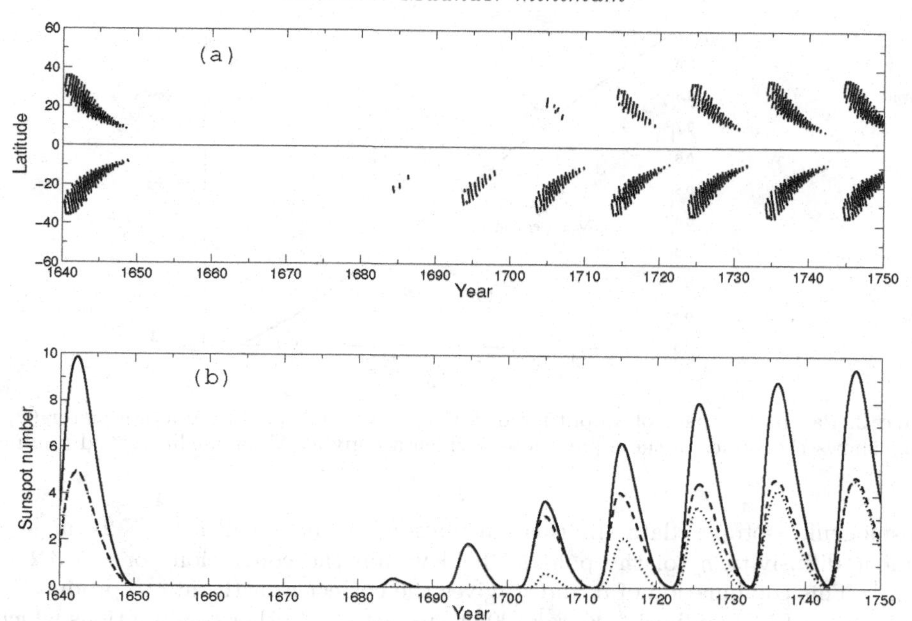

Figure 1. Results covering the Maunder minimum episode. (a) The butterfly diagram. (b) The smoothed sunspot number. The dashed and dotted lines show the sunspot numbers in southern and northern hemispheres, whereas the solid line is the total sunspot number. From Choudhuri & Karak (2009).

we decrease the toroidal field by multiplying it everywhere by 0.8 to stop the eruption for some time. This essentially reduces the strong overlap between two cycles in our model (see figure 13 of Chatterjee *et al.* 2004). After making these changes, we run the model for several cycles without any further change. In the second procedure for reproducing the Maunder minimum, we decrease the amplitude of MC v_0 abruptly to a very low value. After keeping it at low value for few years, we again increase it to the usual value but at different rates in two hemispheres. In the northern hemisphere, it is increased at slightly lower rate than the southern hemisphere. Note that in this case we have varied only v_0 and no other parameters of the model. We have repeated this calculation in the low diffusivity model of Dikpati & Charbonneau (1999) too. Last, we have included the effect of the fluctuations of polar field along with the fluctuations of MC. We have run the model for different values of γs from 0 to 1 at each values of v_0 from a very low value to the average value. Then we find out the critical values of v_0 and the corresponding γ factor for which we get a Maunder-like minimum.

3. Results

First, we discuss the results from the polar field reduction procedure. It is shown in Fig. 1 (see the caption also). In order to facilitate the comparison with the observation data, we have marked the beginning of Fig. 1 to be the year 1640. From this figure we see that the sudden initiation but gradual recovery of Maunder minimum and the north-south asymmetry of sunspot numbers in the last phase have been nicely reproduced. We also find the cyclic oscillation of the poloidal field in the solar wind (shown in figure 2 of Choudhuri & Karak 2009). This oscillation explains the cyclic behavior found

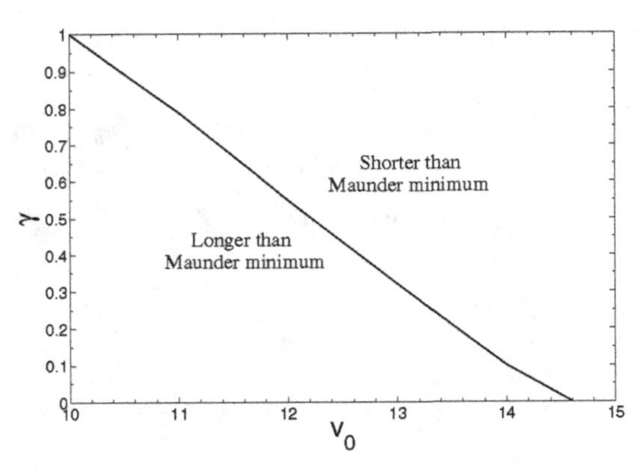

Figure 2. Parameter space of amplitude of MC (v_0) and polar field reduction factor (γ). The line shows the values of these parameters which are giving Maunder-like grand minima.

in cosmogenic isotopes data. In this calculation, we have taken $\alpha = 21$ m s^{-1} and turbulent diffusivity η_p of the poloidal field within the convection zone $= 3.2 \times 10^{12}$ cm^2 s^{-1}. This combination of α and η_p gives the correct growth rate to produce Maunder minimum. In Choudhuri & Karak (2009), we list some other combinations which also reproduce Maunder-like grand minima.

In the second procedure of reducing v_0 alone, we again get results very similar to what are shown in Fig. 1, if v_0 is abruptly made 10 m s^{-1} and then allowed to gain back its strength (see figures in Karak 2010). This procedure also reproduces all the important features of the Maunder minimum remarkably well. We underscore that in the advection-dominated model (e.g. Dikpati & Charbonneau 1999) we do not get this result. This is because the decrease of MC in the advection-dominated model produces more toroidal field and makes the cycle stronger, as analyzed by Yeates *et al.* (2008).

We conclude that it is possible to reproduce the Maunder minimum by decreasing either polar field or v_0 to very low values. However, if we allow both the polar field and v_0 to decrease simultaneously, then it seems possible to reproduce Maunder-like grand minima without making either the polar field or the MC so low. We have found that for each γ there is a particular value of v_0 which can produce a Maunder-like minimum. The line in Fig. 2 shows the combinations of parameters giving a Maunder minimum of appropriate duration. Values of γ and v_0 lying in the lower left of Fig. 2 give grand minima longer than the Maunder minimum, whereas values lying in the upper right give shorter minima. We should mention that Fig. 1 is reproduced just by taking $\gamma = 0.2$ (actually $\gamma_N = 0.0$ and $\gamma_S = 0.4$ is used) keeping v_0 unchanged. However, in this calculation, we have to reduce v_0 to around 13.5 m s^{-1} along with the polar field reduction to 0.2 in both hemispheres. This is because in earlier calculations we had reduced the toroidal field slightly along with the change of polar field. Here we have not done this reduction of the toroidal field. Additionally, here the values of α and η_p are slightly different giving different dynamo growth rate.

4. Conclusion

We have shown that most of the important features of the Maunder minimum can be reproduced quite well by assuming a simple ansatz that the polar field or the amplitude of MC or both decreased significantly at the beginning of Maunder minimum. Because of our lack of knowledge about the physical conditions at the beginning of the Maunder minimum, we cannot say how exactly the Sun was driven to the Maunder minimum. However, we should mention that there are several independent studies (Wang & Sheeley 2003; Miyahara *et al.* 2004; Passos & Lopes 2011) suggesting that the amplitude of MC was weaker during the Maunder minimum. If this happens to be correct, then this study along with several other studies (Chatterjee *et al.* 2004; Chatterjee & Choudhuri 2006; Jiang *et al.* 2007; Yeates *et al.* 2008; Goel & Choudhuri 2009; Karak & Choudhuri 2011) indicate that the solar dynamo actually is diffusion-dominated and not advection-dominated.

References

Chatterjee, P. & Choudhuri, A. R. 2006, *Solar Phys.*, 239, 29
Chatterjee, P., Nandy, D., & Choudhuri, A. R. 2004, *Astron. Astrophys*, 427, 1019
Choudhuri, A. R. 1992, *Astron. Astrophys*, 253, 277
Choudhuri, A. R., Chatterjee, P., & Jiang, J. 2007, *Phys. Rev. Lett.*, 98, 131103
Choudhuri, A. R. & Karak, B. B. 2009, *RAA*, 9, 953
Choudhuri, A. R., Schüssler, M., & Dikpati, M. 1995, *Astron. Astrophys*, 303, L29
Dikpati, M. & Charbonneau, P. 1999, *Astrophys. J.*, 518, 508
Durney, B. R. 1995, *Solar Phys.*, 160, 213
Goel, A. & Choudhuri, A. R. 2009, *RAA*, 9, 115
Hathaway, D. H. 1996, *Astrophys. J.*, 460, 1027.
Jiang, J., Chatterjee, P., & Choudhuri, A. R. 2007, *Mon. Not. Roy. Astron. Soc.*, 381, 1527
Hoyt, D. V. & Schatten, K. H. 1996, *Solar Phys.*, 165, 181
Karak, B. B. 2010, *Astrophys. J.*, 724, 1021
Karak, B. B. & Choudhuri, A. R. 2011, *Mon. Not. Roy. Astron. Soc.*, 410, 1503
Passos, D. & Lopes, I. P. 2011, *JASTP*, 73, 191
Ribes, J. C. & Nesme-Ribes, E. 1993, *Astron. Astrophys*, 276, 549
Wang, Y. -M. & Sheeley, N. R. Jr. 2003, *Astrophys. J.*, 591, 1248
Yeates, A. R., Nandy, D., & Mackay, D. H. 2008, *Astrophys. J.*, 673, 544

The Physics of Sun and Star Spots
Proceedings IAU Symposium No. 273, 2010
D. P. Choudhary & K. G. Strassmeier, eds.

doi:10.1017/S1743921311015717

Evidence for the return meridional flow in the convection zone from latitude motions of sunspots

K.R. Sivaraman[1], H.Sivaraman[2], S.S.Gupta[3] and R.F.Howard[4]

[1]5020, Haven Place, #206, Dublin, CA-94568, USA and formerly of the Indian Institute of Astrophysics, Bangalore, India.
email: kr_sivaraman@yahoo.com

[2]2581, Rivers Bend Circle, Livermore, CA-94550, USA.

[3]Formerly of the Indian Institute of Astrophysics, Kodaikanal, India.

[4]National Solar Observatory, Tucson, Arizona, USA.

Abstract. We have derived the latitude motions of sunspots classified into three area categories using the measures of positions and areas of their umbrae from the white – light images of the Sun for the period 1906 – 1987 from the Kodaikanal Observatory archives. The latitude motions are directed equator – ward in all the three area classes. We interpret that these equator – ward latitude motions reflect the meridional flows at the three depths in the convection zone where the magnetic flux loops of the spots of the three area classes are anchored. We obtain estimates of the anchor depths through a comparison of the rotation rates of the spots in each area class with the rotation rate profiles from helioseismic inversions. The equator – ward flows measured by us thus provide evidence of the return meridional flows in the convection zone as required in the flux transport solar dynamo models. We have done an identical analysis using a similar data set derived from the photoheliogram collections of the Mt.Wilson Observatory for the period 1917 – 1985. There is good agreement between the results from the data sets of the two observatories.

Keywords. Latitude motions, meridional flow, convection zone, flux transport

1. Introduction

The flux – transport model of the solar dynamo (the most successful among solar dynamo models) invokes meridional circulation (one cell each in the north and south hemispheres) consisting of a pole – ward flow on the surface and an equator – ward return flow in the interior, in addition to differential rotation (the Ω – effect) and helical turbulence (the α – effect) in the dynamo mechanism in the ($\alpha - \Omega$) type dynamos (Wang, Sheeley & Nash 1991; Choudhuri, Schüssler & Dikpati 1995; Charbonneau 2005). While the meridional flow on the surface is well known since a long time, the return flow, a key ingredient in the model has not been detected observationally so far (Charbonneau 2005; Dikpati 2005). Studies have shown that the rotation rates measured using sunspots as tracers reflect the rotation rates of the plasma layers in the interior at the respective depths where the foot points of the magnetic flux loops of spots of different ages (or areas) are anchored and not the surface rotation rate (Nesme-Ribes, Ferreira & Mein 1993; Collin *et al.* 1995; Beck 2000; Sivaraman *et al.* 2003). Based on this scenario we have estimated the velocity of meridional flows at three depths in the convection zone by measuring the latitudinal drifts of spots divided into three area classes, assuming that spots of each area class reflect the latitudinal motion at the respective depths where their foot points are anchored, just as their rotation rates do. We matched the rotation rates of the spots in each area class (the same sample of spots divided into the same area classes as was used for deriving the meridional velocities) with helioseismic profiles (plot of rotation rate *vs* depth as $\frac{r}{R_0}$; r, radial distance and R_o solar radius) and obtained estimates of the respective anchor depths in the convection zone.

2. Data and analysis

2.1. *Measurements:*

Our data consists of measures of position (heliographic latitude and longitude) and umbral area of every spot that has appeared on the disc within longitudes $\pm$ 60° from the daily photoheliogram for the period 1906 – 1987 of the Kodaikanal Observatory, with a digitizing pad of spatial resolution of 0.02 mm that translates to $\approx$ 0.02 arc sec on the 20 cm. diameter solar image. Besides, we accessed a similar data set (created from the Mt.Wilson daily photoheliograms covering the period 1917 – 1985) from the NOAA site (ftp: //ftp.ngdc.noaa.gov/ STP/SOLAR-DATA/SUNSPOT-REGION-TILT). The present analysis is on the same lines as the one adopted for determining meridional flow velocities and anchor depths of spot groups (Sivaraman *et al.* 2010).

2.2. *Latitudinal drifts, meridional velocities and anchor depths:*

We computed the latitudinal drift (in deg day^{-1}) of each spot by dividing the drift by the time elapsed (expressed in days) between successive observations. After eliminating spots with latitudinal drifts $\geqslant$ 1.5 deg day^{-1}(equivalent to $\approx$ 200 m s^{-1}) so as to minimize the errors, we had 118760 spots in the Kodaikanal data and 107020 spots in the Mt.Wilson data. To study the variation of latitudinal drifts with latitude we divided the latitude zone + 40° to - 40° into 5° latitude zones and assigned the latitudinal drifts to that 5° zone based on the initial latitudes of the spots. In the next step, we sorted the spots in the 5° zones into three area classes – of umbral areas 0 – 5 μ, 5 – 10 μ and > 10 μ (μ area in millionths of the hemisphere) and computed the mean latitudinal drifts of spots of each area class in all of the 5° zones. We multiplied the latitudinal drifts in deg day^{-1}by the factor of 140.596 cos(λ) to convert to meridional velocities in m s^{-1}. Accordingly, a drift of 0.01 deg day^{-1}would correspond to a velocity of $\approx$ 1.3 m s^{-1} at 15° latitude.

To estimate the anchor depths, we divided 118760 spots (Kodaikanal data) into three area classes (0-5 μ, 5-10 μ and > 10 μ) and computed the sidereal rotation rates of spots in each class and the least square solution for the latitude dependence as in Gupta, Sivaraman & Howard 1999.

We projected the mean rotation rates of spots of each area class one after the other on the rotation rate vs depth ($\frac{r}{R_0}$) profiles at five latitudes (0°, 10°, 20°, 30° and 40°) and read off the depths on the $\frac{r}{R_0}$-axis corresponding to the intercepts . These represent the anchor depths at the five latitudes and their mean , the mean anchor depth of spots of each area class. The internal rotation profiles are from the Global Oscillation Network Group (GONG) data (Antia, 2008, private communication). We repeated the above for the 107020 spots in the Mt. Wilson data.

3. Results and conclusions

We show the variation of latitude drifts with latitude in Figure 1 and the values of latitudinal drifts (deg day^{-1}) and the corresponding meridional velocities (in m s^{-1}) in Table 1 extracted from Figure 1, for the Kodaikanal and Mt.Wilson data.

– Within the latitude zone + 30° to - 30°, the meridional flow is equator – ward in the three area classes (see Figure 1 and Table 1). This flow provides evidence for the

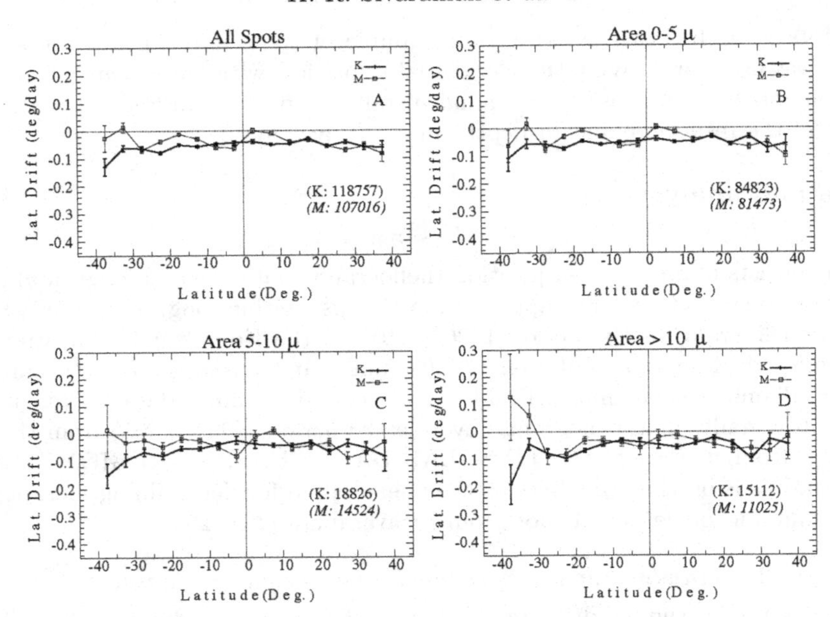

Figure 1. Latitudinal drifts (in deg day^{-1}) averaged over the spots within five degree latitude bins *vs.* latitude from Kodaikanal white – light images for the period 1906 – 1987 (**K** —⊢—) and Mt. Wilson white – light images for the period 1917 – 1985 (**M** ···⊡···). The latitude of the spot at the time of initial observation defines the five-degree latitude zone (bin) to which it is assigned. Negative drifts indicate equator – ward motion in both hemispheres. The number of spots used for determining the latitudinal drifts in the respective area category is shown at the bottom right within brackets in each panel (prefix K stands for Kodaikanal data); similar number in italics with prefix *M* stands for the number of spots of Mt. Wilson data.

Table 1. Latitudinal drift (deg day^{-1}) and the equivalent meridional flow velocity (in m s^{-1}) averaged over 0° – 30° latitude for the northern and southern hemispheres for the three area classes. (in bold: **Kodaikanal data**; in italics: *Mt. Wilson data*). Negative latitudinal drifts represent equator – ward motions. The mean anchor depth of spots of each area class is shown in the last column (see text).

Area class in μ	North deg day^{-1}	North m s^{-1}	South deg day^{-1}	South m s^{-1}	Mean anchor depth ($\frac{r}{R_0}$)
< 5	**-0.042 ± 0.002**	**5.7 ± 0.3**	**-0.055 ± 0.003**	**7.5 ± 0.4**	**0.89**
	-0.034 ± 0.003	*4.6 ± 0.4*	*-0.044 ± 0.003*	*6.0 ± 0.4*	*0.89*
5 – 10	**-0.043 ± 0.004**	**5.8 ± 0.5**	**-0.053 ± 0.004**	**7.2 ± 0.5**	**0.84**
	-0.036 ± 0.006	*4.9 ± 0.8*	*-0.041 ± 0.007*	*5.6 ± 0.9*	*0.85*
>10	**-0.052 ± 0.004**	**7.1 ± 0.5**	**-0.061 ± 0.004**	**8.3 ± 0.5**	**0.78**
	-0.040 ± 0.007	*5.4 ± 0.9*	*-0.056 ± 0.008*	*7.6 ± 1.1*	*0.78*
All spots	**-0.044 ± 0.002**	**6.0 ± 0.3**	**-0.056 ± 0.002**	**7.6 ± 0.3**	Depth cannot
	-0.035 ± 0.002	*4.8 ± 0.3*	*-0.045 ± 0.003*	*6.1 ± 0.4*	be assigned

return meridional flow in the convection zone that has remained undetected so far. This together with the surface pole – ward flow would act as a conveyor belt to transport the magnetic flux for recycling by the dynamo as envisaged in the flux – transport models.

– The equator – ward drift reaches high values in both hemispheres (≈ -0.08 to -0.09 deg day^{-1}or 10 to 12 m s^{-1}) around latitude ± 25° and slows down towards zero drift near the equator and at latitudes ± 35° (see Figure 1). One or two such high values result in increase in the mean velocities particularly for large spots (area > 10 μ). Chou

& Dai (2001) analyzing the images from the Taiwan Oscillation Network for the period 1994 – (2001) with time – distance helioseismology technique noticed that at depths > 10 Mm, a new component of meridional flow diverging from the activity latitudes ($\approx 25°$) appeared in both hemispheres as the solar activity developed during 1998 – 2000. We are of the view that the surge in meridional velocity in the same latitude zone seen in our study is the signature of the divergent flow noticed by Chou & Dai (2001). Our analysis, in addition shows that this divergent flow is stronger for spots of large areas (hence large magnetic fields). In an earlier study it was shown that spot groups show similar equator – ward flow (Sivaraman *et al.* 2010). But the divergent flow was not large enough to cause an overall increase in the velocity as in the case of individual spots. The reason perhaps is that group motions (because the amplitudes are so small) are influenced by the possibly random appearance and disappearance of spots in the group, which shifts the calculated position of the group. Thus there is a cause unrelated to meridional motion that affects group meridional motion. This uncertainty being absent in the case of individual spots, we are able to detect the small and subtle changes in their latitudinal motions.

Acknowledgements

The presentation of this paper in the IAU Symposium 273 was possible due to partial support from the National Science Foundation grant numbers ATM 0548260, AST 0968672 and NASA – Living With a Star grant number 09-LWSTRT09-0039. One of the authors (KRS) wishes to thank the LOC, in particular Dr. Debi Prasad Choudhary and Dr. Cristina Cadavid. We wish to acknowledge the sincere efforts of several generations of observers at the Kodaikanal and Mt. Wilson Observatories who made these observations covering many decades. Without their very dedicated and careful work this study would not have been possible.

References

Beck, J. G.: 2000, *Solar Phys.*191, 47.

Charbonneau, P.: 2005, "Dynamo Models of the Solar Cycle": In *Living Reviews in Solar Physics.* Irsp – 2005 – 2.

Chou, D.-Y. & Dai, D.-C: 2001, *Astrophys. J. Lett.* 559, L175.

Choudhuri, A. R. & Schüssler, M., Dikpati.M.: 1995, *Astron. Astrophys* 303, L29.

Collin, B., Nesme-Ribes, E., Lereoy, B., Meunier, N., & Sokoloff, D.: 1995, *C.R. Acad. Sci. Paris; t. 321, Serie II.b, 111.*

Dikpati, M.: 2005, *Adv. Space Res.* 35, 322.

Gupta, S. S., Sivaraman, K. R., & Howard, R. F.: 1999, *Solar Phys.* 188, 225.

Nesme-Ribes, E., Ferreira, E. N., & Mein, P.: 1993, *Astron. Astrophys* 274, 563.

Sivaraman, K. R., Sivaraman, H., Gupta, S. S., & Howard, R. F.: 2003, *Solar Phys.* 214, 65.

Sivaraman, K. R., Sivaraman, H., Gupta, S. S., & Howard, R. F.: 2010, *Solar Phys.* 266, 247.

Wang, Y.-M., Sheeley, N. R., & Nash, A. G.: 1991, *Astrophys. J.* 383, 431.

The Physics of Sun and Star Spots
Proceedings IAU Symposium No. 273, 2010
D. P. Choudhary & K. G. Strassmeier, eds.

doi:10.1017/S1743921311015729

Observations of density fluctuations in a quiescent prominence

Ken Nakatsukasa
California State University Northridge

Abstract. Density fluctuations in a torsional quiescent prominence were studied with Solar Optical Telescope on board of Hinode satellite. Continuous observations were made in Ca II H line from ~15:00 UT May 03, 2008; the observational duration was ~ 1hr with a cadence time ~ $30s$. The emission intensity along the prominence axis as functions of altitudes and time shows fluctuations in brightness with local peaks in Fourier power spectra, indicating the presence of periods and intervallic displacements to be 3 ~ 11 min and 2 ~ 7 Mm respectively and statistically significant peaks at $368 \pm 63s$ & 620 ± 41 s, and 3.3 ± 0.6 Mm regime. The distance is the line of sight projection. These intensity disturbances may perhaps be caused by density fluctuations originated from compressional waves in which the possible origin could be the combinations of p-mode oscillations induced at the surface and twisting of the flux tube.

Keywords. Sun: prominences.

1. Introduction

Solar prominences are thin elongated clouds made of mostly ionized gas with typical temperatures 4800 ~ 6000K and plasma densities $10^{10}cm^{-3} \sim 10^{11}cm^{-3}$(Zirin 1988). Surrounded is a corona with temperatures and densities higher and lower by order of 2 respectively. They are invisible in the continuum, but may be seen in any strong emission lines. If they are spotted at the solar limb, they appear brighter against the dark space and they are called prominences. They appear dark on the disk due to background radiations from the photosphere; they are called solar filaments. In general, prominences are categorized in two kinds: Quiescent (long-lived) filaments and Active (transient, eruptions, flare-associated) filaments (Zirin 1988).

A comprehensive overview of the oscillations in prominences is given by (Mackay *et al.* 2010). Solar prominences are subject to various types of oscillatory motions. Oscillations are mainly detected from periodic Doppler shifts from spectral lines. Space-Time sliced intensity images laid side by side is another detection technique. In the present paper, study of emission intensity disturbances within a torsional quiescent prominence is shown.

2. Observation

Analysis is based upon Hinode telescope observations of quiescent prominence located at east limb at $\sim -25°$ latitude. The data was taken on May 03, 2008 from 15:00 UT for ~1 hr with a cadence of ~30s. The images consist of 1024 × 1024 pixels in Ca II H (396.8 nm) line with a bandwidth 0.3 nm and the pixel size ~ 80km. The present paper focused on the prominence shown in Fig. 1, in particular, the left flux tube. The detection technique is to obtain averaged emission intensity integrated along the line of sight near the flux tube axis. The results were converted by fast fourier transform for determinations of any intervallic fluctuations. Daily images of full disk H alpha from Solar

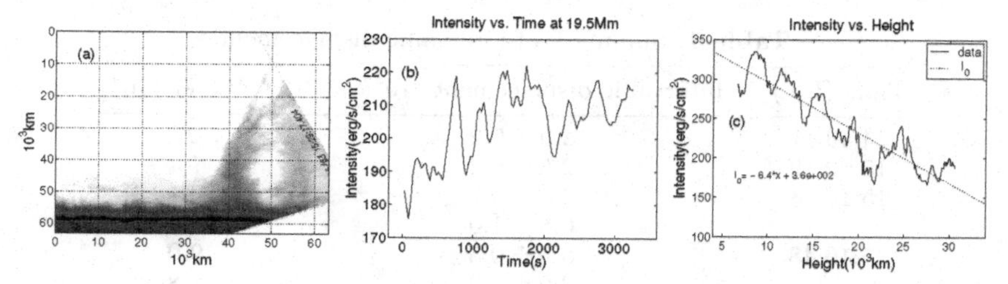

Figure 1. (a)Prominence image at Ca II H line taken on May 03 15:29 UT. The color was inverted and the contrast was enhanced. This image consists of 512 × 512 pixels. (b)Temporal evolution. Distance is approximated from photosphere. (c)Spatial displacement at 14:49:19

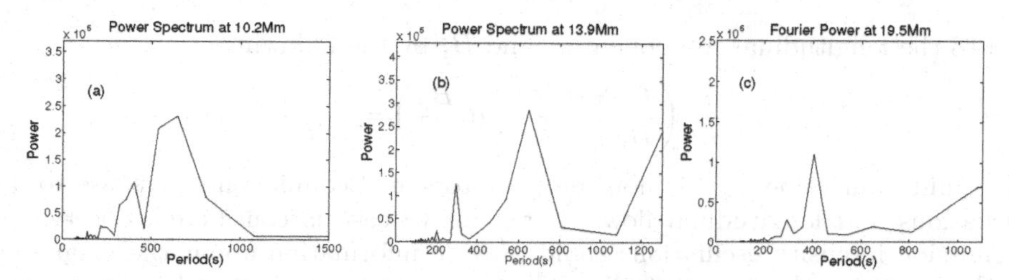

Figure 2. (a)∼(c) Power spectra at different altitudes.

Magnetic Activity Research Telescope (SMART) showed this prominence appeared from west limb around Apr 20, surviving all time before it came to east limb where observations of Hinode took place.

3. Results and analysis

Movies of the prominence showed moving features, partially brighter than ambient, at ∼15:05:08 UT, appearing from near the apex moving in torsional motions toward the left foot point of Fig. 1a and disappearing behind the tube at ∼15:10:18 UT. The line of sight projection speed was determined by tracking the flow; it is estimated 10 ∼ 13 km/s. Several other flows are seen but a much smaller scale. If one supposed these flow directions represent the magnetic field line, then the pitch angle $\tan^{-1}(B_\phi/B_z)$ (in usual cylindrical coordinates with z along the tube axis) seems to be higher at a foot point and decrease at higher altitudes which is seen from movies. Only downflows were observed in view of the left flux tube. Fig. 1 also shows plasma fluids escaping into outer space from near left foot point.

Fig. 1b & 1c show temporal evolutions and spatial displacements at ∼19.7 Mm and 14:49:19 UT respectively(approximated from the photosphere). Fig. 2 shows power spectra of temporal evolutions at three different heights. Table 1 summarizes intervallic displacements measured from power spectra at different time. Fourier power spectra indicate local peaks in periods and intervallic displacements to be 3 ∼ 11min and 2 ∼ 7 Mm respectively and statistically significant peaks at $368 \pm 63s$ & 620 ± 41s, and 3.3 ± 0.6 Mm regime.

Because the intensity and density are correlated, the quasi-periodic and intervallic fluctuations in intensity could be a signature of compressional waves. However, this wave like phenomena did not clearly show signs of drifting, signifying that these intervallic fluctuations might represent near stationary waves rather than propagating. The origin of the stationary-like wave could possibly be due to the twisting of the flux rope. According to Zhugzhda (Zhugzhda 1996), the torsional velocity v_ϕ and the magnetic field B_ϕ are

Table 1. Summary of intervallic displacement

Time (UT)	Intervallic displacement ($10^3\,km$)	Power (10^5)
14:49:17	3.0 ± 0.2	27
15:02:19	2.5 ± 0.2	7.4
15:10:18	4.0 ± 0.7	10
	6.8 ± 1.0	73
15:16:48	3.6 ± 0.5	9.0
	2.0 ± 0.2	7.8
15:28:47	3.5 ± 0.1	55
15:34:20	3.2 ± 0.8	6.3

related to the longitudinal component v_z and B_z by the following.

$$\frac{\partial}{\partial t}\left(\frac{B_\phi}{B_z}\right) + \frac{\partial}{\partial z}\left(v_z \frac{B_\phi}{B_z}\right) = \frac{\partial v_\phi}{\partial B_z} \tag{3.1}$$

If the equilibrium value $B_{\phi,0}$ is non-zero, the torsional components give rise to density perturbations and longitudinal flows. If the above case is considered, observations of such emission intensity oscillations could provide information about the magnetic field strength based from theoretical studies. The moving features discussed earlier are perhaps an indication of mass flows because they were isolated and appear irregularly. As was discussed, only downflows seem to occur in this flux tube. In this sense, the other foot point carries out the upflows and represents continuous supplies of chromospheric materials. This process might be a part of maintenances and stabilities of quiescent prominences.

Other things to note is the intensity I_0 almost decreases linearly with increasing heights. This slow decline in density may be related to its long-lived, continuous mass flows, or combinations of both.

4. Discussion and conclusion

The present paper discovered periods near ~6 min and ~11 min regime. The fact that measurements yield two frequencies might be because the oscillation from twist is accompanied by upward acoustic waves initiated by photospheric motions at the foot point. A number of papers discuss its presence of 5 min global photospheric oscillations generated by p-mode waves. Slow magnetoacoustic waves in coronal loops appear to be created by p-mode (De Mootel *et al.* 2000); likewise, continuous fluctuations in density suggest that these wave like phenomena may also be induced by p-mode oscillations. Shorter and longer periods could be the consequence of p-mode and twist respectively. Although the dominant period is 5 min, Garcia *et al.* have identified 10 modes above ~666 s seen by GOLF (Garcia *et al.* 2001), possibly implying the coupling of other periods as well. However, we emphasize that transverse waves are not due to p-mode oscillations, but by some other means (e.g. microflares) because the oscillations are generally damped and cease after few cycles (Mackay *et al.* 2010). Because of the quasi-periodic fluctuations in intensity, we rather not disregard the approach the there are MHD waves in this prominence.

Acknowledgments

Hinode is a Japanese mission developed and launched by ISAS/JAXA, with NAOJ as domestic partner and NASA and STFC (UK) as international partners. It is operated by these agencies in co-operation with ESA and NSC (Norway). This study was part of a summer research program provided by California State University, Northridge Physics and Astronomy Department.

References

De Moortel, I., Ireland, J., & Walsh, R. W., 2000, *Astrophys. J.*, 355, L23–26

Garcia, R. A., Regulo, C., Turck-Chieeze, S., Berttelo, L., Kosovichev, A. G., Brun, A. S., Counvidat, S., Henney, C. J., Lazrek, M., Ulrich, R. K., & Varadi, F., 2001, *Solar Phys.*, 200, 361–379

Mackay, D. H., Karpen, J. T., Ballester, J. L., Schmieder, B., & Aullanier, G., 2010, *Space Sci. Revs*, 151, 4, 333–399

Zhugzhda, Y. D., 1996, *AIP*, 3, 1, 10–21

Zirin, H., 1988, *Astrophysics of the sun*, Campridge University Press, New York

The Physics of Sun and Star Spots
Proceedings IAU Symposium No. 273, 2010
D. P. Choudhary & K. G. Strassmeier, eds.

doi:10.1017/S1743921311015730

Excitation of magneto-acoustic waves in network magnetic elements

Yoshiaki Kato[1], Oskar Steiner[2], Matthias Steffen[3], and Yoshinori Suematsu[4]

[1]Institute of Space and Astronautical Science, Japan Aerospace Exploration Agency
3-1-1 Yoshinodai, Chuo-ku, Sagamihara, Kanagawa 252-5210, Japan
email: kato.yoshiaki@isas.jaxa.jp

[2]Kiepenheuer-Institut für Sonnenphysik
Schöneckstrasse 6, D-79104 Freiburg, Germany

[3]Astrophysikalisches Institut Potsdam
An der Sternwarte 16, D-14482, Potsdam, Germany

[4]Hinode Science Center, National Astronomical Observatory of Japan
2-21-1 Osawa, Mitaka, Tokyo 181-8588, Japan

Abstract. From radiation magnetohydrodynamic (RMHD) simulations we track the temporal evolution of a vertical magnetic flux sheet embedded in a two-dimensional non-stationary atmosphere that reaches all the way from the upper convection zone to the low chromosphere. Examining its temporal behavior near the interface between the convection zone and the photosphere, we describe the excitation of propagating longitudinal waves within the magnetic element as a result of convective motion in its surroundings.

Keywords. Sun: photosphere, sun: chromosphere, sun: oscillations, Suun: magnetic fields, MHD

1. Introduction

Filtergrams or spectroheliograms in Ca II H and K reveal two main sources of Ca II emission: plages, which spatially coincide with the active regions, and the chromospheric network, which outlines the boundaries of the supergranular velocity field (Simon & Leighton 1964). Both are also the location of small magnetic flux concentrations in the photosphere. The close relationship between Ca II emission and magnetic flux (Skumanich *et al.* 1975; Schrijver *et al.* 1989) suggests that the magnetic field plays a key role for the chromospheric emission. The chromospheric heating must be greatly enhanced in areas where the magnetic flux density is large. The heating source and the dissipation mechanism has remained elusive, but likely candidates are the dissipation of magnetohydrodynamic waves or the direct dissipation of electric currents. With regard to chromospheric heating by MHD-waves, numerous studies have been carried out based on the approximation of slender flux tubes (Herbold *et al.* 1985; Huang *et al.* 1995; Fawzy *et al.* 1998 and Hasan & Ulmschneider 2004 and references therein). They have greatly expanded our understanding of the physics of tube modes, mode coupling, dependency on the excitation mechanism and the tube geometry, shock formation, etc.

Some of these models consider the generation of longitudinal and transverse waves in a flux tube by turbulent motions in the convection zone, where an analytical treatment of turbulence based on the Lighthill-Stein theory of sound generation is used (e.g., Musielak

& Ulmschneider (2001) and references therein). Others take a driving motivated by observations of the motion of photospheric magnetic flux concentrations, suggesting that transverse waves can be generated through the impulse transmitted by granules to magnetic flux tubes, as, e.g., in Choudhuri *et al.* (1993a,b); Hasan *et al.* (2000), and Cranmer & van Ballegooijen (2005).

All the above mentioned simulations impose a given driving, which is either monochromatic or impulsive or derived from a theoretical spectrum of turbulence. The focus of this investigation is the self-consistent excitation of a thick magnetic flux concentration through the ambient convective motion.

2. Numerical method

The simulations were carried out with the CO^{5}BOLD-code (Freytag *et al.* 2002). The code solves the coupled system of the equations of compressible magnetohydrodynamics in an external gravity field and non-local, frequency-dependent radiative transfer in one, two, or three spatial dimensions. The two-dimensional computational domain extends over a height range of 3160 km of which 780 km reach above the mean surface of optical depth unity and the rest below it. The horizontal extension is 11200 km.

The simulation starts with a homogeneous, vertical, unipolar magnetic field of a flux density of 80 G superposed on a previously computed, relaxed model of thermal convection. The magnetic field is constrained to have vanishing horizontal components at the top and bottom boundary but lines of force can freely move in the horizontal direction. The magnetic field quickly starts to concentrate in the intergranular downdrafts by the convective motion. Subsequently, individual flux concentration start to merge. After approximately 100 min, the magnetic field concentrates in a single magnetic 'flux sheet' with a strength of approximately 2000 G near optical depth unity within the flux concentration. It then remains in this state for the following 86 min simulation time. This state, however, is not a static or stationary one—as a consequence of the interaction with the surrounding convective motion, the flux concentration moves laterally, gets distorted, and exhibits internal plasma flow in the course of time. Here we describe the excitation of longitudinal slow modes within the flux sheet.

3. Excitation of magneto-acoustic waves

The top and bottom panels in Fig. 1 show part of the computational domain with the magnetic flux concentration (gray magnetic field lines) in the middle, the temperature field (gray scale), the velocity field (arrows), continuum optical depth unity (black horizontally running contour), and the plasma-beta unity (a black dashed curve) for an arbitrary time instant. They show the full width of the computational domain but only a section of the full height range, which reaches 2300 km below and 830 km above mean optical depth unity. Note that the temperature scale stops at 10 000 K so that the temperature field saturates below $\tau_c = 1$. The snapshot shows warm granular upwellings framed by narrow cool intergranular downflows in the convection zone. It also shows two narrow downflow channels in the close vicinity on both sides but outside of the magnetic flux concentration below the surface of optical depth unity. These 'downflow jets', which were in detailed described by Steiner *et al.* (1998), are a consequence of a baroclinic flow impinging on the magnetic flux concentration from the lateral directions, driven by the radiative cooling at the 'hot walls' of the flux concentration.

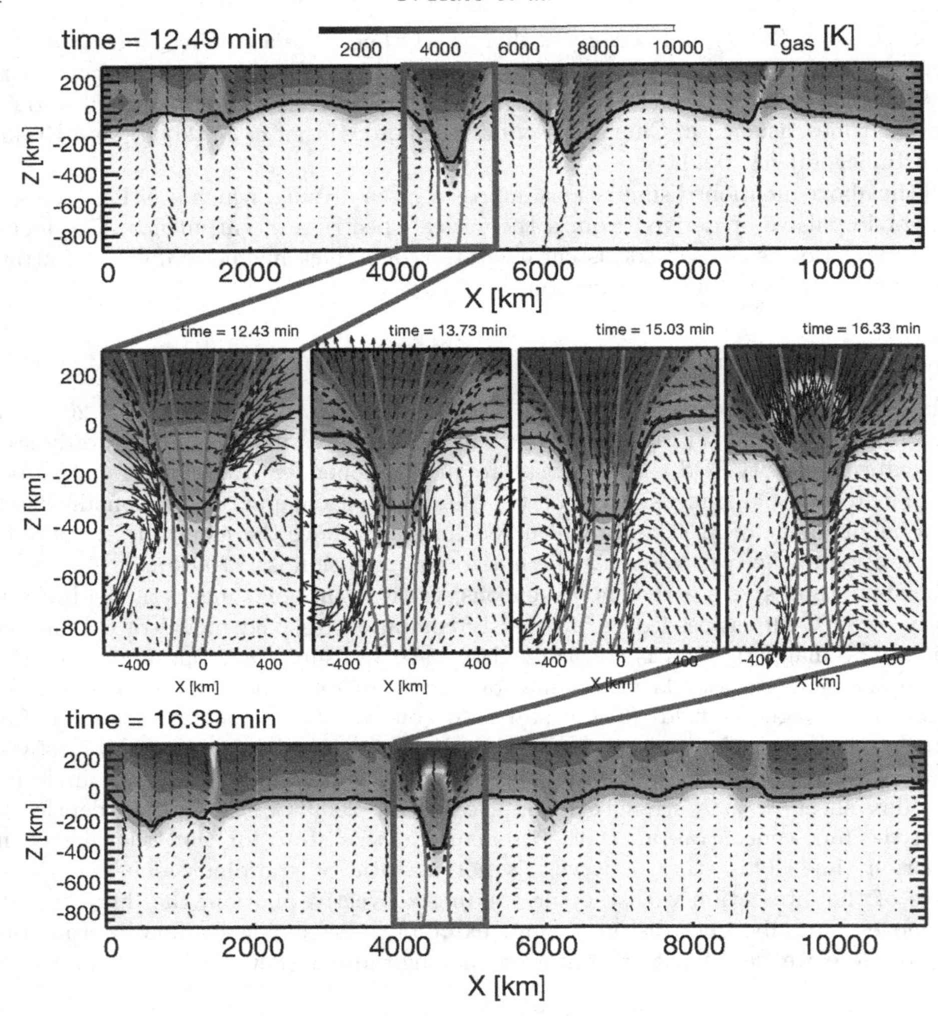

Figure 1. Top and bottom panels: snapshots of a magnetic element showing the full width of the computational domain. The gray-scale indicates the temperature. Gray solid contours indicate magnetic field lines. Black solid curves indicate the optical surface ($\tau = 1$) and black dashed curves indicate the surface of plasma-$\beta = 1$ above which plasma-$\beta < 1$. Arrows indicate the velocity vectors. Middle panels: close-up views of the magnetic element.

In the present simulation we observe that these downflows are far from stationary. They tend to be present most of the time but get transiently enhanced, weakened, or interrupted. Sometimes, the lateral inflow carries a preexisting regular intergranular downflow with it. It then merges with the downflow channel of the flux concentration, which results in a particular strong 'downflow jet'. We identify such transients to be an important source for magneto-acoustic waves within the magnetic flux concentration, in particular a source for longitudinal slow modes. The middle row of Fig. 1 shows a time series of close-ups of such a transient event. The flows in the close surroundings of the magnetic flux concentration generate flows within the magnetic funnel, which leads to an upwardly propagating longitudinal wave and finally to the "bow-shaped" shock front, visible in the last panel of the series.

4. Summary

We have carried out a radiation magnetohydrodynamic simulation of a magnetic flux concentration embedded in the solar atmosphere as representative of network magnetic elements. The analysis of the simulation results focusses on the excitation of magneto-acoustic waves within the magnetic flux concentration. It is shown that convective down-flow events in the close surroundings of the magnetic funnel are responsible for the excitation of propagating longitudinal waves within the magnetic funnel. They steepen to shock waves in the upper photosphere. Presently, we further analyze the simulation data and plan to further elucidate the newly found mechanism in a subsequent paper.

Acknowledgement

O. Steiner gratefully acknowledges financial support and gracious hospitality during his visiting professorship at the National Astronomical Observatory of Japan when part of the work reported herein was carried out. The numerical computations were carried out on NEC SX-9 at JAXA Supercomputer Systems (JSS).

References

Choudhuri, A. R., Auffret, H., & Priest, E. R. 1993a, *Solar Phys.*, 143, 49
Choudhuri, A. R., Dikpati, M., & Banerjee, D. 1993b, *Astrophys. J.*, 413, 811
Cranmer, S. R. & van Ballegooijen, A. A. 2005, *Astrophys. J. Suppl.*, 156, 265
Fawzy, D. E., Ulmschneider, P., & Cuntz, M. 1998, *Astron. Astrophys*, 336, 1029
Freytag, B., Steffen, M., & Dorch, B. 2002, *Astron. Nachr.*, 323, 213
Hasan, S. S., Kalkofen, W., & van Ballegooijen, A. A. 2000, *Astrophys. J.*, 535, L67
Hasan, S. S. & Ulmschneider, P. 2004, *Astron. Astrophys*, 422, 1085
Herbold, G., Ulmschneider, P., Spruit, H. C., & Rosner, R. 1985, *Astron. Astrophys*, 145, 157
Huang, P., Musielak, Z. E., & Ulmschneider, P. 1995, *Astron. Astrophys*, 297, 579
Musielak, Z. E. & Ulmschneider, P. 2001, *Astron. Astrophys*, 370, 541
Schrijver, C. J., Cote, J., Zwaan, C., & Saar, S. H. 1989, *Astrophys. J.*, 337, 964
Simon, G. W. & Leighton, R. B. 1964, *Astrophys. J.*, 140, 1120
Skumanich, A., Smythe, C., & Frazier, E. N. 1975, *Astrophys. J.*, 200, 747
Steiner, O., Grossmann-Doerth, U., Knölker, M., & Schüssler, M. 1998, *Astrophys. J.*, 495, 468

The Physics of Sun and Star Spots
Proceedings IAU Symposium No. 273, 2010
D. P. Choudhary & K. G. Strassmeier, eds.

doi:10.1017/S1743921311015742

Solar flare forecasting using sunspot-groups classification and photospheric magnetic parameters

Yuan Yuan[1,2], Frank Y. Shih[2], Ju Jing[1] and Haimin Wang[1]

[1]Space Weather Research Lab, New Jersey Institute of Technology,
323 MLK Blvd, Newark, New Jersey, United States
email: yy46@njit.edu

[2]Computer Vision Lab, New Jersey Institute of Technology,
323 MLK Blvd, Newark, New Jersey, United States

Abstract. In this paper, we investigate whether incorporating sunspot-groups classification information would further improve the performance of our previous logistic regression based solar flare forecasting method, which uses only line-of-sight photospheric magnetic parameters. A dataset containing 4913 samples from the year 2000 to 2005 is constructed, in which 2721 samples from the year 2000, 2002 and 2004 are used as a training set, and the remaining 2192 samples from the year 2001, 2003 and 2005 are used as a testing set. Experimental results show that sunspot-groups classification combined with total gradient on the strong gradient polarity neutral line achieve the highest forecasting accuracy and thus it testifies sunspot-groups classification does help in solar flare forecasting.

Keywords. Sun: flares, sun: magnetic fields, sunspots

1. Introduction

Sunspot-groups characteristics have long been used in solar flare forecasting and still being used extensively. Contarino *et al.* (2009) studied sunspot-groups parameters (i.e., Zrich class, magnetic configuration, area, morphology of the penumbra), and then performed a flare forecasting campaign based on the results. They claimed that the results obtained by comparing the flare forecasting probability with the number of flares that have actually occurred are quite encouraging. Kasper & Balasubramaniam (2010) found out that the penumbral area, umbral area and irradiance showed promise as possible parameters for predicting solar flares, particularly M-class flares. Qahwaji & Colak (2007) compare the performances of several machine learning algorithm on flare forecasting using classification of sunspot groups and solar cycle data. They found out that Support Vector Machines provide the best performance for predicting whether a classified sunspot group is going to flare.

On the other hand, photospheric magnetic parameters derived form line-of-sight magnetograms are becoming more and more popular in solar flare forecasting. Jing *et al.* (2006) studied the mean value of spatial magnetic gradients at strong-gradient magnetic neutral lines, the length of strong-gradient magnetic neutral lines and the total magnetic energy. They found out there exist statistical correlations between the three parameters of magnetic fields and the flare productivity of solar active regions. Yuan *et al.* (2010) proposed a cascading forecasting approach using total unsigned magnetic flux, length of the strong-gradient magnetic polarity inversion line, and total magnetic energy dissipation. Experimental results show that photospheric parameters is indeed can be used a precursor for solar flares forecasting.

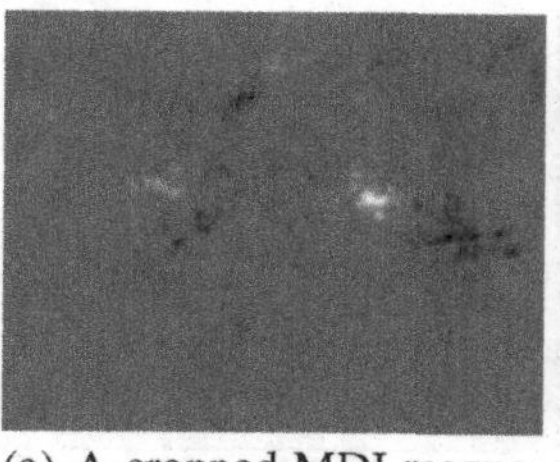

(a) A cropped MDI magnetogram containing active regions

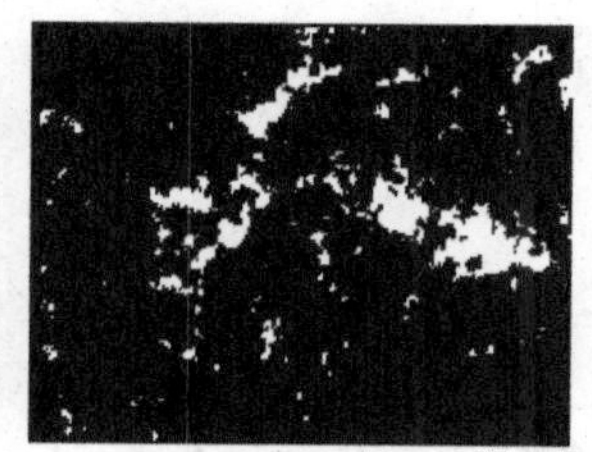

(b) A map indicating the regions used to calculate total unsigned magnetic flux with white color

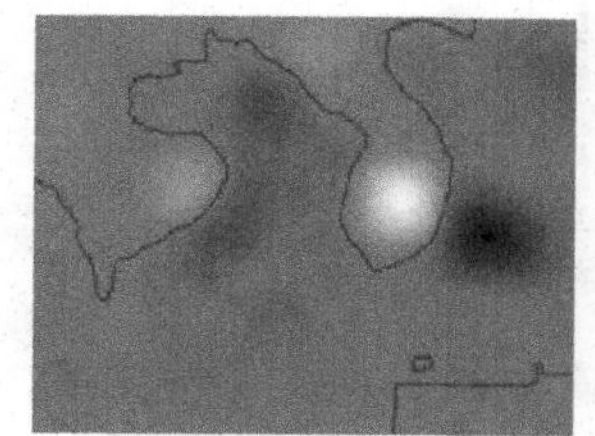

(c) Magnetic polarity inversion line overplotted on a smoothed MDI magnetogram

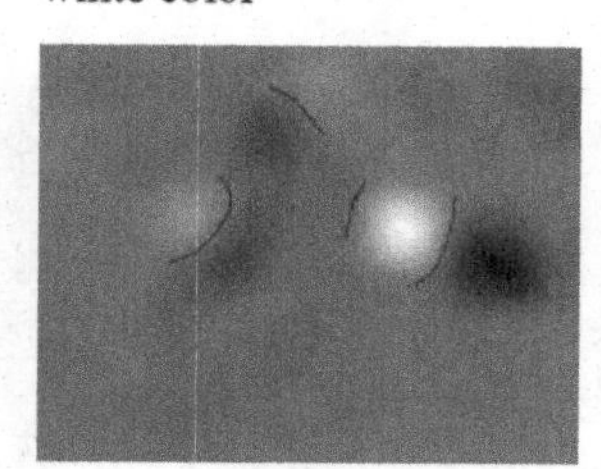

(d) Magnetic polarity inversion line with strong gradient overplotted on a smoothed MDI magnetogram

Figure 1. Illustration of the calculation of total unsigned magnetic flux and total unsigned gradient on strong gradient magnetic polarity inversion line

In this study, aiming to improve the solar flare forecasting performance in our previous study (Song *et al.* 2009) ,we use both sunspot-groups classification and photoshperic magnetic parameters. We view the solar flare forecasting as a classification problem in machine learning field. To forecasting a flare event is usually converted to classify one sample as a flaring sample or a non-flaring sample. Previously, researchers usually adopt support vector machines, such as Qahwaji & Colak (2007), or neural networks, such as Wang *et al.* (2008). The outputs of support vector machines and neural networks are binary labels indicating flaring or nonflaring. However, people sometimes prefer to get a probability instead of a binary label, just like what people get from daily weather reports. In our previous studies by Song *et al.* (2009) and Yuan *et al.* (2010), we have shown that logistic regression, which is a statistical learning method for probability estimation, can be used for flare forecasting. In this paper, the solar flare forecasting is regarded as a classification problem in machine learning field, i.e., flaring population vs. non-flaring population.

2. Dataset

The dataset used in our experiments includes 4913 samples from the year 2000 to 2005, in which 2721 samples from the year 2000, 2002 and 2004 are used as the training set, and the remaining 2192 samples from the year 2001, 2003 and 2005 are used as the testing set. Each sample is a pair of values describing the properties of an active region. A sample composed of a label indicating whether the active region produces a flare or not, a label indicating the classification of the sunspot-groups within the active region, a number indicating the total unsigned magnetic flux within the active region,

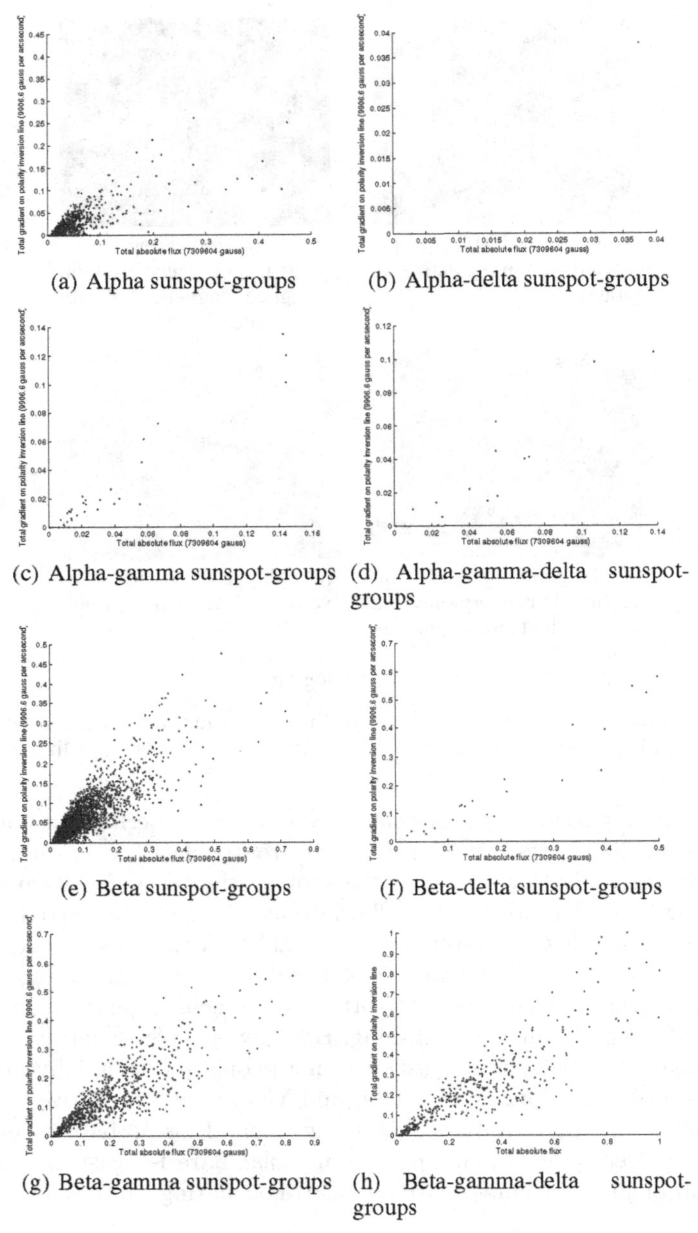

(a) Alpha sunspot-groups (b) Alpha-delta sunspot-groups

(c) Alpha-gamma sunspot-groups (d) Alpha-gamma-delta sunspot-groups

(e) Beta sunspot-groups (f) Beta-delta sunspot-groups

(g) Beta-gamma sunspot-groups (h) Beta-gamma-delta sunspot-groups

Figure 2. Distribution of dataset with respect to different sunspot-groups classifications.

and a number indicating the total gradient of the strong gradient polarity neutral line. Figure 2 illustrates the distribution of our dataset with respect to different sunspot-groups classifications.

Total unsigned magnetic flux is the integration of pixel intensity over the strong magnetic flux region of an active region. In this study, we define strong flux region as the region composed of pixels with intensity greater than median value plus 80 gauss, and pixels with intensity less than median value minus 80 gauss. Total unsigned magnetic flux T_{flux} can be calculated as following:

$$T_{flux} = \int_{B_Z(x,y) \leqslant median-80 \ or \ B_Z(x,y) \geqslant median+80} |B_Z(x,y)| \, dxdy \tag{2.1}$$

where $B_Z(x,y)$ is the intensity of a pixel at location (x,y) of a MDI magnetogram.

The total gradient of the strong gradient polarity neutral line is the integration of the gradient over the pixels whose intensities are zeros and their gradient is greater than a threshold (here we choose 5). The total gradient of the strong gradient polarity neutral line T_{grad} can be calculated as following;

$$T_{grad} = \int_{B_Z(x,y) \equiv 0 \ ,m>5} \sqrt{\left(\frac{\partial B_Z(x,y)}{\partial x}\right)^2 + \left(\frac{\partial B_Z(x,y)}{\partial y}\right)^2} dxdy \tag{2.2}$$

where

$$m = \sqrt{\left(\frac{\partial B_Z(x,y)}{\partial x}\right)^2 + \left(\frac{\partial B_Z(x,y)}{\partial y}\right)^2} > 5 \tag{2.3}$$

Figure 1 contains one sample illustrating the calculation of total unsigned magnetic flux and total unsigned gradient on strong gradient magnetic polarity inversion line. To calculate total unsigned magnetic flux, a binary mask (illustrated as fig. 1(b)) is generated which indicating the regions where magnetic flux is greater than median plus 80 or less than median minus 80. And then the summation of pixel values inside those regions are figured out as total unsigned magnetic flux T_{flux}. To figure out magnetic polarity inversion line, a MDI magnetogram is firstly smoothed with a Gaussian filter with the standard deviation 10 and of size 30 by 30. And then contour lines at height zeros are find out(illustrated as fig. 1(c)). At last, the contour lines with strong gradient is kept (illustrated as fig. 1(d)). The summation of the gradient on strong gradient magnetic polarity inversion line are figured out as the total gradient of the strong gradient polarity neutral line T_{grad}.

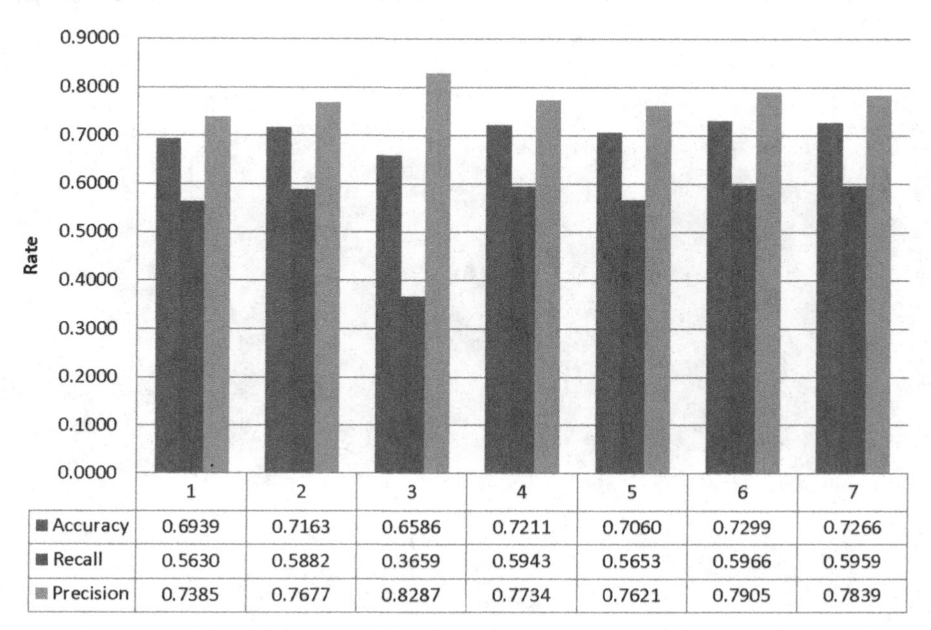

	1	2	3	4	5	6	7
Accuracy	0.6939	0.7163	0.6586	0.7211	0.7060	0.7299	0.7266
Recall	0.5630	0.5882	0.3659	0.5943	0.5653	0.5966	0.5959
Precision	0.7385	0.7677	0.8287	0.7734	0.7621	0.7905	0.7839

Figure 3. Performance evaluation

3. Experimental Results

Figure 3 illustrates the rate of accuracy, recall and precision of the solar flares forecasting method with seven different combinations of input parameters: 1. use T_{flux} alone, 2. use T_{grad} alone, 3. use sunspot-groups classification alone, 4. use T_{flux} and T_{grad} , 5. use T_{flux} and sunspot-groups classification, 6. use T_{grad} and sunspot-groups classification, 7. use T_{flux}, T_{grad} and sunspot-groups classification.

From figure 3, we can see that solar flares forecasting using T_{grad} and sunspot-groups classification achieves best accuracy and recall. Solar flare forecasting using sunspot-groups classification alone achieve best precision. The average performance (measured by accuracy, recall and precision) of solar flare forecasting using T_{grad} and sunspot-groups classification is the best.

Acknowledgment This work is supported by NSF under grants ATM-0716950, ATM-0745744 and NASA under grant NNXO-8AQ90G. The authors thank Dr. Barnes and Dr. Leka for letting us use their data set. Preparation of their data set was funded by NASA LWS TRT contract NNH09CE72C.

References

Contarino, L., Zuccarello, F., Romano, P., Spadaro, D., Guglielmino, S. L., & Battiato, V. 2009, *Acta Geophysica*, 57, 52

Jing, J., Song, H., Abramenko, V., Tan, C., & Wang, H. 2006, *Astrophys. J.*, 644, 1273

Kasper, D. & Balasubramaniam, K. S. 2010, *Bulletin of the American Astronomical Society*, 41, 291

Qahwaji, R. & Colak, T. 2007, *Solar Phys.*, 241, 195

Song, H., Tan, C., Jing, J., Wang, H., Yurchyshyn, V., & Abramenko, V. 2009, *Solar Phys.*, 254, 101

Wang, H. N., Cui, Y. M., Li, R., Zhang, L. Y., & Han, H. 2008, *Advances in Space Research*, 42, 1464

Yuan, Y., Shih, F. Y., Jing, J., & Wang, H.-M. 2010, *Research in Astronomy and Astrophysics*, 10, 785

The Physics of Sun and Star Spots
Proceedings IAU Symposium No. 273, 2010
D. P. Choudhary & K. G. Strassmeier, eds.

doi:10.1017/S1743921311015754

Using SONG to probe rapid variability and evolution of starspots

James E. Neff[1,4], Jon Hakkila[1], Frank Hill[2], Jason Jackiewicz[3], Travis S. Metcalfe[4], Jørgen Christensen-Dalsgaard[4,5], Søren Frandsen[5], Frank Grundahl[5], Hans Kjeldsen[5], Uffe Gråe Jørgensen[6], Per Kjærgaard Rasmussen[6], and Sheng-Hong Gu[7]

[1]Dept. of Physics & Astronomy, College of Charleston, Charleston, SC 29424, USA
email: neffj@cofc.edu
[2]National Solar Observatory, Tucson, AZ, 85719, USA
[3]Astronomy Dept., New Mexico State University, Las Cruces, NM 88003, USA
[4]High Altitude Observatory, NCAR, Boulder, CO 80307, USA
[5]Dept. of Physics & Astronomy, Aarhus University, Ny Munkegade, 8000 Aarhus C, Denmark
[6]Niels Bohr Institute, U. Copenhagen, Juliane Maries Vej 30, DK 2100 Copenhagen, Denmark
[7]Yunnan Astronomical Observatory, Chinese Academy of Sciences, Kunming, China.

Abstract. The Stellar Observations Network Group (SONG) is being developed as a network of 1-meter spectroscopic telescopes designed for and primarily dedicated to asteroseismology. It is patterned after the highly successful GONG project. The Danish prototype telescope will be installed in Tenerife in early 2011. Ultimately we hope to have as many as 8 identical nodes providing continuous high-resolution spectroscopic observations for targets anywhere in the sky. The primary scientific goals of SONG are asteroseismology and the search for Earth-mass exoplanets. The spectroscopic requirements for these programs push the limits of current technology, but the resulting spectrograph design will enable many secondary science programs with less stringent requirements. Doppler imaging of starspots can be accomplished using continuous observations over several stellar rotations using identical instrumentation at each node. It should be possible to observe the evolution of starspot morphology in real-time, for example. We discuss the design and status of the SONG project in general, and we describe how SONG could be used to probe short timescale changes in stellar surface structure.

Keywords. Stars: spots, stars: oscillations, stars: magnetic fields, stars: activity, instrumentation: spectrographs

1. The Stellar Observations Network Group

Carefully designed, dedicated, long-term ground-based (e.g. GONG, BiSON) and space-based observations (e.g. SOHO) have enabled a detailed look inside the the Sun using the techniques of helioseismology. Extending these techniques to stars ("asteroseismology") has long been a dream of the stellar astrophysics community (Christensen-Dalsgaard 2002). Thanks to intensive ground-based observing campaigns and, especially, the CoRoT and Kepler missions, that dream is now becoming a reality (Bedding *et al.* 2010, Chaplin *et al.* 2010, Gilliland *et al.* 2010). The past decade has also seen an explosion of interest in detecting and characterizing extrasolar planets.

GONG ("Global Oscillations Network Group") is a network of 6 extremely sensitive and stable velocity imagers located around the Earth to obtain nearly continuous helioseismology observations. It has been in operation for about 15 years. For at least that long, there have been efforts to develop a similar network for stellar seismology. At last, such a network is being built (Grundahl *et al.* 2008)! It is a Danish-led effort now entitled *SONG* ("Stellar Observations Network Group"). The scientific goals of *SONG* are (1) to study the internal structure and evolution of stars using asteroseismology, and (2) to search for and characterize planets with masses comparable to the Earth in orbit around other stars. It will accomplish the first of these goals using 1-meter telescopes and extremely stable high-resolution spectrographs with an iodine cell. The second goal will be accomplished using lucky imaging for high-cadence microlensing event monitoring and through radial-velocity studies.

2. Design capabilities and status of SONG

The Danish prototype has been fully designed and funded (Grundahl *et al.* 2009). The 1-meter telescope and observatory enclosure is being built by Astelco Systems, GMBH, with delivery at the site in Tenerife expected around the beginning of 2011. The instrumentation and spectrograph enclosure is being built and integrated in Denmark. A second, virtually identical "node" is being developed in China. Preliminary design work has already been completed, and site surveys are being conducted.

A consortium has been formed in the US to actively seek funding for a third node, to be placed in Hawaii (possibly at HAO's Mauna Loa Solar Observatory). Ultimately, we envision an 8-node network (four telescopes in each hemisphere). But even a 3-node network of identical instruments will enable high-resolution spectra to be obtained continuously over a large fraction of the sky.

The telescope is an alt-az design with one Nasmyth focus occupied by two lucky imaging cameras (using a beam splitter for separate red and visual cameras) designed for microlensing planet searches. The coudé focus feeds a spectrograph in a separate, climate-controlled enclosure (a modified shipping container on a stable platform). An iodine cell in the pre-slit area is crucial to measure radial velocities with the precision required for asteroseismology, but it can be removed from the beam for normal spectroscopic observations. The baseline detector is an Andor Technology 2Kx2K CCD with 13.5 micrometer pixels. With the echelle spectrograph design, this will yield 2-pixel resolution of 120,000 with a 1 arc-second slit. The design is optimized for the wavelength range 4814 to 6774 Å, with full wavelength coverage below 5200 Å. The spectrograph design permits the utilization of a larger detector with smaller pixels to achieve greater wavelength coverage and better sampling of the instrumental profile. Much of the data pipeline processing will be accomplished on-site, but the final processing will be performed at the observatory operations center in Denmark. *SONG* is intended to be an "open-source" project – all drawings and in-house software will be available to the community.

3. The need for continuous spectroscopy

The use of Doppler imaging techniques has revealed the surface structure (magnetic and otherwise) of a large number of stars (Berdyugina 2005, Strassmeier 2009). For the most part, these have been based on observations from a single site, with data obtained over several rotations to achieve full phase coverage. Targets are routinely observed for only a few days each observing season. It is very difficult to do otherwise using a single, general purpose facility.

How stable is the structure over several rotations? On what timescales do starspots form, migrate, and decay? These questions are difficult or impossible to answer with an occasional "snapshot" made from a single site. Ideally, we would like to observe a star continuously over several rotations for a single Doppler image, and we would like to combine a series of Doppler images to investigate the structure over the relevant evolutionary timescales. These can be used to probe the nature of stellar dynamos by measuring differential rotation, meridional flows, and starspot behavior on evolutionary and stellar cycle timescales (e.g. stellar "butterfly diagrams"). To accomplish all this, we require a network of high-resolution spectrographs (preferably with identical capabilities) widely distributed in longitude.

Between 1989 and 1998, a global consortium (known as *MUSICOS*) was formed, and tremendous effort was made to arrange for continuous high-resolution spectroscopy using existing facilities (Catala *et al.* (1993)). Several new spectrographs were built and placed at key facilities. Over the decade, 5 major campaigns including 19 scientific programs were carried out. Each of these campaigns interleaved 3 or more scientific programs that all required continuous spectroscopy. Examples include stellar oscillations (e.g., Kennelly *et al.* (1996)), stellar activity (e.g. García-Alvarez *et al.* (2003)), and dynamic circumstellar environments (e.g. Unruh *et al.* (2004)). *MUSICOS* demonstrated that obtaining continuous high-resolution spectroscopy using existing sites requires a heroic effort, is limited by inhomogeneous data sets from non-identical facilities, and can only be done on an ad hoc and infrequent basis. Recent campaigns dedicated to asteroseismology (e.g. Arentoft *et al.* (2008), Bedding *et al.* (2010)) have been more successful, but they also highlight the need for a dedicated, full-time network.

4. Using SONG to observe starspots

The design of *SONG* is driven by the precise radial velocity requirements for asteroseismology. But, as *MUSICOS* demonstrated, the capability of continous high-resolution spectroscopy enables other science. Doppler imaging of starspots is the most obvious. There is a growing overlap of the starspot and stellar oscillation communities, driven by the tremendous data coming from CoRoT and Kepler. Many stars have both oscillations and starspots, which can serve as complimentary probes of their internal structure.

Asteroseismology requires long, uninterrupted observing runs on individual stars using the iodine cell. The network will be dedicated for weeks or even months to a single bright target. During periods of overlapping coverage between sites, or at the beginning and end of each night, individual nodes could be scheduled for other programs. The *SONG* network therefore will be capable of interleaving observations over starspot evolutionary timescales. For stars with stable spots and suitable periods, this coverage will be adequate for Doppler imaging. For most stars, however, continuous coverage with the full *SONG* network for several stellar rotations would be preferable. Short Doppler imaging runs could be scheduled between longer asteroseismology runs, for example.

To obtain such continuous Doppler imaging observations, it is necessary that the stellar activity community become involved in the development of *SONG*. Currently, the top priority is to garner support and funding for the US node, but there is still an opportunity for this community to provide input into the observing and data analysis strategies. Funding the full 8-node *SONG* network is likely to require broad support from the entire stellar astrophysics community. We are planning a workshop (to be held in Charleston, SC during September 2011) to solicit broader community input and support.

Acknowledgements

For more information see http://astro.phys.au.dk/SONG/

References

Arentoft, T. *et al.* 2008, *Astrophys. J.*, 687, 1180
Berdyugina, S.V. 2005, *Living Reviews in Solar Physics*, 2, 8
Bedding, T. R. *et al.* 2010, *Astrophys. J.*, 713, 935
Catala, C. *et al.* 1993, *Astron. Astrophys.*, 275, 245
Chaplin, W. J. *et al.* 2010, *Astrophys. J.*, 713, L169
Cristensen-Dalsgaard 2002, *Reviews of Modern Physics*, 74, 1073
García-Alvarez, D. *et al.* 2003, *Astron. Astrophys.*, 397, 285
Gillilland, R. L. *et al.* 2010, *Pub. Astron. Soc. Pac.*, 122, 131
Grundahl, F., Arentoft, T., Christensen-Dalsgaard, J., Frandsen, S., Kjeldsen, H, & Rasmussen, P. K. 2008, *Journal of Physics Conference Series*, 118(1):012041
Grundahl, F., Cristensen-Dalsgaard, J., Kjeldsen, H., Jørgensen, U. G. Arentoft, T., Frandsen, S., & Rasmussen, P. K. 2009, *ASP-CS*, eds. Dikpati, M. *et al.*, 416, 579.
Kennelly, E. J. *et al.* 1996, *Astron. Astrophys.*, 313, 571
Strassmeier, K. G. 2009, *Astron. Astrophys. Reviews*, 17, 251
Unruh, Y. C. *et al.* 2004, *Mon. Not. Roy. Astron. Soc.*, 348, 1301

The Physics of Sun and Star Spots
Proceedings IAU Symposium No. 273, 2010
D. P. Choudhary & K. G. Strassmeier, eds.

doi:10.1017/S1743921311015766

Optical polarimetry and photometry of young sun-like star LO Peg

J. C. Pandey[1], B. J. Medhi[1], and R. Sagar[1]

[1] Aryabhatta Research Institute of Observational Sciences, Nainital, India -263129
email: jeewan@aries.res.in

Abstract. We have carried out the B,V and R-band polarimetric and V-band photometric study of the star LO Peg. Our analysis reveal that LO Peg is highly polarized among the sun-like stars. The degree of polarization and polarization position angle are found to be rotationally modulated. The levels of polarization observed in LO Peg could be the result of scattering of an anisotropic stellar radiation field by an optically thin circumstellar envelope or scattering of the stellar radiation by prominence-like structures. The long term photometric observations of LO Peg indicate three independent groups of spots are present on the surface of LO Peg.

Keywords. Stars: activity, stars: late-type, stars: spots, techniques: photometric, techniques: polarimetric

1. Introduction

LO Pegasi (LO Peg) is a single, young, K3V-K7V-type and a member of the Local Association (Montes *et al.* 2001; Pandey *et al.* 2005). It is one of the fast-rotating active stars with a period of 0.42 days. LO Peg shows strong Hα and Ca II H and K emission lines (Jeffries *et al.* 1994). Evidence of an intense downflow of material and optical flaring on LO Peg had presented by Eibe *et al.* (1999). Zuckerman *et al.* (2004) have identified LO Peg as a member of a group of 50-Myr-old stars that partially surround the Sun. In this contribution, we have investigated the polarimetric and long term photometric observations of LO Peg.

2. Observations and data reductions

The broad-band B, V and R polarimetric observations of LO Peg have been made in between 2007 October 19 and December 19 using ARIES Imaging Polarimeter (Rautela *et al.* 2004), mounted on the Cassegrain focus of the 104-cm Sampurnanand telescope (ST) of Aryabhatta Research Institute of Observational Sciences, Nainital. Details of polarimetric data reduction are given in Medhi *et al.* 2010 and Pandey *et al.* 2009. For calibration of polarization angle zero-point, we observed highly polarized standard stars. For instrumental polarizations, we observed standard unpolarized stars. Photometric observations in V-band were also taken from ST using 2k$\times$2k CCD (see Pandey *et al.* 2005) during year 2001 and 2002. We have also used the All Sky Automated Survey (ASAS; Pojmanski (2002)) data for our study. We have used only 'A' and 'B' graded data. The ASAS survey has longer span of V-band photometry of 6.5 years. The total time span of the observations is 8 years, which is useful to observe any long term variations as well as migration of spots on the stellar surface.

3. Polarization

We have fold the polarimetric data using the ephemeris HJD = 2448869.93 + 0.42375 E. Top and middle panels of Figure 1 show the variations of degree of polarization

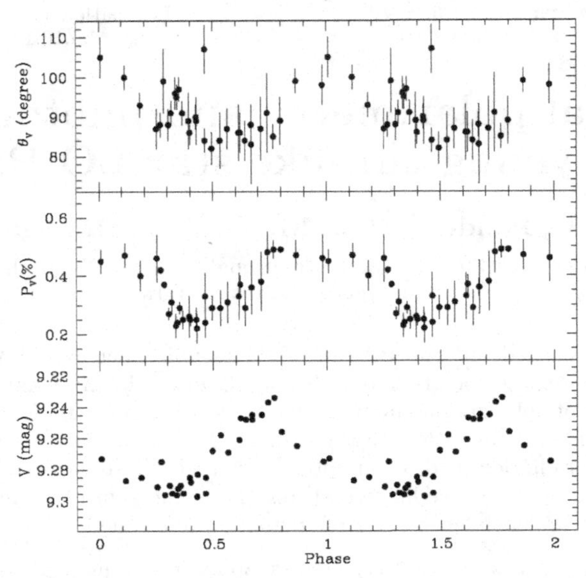

Figure 1. Bottom Panel: V-band light variation, Middle panel: V-band polarization variation and Top panel: Position angle (θ_V) variation with rotational phase of the star LO Peg. The data are folded with rotational period of 0.42 d

(P_V) and position angle (θ_V) in V-band. Both are found to be rotationally modulated. The average values of degree of polarization in B, V and R bands are found to be $0.387 \pm 0.004\%$, $0.351 \pm 0.004\%$ and $0.335 \pm 0.003\%$, respectively. However, the average values of polarization position angle are found to be $88° \pm 1°$, $91° \pm 1°$ and $91° \pm 1°$ in B, V and R band, respectively. The nearly similar values of polarization position angle in each band indicates that the scattering geometry is identical at all wavelengths. Below, we investigate the origin of polarization in LO Peg.

LO Peg is located at a distance of 25 pc, and therefore has a negligible reddening. It is quite natural to assume that the observed polarization in LO Peg is not foreground in origin. Further, the observed polarization in any distant star located near the Galactic plane may have small negligible interstellar component. Further, the existence of a time-dependent polarization is a well-established criterion for intrinsic polarization. Huovelin & Saar 1991 showed that the B, V and R-band polarization due magnetic filed should be less than 0.16% for a K-type dwarf having 2.7 k Gauss magnetic filed and total spot area of $\approx 24\%$. The observed values of polarization in LO Peg are more than 0.16%. Therefore, the polarization in LO Peg could not be magnetic in origin. The observed polarization in LO Peg was found to decrease towards longer wavelength. This could be due to either selective absorption by circumstellar dust, which grows towards shorter wavelengths or wavelength-dependent albedo, which decreases towards longer wavelengths resulting less scattering and thus polarization. We have also investigated that whether the polarization in LO Peg due to cool prominence like structures which are located at the co-rotation radius (r_c). Pfeiffer(1978) showed that polarization due prominence like structure is $P = N\sigma f/(r_c^2 + N\sigma f)$. Here, N is number of scatterer, f is angular scattering function and σ is scattering cross-section. Using above relation, the amount of scatterers are determined to be $\sim 10^{48}$. If the cloud contains mostly hydrogen, this amounts to about $\sim 10^{24}$ g, which is only an order larger than the mass-loss rate (2.0×10^{-11} $M_\odot$ yr^{-1}).

Variation of V-band magnitude is shown in bottom panel of Fig. 1. Rotational modulation is clearly visible. It appears that the variation in V-band light is correlated with the

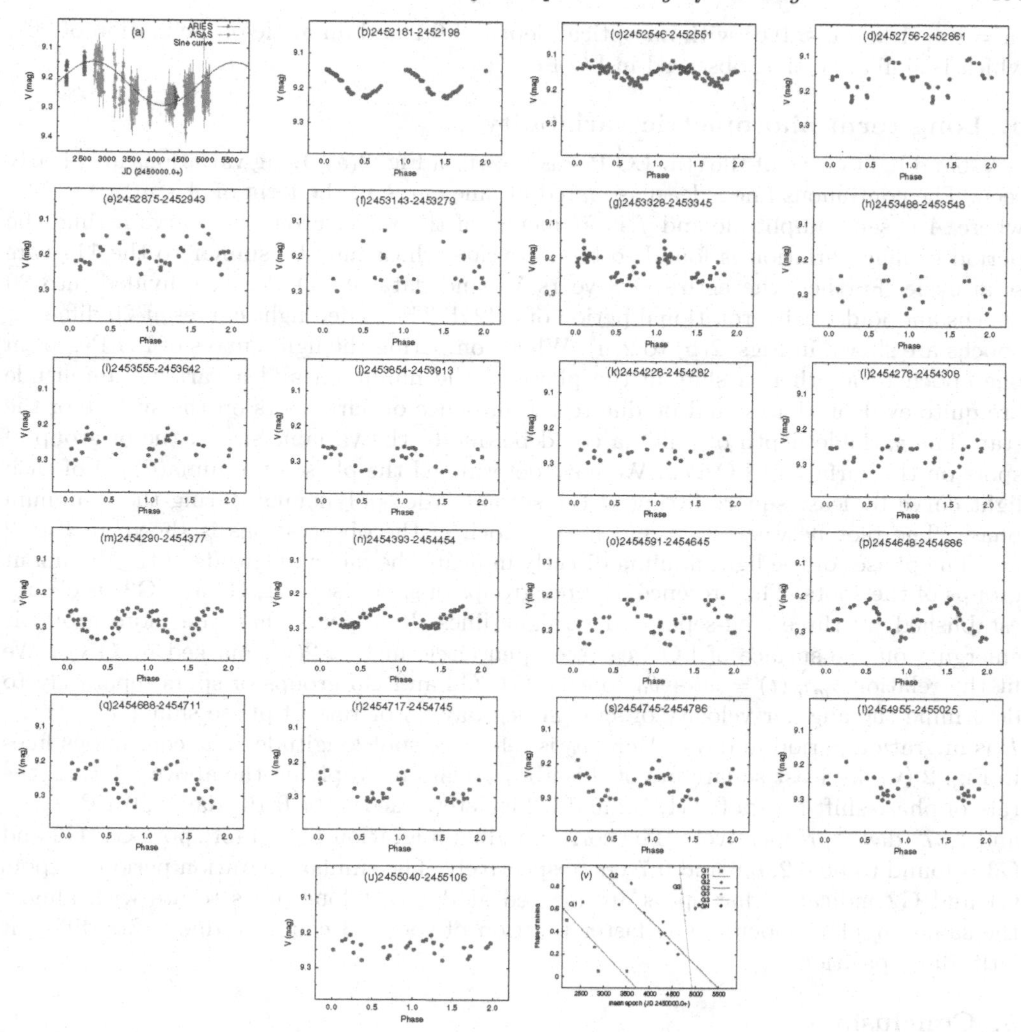

Figure 2. The top-first panel is V-band light curve as function of JD. From top-second to bottom-third panels are light curves as a function of rotational phase at 20 different epochs in chronological order. The epochs are mentioned at top of each diagram. The bottom forth panel shows a plot between minima of phase and mean epoch of observations.

V-band polarization. The significance of correlation has been calculated by determining the linear correlation coefficient r between the magnitude and the degree of polarization. The values of r between V-band magnitude and P_V, and V-band magnitude and θ_V are found to be 0.47 and 0.45 with the corresponding probability of no correlation being 0.008 and 0.013, respectively. Star was brighter when P_V was higher. This implies that the surface brightness inhomogeneity seems to play a major role in the polarization variability of LO Peg. Inhomogeneities in the distribution of both the material in the circumstellar envelope and brightness on the surface of the illuminating star can provide the necessary asymmetry, and hence produces a net polarization in the integrated light. The mechanism which can produce periodic linear polarization variability is Rayleigh or Thomson scattering in a non-spherically symmetric distribution of circumstellar material. Recently, Al-Malki *et al.* (1999) showed that a $\sim$ 0.5 % of polarization can be achieved from an anisotropic light source scattered by a Maclaurian spheroidal envelope (a = b = 1 and c $\approx$ 0.6) with an optical depth of 0.1 and an angle of inclination of 60°, which is similar to that observed in LO Peg.

4. Long term photometric variability

V-band light curve of the star LO Peg is shown in Fig. 2(a). Long variations are clearly seen. The continuous line is least -squared fit sine curve in the form of $A\ sin(2\pi fx + \phi)$, where A is semi-amplitude and f is frequency of the wave. From sine curve fitting, the period of long variation is found to be $\sim$ 8 years which may be similar to the 11-years solar-cycle. Further, the entire $\sim$ 8 years V-band data of LO Peg are divided into 20 epochs and folded with rotational period of 0.42 d. The folded light curves at 20 different epochs are shown in Figs. 2(b) to 2(u). When comparing the light curves of LO Peg from one epoch to another, a shift in the phase of the minimum and a variable amplitude are quite evident. This could be due to the presence of dark spots on the surface of the star. The variable depth of minima could be due to the variable size of the of group of spots on the surface of LO Peg. We have determined the phase of minima(θ_{min}) of each light curve by least square fitting of the second order polynomial during the minimum phase. The plot between θ_{min} and mean epoch of the observations is shown in Fig. 2 (v). The phases of the light minima directly indicate the mean longitude of the dominant groups of the spots. The presence of three groups of spots (say G1, G2 and G3) is clearly established by three well-separated straight lines. It appears that one more group is emerging on the surface of LO Peg (see open circle in Fig. 2(v), marked by GN) . We fit the relation $\theta_{min}(t) = \omega t + \theta_0$ to each G1, G2 and G3 groups of spots separately to determine the angular velocity of spot(in deg day^{-1}) or rate of phase shift ($\omega = 2\pi/P$, P is migration period in days). Here, θ_0 is reference spot longitude. The continuous lines in Fig. 2 (v) are least square fits of this above relation. Applying the above relation, the rate of phase shift for spots G1, G2 and G3 is determined to be 0.16° day^{-1}, 0.15° day^{-1} and 1.37° day^{-1}, respectively. The corresponding migration period for spots G1, G2 and G3 is found to be 6.2, 6.6 and 0.7 yrs, respectively. The similar migration period of spots G1 and G2 indicates that spots are located at different longitudes rotate with almost the same angular velocity. The faster rotation of spot G3 could be due to its different latitudinal position.

5. Conclusions

We conclude that the high values of polarization observed in LO Peg require either a spheroidal envelope with an optical depth of 0.1 or a clumpy material (e.g. solar prominence-like structures) of mass of the order of $\sim 10^{10}$ $M_{\odot}$. From long term photometric data, we established that three groups of spots are present on the surface of LO Peg, out of which one spot is migrating faster than other two.

Acknowledgments. Many thanks to IAU for financial support to attend the symposium and ASAS for the V-band data.

References

Al-Malki M. B., Simmons J. F. L., Ignace R., Brown J. C., & Clarke D., 1999, *Astron. Astrophys*, 347, 919

Eibe M. T., Byrne P. B., Jeffries R. D., & Gunn A. G., 1999, *Astron. Astrophys*, 341, 527

Huovelin J., & Saar S. H., 1991, *Astrophys. J.*, 374, 319

Jeffries R. D. *et al.* 1994, *Mon. Not. Roy. Astron. Soc.*, 270, 153

Medhi B. J., Maheswar G., Pandey J. C., & Sagar R., 2010, *Mon. Not. Roy. Astron. Soc.*, 403, 1577

Montes D., *et al.* 2001, *Mon. Not. Roy. Astron. Soc.*, 328, 45

Pandey J. C., Singh K. P., Drake, S. A., & Sagar R. 2005, *Astron. J.*, 30, 490

Pandey J. C., Medhi B. J., Sagar R., & Pandey A. K. 2009, *Mon. Not. Roy. Astron. Soc.*, 30, 490

Pfeiffer R. J., 1979, *Astrophys. J.*, 232, 181

Pojmanski G. 2002, *Acta Astronomica*, 52,397

Rautela B. S., Joshi G. C., & Pandey J. C., 2004, *Bull. Astron. Soc. India*, 32, 159

Zuckerman B., Song I., & Bessell M. S., 2004, *Astrophys. J.*, 613, L65

The Physics of Sun and Star Spots
Proceedings IAU Symposium No. 273, 2010
D. P. Choudhary & K. G. Strassmeier, eds.

doi:10.1017/S1743921311015778

Surface evolution in stable magnetic fields: the case of the fully convective dwarf V374 Peg

K. Vida[1], K. Oláh[1], and Zs. Kővári[1]

[1]Konkoly Observatory of the Hungarian Academy of Sciences
H-1121 Budapest, Konkoly Thege Miklós str. 15-17.
email: vidakris@konkoly.hu

Abstract. We present $BV(RI)_C$ photometric measurements of the dM4-type V374 Peg covering ~ 430 days. The star has a mass of $\sim 0.28 M_{\rm Sun}$, so it is supposed to be fully convective. Previous observations detected almost-rigid-body rotation and stable, axisymmetric poloidal magnetic field. Our photometric data agree well with this picture, one persistent active nest is found on the stellar surface. Nevertheless, the surface is not static: night-to-night variations and frequent flaring are observed. The flares seem to be concentrated on the brighter part of the surface. The short-time changes of the light curve could indicate emerging flux ropes in the same region, resembling to the active nests on the Sun. We have observed flaring and quiet states of V374 Peg changing on monthly timescale.

Keywords. Magnetic fields, techniques: photometric, stars: activity, stars: flare, stars: individual (V374 Peg), stars: late-type, stars: spots, stars: magnetic fields

---

1. Introduction

V374 Peg is an M4 dwarf (Reid, Hawley, & Gizis 1995) whose X-ray emission has been detected by the ROSAT satellite (Bade *et al.* 1998). Greimel & Robb (1998) and Batyrshinova & Ibragimov (2001) presented photometry in R and $UBVRI$ filters, both paper report intense and frequent flares on the star. Recently, Korhonen *et al.* (2010) published spectroscopic observations of V374 Peg (along with some of the photometry presented in this paper), which showed intensive emission in the Hα line.

A reason why this object is especially interesting is the mass of the star: it is $0.28 M_{\rm Sun}$ (Delfosse *et al.* 2000), which places it just below the theoretical limit of full convection ($0.35 M_{\rm Sun}$ according to Chabrier & Baraffe 1997). Above this limit stars have a radiative core with a convective envelope and are supposed to sustain a solar-type $\alpha\Omega$ dynamo (Parker 1955; Babcock 1961; Leighton 1969). Below $0.35 M_{\rm Sun}$ stars are fully convective, but the origin of their magnetic fields is unclear in details. Küker & Rüdiger (2005) and Chabrier & Küker (2006) showed, that these kind of stars can produce large-scale, non-axisymmetric fields using α^2-dynamo, if they rotate rigidly. On the other hand, they can have axisymmetric, poloidal fields, given that they have strong differential rotation (Dobler, Stix, & Brandenburg 2006). As spectropolarimetric observations of Donati *et al.* (2006) and Morin *et al.* (2008) have showed V374 Peg has a very weak differential rotation, but at the same time the star has stable, axisymmetric, poloidal magnetic field, which doesn't fit to the current theoretical models.

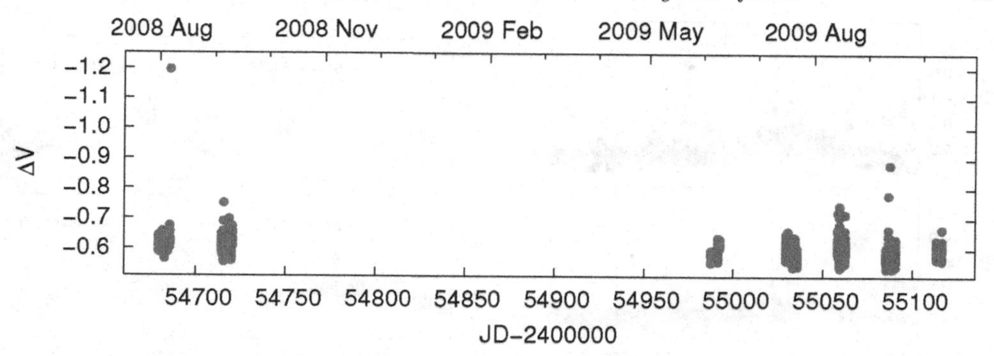

Figure 1. V light curve of V374 Peg. The deviating points are flares.

In this paper we study the behaviour of the photosphere of V374 Peg on a yearly timescale.

2. Observations

We present photometric data obtained by the 1m RCC telescope of the Konkoly Observatory at Piszkéstető Mountain Station. The telescope is equipped with a 1300×1300 Princeton Instruments CCD. The $BV(RI)_C$ light curves have been obtained between 2008 July 31 and 2009 October 12 in seven observing runs. The observing runs are separated by about 3 weeks. Data reduction and aperture photometry was done using standard IRAF† routines. The resulting light curve in V passband is plotted in Fig. 1. For the phased light curves we used the following ephemeris:

$$\mathrm{HJD} = 2453601.786130 + 0.445679 \times E,$$

where the period was determined for the whole photometric observation using the SLLK method described by Clarke (2002). This algorithm phases the datasets with different periods and chooses the period giving the 'smoothest' light curve as the correct one, giving a more reliable result than Fourier-analysis for quasi-periodic, non-sinusoidal light curves.

3. Analysis of the data

When looking at the light curve (Fig. 1) the first impression is that – except the flares – there is no change in the mean brightness level, i.e. the total spottedness remains the same. A closer look reveals that during our observations two stable active regions could be found on the stellar surface: one at ~ 0.0 and another one at ~ 0.6 phase (see Fig. 2). For easier visualization we have made a spotmodel (see Fig. 3) using the SML software (Ribárik *et al.* 2003) supposing a surface temperature of 3000K, a spot temperature of 2800K (Morin *et al.* 2008), and circular spot shape. The overall shape of the phased light curve remains the same during our ~ 430 day-long time-series, which means, that these active regions do not change their size and position in appreciable extent. Thus, from our

† IRAF is distributed by the National Optical Astronomy Observatory, which is operated by the Asso- ciation of Universities for Research in Astronomy, Inc., under cooperative agreement with the National Science Foundation.

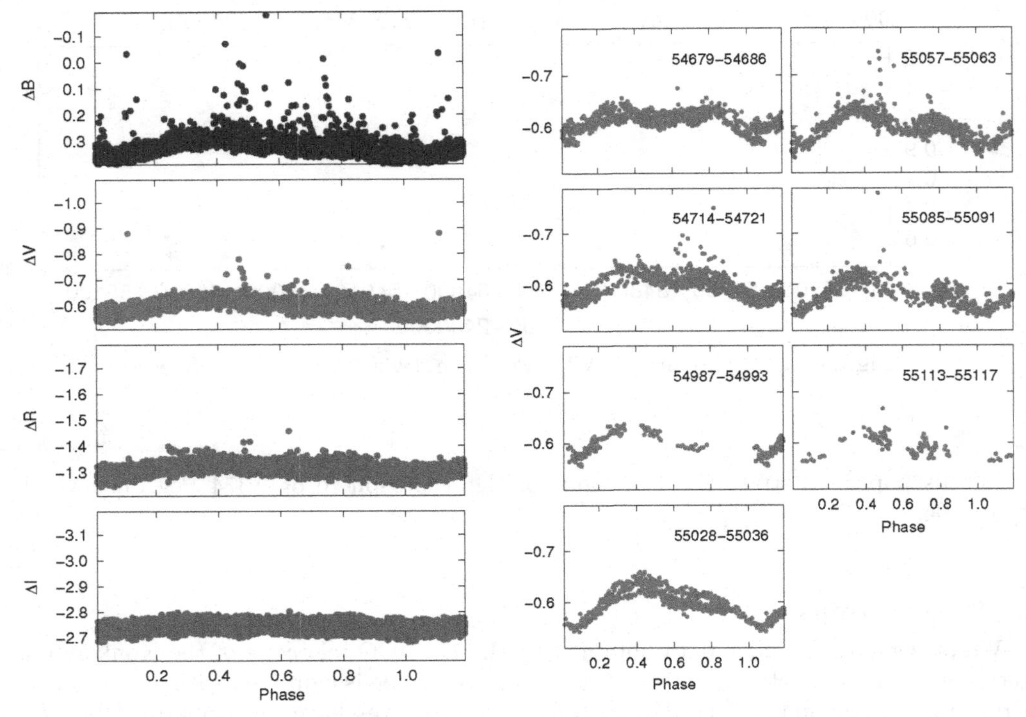

Figure 2. *Left:* Phased $BV(RI)_C$ light curves of V374 Peg showing all the observations. Frequent and intensive flares can be seen in B, V, and R_C bands. The flares tend to be concentrated around phase 0.5. *Right:* Phased V light curves from different observing runs. The star shows active (flaring) and less active phases changing on monthly timescale.

photometry we can confirm the finding of Donati *et al.* (2006) and Morin *et al.* (2008): the light curve of V374 Peg is stable on the timescale of a year.

However, the light curve is not constant at all: small changes can be seen on a nightly timescale (see Fig. 3). These variations are mainly seen around the active region at 0.6 phase. It is possible, that these variations are caused by newly-born and decaying spots in the same active region, similarly to the emerging flux ropes in active nests on the surface of the Sun. We can conclude, that the location and size of the two active regions are stable, but the exact spot configuration is changing very rapidly.

Looking at Fig. 2, it can be clearly seen that V374 Peg has active and less active states. During the first two observing runs frequent flaring could be observed, while in the next two runs the light curve is free of flares, with the spot configuration and the rapid nightly variations remaining the same as before (note, that there is a gap of $\sim$ 260 days between the first two and the subsequent observing runs). During the following two runs the star is more active again, and in the last run the phase coverage of the light curve is very poor, so no exact conclusions can be drawn from that one. It seems that the flaring and non-flaring phases on V374 Peg vary on monthly timescale.

The fact, that the nightly variation of the light curve occurs around the same phase ($\sim$ 0.6) as the bulk of the flares, suggests that there is a connection between them. The frequent flares can be a result of reconnection between the emerging flux ropes of the active nest around phase 0.6.

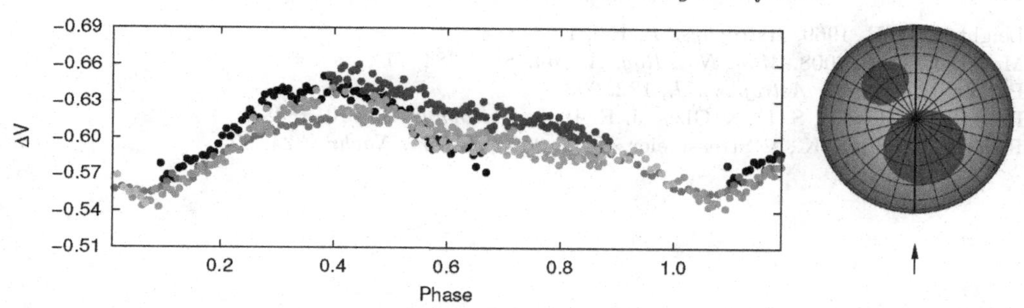

Figure 3. *Left:* Phased V light curve between JDs 2455029 and 2455036. The consecutive nights are plotted with different shades (getting lighter with time). The overall shape of the light curve remains the same, but nightly changes can be seen, especially at the location of the spot around 0.6 phase. *Right:* Spot model of V374 Peg from polar view. Arrow at the bottom indicates the line of sight at phase zero.

Our observations have raised some interesting questions. Is there any periodicity in the changing of the flaring/non-flaring state of V374 Peg? If yes, what is the underlying cause of this behaviour? Until now, all the studies have noted, that the magnetic configuration of the object seems to be stable. On what timescale are the large-scale changes happening? Hopefully these questions can be answered by further photometric monitoring and by more detailed modelling of the object.

Acknowledgement

The authors are supported by the Hungarian Science Research Programme (OTKA) grants K-68626, K-81421, and the "Lendület" Program of the Hungarian Academy of Sciences.

References

Babcock, H. W. 1961, *Astrophys. J.*, 133, 572

Bade, N., *et al.* 1998, *Astron. AstrophysSuppl.*, 127, 145

Batyrshinova, V. M. & Ibragimov, M. A. 2001, *Astron. Lett.* 27, 29

Chabrier, G. & Baraffe, I. 1997, *Astron. Astrophys*, 327, 1039

Chabrier, G. & Küker, M. 2006, *Astron. Astrophys*, 446, 1027

Clarke, D. 2002, *Astron. Astrophys*, 386, 763

Delfosse, X., Forveille, T., Ségransan, D., Beuzit, J.-L., Udry, S., Perrier, C., & Mayor, M. 2000, *Astron. Astrophys*, 364, 217

Dobler, W., Stix, M., & Brandenburg, A. 2006, *Astrophys. J.*, 638, 336

Donati, J.-F., Forveille, T., Cameron, A. C., Barnes, J. R., Delfosse, X., Jardine, M. M., & Valenti, J. A. 2006, *Science*, 311, 633

Greimel, R. & Robb, R. M. 1998, *Information Bulletin on Variable Stars*, 4652, 1

Korhonen, H., Vida, K., Husarik, M., Mahajan, S., Szczygiel, D., & Oláh, K. 2010, *Astron. Nachr.*, 331, 772

Küker, M. & Rüdiger, G. 2005, *Astron. Nachr.*, 326, 265

Leighton, R. B. 1969, *Astrophys. J.*, 156, 1
Morin, J., *et al.* 2008, *Mon. Not. Roy. Astron. Soc.*, 384, 77
Parker, E. N. 1955, *Astrophys. J.*, 122, 293
Reid, I. N., Hawley, S. L., & Gizis, J. E. 1995, *Astrophys. J.*, 110, 1838
Ribárik, G., Oláh, K., & Strassmeier, K. G. 2003, *Astron. Nachr.*, 324, 202

The Physics of Sun and Star Spots
Proceedings IAU Symposium No. 273, 2010
D. P. Choudhary & K. G. Strassmeier, eds.

doi:10.1017/S174392131101578X

A nonlinear model for rotating cool stars

Sydney A. Barnes[1]

[1]Lowell Observatory, 1400 W. Mars Hill Road, Flagstaff, AZ 86001, USA
email: barnes@lowell.edu

Abstract. A simple nonlinear model is introduced here to describe the rotational evolution of main sequence cool (FGKM) stars. It is formulated only in terms of the ratio of a star's rotation period, P, to its convective turnover timescale, τ, and two dimensionless constants which are specified using solar- and open cluster data. The model explains the origin of the two sequences, C/fast and I/slow, of rotating stars observed in open cluster color-period diagrams, and describes their evolution from C-type to I-type through the rotational gap, g, separating them. It explains why intermediate-mass open cluster stars have the longest periods, while higher- and lower-mass cool stars have shorter periods. It provides an exact expression for the age of a rotating cool star in terms of P and τ, thereby generalizing gyrochronology. The possible range of initial periods is shown to contribute upto 128 Myr to the gyro age errors of solar mass field stars. A transformation to color-period space shows how this model explains some detailed features in the color-period diagrams of open clusters, including the shapes and widths of the sequences, and the observed number density of stars across these diagrams.

Keywords. Convection, methods: analytical, stars: evolution, stars: late-type, stars: rotation

---

1. Introduction

This paper is a condensed version of Barnes (2010), which introduces a simple nonlinear model to describe the rotational evolution of cool stars, and solves it to provide an improved understanding of their spin-down. The convective turnover timescale plays a key role in this model. The theoretical context for this work can be traced to Parker (1958), Schatzman (1962), Weber & Davis (1967), Mestel (1968), Kawaler (1988), MacGregor & Brenner (1991), Chaboyer et al. (1995), and Noyes *et al.* (1984).

The immediate context for this work is provided by Barnes & Kim (2010), who showed that the fast/C- and slow/I limits of stellar rotation, exemplified by two corresponding sequences of stars in open cluster color-period diagrams, (identified and named by Barnes 2003), can be described by

$$\frac{dP}{dt} = \begin{cases} {}^{k_C P}/_{\tau}, & \text{for early times/C sequence} \\ {}^{\tau}/_{k_I P}, & \text{for late times/I sequence,} \end{cases} \tag{1.1}$$

where P, t, τ are, respectively, the rotation period, age, and convective turnover timescale in cool stars, and $k_I = 452 Myr/d$ and $k_C = 0.65 d/Myr$ are two dimensionless constants. This work, described in detail in Barnes (2010), combines the two tines of that relationship into the *period evolution equation*,

$$\frac{dP}{dt} = \left\{ \frac{k_I P}{\tau} + \frac{\tau}{k_C P} \right\}^{-1}, \tag{1.2}$$

and summarizes its consequences for open cluster color-period diagrams.

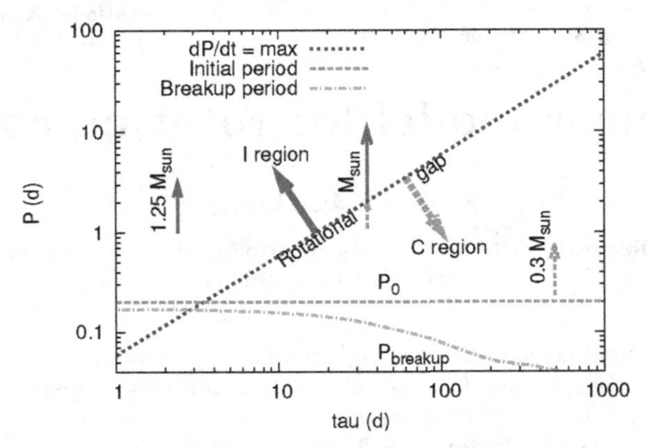

Figure 1. The $\tau - P$ plane is split into two regions C/(lower-right) and I/(upper left) by the straight line $P = \tau/\sqrt{k_I k_C}$ (diagonal) where dP/dt is maximum. Note (1) that this line is necessarily cut by the initial period line, and (2) the locations of the individual stellar trajectories.

2. Consequences for the $\tau - P$ plane

dP/dt approaches zero for both $P/\tau \to 0$ and $P/\tau \to \infty$, and has a maximum of $0.02d/Myr$ when $P = \tau/\sqrt{k_I k_C}$. Consequently, as displayed in Fig. 1, the $\tau - P$ plane is split into two regions, lower-right/C and upper-left/I by the diagonal (Rotational gap) line, where the model predicts a low number density of stars.

3. Generalized gyro age

Equation (1.2) can be immediately integrated to provide the gyro age, t, of a star:

$$t = \frac{\tau}{k_C} ln\left(\frac{P}{P_0}\right) + \frac{k_I}{2\tau}(P^2 - P_0^2), \tag{3.1}$$

where t is returned in Myr when P and τ are specified in d. The gyro ages are thus generalized with respect to the original ones of Barnes (2007), where only the I-type stars were considered. Initial period variations can be shown to contribute an additional error of upto 128 Myr for a solar-mass star.

4. Solution in $\tau - P/P_0$ plane

The solutions to the period evolution equation are conveniently displayed in the $\tau - P/P_0$ plane (Fig. 2) (see Barnes 2010 for the $t - P$ plane), and are specified implicitly by

$$\tau = \frac{k_C t \pm \sqrt{(k_C t)^2 - 2k_I k_C (P^2 - P_0^2) ln(x)}}{2ln(P/P_0)}. \tag{4.1}$$

Isochrones for specified ages - 100 Myr and 1 Gyr are displayed - have a higher-mass branch where periods increase with τ, and a lower-mass branch where they decline with τ. Consequently, in the context of this model, open cluster stars are expected to have a maximum rotation period at an intermediate mass, and short rotation periods for both higher- and lower-mass stars. The 100 Myr $\to$ 1 Gyr trajectories of $1.25M_\odot$, $1M_\odot$, and $0.3M_\odot$ models are also indicated.

5. Rotational gap

The $\tau - P$ plane of Fig. 1 is easily transformed into the color-period plane of observations, and this transformation warps the diagonal line of maximum dP/dt in the former

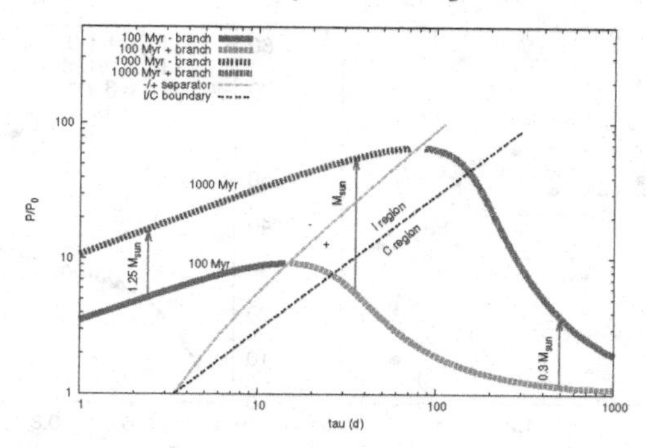

Figure 2. The solution to the period evolution equation in the $\tau - P$ plane. Note the ascent with τ of the higher-mass branch and the descent of the lower-mass branch, and the 100 Myr→1 Gyr trajectories for illustrative stars. (See Barnes 2010 for the dual $t - P$ plane.)

into the curved diagonal line marked in Fig. 3 for $B - V$ color. (Other colors in the set $[UBVRIJHK]$ can be obtained by using the table provided in Barnes & Kim 2010.) As expected from the theory, we observe a remarkable paucity of stars in the vicinity of this (Rotational gap) line in both panels, the left displaying the young (100-150 Myr) Pleiades (Hartman *et al.* 2010) and M35 (Meibom *et al.* 2009) open clusters, and the right displaying the older (550 Myr) M37 open cluster (Hartman *et al.* 2009). The division of rotating stars into fast/C- and slow/I sequences proposed by Barnes (2003) is also clearly visible in both panels.

6. Isochrones

One can also calculate isochrones using an appropriate range of initial periods, and we display two panels corresponding to (1) the classic Hyades/Coma Ber (600 Myr) data of Radick *et al.* (1987; 1990) and (2) the Solar datum/age.

The left panel shows the fidelity with which the position and dispersion of the observations can be accounted for, including the downturn for the coolest stars, which shows them to be of C-type. Note the absence of stars at the location of the rotational gap. The right panel shows the convergence of periods expected at solar age. It also displays

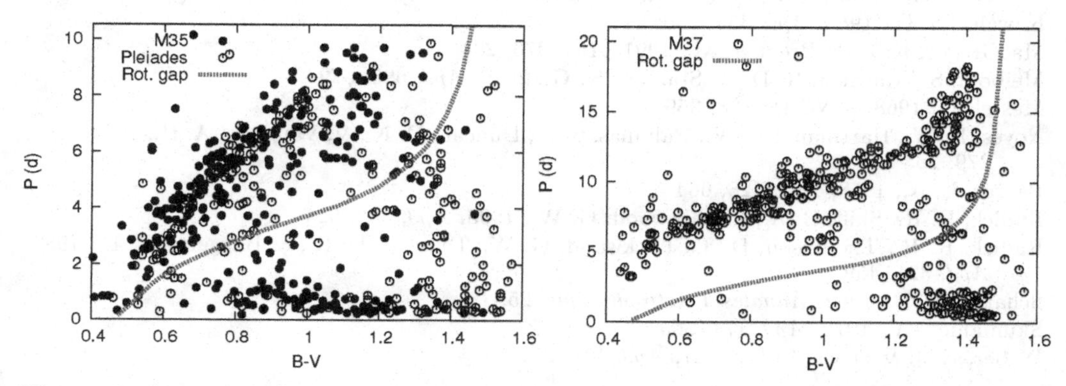

Figure 3. A paucity of stars is expected and observed at the C/I boundary (rotational gap) region (dotted pink lines) because dP/dt is at a maximum here. The fast/C- and slow/I sequences proposed by Barnes (2003) are clearly visible. **Left:** The (100-150 Myr) Pleiades/M35 rotation period data. **Right:** The (550 Myr) M37 period data.

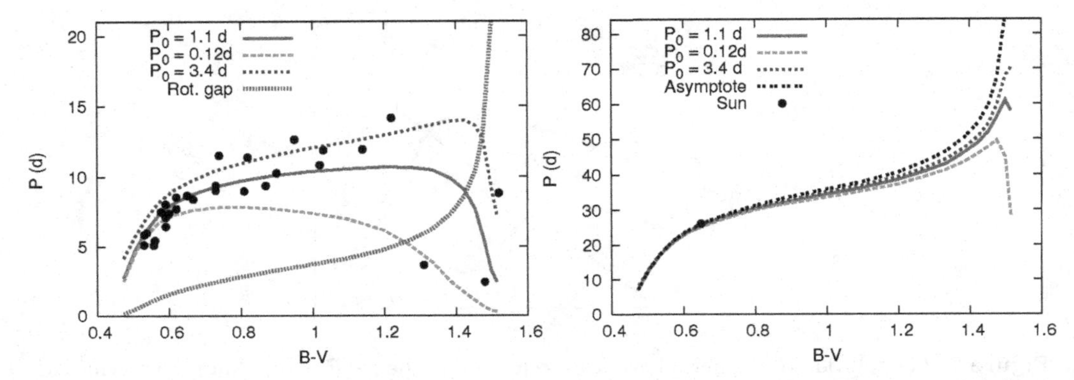

Figure 4. Left: Isochrones for 600 Myr, overlaid on the Hyades-Coma Ber rotation period data. The model reproduces the tight rotational sequence for F- and G stars, the increasing dispersion for K stars, and the fast C-type rotation for the M stars. (The dotted pink line is the calculated rotational gap.) **Right:** The Solar datum is overlaid on the theoretical $B-V$ Color-Period isochrones for solar age, 4.57 Gyr. Note the very tight I sequence expected for all masses but the lowest. The asymptotic relationship $P \to \sqrt{t} \times \sqrt{2\tau/k_I}$ (dotted black line) is also displayed.

the asymptotic relationship $P \to \sqrt{t} \times \sqrt{2\tau/k_I}$, showing the connection to the work of Skumanich (1972), and the mass dependence expected for late times.

SAB gratefully acknowledges the contributions of Y.-C. Kim of Yonsei University, Korea, to the numerics undergirding this work. The presentation of this paper at IAU Symp. 273 was possible due to partial support from the National Science Foundation grant numbers ATM 0548260, AST 0968672 and NASA - Living With a Star grant number 09-LWSTRT09-0039.

References

Barnes, S. A., 2003, *ApJ*, 586, 464

Barnes, S. A., 2007, *ApJ*, 669, 1167

Barnes, S. A., 2010, *ApJ*, in press

Barnes, S. A. & Kim, Y.- C., 2010, *ApJ*, in press

Chaboyer, B., Demarque, P. D., & Pinsonneault, M. H., 1995,*ApJ*, 441, 876

Hartman, J. D., Bakos, G. A., Kovacs, G., & Noyes, R. W., 2010, *MNRAS*, in press

Hartman, J. D., Gaudi, B. S., Pinsonneault, M. H., Stanek, K. Z., Holman, M. J., McLeod, B. A., Meibom, S., Barranco, J. A., & Kalirai, J. S., 2009, *ApJ*, 691, 342

Kawaler, S. D., 1988, *ApJ*, 333, 236

MacGregor, K. B. & Brenner, M., 1991, *ApJ*, 376, 204

Meibom, S., Mathieu, R. D., & Stassun, K. G., 2009, *ApJ*, 695, 679

Mestel, L., 1968, *MNRAS*, 138, 359

Noyes, R. W., Hartmann, L. W., Baliunas, S. L., Duncan, D. K., &, Vaughan, A. H., 1984, *ApJ*, 279, 763

Parker, E. N., 1958, *ApJ*, 128, 664

Radick, R. R., Skiff, B. A., & Lockwood, G. W., 1990, *ApJ*, 353, 524

Radick, R. R., Thompson, D. T., Lockwood, G. W., Duncan, D. K., & Baggett, W. E., 1987, *ApJ*, 321, 459

Schatzman, E., 1962, *Annales d'Astrophysique*, 25, 18

Skumanich, A., 1972, *ApJ*, 171, 565

Weber, E. J. & Davis, L. L., 1967, *ApJ*, 148, 217

The Physics of Sun and Star Spots
Proceedings IAU Symposium No. 273, 2010
D. P. Choudhary & K. G. Strassmeier, eds.

doi:10.1017/S1743921311015791

The dependence of maximum starspot amplitude and the amplitude distribution on stellar properties

Steven H. Saar[1], Michelle Dyke[1,2], Søren Meibom[1] and Sydney A. Barnes[3]

[1]Smithsonian Astrophysical Obs.,
60 Garden Street, Cambridge,MA 02138, USA
email: saar@cfa.harvard.edu, smeibom@cfa.harvard.edu

[2]Yale Univ. New Haven, CT USA

[3]Lowell Obs., 1400 W. Mars Hill Road, Flagstaff, AZ 86011
email: barnes@lowell.edu

Abstract. We combine photometric data from field stars, plus over a dozen open clusters and associations, to explore how the maximum photometric amplitude ($A_{\rm max}$) and the distribution of amplitudes varies with stellar properties. We find a complex variation of $A_{\rm max}$ with inverse Rossby number Ro^{-1}, which nevertheless can be modeled well with a simple model including an increase in $A_{\rm max}$ with rotation for low Ro^{-1}, and a maximum level. $A_{\rm max}$ may then be further affected by differential rotation and a decline at the highest Ro^{-1}. The distribution of $A_{\rm spot}$ below $A_{\rm max}$ varies with Ro^{-1} : it peaks at low $A_{\rm spot}$ with a long tail towards $A_{\rm max}$ for low Ro^{-1}, but is more uniformly distributed at higher Ro^{-1}. We investigate further dependences of the $A_{\rm spot}$ distributions on stellar properties, and speculate on the source of these variations.

Keywords. Stars: spots, stars: rotation, stars: magnetic fields, stars: late-type, stars: evolution

1. Introduction and observations

A number of extensive photometric studies of open clusters and associations have been completed recently. Since the major compilation and study of spot amplitudes $A_{\rm spot}$ was by Messina *et al.* (2001), it timely to revisit the dependence of $A_{\rm spot}$ (indicative of starspots) on other stellar properties, informed by the new data and other recent results (e.g., on differential rotation). We have collected photometric V amplitude and rotation period $P_{\rm rot}$ data from a large number of sources (Table 1) and made new fits to the light curves for two clusters (M35, NGC 3532). The total data set includes over 1200 stars with ages ranging from 10 Myr to roughly solar. We include only single stars or well separated binaries to avoid complications due to tidal/magnetic interactions, and difficulties in assigning amplitudes to a particular component. We also restrict ourselves to V data, which are more numerous (sadly, eliminates some recent cluster surveys).

2. Analysis and models

Based on Messina *et al.* (2001) we expect $A_{\rm spot}$ to depend on both rotation and mass. A simple, physically motivated parameterization for this is the inverse Rossby number $\mathrm{Ro}^{-1} = \tau_C \Omega$ (where τ_C is the convective turnover timescale); the mean-field $\alpha\Omega$ dynamo

Table 1. Sources of $A_{\rm spot}$ data

Source	$N_{\rm star}$	Age [Myr]	reference
β Pic assoc.	17	10	Messina *et al.* (2010b)
Tuc/Hor assoc.			
Col assoc.	52	30	Messina *et al.* (2010b)
Car assoc.			
NGC 2391	16	40	Patten & Simon(1996)
NGC 2602	29	40	Barnes *et al.* (1999)
IC 4665	8	40	Allain *et al.* (1996)
α Per	42	70	Messina *et al.* (2001)[1]
Pleiades	57	100	Messina *et al.* (2001)[1]
AB Dor group	32	100	Messina *et al.* (2010b)
M35	218	150	Meibom *et al.* (2009)
M11	8	215	Messina *et al.* (2010a)
NGC 3532	73	300	Barnes(2003)[2]
M37	504	550	Hartman *et al.* (2009)
Coma	39	600	Collier Cameron *et al.* (2009)
Hyades	21	600	Messina *et al.* (2001)[1]
field	99	300-4000	Messina *et al.* (2001)[1]

Notes: [1]Data compilation (see references for original source); [2]$P_{\rm rot}$ source, we measure $A_{\rm spot}$ for the first time.

number is proportional to Ro^{-2}, and Ro^{-1} figures prominently in stellar activity and rotational evolution (e.g., Barnes & Kim 2010). We take τ_C from Gunn *et al.* (1998).

Since $A_{\rm spot}$ is a relative measure, yielding the peak *difference* in spot coverage over rotation, we expect that the maximum $A_{\rm spot}$ seen at a given Ro^{-1}, $A_{\rm max}$, will be the most indicative diagnostic of the true strength of dynamo-driven spot generation at that rotation rate. We study $A_{\rm max}$ (Fig. 1), the binned $A_{\rm spot}(\mathrm{Ro}^{-1})$ distributions, and moments thereof (Fig. 2).

If $A_{\rm max}$ behaves like other magnetic activity diagnostics (e.g., X-ray emission), we expect to see an increase in $A_{\rm max}$ with rotation at lower Ro^{-1}, a possible maximal "saturated" spottedness level above some critical Ro^{-1}, and a possible decrease in $A_{\rm max}$

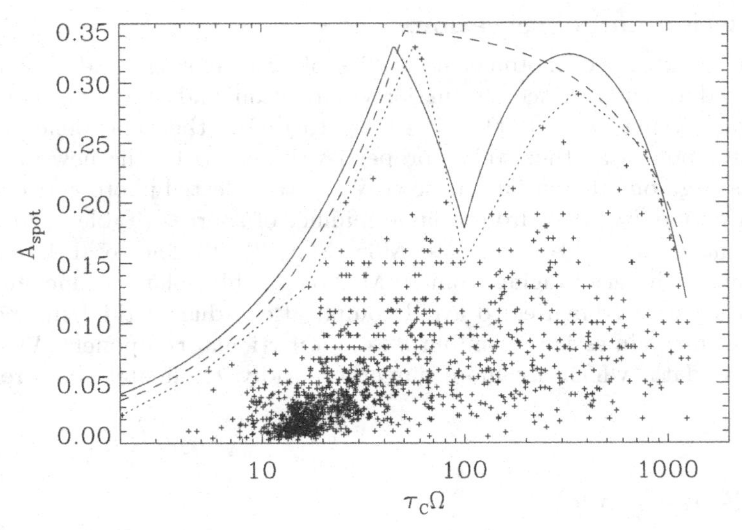

Figure 1. $A_{\rm spot}$ vs. Ro^{-1} for our sample (Table 1). The dashed line connects local $A_{\rm spot}$ maxima ($A_{\rm max}$); the dashed and solid lines gives our approximate models (1 and 2, respectively) for $A_{\rm max}$ (see discussion).

again at very fast rotation rates (e.g., analogous to the "super-saturation" seen in coronae; e.g., Randich 1998). Causes for the latter phenomena are much debated, and part of the reason for such a study. Saturation may result from maximum surface coverage by fields/spots, Coriolis force-driven concentration of spots at poles, maximum dynamo output limited by back-reaction of magnetic fields on flows. A decline at the highest Ro^{-1} may be due to a continuation of the above, or simply due to lack of additional surface area to permit higher levels of inhomogeneity.

Connecting the highest observed $A_{\max}(\mathrm{Ro}^{-1})$ (Fig. 1, dotted line) does indeed show a general rise in $A_{\max}$ at lower Ro^{-1}, and a decrease at the highest Ro^{-1}. For intermediate values, however, the situation is less clear. One *could* more coarsely follow $A_{\max}(\mathrm{Ro}^{-1})$, and connect $A_{\max}$ high points at $\mathrm{Ro}^{-1} \approx 60$ and 350, making a fairly smooth, though gently declining "saturated" $A_{\max}$ level. Alternatively, connecting $A_{\max}$ as indicated in Figure 1 shows a large "wedge" removed, with local minimum at $\mathrm{Ro}^{-1} \approx 100$.

We have tried to construct simple models to match both of these scenarios. The first ("no-wedge") $A_{\max}(\mathrm{Ro}^{-1})$ curve is well fit by:

$$A_{\max} = \min\{a_1 \mathrm{Ro}^{-b_1}; c_1(2 - e^{d_1/\mathrm{Ro}})\} \qquad \textbf{[Model 1]}$$

where a_1, b_1, c_1, and d_1 are adjustable constants. This model contains both a power law increase in $A_{\max}$, limited by a gradually decaying "saturation" level (the exponential term). The $A_{\max}(\mathrm{Ro}^{-1})$ curve with the "wedge" removed, can, surprisingly be fit by adding only one additional parameter:

$$A_{\max} = \min\{a_2 \mathrm{Ro}^{-b_2}; c_2(2 - e^{d_2/\mathrm{Ro}})\} - f_2 \Delta\Omega \qquad \textbf{[Model 2]}$$

where a_2, b_2, c_2, d_2, and f_2 are adjustable constants, and $\Delta\Omega$ is the empirical relationship between surface differential rotation (DR) and Ro^{-1} for single dwarfs (Saar 2009; 2010, this volume). The latest fits for this yield $\Delta\Omega \propto \mathrm{Ro}^{-1.00}$ for $\mathrm{Ro}^{-1} < 100$ and $\Delta\Omega \propto \mathrm{Ro}^{1.28}$ for $\mathrm{Ro}^{-1} > 100$ (Saar 2010, this volume).

We also binned the $A_{\max}$ data into 12 bins of equal numbers in each (100 stars). We also studied the resulting distributions in each bin. Maxima and moments of these distributions are shown in Fig. 2.

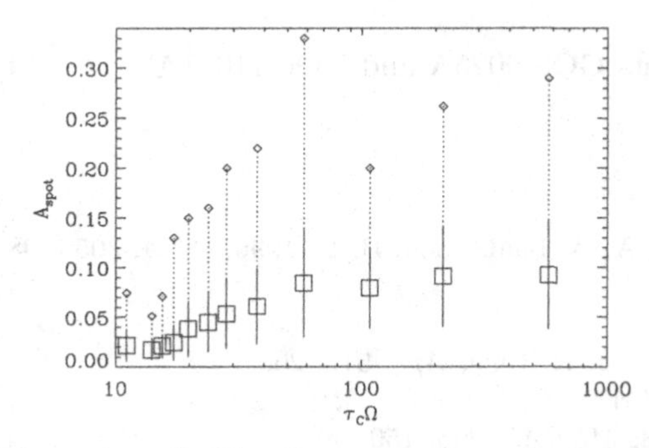

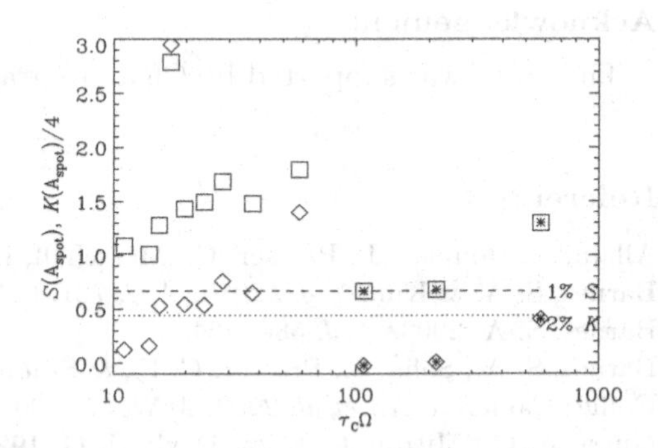

Figure 2. [Left]: The mean (boxes), $\pm\sigma$ width (solid), and maximum (diamonds) of the $A_{\mathrm{spot}}(\mathrm{Ro}^{-1})$ distribution in each of 12 bins with 100 stars in each. All three increase fairly steadily until the coronal saturation point ($\mathrm{Ro}^{-1} \approx 100$). The mean A_{spot} also saturates here, while $A_{\max}$ and the distribution σ retreat, only to resume increasing towards higher Ro^{-1}. [Right]: Skewness S (boxes) and excess kurtosis $K/4$ (diamonds) of the $A_{\mathrm{spot}}(\mathrm{Ro}^{-1})$ distributions in 100 star bins. Similar to the distribution σ width (at left), S and K both increase with Ro^{-1} below saturation ($\mathrm{Ro}^{-1} \approx 100$), whereafter they sharply decline (symbols marked with *). The 1% significance level for S (dashed) and 2% level for K (dotted) are also indicated.

3. Results and conclusions

We interpret the success of the added DR term as arising from the effect of strong shear on spot groups, smearing out the spot concentrations (and shredding the spots themselves), inevitably leading to more uniform spatial distributions and lower $A_{\rm max}$. The shearing effect of DR reaches a peak in a range of rapid rotators near the start of the "saturated activity" branch (where, e.g., $L_X/L_{\rm bol}$ reaches a constant maximum of $\approx 10^{-3}$). The sharp increase in $A_{\rm max}$ for $\rm Ro^{-1} > 100$ (Fig. 1) echoes the Messina *et al.* (2001) result, which we suggest is due to decreasing rotational shear in these stars, permitting larger spots, spot groups, and thence larger $A_{\rm max}$.

There are also a significant changes in $A_{\rm spot}$ *distribution* around $\rm Ro^{-1} \sim 100$: it becomes narrower (Fig. 2, left), and markedly less skewed and more Gaussian (excess kurtosis $K \rightarrow 0$; Fig. 2, right) even while $\langle A_{\rm spot}\rangle$ stays roughly constant (Fig. 2, left). These changes, occurring precisely at the local $A_{\rm max}$ minimum, lends weight to the "wedged" interpretation of $A_{\rm max}$. Again, high shear should tend to smear out spot groups and structures, making distributions more homogeneous and reducing their moments.

Modeling and further observations are needed to help understand the physical processes underlying these results. Stronger Coriolis forces at high $\rm Ro^{-1}$ will drive increased flux emergence towards the poles (e.g., Schüssler & Solanki 1992), but spots are still observed to appear at all latitudes (e.g., Strassmeier 2009), suggesting a source of flux/spots nearer the surface. The implied increase in importance of a convection zone-based dynamo at higher $\rm Ro^{-1}$ (vs. a tachocline-driven one) is probably an important factor driving changing spot distributions, as are altered velocity fields (e.g., DR, meridional flows). Is there any significance to the sharp spike in S and K near $\rm Ro^{-1} \approx 17$? Is there any relationship between coronal "super-saturation" and the drop in $A_{\rm max}$ at high $\rm Ro^{-1}$? Do true spot areas continue increasing, even as $\langle A_{\rm spot}\rangle$ saturates and $A_{\rm max}$ declines? Indeed, *does* $A_{\rm max}$ decline sharply? (The drop is not apparent in the binned data; Fig. 2, left.) More $A_{\rm spot}$ data at high $\rm Ro^{-1}$, and measurements of absolute spot coverage using molecular bands (e.g., ONeal *et al.* 1996; Saar *et al.* 2001) could be used to probe these issues. We are also beginning to explore some of these questions with simple models.

Acknowledgement

This work was supported by Chandra grants GO8-9025A and GO0-11041A.

References

Allain, S., Bouvier, J., Prosser, C., Marschall, L. A., & Laaksonen, B. D. 1996, *A&A*, 305, 498
Barnes, S. A. & Kim, Y.-C. 2010, *ApJ*, 721, 675
Barnes, S. A. 2003, *ApJ*, 586, 464
Barnes, S. A., Sofia, S., Prosser, C. F., & Stauffer, J. R. 1999, *ApJ*, 516, 263
Collier Cameron, A., *et al.* 2009, *MNRAS*, 400, 451
Gunn, A. G., Mitrou, C. K., & Doyle, J. G. 1998, *MNRAS*, 296, 150
Hartman, J. D., *et al.* 2009, *ApJ*, 691, 342
Meibom, S., Mathieu, R. D., & Stassun, K. G. 2009, *ApJ*, 695, 679
Messina, S., Desidera, S., Turatto, M., Lanzafame, A. C., & Guinan, E. F. 2010, *arXiv:1004*.1959
Messina, S., Parihar, P., Koo, J.-R., Kim, S.-L., Rey, S.-C., & Lee, C.-U. 2010, *A&A*, 513, A29
Messina, S., Rodonò, M., & Guinan, E. F. 2001, *A&A*, 366, 215
O'Neal, D., Saar, S. H., & Neff, J. E. 1996, *ApJ*, 463, 766
Patten, B. M. & Simon, T. 1996, *ApJS*, 106, 489

Randich, S. 1998, 10th Cool Stars, Stellar Systems, and the Sun, *ASP Conf. Ser.*, 154, 501
Saar, S. H. 2009, Astronomical Society of the Pacific Conference Series, *ASP Conf. Ser.*416, 375
Saar, S. H., Peterchev, A., O'Neal, D., & Neff, J. E. 2001, 11th Cambridge Workshop on Cool Stars, Stellar Systems and the Sun, *ASP Conf. Ser.*, 223, 1057
Schüssler, M. & Solanki, S. K. 1992, *A&A*, 264, L13
Strassmeier, K. G. 2009, *A&A Rev.*, 17, 251

The Physics of Sun and Star Spots
Proceedings IAU Symposium No. 273, 2010
D. P. Choudhary & K. G. Strassmeier, eds.

doi:10.1017/S1743921311015808

The effects of star spots on transit photometry

John R. Hodgson II and Damian J. Christian

Physics and Astronomy Department, California State University, Northridge
18111 Nordhoff Street, Northridge, California 91330-8268, United States of America
email: John.Hodgson.71@my.csun.edu
Damian.Christian@csun.edu

Abstract. We have undertaken an observational program to photometrically monitor several transiting planet host stars. The Rabus *et al.* result for TrES-1 showed the dramatic effects star spots can have on transit photometry. We will investigate the effects of spots on transit light curves and estimates of planetary radii. The observed spot patterns will be used to derive the rotational periods of our sample. Our sample includes several of the newly discovered transiting ESPs from the SuperWASP, HAT, TrES, and Kepler projects.

Keywords. Stars: starspots, planetary systems, techniques: photometric

1. Introduction

Currently, the discovery of new extra-solar planets is occurring nearly every day. Approximately 20% of these are 'transiting' extra-solar planets (TESPs) found through photometric observations of their host stars. Precision photometry combined with radial velocity measurements constrain the values of the planetary radii in addition to the calculated inclination angle of the planetary system. Variations in the photometry of a planetary system is often due to stellar activity. While these variations are normally detrimental to establishing the physical parameters of the planetary system, they can be used to gain information about the host stars.

Stellar activity is a prominent problem in deducing the planetary parameters of late type stars. The occultation of stellar activity by a planet changes the shape of the transit light curve and can lead to incorrect estimates of the planet's radius. An example of this occurred during observations of the active K0V star TrES-1 and was reported by (Rabus *et al.* 2009, Rabus *et al.* 2009). Their report stimulated a search for other instances of stellar activity on Tres-1 and other stars with known transiting extra-solar planets. Additionally, planetary transits may be used to infer physical properties of stellar spots, and the multiple transits easily available from space-based missions, such as CoRoT and Kepler have been used for these studies (Silva-Valio *et al.* 2010).

2. Overview

We have undertaken an observational program to photometrically monitor several transiting planet late-type host stars that are expected to have starspots. We are interested in investigating how starspots effect the transit depth and derived planetary radii. Additionally, information on stellar rotation rates can be determined from stellar activity present over multiple transits. In the current paper, we present our initial findings for this project.

Table 1. Known characteristics of targets selected for follow-up photometry.

Target Name	Spectral Type	Stellar Magnitude [V]	Stellar Radius [R_{Sun}]	Rotational Period [days]	Planet Radius [R_J]	Planetary Period [days]	Inclination [degrees]
WASP-10	K5	12.7	0.783	11.9 ± 0.2	1.08	3.093	86.8
Kepler-6	G1	13	1.391	≈ 6	1.323	3.234	86.8
Kepler-7	F8\G0	13.9	1.843		1.478	4.886	86.5
Kepler-8	F7\9	13.9	1.486		1.419	3.523	84.07
TrES-1	K0V	11.79	0.82	$40.2^{+22.9}_{-14.6}$	1.081	3.030	86.4

Table 1 contains an overview of some of our primary targets that have been selected for follow-up photometry.

Below is a list of the properties we will be looking for.

Planetary Radii. The radii of exoplanets is found from the change in the host star's intensity during a transit. When a starspot is occulted by the transiting planet, a decrease in the transit depth results which can cause an underestimate of the planet's radius. Similarly, when a plage is occulted by a transiting planet, an increase in the transit depth results which can cause an increase in the calculated planetary radii.

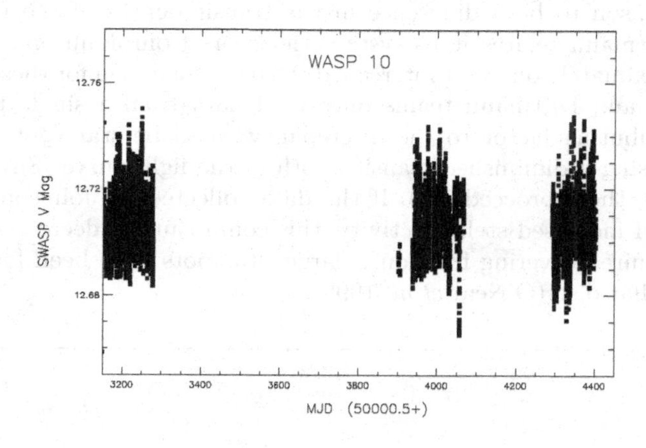

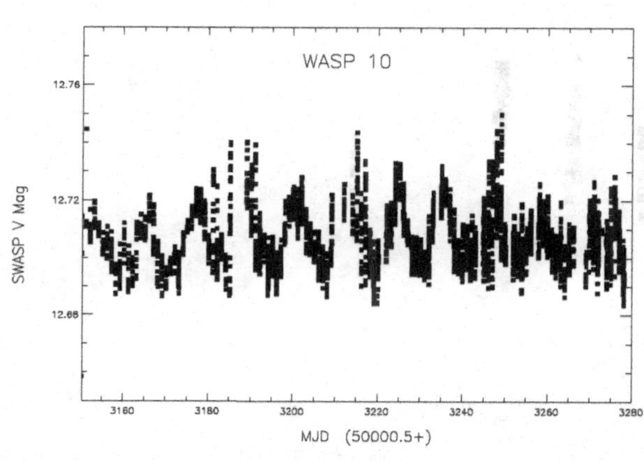

Figure 1. SuperWasp lightcurve of WASP-10 data from 2004 to 2007 (1 January 2004 is JD 2453005.5). The lower image is an expanded view of the smoothed 2004 WASP-10 data showing the 11.9 day period.

Stellar Rotation. The observed occultations of a star spot over multiple transits can yield the rotation rate of the host star. Given the radius of the star, one only needs to measure the distance the spot has migrated across the stellar disk and the time between the transit measurements (Silva-Valio 2008). Assuming no change in the spot's area, variations in the modulations caused by the spot can be used to measure the inclination of the stellar rotational axis.

3. Discussion

In the first phase of our program we have analyzed photometric data from the SuperWASP and Kepler archives and searched for possible stellar rotation periods. We have focused on the stars in Table 1. We have determined a rotational period for WASP-10 of 11.9 days (see Figure 1) in agreement with Christian *et al.* (2008) and more recently with Maciejewski *et al.* (2010), and determine a suggested rotational period of approximately 6 days for Kepler-6 (Figure 2). These specific cases are discussed below.

WASP-10. Currently, the radii of WASP-10b and its K5 host star are under debate. Christian *et al.* (2008) and Dittman *et al.* (2009) both find the radius of WASP-10b to be 1.28 R_J, but Johnson *et al.* (2009) found a radius of 1.08 R_J. This variation is explained by Johnson to be a difference in the transit depths of the normalized light curves, while Dittmann relates it to systematic errors from Johnson's optical system. Intervals of approximately one year interceded the data collection for these measurements and the Johnson and Dittmann teams only used data from a single transit for their findings. A contributing factor to the discrepancy could be star spots located off the transit chord causing a diminished transit depth in the light curve (Silva-Valio (2008); Valio; Boise 2010, these proceedings). If the data collected by Johnson *et al.* occurred during a period of increased stellar activity, this could cause a decreased transit depth. Although the required covering fraction is large, starspots have been found to cover up to 40% of the stellar disc (O'Neal *et al.* 1998).

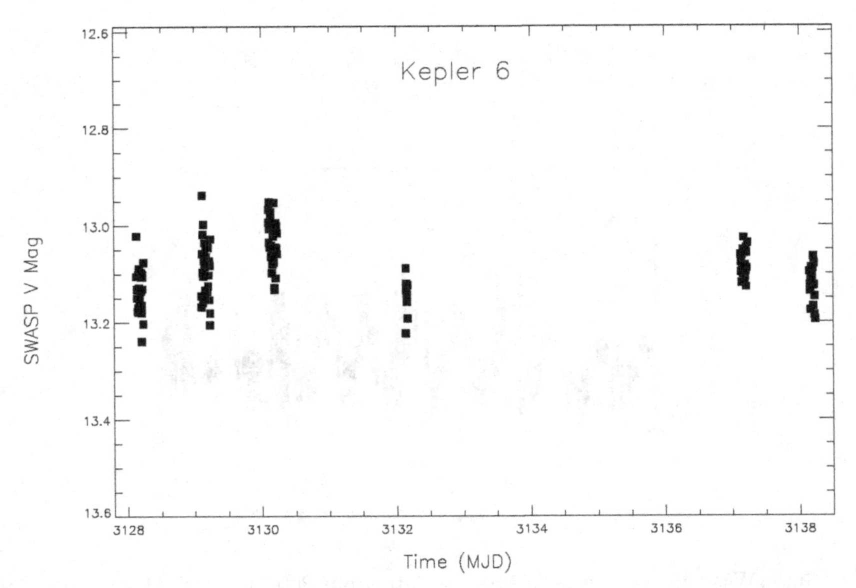

Figure 2. SuperWASP light curve of Kepler-6 from May 2004 (MJD 53128.5 corresponds to 03 May 2004).

Kepler. Data on the Kepler targets have only recently been acquired by the Kepler mission. However, a review of the SuperWASP light curves for the Kepler targets has shown variation in some of the light curves and a rotational period of approximately 6 days for Kepler-6.

4. Future

We will continue to gather and analyze data on our current targets to establish activity cycles and rotational periods, and then investigate individual transits to derive both planetary and stellar parameters. Further interrogations of the SuperWASP and Kepler archives are under way and new targets being considered.

Acknowledgements

We thank the SuperWASP and Kepler projects for data access. We would also like to thank the CSUN Physics and Astronomy Department and the IAU 273 LOC for support of this project.

References

Bakos, G. À.; Torres, G.; Pàl, A.; Hartman, J.; Kovàcs, Gza; Noyes, R. W.; Latham, D. W.; Sasselov, D. D.; Sip:ocz, B.; Esquerdo, G. A.; and 14 coauthors 2010, *Astrophys. J.*, 710, 1724

Basri, G., Walkowicz, L., Batalha, N. and 10 coauthors 2010, *Astron. J.*, 141, 20

Charbonneau, D., Brown, T., & Latham, D. 2000, *Astrophys. J.*, 529, L45

Christian, D. J. & Gibson, N. J. 2008, *Mon. Not. R. Astron. Soc.*, 392, 1585

Dittman, J. A. & Close, L. M. 2009, *Astrophys. J.*, 701, 756

Dittman, J. A. & Close, L. M. 2010, *Astrophys. J.*, 717, 235

Fares, R. & Donati, J. F. 2010, *Mon. Not. R. Astron. Soc.*, 406, 409

Johnson, J., Winn, J., Cabrera, N. 2009. *Astrophys. J.*, 692, 100

Maciejewski, G., Dimitrov, D., Neuhäuser, R. 2010. *Mon. Not. R. Astron. Soc.* In Press (arXiv:1009.4567v1)

O'Neal, D., Saar, S., Neff, J. 1998. *Astrophys. J.*, 501, L73

Pollacco, D. L. & Skillen, I. et al. 2006, *Pub. Astron. Soc. Paci.*, 118, 1407

Rabus, M. & Alonso, R. 2009, *Astron. Astrophys*, 494, 391

Schneider, J. 2000, *ASPCS*, 212, 284

Schneider, J. 2010, *The Extrasolar Planets Encyclopedia*, retrieved from www.exoplanet.eu/catalog-transit.php

Silva-Valio, A. 2008, *Astrophys. J.*, 683, 179

Silva-Valio, A., Lanza, A. F., Alonso, R., Barge, P. 2010 *Astron. Astrophys*, 510, 25

Physics of the Sun and Star spots
Proceedings IAU Symposium No. 273, 2010
D. P. Choudhary and K. G. Strassmeier, eds.

doi:10.1017/S174392131101581X

Structure of sunspots observed with Hinode Solar Optical Telescope

Debi Prasad Choudhary[1], Gordon A. MacDonald[2], Na Deng[3]& Shimizu Toshifumi[4]

[1,2,3]Department of Physics and Astronomy, California State University Northridge, 18111 Nordhoff St., Northridge Ca, USA, 91330,
[4] Institute of Space and Astronautical Science, Japan Aerospace Exploration Agency (ISAS/JAXA) 3-1-1 Yoshinodai, Sagamihara, Kanagawa 229-8510, Japan

Abstract. We studied 16 sunspots with different sizes and shapes using the observations with the Hinode Solar Optical Telescope. The ratio of G-band and CaII H images reveal rich structures both within the umbra and penumbra of most spots. The striking features are the compact blob at the foot point of the umbra side of the penumbral fibrils with disk center-limb side asymmetry. In this paper, we present properties of these features using the spectropolarimetry and images in G-band, CaII and blue filters. We discuss the results using the contemporary models of the sunspots.

Keywords. Sunspots, G-band, Umbral dots, bright points

---

1. Introduction

"Clearly, observations must point the way, because it is evident now that the sunspot is too complicated a structure to be the product of a single overwhelming theoretical effect. The sunspot results from the conjunction of several effects, and we can spend a lot of time guessing what those effects might be without getting anywhere".
Eugene Parker (Personal communication 2010)

We analyzed the sunspots observations obtained with Solar Optical Telescope (SOT) onboard Hinode to look for the signatures that are relevant to identify theoretical effect of different models. A sunspot may be made up of individual flux tubes pressed together or a monolithic block of magnetized plasma immersed in the solar convective zone Parker (1979), Meyer, Schmidt & Weiss (1977). High spatial resolution observations of sunspot formation show that individual magnetic knots appear on the solar surface that get collected to form larger bodies perhaps by mutual hydrodynamic attraction Parker (1978). The question still remains that after the knots are collected together to form a sunspot, do they still retain their individuality? In other words, are there gaps between the individual flux tubes that contain plasma devoid of magnetic field? Existence of field free plasma below the visible layer of the sunspots distinguish the two types of sunspots models del Toro Iniesta (2001). The physical processes leading to the stability and decay of the sunspot would depend largely on its subphotospheric structure. The time series of high spatial resolution Hinode observations of sunspots, without the interruption of atmospheric seeing, provides opportunity to study the finer structures and their temporal evolution governed by the subphotospheric dynamics. The G-band images serve as a flux tube diagnostic tool due to their association with magnetic concentrations, with efficient heat transport leading to the weakening of molecular band and shifting of optical depth Ishikawa *et al.* (2007), Sanches Almeida (2001). The Ca II H images provides information on the layer slightly higher than the height of G-band formation, the combined

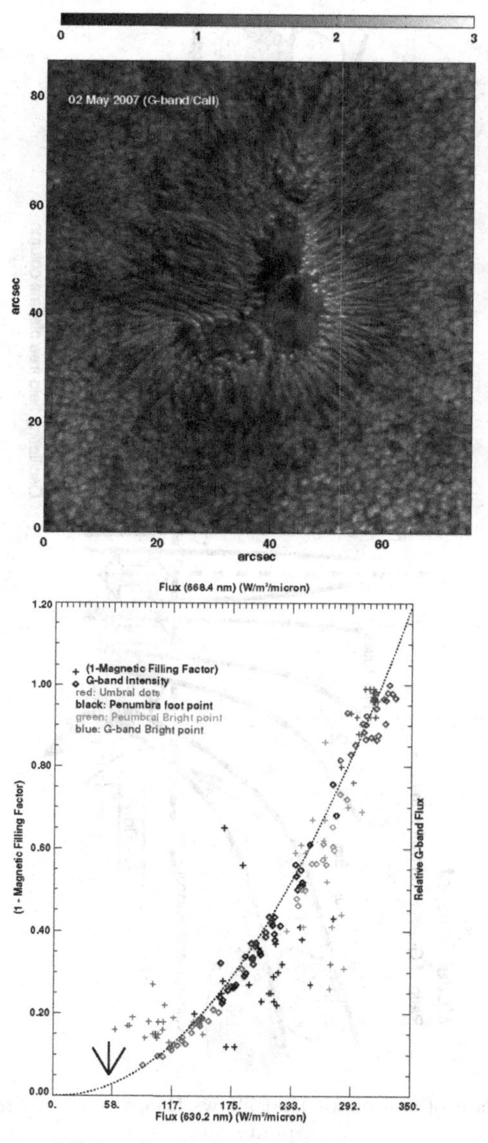

Figure 1. (Right) (a) Ratio of G-band and CaII image of the sunspot. The suncenter is to the right. The umbral dots are seen enhanced. The disk center is towards right of the sunspot. (Left) (b) Magnetic Filling factor and G-band intensity of sunspot bright points. The left and bottom axis labels is for the magnetic filling factor. The Top and Right axis labels are for the G-band intensity. The bottom arrow indicates that the actual value of the magnetic filling factor of the umbral bright point may be higher.

understanding of which would provide new understanding of sunspot magnetic structure Carlsson *et al.* (2007).

2. Data Analysis

We use the Broadband Filter Imager (BFI) and Spectropolarimeter (SP) observations obtained with SOT Tsuneta *et al.* (2008). After initial processing of the level0 data,

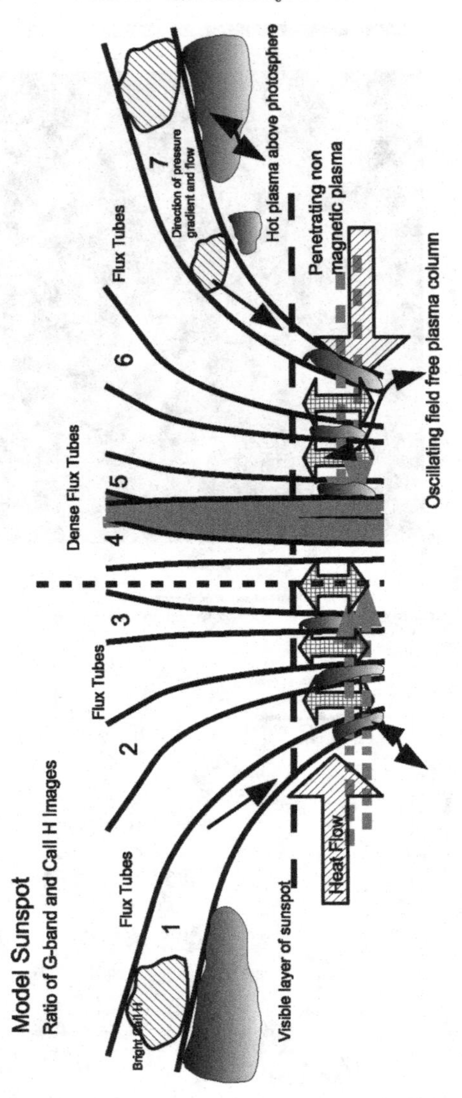

Figure 2. The cartoon of a sunspot that shows theoretical effects of model umbra and penumbra.

flux calibration was performed using the measurements of solar radiance through filters made by the detectors on ground and space Shimizu *et al.* (2007). The ground based measurements were performed by recording the the full field of view images solar disk center by the CCD. The space measurements are made by using the synoptic images that measure the intensities at the solar disk center. The comparison of ground and space measurements show consistent results. The ratio of the quiet region and the zero air mass solar flux was used to convert sunspot images into solar flux units. The localized bright features in the sunspot that could be successfully fitted with a two dimensional Gaussian function were selected from the co-aligned flux calibrated images. The flux contained in the half width of the Gaussian surface was measured as the intensity of the bright points. The magnetic filling factor was obtained by using inversion codes to the SP observations.

3. Discussion and Conclusions

Figure 1 (a) shows the example of the ratio of G-band and CaII H images of a sunspot observed on 02 May 2007. The important features of the ratio images are: (1) The inner penumbra of the ratio images are brighter compared to the outer part where the CaII H signal is higher compared to the G-band. There is a sharp boundary at the middle of penumbra dividing bright inner and dark outer area. (2) The foot points of the penumbra, which are the locations of the peripheral umbral dots, are brighter compared to the inner umbral dots. The disk-centerward penumbral boundary has brighter points than their limb side counterparts. (3) The umbral dots are not uniformly distributed in the umbra. (4) The umbral dots, closer to the penumbra, are associated with elongated structures directed inwards. (5) Most umbral dots are not circular but elongated elliptical shape. The light bridges consists of compact bright points that have elongated structures. Figure 1 (b) shows the G-band flux and magnetic filling factor as a function of continuum intensity in BFI red filter through at 6685 Å and 6302.5 Å. The solid curve represent the quadratic function of continuum intensity that follows the trend of both G-band flux and filling factor.

We interpret these observations using the theoretical models for sunspot structure below and above the visible layer with forest of field free gaps between the flux tubes Parker (1979), Spruit & Scharmer (2006). Figure 2 describes the physical processes leading to the observed sunspot bright point. The figure shows seven flux tubes in a bundle embedded in field free plasma expanding above photosphere as the gas pressure drops. The field free plasma penetrates in the gap between the flux tubes as shown by gray arrows that transport heat from non magnetic convective zone. The penetrating field free plasma between the flux tubes oscillates, represented by hatched columns with double arrow, due to longitudinal overstability Parker (1979) and heats the upper layers represented by shaded columns, dissociate the molecules in them and produce umbral dots by altering the optical depth to expose the deeper layers. The inclinaed flux tubes result viewing of larger part of hot flux tube wall Rutten *et al.* (2001). The oscillating plasma columns themselves do not appear at the visible surface as blocked by the overlying magnetic canopy. The penumbral flux tubes that are at the periphery of the sunspot get extra heating by direct contact with the convective zone hot plasma represented by hatched side arrows leading to the brighter penumbral foot points. The penumbral flux tubes are highly inclined which contribute to the G-band brightness with limb-disk side asymmetry. As the matter in the flux tube gets heated it expands leading to downward drop in pressure resulting directional flow indicated by arrows in flux tube 1 and 7 Degenhardt & Lites (1993). The umbra-ward motion of penumbral grains may also be due to this effect. In some locations of the umbra the flux tubes are so dense that there is not much gap between them for the oscillating field free plasma represented by shaded flux bundle between the 4 and 5 flux tube. At these locations there are few umbral dots leading to non uniform distribution in the umbra. In the penumbra, the field free plasma would heat localized regions and produce the bright points represented by shaded and hatched region near flux tube 1 and 7 and produce hotter flux tube plasma in the outer penumbra where the plasma is tenuous due to the expanding flux tubes.

4. Acknowledgements

This work was supported by NSF grant ATM-0548260 at California State University Northridge. The data was obtained using Hinode Solar Optical Telescope. Hinode is a Japanese mission developed and launched by ISAS/JAXA, with NAOJ as domestic partner and NASA and STFC (UK) as international partners. It is operated by them in co-operation with ESA and NSC (Norway).

References

Carlsson, M. *et al.* 2007, *Pub. Astron. Soc. Japan*, 59, S663
Degenhardt, D. & Lites, B. W. 1993, *Astrophys. J.*, 416, 875
del Toro Iniesta, J. C. *ASP Conference Series*, 248, 35
Ishikawa, R., Tsuneta, S., Kitakoshi, Y., Katsukawa, *et al.*, *Astron. Astrophys*, 472, 911
Meyer, F., Schmidt, H. U., & Weiss, N. O. 1977, *Mon. Not. Roy. Astron. Soc.*, 179, 741
Parker, E. N. 1978, *Astrophys. J.*, 222, 357
Parker, E. N. 1979, *Astrophys. J.*, 230, 905
Parker, E. N. 1979, *Astrophys. J.*, 234, 333
Rutten, R. J. *et al.* 2001, *Astron. Soc. Pac. Conf. Series*. 236, 445
Sanches Almeida, J. *et al.* 2001, *Astrophys. J.*, 555, 978
Shimizu, T, Kubo, M., Tarbell, T. D. and 10 coauthors, 2007, *ASP Conference Proceedings*, 369, 51.
Spruit, H. C. & Scharmer, G. B. 2006, *Astron. Astrophys*, 447, 343
Tsuneta, S., Ichimoto, K., Katsukawa, Y., *et al.* 2008, *Solar Phys.*, 249, 113.

Physics of Sun and Star Spots
Proceedings IAU Symposium No. 273, 2010
Debi Prasad Choudhary

doi:10.1017/S1743921311015845

Laboratory simulation of solar magnetic flux rope eruptions

S. K. P. Tripathi[1] and W. Gekelman[2]

Physics & Astronomy, University of California at Los Angeles,
Los Angeles, California 90095, USA
emails: [1]tripathi@physics.ucla.edu and [2]gekelman@physics.ucla.edu

Abstract. A laboratory plasma experiment has been constructed to simulate the eruption of arched magnetic flux ropes (AMFRs e.g., coronal loops, solar prominences) in an ambient magnetized plasma. The laboratory AMFR is produced using an annular hot LaB_6 cathode and an annular anode in a vacuum chamber which has additional electrodes to produce the ambient magnetized plasma. Two laser beams strike movable carbon targets placed behind the annular electrodes to generate controlled plasma flows from the AMFR footpoints that drives the AMFR eruption. The experiment operates with a 0.5 Hz repetition rate and is highly reproducible. Thus, time evolution of the AMFR is recorded in three-dimensions with high spatio-temporal resolutions using movable diagnostic probes. Experimental results demonstrate outward expansion of the AMFR, release of its plasma to the background, and excitation of fast magnetosonic waves during the eruption.

Keywords. magnetic flux rope, prominence, coronal loop, coronal mass ejections, flares

1. Introduction

Arched Magnetic flux ropes (AMFRs) are arched magnetic structures that confine plasma and carry electrical current along its curved axis. The current generates twist in the AMFR magnetic field. Coronal loops, prominences, and filaments are main examples of solar AMFRs that have both footpoints anchored underneath the photosphere (Lang 2001). Solar AMFRs are observed to remain stable for many Alfvén transit times (time required for the Alfvén wave to travel from one footpoint to the other) in traditional astronomical and spacecraft observations (Abbot 1911). The persistent appearance of solar AMFRs for such a long duration (lasting days to weeks) implies that the current carried by solar AMFRs during this stable phase is well below the current threshold required for the kink instability. The stable phase of the solar AMFRs ends when instabilities that significantly affect the force balance of the AMFRs appear. The kink instability is a prime example of one such instability (Hood 1979). Several other possibilities include sausage instability, interaction of the AMFR with another AMFR, and changes in the ambient plasma conditions (Antiochos 1998 and Hansen 2004). Following the appearance of the instability, the AMFR erupts and frequently evolves into energetic events such as solar flares and coronal mass ejections (CMEs) [*e.g.*, Dennis & Schwartz (1989), Chen (1996), Cremades & Bothmer (2004), and Hildner (1975)].

Technological developments in solar observations have drastically improved the measurement of the global properties of the solar AMFRs. However, it is still not feasible to directly measure the internal magnetic field and other important plasma parameters of the solar AMFRs with a good spatial and temporal resolutions. This severely limits our capability to experimentally test the validity of the theoretical models and identify the important microphysics that may not be captured in numerical simulations. Laboratory

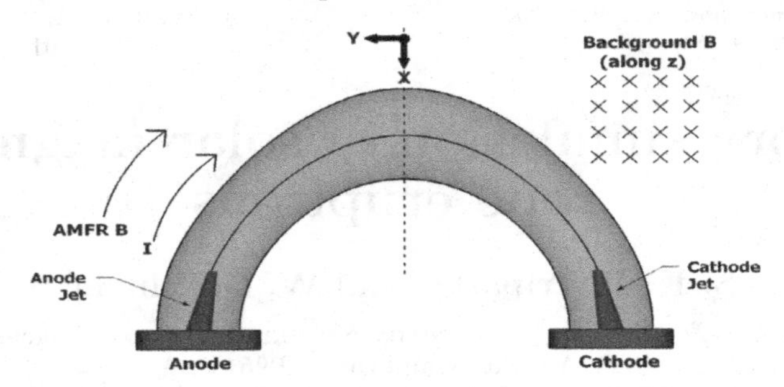

Figure 1. Schematic showing the experimental arrangement for simulating the solar AMFR eruptions. Laser generated plasma jets, relative directions of the AMFR and background magnetic fields, AMFR current I, Cartesian coordinates xyz, and both electrodes are indicated.

experiments that produce carefully scaled models of actual solar AMFRs have started playing an important role in bridging this gap (Alexander 2007, Tripathi, Bellan, & Yun 2007, and Tripathi & Gekelman 2010).

The conventional laboratory technique for producing an AMFR utilizes a magnetized arc plasma source which produces an AMFR containing several kilo-ampere electrical current (Hansen 2004). The current builds up within few microseconds and greatly exceeds the threshold for the kink instability. Hence, the stable phase of the AMFRs is not observed in these experiments. Moreover, the AMFRs in conventional experiments evolve in a vacuum environment unlike actual solar environment where magnetized plasma exists in the ambiance. Thus, a non-conventional approach is needed to simulate the impulsive eruption of a stable solar AMFR. In this article, we describe a new laboratory plasma experiment that utilizes three high-repetition-rate plasma sources to generate the AMFR, an ambient plasma, and controlled plasma flows from the AMFR footpoints.

2. Experimental setup

The AMFR is produced in a 4.0 m long (1.0 m diameter) vacuum chamber that has electromagnets to produce a 25 G magnetic field along its axis. A hot LaB_6 cathode (20 cm × 20 cm, 1700 °C) and a wire-mesh anode are mounted at one end of the chamber to produce an ambient argon plasma (neutral gas pressure = 5.0×10^{-4} torr, plasma density $n = 2.0 \times 10^{12}$ cm^{-3}, electron temperature $T_e \sim 4$ eV, pulse width = 20 ms, repetition rate = 0.5 Hz). A Cartesian coordinate system xyz with the z direction oriented along the axis of the vacuum chamber (towards the rectangular cathode) is used to delineate the relative position of the AMFR plasma source in the ambient plasma. As shown in Fig. 1, the symmetry plane of the AMFR is a xy plane at $z = 0$ cm. The AMFR is produced using an additional set of an annular disk-shaped anode and a hot LaB_6 cathode (7.6 cm outer and 1.3 cm inner diameter) mounted inside the vacuum chamber on two radially movable shafts. The AMFR footpoints are separated by 23 cm. Each AMFR electrode is surrounded by an electromagnet to generate an arched magnetic field ($B \sim 1$ kG at the AMFR footpoints). The AMFR magnetic field makes a 90^0 angle from the magnetic field of the ambient plasma. Using this arrangement, highly reproducible AMFRs ($n \sim 10^{13}$ cm^{-3}, $\delta n/n < 0.005$, $T_e \sim 4$ eV, pulse width = 2.0 ms, I = 42 A) are produced in the afterglow of the every ambient plasma pulse. During each pulse, a stable AMFR is formed ~ 150 μs after application of the voltage on the AMFR electrodes. Figure 2(a) shows a fast camera image of the stable AMFR. The persistent appearance of the images and a stationary density profile characterize the stable phase of the AMFR.

Two infrared laser beams (1064 nm, $E \sim 0.8$ J/pulse) are triggered in the middle of the AMFR pulse to ablate two carbon rods placed behind the annular electrodes. The

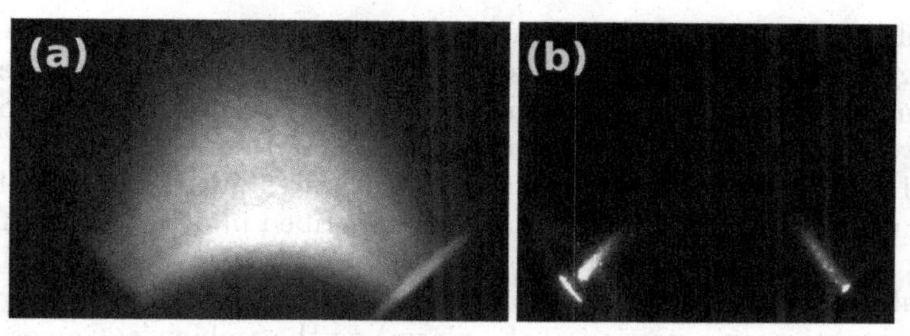

Figure 2. (a) An unfiltered fast camera image of the stable argon AMFR before the laser trigger and (b) a 460 nm narrow-passband filtered image of the carbon jets at 0.24 μs after the laser trigger. The narrow-passband filter transmits radiation from the carbon jets but blocks visible radiation from the argon AMFR.

laser-ablated carbon generates plasma jets from both AMFR footpoints. Following the laser trigger, both rods are moved by a small step using computer-controlled motors to ensure that a fresh target surface is ablated at every shot and a good reproducibility is maintained in the experiment. A fast camera image of the laser jets [see Fig. 2(b)] in the AMFR was acquired using a 460 nm passband filter which shows that the jets are composed of C^{++} ions. The jets propagate with a supersonic velocity ($v \sim 5.0 \times 10^4$ m/s) in the AMFR (ion acoustic speed $c_s \sim 5.0 \times 10^3$ m/s). The AMFR current following the appearance of the jets increases to $\sim$ 600 A which is above the kink instability threshold. Since the experiment is highly reproducible and runs at 0.5 Hz repetition rate, computer-controlled movable probes are employed to collect the magnetic field and the density data with a high spatio-temporal resolutions ($\Delta t = 20$ ns, $\Delta x = 1$–5 mm).

3. Results and Discussion

The density profile of the AMFR is presented in Fig. 3 at four distinct times to show the evolution of the AMFR following the laser trigger ($t = 0$ μs). The profile has been

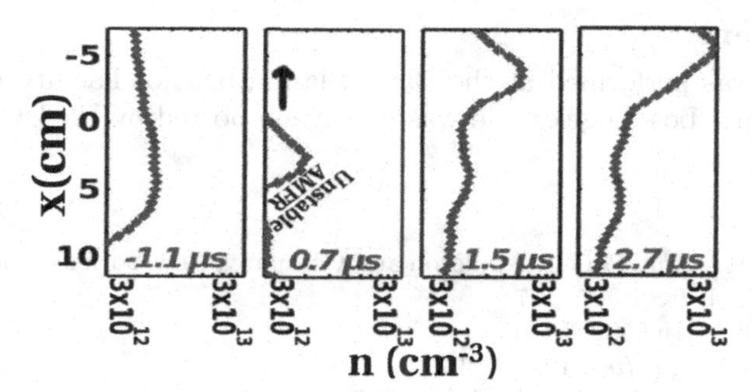

Figure 3. The AMFR density profiles $n(x)$ are shown at four distinct times. The outward motion of the AMFR can be seen as motion of the density peak along the $-x$ direction.

measured along the dashed x line shown in Fig. 1 at $y = 0$ cm and $z = -5$ cm. Collection of the data in the $z = 0$ cm plane is avoided since the laser beams propagate in this plane and they can easily destroy miniature diagnostic probes. As stated earlier, the laser generated jets significantly increase the AMFR current making it kink unstable. Moreover, the stable phase of the AMFR is maintained due to the balance of the outward hoop force and the inward forces including the magnetic tension force (Krall *et al.* 2000). The increase in the AMFR current significantly enhances the outward hoop force and the AMFR expands in the outward direction. The $t = -1.1$ μs plot represents the stationary density profile of the stable AMFR. The AMFR becomes extremely dynamic following the laser trigger. The $t = 0.7\ \mu s$ plot shows that the AMFR plasma is released to its ambiance immediately following the appearance of the laser jets and it becomes thinner. The other two plots clearly show the outward motion of the AMFR. In our recent paper (see Tripathi & Gekelman 2010), excitation of fast waves in the ambient plasma during the eruption of the laboratory AMFR has been described in detail alongwith changes in the magnetic structure of the AMFR.

These results demonstrate the important characteristics of solar AMFR eruptions such as its outward motion and ejection of the plasma to the ambiance. Camera images also show that the laboratory AMFRs resemble the structure of the solar AMFRs. Comparison of the plasma parameters of the laboratory and solar AMFRs are presented in Table 1. A MHD scaling of the solar AMFR parameters to the laboratory parameters requires

Table 1. Comparison of solar and laboratory plasma parameters

	Lower Corona	**Coronal loops**	**Laboratory AMFR**
Density n (cm^{-3})	$\sim 10^9$	$10^9 - 10^{11}$	$\sim 10^{13}$
Temperature T (K)	$\sim 10^6$	10^6–10^7	$\sim 10^5$
Length L (m)	$\sim 10^8$–10^9	$\sim 10^7$	0.51
Magnetic field B (G)	~ 10	20	1000
Plasma β		10^{-2}–10^{-1}	10^{-3}–10^{-1}
Lundquist number S		10^{13}	10^2–10^4

that a large Lundquist number and low plasma β (ratio of plasma pressure and magnetic field pressure) are retained in the experiment. The results indicate that the laboratory AMFR has appropriate values to simulate the actual solar AMFRs. High resolution measurements of the magnetic field, density and electron temperature are planned in near future to further explore the details of the AMFR eruption in our experiment.

Acknowledgment

This experiment was performed at the Basic Plasma Science Facility (BaPSF) at University of California, Los Angeles and was jointly supported by US DOE and NSF.

References

Abbot, C. G. Sep 1911, *The Sun* (D. Appleton and Company, New York and London), p. 128-182
Alexander, D. 2007, *Ap&SS* 307, 197
Antiochos, S. K. 1998, *ApJ*,502, L181
Chen, J. 1996, *J. Geophys. Res.* 101, 27499
Cremades H. & Bothmer, V. 2004, *A&A* 422, 307
Dennis, B. R. & Schwartz, R. A. 1989, *Solar Phys.* 1989, 121, 75
Hansen, J. F. & Bellan, P. M. 2001, *ApJ*, 563, L183
Hansen, J. F., Tripathi, S. K. P., & Bellan, P. M. 2004, *Phys.Plasmas* 11, 3177
Hildner, E. *et al.* 1975, *Solar Phys.* 42, 163
Hood, A. W. & Priest, E. R. 1979, *Solar Phys.*, 64, 303
Krall, J., Chen J., & Santoro R. 2000, *ApJ*, 539, 964
Lang, K. R. 2001, *The Cambridge Encyclopedia of the Sun* (Cambridge University press), 1st Ed. p. 106-143
Tripathi, S. K. P., Bellan, P. M., & Yun, G. S. 2007, *Phys. Rev. Lett.*, 98, 135002
Tripathi, S. K. P. & Gekelman, W. 2010, *Phys. Rev. Lett.*, 105, 075005

Physics of Sun and Star Spots
Proceedings IAU Symposium No. 273, 2010
A. C. Editor, B. D. Editor & C. E. Editor, eds.

doi:10.1017/S1743921311015857

Microwave Depolarization above Sunspots

Jeongwoo Lee[1] and Stephen M. White[2]

[1]Physics Department, New Jersey Institute of Technology,
Newark, NJ 07102, U.S.A.
email: leej@njit.edu

[2]AFRL, Space Vehicles Directorate, Kirtland AFB,
Albuquerque, NM 87117, U.S.A.
email: afrl.rvb.pa@hanscom.af.mil

Abstract. Microwave emissions from sunspots are circularly polarized in the sense of rotation (right or left) determined by the polarity (north or south) of coronal magnetic fields. However, they may convert into unpolarized emissions under certain conditions of magnetic field and electron density in the corona, and this phenomenon of depolarization could be used to derive those parameters. We propose another diagnostic use of microwave depolarization based on the fact that an observed depolarization strip actually represents the coronal magnetic polarity inversion line (PIL) at the heights of effective mode coupling, and its location itself carries information on the distribution of magnetic polarity in the corona. To demonstrate this diagnostic utility we generate a set of magnetic field models for a complex active region with the observed line-of-sight magnetic fields but varying current density distribution and compare them with the 4.9 GHz polarization map obtained with the Very Large Array (VLA). The field extrapolation predicts very different locations of the depolarization strip in the corona depending on the amount of electric currents assumed to exist in the photosphere. Such high sensitivity of microwave depolarization to the coronal magnetic field can therefore be useful for validating electric current density maps inferred from vector magnetic fields observed in the photosphere.

Keywords. Sun: corona, Sun: magnetic fields, Sun: radio radiation, polarization

---

1. Introduction

Magnetic fields in the sun and stars are usually measured from polarized spectral lines typically formed in the photosphere and the chromosphere (e.g. Zirin 1988, Dalgarno & Layzer 1987). At greater coronal heights ($\sim 10^5$ km), however, the plasma is dilute and fully ionized, and fewer lines sensitive to magnetic field are available (Lin *et al.* 2000). Instead strong radiations exist at radio wavelengths and it is desirable to make use of their polarization to measure the magnetic field at those heights (Dulk & McLean 1978). The coronal radio emissions appear in two natural modes: X mode which is polarized in the same sense that the electron gyrates about the magnetic field and O mode, polarized in the opposite sense. They are circularly polarized everywhere except for a narrow range of propagation angles orthogonal to the local magnetic field, and thus observed as either right or left-hand circular polarized (RCP and LCP hereafter) depending on the orientation of ambient magnetic field (Ratcliffe 1959). The X and O modes generally propagate independently of one other, and hence the polarization is unchanged during propagation. However, coupling between the two modes may occur when they pass through the region of magnetic fields orthogonal to the line of sight, called quasi-transverse (QT) layer (Kakinuma & Swarup 1962). One particular outcome of the mode coupling is the complete depolarization of the emission which depends on the frequency, magnetic field, and density (Cohen, 1960). Many studies of mode coupling have thus far been focused on the use of specific relationship between density and magnetic field required for depolarization

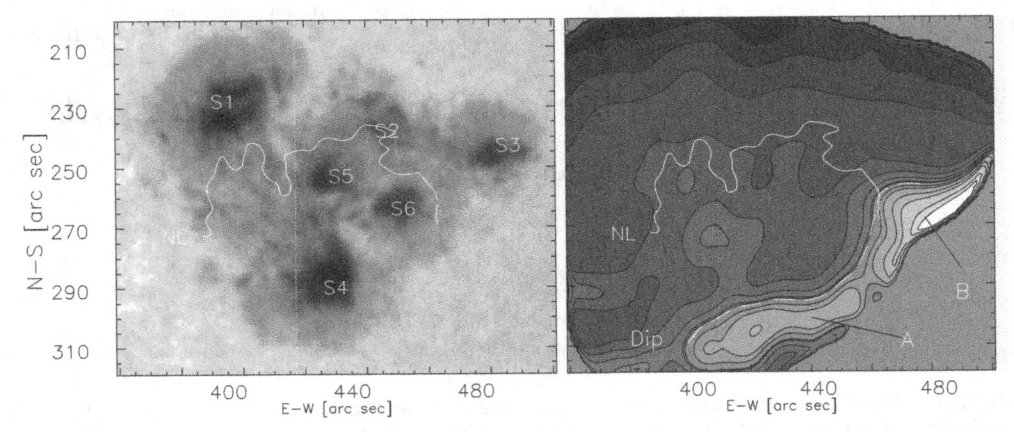

Figure 1. A white-light picture of a complex active region, AR 6615, obtained at the BBSO and a map of the degree of circular polarization taken with the VLA at 4.9 GHz. The depolarization lines, $V_{\rm obs} = 0$, (dashed line) is noticeably shifted from the PIL of longitudinal photospheric magnetic fields (solid line). Thus, in this case, the observed sense of circular polarization does not well correspond to the magnetic polarity in the photosphere. The (x, y) coordinates are geocentric coordinates in the west and the south, respectively, and x and y increase toward the west and the south, respectively. (from Lee *et al.* 1998)

in deriving one of these quantities (e.g., Kundu and Alissandrakis 1984, Alissandrakis & Chiuderi Drago 1994, Alissandrakis *et al.* 1996, Ryabov *et al.* 1999). In this paper, we instead focus on the location of the depolarization strip as an important constraint on models for the large scale coronal magnetic field structure.

2. Depolarization of Microwave Emissions

Figure 1 shows the optical continuum image and a polarization map obtained for a complex active region AR 6615 from the Big Bear Solar Observatory (BBSO) and the Very Large Array (VLA), respectively. The radio polarization map was created from the conventional CLEAN images at 4.9 GHz in which the white (black) regions are dominated by RCP (LCP). According to the longitudinal magnetogram (Fig. 2), the three northernmost spots (labeled S1–S3) are of negative (going into the Sun) polarity while the other three spots (S4–S6) are of positive (going out of the Sun) polarity and the bipolarity in this active region is generally north-south. Nonetheless, the negative polarization prevails over most of the active region so that the radio map is dominated by one polarity although the photospheric fields are bipolar. This phenomenon is expected to occur when the inversion line shifts so far to one side (southward in the present case) that the QT surface covers the southern half of the active region. In that case, waves emitted from the upgoing magnetic fields there start with RCP and then reverse their polarization to LCP on passing through the overlying QT surface. The waves emitted from the northern half of the active region start with LCP and never encounter any QT surface to remain as LCP all the way. Consequently most of the active region appears to be LCP. As a result, the only two emission regions which are not LCP are seen in the south end (region A) and on the western side (region B). The $V_{\rm obs} = 0$ line seen in region A is a long way from the photospheric PIL and the QT surface should be at a considerable height, where the density and magnetic field strength could be low enough to allow strong coupling to occur. We can thus conclude that the $V_{\rm obs} = 0$ line in region A is due to depolarization. However, the corresponding line in region B cannot be understood

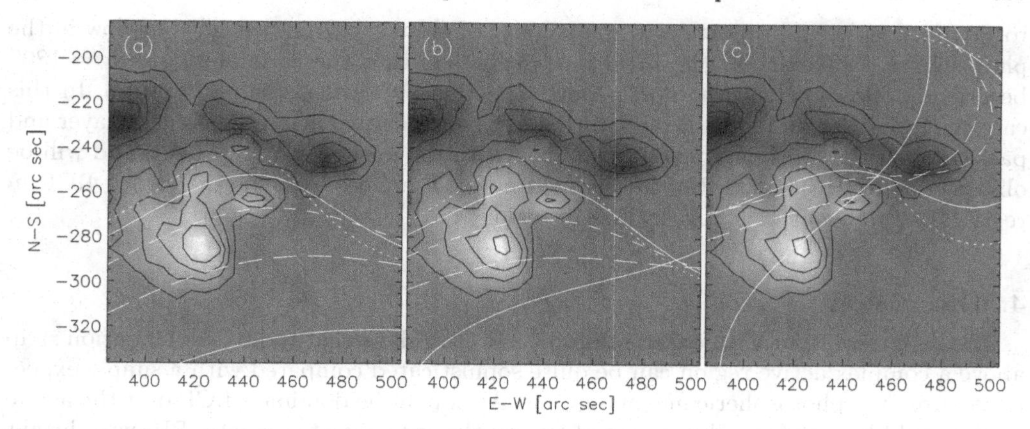

Figure 2. The coronal magnetic PILs predicted by the three magnetic field models (see text for description) overlaid on the photospheric magnetogram. The PILs correspond to heights of $h = 30''$ (dotted lines), 60", $100''$, $140''$, and $180''$ (solid lines).

in the same way. This location is rather close to the photospheric PIL, and the waves emitted in region B would encounter the QT surface at a much lower height where the mode coupling is likely to be weak. Therefore the strong RCP seen in the western part of this active region was of curiosity and has remained to be explained (Lee *et al.* 1998). In the next we check whether a specific magnetic field configuration can explain the RCP in region B.

3. Magnetic Field Models for the QT layers

We create three magnetic field models for this active region using combinations of six point charges placed at the center locations of S1–S6 (Fig. 1). The first model has all potential field charges with their relative strengths and depths chosen to reproduce the photospheric longitudinal magnetogram as closely as possible. The second and third models are created by adding charges of linear force-free fields only at the locations of S2, S3, S5, and S6 so that we have local electric currents confined in those regions with strengths gauged by the force-free parameters, $\alpha = -2.1 \times 10^{-5}$ km^{-1} and $\alpha = -2.9 \times 10^{-5}$ km^{-1}, respectively. We must note that this is not a standard procedure for constructing the nonlinear force-free field, but should be regarded as a tentative experiment to quickly check how the magnetic field deviates from the potential field configuration in the presence of local currents. In Figure 2 we plot the coronal PILs projected into the sky plane predicted by each model for heights of $h = 30'', 60'', 100'', 140''$, and $180''$, respectively, over the longitudinal magnetogram of the Mees Solar Observatory (MSO). Here the coronal PIL means the points on a coronal plane of height h where the line-of-sight magnetic field vanishes. A set of these coronal PILs at selected heights is shown in Figure 2 as way to visualize the shape of the QT surface. The potential field model shown in the leftmost panel predicts a monotonic shift of the magnetic neutral line to the south as height increases. This explains the depolarization strip in region A, but not the strong RCP in region B, because emissions from S6 then pass through the overlying region with the negative magnetic polarity to be observed in LCP. The other two models also predict the general shift of the negative magnetic polarity toward southward, although the predictions differ in the slope of the QT surface with respect to the line of sight. A noticeable difference between the models appears in region B. Here, due to the presence of electric currents, the coronal PIL rotates counterclockwise with height. The degree of rotation increases in proportion to the amount of electric current assumed to flow in the photosphere. In particular the third model predicts that the coronal PIL at $h = 180''$ be oriented almost along the north-south direction above the spots S3 and S6. In this case the radiations in region B are emitted from the positive magnetic polarity layer and pass through the overlying regions also filled with the positive polarity fields and will be observed as RCP. The third magnetic field model therefore explains the strong RCP in region B as well as the depolarization in region A.

4. Discussion

The high resolution VLA observations of AR 6615 show that radio depolarization strip above a complex active region can be quite sophisticated compared with a simple expectation from the photospheric magnetograms. Although the dominant LCP over the active region could be understood in terms of the southward shift of magnetic PIL with height and mode coupling in the high corona, the strong RCP seen in the western side of this active region was a puzzle as pointed out in the previous study (Lee *et al.* 1998). A simple modeling presented in this paper oers an alternative interpretation that due to strong currents flowing in the western part of the active region the coronal magnetic PIL rotates with height counterclockwise so that the RCP emission from the region could remain as RCP all the way to the observer. Note that this diagnostic power increases with the number of observing frequencies. With radio polarization maps available at multifrequencies, we can even trace the change of PIL orientation with height, as the depolarization strip appears close to the photospheric PIL at higher frequencies (low corona) and moves gradually away from it at lower frequencies (high corona). This procedure will reveal the location of the QT layer as a function of coronal height, which, in turn, depends on the magnetic field and currents underneath. Microwave polarization maps may thus serve as a powerful diagnostic for the large scale field structure in the outermost active region corona where no other radiation is so sensitive to magnetic field.

The presentation of this paper in the IAU Symposium 273 was possible due to partial support from the NSF grants ATM 0548260, AST 0968672 and NASA LWS grant 09-LWSTRT09-0039. J.L. has been supported by NSF grant AST-0908344.

References

Alissandrakis, C. E., Borgioli, F., Chiuderi Drago, F., Hagyard, M., & Shibasaki, K. 1996, *Solar Phys.*, 167, 167

Alissandrakis, C. E., Chiuderi-Drago, F. 1994, *ApJ*, 428, 73

Cohen, M. H. 1960, *ApJ*, 131, 664

Dalgarno, A. & Layzer, D. 1987, Spectroscopy of astrophysical plasmas, Cambridge U. Press, Cambridge and New York

Dulk, G. A. & McLean, D. J. 1978, *Solar Phys.*, 57, 279

Kakinuma, T. & Swarup, G. 1962, *ApJ*, 136, 975,

Kundu, M. R. & Alissandrakis, C. E. 1984, *Solar Phys.*, 94, 249

Lee, J., White, S. M., Kundu, M. R., Mikic, Z., & McClymont, A. N. 1998, *Solar Phys.*, 180, 193

Lin, H., Penn, M. J., & Tomczyk, S. 2000, *ApJ*, 541, L83

Ratcliffe, J. A. 1959, *The Magneto-ionic Theory and its Applications to the Ionosphere*, Cambridge U. Press, Cambridge

Ryabov, B. I., Pilyeva, N. A., Alissandrakis, C. E., Shibasaki, K., Bogod, V. M., Garaimov, V. I., & Gelfreikh, G. B. 1999, *Solar Phys.*, 185, 157

Zirin, H. 1988, *Astrophysics of the Sun*, Cambridge U. Press, Cambridge and New York

Physics of Sun and Star Spots
Proceedings IAU Symposium No. 273, 2010
A.C. Editor, B.D. Editor & C.E. Editor, eds.

doi:10.1017/S1743921311015869

Damping and the period ratio $P_1/2P_2$ of non-adiabatic slow mode

N. Kumar[1] and A. Kumar[2]

[1]Department of Mathematics, M.M.H. College, Ghaziabad 201009, Uttar Pradesh, India
[2]Department of Mathematics, Vishveshwarya Institute of Engineering and Technology, Dadri, G. B. Nagar 203207, Uttar Pradesh, India

Abstract. We investigate the combined effects of thermal conduction, compressive viscosity and optically thin radiative losses on the period ratio, $P_1/2P_2$, (P_1 is the period of the fundamental mode and P_2 is the period of its first harmonic) of a slow mode propagating one dimensionally. We obtain the dispersion relation and solve it to study the influence of non-ideal effects on the period ratio. The dependence of period ratio on thermal conductivity, compressive viscosity and radiative losses has been shown graphically. It is found that the effect of thermal conduction on the period ratio is negligible while compressive viscosity and radiation have sufficient effects for small loops and large loops respectively.

1. Introduction

Coronal loop oscillations have recently become a subject of considerable observational and theoretical interest. Since the launch of SoHo and TRACE, many examples of both standing and propagating waves have been detected in a variety of solar structures. There are observational evidences of slow modes occurring as propagating waves (DeForest & Gurman 1998; Ofman *et al.* 1997, 1999; Robbrecht *et al.* 2001; De Moortel *et al.* 2002a,b; McEwan & De Moortel 2006). The standing slow mode oscillations in solar corona has been detected (Kliem *et al.* 2002; Ofman & Wang 2002; Wang *et al.* 2003, 2004). We are interested in the detection of multiperiods in loops. Multiperiods have been first reported in standing fast waves (Verwichte *et al.* 2004; Van Doorsselaere *et al.* 2007) and now very recently in slow modes (Srivastava & Dwivedi 2010). Since higher harmonics have lower wavelengths, they carry more detailed information about a structure and are more influenced by chromospheric structure or the gravitational scale height. Andries *et al.* (2005a) studied the ratio P_1/P_2 of the fundamental oscillation period, P_1, and its first harmonic, P_2, of a kink mode oscillation, showing that this ratio falls below 2. Srivastava & Dwivedi (2010) reported the period ratio of slow mode P_1/P_2=1.54 and 1.84. The observed tendency of period ratio, $P_1/2P_2$, to be less than unity has led to a number of researchers to assess the influence of various physical effects such as longitudinal and transverse density structuring, wave dispersion and gravitational stratification on the period ratio (Andries *et al.* 2005b; McEwan *et al.* 2006, 2008).

The effect of damping on the period ratio of slow mode has been studied recently by Macnamara & Roberts (2010). They discussed the role of thermal conduction and compressive viscosity but have not included optical thin radiation to study the effects on period ratio. So in this paper, we aim at investigating the joint effects of radiation, thermal conduction and compressive viscosity on the period ratio of non-adiabatic slow modes.

2. Model Equations and Dispersion Relation

We model a single coronal loop tied at footpoints located in photosphere. We suppose that the wavelengths are much smaller than the gravitaional scale height i.e. gravitaional effects are neglected. We take the longitudinally propagating waves as purely one dimensional wave. The basic linear MHD equations describing plasma motion in 1D are (Macnamara & Roberts, 2010)

$$\frac{\partial \rho}{\partial t} + \rho_0 \frac{\partial v_z}{\partial z} = 0, \tag{1}$$

$$\rho_0 \frac{\partial v_z}{\partial t} + \frac{\partial p}{\partial z} = \frac{4}{3}\nu \frac{\partial^2 v_z}{\partial z^2}, \tag{2}$$

$$\frac{\partial p}{\partial t} - \frac{\gamma p_0}{\rho_0}\frac{\partial \rho}{\partial t} - (\gamma-1)\kappa_{\parallel}\frac{\partial^2 T}{\partial z^2} + (\gamma-1)(L+\rho_0 L_\rho)\rho + (\gamma-1)\rho_0 L_T T = 0, \tag{3}$$

$$\frac{p}{p_0} = \frac{\rho}{\rho_0} + \frac{T}{T_0}. \tag{4}$$

Here ρ, p, v and T represent perturbed density, pressure, velocity and temperature respectively whereas ρ_0, p_0 and T_0 represent equilibrium density, pressure and temperature respectively. ν is the coefficient of compressive viscosity of the form $\nu = \nu_0 T^{5/2}\mathrm{kgm^{-1}s^{-1}}$ with $\nu_0 = 10^{-17}$. γ is the ratio of specific heats. Equation (3) is linearized energy equation and present form is due to non-ideal effects (radiation losses, thermal conduction and heating). We take thermal conduction to act purely along the z-axis setting $\kappa_{\parallel} = 10^{-11}T^{5/2}\mathrm{Wm^{-1}K^{-1}}$ as thermal conduction is strongly suppresed across a magnetic field (Spitzer, 1962). $L(\rho, T)$ is the net heat- loss function per unit mass and the time having the form $L(\rho, T) = \chi\rho T^\alpha - h$, where χ and α are piecewise continuous functions depending on the temperature (Hildner 1974; Carbonell *et al.* 2004). The heating term h is assumed fixed that maintains the equilibirium temperature without contributing to the linearized perturbation equations. L_T and L_ρ are the partial derivatives of heat-loss function with respect to temperature and density respectively i.e. $L_T = (\partial L/\partial T)_\rho, L_\rho = (\partial L/\partial \rho)_T$.

Fourier analysing equations (1)–(4) as exp $(i\omega t - k_z z)$, we obtain the following dispersion relation

$$\omega^3 - i\left(\frac{4}{3}\frac{\nu c_s^2}{\gamma p_0} + \frac{(\gamma-1)\kappa_{\parallel}T_0}{p_0}\right)k_z^2\omega^2 - \left(c_s^2 + \frac{4}{3}\frac{\nu c_s^2}{\gamma p_0}\frac{\gamma(\gamma-1)\kappa_{\parallel}T_0}{\gamma p_0}k_z^2\right)k_z^2\omega$$
$$+ i\frac{(\gamma-1)\kappa_{\parallel}T_0}{\gamma p_0}c_s^2 k_z^4 + (\gamma-1)\rho_0 L_T\left(-\frac{4}{3}\frac{k_z^2\nu\omega}{p_0\rho_0} - i\frac{\omega^2}{p_0} + i\frac{k_z^2}{\rho_0}\right)T_0$$
$$-i(\gamma-1)(L+\rho_0 L_\rho)k_z^2 = 0. \tag{5}$$

If we solve dispesion relation (5) for complex freequency $\omega = \omega_r + i\omega_i$ with $k_z L = \pi/2$ and π to obtain $\omega_1 = \mathrm{real}(\omega)$ and $\omega_2 = \mathrm{real}(\omega)$ respectively, the fundamental period, P_1 can be obtained as $P_1 = 2\pi/\omega_1$ and period P_2 for first overtone as $P_2 = 2\pi/\omega_2$. So, $\frac{P_1}{2P_2} = \frac{\omega_2}{2\omega_1}$.

We now introduce the dimensionless parameters namely thermal ratio, d, radiation ratio, r (De Moortel $ Hood, 2004) and viscosity measure ϵ as

$$d = \frac{(\gamma-1)\kappa_{\parallel}T_0\rho_0}{\gamma^2 p_0^2\tau} = \frac{1}{\gamma}\frac{\tau_s}{\tau_{cond}}, \quad r = \frac{(\gamma-1)\tau\rho_0^2\chi T_0^\alpha}{\gamma p_0} = \frac{\tau_s}{\tau_{rad}}, \quad \epsilon = \frac{4}{3}\nu\frac{c_s}{\gamma p_0 L}.$$

where τ_s is sound travel time. The ratios d and r are expressed in terms of time scales τ because most observed waves have a prescribed period, which we take as τ rather than a prescribed loop length. Using standard coronal values for all the variables $T_0 = 10^6 K$, $\rho_0 = 1.67 \times 10^{-12}\mathrm{kg\,m^{-}3}$, $\kappa_{\|} = 10^{-11}\mathrm{T}_0^{5/2}\mathrm{Wm^{-1}deg^{-1}}$, $\tilde{\mu} = 0.6$, $\mathrm{R} = 8.3 \times 10^3\mathrm{m^2s^{-1}deg^{-1}}$, $\gamma = 5/3$, $\tau = 300\mathrm{s}$. gives a value $d = 0.025$ for thermal ratio and $r = 0.06$ for radiation ratio. If we set $\Omega = \omega/k_z c_s$, the dispersion relation (5) in non-dimensional parameters becomes

$$\Omega^3 - i(V + \gamma D + \alpha\gamma R)\Omega^2 - (1 + \gamma VD + \alpha\gamma VR)\Omega + iD - i(2-\alpha)R = 0, \quad (6)$$

where

$$V = \epsilon k_z L, \quad D = dk_z L \quad \text{and} \quad \mathrm{R} = \frac{\mathrm{r}}{\mathrm{k_z L}}.$$

When $\epsilon = d = r = 0$, then $\Omega = 0$ or $\Omega = \pm 1$. Therefore $\omega = k_z c_s$ is a solution of equation (6). The period ratio $P_1/2P_2 = \omega_2/2\omega_1 = [(\pi c_s/L)/(2\pi c_s/2L)] = 1$.

3. Results and Discussion

We solve dispersion relation (6) to find out the period ratio $P_1/2P_2$ of non-adiabatic slow mode in order to discuss how thermal conductivity, viscosity and radiative losses bring about a shift in the period ratio from unity. In the absence of thermal conduction and compressive viscosity, the variation of period ratio $P_1/2P_2$ with radiation parameter r is shown in figure 1(a). Figure 1(a) depicts the behaviour of period ratio with radiation

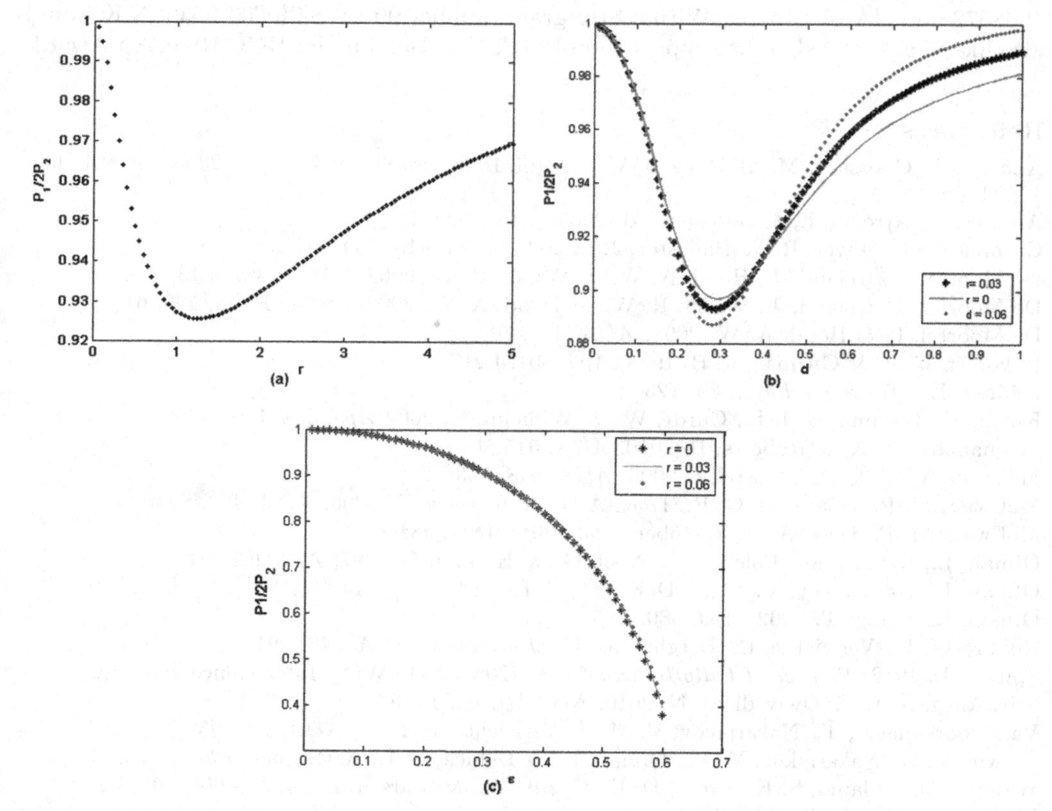

Figure 1. Period ratio as a function of (a) radiation parameter r, (b) thermal conduction parameter d for different values of radiation parameter r, and (c) compressive viscosity parameter ϵ for different values of radiation parameter r.

ratio. The period ratio is unity when $r = 0$ as expected and decreases uniformly to a minimum value 0.9257 at $r = 1.25$. Thereafter the period ratio increases and moves towards unity for sufficiently large values of r.

Figure 1(b) shows the variation of period ratio with thermal conduction parameter d for three different values of radiation parameter $r = 0.0, 0.03$ and 0.06. The period ratio decreases to minimum values 0.8669, 0.8832 with the increaes in d up to 0.29 for $r = 0$ and $r = 0.03$ respectively. For $r = 0.06$, the minimum value of peroiod ratio 0.8873 is found at $d = 0.28$. After attaining minimum value, period ratio tends to increase and returns to unity for sufficiently large d. It is interesting to note that the period ratio is almost same at $d = 0.41$ for all values of radiation parameter. For radiation parameter $r = 0.06$, the period ratio is less than the period ratios for r=0.0 and 0.03 upto the values of thermal measure d from 0.0 to 0.41 whereas the period ratio is higher than those for $r = 0.0$ and 0.03 for the values of d from 0.41 onwards.

Figure 1(c) depicts the departure of period ratio from unity as a function of compressive viscosity for different values of radiation parameter. It is observed that when we consider the compressive viscosity together with radiation ratio, the effect of small values of radiation parameter does not influence the departure of period ratio from unity.

Acknowledgements

The presentation of this paper in the IAU Symposium 273 was possible due to partial support from the National Science Foundation grant numbers ATM 0548260, AST 0968672 and NASA - Living With a Star grant number 09-LWSTRT09-0039. N.K would also like to acknowledge the support from UGC, New Delhi under UGC Research Award.

References

Andries, J., Goossens, M., Hollweg, J. V.,Arregui, I., & Van Doorselaere, T. 2005a, *A &A*, 430, 1109
Andries, J., Arregui, I., & Goossens, M. 2005b, *ApJ*, 624, L57
Carbonell, M., Oliver, R., & Ballester, J. L. 2004, *A&A*, 415, 739
De Moortel, I., Ireland, J., Hood, A. W., & Walsh, R. W. 2002a, *A&A*, 387, L13
De Moortel, I., Ireland, J., Walsh, R. W., & Hood, A. W. 2002b, *Solar Phys.*, 209, 61
De Moortel, I., & Hood, A. W. 2004, *A&A*, 415, 705
DeForest, C. E. & Gurman, J. B. 1998, *ApJ*, 501, L217
Hildner, E. 1974 *Solar Phys.*, 35, 123
Kliem, V., Dammasch, I. E., Curdt, W., & Wilhelm, K. 2002 *ApJ*, 568, L61
Macnamara, C. K. & Roberts, B. 2010, *A&A*, 515, 41
McEwan, M. P. & De Moortel, I. 2006, *A&A*, 448, 763
McEwan, M. P., Donnellly, G. R., Diaz, A. J., & Roberts, B. 2006, *A&A*, 460, 893
McEwan, M. P., Diaz, A. J., & Roberts, B. 2008, *A&A*, 481, 819
Ofman, L., Romoli, M., Poletto, G., Noci, G., & Kohl, J. L. 1997, *ApJ*, 491, L111
Ofman, L., Nakariakov, V. M., & DeForest, C. E. 1999, *ApJ*, 514, 441
Ofman, L., Wang, T. 2002, *ApJ*, 580, L85
Robbrecht, E., Verwichte, E., Berghmans, D. *et al.* 2001, *A&A*, 370, 591
Spitzer, L. 1962, *Physics of Fully Ionized Gases* (New York: Wiley Interscience 2nd edn.)
Srivastava, A. K. & Dwivedi, B. N. 2010, *New Astron.*, 15, 8
Van Doorsselaere, T., Nakariakov, V. M., & Verwichte, E. 2007, *A&A*, 473, 959
Verwichte, E., Nakariakov, V. M., Ofman, L., & Deluca, E. E. 2004, *Solar Phys.*, 223, 77
Wang, T. J., Solanki, S. K., Innes, D. E., Curdt, W., & Marsch, E. 2003, *A&A*, 402, L17
Wang, T. J. & Solanki, S.K. 2004, *A&A*, 421, L33

Physics of Sun and Star Spots
Proceedings IAU Symposium No. 273, 2010
A. C. Editor, B. D. Editor & C. E. Editor, eds.

doi:10.1017/S1743921311015870

Pre-Eruption Magnetic Configurations in the Active-Region Solar Photosphere

Manolis K. Georgoulis

Research Center for Astronomy and Applied Mathematics (RCAAM),
Academy of Athens, 4 Soranou Efesiou Street, Athens, Greece, GR-11527
email: manolis.georgoulis@academyofathens.gr

Abstract. Most solar eruptions occur above strong photospheric magnetic polarity inversion lines (PILs). What overlays a PIL is unknown, however, and this has led to a debate over the existence of sheared magnetic arcades vs. helical magnetic flux ropes. We argue that this debate may be of little meaning: numerous small-scale magnetic reconnections, constantly triggered in the PIL area, can lead to effective transformation of mutual to self magnetic helicity (i.e. twist and writhe) that, ultimately, may force the magnetic structure above PILs to erupt to be relieved from its excess helicity. This is preliminary report of work currently in progress.

Keywords. Sun: coronal mass ejections (CMEs), Sun: flares, Sun: magnetic fields

1. Introduction

Solar eruptions, that is, *solar flares* and *coronal mass ejections* (CMEs) are enigmatic dynamical manifestations in the solar magnetized atmosphere. The two phenomena do not always occur in pairs: not all flares are eruptive, that is, associated to CMEs, as much as not all CMEs are flare-associated (Yashiro *et al.*, 2004; Wang & Zhang 2007).

In solar active regions (ARs) there are mainly two types of eruptive photospheric magnetic configurations: cases of intense magnetic flux emergence/cancellation and cases of intense magnetic polarity inversion lines (PILs; see Forbes *et al.* [2006] for a review). Both involve intense photospheric flows: flux emergence is accompanied by flows carried from the sub-photosphere while PILs invariably exhibit strong shear flows. Often, one sees a combined situation with eruptions occurring above strong, *developing* PILs where flux persistently accumulates. In the quiet Sun, eruptions (CMEs) typically occur above extended, but weaker, PILs almost invariably outlined by unstable filaments in the low corona. Thus PILs are a prerequisite of most solar eruptions. For ARs, the development of strong PILs is long-known to lead to *at least* one *eruptive* flare before the PIL fades and the AR disappears from the disk (e.g. Hagyard *et al.* 1984). This is not the case with intense flux emergence that may or may not lead to eruptions.

Our inability to measure the coronal magnetic field has led to a debate over what magnetic structure(s) above PILs can actually trigger eruptions. The most obvious such structure is a *sheared magnetic arcade*, stretched due to the (observed) photospheric shear flows. Subject to strong shear and suitable topological conditions higher in the corona a sheared arcade may give rise to a *breakout* eruption (Antiochos *et al.* 1999). Another candidate structure is a *helical flux rope* (Rust & Kumar 1996) that is either emerged from the sub-photosphere or is formed prior to the eruption. A flux rope can give rise to more than one eruption scenarios (Forbes *et al.* 2006). A defining characteristic of flux ropes is their *magnetic helicity* (i.e., their internal twist and writhe) that, in some cases (helical kink instability) is the main driving force of eruptions. On the contrary, sheared arcades do not need helicity to trigger eruptions (Phillips *et al.* 2005).

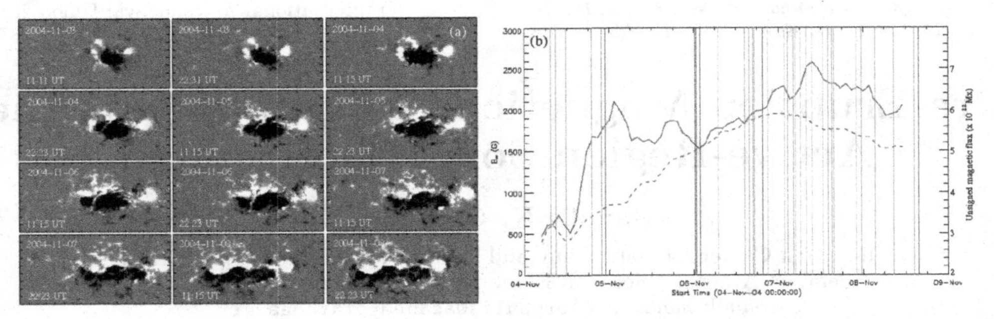

Figure 1. PIL evolution by magnetic flux accumulation and eruptive activity in NOAA AR 10696 over a 5-day period in 2004 November. (a) Timeseries of the photospheric magnetic configurations extracted from full-disk SoHO/MDI data. Tic mark separation in all images is 10″ (adapted by Georgoulis 2008). (b) Respective timeseries of the B_{eff}-values (left ordinate; solid curve) and the unsigned magnetic flux (right ordinate; dashed curve) in the AR. The vertical lines indicate the onset times of several flares, many of which were eruptive.C-, M-, and X-class flares are indicated by green, blue, and red, respectively.

2. Increasing PIL strength and implications

Why do strong PILs *always* host major eruptions? The free magnetic energy, consistently built-up due to electric currents as the PIL evolves but also consistently dissipated in small-scale, local reconnection episodes (EUV and soft X-ray "flickering") may not be the sole physical reason forcing strong PILs to always erupt. We argue that this is achieved by the coupling of free magnetic energy and *magnetic helicity*. This is *not* the same as claiming that a helical flux rope always pre-exists an eruption. Total helicity is roughly conserved in case of magnetic reconnection (e.g., Berger 1999). Free magnetic energy, on the other hand, cannot be totally released in the presence of helicity: theoretically, it can reach a minimum corresponding to a linear force-free magnetic structure with the prescribed helicity (Taylor 1974). The only option for the structure to relax is to bodily expel its excess helicity; this is a proposed CME mechanism (Low 1994). But how does helicity accumulate in the PIL area even if a flux rope is not physically present?

Aiming to quantify the PIL strength, Georgoulis & Rust (2007) introduced the *effective connected magnetic field strength* (B_{eff}), a measure of the magnetic connectivity in an AR. If $\Phi_{i,j}$ and $\mathcal{L}_{i,j}$ are the flux and length, respectively, connectivity matrix elements between the i ($i = 1, ..., p$) positive-polarity and the j ($j = 1, ..., n$) negative-polarity photospheric flux partitions, then $B_{eff} = \sum_{i=1}^{p} \sum_{j=1}^{n} (\Phi_{i,j}/\mathcal{L}_{i,j}^2)$. Clearly, large B_{eff}-values favor short connections in the AR. Developing PILs give rise to larger B_{eff}-values, tale-telling of eruptive ARs. In Figure 1 we show an example of the developing PIL in NOAA AR 10696 with the respective increase of B_{eff} and a resulting series of intensifying eruptive activity in the AR.

But does increasing B_{eff} imply increasing helicity even without a flux rope? This is possible, indeed: Georgoulis & LaBonte (2007) expressed the relative magnetic helicity H_m of an AR in the linear force-free approximation as $H_m = 8\pi \mathcal{F}_\ell d^2 \alpha E_p$, where d is the size element, α is the (constant) force-free parameter, E_p is the minimum (current-free) magnetic energy, and $\mathcal{F}_\ell$ is a known dimensionless parameter. If we now assume that there is a *single* magnetic flux tube with flux content Φ one can show that $\mathcal{F}_\ell E_p \simeq A\Phi^{2\delta}$ (details included in Georgoulis (2010)), where A is a constant and δ is a scaling index, with $1 < \delta \leqslant 1.2$. Then, magnetic helicity H_m can be written as $H_m = 8\pi d^2 \alpha A \Phi^{2\delta}$.

Now recall that H_m in an isolated flux tube reflects its twist T and writhe W, thus $H_m \sim (T + W)\Phi^2$ (Moffatt & Ricca 1992). Combining this and the above expression for

H_m we can solve for the ratio $K \sim (W/T)$ between writhe and twist as follows:

$$K = \lambda \frac{W}{T} = \frac{8\pi d^2}{L} A \Phi^{2(\delta-1)} - \lambda \ . \tag{2.1}$$

Notice that K does *not* depend on the force-free parameter α. Instead, one needs the flux Φ, the footpoint separation L, and the geometrical parameter λ that reflects the coronal shape of the flux tube axis and connects T, L, and α via the equation $T = \lambda(\alpha L)$, assuming a thin coronal tube (Longcope & Klapper 1997). Georgoulis (2010) determines the extremes of λ: $\lambda \in [1/(4\pi), 1/8]$, from a highly eccentric elliptical flux tube to a semi-circular flux tube. For a single flux tube $B_{eff} = \Phi/L^2$, so Eq. (2.1) gives

$$K = \frac{8\pi d^2}{L^{5-4\delta}} A B_{eff}^{2(\delta-1)} - \lambda \ . \tag{2.2}$$

Therefore, large B_{eff}-values, hence intense PILs, favor $K > 0$. However, $K > 0$ means that the twist and writhe have the same sign, which is a *necessary* condition for the helical kink instability (Rust & LaBonte 2005).

3. Discussion and Conclusion

We showed that increasing PIL strength, via persistent flux accumulation along it, implies directly an increasing likelihood of triggering of the helical kink instability in the PIL, at least in case of a single flux tube.

The fundamental requirement of the helical kink instability, however, is the existence of strong twist (self-helicity) that is not evident in case of a sheared arcade. Instead, a sheared arcade includes substantial mutual helicity (Régnier *et al.*, 2005; Démoulin *et al.* 2006). There is only one way to transform mutual into self helicity and this is via magnetic reconnection that roughly conserves the total helicity. Notice from Eq. (2.1) how the kink instability with large writhe favors short footpoint distances ($K \propto 1/L$) in the unstable flux tubes which aligns with the small-scale nature of magnetic reconnection.

We argue, and show in Georgoulis (2010), that the small-scale flickering of numerous magnetic reconnection events always observed in PILs effectively transforms large amounts of the shear-induced mutual helicity into self helicity. Small-scale kinks form and reconnect with the overlying structure (the sheared arcade), transferring their self-helicity to it. The larger structure evolves through increasingly unstable states and finally erupts to shed its excess helicity. The eruption can happen when overlaying conditions are suitable (i.e. minimal confinement), when the overall twist exceeds the kink-instability threshold (large-scale kink), or when a locally initiated sequence of magnetic reconnections causes such intense energy dissipation that the system cannot stay confined.

Notice how this general scenario (1) encompasses flux cancellation models, refining them with a resulting inverse cascade of magnetic helicity from smaller to larger scales, (2) does not remand a pre-existing flux rope, but may allow its formation prior to eruption, and (3) explains the necessity of strong PILs to always erupt.

Figure 2 shows an illustrative example of calculating $K + \lambda$ via Eq. (2.1). A series of roughly coaligned soft X-ray images of the AR are also provided (Figures 2a-c). Notice that candidate locations for the small-scale helical kink instability are *exclusively* along the flux-massive PIL. When the number and cumulative flux of these candidate locations maximized - creating maximum-likelihood conditions for the helical kink instability - a major eruption associated with a X2.3 flare was triggered in the AR. The eruption started locally, from a magnetic structure precisely within the candidate unstable locations and resembled an unstable kink in X-rays (Figure 2c).

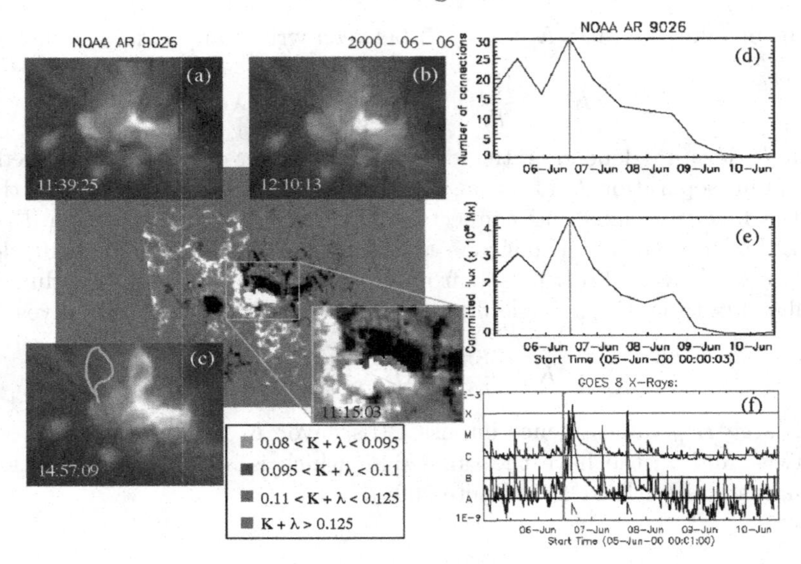

Figure 2. Calculation of candidate kink-unstable connections in NOAA AR 9026 on 2000 June 6. The small-scale connections satisfying $K > 0$ are shown in different colors; from green (weakest; $\lambda \simeq 0.08$) to red (strongest; $\lambda \simeq 0.125$). The X-ray images of the AR (a-c) sample pre-eruption instances. An apparently small-scale kink-unstable structure initiated the eruption (c); a hand-drawn cyan curve next to the loop exemplifies its shape. On the right, the plots show the number of candidate kink-unstable locations (d), their cumulative flux (e), and the GOES X-ray plot (f). The red vertical line in these plots indicates the magnetogram time (11:15 UT).

Acknowledgements

I gratefully acknowledge financial support by the Organizers to attend the Symposium.

References

Antiochos, S. K., DeVore, C. R., & Klimchuk, J. A. 1999, *ApJ*, 510, 485
Berger, M. A., 1999, *Plasma Phys. Contr. Fusion*, 41, B167
Démoulin, P., Pariat, E., & Berger, M. A. 2006, *Solar Phys.*, 233, 3
Forbes, T. G., *et al.* 2006, *Space Sci. Revs*, 123, 251
Georgoulis, M. K. 2008, *Geophys. Res. Lett.*, 35, L06S02, doi:10.1029/2007GL032040
Georgoulis, M. K. 2010, *Solar Phys.*, in preparation
Georgoulis, M. K. & LaBonte, B. J. 2007, *ApJ*, 671, 1034
Georgoulis, M. K. & Rust, D. M. 2007, *ApJ*, 661, L109
Hagyard, M. J., Teuber, D., West, E. A., & Smith, J. B. 1984, *Solar Phys.*, 91, 115
Longcope, D. W. & Klapper, I. 1997, *ApJ*, 488, 443
Low, B. C. 1994, *Phys. Plasmas*, 5, 1684
Moffatt, H. K. & Ricca, R. L. 1992, *Proc. Math. Phys. Sci.*, 439, 411
Phillips, A. D., Macneice, P. J., & Antiochos, S. K. 2005, *ApJ*, 624, L129
Régnier, S., Amari, T., & Canfield, R. C. 2005, *A&A*, 442, 345
Rust, D. M. & Kumar, A. 1996, *ApJ*, 464, L199
Rust, D. M. & LaBonte, B. J. 2005, *ApJ*, 622, L69
Taylor, J. B. 1974, *Phys. Rev. Lett.*, 19, 1139
Wang, Y. & Zhang, J. 2007, *ApJ*, 665, 1428
Yashiro, S., Gopalswamy, N., Michalek, G., St. Cyr, O. C., Plunkett, S. P., Rich, N. B., & Howard, R. A. 2004, *J. Geophys. Res.*, 109(A7), A07105

Discussion

NAME: DIALOG

Author Index

Subject Index